Autism and Child Psychopathology Series

Series Editor
Johnny L. Matson, Department of Psychology, Louisiana State University, Baton Rouge, LA, USA

Brief Overview

The purpose of this series is to advance knowledge in the broad multidisciplinary fields of autism and various forms of psychopathology (e.g., anxiety and depression). Volumes synthesize research on a range of rapidly expanding topics on assessment, treatment, and etiology.

Description

The **Autism and Child Psychopathology Series** explores a wide range of research and professional methods, procedures, and theories used to enhance positive development and outcomes across the lifespan. Developments in education, medicine, psychology, and applied behavior analysis as well as child and adolescent development across home, school, hospital, and community settings are the focus of this series. Series volumes are both authored and edited, and they provide critical reviews of evidence-based methods. As such, these books serve as a critical reference source for researchers and professionals who deal with developmental disorders and disabilities, most notably autism, intellectual disabilities, challenging behaviors, anxiety, depression, ADHD, developmental coordination disorder, communication disorders, and other common childhood problems. The series addresses important mental health and development difficulties that children and youth, their caregivers, and the professionals who treat them must face. Each volume in the series provides an analysis of methods and procedures that may assist in effectively treating these developmental problems.

Pamela McPherson
Editor

Handbook of Treatment Planning for Children with Autism and Other Neurodevelopmental Disorders

Springer

Editor
Pamela McPherson
Department of Psychiatry and Behavioral Medicine
Louisiana State University Health Sciences Center
Shreveport, LA, USA

ISSN 2192-922X ISSN 2192-9238 (electronic)
Autism and Child Psychopathology Series
ISBN 978-3-031-06119-6 ISBN 978-3-031-06120-2 (eBook)
https://doi.org/10.1007/978-3-031-06120-2

This Springer imprint is published by the registered company Springer Nature Switzerland AG
The registered company address is: Gewerbestrasse 11, 6330 Cham, Switzerland

Notes on Language Use

Words matter and language matters. Medicine has its' own terminology to standardize communication among medical professionals with the goal of promoting quality patient care. An individual's experience and preference also matter. Medical terminology may fail to respect lived experience. In the clinical setting, respectful inquiry should be made of an individual's preference as using preferred language upholds the fundamental principles of medical ethics – beneficence, nonmaleficence, autonomy, and justice. While recognizing that person-first language may not be preferred by all individuals, it is the prevailing convention in medicine at the time this volume was written and is used by most authors in this volume. Still, we recognize and respect that some prefer identity-first language. In addition, the definitions of disability, developmental disability, medical complexity, and special needs can vary by use and, sometimes, legal statute. Chapter authors use terminology which best captures their subject matter. When referring to a specific study, the terminology of the study is followed.

Acknowledgments

Before we venture into the world, our first teachers are our families. Words cannot express the gratitude that I hold for my family. Their enduring support of my professional and personal endeavors and especially during the process of preparing this book for publication has been uplifting and grounding. Because actions speak louder than words, I will strive to nurture each of these cherished relationships

During a medical career, our patients become our teachers and remain our most important teachers. The trust that a parent places in us to care for their child is sacred and an immense privilege. Parents and children instruct us and inspire us to be better carers and better people. Seeing the hope and relief a parent feels when they are heard rallies us and weathering their criticism when we fall short reminds us that we don't have all the answers and that lived experience has much to teach us. The best medical care is a partnership of trust between those providing and receiving care with mutual respect and learning.

In our professional lives, if we are fortunate, we have beloved mentors and mentees who become friends and colleagues. I have been wonderfully blessed to work with incredible professionals in many settings. Instructors in clinical programs and colleagues have been and remain valued teachers. I thank all who guided me on my professional journey, including some who assisted with this book, including Dr. Johnny Matson who offered me the opportunity to edit this volume. The experience of asking each colleague my most pressing questions about their field has been incredible and I thank each of you for your contributions to this volume and the expert care you provide to children each day.

Introduction

Teamwork promotes success. None of us can achieve the successes we envision for our patients if we work alone. This is especially true for those of us treating children living with autism spectrum disorders or other intellectual/developmental disabilities. It is critical to children's health that partnerships be forged with the family, across specialties, and with community partners to address medically complex needs. Ideally, a patient-centered, pediatric medical home links a child and family to their network of providers; however, this is not the reality for most children living with disability and the burden falls on families to coordinate care. Medical providers also experience the burden and frustrations of coordinating care. Obtaining a subspecialty consultation or special education services can take months. Our patients cannot wait, and neither can we. We have questions for our colleagues that we want answered now. This book provides answers to the questions you would ask your colleagues if only you have them on speed dial.

This volume of Dr. Johnny Matson's Autism and Child Psychopathology Series provides valuable clinical insights into the contributions of 20 subspecialists for the care of children living with disability from perinatal considerations to navigating the emerging adult's transition to adult healthcare services. If you have decades of experience, this book will provide a view of what has changed since your days in training, and if you are in the early years of your career, you will find the wisdom of seasoned experts to be priceless. The clinical insights shared in this volume will allow you to make goal-directed referrals, prepare families for consultation with specialists, and begin evaluations or interventions while awaiting specialist assessments. In short, the goal of this volume is to provide the valuable knowledge that will enhance your practice, allow you to communicate more effectively with families and colleagues, and help you to achieve the success you envision for your patient.

This volume is arranged in two sections. The first provides an overview of treatment planning with a focus on the medical home model, opening with the history and basic elements of individualized treatment planning. From the rise of the overarching biopsychosocial to the evolving medical home and wraparound models of care, the reader is guided through these basic models of care delivery. The foundations of treatment planning and care coordination including engaging children and parents, goal setting, and progress monitoring are detailed for you. Section I includes chapters detailing family as well as systems perspectives to promote comprehensive treatment planning and concludes with comprehensive recommendations for progress monitoring.

Section II provides chapters with in-depth clinical insights from 20 specialists. Each chapter explores the specialist's scope of practice highlighting the conditions treated, common procedures, recommendations for referrals, and tips for preparing families for appointments. The needs of children and their families are explored from the perinatal period, across childhood and adolescence, to the transition to adult care across community and hospital settings. Section II is designed to promote efficient, goal-directed communication between specialists in order to promote successful care coordination and advance the goals set by our patients for health and wellness.

Contents

Contributors

Heather Anderson Louisiana State University Health Sciences Center, School of Allied Health, Shreveport, LA, USA

Junaid Ansari Department of Neurology, Louisiana State University Health Sciences Center, Shreveport, LA, USA

Ramachandra Bhat Department of Pediatrics, Neonatal Perinatal Medicine, Louisiana State University, Shreveport, LA, USA

Erna Imperatore Blanche Chan Division of Occupational Science & Occupational Therapy, Ostrow School of Dentistry, University of Southern California, Los Angeles, CA, USA
Therapy West, Los Angeles, CA, USA

Adam T. Blancher Louisiana State University – Shreveport, Shreveport, LA, USA

Stefanie C. Bodison Department of Occupational Therapy, College of Public Health and Health Professions, University of Florida, Gainesville, FL, USA

Andrew P. Bourgeois Department of Anesthesiology, Louisiana State University Health Sciences Center, Shreveport, LA, USA

Megan Callahan Department of Psychology, Louisiana State University, Baton Rouge, LA, USA

Donald L. Chi School of Dentistry and School of Public Health, University of Washington, Seattle, WA, USA

Zachary M. Connelly Department of Urology, Louisiana State University Health Shreveport, Shreveport, LA, USA

Elyse M. Cornett Department of Anesthesiology, Louisiana State University Health Sciences Center, Shreveport, LA, USA

Jennifer Creedon Louisiana State University School of Medicine, New Orleans, LA, USA

Maria Gabriela Dominguez Garcia Department of Pediatrics, Neonatal Perinatal Medicine, Louisiana State University, Shreveport, LA, USA

Leah I. Stein Duker Chan Division of Occupational Science & Occupational Therapy, Ostrow School of Dentistry, University of Southern California, Los Angeles, CA, USA

Michaelann Gabriele Center for Developing Kids, Inc., Pasadena, CA, USA

Anjalee Galion Department of Pediatrics, Division of Pediatric Neurology, University of California Irvine – Children's Hospital of Orange County, Orange, CA, USA

Amanda Glinky Children's Hospital New Orleans, New Orleans, LA, USA

Alexander Gomelsky Department of Urology, Louisiana State University Health Shreveport, Shreveport, LA, USA

Neslihan Gungor Department of Pediatrics, Endocrinology Division, Louisiana State University Health Sciences Center, Shreveport, LA, USA

Sandra L. Hayes Louisiana State University Health Sciences Center, School of Allied Health, Shreveport, LA, USA

Karen Johal Department of Pediatrics, Endocrinology Division, Louisiana State University Health Sciences Center, Shreveport, LA, USA

Ryan Jones Louisiana State University Health Sciences Center, New Orleans, LA, USA

Alan D. Kaye Department of Anesthesiology, Louisiana State University Health Sciences Center, Shreveport, LA, USA

Department of Anesthesiology, Louisiana State University Health Sciences Center, New Orleans, LA, USA

Colleen Kraft Children's Hospital Los Angeles, Los Angeles, CA, USA

Kelvin Lee AdventHealth for Children, AdventHealth Medical Group, Orlando, FL, USA

Shabih Manzar Department of Pediatrics, Neonatal Perinatal Medicine, Louisiana State University, Shreveport, LA, USA

Michael Marble University of New Mexico Health Sciences Center, Department of Pediatrics, Division of Pediatric Genetics, Albuquerque, NM, USA

Nanette Massey Louisiana State University – Shreveport, Shreveport, LA, USA

Monika Mathur Department of Pediatrics, Division of Pediatric Neurology, University of California Irvine – Children's Hospital of Orange County, Orange, CA, USA

Dexter A. Matrana Division of Allergy/Immunology, Department of Pediatrics, Louisiana State University Health Sciences Center, Children's Hospital New Orleans, New Orleans, LA, USA

Johnny L. Matson Department of Psychology, Louisiana State University, Baton Rouge, LA, USA

Michael M. McGill Louisiana State University Health Sciences Center, School of Allied Health, Shreveport, LA, USA

Shawn E. McNeil Department of Psychiatry and Behavioral Medicine, LSU Health Shreveport – School of Medicine, Shreveport, LA, USA

Pamela McPherson Department of Psychiatry and Behavioral Medicine, Louisiana State University Health Sciences Center, Shreveport, LA, USA

Thomas Megerian Department of Pediatrics, Division of Pediatric Neurology, University of California Irvine – Children's Hospital of Orange County, Orange, CA, USA

Maroun J. Mhanna Department of Pediatrics, Neonatal Perinatal Medicine, Louisiana State University, Shreveport, LA, USA

Christopher B. Min Department of Pediatrics, Division of Pediatric Neurology, University of California Irvine – Children's Hospital of Orange County, Orange, CA, USA

Sarah Murnane School of Allied Health Professions, LSU Health Shreveport, Shreveport, LA, USA

Bonnie Nakasuji Therapy West, Los Angeles, CA, USA

Chikezie N. Okeagu Department of Anesthesiology, Louisiana State University Health Sciences Center, New Orleans, LA, USA

Tomas Paneque Department of Urology, Louisiana State University Health Shreveport, Shreveport, LA, USA

Jessica Picone Children's Hospital New Orleans, New Orleans, LA, USA

Rebecca Pistorius Willis-Knighton Health System, Shreveport, LA, USA

Andres J. Pumariega Division of Child and Adolescent Psychiatry, Department of Psychiatry, University of Florida College of Medicine, Gainesville, FL, USA

Marcia Rankine Department of Pediatrics, Endocrinology Division, Louisiana State University Health Sciences Center, Shreveport, LA, USA

Syed Ali Raza Department of Neurology, Louisiana State University Health Sciences Center, Shreveport, LA, USA

Devin S. Reed Department of Anesthesiology, Louisiana State University Health Sciences Center, New Orleans, LA, USA

Erin Reuther Children's Hospital New Orleans, New Orleans, LA, USA

Rosario M. Riel-Romero Department of Neurology, Louisiana State University Health Sciences Center, Shreveport, LA, USA

Joshua Sanderson Louisiana State University School of Medicine, New Orleans, LA, USA

Kristen Sandoz Louisiana State University Health Sciences Center, New Orleans, LA, USA

Noeen Sarfraz Louisiana State University Health Sciences Center, Department of Psychiatry and Behavioral Medicine, Shreveport, LA, USA

Hannah Scott Louisiana State University School of Medicine, New Orleans, LA, USA

Emma Simmons University of California Medical School, Riverside, CA, USA

Elizabeth R. Smith Division of Allergy/Immunology, Department of Pediatrics, Louisiana State University Health Sciences Center, Children's Hospital New Orleans, New Orleans, LA, USA

Jennifer Stanfill Department of Anesthesiology, Louisiana State University Health Sciences Center, New Orleans, LA, USA

Shavonne Temple Louisiana State University Health Sciences Center, Shreveport, LA, USA

Celeste Tevis Department of Psychology, Louisiana State University, Baton Rouge, LA, USA

Rebecca Wallace Children's Hospital New Orleans, New Orleans, LA, USA

Luke A. Wall Division of Allergy/Immunology, Department of Pediatrics, Louisiana State University Health Sciences Center, Children's Hospital New Orleans, New Orleans, LA, USA

Nitin Walyat Department of Pediatrics, Neonatal Perinatal Medicine, Louisiana State University, Shreveport, LA, USA

Barzanna A. White Caddo Parish Schools, Shreveport, LA, USA

Nicole A. Withrow University of Northern Colorado, Nutrition and Dietetics, Greeley, CO, USA

Michelle M. Yetman Louisiana State University – Shreveport, Shreveport, LA, USA

Russell Zwiener Children's Hospital New Orleans, New Orleans, LA, USA

Part I

Overview of Treatment Planning

1 Treatment Planning for Children with Autism and Developmental Disability

Pamela McPherson and Noeen Sarfraz

Abstract

The need for formal, patient-centered treatment planning is relatively new to the field of heath care, first becoming common in the 1960s. Individual treatment planning is necessary to ensure that goals for a patient's care are identified, prioritized, fully addressed, and memorialized for future reference. Children with autism and developmental disability commonly have complex needs addressed across multiple systems. Quality care for children with complex needs requires formal treatment planning. Models of treatment planning including the biomedical, biopsychosocial, and medical home model are reviewed in this chapter. The process of working with a child and family to develop measurable goals and implement goals is detailed. Service plans in nonmedical setting are also reviewed in order to provide a comprehensive description of treatment planning for children with autism and developmental disability.

Keywords

Treatment plan · Service plan · Care plan · Autism spectrum disorders · Developmental disability · Intellectual disability · Medical home model · Shared plan of care · Biopsychosocial model · Care coordination

Introduction

Children with autism and developmental disability commonly receive services from a complex network of medical, community, and at home care providers, each with written or unwritten ideas regarding the purpose and goals of treatment. Formal treatment or care planning is the process used by health professionals in conjunction with patients and their family to reach diagnostic agreement and identify and detail the treatment interventions aimed at ameliorating the target symptoms of medical conditions. Individual treatment planning is necessary to ensure that goals for each youth are identified, prioritized, fully addressed, and memorialized for future reference. A written plan may take many forms; a sentence or two at the end of a clinician's note, the problem-oriented SOAP note, the formal treatment plan that is common in

P. McPherson (✉) · N. Sarfraz
Department of Psychiatry and Behavioral Medicine, Louisiana State University Health Sciences Center, Shreveport, LA, USA
e-mail: Pam.mcpherson@lsuhs.edu

P. McPherson (ed.), *Handbook of Treatment Planning for Children with Autism and Other Neurodevelopmental Disorders*, Autism and Child Psychopathology Series,
https://doi.org/10.1007/978-3-031-06120-2_1

behavioral health settings, the care plan used by in-home providers, and the individualized education plan, to name a few. For a child with chronic and complex needs, a parent may be juggling multiple plans. When multiple providers are involved, a medical home promotes care coordination – the active and ongoing integration of the treatment plans of each provider. Officially endorsed by the American Academy of Pediatrics in 1992, the Patient or Family Centered Medical Home is a framework designed to coordinate all caregivers as a team with the shared goal of providing high quality, comprehensive care ("The Medical Home", 2002).

Since the end of World War II, access to high-quality medical and mental health care has been a primary societal concern (Blendon & Benson, 2001). Historically, medical and mental health service delivery systems operated in parallel, only becoming more integrated in the last 30 years. As detailed in Chap. 2, the medical home model for pediatric care was first proposed in 1967 to address the fragmentation of medical care, which was particularly common for children with special needs (Sia et al., 2004). A the same time, mental health care was shifting from an institutional to community-based care model (Grella & Grusky, 1989; Morell, 1979). As Hawaii was implementing the medical home model during the 1980s, the mental health community adopted the *system of care* model implemented through a wraparound approach (Stroul & Friedman, 1986; VanDenBerg et al., 2003). Both the medical home and the system of care are national models that are patient- and family-centered, evidence-based, culturally responsive, and strive to provide comprehensive and coordinated care. Over the past decade, increasing attention has focused on integrating the models (Croghan & Brown, 2010). Integrating care requires understanding the treatment planning practices of medical and behavioral health care.

Treatment Planning

The need for formal, patient-centered treatment planning is relatively new to the medical community and until the 1960s was uncommon. Historically, beneficence was valued over patient autonomy, resulting in details of a diagnosis or treatment plan being withheld from the patient (Weiss, 1985). With the law coining *informed consent* in 1957 and bioethicists advancing the concept of patient autonomy, the concept of shared decision-making in treatment planning has evolved over the last 50 years to become the standard of care (Will, 2011). Professional ethics, the law, third-party payors, and the patient advocacy movement have informed the treatment planning process with the aim of creating a living document that focuses treatment on realistic, agreed-upon goals. Patients and their families benefit from the deeper understanding of treatment needs that the process promotes. Treatment providers and reviewers benefit from a clear, standardized record of the care needs and interventions. A sound treatment plan is a win for all.

Treatment Planning in General Medical Care: The Biomedical Model

The 1910 Flexner Report on Medical Education is credited with firmly establishing the biomedical model as the standard for medical practice (Duffy, 2011; Flexner, 2002). Funding for medical research and the standardization of medical education with the goal of promoting of a scientific approach to the treatment of disease are laudable successes fueled by the Flexner Report, yet even at the time, Sir William Osler and others recognized limitations of the biomedical model. Perhaps because he was not a physician but an educator, Flexner detailed the similarities between the scientific method and medical decision-making with no attention to the subtleties of the physician-patient relationship (Marcum, 2019). In practice, the biomedical model reduces the experience of the patient to a sentence or two documented as the subjective chief complaint. In general medical care, the most common method of documenting treatment is the problem-oriented medical note following the SOAP format – subjective, objective, assessment, and plan (Belden et al., 2017; Weed, 1968). For each patient contact, treatment planning is captured in the plan section. The plan includes diagnostic tests and/or procedures and management objectives including specialty referrals. In addition, the goals and the expected course of treatment are documented in the plan section. Ideally, the plan should note medical decision-making rationales. The goal of the plan section of the SOAP note is to be succinct while synthesizing all relevant data. While most sections of electronic health records provide a robust structure to capture elements of the subjective and objective assessments, little or no formatting is typically offered for the plan section. This flaw requires busy clinicians to create plans de novo with each patient contact toggling between the current and past notes to review progress and update the plan burdening clinician memory and subtracting valuable time from patient care. New generations of electronic documentation hold the promise of a reduced documentation burden, an expanded plan, and improved integration of the treatment plans of multiple providers (O'Donnell et al., 2020).

The Biopsychosocial Model

The limitations of the biomedical model are addressed by internist and psychiatrist, George Engel's biopsychosocial model (Engel, 1977). The biopsychosocial model recognizes the contributions of an individual's psychology and place in society as contributors to disease and as important considerations for treatment planning. The merits of the biopsychosocial approach as a model have been widely debated with challenges that the approach is not truly a model because it cannot be tested and it is too vague as the pertinent biological, psychological, and social elements are not clearly defined (Bolton & Gillett, 2019; Farre & Rapley, 2017). Smith (2020) has countered criticisms citing the standardized patient-centered interview as the evidence-based method defining the model's elements. Recent advances in the study of subjective well-being and the linkage social determinants of health to biomarkers support the biopsychosocial model (Karunamuni et al., 2020). Patient interviews have been developed to capture social determinants of health. A patient-centered interview for chronic illness assessment, called the History and Physical 360, has shown promise in a randomized clinical trial at four medical schools. The H & P 360 includes physician review of eight domains: perception of health, goals/priorities, psychosocial concerns, behavioral health, relationships, resources, functional status, and biomedical factors, with prompts to address concerns as part of the treatment plan (Kirley et al., 2020). A youth-centered psychosocial interview for adolescents, the HEEADSSS (*H*ome environment, *E*ducation and employment, *E*ating, peer-related *A*ctivities, *D*rugs, *S*exuality, *S*uicide/depression, and *S*afety from injury and violence), is available as an app, with the HEEADSSS 3.0 including internet safety and social media questions (Doukrou & Segal, 2018; Goldenring & Cohen, 1988). Through the diagnostic interview process, the biopsychosocial model informs treatment planning in behavioral health care and the medical home model.

Treatment Planning in Behavioral Health Care

In behavioral health, the process of treatment planning was formalized in the 1970s, largely driven by managed care and accrediting bodies (Maruish, 2019). Specific components of the mental health treatment plans are dictated by professional guidelines and payors. Typically the treatment plan includes patient demographic information, a brief case summary, diagnosis(es), target symptoms, goals, objectives, interventions with the party responsible for implementation, and criteria for discharge planning. Because a patient may not manifest all symptoms of a given diagnosis, target symptoms are identified. A target symptom is the symptom of the diagnosis that the proposed intervention will address. Expected changes are documented as goals and/or objectives. Person-centered goals reflect the patient's desired outcomes in a brief statement. Goal setting creates hope and establishes shared priorities for the patient, parents, and the mental health provider. Setting goals has been linked to improved treatment engagement

(Cairns et al., 2019). Objectives detail how goals will be achieved, for example, the skills to be learned with a timeline for mastery. Objectives should be *S.M.A.R.T.* – specific, measurable, achievable, realistic/relevant, and timed (Bovend'Eerdt et al., 2009; Doran, 1981). An intervention is the specific type of therapy or modality used to achieve the objective, such as cognitive behavior therapy, applied behavioral analysis, or psychotropic medication. A behavioral health treatment plan is a dynamic document that is reviewed at regular intervals and updated as circumstances indicate and payors demand.

Treatment Planning in the Medical Home

Routine pediatric care and pediatric subspecialty care is documented using the SOAP format; however, additional documentation is required to capture the complex needs of children with autism spectrum disorders (ASD) and conditions meeting the criteria for special healthcare needs including intellectual or developmental disabilities (IDD). In addition, children with ASD and/or IDD often meet criteria for medical complexity if they have conditions impacting multiple organ systems and causing functional limitations necessitating high resource utilization (Croen et al., 2017; Kuo & Houtrow, 2016). The National Youth Survey of 3- to 17-year-olds found an 18% prevalence of developmental disability and approximately 20% of the sample meeting criteria for special healthcare needs (Coller et al., 2020; Zablotsky et al., 2019). While states may have their own definition of children with special healthcare needs, the federal Maternal and Child Health Bureau's Division of Services for Children With Special Health Care Needs work group defined children with special needs as

> those who have or are at increased risk for a chronic physical, developmental, behavioral, or emotional condition and who also require health and related services of a type or amount beyond that required by children generally (McPherson et al., 1998, p. 138).

Increasingly, primary care providers treat children requiring complex and highly coordinated care in a medical home. About 43% of children with special healthcare needs receive care in a medical home (Child and Adolescent Health Measurement, 2017). For children with complex, chronic conditions receiving care in a medical home, treatment planning is documented in a shared plan of care (SPoC). A SPoC

> focuses on a comprehensive compiling of the information, in partnership with the family, needed to support coordination of care for the multiple needs of an individual child or youth, and his or her family (Wirth & Kuznetsov, 2016, p. 1).

The National Academy for State Health Policy and the Lucile Packard Foundation for Children's Health have published standards for the SPoC (National Care Coordination Standards for Children and Youth with Special Health Care Needs, 2020). The SPoC integrates the child and family goals with the clinical recommendations of multiple professionals to map the actions expected to promote progress toward identified goals (American Academy of, 2014; McAllister, 2014b). A SPoC includes demographic information, a contact list for the medical home neighborhood members, a medical summary, child/family goals, clinical treatment goals, actions or strategies to achieve the goals, and signatures of participants in the process (McAllister, 2014b). The aim of the SPoC process is to empower the treatment team and family partners to:

- Engage in discussion, assessment, priority setting, and declaration of shared goals
- Build trusting relationships with effective communication and collaboration
- Develop and use a shared plan of care
- Enhance the child and family experience of a culturally effective care partnership
- Improve child health and family outcomes
- Enhance the clinical team's experience of providing care (McAllister, 2014a, p. 3)

Treatment planning for a child with complex needs requires the thoughtful construction of a shared plan of care (SPoC) with careful attention to engaging the family in the process in order to promote the successful plan implementation. Engaging the family is a dynamic process influenced by motivation and readiness as well as familial and cultural factors (Gentles et al., 2019). A goal of engagement is to increase activation defined as the knowledge, skill, and confidence to manage chronic illness (Hibbard & Greene, 2013; Hibbard et al., 2004). Increased parental activation, as measured with the Patient Activation Measure-Developmental Disability, is associated with lower stress and greater parental efficacy in parents of children with ASD (Ruble et al., 2018). Engagement and activation are promoted through a discussion of the child's diagnosis that is guided by the parent's capacity to receive information and is responsive to the emotion evoked by the conversation (Berkey et al., 2018). As discussed in Chap. 3, becoming a parent is a dynamic process. The Pistorius' Model detailed in Chap. 3 captures the complexity of parental development when medical complexity is an issue. It is critical that professionals strive to understand and respect the lived experience of parents and our patients. When parents and patients are heard, engagement and activation are enhanced. After hearing concerns and responding to the parent's initial questions, an appointment should be made to complete a SPoC.

Elements of the SPoC

A SPoCSPoC elements is a dynamic document that is organized according to the needs of the patient, family, and the pediatric practice or healthcare system. Medical home accrediting bodies allow for flexibility in SPoC design, with elements of the SPoC typically including family information, a medical summary with diagnoses, negotiated actions to meet goals, and a list of team members. The National Institute for Children's Health Quality (2015) has developed an interactive template for constructing a SPoC that allows for customization. The family or caregiver should be given a copy of the SPoC and asked to sign release of information forms to facilitate the distribution of the plan. See Table. 1.1.

SPoC: The Medical Summary

The SPoC medical summary notes demographic information, the neighborhood map, an overview of medical history, and active conditions including specialized equipment or technological needs, 504 or individualized educational plans, and social needs. A medical summary may be documented as a list of needs or as a

Table 1.1 Ten underlying principles for a shared plan of care

A successful shared plan of care occurs when:
1. Children, youth, and families are actively engaged in their care
2. Communication with and among their medical home team is clear, frequent, and timely
3. Providers/team members base their patient and family assessments on a full understanding of child, youth and family needs, strengths, history, and preferences
4. Youth, families, healthcare providers, and their community partners have strong relationships characterized by mutual trust and respect
5. Family-centered care teams can access the information they need to make shared, informed decisions
6. Family-centered care teams use a selected plan of care characterized by shared goals and negotiated actions; all partners understand the care-planning process, their individual responsibilities, and related accountabilities
7. The team monitors progress against goals, provides feedback, and adjusts the plan of care on an ongoing basis to ensure that it is effectively implemented
8. Team members anticipate, prepare, and plan for all transitions (e.g., early intervention to school; hospital to home; pediatric to adult care)
9. The plan of care is systematized as a common, shared document; it is used consistently by every provider within an organization, and by acknowledged involved providers across organizations
10. Care is subsequently well coordinated across all involved organizations/systems

From McAllister, 2014; Used with permission. McAllister (2014a), p. 3

more detailed narrative. Both promote the communication of needs among team members listed on the neighborhood map. A list has the advantages of being easy to compile, review, and update, while a narrative captures the interwoven complexities faced by the family. A narrative detailing the family's lived experience can be a helpful addition to the SPoC. Understanding complex family dynamics assists team members in identifying the most productive approaches to care and navigating challenges successfully. Parents are experts about their child and appreciate collaborating with medical experts as equals. Both parties have responsibilities with regard to the medical summary. The healthcare team should update medical information regularly, include the family's expertise in the development of the summary, inquire and be observant for barriers to engagement and activation, and provide a copy of the summary to the family. The family's responsibility is to provide updated information as it becomes available, to be alert to barriers to engagement and activation, and to ask questions.

An emergency information form (EIF) may be included in the medical summary. Emergency situations may pose special risks for children with complex health needs. Emergency preparedness should be discussed with families and an EIF developed if necessary. The EIF details the care recommended in the event of a medical emergency or natural disaster with critical information for emergency care providers including demographic and medical information with a problem list and recommended interventions. An EIF may include a description of baseline functioning, any needed antibiotic prophylaxis, procedures to avoid, accommodations that may be calming or necessary for certain procedures, and details regarding equipment or technology that is in use. The American Academy of Pediatrics and the American College of Emergency Physicians (2010) have collaborated on the development of sample EIFs, an interactive EIF electronic template, and policy recommendations. While medical summaries may vary from practice to practice, a standardized EIF is recommended to optimize patient care (American College of Emergency, 2010). The EIF may also include advance directives and note who has medical decision-making capacity.

For children who wander, are at risk of falling, and have self-injurious behaviors (SIB) or mental health needs, a safety plan may be developed as part of the medical summary. Wandering or becoming lost occurs in 25% of children with ASD or ID and 38% of children with both (Rice et al., 2016). An increased incidence of SIB has been reported in children with autism, intellectual disability, and sensory disorders as well as genetic syndromes including Lesch-Nyhan, Niemann-Pick, Smith Magenis, fragile X, Cornelia de Lange, and Prader-Willi (Fletcher et al., 2016). When SIB is diagnosed, a psychologist or other mental health professional conducts a functional behavioral assessment to identify risk factors for SIB and develop a behavioral safety plan to reduce self-injury and minimize the need for restrictive measures (Vollmer et al., 2018). A safety plan identifies SIB warning signs and triggers, calming/coping strategies to prevent or reduce SIB, suggestions for improving environmental safety, and emergency contacts. If the safety plan is developed to prevent suicidal behaviors, it will also include reasons for living. If restraint has been approved, the safety plan will detail the equipment or procedure to be used. Safety plans for self-harming behaviors have been extensively reviewed elsewhere (Berk, 2019; Cassidy et al., 2020; Higgins et al., 2016).

SPoC: Goal Setting

Goal settingenvisions a better future. Goals are the outcomes sought through treatment for the management of illness, improved functionality, and enhanced quality of life (Elwyn & Vermunt, 2020). The goals section of the SPoC documents the parent, child, and clinical goals reached thorough the shared decision-making process. Shared goal setting increases parental confidence in carrying out treatment tasks and promotes greater parental engagement in treatment (An et al., 2019), and children who are active participants in goal setting show greater self-efficacy in achieving goals (Costa et al., 2017). Clinical goals should be evidence-based and framed in language that respects the personal goals of the parents and child. Again, respect for the lived experience of our patients and families is critical for engagement and, frankly, a critical learning opportunity for providers. In day-to-day practice, it is our patients and their families who keep us connected to the community and energized to provide and demand the high-quality care our patients need. Setting goals together keeps us and the family engaged and promotes better outcomes. Goal setting theory suggests that successful goals should be specific and challenging but within the child's ability (Locke & Latham, 2019). Prior to the SPoC team meeting, the family should be asked to reflect on possible goals and may be given prompts to promote goals setting. See Table 1.2. Motivational interviewing techniques, visualizing goals on a goal board, and encouraging parents and youth to share stories of achieving goals in group settings are strategies that promote successful goal setting (Elwyn & Vermunt, 2020; Lee, 2015).

While acute health conditions are commonly addressed by the healthcare provider identifying the problem and using clinical decision-making skills to prescribe a solution, chronic health conditions are better treated through a team approach with the child, parents, and healthcare providers developing shared treatment goals. The team approach promotes the identification of goals for optimizing quality of life for the child and family (Mold, 2017; Sia et al., 2004). The World Health Organization defines quality of life (QoL) as *individuals' perceptions of their position in life in the context of the culture and value systems in which they live and in relation to their goals, expectations, standards and concerns* (WHOQOL Group, 1998, p. 1570). WHO health-related QoL (HRQoL) is component of QoL that refers to a child's *overall quality of life that can be clearly shown to affect health—either physical or mental* (CDC, 2018). There are many validated measures for assessing QoL and HRQoL, with children as young as 4 years self-reporting (Wallander & Koot, 2016). A patient's reported QoL may differ significantly from parental report of their child's QoL and from TD peers (Arias et al., 2018; Sentenac et al., 2020). Youth with ASD report decreased QoL compared to TD peers, largely driven by difficulty with uncertainty and socialization (Adams et al., 2019; Arias et al., 2018; Stokes et al., 2017). As with child QoL, understanding family and carer QoL promotes the identification of motivations for achieving goals. The Family QoL Survey: Main Caregivers of People with Intellectual or Developmental Disability and the Beach Center Family QoL Survey are validated family outcome measures that show utility in the assessment of early childhood intervention services outcomes (Bhopti et al., 2016). The importance of carer QoL to child health outcomes is an area of growing interest as carer QoL captures the physical, mental, and social challenges that are part of the lived experience of parenting that directly impact activation (Leontine et al., 2020). It is important to consider parent and child QoL reports as disagreement does occur and has important implications for tailoring goals to meet both parental and child needs and wishes (Lee et al., 2019). QoL for each child, parent, and family will have a unique significance that should be explored and reflected in the SPoC goals. Children's goals are likely to be motivated by a desire for an activities and inclusion, while parents are more often related to maximizing a child's school performance and independence (Costa et al., 2017). With creativity and knowledge of the child's medical home neighborhood map, child and parent goals can be woven into overall goals. For example, a child may have the goal of making friends at school and the parent a goal of improved academic performance. Social and academic

Table 1.2 Questions to facilitate goal setting with parents

What is your vision for your child's future?
What concerns you the most about your child's health?
If one thing could be changed, what would you want that to be?
Who can help you make this change?
What barriers do you see to making this change?
What other concerns do you have?

goals might be met by enrolling in an afterschool homework group or implementing the accommodation of a study buddy at school.

SPoC: Negotiated Actions

Negotiated actions are objectives or strategies for achieving goals. Negotiated actions are documented in the SPoC indicating the accountable person and the time frame for completion of each action. Negotiated actions may be completed by the child, family member, a professional, or another party. The *SMARTER*criteria should be considered when discussing negotiated actions. See Table 1.3. Ogbeiwi (2021) has reviewed the history of goal setting and the SMART criteria in detail. The application of the SMARTER criteria to the SPoC process promotes negotiated actions that have the necessary specificity to allow each team member to envision the process and outcomes. The discussion of negotiated actions can be started by asking *What would a movie of the action and outcome look like?*Measurable actions allow progress monitoring and increase accountability, which are increasingly demanded by third-party payors. Parents, children, or providers may have aspirational hopes for progress that could derail the SPoC if actions are not reframed to be achievable. Relevant, realistic actions promote progress toward goals and prioritize goals that improve patient, carer, and family QoL. Time frames for actions should be guided by family priorities and resource availability. Engagement empowers families to be active participants in decision-making and is associated with better health status, improved satisfaction with care, and improved treatment compliance (Cené et al., 2016). For negotiated actions to be rewarding and engage participation, understanding and supporting parent and child motivations is important. Motivational mastery is the multifaceted drive that children possess to explore and master their environment with mastery serving as its own reward (Gilmore & Cuskelly, 2011; White, 1959). For each child, motivational mastery may differ across developmental domains. Motivational mastery predicts task mastery independent of cognitive competence in children with disability (Miller et al., 2014; Morgan et al., 2017; Salavati et al., 2018). To frame moderately challenging negotiated actions for a child, motivation and competence or ability should be considered to promote success that avoids frustrations that would impede to process. In practice, negotiated actions should be at an individual's learning or instructional level not at a frustrational level if independence is to be achieved. The concept of mental work load and learning levels comes from educational psychology and has been applied to online learning and training astronauts (Loderer et al., 2020; Said et al., 2020).

According to goal setting theory, goals serve to motivate and guide a patient or family to achieve desired outcome (Bandura, 1988; Locke & Latham, 1990). Goal setting theory can inform the SPoC process in two important ways. First, the SPoC process should seek to identify internal motivations of the child and family in order to minimize the need for external rewards or pressures and increase goal attainment. For children with ASD, this requires close attention to verbal

Table 1.3 SMARTER negotiated actions

Specific	The negotiatied action creates a clear picture of the task or expected outcome and assigns responsibility for completion
Measurable	The negotiated action states how progress or success will be measured
Achievable	The negotiated action is realistic given the patient's and family's skill set and the resources available
Relevant	The negotiated action is necessary to achieve patient, family, or clinical goals
TimeBound	The time frame for completion of the negotiated action is clear and realistic
Engaging	The negotiated actions support and sustain child, parent, and clinician engagement in progress toward goals
Rewarding	The negotiated action recognizes internal and external motivations to promote ownership of the task

Adapted from MacLeod (2012)

and nonverbal information processing. For many children with ASD, some aspects of communication for others may be undervalued during thought processes which can lead to misperception of goals or actions and decreased motivation. Nonverbal information is particularly at risk of being undervalued. So for a child requiring medication, dramatic warnings regarding the effect of missing medication may have the desired impact on the parent, but the warnings may scare the child. A child's compliance maybe improved with reminders of what the medication makes possible. In setting goals, the SPoC process should identify motivations and link the steps to goal attainment to those motivating factors; in other words, identify the prize and keep eyes on the prize. Second, goals, or more accurately the steps toward attaining goals, serve as a guide. Steps must be attainable with minimal frustration if they are to be successful. This requires that we initiate a discussion about the tasks that our patients, their families, and medical home team members find frustrating. All team members must strive to work at optimal levels. Minimizing frustrations allows each team member to work smarter, not harder.

SPoC: Coping Plan

Within the SPOC framework, a coping plan may be written to address specific challenging situations. A coping plan may describe setting-specific, individualized patient care strategies to assist a patient in successfully navigating a setting or task or provide strategies for addressing challenges in completing negotiated actions. Patient care strategies might address communication tactics, physical supports, calming techniques, or other child-specific interventions to facilitate medical care, such as cooperating with a medical procedure. Coping plans accommodating a child's needs have been shown to *reduce stress, improve communication, and enhance workflow* during surgical procedures (Gettis et al., 2018, p. 128). To promote coping plan use on a surgical unit, Wittling et al. (2018) describe the development and staff education on an electronic health record-based coping plan for surgical patients with ASD, a strategy that resulted in improved patient satisfaction scores. If a SPoC and EIF are in place, they should inform the coping plan.

The second type of coping plan addresses anticipated challenges completing negotiated actions that call for behavioral change, the so-called intention-behavior gap, a phenomenon whereby people do not act in a manner in which they had wanted or intended (Sheeran & Webb, 2016). Approximately half of intentions are translated into actual actions. Therefore, even when negotiated actions are SMARTER and self-efficacy is high, the best of intentions may not result in the desired behavioral changes. Anticipating this gap when negotiating actions improves outcomes (Kwasnicka et al., 2013). Clinicians should normalize and assist the family to anticipate challenges initiating tasks, sustaining tasks, and completing tasks (Sheeran & Webb, 2016). Parental support of the child and community-based skills coaching increase self-regulation and goal attainment in children (Brown et al., 2020; Keenan et al., 2014). Parental support for goal attainment may be offered through education/skill building groups, individual coaching, and peer mentoring (Bray et al., 2017; Ogourtsova et al., 2019). Technology-based interventions are increasingly being used to support education and goal attainment (Chia et al., 2018).

SPoC: Progress Monitoring

SMARTER goals are measurable goals. The treatment plan or SPoC should indicate how progress toward goals will be measured and the frequency of monitoring. Routine outcome monitoring improves patient outcomes, enhances engagement, and promotes the early identification of challenges and implementation of course corrections if necessary (Maruish, 2019). Monitoring data is increasingly being requested by third-party payors and can be used to support treatment interventions by providers challenging coverage denials by payors. Data from progress monitoring is an important quality of care indicator used by care systems to justify to expenses associated with the medical home model.

When possible, standardized instruments should be used for treatment monitoring. Familiarity with progress monitoring instruments facilitates the development of measurable goals and supports the selection of instruments with a focus on goal achievement rather than selection based on diagnosis (Jacob et al., 2015). This aligns with the aims of personalized medicine. Selecting monitoring instruments based on SPoC goals recognizes that while available measures have limitations, they can provide valuable data if interpreted with knowledge of limitations. The Autism Speaks Autism Treatment Network recommends an idiographic approach to monitoring target symptoms. In this system, frequency, duration, and severity of target symptoms are tracked using a Likert scale with caregivers noting of medication changes and life events for provider consideration (ATN/AIR-P Autism and Medication: Safe and Careful Use Kit, 2012). Medical specialists commonly advise the use of patient diaries to monitor symptoms and outcomes for sleep disorders, epilepsy, gastrointestinal disorders, behavior patterns, and other conditions. Progress monitoring is reviewed in detail in Chap. 5.

Navigating Challenges in Treatment Planning

The challenges of treatment planning are diverse and vary across individuals and systems. From an ecological perspective, challenges may lie with the patient, parent, provider, healthcare system, or healthcare oversight/payment structures. Common across these groups are financial pressures, time constraints, knowledge gaps, information overload, and decision-making fatigue. The fiscal realities of operating a medical home and completing the SPoC or even completing a treatment plan may overwhelm small practices (Lieberthal et al., 2017). Families with copays or paying out of pocket may question the value of treatment planning sessions. A patient-centered medical home in Maryland reported success with an alternate compensation model, where physicians were held responsible for their own expenses, e.g., medical assistant salaries, which then allowed physicians to reallocate a greater proportion of funds to matters of care coordination with the PCMH (O'Malley et al., 2015). While treatment planning should be efficient, fiscal realities remain to be addressed multiple levels.

Time constraints are a reality of life. Families commonly cite time pressures as a stressor (Whiting, 2014). O'Malley et al. (2015) found that providers cited the pressure of completing nonclinical tasks, e.g., patient scheduling, as a major challenge. The authors noted intra-team communication and formal teamwork training to streamline tasks and save time. On-site for convenience, e.g., using practice coaches, is particularly efficient. Authors also highlight safe workplace culture, where staff supports one another, as a significant contributor to enhanced teamwork and improved care for vulnerable populations. Finally, the merit in organizing formal leadership training sessions to improve teamwork was suggested.

Each member of the medical home team brings valuable knowledge to treatment planning sessions. Knowledge in the form of lived experience, technical skills, and medical expertise must be synthesized through the lens of the individual patient in order to craft a treatment plan. In order for patients and families to apply their lived experience and knowledge, clinician must listen while nimbly using our knowledge to tailor assessments and interventions to life stage and individual needs (Aman, 2005). Education to care for persons with IDD is a common deficit across medical professions (Auberry, 2018; Williamson et al., 2017). In a 2020 study, fewer than 35% of family and internal medicine residents received any disability-focused education in medical school and only 11.2% in residency (Stillman et al., 2021). In addition, didactics without practice typically does not translate into provider behavioral change or better patient outcomes (Adirim et al., 2021). Knowledge gaps in professionals are not due to a lack of interest. Stillman et al. (2021) reported that 27% of residents would consider a fellowship focusing on the care of people with IDD if available.

Knowledge is critical, but good communication is necessary if knowledge is to be useful. In addition, accommodating communication needs is an American with Disabilities Act requirement. Professionals commonly report that communication with patients with IDD is a skill deficit (Agaronnik et al., 2019; Novak et al., 2019). Solutions to communication barriers include prioritizing building rapport with patients and parents/caregivers, operationalizing a family-centered approach to treatment planning, and, finally, employing communication aids to facilitate communication (Sharkey et al., 2016). The National Joint Committee on the Communication of Persons with Severe Disabilities has issued guidance for best practices for assessing and accommodating communication needs (Brady et al., 2016). Another excellent resource for professional development is The Vanderbilt Kennedy Center for Excellence in Developmental Disabilities (www.IDDtoolkit.org/).

As clinicians, we must watch for information overload and decision-making fatigue in the families we treat, peers, and ourselves. Attending to information, analyzing information, and making decisions require a significant amount of energy, more commonly called work. Our ability to maintain a high workload is multidetermined with the cognitive and physical demands of a task, emotional and physical well-being of the individual, and motivation contributing to outcomes. There are significant physical and mental costs to maintaining a high workload without adequate time for recuperation. Parents of children with medical complexity and medical professionals are at increased risk of physical and mental issues due to high workload demands (Gallagher & Hannigan, 2014; Patel et al., 2018; Whiting, 2014). Collecting information efficiently through the use of questionnaires, electronic communications, and the skills of paraprofessionals and providing information in multiple formats that can be reviewed at leisure can decrease information overload and decision-making fatigue.

Social determinants of health (SDoH)contribute to overload and are significant drivers of poor health. SDoH include race, ethnicity, and socioeconomic status, each of which influences a person's experience interfacing with healthcare systems (Zuckerman et al., 2015). It has been estimated that SDoH can account for up to 40% of health outcomes (Booske et al., 2010). SDoH is an active area of research in the field of ASD and DD (Iadarola et al., 2019; Johnson, 2020; Zuckerman et al., 2015). Parents of children with ASD have stressed the necessity of services that are culturally and linguistically appropriate, the importance of providing written diagnostic and treatment information to assist in navigating services, and connections to other families with similar lived experience (Stahmer et al., 2019). Johnson examines the intersection of bias, structural racism, and SDoH with healthcare inequities, making a point to recommend that healthcare personnel consciously consider SDoH in treating their patients (Johnson, 2020; MacIntyre, 2020). It is clear that providers will first need to identify, and then mitigate, the effects of their implicit bias when treating patients and interacting with their families. Johnson admonishes providers to advocate for systems-level changes, i.e., policies that help this population attain fair housing, healthy groceries, and the like. In a brief detailing screening for SDoH in populations with complex needs, Thomas-Henkel and Schulman use the *Transforming Complex Care (TCC)*initiative, which ran from 2016 to 2018, to demonstrate effective ways clinicians have implemented the use of SDoH assessment tools and collection of patient-level SDoH data (e.g., transportation, housing, etc.) (Thomas-Henkel & Schulman, 2017).

A sometimes overlooked component of navigating challenges in treatment planning for ASD and DD populations is at the legislative and policy levels, but it is certainly critical for winning the long game (Alderwick & Gottlieb, 2019). While it is encouraging that healthcare policymakers are increasingly using language related to SDoH in healthcare policy, there is still confusion regarding the exact nature of SDoH, and since SDoH is a multisector entity, best practices regarding the role of the healthcare sector in incorporating SDoH in routine, day-to-day patient care remain to be fully defined. An important example of families and clinicians collaborating with lawmakers in such a way to improve care for ASD and DD populations was the creation of Medicaid waivers, which increased access to care by waiving certain requirements for qualified children.

Care Coordination

Care coordination ensures that the SPoC is implemented and updated as necessary to promote patient health and improve the family/patient healthcare experience with the added goal of reducing costs (Roman et al., 2020). Care coordination also seeks to identify and address psychosocial factors and healthcare disparities (Mattson et al., 2019); thus, care coordination is family-centered but serves the needs of providers and payors as well (Schor, 2019). Federal and state agencies have endorsed care coordination as a critical service for children with chronic health conditions (Barbara Wirth et al., 2018). While there is not a universally accepted definition of care coordination, the American Academy of Pediatrics definition is representative, stating:

> Care coordination is a cross-cutting system intervention that is the deliberate organization of patient care activities between ≥2 participants (including the patient) involved in a patient's care to facilitate the appropriate delivery of health care services (American Academy of, 2014, p. e1452).

The SPoC serves as a road map for care coordination, but the driving force of care coordination is relationship building. A care coordinator seeks to address the fragmentation of services commonly faced by children with special needs through connecting providers within and across systems of care. This is in contrast to case managers who are agency specific. The Care Coordination Measurement Tool has been widely used for assessing the efficacy of care coordination activities, capturing resources needed for care coordination (Antonelli & Antonelli, 2004). Roman et al. (2020) reviewed care coordination service use by over 2600 children grouped into 13 diagnostic categories and reported that children with ASD had the highest service utilization in sectors (i.e., education, financial, medical/dental, behavioral health, advocacy, legal services, transition to adult health services, securing interagency coordination, access to social connections, basic needs, and basic information). In addition, 41% of the families required services in three or more sectors indicating that care coordinators must have the ability to move nimble across sectors. The study found that diagnosis predicted the sectors involved, highlighting the value of care coordinator specialization. Specialized care coordinators have been employed to address the needs of children with diverse diagnoses, including children with ASD, maxillofacial disorders, juvenile rheumatoid arthritis, muscular dystrophy, and behavioral health disorders (Normile & VanLandeghem, 2018; Shing et al., 2018).

Like other aspects of the medical home, guidance has been developed for care coordination. The Boston Children's Hospital has a curriculum for care coordination (An Interprofessional Resource to Effectively Engage Patients and Families in Achieving Optimal Child Health Outcomes, 2019). The National Academy for State Health Policy and the Lucile Packard Foundation for Children's Health have published guidance for care coordination which addresses the training necessary for care coordinators who may be licensed or unlicensed providers or a family member (National Care Coordination Standards for Children and Youth with Special Health Care Needs, 2020).

Other Types of Service Plans

In addition to a SPoC and other medical, dental, or mental health plans, children with ASD and other IDD typically receive other supports and services including home- and community-based services, special educational services, and in some cases legal services which are documented in a written plan. In some cases, information from the medical home team may be sought to inform the plan. It behooves the medical home team to review other plans in order to understand the range of services received and the child's interaction with the community.

Home- and Community-Based Services

Youth may receive services from community providers with a written plan specifying the details of the service. Supports and services may be accessed through private insurance or government programs like Medicaid. While the federal government provides funding to the states for supports and services, the administration of programs for persons with developmental disabilities is subject to state interpretation of federal guidelines. Services typically include home- and community-based services waivers. While waivers vary by state, they typically cover case management, in-home support staff, and respite services. Once qualified for a state program, a medical order or certification of need from the medical provider may be required to access a specific service. The community agency providing the service will conduct a needs assessment then formulate a plan of care with the patient and family. An in-home support services individualized plan of care typically includes services to be rendered, the discipline rendering the service, and frequency and duration of the service. States issue guidelines for plan content and the periodic review of the plan of care.

School-Based Services

For most of the year, children spend the majority of their waking hours at school where it is expected that children will benefit from a free and appropriate public education and related services to meet their unique needs. The Individuals with Educational Disabilities Act (IDEA) defines a number of plans used in educational settings including Individualized Family Service Plan (IFSP), Individualized Educational Plan (IEP), Individualized Transition Plan (ITP), 504 Accommodation Plan (IAP), and Individualized Health Plan (IHP). An IFSP defines early intervention services addressing a disability in physical, cognitive, communication, social, or emotional development in children under 3 years of age. For children 3 and over, an IEP details how a free and appropriate public education is provided to those qualifying for special education services. When children with an IEP turn 16, a plan for transitional services (ITP) is added to the IEP outlining the actions to support adult living and community involvement. Children who do not qualify for services under IDEA may receive services under Section 504 of the Rehabilitation Act with accommodations documented in a 504 Plan or IAP. Children who may require medical care or medication during the school day will have an IHP detailing the care to be provided routinely or in an emergency. In 2019–2020, about 14% of all public school students received individualized services under an individualized education plan (IEP) and an additional 2.3% received accommodations under a Section 504 or individualized accommodation plan (IAP) (Lee, 2018; Students with Disabilities, 2021). During the 2019–2020 school year, 11% of students received special education services for autism, 7% for developmental delay, and 6% for intellectual disability (Students with Disabilities, 2021). For additional information regarding the accommodation of special needs in school settings, see Chap. 12.

Court-Based Services

Millions of children face proceedings in juvenile or family courts each year with documentation in court records, probation plans, foster care plans, and custody or guardianship court orders, among others. Rutten et al. (2017) reported an ASD prevalence of 2.3–15% among youth facing delinquency charges. Nearly 10% of justice involved youth qualify for an intellectual disability diagnosis (Thompson & Morris, 2016). Court involved youth will have probation plans that may include court-ordered treatment or supervision requirements that should be informed by the medical home team. Family court oversees children in need of care. Approximately 32% of youth in foster care have a disability and about 10% qualify as medically fragile (Seltzer et al., 2019; Slayter, 2016). Federal law requires that children in foster care have written case plans that include reunification plans and transition plans. Case plans are required to document medical needs and required treatments and can provide important information to the medical home team about custody, living arrangements, and who may make medical decisions for the child in care. In addition to juvenile or family court involvement, a considerable number of children live with one parent, a grandparent, extended family, or a nonrelative requiring legal documentation of guardianship or custody status. Custody or guardianship papers should be part of the treatment record and SPoC.

Summary

Parents of children with complex medical needs become adept at navigating numerous systems of care. The SPoC and the treatment plans for specific services map the journey and facilitate communication within and across systems. Just as individual treatment plans are living documents that are updated according to patient needs, the process of treatment planning evolves. The process evolves in response to legislation, payor demands, medical demands, and the wisdom gained from the lived experience of patients and families. Ultimately, treatment planning for children with complex needs must be addressed on many levels simultaneously. Patients, families, and treatment providers must consider individual needs during treatment planning while informing system and community needs including federal, state, and local policy development (Maree et al., 2020). Medical informatics holds the promise of integrating the diverse and complex needs of all users into a robust, dynamic treatment plan. As information mounts and systems change, we must not lose sight of the fact that treatment planning is about relationships and good communication in the service of a patient's health and well-being.

References

Adams, D., Clark, M., & Keen, D. (2019). Using self-report to explore the relationship between anxiety and quality of life in children on the autism spectrum. *Autism Research, 12*(10), 1505–1515.

Adirim, Z., Sockalingam, S., & Thakur, A. (2021). Postgraduate medical training in intellectual and developmental disabilities: A systematic review. *Academic Psychiatry, 45*(3), 371–381.

Agaronnik, N., Campbell, E. G., Ressalam, J., & Iezzoni, L. I. (2019). Communicating with patients with disability: Perspectives of practicing physicians. *Journal of General Internal Medicine, 34*(7), 1139–1145.

Alderwick, H., & Gottlieb, L. M. (2019). Meanings and misunderstandings: A social determinants of health lexicon for health care systems. *The Milbank Quarterly, 97*(2), 407–419. https://doi.org/10.1111/1468-0009.12390

Aman, M. G. (2005). Treatment planning for patients with autism spectrum disorders. *Journal of Clinical Psychiatry, 66*, 38.

American Academy of, P. (2014). Council on Children with Disabilities, Medical Home Implementation Project Advisory Committee. Patient-and family-centered care coordination: A framework for integrating care for children and youth across multiple systems. *Pediatrics, 133*(5), e1451–e1460.

American Academy of, P, Committee on Pediatric Emergency, M, Council on Clinical Information, T, American College of Emergency, P, & Pediatric Emergency Medicine, C. (2010). Emergency information forms and emergency preparedness for children with special health care needs. *Pediatrics, 125*(4), 829–837.

American College of Emergency, P. (2010). Emergency information form for children with special health care needs. *Annals of Emergency Medicine, 56*(3), 315.

An interprofessional resource to effectively engage patients and families in achieving optimal child health outcomes. (2019). Retrieved from https://medicalhomeinfo.aap.org/tools-resources/Documents/PCCC%202nd%20Edition/Full%20Pediatric%20Care%20Coordination%20Curriculum.pdf

An, M., Palisano, R. J., Yi, C.-H., Chiarello, L. A., Dunst, C. J., & Gracely, E. J. (2019). Effects of a collaborative intervention process on parent empowerment and child performance: A randomized controlled trial. *Physical & Occupational Therapy in Pediatrics, 39*(1), 1–15.

Antonelli, R. C., & Antonelli, D. M. (2004). Providing a medical home: The cost of care coordination services in a community-based, general pediatric practice. *Pediatrics, 113*(Supplement 4), 1522.

Arias, V. B., Gómez, L. E., Morán, M. L., Alcedo, M., Monsalve, A., & Fontanil, Y. (2018). Does quality of life differ for children with autism spectrum disorder and intellectual disability compared to peers without autism? *Journal of Autism and Developmental Disorders, 48*(1), 123–136. https://doi.org/10.1007/s10803-017-3289-8

ATN/AIR-P Autism and medication: Safe and careful use kit. (2012). In A. S. ATN (Ed.), (p. 33).

Auberry, K. (2018). Intellectual and developmental disability nursing: Current challenges in the USA. *Nursing: Research and Reviews, 8*, 23–28.

Bandura, A. (1988). Self-regulation of motivation and action through goal systems. In *Cognitive perspectives on emotion and motivation* (pp. 37–61). Springer.

Belden, J. L., Koopman, R. J., Patil, S. J., Lowrance, N. J., Petroski, G. F., & Smith, J. B. (2017). Dynamic electronic health record note prototype: Seeing more by showing less. *The Journal of the American Board of Family Medicine, 30*(6), 691. https://doi.org/10.3122/jabfm.2017.06.170028

Berk, M. (2019). *Evidence-based treatment approaches for suicidal adolescents: Translating science into practice*. American Psychiatric Pub.

Berkey, F. J., Wiedemer, J. P., & Vithalani, N. D. (2018). Delivering bad or life-altering news. *American Family Physician, 98*(2), 99–104.

Bhopti, A., Brown, T., & Lentin, P. (2016). Family quality of life: A key outcome in early childhood intervention services—A scoping review. *Journal of Early Intervention, 38*(4), 191–211.

Blendon, R. J., & Benson, J. M. (2001). Americans' views on health policy: A fifty-year historical perspective. *Health Affairs, 20*(2), 33–46.

Bolton, D., & Gillett, G. (2019). *The biopsychosocial model of health and disease: New philosophical and scientific developments*. Springer Nature.

Booske, B. C., Athens, J. K., Kindig, D. A., Park, H., & Remington, P. L. (2010). *Different perspectives for assigning weights to determinants of health*. University of Wisconsin: Population Health Institute.

Bovend'Eerdt, T. J. H., Botell, R. E., & Wade, D. T. (2009). Writing SMART rehabilitation goals and achieving goal attainment scaling: A practical guide. *Clinical Rehabilitation, 23*(4), 352–361.

Brady, N. C., Bruce, S., Goldman, A., Erickson, K., Mineo, B., Ogletree, B. T., ... Siegel, E. (2016). Communication services and supports for individuals with severe disabilities: Guidance for assessment and intervention. *American Journal on Intellectual and Developmental Disabilities, 121*(2), 121–138.

Bray, L., Carter, B., Sanders, C., Blake, L., & Keegan, K. (2017). Parent-to-parent peer support for parents of children with a disability: A mixed method study. *Patient Education and Counseling, 100*(8), 1537–1543.

Brown, D. M., Arbour-Nicitopoulos, K. P., Martin Ginis, K. A., Latimer-Cheung, A. E., & Bassett-Gunter, R. L. (2020). Examining the relationship between parent physical activity support behaviour and physical activity among children and youth with autism spectrum disorder. *Autism, 24*(7), 1783–1794.

Cairns, A. J., Kavanagh, D. J., Dark, F., & McPhail, S. M. (2019). Goal setting improves retention in youth mental health: A cross-sectional analysis. *Child and Adolescent Psychiatry and Mental Health, 13*(1), 31.

Cassidy, S. A., Robertson, A., Townsend, E., O'Connor, R. C., & Rodgers, J. (2020). *Advancing our understanding of self-harm, suicidal thoughts and behaviours in autism*. Springer.

CDC. (2018). HRQOL Concepts. Retrieved from https://www.cdc.gov/hrqol/concept.htm

Cené, C. W., Johnson, B. H., Wells, N., Baker, B., Davis, R., & Turchi, R. (2016). A narrative review of patient and family engagement: The "foundation" of the medical home. *Medical Care, 54*(7), 697.

Chia, G. L. C., Anderson, A., & McLean, L. A. (2018). Use of technology to support self-management in individuals with autism: Systematic review. *Review Journal of Autism and Developmental Disorders, 5*(2), 142–155.

Child and Adolescent Health Measurement, I. (2017). *2016–2017 National Survey of Children's Health (NSCH) data query*. Data Resource Center for ChildandAdolescent Health [Internet].

Coller, R. J., Berry, J. G., Kuo, D. Z., Kuhlthau, K., Chung, P. J., Perrin, J. M., ... Thompson, L. R. (2020). Health system research priorities for children and youth with special health care needs. *Pediatrics, 145*(3), 1–12.

Costa, U. M., Brauchle, G., & Kennedy-Behr, A. (2017). Collaborative goal setting with and for children as part of therapeutic intervention. *Disability and Rehabilitation, 39*(16), 1589–1600. https://doi.org/10.1080/09638288.2016.1202334

Croen, L. A., Shankute, N., Davignon, M., Massolo, M. L., & Yoshida, C. (2017). Demographic and clinical characteristics associated with engagement in behavioral health treatment among children with autism spectrum disorders. *Journal of Autism and Developmental Disorders, 47*(11), 3347–3357.

Croghan, T. W., & Brown, J. D. (2010). *Integrating mental health treatment into the patient centered medical home*. Agency for Healthcare Research and Quality.

Doran, G. T. (1981). There'sa SMART way to write management's goals and objectives. *Management Review, 70*(11), 35–36.

Doukrou, M., & Segal, T. Y. (2018). Fifteen-minute consultation: Communicating with young people—How to use HEEADSSS, a psychosocial interview for adolescents. *Archives of Disease in Childhood-Education and Practice, 103*(1), 15–19.

Duffy, T. P. (2011). The Flexner Report--100 years later. *The Yale Journal of Biology and Medicine, 84*(3), 269–276.

Elwyn, G., & Vermunt, N. P. C. A. (2020). Goal-based shared decision-making: developing an integrated model. *Journal of patient experience, 7*(5), 688-696.

Engel, G. L. (1977). The need for a new medical model: A challenge for biomedicine. *Science, 196*(4286), 129–136.

Farre, A., & Rapley, T. (2017). The new old (and old new) medical model: Four decades navigating the biomedical and psychosocial understandings of health and illness. *Healthcare (Basel, Switzerland), 5*(4), 88. https://doi.org/10.3390/healthcare5040088

Fletcher, R., Barnhill, J., & McCarthy, S.-A. (2016). *Diagnostic manual-intellectual disability (DM-ID): A clinical guide for diagnosis of mental disorders in persons with intellectual disability* (2nd ed.). NADD.

Flexner, A. (2002). Medical education in the United States and Canada. *Bulletin of the World Health Organization, 80*, 594–602.

Gallagher, S., & Hannigan, A. (2014). Depression and chronic health conditions in parents of children with and without developmental disabilities: The growing up in Ireland cohort study. *Research in Developmental Disabilities, 35*(2), 448–454.

Gentles, S. J., Nicholas, D. B., Jack, S. M., McKibbon, K. A., & Szatmari, P. (2019). Parent engagement in autism-related care: A qualitative grounded theory study. *Health Psychology and Behavioral Medicine, 7*(1), 1–18.

Gettis, M. A., Wittling, K., Palumbo-Dufur, J., McClain, A., & Riley, L. (2018). Identifying best practice for healthcare providers caring for autistic children perioperatively. *Worldviews on Evidence-Based Nursing, 15*(2), 127–129.

Gilmore, L., & Cuskelly, M. (2011). Observational assessment and maternal reports of motivation in children and adolescents with Down syndrome. *American Journal on Intellectual and Developmental Disabilities, 116*(2), 153–164.

Goldenring, J. M., & Cohen, E. (1988). Getting into adolescent heads. *Contemporary Pediatrics, 5*(7), 75–90.

Grella, C. E., & Grusky, O. (1989). Families of the seriously mentally ill and their satisfaction with services. *Psychiatric Services, 40*(8), 831–835.

Hibbard, J. H., & Greene, J. (2013). What the evidence shows about patient activation: Better health outcomes and care experiences; fewer data on costs. *Health Affairs, 32*(2), 207–214.

Hibbard, J. H., Stockard, J., Mahoney, E. R., & Tusler, M. (2004). Development of the patient activation measure (PAM): Conceptualizing and measuring activation in patients and consumers. *Health Services Research, 39*(4 Pt 1), 1005–1026. https://doi.org/10.1111/j.1475-6773.2004.00269.x

Higgins, A., Doyle, L., Morrissey, J., Downes, C., Gill, A., & Bailey, S. (2016). Documentary analysis of risk-assessment and safety-planning policies and tools in a mental health context. *International Journal of Mental Health Nursing, 25*(4), 385–395.

Iadarola, S., Pérez-Ramos, J., Smith, T., & Dozier, A. (2019). Understanding stress in parents of children with autism spectrum disorder: A focus on under-represented families. *International Journal of Developmental Disabilities, 65*(1), 20–30. https://doi.org/10.1080/20473869.2017.1347228

Jacob, J., Edbrooke-Childs, J., Law, D., & Wolpert, M. (2015). Measuring what matters to patients: Using goal content to inform measure choice and development. *Clinical Child Psychology and Psychiatry, 22*(2), 170–186. https://doi.org/10.1177/1359104515615642

Johnson, T. J. (2020). Intersection of bias, structural racism, and social determinants with health care inequities. *Pediatrics, 146*(2). https://doi.org/10.1542/peds.2020-003657

Karunamuni, N., Imayama, I., & Goonetilleke, D. (2020). Pathways to well-being: Untangling the causal relationships among biopsychosocial variables. *Social Science and Medicine, 272*, 112846.

Keenan, S., King, G., Curran, C. J., & McPherson, A. (2014). Effectiveness of experiential life skills coaching for youth with a disability. *Physical and Occupational Therapy in Pediatrics, 34*(2), 119–131.

Kirley, K., Hayer, R., Khan, T., Johnson, E., Sanchez, E. S., Kosowicz, L., … Williams, B. C. (2020). Expanding the traditional history and physical examination to address chronic diseases and social needs: A multisite randomized control trial of 4 medical schools. *Academic Medicine, 95*(11S).

Kuo, D. Z., & Houtrow, A. J. (2016). Recognition and management of medical complexity. *Pediatrics, 138*(6), e20163021.

Kwasnicka, D., Presseau, J., White, M., & Sniehotta, F. F. (2013). Does planning how to cope with anticipated barriers facilitate health-related behaviour change? A systematic review. *Health Psychology Review, 7*(2), 129–145.

Lee, J. (2015). Helping your patients set achievable goals. *AADE in Practice, 3*(5), 42–45. https://doi.org/10.1177/2325160315598289

Lee, A. (2018). Thousands of sublic schools aren't using 504 plans, new analysis suggests.

Lee, Y.-C., Yang, H.-J., Lee, W.-T., & Teng, M.-J. (2019). Do parents and children agree on rating a child's HRQOL? A systematic review and meta-analysis of comparisons between children with attention deficit hyperactivity disorder and children with typical development using the PedsQLTM. *Disability and Rehabilitation, 41*(3), 265–275.

Leontine, W., de Nijs, P. F. A., Duvekot, J., Greaves-Lord, K., Hillegers, M. H. J., Brouwer, W. B. F., & Hakkaart-van Roijen, L. (2020). Children with an autism spectrum disorder and their caregivers: Capturing health-related and care-related quality of life. *Journal of Autism and Developmental Disorders, 50*(1), 263–277.

Lieberthal, R. D., Payton, C., Sarfaty, M., & Valko, G. (2017). Measuring the cost of the patient-centered medical home: A cost-accounting approach. *The Journal of Ambulatory Care Management, 40*(4), 327–338. https://doi.org/10.1097/JAC.0000000000000196

Locke, E. A., & Latham, G. P. (1990). *A theory of goal setting & task performance*. Prentice-Hall, Inc..

Locke, E. A., & Latham, G. P. (2019). The development of goal setting theory: A half century retrospective. *Motivation Science, 5*(2), 93.

Loderer, K., Pekrun, R., & Lester, J. C. (2020). Beyond cold technology: A systematic review and meta-analysis on emotions in technology-based learning environments. *Learning and Instruction, 70*, 101162.

Macintyre, A. K., Torrens, C., Campbell, P., Maxwell, M., Pollock, A., Biggs, H., ... McLean, J. (2020). Socioeconomic inequalities and the equity impact of population-level interventions for adolescent health: An overview of systematic reviews. *Public Health, 180*, 154–162. https://doi.org/10.1016/j.puhe.2019.11.008

MacLeod, L. (2012). Making SMART goals smarter. *Physician Executive, 38*(2), 68–72.

Marcum, J. A. (2019). Professing clinical medicine in an evolving health care network. *Theoretical Medicine and Bioethics, 40*(3), 197–215. https://doi.org/10.1007/s11017-019-09492-x

Maree, P., Hughes, R., Radford, J., Stankovich, J., & Van Dam, P. J. (2020). Integrating patient complexity into health policy: A conceptual framework. *Australian Health Review, 45*(2), 199–206.

Maruish, M. E. (2019). *Essentials of treatment planning*. John Wiley & Sons.

Mattson, G., Kuo, D. Z., & Committee on Psychosocial Aspects of Child and Family, H. (2019). Psychosocial factors in children and youth with special health care needs and their families. *Pediatrics, 143*(1), 1–14.

McAllister, J. (2014a). *Achieving a shared plan of care for children and youth with special health care needs: An implementation guide*. Lucille Packard Foundation for Children's Healthcare.

McAllister, J. W. (2014b). *Achieving a shared plan of care with children and youth with special health care needs*. Lucille Packard Foundation for Children's Healthcare.

McPherson, M., Arango, P., Fox, H., Lauver, C., McManus, M., Newacheck, P. W., ... Strickland, B. (1998). A new definition of children with special health care needs. *Pediatrics, 102*(1 Pt 1), 137–140. https://doi.org/10.1542/peds.102.1.137

Medical Home Initiatives for Children With Special Needs Project Advisory Committee. (2002). The medical home. *Pediatrics, 110*(1), 184. https://doi.org/10.1542/peds.110.1.184

Miller, L., Ziviani, J., & Boyd, R. N. (2014). A systematic review of clinimetric properties of measurements of motivation for children aged 5–16 years with a physical disability or motor delay. *Physical and Occupational Therapy in Pediatrics, 34*(1), 90–111.

Mold, J. (2017). Goal-directed health care: Redefining health and health care in the era of value-based care. *Cureus, 9*(2), e1043. https://doi.org/10.7759/cureus.1043

Morell, B. B. (1979). Deinstitutionalization: Those left behind. *Social Work, 24*(6), 528–532. https://doi.org/10.1093/sw/24.6.528

Morgan, G. A., Józsa, K., & Liao, H.-F. (2017). Introduction to the HERJ special issue on mastery motivation: Measures and results across cultures and ages. *Hungarian Educational Research Journal (HERJ), 7*(2), 5–14.

National care coordination standards for children and youth with special health care needs. (2020). Retrieved from https://www.nashp.org/national-care-coordination-standards-for-children-and-youth-with-special-health-care-needs/

NICHQ. (2015). NICHQ's care plan template.

Normile, B., & VanLandeghem, K. (2018, June). *State strategies to advance medical homes for children and youth with special health care needs*. National Academy for State Health Policy.

Novak, I., Morgan, C., McNamara, L., & Te Velde, A. (2019). Best practice guidelines for communicating to parents the diagnosis of disability. *Early Human Development, 139*, 104841.

O'Donnell, H. C., Suresh, S., & Council on Clinical Information, T. (2020). Electronic documentation in pediatrics: The rationale and functionality requirements. *Pediatrics, 146*(1), 00.

O'Malley, A. S., Gourevitch, R., Draper, K., Bond, A., & Tirodkar, M. A. (2015). Overcoming challenges to teamwork in patient-centered medical homes: A qualitative study. *Journal of General Internal Medicine, 30*(2), 183–192.

Ogbeiwi, O. (2021). General concepts of goals and goal-setting in healthcare: A narrative review. *Journal of Management & Organization, 27*(2), 324–341.

Ogourtsova, T., O'Donnell, M., De Souza Silva, W., & Majnemer, A. (2019). Health coaching for parents of children with developmental disabilities: A systematic review. *Developmental Medicine and Child Neurology, 61*(11), 1259–1265.

Patel, R. S., Bachu, R., Adikey, A., Malik, M., & Shah, M. (2018). Factors related to physician burnout and its consequences: A review. *Behavioral Sciences, 8*(11), 98.

Rice, C. E., Zablotsky, B., Avila, R. M., Colpe, L. J., Schieve, L. A., Pringle, B., & Blumberg, S. J. (2016). Reported wandering behavior among children with autism spectrum disorder and/or intellectual disability. *The Journal of Pediatrics, 174*, 232–239.

Roman, S. B., Dworkin, P. H., Dickinson, P., & Rogers, S. C. (2020). Analysis of care coordination needs for families of children with special health care needs. *Journal of Developmental and Behavioral Pediatrics, 41*(1), 58–64.

Ruble, L., Murray, D., McGrew, J. H., Brevoort, K., & Wong, V. W. (2018). A preliminary study of activation, stress, and self-management of parents of children with autism spectrum disorder. *Journal of Child and Family Studies, 27*(3), 825–834.

Rutten, A. X., Vermeiren, R., & Van Nieuwenhuizen, C. (2017). Autism in adult and juvenile delinquents: A literature review. *Child and Adolescent Psychiatry and Mental Health, 11*(1), 1–12.

Said, S., Gozdzik, M., Roche, T. R., Braun, J., Rössler, J., Kaserer, A., … Tscholl, D. W. (2020). Validation of the raw national aeronautics and space administration Task Load Index (NASA-TLX) questionnaire to assess perceived workload in patient monitoring tasks: Pooled analysis study using mixed models. *Journal of Medical Internet Research, 22*(9), e19472.

Salavati, M., Vameghi, R., Hosseini, S. A., Saeedi, A., & Gharib, M. (2018). Comparing levels of mastery motivation in children with cerebral palsy (CP) and typically developing children. *Medical Archives, 72*(1), 41.

Schor, E. L. (2019). Ten essential characteristics of care coordination. *JAMA Pediatrics, 173*(1), 5–5. https://doi.org/10.1001/jamapediatrics.2018.3107

Seltzer, R. R., Raisanen, J. C., Williams, E. P., Da Silva, T., Donohue, P. K., & Boss, R. D. (2019). Exploring medical foster care as a placement option for children with medical complexity. *Hospital Pediatrics, 9*(9), 697–706.

Sentenac, M., Rapp, M., Ehlinger, V., Colver, A., Thyen, U., & Arnaud, C. (2020). Disparity of child/parent-reported quality of life in cerebral palsy persists into adolescence. *Developmental Medicine and Child Neurology, 63*(1), 68–74.

Sharkey, S., Lloyd, C., Tomlinson, R., Thomas, E., Martin, A., Logan, S., & Morris, C. (2016). Communicating with disabled children when inpatients: Barriers and facilitators identified by parents and professionals in a qualitative study. *Health Expectations, 19*(3), 738–750.

Sheeran, P., & Webb, T. L. (2016). The intention–Behavior gap. *Social and Personality Psychology Compass, 10*(9), 503–518.

Shing, J. Z., Liu, T., & Valdez, R. (2018). Changes in care coordination and health insurance in the population of US children with muscular dystrophy, 2005–2006 and 2009–2010. *International Journal of Care Coordination, 21*(4), 140–152.

Sia, C., Tonniges, T. F., Osterhus, E., & Taba, S. (2004). History of the medical home concept. *Pediatrics, 113*(Supplement 4), 1473–1478.

Slayter, E. (2016). Youth with disabilities in the United States child welfare system. *Children and Youth Services Review, 64*, 155–165.

Smith, R. C. (2020). Making the biopsychosocial model more scientific—Its general and specific models. *Social Science and Medicine, 272*, 113568.

Stahmer, A. C., Vejnoska, S., Iadarola, S., Straiton, D., Segovia, F. R., Luelmo, P., … Bronstein, B. (2019). Caregiver voices: Cross-cultural input on improving access to autism services. *Journal of Racial and Ethnic Health Disparities, 6*(4), 752–773.

Stillman, M. D., Ankam, N., Mallow, M., Capron, M., & Williams, S. (2021). A survey of internal and family medicine residents: Assessment of disability-specific education and knowledge. *Disability and Health Journal, 14*(2), 101011.

Stokes, M. A., Kornienko, L., Scheeren, A. M., Koot, H. M., & Begeer, S. (2017). A comparison of children and adolescent's self-report and parental report of the PedsQL among those with and without autism spectrum disorder. *Quality of Life Research, 26*(3), 611–624. https://doi.org/10.1007/s11136-016-1490-4

Stroul, B. A., & Friedman, R. M. (1986). *A system of care for severely emotionally disturbed children & youth*. CASSP Technical Assistance Center, Georgetown University Child Development Center.

Students with Disabilities. (2021). Retrieved from https://nces.ed.gov/programs/coe/indicator/cgg

Thomas-Henkel, C., & Schulman, M. (2017). *Screening for social determinants of health in populations with complex needs: Implementation considerations*. Center for Health Care Strategies.

Thompson, K. C., & Morris, R. J. (2016). Juvenile delinquency and disability. In *Juvenile delinquency and disability* (pp. 31–39). Springer.

VanDenBerg, J., Bruns, E., & Burchard, J. (2003). History of the wraparound process. *Focal Point, 17*(2), 4–7.

Vollmer, T. R., Deshais, M., Sloman, K. N., & Borrero, C. S. W. (2018). Behavioral assessment of self-injury. In *Handbook of childhood psychopathology and developmental disabilities assessment* (pp. 263–288). Springer.

Wallander, J. L., & Koot, H. M. (2016). Quality of life in children: A critical examination of concepts, approaches, issues, and future directions. *Clinical Psychological Review, 45*, 131–143. https://doi.org/10.1016/j.cpr.2015.11.007

Weed, L. L. (1968). Medical records that guide and teach (concluded). *Yearbook of Medical Informatics, 212*, 1.

Weiss, G. B. (1985). Paternalism modernised. *Journal of Medical Ethics, 11*(4), 184–187.

White, R. W. (1959). Motivation reconsidered: The concept of competence. *Psychological Review, 66*(5), 297.

Whiting, M. (2014). Children with disability and complex health needs: The impact on family life. *Nursing Children and Young People, 26*(3), 26–30.

WHOQoL Group, T. W. (1998). The World Health Organization quality of life assessment (WHOQOL): Development and general psychometric properties. *Social Science and Medicine, 46*(12), 1569–1585.

Will, J. F. (2011). A brief historical and theoretical perspective on patient autonomy and medical decision making: Part II: The autonomy model. *Chest, 139*(6), 1491–1497. https://doi.org/10.1378/chest.11-0516

Williamson, H. J., Contreras, G. M., Rodriguez, E. S., Smith, J. M., & Perkins, E. A. (2017). Health care access for adults with intellectual and developmental disabilities: A scoping review. *OTJR: Occupation, Participation and Health, 37*(4), 227–236.

Wirth, B., & Kuznetsov, A. (2016). *Shared plan of care: A tool to support children and youth with special health care needs and their families*. National Center for Medical Home Implementation.

Wirth, B., VanLandeghem, K., Normile, B., Kim, E., & Handler, M. (2018). State strategies for shared plans of care to improve care coordination for children and youth with special health care needs.

Wittling, K., Dufur, J. P., McClain, A., & Gettis, M. (2018). Behavioral coping plans: One inter-professional team's approach to patient-centered care. *Journal of Pediatric Nursing, 41*, 135–139.

Zablotsky, B., Black, L. I., Maenner, M. J., Schieve, L. A., Danielson, M. L., Bitsko, R. H., … Boyle, C. A. (2019). Prevalence and trends of developmental disabilities among children in the United States: 2009–2017. *Pediatrics, 144*(4), e20190811.

Zuckerman, K. E., Lindly, O. J., Sinche, B. K., & Nicolaidis, C. (2015). Parent health beliefs, social determinants of health, and child health services utilization among US School-age children with autism. *Journal of Developmental and Behavioral Pediatrics, 36*(3), 146–157.

Medical Home for Children with Autism Spectrum Disorder and Other Developmental Disabilities

2

Colleen Kraft

Abstract

Pediatrics embraces the ongoing connection, a medical home, starting with the birth of the child, the early visits with new parents, and adventures from preschool through adolescence. The concept of the Medical Home, a structure and process of care that prioritizes this special relationship, is even more important when a child has a developmental disability. The Medical Home Model promotes care delivered by a team (pediatrician, advance practice professionals, nurses, medical assistants, and administrative staff) that can coordinate preventive care, immunizations, urgent sick visits, developmental surveillance and screening, and management of chronic conditions in conjunction with a specialist if needed. This chapter guides professionals through the setup of a medical home practice.

Keywords

Medical home · Autism spectrum disorder · Developmental disability · Care coordination · Toolbox

C. Kraft (✉)
Children's Hospital Los Angeles, Los Angeles, CA, USA
e-mail: ckraft@chla.usc.edu

The Medical Home Concept

At the heart of primary care pediatrics is the relationship between the child, the family, and the pediatric clinician. Pediatrics embraces the ongoing connection, a medical home, starting with the birth of the child, the early visits with new parents, and adventures from preschool through adolescence. The concept of the Medical Home, a structure and process of care that prioritizes this special relationship, is even more important when a child has a developmental disability. The American Academy of Pediatrics (AAP)defines the Medical Home as care that is "accessible, continuous, coordinated, comprehensive, family-centered, compassionate, and culturally effective" (American Academy of Pediatrics, 2002, p. 184). The Medical Home Model promotes care delivered by a team (pediatrician, advance practice professionals, nurses, medical assistants, and administrative staff) that can coordinate preventive care, immunizations, urgent sick visits, developmental surveillance and screening, and management of chronic conditions in conjunction with a specialist if needed. The public's high level of trust in healthcare professionals has been repeatedly affirmed by Gallup and the Wellcome Global Monitor (Belseck, 2019). Families and children trust pediatricians and their Medical Home team; it's the first place they call when they need advice about their child. The pediatric

P. McPherson (ed.), *Handbook of Treatment Planning for Children with Autism and Other Neurodevelopmental Disorders*, Autism and Child Psychopathology Series,
https://doi.org/10.1007/978-3-031-06120-2_2

team and the family provide continuity from hospital to home, with transitions from childcare to preschool to school, and between specialists and therapy. The continuity and trusted relationship make the Medical Home an ideal locus of care for children with developmental disabilities (Bright Futures Steering, 2006).

Why the Medical Home Should Be the Locus of Care for Kids with Autism Spectrum Disorders and Other Developmental/Behavioral Disabilities

Growing Numbers of Children Require Specialized Medical Care

Many factors support the need for the pediatric Medical Home to be the convener of care for children with developmental and behavioral disabilities. There is recognition that the number of children with these disorders is increasing, likely due to better awareness on the part of pediatricians as well as the public. The prevalence of autism spectrum disorder (ASD) in 2007 estimated by the Centers for Disease Control and Prevention (CDC) was 1:150 for children aged 8 years, based on 2000-study-year-data. In 2018, the prevalence increased to 1:59 for children aged 8 years, based on 2014-study-year-data. The prevalence of intellectual disability (ID) ranges from 1.58% to 3% and developmental disability (DD) from 6.99% to 16.24%, with variation due to the limitation of health surveillance data collection, among other factors (Krahn, 2019). In addition, children living with disabilities commonly experience multi-morbidity, defined as two or more comorbid medical conditions, as well as medical complexity. Children with medical complexity are "the most medically fragile and have the most intensive health care needs" (Cohen et al., 2011, p. 259). Prevalence estimates for children living with medical complexity range from 1.5% to 13.2% (Justin et al., 2021). With the prevalence of children requiring specialized pediatric care increasing, the Medical Home offers the most comprehensive care model and has been endorsed widely by professional medical organizations since endorsed by the American Academy of Pediatrics in 1992 (Dadashzadeh, 2019).

Pediatricians Are Uniquely Situated to Identify Care Needs

Developmental surveillance and screening are embedded in well-child care (Lipkin & Macias, 2020). Preventive visits include assessing appropriate milestones and answering family questions about their child's speech, gross and fine motor skills, and behavior. In addition to assessing the child's development, the pediatrician offers parents anticipatory guidance, featuring advice to assist parents to encourage the next skills that their child should acquire. These conversations allow the family and physician to form a partnership designed to identify each child's unique developmental needs.

Newborn Screening

Newborn screening is a public health innovation that dates back to the 1960s with the goal of identifying infants at risk for treatable genetic disorders (Newborn Screening Authoring, 2008). After the first 24 hours of life, a few drops of blood are taken from a baby's heel and placed on a card that is sent to the state's newborn screening lab. Infants are screened for genetic conditions which are treatable but not clinically evident in the newborn period. Results of the screening test are sent to the infant's medical home as well as the hospital where they were born.

Newborn screening started with the discovery that the amino acid disorder phenylketonuria (PKU) could be treated by changes in the child's diet and that removal of phenylalanine from the diet was required for that child's optimal development (Berry, 2015). Infants with PKU appear normal at birth but are unable to metabolize the essential amino acid phenylalanine, resulting in irreversible intellectual disability. Since that time, a number of additional tests have been added to the newborn screening panel. Infants with conditions such as congenital hypothyroidism, galactosemia, or congenital adrenal hyperplasia and up to 80 other inborn errors of metabolism can be identified. These children can then undergo specific testing and treatment for these disorders (kelly et al., 2016).

Screening programs are run by state or national public health programs with the goal of screening all infants born in the jurisdiction for a defined panel of treatable disorders. The number of diseases screened for is set by each district and can vary greatly, creating disparity between and sometimes within countries. In addition to blood samples, bedside hearing screening using automated auditory brainstem response and congenital heart defects using pulse oximetry are included in some state newborn screening programs. Infants who screen positive undergo further testing to determine if they are truly affected with a disease or if the test result was a false positive. Most state health departments link with a geneticist who can help guide follow up for additional testing and coordination with the infant's medical home (Newborn Screening Authoring, 2008).

Screening for ASD

Developmental screening involves using a validated tool to detect any concerns. Some of these screening tools include a component that specifically looks for autism. Examples of these tools are listed in Table 2.1.

Table 2.1 ASD screening tools

Screening instrument	Description
The Ages and Stages Questionnaires, Third Edition (ASQ-3)	This is a developmental screening tool that asks parents to report on their child's skills in the domains of communication, gross motor, fine motor, personal-social, and problem-solving. Questionnaires are available to pinpoint progress in children ages 1 month to 5½ years (Agarwal et al., 2020)
Modified Checklist for Autism in Toddlers (MCHAT)	The M-CHAT was developed to screen for autism in children ages 16–30 months, with a follow-up screen (M-CHAT-R/F) to be used when the initial screen is concerning (Robins et al., 2001)
Parents' Evaluation of Developmental Status (PEDS)	This is a surveillance and screening tool for children ages 0–8 years. This tool elicits and addresses parents' concerns about their child's development and behavioral health (Glascoe, 1998)
The Survey of Well-being in Young Children/ Parent's Observations of Social Interaction (SWYC/POSI)	The SWYC includes questions relating to a child's developmental milestones, emotional/behavioral health, and family context. The Parent's Observations of Social Interactions (POSI) screens for autism-specific behaviors (Perrin et al., 2021)

The Medical Home Can Best Respond to Initial Concerns or a Positive Screen and Coordinate Care

The Medical Home is usually the first place where a parent might communicate a concern about their child. Traditionally, the response has been referral to a pediatric specialist, including Developmental/Behavioral Pediatrics, Child and Adolescent Psychiatry, Child Psychology, or Child Neurology. Due to small numbers of these specialists, there can be a 6- to18-month wait time for an appointment, which is concerning to both clinicians and families. While the age of first concern is often before the age of 2, the average age at which an ASD is diagnosed is generally after 5 years with girls diagnosed at a later age (McDonnell et al., 2020). The medical home model addresses system issues which pose a barrier to early identification of ASD including inefficient systems of care, family engagement, provider attitudes, and organizational culture while promoting facilitators, including strong stakeholder relationships and well-coordinated care (Broder Fingert et al., 2019).

Provider attitudes and organizational culture are shifting to empower pediatricians in the early identification of ASD. In 2020, the American

Academy of Pediatrics (AAP) published an updated clinical report on autism spectrum disorder, *Identification, Evaluation, and Management of Children With Autism Spectrum Disorder*, which provides pediatricians with a framework for the identification/evaluation and management of at-risk children based on the available evidence (Hyman et al., 2020). The reality is that pediatric clinicians can and should have an active role that extends beyond screening and to the child's evaluation. Pediatricians are a trusted partner with the child's family, and an active partnership in evaluating the child's core deficits, medical and behavioral comorbidities, and connection to services facilitates the best results for the child.

Programs such as Project Echo Autism are actively training primary care clinicians in the early identification, diagnosis, and comprehensive care of children with autism. The Project Echo Autism model involves a "hub" of pediatric specialists, therapists, psychologists, and family members, and "spokes" of primary care practices interested in diagnosing and caring for these children. The teams meet on a regular basis with sharing of didactic topics and the "spoke" practices presenting a patient with developmental concerns. The "hub" team provides ongoing mentorship for "spoke" practices. In time, primary care clinicians become competent in the diagnosis and management of children with autism, as noted in the Project Echo Autism program evaluation (Mazurek et al., 2017).

Role of the Medical Home in the Diagnostic Journey for Children with Developmental Delays

Medical considerations are important in the differential diagnosis of developmental delay. Physical exam, including height, weight, body mass index, head circumference, and other vital signs, may suggest a genetic syndrome. History of seizure activity or specific concerning physical exam findings might elicit studies such as an MRI or EEG, or a referral to Genetics or Neurology. Audiology evaluation is often recommended to rule out hearing loss as a factor.

The pediatric clinician can respond to a parent concern with many resources, the first being the local Early Intervention program, serving children from birth until age 3. A child can be referred if there is a suspected delay. A diagnosis is not necessary, and families do not need a referral to connect with Early Intervention. The outcome of this evaluation may result in an Individual Family Services Plan (IFSP), a living tool that is shared with the pediatric clinician. This document outlines goals and timelines for the child's progress with his developmental concern and can be reviewed at subsequent well visits or chronic care management visits in the Medical Home. The pediatric clinician may know the therapists working with the child and family and offer valuable guidance to the Early Intervention team.

Once the child turns 3 years of age, her concerns are addressed by the local school system. She may be eligible for an Individualized Education Plan (IEP) that addresses her disability and how it may impact her education. After age 3, services such as speech therapy may be provided by the school. Pediatric clinicians have the advantage of awareness of the local schools and their staff. As these clinicians settle into a community, familiarity with local school resources is important for all their patients, particularly those children with special healthcare needs. Local school staff can serve as a powerful partner in optimizing the development and education for these children. Chapter 12 explores educational services for children living with ASD and IDD in greater detail.

Telehealth has provided additional options for following the family journey and the child's progress. During the evaluation stage, families value frequent contact; telehealth can facilitate discussion and shared decision-making with a family without requiring the child to physically travel to the office. The use of telehealth for the assessment and diagnosis of ASD is promising, including real-time and store-and-forward models (Alfuraydan et al., 2020). Recently, telehealth has expanded dramatically due to changing federal regulations and insurance policies driven by the COVID-19 pandemic. A team at Vanderbilt University has developed the TELE-ASD-PEDS, a telemedicine-based tool for the evaluation of ASD in toddlers under 36 months and offers webinars, administration guidelines, checklists, and assessment supports (Corona et al., 2020). A full exploration of the uses of telehealth is beyond the scope of this chapter but has been explored by Solomon and Soares (2020).

Coordinating Referrals to Therapy Evaluations Before the Specialist Visit? Yes!

Interventions for children with suspected ASD/DD can be initiated prior to a pediatric subspecialist diagnosis. The Medical Home initiating evidence-based therapeutic evaluations and treatment constitutes a changing approach to children with developmental concerns. It is far more child- and family-centered and provides solutions along the diagnostic journey. A child with suspected ASD goes from "waiting in line" for the full subspecialist evaluation and diagnosis to obtaining the necessary assessments and starting therapy, even while "waiting in line." Important interventions include speech and language therapy, social communication and skill building, and sensory processing.

Speech and language therapy is necessary to initiate and reinforce communication throughout the child's day. For children who do speak, the therapy often directs nonfunctional or echolalic speech to a practical method of communication. For the 30% of children with ASD who do not speak, therapy provides alternatives to speech (such as signing or voice output devices) (Rose et al., 2016). Picture Exchange Communications Systems can serve as a bridge for a child who has evolving receptive language to a child who will eventually develop speech. Speech-language pathologists (SLPs) work to prevent, assess, diagnose, and treat speech, language, social communication, cognitive-communication, voice, fluency, and swallowing disorders in children and adults (American Speech Language Hearing Association, 2015). Speech services begin with initial screening for communication disorders and continue with assessment and diagnosis. Then options for providing treatment of the identified problem and follow-up are implemented. Some of the areas in which SLPs are helpful include:

- Cognitive aspects necessary for communication (e.g., attention, memory, problem-solving, executive function)
- Expressive speech (phonation, articulation, fluency, resonance, and voice including aeromechanical components of respiration)
- Language (phonology, morphology, syntax, semantics, and pragmatic/social aspects of communication) including comprehension and expression in oral, written, graphic, and manual modalities; language processing; pre-literacy and language-based literacy skills, phonological awareness
- Alternative and augmentative communication, for children who may be nonverbal or with severe language and communication impairments
- Swallowing or other upper aerodigestive functions such as infant feeding and aeromechanical events (evaluation of esophageal function is for the purpose of referral to medical professionals)
- Voice (hoarseness, dysphonia), poor vocal volume (hypophonia), abnormal (e.g., rough, breathy, strained) vocal quality
- Sensory awareness related to communication, swallowing, or other upper aerodigestive functions

Behavioral Therapy and Autism

When constructing a shared plan of care with a family, it is critical to identify target symptoms for intervention before choosing a specific therapy. Selecting interventions targeting behaviors of greatest concern to the patient and family increases compliance and improves outcomes (Hyman et al., 2020). Some of the behaviors that cause social impairment for children with ASD are listed in Table 2.2.

Table 2.2 Behaviors that cause social impairment for children with ASD

Lack of eye contact
Lack of interest in others
Difficulty understanding others' feelings
Not liking being touched
Delayed speech
Unusual or inappropriate body language
Repeating words or phrases without communicative intent
Repetitive movements like rocking back and forth
Limited topic of interest
Tendency to strictly stick to routines
Being upset due to changes in their routine or environment
Hyper- or hyporeactivity to sensory input
Clumsiness, odd ways of moving
Being obsessive about unusual object

Autism-Specific Behavioral Therapy

The most generally successful approach for children with autism is behavioral therapy. Often applied behavior analysis is mentioned as a the "gold standard" behavioral intervention, but there are many different approaches (Lindgren & Doobay, 2011). Knowing what's available, who performs these therapies, and how your patients have responded becomes key information in the management of these children in the Medical Home. One challenge is determining which type of behavioral therapy matches your patient's needs (Myers & Johnson, 2007). Current evidence-based guidelines support the idea that ongoing, intensive therapy, especially early on after the diagnosis of autism, clearly benefits children (Reichow et al., 2018; Volkmar et al., 2014). Often more than one behavioral therapy is used with a specific child. A description of behavioral therapies for children with autism follows:

Applied Behavior Analysis (ABA)

This therapy is the most-researched intervention for autism and has been used for more than 50 years. It is a highly structured, scientific approach that teaches play, communication, self-care, academic, and social living skills and reduces problematic behaviors. A lot of research shows that it improves outcomes for children with autism.

ABA involves a therapist breaking down skills into component parts and teaching them to a child through repetition, reinforcement, and encouragement. With ABA, a therapist observes a child's abilities and defines what would benefit him, particularly when a child is not interested in learning these skills. For example, if a child is not interested in greeting others or in learning toilet training, an ABA therapist might focus on those skills anyway, because she recognizes their long-term value long before a child can.

ABA is the usual starting point for children with the most severe behaviors. Therapists recommend as many as 40 hours a week of therapy, often in a full-time, classroom-based program. Many parents are eventually able to learn some of the ABA skills and use them in ongoing home-based therapy. As skills improve and children begin to make friends and improve socially, ABA often continues to play a useful role in refining skills and addressing new challenges.

Verbal Behavior Therapy (VBT)

This type of applied behavior therapy teaches non-vocal children how to communicate purposefully. Children learn how we use words functionally – to get a desired response. It's not enough for a child to know that a cookie is called a cookie or to point to a cookie that he wants. VBT seeks to move children beyond labeling, a first step of learning language, and gesturing to vocalizing their requests – "I want a cookie."

In a typical session, the therapist will present stimuli, such as food, activities, or toys, based upon a child's preferences. The therapist uses stimuli that will attract a child's interest – a cookie in the kitchen or a swing on the playground. Children are encouraged through repetition to understand that communication produces positive results; they get what they want because they use language to ask for it.

Cognitive Behavioral Therapy (CBT)

Cognitive behavioral therapy is often recommended for children with higher intellectual functioning and milder symptoms of autism. Cognitive behavioral therapy aims to define the triggers of particular behaviors, so that a child starts to recognize those moments himself. Through practice, a therapist introduces practical responses. In other words, kids learn to see when they are about to head down a habitual behavioral or mental path ("I get scared riding the bus.") and to practice something different instead ("I'm going to do that relaxation exercise I was taught before I get on the bus and while I am riding"). CBT helps with concerns common to autism, such as being overly fearful or anxious.

Other behavioral models for autism focus more on developing skills a child already has and working on their deficiencies in subtler ways.

Developmental and Individual Differences Relationship (DIR) Therapy

DIR therapy (also called Floortime). With this therapy, a therapist and parents connect with children through activities that the child enjoys. It relies on a child having the motivation to interact with others. The therapist follows a child's lead in working on new skills.

Therapists then work with parents to engage their child, expand their "circles of communication," and meet the child at their developmental level and build on their strengths.

DIR therapy aims to help the child reach six key milestones that contribute to emotional and intellectual growth:

- Self-regulation and interest in the world
- Intimacy or engagement in relationships
- Two-way communication
- Complex communication
- Emotional ideas
- Emotional thinking

Therapists teach parents how to direct their children into more and more complex interactions. This process, called "opening and closing circles of communication," is central to the DIR approach.

Relationship Development Intervention (RDI)

RDI is a family-centered approach to treat autism focusing on defined emotional and social objectives meant to establish more meaningful relationships. This includes the ability to form an emotional bond and share experiences. It is commonly used with parents trained by RDI consultants. Goals are set to develop skills related to interpersonal engagement, such as empathy and overall motivation to engage with others. RDI breaks its various objectives down into step-by-step paths adults use to prompt development, such as building eye contact or back-and-forth communication. RDI builds on the idea that "dynamic intelligence" is key to improving quality of life for individuals with autism.

Dynamic intelligence means the ability to think flexibly:

- Understand different perspectives
- Cope with change
- Integrate information from multiple sources (e.g., sights and sounds)

There are six objectives of RDI:

1. Emotional referencing: the ability to learn from the emotional and subjective experiences of others
2. Social coordination: the ability to observe and control behavior to successfully participate in social relationships
3. Declarative language: the ability to use language and nonverbal communication to express curiosity, invite others to interact, share perceptions and feelings, and coordinate your actions with others
4. Flexible thinking: the ability to adapt and alter plans as circumstances change

5. Relational information processing: the ability to put things into context and solve problems that lack clear cut solutions and have no “right and wrong” solutions
6. Foresight and hindsight: the ability to think about past experiences and anticipate future possibilities based on past experiences

Social Skills Groups

Social skills groups help children engage in pragmatic language and manage real-world difficulties with peers. While observational studies show them to be effective, less research supports their success so far (Gates et al., 2017). Because children with autism are usually more comfortable talking and interacting with adults than with peers, social skills groups bring out difficulties that come up when being with peers. In such groups, the leader sets up specific situations that mimic real life and guides a child to develop appropriate behaviors. The groups often use text and pictures to demonstrate social skills. Social scripts give children the specific language to manage difficult situations.

Occupational Therapy for Sensory Processing

A child with autism often has symptoms that include difficulty processing sensory information such as textures, sounds, smells, tastes, bright lights, and movement. Combined with communication impairments, these sensory difficulties can make ordinary situations feel overwhelming and can interfere with the daily function of a child and their families.

Sensory integration therapy, as practiced by occupational therapists, uses play activities in ways designed to change how the brain reacts to touch, sound, sight, and movement. The use of items such as swings, auditory blocks, or weighted vests allows a child to find a calming alternative to a stressful situation or negative behavior (Schoen et al., 2019). This activity is important to build a child’s self-regulation skills.

Medical Specialty Referrals

Medical specialty referrals are often required. Medical multi-morbidity is common in children with autism. Gastrointestinal problems are reported in more than 50% of children with autism (Celia et al., 2016). Selective eating and subsequent chronic constipation are common. Assess the child’s eating, stooling pattern, and periods of irritability and you may well identify a pattern! Sleep disorders are just as common in children with autism (McPherson et al., 2020). Sleep problems include insomnia, bedtime resistance, and frequent awakening at night. In some children, parasomnia can be problematic. A pediatric clinician can advise the family on sleep routines, which are often comforting to the child. Recognizing sensory triggers that reduce sleep (constipation/abdominal pain, auditory or visual sensitivity) will allow the pediatric clinician to suggest interventions to address these concerns. The use of melatonin may help with initiating sleep as well.

Care Coordination and the Medical Home

Care coordination is a unique feature of the Medical Home, one which highlights the importance of this locus of care for children with developmental concerns (Fig. 2.1).

Most primary care practices utilize some form of care coordination. In larger systems, care managers may be hired to maintain a registry, or list, of children with special healthcare needs. These care managers will call and interact with families on a regular basis to assess a child’s status and needs, set goals, and plan care as well as the next monitoring visit. For a child with autism, this care manager visit may include reviewing the child’s IEP or IFSP; functional status with eating, stooling, sleep; and functional areas of improvement or regression. Then, in conjunction with the child’s primary care clinician, the team can brainstorm solutions, resources, and next steps with the family.

Care Coordination Definition:

Pediatric care coordination is a patient- and family-centered, assessment-driven, team-based activity designed to meet the needs of children and youth while enhancing the caregiving capabilities of families. Care coordination addresses interrelated medical, social, developmental, behavioral, educational, and financial needs to achieve optimal health and wellness outcomes.

Defining Characteristics of Care Coordination:

1. Patient- and family-centered
2. Proactive, planned, and comprehensive
3. Promotes self-care skills and independence
4. Emphasizes cross-organizational relationships

Care Coordination Competencies:

1. Develops partnerships
2. Communicates proficiently
3. Uses assessments for intervention
4. Is facile in care planning skills
5. Integrates all resource knowledge
6. Possesses goal/outcome orientation
7. Takes an adaptable and flexible approach
8. Desires continuous learning
9. Applies team-building skills
10. Is adept with information technology

Care Coordination Functions:

1. Provides separate visits and care coordination interactions
2. Manages continuous communications
3. Completes/analyzes assessments
4. Develops care plans with families
5. Manages/tracks tests, referrals, and outcomes
6. Coaches patients/families
7. Integrates critical care information
8. Supports/facilitates care transitions
9. Facilitates team meetings
10. Uses health information technology

Delivery of Family-Centered Care Coordination Services Includes:

(Antonelli, McAllister, & Popp, 2009)

Antonelli, R., McAllister, J., & Popp, J. (2009). Making care coordination a critical component of the pediatric health system: a multidisciplinary framework. *110.*

Fig. 2.1 A Framework for high-performing pediatric care coordination. (Antonelli et al., 2009)

In a smaller practice, the care coordinator is usually a staff member well known to the family, a nurse or medical assistant, or sometimes even an administrative staff member. Not uncommonly, the care coordinator is a parent of a child with a special healthcare need. This is advantageous for the family as well as the practice, as these parents can be a crucial link between families and services for their children. The notion of a "Parent Navigator" has been endorsed by national organizations such as Family Voices, or state/local organizations such as Medical Home Plus. It is the function of care coordination to assess the child's and family's strengths and concerns to the primary care clinician, who can then take the time during the visit to address these concerns during the follow-up visit. The ongoing communication and collaboration between the

Medical Home team and the child/family becomes a foundation of support for children with developmental disabilities.

The Medical Home "Toolbox"

A Medical Home "Toolbox" for working with families of children with suspected delay could contain resources that would aid in the structure and communication in these follow-up visits. The Toolbox could contain:

1. Information on local speech, occupational, behavior, and physical therapists, including names, services, contact information, and insurances accepted.
2. Childcare centers where your patients spend their day.
3. Local school systems and Early Childhood Special Education (ECSE) classes. Often there is one elementary school within a district where ESCE class is held for children who are found eligible for this service.
4. A list of local and national family advocacy groups.
5. Templated letters for your families (see sample letters at the end of this chapter.), including:
   (a) An Introduction to the local Early Intervention System, where your patient's name and developmental concern can be communicated.
   (b) A request for a Child Study to the local school system. This request, signed and dated by the family and presented to the school, starts the legal process for a school system to evaluate the child for Special Education services.
6. Templated information for your Electronic Health Record, including:
   (a) Therapy prescriptions, where your patient's name, requested therapy evaluation and treatment, and a checklist of possible diagnoses to print or send electronically to a family through a patient portal
   (b) Letters of Medical Necessity for insurance authorization (see sample letter at the end of this chapter)
   (c) Letters for Appeal to insurance for therapy, specialist, or medication denials
   (d) Structured documentation for your EHR during the patient visit. This might include:
      (i) Child's diagnosis, or domains of concern if no diagnosis yet
      (ii) Family concerns
         1. Medical
         2. Developmental, including behavior
         3. Functional, including diet and sleeping
      (iii) Present therapies and therapists
      (iv) Review of progress made against the goals stated in the IFSP or IEP
      (v) Other services the child is receiving (home nursing, DME, SSI, etc.)
      (vi) Pending services and barriers (insurance authorization, wait times for schools or specialists)
      (vii) Care plan, including goals and follow-up for the next visit

This list will be informed by your patients and families. As a pediatric clinician follows up with their patient, they will learn which therapies have been helpful or effective, which childcare centers and preschools work with Early Intervention, what services are offered through local school systems, and how the effective the transition process works between Early Intervention and School.

Autism Spectrum Disorder: The Diagnosis and Ongoing Care

Can ASD be diagnosed in the Medical Home? In the 2020, the American Academy of Pediatrics (AAP) updated clinical report on *Identification, Evaluation, and Management of Children With Autism Spectrum Disorder* stated:

> Although most children will need to see a specialist, such as a developmental-behavioral or neurodevelopmental pediatrician, psychologist,

> neurologist, or psychiatrist, for a diagnostic evaluation, general pediatricians and child psychologists comfortable with application of the DSM-5 criteria can make an initial clinical diagnosis (p. 13).

Most pediatric clinicians recognize the DSM-5 features in their patients that support ASD diagnostic criteria. These include persistent deficits in social communication and interaction across multiple contexts; restrictive and repetitive patterns of behavior, interests, and activities; and symptoms causing clinically significant impairment in social, occupational, or other important areas of functioning which were present in the early developmental period. Yet, many primary care clinicians fear making the ASD diagnosis. A study of Canadian pediatricians' experience with assessing and diagnosing ASD highlighted the importance of the increasing access to high-quality training on diagnosis, providing toolkit resources, and increasing reimbursement (Penner et al., 2017). Diagnostic uncertainty has been reported as more common for children with higher adaptive behavior scores, a mild-moderate ADOS score, high stress home environments, or parents with limited grasp of developmental milestones (McDonnell et al., 2019). A pilot project of the diagnosis of ASD by pediatricians as compared to traditional referrals to a psychologist found time to the diagnosis of ASD 44% faster and less costly with a 93% diagnostic agreement between pediatricians and psychologists (Ahlers et al., 2019).

ASD is recognized as being both common and complex. The diagnosis of ASD bodes a family trajectory familiar to any parent of a child with a chronic health condition. That trajectory may include multiple physician and therapy appointments, risk of at least one parent exiting the workforce, and concern about a child's lifelong chance for success. Thus, it is understandable that a pediatric clinician would be hesitant to make the diagnosis. However, once a diagnosis is made, this same clinician can help open the door to life-changing services for the child and family.

Chronic care management for children with ASD includes many of the features highlighted in the care of children before the diagnosis. Regular visits, either in person or via telehealth, can serve as a forum for discussion of the child's services, progress, new concerns, and new referrals. These visits should review:

1. The child/family team members, including:
   (a) Early Intervention Services
   (b) School and Early Childhood Special Education
   (c) Local therapists and therapy groups
   (d) Psychologist
   (e) Behavior analysts and interventionists
   (f) Occupational therapist
   (g) Speech therapist
   (h) Nutrition
   (i) Family advocacy groups
   (j) Family navigators; the savvy parents in your practice
2. Recognition of and addressing the child's multi-morbidity with ASD. These may include conditions that a primary care clinician may feel confident in managing, such as:
   (a) ADHD
   (b) Anxiety
   (c) Aggression
   (d) Sleep concerns
   (e) Gastrointestinal concerns
   (f) Allergy/immunology
   (g) Seizures
3. Collaborative care with pediatric subspecialists. There may be conditions that require further subspecialty referral, particularly in the fields of Child and Adolescent Psychiatry and Neurology.

Pediatric Psychiatry Collaboratives

Pediatric Psychiatry Collaboratives (PPC) are regional systems of behavioral health services that provide timely access to psychiatric consultation to the pediatric Medical Home. The Massachusetts Child Psychiatry Access Project (https://www.mcpap.com/) was the original program designed to facilitate a working relationship between primary care clinicians and child and adolescent behavioral health resources. The program started in 2003 and is now funded by the

Massachusetts Department of Mental Health and in part by major commercial insurance companies in Massachusetts. Each MCPAP "team" is staffed with two full-time child and adolescent psychiatrists, independently licensed behavioral health clinicians, resource and referral specialists, and program coordinators. A primary care practice joining MCPAP will access education to improve the practice competency and comfort with screening, identification, and assessment of pediatric behavioral health disorders, eventually treating mild to moderately affected children according to current evidence-based practices. There is also assistance in making effective referrals and coordinating care for patients who need community-based specialty behavioral health services (Walter et al., 2018).

Recently, 21 states were awarded funding from HRSA (Health Resources and Services Administration of the Maternal Child Health Bureau) in the Pediatric Mental Health Care Access Program. This program, designed to spread the innovation started with the Massachusetts program, promotes behavioral health integration into pediatric primary care using telehealth. State or regional networks of pediatric mental health teams provide teleconsultation, training, technical assistance, and care coordination for pediatric primary care clinicians to diagnose, treat, and refer children with behavioral health conditions. This resource, if available in the state where the Medical Home is located, could provide assistance with the common behavioral health comorbidities seen in children with ASD (Stein et al., 2019).

Referrals Beyond the Medical Home Team

Programs exist to help the Medical Home with the additional expertise needed to care for children with ASD in a comprehensive manner. An important function of the Medical HomeFamily Support Organizations is to provide resources to families for information, education, and support. Here are some great resources for families in Table 2.3.

Medical Home: Coordinated, Family-Centered Care That Improves Outcomes

The concept of Medical Home for Children with Special Health Care Needs is a priority for the American Academy of Pediatrics; children with medical complexity have found their champions for coordinated care in the Medical Home. Children with ASD deserve no less; care for these children requires a comprehensive view that starts with developmental surveillance and screening and continues with coordination of diverse and sometimes eclectic services. The Medical Home can support families as well as be informed of local resources by the community that educates, treats, and employs individuals with ASD. An ongoing chronic care management plan for children with ASD is the strategy to improve the social, educational, medical, and functional outcomes for these children.

Table 2.3 Resources for families and professionals

Resource	Description
The American Academy of Pediatrics Resources for Autism Spectrum Disorder www.aap.org/autism	The AAP has several practical resources for pediatricians caring for children with autism. The webpage provides links to the Autism Toolkit, screening resources, policy statements, and training available to help manage these children in the pediatric Medical Home
Autism Highway http://autismhwy.com/	Started by a woman whose son was diagnosed with autism, Autism Highway is both informative and fun. The website is easy to navigate and it provides an extensive list of autism-related events and specialists. In addition, Autism Highway includes interactive games for kids
Autism Navigator https://autismnavigator.com/	Autism Navigator is a collection of web-based tools and courses developed to bridge the gap between science and community practice. They have integrated the most current research into an interactive web platform with video to illustrate effective evidence-based practice. The video clips come from the rich library of video from federally funded research projects at the Autism Institute at Florida State University
Autism Speaks https://www.autismspeaks.org	Autism Speaks is the largest autism research organization in the United States. It sponsors autism research and conducts awareness and outreach activities aimed at families, governments, and the public. It is the family support and advocacy organization that sponsors the Autism Treatment Network. The website provides a comprehensive resource guide for all states. The 100 Day Kit for Newly Diagnosed Families of Young Children was created specifically for families of children ages 4 and under. Visit Autism Speaks to see their comprehensive listing of autism websites for families
The Autism Society https://www.autism-society.org/	The Autism Society is a grassroots autism organization with local chapters working to increase public awareness about the day-to-day issues about people across the spectrum, advocate for appropriate services for individuals of every age, and provide the latest information regarding treatment, education, research, and advocacy
Autism Treatment Network https://www.autismspeaks.org/autism-treatment-network-atn	The Autism Treatment Network of Autism Speaks is a comprehensive repository of research and evidence-based information on many aspects of ASD treatment. Clinicians and families can search for the best treatment for comorbidities such as sleep or GI disorders, new technology, ongoing clinical trials, safety, advocacy, and education. Guides to Understanding Autism for grandparents, siblings, and other community members are available as well
Project ECHO Autism https://echoautism.org/	Project ECHO (Extension for Community Healthcare Outcomes) Autism is a collaborative learning model that links specialist care teams ("hubs") with clinicians practicing in local communities ("spokes"). It is hosted in a teleconsultation format involving a mix of didactic presentations and in-depth case consultations. The sessions are held at an arranged time for the "Hub" and "Spoke" teams, and both teams benefit from the ongoing interaction. As a Medical Home team, a pediatric clinician and her staff can learn about ASD diagnosis, specific therapies, and tips in caring for these children in a primary care setting. Some Medical Home teams have been trained in the STAT Autism tool and are using this to aid in the diagnosis of children with ASD in primary care practice
Sesame Street and Autism: See Amazing in All Children https://autism.sesamestreet.org/	Sesame Workshop created Sesame Street and Autism: See Amazing in All Children, a nationwide initiative aimed at communities with children ages 2–5. Developed with input from parents, people who serve the autism community, and people with autism, See Amazing in All Children offers families ways to overcome common challenges and simplify everyday activities. The project also fosters an affirming narrative around autism for all families and kids
Wrightslaw Special Education Law and Advocacy https://www.wrightslaw.com/	A great site for accurate, reliable information about special education law, education law, and advocacy for children with disabilities. Early intervention is invaluable because it links parents to services in the community, but it can be hard to find services without a long waiting list. Families can search on their own for providers using the Wrightslaw Yellow Pages for Kids With Disabilities

SAMPLE EARLY INTERVENTION REFERRAL LETTER

Practice name/logo/contact information

Date:

To (Name) Early Intervention Program:

________________(DOB)______is followed by me for general pediatric care. This child's family and I have concerns about this child's:

___ Speech and language development

___ Gross motor development

____Fine motor development

____Personal social and problem-solving skills

____Atypical behavior

Thank you for addressing these concerns with my patient. Please send your report to the (fax or email) above, and feel free to reach out to my office with questions.

Sincerely,

____

SAMPLE LETTER OF MEDICAL NECESSITY

Your practice address/logo

Date:

RE: Patient XXXXX and DOB

Reference Number: [Number on insurance denial information]

To whom it may concern:

XXXXX is followed by me for general pediatric care. He has autism spectrum disorder and has had difficulty with __(concern). This translates into regression with attending to tasks, overall behavior, eating and sleeping.

_____ therapy has been recommended with a specific focus on _____ therapy for this child. I am writing a letter of medical necessity for XXXXX to be evaluated and treated by Dr. YYYY at (location) for _concern__.

This request is to specifically provide task-oriented therapeutic activities designed to significantly improve, develop or restore physical functions lost or impaired as a result of a disease, or injury. XXXX has Autism Spectrum Disorder which has resulted in regression in behavior and overall physical functions.

In my professional opinion, XXXX's condition can improve significantly based on objective measures within 1 month of the date that therapy begins. He will need a professional evaluation by Dr. YYYY, who is a specialist in this field. The goal is that therapy services proposed will lead to the establishment of a safe and effective maintenance program that will be performed by the member without ongoing skilled therapy services.

This is necessary to care for XXXX's specific deficits due to his Autism Spectrum Disorder. Therapy will be provided in accordance with an ongoing, written plan of care that is reviewed with and approved by the treating physician in accordance with applicable state laws and regulations.

The first step is to allow XXXXX the evaluation with the right therapist to create the evaluation and plan. The Therapy Plan of Care should be of such sufficient detail and include appropriate objective and subjective data to demonstrate the medical necessity of the proposed treatment.

Thank you for this consideration. It is my goal to assist my patient and his family to find the right therapist for evaluation and treatment that will results in XXXX's improved function and behavior. Given these specific deficits due to autism spectrum disorder, I am recommending that XXXXX be allowed evaluation and treatment for his __(concern)__ by Dr. YYYY.

Sincerely,

Your signature

____

SAMPLE LETTER REQUESTING SPECIAL EDUCATION SERVICES FOR A CHILD WITH ASD

Date:

ATTN: Principal, Special Education Coordinator

Dear: I am the parent of __________, date of birth _________, who is currently enrolled at ____________School in the ___ grade. My child has a developmental disability and I am concerned about [his/her] educational progress.

I am writing to make a formal request for assessment for special education services as allowed under the Individuals with Disabilities Act (IDEA). I am requesting that ______ be given a comprehensive assessment by the school district in all areas of suspected disability, and that an IEP eligibility meeting be scheduled for [him/her.], as [He/She] may be eligible for special education assistance.

Some of my concerns are stated below.

1. Speech and Language: (elaborate on your concerns)
2. Social Reciprocity:
3. Fine Motor Skills:
4. Gross Motor Skills
5. Problem Solving:

I specifically request that the school district conduct the following evaluations of my [daughter/son], __________ but that they not be limited to:

- An evaluation by a Behavioral Specialist, School Psychologist and Speech Pathologist specifically trained to identify and assess children with Autism Spectrum Disorder.
- A Functional Behavioral Assessment (FBA) to determine the triggers and function of my [son's/daughter's] behavioral difficulties in order to develop a Positive Behavioral Support Plan (BSP) and goals to enable [him/her] to learn to replace difficult behaviors with acceptable behaviors so that [he/she] may make educational progress.

I look forward to receiving an Assessment Plan within 15 days for my review and consent so that the evaluations can proceed. Even more so, I look forward to these evaluations being completed promptly and an IEP eligibility meeting within 60 days to discuss the results and plan for my child's supported education.

Also, please ensure that I get copies of the assessment reports at least 5 days before the IEP eligibility meeting pursuant to the Family Educational Rights and Privacy Act (FERPA) so that I will have adequate time to review them and prepare any questions I may have for the team.

Sincerely,

Parent Signature

[Note: Under federal law, the parent's signature (not the physician's) is required on the request letter in order to secure FERPA protections.]

INSTRUCTIONS FOR COMPLETING THE REQUEST FOR SPECIAL EDUCATION SERVICES

- Fill out above letter with your child's info. Any [ ] mean that information needs to be added or removed, depending on what is pertinent.
- Send the letter either certified mail, return receipt requested OR hand deliver and request date and time stamp on both your copy and the copy dropped off
- Send to BOTH school district AND school principal at the local elementary school, whether or not you attend that school
- Keep all written documentation in a binder for easy access
- Call our office with any questions regarding process, especially if something doesn't sound right or you do not agree with the recommendations of the school district
- Do not take no for an answer! You are your child's best advocate and you know what is best for your child! Fight for your child's right to an education!!

References

Agarwal, P. K., Xie, H., Rema, A. S. S., Rajadurai, V. S., Lim, S. B., Meaney, M., & Daniel, L. M. (2020). Evaluation of the Ages and Stages Questionnaire (ASQ 3) as a developmental screener at 9, 18, and 24 months. *Early Human Development, 147*, 105081.

Ahlers, K., Gabrielsen, T. P., Ellzey, A., Brady, A., Litchford, A., Fox, J., ... Carbone, P. S. (2019). A pilot project using pediatricians as initial diagnosticians in multidisciplinary autism evaluations for young children. *Journal of Developmental and Behavioral Pediatrics, 40*(1), 1–11.

Alfuraydan, M., Croxall, J., Hurt, L., Kerr, M., & Brophy, S. (2020). Use of telehealth for facilitating the diagnostic assessment of autism spectrum disorder (ASD): A scoping review. *PLoS One, 15*(7), e0236415.

American Academy of, P. (2002). Medical home initiatives for children with special needs project advisory committee. American academy of pediatrics. *The Medical Home Pediatrics, 110*(1 Pt 1), 184–186. PMID: 12093969.

Antonelli, R., McAllister, J., & Popp, J. (2009). Making care coordination a critical component of the pediatric health system: A multidisciplinary framework. *Commonwealth Fund, 110*. https://www.commonwealthfund.org/publications/fund-reports/2009/may/making-care-coordination-criticalcomponent-pediatric-health.

Belseck, N. (2019). How the world really feels about health and science. *Medical Chronicle, 2019*(10), 10.

Berry, S. A. (2015). Newborn screening. *Clinics in Perinatology, 42*(2), 441–453.

Bright Futures Steering, C. (2006). Medical Home Initiatives for Children with Special Needs Project Advisory Committee identifying infants and young children with developmental disorders in the medical home: An algorithm for developmental surveillance and screening. *Pediatrics, 118*(1), 405–420.

Broder Fingert, S., Carter, A., Pierce, K., Stone, W. L., Wetherby, A., Scheldrick, C., ... Ibañez, L. (2019). Implementing systems-based innovations to improve access to early screening, diagnosis, and treatment services for children with autism spectrum disorder: An autism spectrum disorder pediatric, early detection, engagement, and services network study. *Autism, 23*(3), 653–664.

Celia, T., Freysteinson, W. W., & Frye, R. E. (2016). Concurrent medical conditions in autism spectrum disorders. *Pediatric Nursing, 42*(5), 230–234.

Cohen, E., Kuo, D. Z., Agrawal, R., Berry, J. G., Bhagat, S. K. M., Simon, T. D., & Srivastava, R. (2011). Children with medical complexity: An emerging population for clinical and research initiatives. *Pediatrics, 127*(3), 529–538.

Corona, L., Hine, J., Nicholson, A., Stone, C., Swanson, A., Wade, J., ... Warren, Z. (2020). *TELE-ASD-PEDS: A telemedicine-based ASD evaluation tool for toddlers and young children*. Vanderbilt University Medical Center.

Dadashzadeh, S. (2019). The role of patient-centered medical home recognition in addressing public health priorities. *The Health Care Manager, 38*(2), 124–145.

Gates, J. A., Kang, E., & Lerner, M. D. (2017). Efficacy of group social skills interventions for youth with autism spectrum disorder: A systematic review and meta-analysis. *Clinical Psychology Review, 52*, 164–181.

Glascoe, F. P. (1998). *Collaborating with parents: Using parents' evaluation of developmental status to detect and address developmental and behavioral problems*. Ellsworth & Vandermeer Press.

Hyman, S. L., Levy, S. E., & Myers, S. M. (2020). Identification, evaluation, and management of children with autism spectrum disorder. *Pediatrics, 145*(1), p. 13.

Justin, A. Y., McKernan, G., Hagerman, T., Schenker, Y., & Houtrow, A. (2021). Identifying children with medical complexity from the national survey of children's health combined 2016–17 data set. *Hospital Pediatrics, 11*(2), 192–197.

Kelly, N., Makarem, D. C., & Wasserstein, M. P. (2016). Screening of newborns for disorders with high benefit-risk ratios should be mandatory. *The Journal of Law, Medicine and Ethics, 44*(2), 231–240.

Krahn, G. L. (2019). A call for better data on prevalence and health surveillance of people with intellectual and developmental disabilities. *Intellectual and Developmental Disabilities, 57*(5), 357–375.

Lindgren, S., & Doobay, A. (2011). *Evidence-based interventions for autism spectrum disorders*. The University of Iowa.

Lipkin, P. H., & Macias, M. M. (2020). Promoting optimal development: Identifying infants and young children with developmental disorders through developmental surveillance and screening. *Pediatrics, 145*(1), e20193449. https://doi.org/10.1542/peds.2019-3449

Mazurek, M. O., Brown, R., Curran, A., & Sohl, K. (2017). ECHO autism: A new model for training primary care providers in best-practice care for children with autism. *Clinical Pediatrics, 56*(3), 247–256.

McDonnell, C. G., Bradley, C. C., Kanne, S. M., Lajonchere, C., Warren, Z., & Carpenter, L. A. (2019). When are we sure? Predictors of clinician certainty in the diagnosis of autism spectrum disorder. *Journal of Autism and Developmental Disorders, 49*(4), 1391–1401.

McDonnell, C. G., DeLucia, E. A., Hayden, E. P., Penner, M., Curcin, K., Anagnostou, E., ... Liu, X. (2020). Sex differences in age of diagnosis and first concern among children with autism spectrum disorder. *Journal of Clinical Child and Adolescent Psychology*, 1–11.

McPherson, P., Kaushal, M., & Kothapalli, V. (2020). The treatment of dually diagnosed individuals with sleep disturbances and intellectual disabilities. In *Handbook of dual diagnosis* (pp. 613–645). Springer International Publishing.

Myers, S. M., & Johnson, C. P. (2007). Management of children with autism spectrum disorders. *Pediatrics, 120*(5), 1162–1182.

Newborn Screening Authoring, C. (2008). Newborn screening expands: Recommendations for pediatricians and medical homes—Implications for the system. *Pediatrics, 121*(1), 192–217.

Penner, M., King, G. A., Hartman, L., Anagnostou, E., Shouldice, M., & Hepburn, C. M. (2017). Community general pediatricians' perspectives on providing autism diagnoses in Ontario, Canada: A qualitative study. *Journal of Developmental and Behavioral Pediatrics, 38*(8), 593.

Perrin, E. C., Sheldrick, R. C., Visco Z., & Mattern K. *The survey of well-being of young children (SWYC) user's manual*. Available at: https://www.floatinghospital.org/The-Survey-ofWellbeing-of-Young-Children/ManualTraining-Resources. Accessed 4 Apr 2021.

Reichow, B., Hume, K., Barton, E. E., & Boyd, B. A. (2018). Early intensive behavioral intervention (EIBI) for young children with autism spectrum disorders (ASD). *Cochrane Database of Systematic Reviews*.

Robins, D. L., Fein, D., Barton, M. L., & Green, J. A. (2001). The modified checklist for autism in toddlers: An initial study investigating the early detection of autism and pervasive developmental disorders. *Journal of Autism and Developmental Disorders, 31*(2), 131–144.

Rose, V., Trembath, D., Keen, D., & Paynter, J. (2016). The proportion of minimally verbal children with autism spectrum disorder in a community-based early intervention programme. *Journal of Intellectual Disability Research, 60*(5), 464–477.

Schoen, S. A., Lane, S. J., Mailloux, Z., May-Benson, T., Parham, L. D., Smith Roley, S., & Schaaf, R. C. (2019). A systematic review of Ayres sensory integration intervention for children with autism. *Autism Research, 12*(1), 6–19.

Solomon, D., & Soares, N. (2020). Telehealth approaches to care coordination in autism spectrum disorder. In *Interprofessional care coordination for pediatric autism spectrum disorder* (pp. 289–306). Springer.

Speech–Language Pathologists. ASHA.org. American Speech–Language–Hearing Association. Retrieved 6 Apr 2015.

Stein, B. D., Kofner, A., Vogt, W. B., & Yu, H. (2019). A national examination of child psychiatric telephone consultation programs' impact on children's mental health care utilization. *Journal of the American Academy of Child and Adolescent Psychiatry, 58*(10), 1016.

Volkmar, F., Siegel, M., Woodbury-Smith, M., King, B., McCracken, J., & State, M. (2014). Practice parameter for the assessment and treatment of children and adolescents with autism spectrum disorder. *Journal of the American Academy of Child and Adolescent Psychiatry, 53*(2), 237–257.

Walter, H. J., Kackloudis, G., Trudell, E. K., Vernacchio, L., Bromberg, J., DeMaso, D. R., & Focht, G. (2018). Enhancing pediatricians' behavioral health competencies through child psychiatry consultation and education. *Clinical Pediatrics, 57*(8), 958–969.

3 The Family Perspective: Treatment Planning for Children with Autism Spectrum Disorder, Developmental Disabilities, and Medical Complexity

Rebecca Pistorius

For Seth

Abstract

Vulnerable children with medically complex disorders commonly require an intensive level of services, and deserve to have the care they need. This chapter combines my professional and personal perspectives as a physician-parent of a child with medical complexity, and focuses on the family viewpoint and experience of the treatment of children with medical complexity, including autism spectrum disorder (ASD) and developmental disabilities (DD). The purpose of this chapter is to provide guidance both for families and for healthcare providers on the multifaceted and often unique challenges of caring for children with medical complexity, and demonstrates how providers can facilitate parent and caregiver involvement in the medical home team as partners in shared decision-making. Maximizing the available treatments and resources will lead to better outcomes for the child and for the family as a whole.

R. Pistorius (✉)
Willis-Knighton Health System, Shreveport, LA, USA
e-mail: rpistoriu@wkhs.com

Keywords

Autism spectrum disorder · Developmental disability · Family · Treatment planning · Pistorius model

Introduction

A Lesson to Remember

A wise medical school professor told me, "Every time you see a disease you think is fascinating, always remember there is a mother somewhere crying." The truth of those words has remained with me, guiding the care I give my patients, and offering me solace as a parent and greater compassion as a physician. These words embody the humanity and the attitude of humility that healthcare providers must adopt to truly form a healing partnership with families. Vulnerable children with medically complex disorders who commonly require intensive services, and deserve to have the care they need. It is from this human aspect, from the perspective of a physician-parent of a child with medical complexity, that I contribute this chapter on the family viewpoint and experience of the treatment of children with medical complexity, including autism spectrum disorder and developmental disabilities.

P. McPherson (ed.), *Handbook of Treatment Planning for Children with Autism and Other Neurodevelopmental Disorders*, Autism and Child Psychopathology Series,
https://doi.org/10.1007/978-3-031-06120-2_3

A Change in Perspective

Physician-parents of medically complex children bring an important perspective to the medical home team and a critical voice for families. A pediatric hospitalist, Dr. Amy R. L. Rule, describes her initial viewpoint as a physician toward parents' approaches to obtaining care for their medically complex children, and how her viewpoint evolved over time. She reflects that early in her career, she had a tendency to read these parents as demanding and wanting to dictate the medical care for their children, which frustrated and even annoyed her. She acknowledges that she did not understand their perspectives and was unable to empathize with these parents until she herself had a child with a rare and medically complex disease. It was only then that she was able to relate and understand the necessity of their proactive and highly involved approaches to obtaining health care for their children. She realized these parents were appropriately invested in seeking care for their children whom they had created and for whom they were fully responsible in all aspects of their health and well-being. She recognized that they were motivated by love for their children to fight to provide them with the best possible opportunities and outcomes (Rule, 2018).

Dr. Rule's personal story provides insight into the challenges that parents and caregivers often face in obtaining appropriate care for their children, and the need for empathy and compassion on the part of healthcare providers. Parents and caregivers should be recognized as experts when it comes to their children, and we, as clinicians, must recognize and respect the ways in which they are making themselves vulnerable by sharing their feelings, perspectives, and often very personal aspects of their lives in order to advocate for their children to receive the level of care they need (Carbone et al., 2010).

Overview

The purpose of this chapter is to provide information and guidance both for families and for healthcare providers as they navigate the complex challenges of caring for children with medical complexity, including autism spectrum disorder (ASD) and developmental disabilities (DD), and to demonstrate how providers can bring parents and caregivers into the medical home team as partners in shared decision-making. Providers must consider the needs of the family as a whole to optimize outcomes for the child. It is essential for providers to seek out, listen to, and value the perspectives of parents and caregivers to determine the most appropriate and individualized treatment planning for each child and to construct a meaningful shared plan of care (SPoC) (Wirth & Kuznetsov, 2016). An integral part of fulfilling this objective with the best results for both the child and their caregivers is for clinicians to embrace the medical home model as a holistic, interdisciplinary approach to treatment as each family progresses through their unique journey. Families do not have a specific rule book for raising children with medical complexity. It is the providers' responsibility to provide up-to-date, accessible information and to coordinate appropriate resources and supports that will maximize the child's abilities (NAMI, 2016). Parents and caregivers must have a shared role in decision-making and creating a SPoC, and providers should monitor patient and family satisfaction with treatment throughout their care (Carbone et al., 2010). The treatment approach should focus on individualized goals that are realistic, yet hopeful. As caring for a child with medical complexity can bring unique and even unexpected challenges, parents and caregivers will need support as well while they are on their own journeys of learning about and accepting their child for the person he or she is and will be (NAMI, 2016).

This chapter will start at the beginning of the family's experiences, when they first notice the initial red flags in their child's development and will provide information about the diagnostic process. Many resources on the stages of child development are available in books and online, but this chapter will explore commonly experienced parental stages of development that complement progression through the child's developmental stages. In addition, this chapter will describe the progression of parents and caregivers toward becoming experts in caring and advocating for their medically complex children.

Information for Parents and Caregivers

Every child is unique and every family's journey is distinct. As medical professionals must strive to understand each family, families can benefit from insight into the medical world. Families can best lead and participate in their child's care if they understand the red flags, diagnostic process, treatment options, how to zero in on specific needs, and how to prepare for the future. Providing the information that parents need to make decisions is a critical role of the medical home team.

Initial Red Flags: *"My Child Versus the Typical Trajectory of Development"*

Even before a child is born, the family often has begun to develop dreams and goals for what their child will look like, his or her personality and interests, where he or she will go to school, and what life will be like in general with that child in their lives. Sometimes prenatal testing can indicate genetic or structural abnormalities, but often, there is no indication at birth that a child later may be diagnosed with a DD or ASD. For many families, the first evidence of a problem is a gut feeling that something is not right which leads to searching the internet and, finally, honing in on developmental milestones, then panic on identifying a delay. This is that moment that healthcare providers must approach with humanity and humility in offering expert knowledge and rendering skilled care.

Typical developmental milestones are in the areas of gross and fine motor skills and speech acquisition and communication. Developmental milestone screening occurs during well-child pediatric appointments starting in infancy. There are many books, online resources, and apps that provide parents with the "typical trajectory" of child development month by month. Parents may be the first to realize their child is not meeting these milestones such as walking or talking. Some parents or caregivers may notice a regression of previously acquired milestones, such as a child whose speech development seemed to be on track when suddenly they notice their child has stopped talking. In other cases, especially if it is the first-born child, the parents may not notice the presence of delayed milestones, and it may be the pediatrician who first suspects there is a problem. Understanding a family's journey cannot be captured by a milestone checklist. The skill of taking a developmental history that fully informs treatment planning is an art that should not be overlooked.

Developmental Disability: *"What Are You Saying About My Child?"*

Language is often imperfect. This is especially true of the literature surrounding medically complex children. "Developmental disability" is a broad term that can be used when describing a range of conditions and diagnoses and does not refer to any one specific condition. Developmental disabilities refer to a physical or cognitive impairment that can interfere with any aspect of development, such as gross motor, fine motor, speech and language, mobility, intellectual function, or emotional function. Although some developmental disabilities are evident in the first few months of age, others may not be recognized until the early school years. It can be difficult to diagnose specific disabilities when children are very young, so *developmental disability* (DD) is often used as a nonspecific umbrella term by clinicians indicating functional delays without a clear underlying medical disorder. The DD diagnosis is an important one because it can qualify a child for early intervention services (Ganjiwale et al., 2016). In addition to the medical use of "developmental disability," federal and state laws often have specific definitions. Assisting families to navigate the complex language of the systems touching the lives of medically complex children is an important role of the medical home care coordinator (Normile & VanLandeghem, 2018).

IDEA and Early Childhood Intervention Services: *"My 3-Year-Old Can Go To School?"*

The Individuals with Disabilities Education Act (IDEA) requires that early intervention services are available in every state, as early intervention is recognized as critical for the best outcomes for children. While the response to early childhood education services can vary, a referral for early intervention is still worthwhile and an invaluable resource to parents. Sometimes a specific diagnosis that explains the developmental delays is clear from the beginning, but for other children, more testing and time may be needed before a definitive diagnosis can be determined. For some children, a specific diagnosis is never found, despite all efforts, but it is important for parents and caregivers to know that addressing the child's symptoms with early and targeted intervention is still beneficial and recommended (IDEA, 2004). Schools address a child's needs based on an exceptionality as defined by IDEA, state regulations, and school district assessments. Chapter 12 explores educational interventions.

The Diagnostic Process: *"We Have to See Another Specialist?"*

ASD can be recognized as early as 18–24 months of age, but many children are not diagnosed until years later. Early warning signs can be seen with deficits in eye contact, language, adaptability, and social interactions. Typically, a child's pediatrician performs developmental monitoring as well as developmental screening as recommended by the American Academy of Pediatrics (AAP) at well-child visits for 9 months, 18 months, and 30 months of age. Screening specifically for ASD is recommended at the 18- and 24-month visits. If a child is not meeting developmental milestones at these visits, often the child will be referred by his or her pediatrician to a specialist at that time for further and more comprehensive evaluation. Depending on the area of concern with the child, they may be referred to a specialist such as a developmental pediatrician, child psychiatrist or psychologist, or for evaluation by physical, occupational, and/or speech therapy, to help clarify the source of the delay and to determine if early intervention services are needed (https://www.cdc.gov/ncbddd/autism/screening.html).

DSM-V Diagnostic Criteria for Autism Spectrum Disorder

ASD is diagnosed using the *Diagnostic and Statistical Manual of Mental Disorders*, 5th edition (DSM-5) (American Psychiatric Association, 2013). DSM-5 criteria for the diagnosis of ASD include the presence of persistent deficits in two major areas: (1) lack of reciprocity in social communication and interaction and (2) limited, often repetitive interests, activities, and behavior which may include stereotyped speech or movements, sensory sensitivities, and rigid adherence to routines with high resistance to transitions or changes in the routine. Language and intellectual impairments may or may not be present. These symptoms must be present from childhood and cause impairment in daily life, but it is important to note that the presentation of individual children varies greatly. The diagnostic process for ASD typically involves standardized testing materials and evaluation methods such as questionnaires, interviews with the parents or caregivers, and clinical observation of the child from an experienced professional. Prognosis is highly correlated with the presence of intellectual disability, language impairment (especially if it persists after age 5), and comorbid mental health issues (American Psychiatric Association, 2013). Earlier detection of ASD and prompt intervention can greatly improve the child's long-term prognosis (Volkmar et al., 2014).

Diagnostic Considerations: The Evaluation of Differential Diagnoses and Comorbidities

It is important to distinguish autism spectrum disorder from other developmental disorders and to determine if ASD is comorbid with a DD during the diagnostic process. ASD is diagnosed when a child meets the DSM 5 criteria for this disorder and when other disorders have been ruled out. Specific disorders which frequently have overlapping features with ASD include sensory disorders (e.g., deafness), sensory processing disorders, speech and language disorders, intellectual disability, reactive attachment disorder, obsessive-compulsive disorder, selective mutism, childhood-onset schizophrenia, and other organic disorders. In particular, as many children with ASD have symptoms related to communication and cognitive factors, it may be difficult to distinguish these types of diagnoses from ASD itself. In some cases, comorbidities may be more easily overlooked in children with a diagnosis of ASD. Common comorbidities of ASD are anxiety, difficulty with attention, intellectual disability, deficits in verbal communication, and behavioral issues. In patients with ASD, about 50% have profound intellectual disability (ID), 35% have mild to moderate ID, and 20% have intelligence quotients (IQs) in the normal range. As not all children with ASD have intellectual disability, it is important to note ID as a comorbidity when present. Prior to the DSM 5, diagnoses of autism disorder and Asperger's disorder were made. Under the older classification system, children who were classified as having autistic disorder usually have more deficits in verbal communication whereas those with Asperger's disorder often have more impairment in nonverbal than with verbal communication. Children with ASD who meet criteria for ADHD can be diagnosed with both disorders. Behavioral issues commonly seen with ASD include hyperactivity, obsessive and compulsive symptoms, aggression toward self or others, stereotypies, tics, and mood or affective problems which may be due to difficulty with emotional regulation. Symptoms of emotional dysregulation can include depression, anxiety, mood swings, or an inappropriate emotional response for a given situation such as over- or under-reactivity. Depression is a common comorbidity for children with Asperger's disorder. Children with ASD and other developmental disorders may be at greater risk for bullying which can intensify negative emotions and behaviors. These disorders may be diagnosed as comorbidities as appropriate based on DSM 5 symptom criteria (Volkmar et al., 2014).

Recommendations for the Treatment of Autism Spectrum Disorder

The American Academy of Child and Adolescent Psychiatry published practice parameters for the assessment and treatment of children and adolescents with ASD. The guidelines first were published in 1999 and then were updated in 2014 based on thousands of new research articles and a revision of the diagnostic criteria in the DSM 5 (AACAP, 2009). It may be helpful for parents and caregivers to be familiar with these parameters so they will know what to expect during the diagnostic and treatment process while keeping in mind that the parameters are general guidelines for clinicians rather than an official standard of care (Volkmar et al., 2014). The treatment of ASD is beyond the scope of this chapter. In general, treatment should be evidence-based, the symptoms targeted by the treatment clearly identified, and outcomes systematically tracked. There are many excellent resources to inform patients, their parents or caregivers, and their siblings. For a summary of resources, see Table 3.1.

For Providers: The Provider-Family Relationship – *"I'm an Expert About My Child Too Because I'm the Parent. Let's Work Together"*

Research has supported the idea that we as healthcare providers do not always adequately address what parents and caregivers want in the treatment of their children with ASD and other

Table 3.1 Resources for parents, caregivers, and families

Resources	Description
National Alliance for the Mentally Ill nami.org	NAMI provides an overview of treatment approaches available for mental health disorders including ASD. NAMI contains a plethora of information and support for patients as well as their family members. The resources on this website also include more personal supports such as a platform for individuals to share their stories to inspire hope for patients and their families, online discussion groups, NAMI Connection (a support group for individuals with mental illness), and NAMI Family Support groups to support their caregivers and other family members (NAMI, 2016)
The ARC Alliance thearcalliance.org	The ARC Alliance provides helpful descriptions of common symptoms of developmental disabilities, information on prevalence, and includes checklists of red flags for parents to monitor for speech and motor milestones as well as symptoms of autism. The ARC also includes a description of early intervention services for children from birth to age 3, which include play therapy, vision services, education, occupational therapy, speech therapy, physical therapy, social work, and nutrition services (The Arc Alliance, 2017)
Autism Speaks Website autismspeaks.org	Autism Speaks has extensive resources and support for family members of children with autism, including specific guides for parents, siblings, and grandparents. Personal stories from parents are shared to provide practical tips from their own experiences with their family journey with autism. In particular, there are tips for parents whose child has just been diagnosed. This site offers support and solidarity for parents and families as they adjust to the realities of a different parenting journey than they expected (Autism Speaks Website, 2021)
University of Michigan Medicine mottchildren.org	The University of Michigan Medicine website features Your Child Development and Behavioral Resources with extensive information on developmental delays. It describes how developmental delays are recognized, potential causes of delays, and options for early intervention. It includes information on the role of the school system in evaluations, the special education system, Individualized Education Plans (IEPs), transition planning into adulthood, the Individuals with Disabilities Education Act (IDEA), and legal information including parental rights. Finally, there are multiple recommendations for books, organizations, and other resources to equip parents with more understanding and approaches to parenting a child with developmental disabilities (University of Michigan Medicine, C. S. Mott Children's Hospital Website, 2021)
American Academy of Pediatrics aap.org	The American Academy of Pediatrics (AAP) provides a wealth of information on parenting that includes articles on parental mental health, parenting styles, tips for parenting children of various ages, and how to navigate conflicts as a parent. The AAP operates a companion site for families at healthychildren.org (American Academy of Pediatrics Website, 2021)

developmental disabilities. Some parents have reported that what they find to be helpful and unhelpful in treatment and what they want in the provider-family relationship sometimes conflicts with what providers think they need. Todorow et al. report that children diagnosed with autism spectrum disorder often have more physical and mental health needs than their peers with typical development, and even compared with other children with special healthcare needs. Often, these children require specialty services over the course of their development, including neurology, psychiatry or psychology, gastroenterology, endocrinology, allergy and immunology, and physical, speech, and occupational therapy services. The higher level of need underscores the importance of a high-quality medical home model for children with medical complexity. Patients with a higher-quality, family-centered medical home are found to have fewer unmet needs in comparison with children who receive typical primary care only or care in a less responsive medical home setting (Todorow et al., 2018).

A qualitative study by Carbone et al. (2010) compares discrepancies in the perspectives of pediatricians and families on care within a medical home model to meet the needs of children with autism spectrum disorder. Parents recognized the importance of early screening, diagnosis, and treatment interventions in the overall trajectory of their child's development. Some expressed frustration that their pediatricians did not respond soon enough when they raised concerns about their children's development. Having a pediatrician as a collaborator who promotes open communication can assist with creating a solid and trusting partnership between the family and the pediatrician. Parents often reported that the medical home model did not offer the level of care coordination and expertise required. They identified themselves as the primary care coordinators for the majority of care for their children's special needs. Many parents felt there was a lack of recognition from healthcare providers that they were experts on their children and did not feel valued as partners in shared decision-making with regard to the care of their children. Some parents perceived a lack of concern, poor listening, and dismissive attitudes at times from providers. Some important concepts such as puberty and transitioning to adulthood were not discussed with parents at all. Other families also reported frustration that their pediatricians were not knowledgeable or receptive to complementary and alternative medicine interventions as treatment options. Families reported often turning to other parents, friends, and the internet instead of to their providers for additional information and support (Carbone et al., 2010).

Early experiences with medical care can establish the tone between parents and healthcare providers for years to come. Parents in the Carbone study expressed that the negative experiences with their providers greatly affected later interactions with them as well. Parents often felt frustration, anger, and feeling isolated, fatigued, and overwhelmed due to so many aspects of care coordination falling upon them, as well as additional stress from financial strain (Carbone et al., 2010). Bringing humanity to patient care requires physicians to ask how having a child with special needs has affected the child's siblings and parents and to recognize the emotional burdens families face including grief for the loss of or change in the dreams and goals they originally had for their child. Families reported needing support for their child's challenging behaviors, and awareness of the potential for caregiver burnout. When needs for support are not met, parent and caregiver stress can result in poorer outcomes for the child (Van Orne et al., 2018).

From the pediatricians' perspectives, they reported obstacles of time constraints, lack of training and familiarity with prescribing psychotropic medications, a shortage of resources, and a limited knowledge of what community resources were available. Recommendations for improving patient advocacy in medical homes included improving pediatricians' education on autism spectrum disorder, scheduling longer appointments, and an interdisciplinary care model with better identification of resources, especially mental health resources to address behavioral challenges and for prescribing psychotropic medications. Other recommendations were for more extensive collaboration with specialists, schools, and ancillary providers as needed. Parents who were able to participate in family support groups and access respite care found these resources very helpful (Carbone et al., 2010).

For Providers: What Your Patients Want You to Know

The American Academy of Child and Adolescent Psychiatry (AACAP) published a Clinical Perspective on ten ways to support a person diagnosed with autism spectrum disorder. The *Ten Tips to Support Me* was endorsed by a group of youth diagnosed with Asperger syndrome and by the Board of Families from the Gipuzkoa Autistic Society. Although autism spectrum disorder encompasses a broad spectrum of presentations and personalities, these tips can apply to many people with this diagnosis. These *Tips* reflect the motto of The Autistic Self Advocacy Network https://autisticadvocacy.org/ "Nothing about us without us." The ASAN is a leading voice for the autistic self-advocacy movement.

Living with ASD, I/DD, and Medical Complexity: "Help Us By Respecting Us and Supporting Better Quality of Life"

Preferred Language Matters

It is important for clinicians to know about preferred language and how to ask both the patient and their parents or caregivers how the patient would like to be described. Discussing a patient's disability is another opportunity to practice humanity and show sensitivity and empathy. The National Alliance on Mental Illness (NAMI) and the National Center on Disability and Journalism (NCDJ) emphasize that language that describes a person's diagnosis should put the person first in a way that is respectful of the person and reduces stigma. The National Center on Birth Defects and Developmental Disabilities of the Centers for Disease Control & Prevention provides a guide for people-first language, which avoids identifying an individual by his or her disability. In most cases, this entails placing the reference to the disability after a reference to a person, as in "a person with a disability," or "a person living with a disability," rather than "the disabled person" (NAMI, 2016; National Center on Disability and Journalism, 2018).

The use of the term "developmental disability" is accepted, but referring to the specific diagnosis is always preferred, if possible. The NCDJ cautions against general use of the term "disability," which is no longer generally used or accepted. The prefix "dis" means "not," which labels a person as "not able." In the 1990s, the term "differently abled" emerged to take the place of terms "disabled," "mentally retarded," and "handicapped." This term is controversial and is not considered to be acceptable. "Handicapable" is a similar, but rarely used term. "Person with a disability" is often preferred instead (National Center on Disability and Journalism, 2018).

The NCDJ recommends "nondisabled," "does not have a disability," or "is not living with a disability" to describe typically developing individuals instead of "healthy," "whole," "normal," or "able-bodied." Terms such as "normal" and "abnormal" are considered inappropriate when describing a person. The term "neurotypical" has gained popularity in recent years and refers to an individual within the normal variation of the population (National Center on Disability and Journalism, 2018).

It is important not to assume that every person who has a disability has a lesser quality of life, is suffering, or considers himself or herself to be "afflicted," "stricken with," or a "victim," all of which imply that the person should be pitied. Other terms to avoid are "invalid," "disorder," "defect," "defective," "deformed," "deformity," "disfigurement," "disfigured," "crippled," "lame," or "damaged," as they imply the person is deficient, inferior, or incomplete. Alternative descriptions consistent with person-first language are "has," "lives with," or "has been diagnosed with." It is recommended to use the term "condition" instead of "disorder" and to specify the name of the specific diagnosis whenever possible (National Center on Disability and Journalism, 2018).

As mentioned previously, prior to the fifth edition of the *American Psychiatric Association's Diagnostic and Statistical Manual of Mental Disorders* (DSM-V) in 2013, the now integrated diagnosis of autism spectrum disorder formerly had subtypes of Asperger's syndrome, autism disorder, and childhood disintegrative disorder. For persons with this diagnosis, some prefer the term "autistic" or "autistic person," while others dislike the use of "autistic" as an adjective. Some prefer "a person with autism" instead. Many patients diagnosed with autism spectrum disorder have verbal limitations. In the past, people without expressive verbal communication were referred to as "dumb" or "mute," but these terms imply that the person has no ability to communicate. It is better to state how the person communicates (e.g., American Sign Language, augmentative or alternative communication). It is important for clinicians to ask the patient and family about their preferred language, rather than making assumptions on their preferences. For people who have experienced a traumatic brain injury, using the term "brain damaged" is not appropriate. Instead, "person with a (traumatic) brain injury" is preferred. According to the National Down Syndrome Society, preferred terms for patients with Down syndrome are "has/is living with/a person with Down syndrome" and include "cognitive disability" and "intellectual disability" (National Down Syndrome Society, 2021). Cerebral palsy is a broad term for various congenital neurological manifestations that affect movement and coordination due to early injuries in the brain. Again, person-first language is recommended (National Center on Disability and Journalism, 2018).

However, it is important to recognize that not all people with disabilities favor people-first language. In contrast with "people-first language," "identity-first language" (e.g., "disabled," "autistic," etc.) begins with the disability. Identity-first language developed from the concept that some people consider their disabilities to be integral and inseparable parts of who they are. This method is preferred in Great Britain and has been adopted by some disability activists in the United States as well. If unable to ask the person, it is recommended to use people-first language. Other terms have been more recently created, including "diversabled" and "diversability" to highlight the aspect of diversity as a part of disabilities, which are not widely used as of now (National Center on Disability and Journalism, 2018).

During the twentieth century, the word "retarded" came into use to describe a person's level of intelligence. With the beginning of IQ testing, additional terms were even more pejorative and now are considered offensive. With Rosa's Law in 2010, "intellectual disability" replaced the term "mental retardation." The terms "high functioning" and "low functioning" often are used to describe people with various conditions but often are considered derogatory, as they may trivialize or discount the person's abilities. An example of an alternative description is "a person who lives with minimal assistance" or "a person who lives with no special assistance." It is better to refer to the actual diagnosis and to specifically describe the areas in which the person excels, as well as areas of challenge (National Center on Disability and Journalism, 2018).

The designation "special needs" emerged at the beginning of the twentieth century during the development of educational programs for people with a range of disabilities. Many people now view the term "special" as offensive and stigmatizing, so this term is best used only in conjunction with formal programs with this designation. "Functional needs" in the context of the person's specific disability is preferred to "special needs" (National Center on Disability and Journalism, 2018).

Stages of Parental Development

Since the 1950s, the concept of "parent development" has been a recognized area of study (Freidman, 1957; Galinksy, 1987). For providers, it is important to determine where a family is in the process of parental development (at baseline and monitoring at intervals over time), and how the provider can serve as a better partner in this journey. The parental development stages are a general guideline for all parents but should be considered with more flexibility for parents of children with medical complexity, as the timeline of stages may need to be adapted based on the developmental progress of the child and the needs of the family (Table 3.2).

Table 3.2 Stages of parenting with modifications for parents and caregivers of children with medical complexity

Freidman parenting stages	Galinsky parenting stages and tasks	Potential challenges and modifications of stages (Pistorius)
Learning cues to accurately interpret an infant's needs	*Image-making: pregnancy to birth* Parental tasks: preparing for changes in current relationships, forming images of how the child will be and how they will be as parents	• Change in initial expectations, dreams, and goals • Feelings of frustration, confusion, anger, grief
Learning to accept a toddler's growth and development and corresponding parental loss of control	*Nurturing: 18–24 months* Parental tasks: balancing needs of the child with their own; lovingly caring for the child	• May be prolonged with delays in the child's development if more reliant on caregivers compared with typically developing children • May be more difficult for parents/caregivers to balance their own needs with those of their child if their child requires more specialized care and attention than TD peers
Learning to separate and allow independent development during childhood	*Authority: ages 2–4 or 5* Parental tasks: developing rules and expectations for behavior, helping the child learn and follow those rules	• Process of developing effective discipline may be extended and potentially more challenging for parents or caregivers to help their child learn and follow rules if any intellectual or behavioral problems are present
Learning to accept rejection from a child while balancing an appropriate level of support (being present when needed without intrusion or desertion)	*Interpretive: age 5 or 6 to age 12* Parental tasks: teaching the child about the world, helping the child identify authority figures, teaching morals and values to guide behavior and socialization	• For children who have difficulty with social reciprocity, it may be harder to learn how to interpret and navigate relationships appropriately • If the child is more concrete (i.e. has a more literal interpretation of the world rather than being able to engage in abstract thinking), he or she may have more difficulty applying and internalizing morals and values
Learning to create a new, independent life while supporting a teenager who is developing his or her own identity	*Interdependent: teenage years* Parental tasks: monitoring and guiding behavior, allowing the child to have more autonomy while the parent maintains authority	• The process of graduated autonomy may require modification depending on the teenager's degree of impulsivity and understanding of the world • Safety is an important concern based on the individual teenager's behavior and increase in relative vulnerability compared with peers

(continued)

Table 3.2 (continued)

Freidman parenting stages	Galinsky parenting stages and tasks	Potential challenges and modifications of stages (Pistorius)
	Departure: late teenage years or early adult years Parental tasks: general reflection (evaluating themselves as parents including identification of their successes with raising their children and also reflecting on ways they would have parented differently in retrospect)	• May require modification based on the functional abilities of the older teenager or young adult (e.g., if unable to live independently, may need to stay in home with caregivers or transition to an assisted living situation.) • Parental self-evaluation may differ between parents/caregivers of children with medical complexity if the child did not fulfill parental expectations and goals or is unable to function independently. In this situation, parents/caregivers may be more likely to have feelings of guilt or grief. As a result, it will be important for parents/caregivers to: Reflect honestly and realistically on their parenting Note their parental achievements as well as those of their child Balance the negative with the positive Recognize factors over which they had no control Recognize their child as valuable and as having a unique purpose in life no matter what his or her abilities Develop acceptance that they have done their best with the information and resources available to them

General: Parents/caregivers may experience more overlap between stages and different time frames for progressing through the stages

Proposed Stages for Parent Development for Families with a Child with Medical Complexity

Parents and caregivers of children with ASD or medical complexity may have an experience that differs from the stages of parental development detailed by Friedman and Galinsky. Acknowledging the spectrum of paths that good parents may take in their parenting journey is important to empower parents and inform professionals. Here, modifications to Friedman and Galinsky's stages are introduced to better align with the parent development in families with a child with medical complexity. For example, Galinsky's image-making stage of how their child will be and how their family relationships will change may be very different than their initial expectations. A greater range of emotions may be experienced by families with a medically complex child with a corresponding process of frustration, confusion, anger, and even grief as their child is noted to be developing differently than same-age peers.

Jolene Philo's book, *Different Dream Parenting: A Practical Guide to Raising a Child with Special Needs*, introduces the concept of how parents' and caregivers' dreams and goals sometimes end up having to change from what they had expected in the beginning (Philo, 2011). Olshanksy (1962) described this complex process as *chronic sorrow* (p. 190). While research points to mothers experiencing more chronic sorrow than fathers, additional research is needed to determine if this difference is related to care taking roles, personality factors, or some other variable (Coughlin & Sethares, 2017). Helping parents identify emotions related to the letting go of expectations and treating parents with respect and as experts about their child can help parents to more fully engage with the medical home.

Other parental stages such as the nurturing stage of toddlerhood described by Galinsky and the stage of growing autonomy on the part of the toddler described by Freidman may be prolonged if the child takes longer to progress with development and is more reliant on their caregivers than typically developing children of the same age. Parents and caregivers may also find it more difficult to balance their own needs with those of their child if their child requires more specialized care and attention than other children. Care needs may make finding childcare difficult or parents may feel reluctant to take time for themselves contributing to poorer health and mental health (Küçük & Alemdar, 2018; Lee et al., 2017).

The parental task of developing effective discipline in Galinsky's authority stage may also be extended. General rules and traditional wisdom about limit setting often requires child-specific modification. Parents or caregivers may face challenges helping their child learn and follow rules when intellectual or behavioral problems are present. Parents often report high stress around limit-setting and discipline (Shawler & Sullivan, 2017). The medical home team should consider referral to a mental health professional to help parents develop effective strategies for guiding their child's behavioral development.

The later parental development stages may also need modification when parenting a child with special needs. Parents and caregivers may experience more overlap between stages as well as different time frames for progressing through the stages. For example, during Galinsky's interpretive stage, children who have difficulty with social reciprocity may have a harder time learning how to interpret and navigate relationships appropriately. If the child has difficulty thinking abstractly and has a concrete, or literal, style of thinking, there may be more difficulty applying and internalizing morals and values. The graduated autonomy described in the interdependent stage may need to be modified depending on the teenager's degree of impulsivity and understanding of the world. While safety is an important concern at all stages, the increasing independence of the teenagers may create new vulnerabilities for the teen and challenges for parents.

For emerging adults with medical complexity, Galinsky's departure stage may be altered based on the functional abilities of the older teenager or young adult, and for parents, sorrow may reemerge. For some families, the young adult with special needs may not be able to live independently and may either need to stay at home with the caregivers or transition to an assisted living situation. The parental tasks of self-evaluation as a parent may differ for those with children who have special needs if the child did not fulfill parental expectations and goals or is unable to function independently. In this situation, parents and caregivers may be more likely to have feelings of guilt or grief, but it is important for them to reflect honestly and realistically on their parenting and to note their parental achievements as well as those of their child. During this evaluation process, it will be helpful for parents and caregivers to balance the negative with the positive, to recognize factors over which they had no control, to recognize their child as valuable and as having a unique purpose in life no matter what his or her abilities, and to develop acceptance that they have done their best with the information and resources that were available to them.

Professionals are cautioned to examine biases around the concepts of parenting and family. Like the development of medically complex children, the stages of development that their parent's experience should be thoughtfully assessed with interventions offered to maximize well-being and to promote full engagement with the medical home.

What Is Good Parenting?

Parenting is not a one-sided process directed toward the children. During the child-rearing years, parents and caregivers grow tremendously in their knowledge and wisdom about their child but also about themselves as they gain experience through their roles as parents. It is unrealistic to expect perfection from ourselves as parents. The process of learning to be a good parent is an important one that the medical home should support. In 1953, D. W. Winnicott introduced the concept of the "good enough mother" as one who is fully responsive to her infant's needs early on and then allows the child to experience graduated degrees of frustration as he or she gets older. This gradual process of introducing the child to the external world facilitates the independence that is essential for the child to separate from the mother in a way that promotes healthy development (Winnicott, 2016). The "good enough mother" does not have to be perfect but must be good enough to meet her child's needs, which often are more involved and demanding for a medically complex child. The medical home can provide critical information to parents on meeting their child's needs.

The quality of early parenting is important in determining the trajectory of children's short-term and long-term success in becoming well-adjusted, developing self-efficacy, and for the child with medical complexity, parenting comes with additional responsibilities for their child's health. Evidence shows that the parental behaviors required while raising a child must evolve based on the child's developmental stages: poor parenting skills have a greater association with psychopathology in children, as has been demonstrated in some cases of abuse (Rutter, 1998; Woodhouse et al., 2018). On the other hand, exposure to good parenting skills can help provide children with the best possible set of skills for emotional regulation, with corresponding positive effects. This does not mean that parenting alone can "fix" the diagnosis of ASD or any other diagnosis, but good parenting can help to provide children with the best outcomes possible (Mrazek et al., 1995).

As it is difficult to define "good parenting," there is not a bright line definition. Good parenting is dependent on context and child specific needs. Research into the elements of good parenting necessarily focuses on the parents; however, research must be examined critically and thoughtfully when considering the parents of a medically complex child. The approaches of Mrazek et al. (1995) and Eve et al. (2014) are offered as examples. Mrazek et al. (1995) identified the core aspects of parenting as emotional availability, control, psychiatric disturbance, knowledge base, and commitment. Emotional availability pertains to the amount of emotional warmth the parent or caregiver provides to the child. Control refers to the level of flexibility and permission from the parents or caregivers. Psychiatric disturbance describes the potential situation in which the parent or caregiver themselves has a psychiatric disorder and the severity of symptoms displayed in front of the child. Knowledge base is the parent or caregiver's basic understanding of the child's needs and physical and emotional development. Commitment refers to the parent or caregiver's ability to appropriately follow through with their responsibilities of the care of their child (Mrazek et al., 1995). Interestingly, this listing does not consider the parent's overall health, child's medical complexity, or social determinants of health, which contribute significantly to a parent's ability to achieve these core aspects.

Eve et al. (2014) compared feedback between professionals to clarify the concept of good parenting. Professionals included social workers, community services employees, psychologists, lawyers, and magistrates who had experience with conducting parenting capacity assessments. Six common themes emerged for qualities of good parenting: insight into their role as a parent, including knowing their child and being able to acknowledge their own limitations and when and how to get support as needed; being able and motivated to be a parent and do things for their children, acknowledge responsibility, and make changes when needed; considering and addressing both short-term and long-term needs of their child; protecting and making sacrifices for their children, and placing their child's needs before their own; developing a stable, healthy parent-child bond; and appropriately balancing consistent parenting routines and boundaries with the ability to be flexible when needed. Insight is related to "meta-parenting" which includes objective assessment of a parenting situation beginning with anticipating, correct assessment of the situation and necessary problem-solving and interventions, and general reflection, with the goal of improving one's parenting efficacy (Eve et al., 2014). Again, these themes fail to accommodate challenges of parenting medically complex children or recognize that a range of parenting styles can be good enough.

Parenting Styles

Parenting styles vary by culture, geography, and even over time in the same family. Still, there are some important generalizations about parenting style to consider. Baumrind (1971) studied parenting styles as they related to control, from which developed the concepts of authoritative, authoritarian, and permissive parenting. Parental control can be separated into three components of parenting style: (1) acceptance versus rejection; (2) psychological control versus autonomy; and (3) firm control versus lax control (Baumrind, 1967). An authoritative parenting style is considered to be superior to authoritarian and permissive approaches. Authoritarian patients place a high level of rigid demands on their children, often with strict punishments for disobedience, and provide a low degree of warmth and nurturing. Permissive parents are overly indulgent with low expectations, make few demands on their children without firm limits on their behavior, and often have blurring of the boundaries in the parent-child relationship. In contrast, authoritative parents provide a balance of firm, yet reasonable demands, appropriate discipline, and are kind and warm toward their children (Mrazek et al., 1995).

How the concepts of authoritative, authoritarian, and permissive parenting apply to families with medically complex children is in need of further study. Dyches et al. (2012) found positive parenting associated with better functional outcomes for children, including children being more prosocial, mature, empathic, and responsible, with enhanced cooperation and self-regulation skills, as well as superior language abilities and overall academic performance, with lower levels of maladaptive social behaviors. In contrast, a lack of positive parenting was more likely to result in depression, anxiety, and antisocial or aggressive behaviors during childhood. However, parents may default to more authoritarian or permissive parenting styles when under stress which in turn is associated with poorer executive functioning in children (Hutchison et al., 2016). Any parent knows that stress impacts our parenting style. Addressing stress through individual and group interventions as well as through broader interventions targeting social determinants of health shows promise to support parents to be the best version of themselves (Pierron et al., 2018).

Protective and Risk Factors

Based on clinical observation, the way in which parents relate to their children and the parent-child bond remain relatively stable over time. The presence of protective and risk factors can affect the quality of parenting and attachment and are an important point of intervention. Parenting abilities often originate from the parent's own genetic predisposition and early experiences of being parented in their family of origin. Intergenerational attachment patterns have been studied and have shown that attachment patterns between mother and child, in particular, often are similar to those the mother had with her own mother. The temperament of the individual child will play a role in the degree of parenting difficulty as well. In every parenting or caregiver situation, although the knowledge and skills needed to care for a child will change as the child progresses through each developmental stage, it is likely that parenting abilities will improve with experience over time. However, negative and unexpected life stressors in the family may negatively impact parenting. How difficult a child is to care for and the resources available to the family are very important components that can impact parenting. Social support such as extended family members who are available to help with childcare and parental self-care (including mental health treatment when needed) can be protective factors (Mrazek et al., 1995).

Parental involvement with a shared plan of care focusing on functional outcomes is a protective factor promoting better family relationships and overall child functioning and behavior.

Ventola et al. (2017) emphasized the importance of parental involvement for children with ASD to achieve improved outcomes including decreased internalizing and externalizing symptoms in children and in parents' symptoms of anxiety and depression. Szatmari et al. (2021) examined five domains of functioning (communication, socialization, activities of daily living, internalizing, and externalizing) in the middle childhood years for children with ASD and investigated corresponding attributes that led to proficiency or growth in these areas. Attributes associated with doing well were language ability, family functioning, and household income, with the later acting as a maker for multiple social determinants of health. Interestingly, cognitive ability was not a significant factor. The strong impact of family functioning indicates that improving the level of family functioning as well as other modifiable factors can lead to better outcomes for children with ASD.

The parental characteristic of optimism is a positive factor associated with an ability to reframe challenges that results in an improved sense of self-efficacy in parents of children with developmental disability (Minnes et al., 2015). A focus on the positive aspects of parenting children with intellectual disabilities rather than adopting a traditional focus on the negative aspects related to their care has been shown to benefit the parent-child relationship, increase parent enjoyment and fulfillment in their caretaking role, and foster personal growth despite difficult circumstances (Beighton & Wills, 2017) (Table 3.3).

When parents and caregivers are able to find meaning and goodness in this role of caring for their child with ID, they often experience additional benefits of better personal health and fewer symptoms of depression than when they have more negative perspectives of the situation. Due to the benefits both to the patients and their parents or caregivers in appreciating the positive aspects of their individual situations, this may be an area for healthcare providers to assist in developing therapeutic interventions to increase positivity for the families with whom they work. The concept of "meaning-focused coping" describes a method of cognitive reappraisal for stressful events, such as caring for a child with ID, in which the positive aspects of the situation are identified and focused upon so that any emotional and psychological burden related to the situation are diminished, and the parent or caregiver is able to find meaning in the hardship, cope better with the stressors, and shift their priorities to focus on what matters most. As mentioned earlier in the parenting stage of imagery, parents generally

Table 3.3 Positive aspects of parenting children with medical complexity

Appreciation of the child	Parental benefits	Relationship changes	Change in priorities and purpose
• Child is a source of joy/happiness • Sharing love with the child • Child's accomplishments (triumph over adversity) • Positive effect of the child on others (acceptance from others, raising awareness of disabilities)	• Pleasure/satisfaction in providing care for the child • Child provides a challenge or opportunity to learn and develop • Led to the development of new skills, abilities, or career opportunities • Increased personal strength or confidence; fighting for their child and developing inner strength • Sense of accomplishment in having done one's best for the child	• Strengthened family and/or marriage • More meaningful relationships (closer relationships, more empathy for others in similar situations) • Expanded social and community networks	• Led to a new or increased sense of purpose in life • Becoming a better person (more compassionate, less selfish, more tolerant) • Increased spirituality and faith • Changed perspective on life (clarified what is important in life, more aware of the future) • Making the most of each day, living life at a slower pace, greater appreciation of simple things

Adapted from Beighton and Wills (2017), p. 328, Hastings and Taunt (2002)

assume their child will follow a typical pattern of development, but if their child follows a different, unexpected trajectory of development, there is often a loss of that anticipated image they had of the way their child would be, what his or her life would be like, and what their own lives would be like as parents and as a family. The process of coming to terms with the shift in their reality and its associated loss is likely to be accompanied by feelings of unfairness. With meaning-focused coping, the parent or caregiver must adjust their dreams and plans, modify priorities and goals as needed, find new meaning and purpose in their situation, and work toward acceptance (Beighton & Wills, 2017).

Protective and risk factors for parents or caregivers may be identified in themselves or by a clinician who is working with the family. The identification of these factors is critical to the success of the child and the family, and various intervention strategies may be required based on the individual needs of that family to create a better outcome for all. Clinicians should ask about any problems the child may be having, conflict in the home, mental illness of anyone in the home, recent stressors, appropriateness of childcare, adverse childhood experiences, and available social support for the family. Clinicians should be aware of signs of abuse or neglect but also should take note of rejection from the parents or caregivers, as this can be another red flag in the family relationships. It is important to assess key factors at home including the emotional climate, parental expectations of the child, degree of support for the child in general as well as for his or her need for graduated autonomy, and the presence of any negative attributions toward the child. In the category of "adequate parenting," problems and stressors would be addressed in a timely and appropriate manner, and there would not be any ongoing problems. Both average and exceptional parenting would be in this category (Mrazek et al., 1995).

Quality of Life

The World Health Organization (WHO) has defined quality of life as "an individual's perceptions of their position in life in the context of culture and value systems in which they live, and in relation to their goals, expectations, standards, and concerns" (WHOQOL, 1994, p. 41–57). The WHO developed a quality of life (QOL) scale, the WHOQOL, to measure QOL for patients with a range of diseases, severities, ages, cultural, and socioeconomic groups across 15 countries. The US version of the WHOQOL was evaluated by Bonomi et al. (2000) and was found to have good internal consistency, reliability, and construct validity for measuring QOL in US adults. There are dozens of instruments for measuring health-related QOL in children and adolescents including general measures and disorder-specific measures (Wallander & Koot, 2016).

Quality of Life for Children with ASD and Developmental Disabilities

Measuring QOL for children with ASD/IDD can inform the shared plan of care and medical home practice to improve outcomes for children and families. Williams et al. (2020) investigated the QOL of children with ASD/IDD, defining QOL as life satisfaction in important areas and maintaining physical and emotional well-being. Parental ratings of the child's level of functioning and independence as well as the child's overall level of participation and estimated quality of life were tracked. The potential for a range of physical comorbidities to affect quality of life was also considered. Lower quality of life was associated with a greater level of dependence for daily activities and with poorer eye contact, but communication and mobility were not correlated with a lower quality of life. Better quality of life was associated with increased participation and the rewards of increased emotional well-being and social connectedness gained from participation. Potential areas to target for enhanced participation are art, sports, and recreational activities that are directed toward the child's interests and abilities. Having the opportunity to choose and to have some control over the activity can improve the child's engagement, enjoyment, and sense of self-efficacy. Williams et al. (2020) recognized a "disability paradox" (p. 98) in which a child's impairment does not have to result in poor quality of life, and demonstrated that activities can be modified as needed instead of having to be avoided or restricted. Interventions that maximize a child's level of independent functioning and communication, and those which encourage meaningful social interactions correspond with improved quality of life.

Quality of Life for Parents and Caregivers

The physical and emotional demands for parents and caregivers of medically complex children may increase beyond the level of care that is required for typically developing children. Most people with developmental disabilities need some level of lifelong support, which most often is provided by their parents and extends the parenting role indefinitely. The complex caregiving and prolonged responsibilities of parenting children in these situations can negatively affect parents' general health, mental health, and marital relationships (Namkung et al., 2015; Fairfax et al., 2019). Perceptions of quality of life of the parents or caregivers may decrease as a result of the increased stress of taking care of a medically complex child. Self-care must become a priority for parents and caregivers in order to prevent frustration and burnout. While some families adjust well to the challenges of parenting a child with a disability, it may be more difficult for other families to make these physical and mental adjustments. Learning to manage stress appropriately and developing positive coping skills will be essential, not only for the benefit of the caregiver but also for the child and the family as a whole (Ganjiwale et al., 2016).

While the relationship between parents may be strained when caring for their medically complex child, the Wisconsin Longitudinal Study found no significant increase in the risk or timing of divorce for parents who had a child with a developmental disability (Namkung et al., 2015). Maintaining strong, supportive relationships with others, including one's spouse, extended family, friends, and society in general (including other parents of children with developmental disabilities and self-help groups) is essential to maintaining a good QOL. Social support that allows for open communication and family cooperation in the care of the child and the household was found to be very helpful for parents and caregivers. Strategies that support parents' financial stability and the ability to participate in relaxing and enjoyable activities also are important to QOL (Ganjiwale et al., 2016).

Ganjiwale et al. (2016) observed that QOL can be improved by teaching parents and caregivers coping strategies including problem-focused coping (utilizing active coping and planning, including religion and other types of support), active emotional coping ("venting," use of positive reframing, humor, acceptance, and emotional support), and limiting avoidant emotional coping (such as distraction, denial, self-blame, behavioral disengagement, and substance use). The most effective coping strategies varied by the child's diagnoses. Parents and caregivers of children with cerebral palsy, epilepsy, and learning disabilities tended to utilize the "active emotional" style of coping. Those whose children had Down syndrome tended to use the "problem-focused coping." Parents of children with ADHD used the "avoidant emotional" coping style more frequently. Parents of children with intellectual disabilities tended to use both problem-focused and active emotional coping strategies on an equal basis. Both of these coping styles were associated with higher self-reports of quality of life. Parents reported that having realistic and positive perspectives and acceptance of their child's disability had positively impacted their coping abilities and had resulted in a more optimistic and hopeful outlook on the future for their children and for themselves (Ganjiwale et al., 2016). Similarly, Fairfax et al. (2019) found an association between coping methods and psychological QoL for parents or caregivers raising a child with a disability or chronic illness. Adaptive coping strategies including problem-focused coping, stress management techniques, flexibility, and cognitive reappraisal strategies were positively associated with psychological QoL, while maladaptive coping methods were negatively associated with QoL. These findings support targeted treatment interventions for developing caregiver coping strategies to improve the lives of both the caregivers and their children. These findings emphasize the need for healthcare providers to be aware of their patients' family dynamics. By having better knowledge of how parents are responding to and coping with the stress of caregiving, the providers can then offer family interventions targeting parental education, parenting skills and management training, problem-solving, coping skills, community support, and family cohesion as needed to improve the QOL of the family unit and promote the best outcomes for the child (Puka et al., 2018).

Instilling Hope: *"What Is On the Horizon?"*

With some diagnoses, general prognosis is known. However, with the spectrum of developmental delay and intellectual disability commonly seen in patients with autism spectrum disorder and in other children with medical complexity, it is not always possible to predict what a child will be able to accomplish and what symptoms may be lifelong limitations. It is essential for clinicians to meet the patient and the family where they are by addressing each symptom to the best of their ability, and providing the necessary education, instructions, resources, and referrals to address these symptoms. As the patient becomes an adolescent and gets closer to adulthood, it is important to provide information on transitioning to adulthood. Further exploration of the transition to adulthood can be found in the final chapter of this book.

Raun Kaufman, whose family founded The Autism Treatment Center of America in 1983, continues to perpetuate hope through his personal and professional experiences with autism spectrum disorder. His message highlights the importance of hope for patients, families, and clinicians. He writes that although there are no guarantees, families should never have to accept a message of discouragement as the certain reality for their children. Instead, Kaufman gives a message that hope can only lead to beneficial rather than negative results. He points out the value of families being open to possibilities in their child's future as a necessity for improved outcomes, even when others have diminished expectations, and not to assume that these limitations must be reality. His essential message to parents and caregivers of children with autism is that hope can motivate patients and families into action, which is essential for positive outcomes (Kaufman, 2021).

Parents of children with ID most frequently described their children in terms of the joy they brought to their families, pride in their children as unique individuals, and from a perspective of hope. Making room for the positive along with the challenges inherent in caring for a medically complex child can lead to resiliency and hope for families that their child will still be able to make progress and reach goals in their lives (Beighton & Wills, 2017). Hope for families in these situations has been described as "a complex intangible in the healing and coping process" (Kauser et al., 2003: 35). Extensive interviews of parents and caregivers with medically complex children clearly illustrate that the emotions expressed by families indicate how much they love their children (Beighton & Wills, 2017).

Determining prognosis for children with autism can be difficult, as every child is unique. Early diagnosis, lower severity of symptoms, showing joint attention, greater cognitive abilities, and functional play skills all are associated with better outcomes. A major shared goal of families and clinicians should be maximizing the quality of life for a child with autism. This objective requires that the SPoC be regularly updated to ensure an interdisciplinary, collaborative approach that utilizes social, communication, cognitive, skill acquisition, and behavioral interventions to improve participation in all aspects of life. Treatment approaches may consist of behavioral (including ABA therapy), medical, pharmacological, and nutritional interventions. In addition to currently available treatments, substantial research is being conducted on an ongoing basis for treatments to help improve symptoms, decrease level of disability, and enrich the child's quality of life. Maximizing the available treatments and resources will lead to better outcomes for the child and for the family as a whole (CDC, 2019).

References

American Academy of Child and Adolescent Psychiatry. (2009). Practice parameter on the use of psychotropic medication in children and adolescents. *Journal of the American Academy of Child & Adolescent Psychiatry, 48*(9), 961–973.

American Academy of Pediatrics Website. (2021). Screening Technical Assistance & Resource Center: Developmental screening. https://www.aap.org/en-us/advocacy-and-policy/aap-health-initiatives/Screening/Pages/Early-Childhood-Development.aspx

American Psychiatric Association, American Psychiatric Association, & DSM-5 Task Force. (2013). *Diagnostic and statistical manual of mental disorders: DSM-5* (5th ed.). American Psychiatric Association.

Autism Speaks Website. (2021). https://www.autismspeaks.org/

Baumrind, D. (1967). Child care practices anteceding three patterns of preschoool behaviour. *Genetic Psychology Monographs, 75*(1), 43–88.

Baumrind, D. (1971). Current patterns of parental authority. *Developmental Psychology, 4*, 1–103.

Beighton, C., & Wills, J. (2017). Are parents identifying positive aspects to parenting their child with an intellectual disability or are they just coping? A qualitative exploration. *Journal of Intellectual Disabilities, 21*(4), 325–345.

Bonomi, A. E., Patrick, D. L., Bushnell, D. M., & Martin, M. (2000). Validation of the United States' version of the world health organization quality of life (WHOQOL) instrument. *Journal of Clinical Epidemiology, 53*(1), 1–12.

Carbone, P. S., Behl, D. D., Azor, V., & Murphy, N. A. (2010). The medical home for children with autism spectrum disorders: Parent and pediatrician perspectives. *Journal of Autism and Developmental Disorders, 40*(3), 317–324.

Center for Disease Control (CDC). (2019). Autism spectrum disorder: Treatment and intervention services for autism spectrum disorder.

Centers for Disease Control and Prevention, National Center on Birth Defects and Developmental Disabilities (NCBDDD). (2020). Screening and diagnosis of autism spectrum disorder. https://www.cdc.gov/ncbddd/autism/screening.html

Coughlin, M. B., & Sethares, K. A. (2017). Chronic sorrow in parents of children with a chronic illness or disability: An integrative literature review. *Journal of Pediatric Nursing, 37*, 108–116.

Dyches, T. T., Smith, T. B., Korth, B. B., Roper, S. O., & Mandleco, B. (2012). Positive parenting of children with developmental disabilities: A meta-analysis. *Research in Developmental Disabilities, 33*(6), 2213–2220.

Eve, P. M., Byrne, M. K., & Gagliardi, C. R. (2014). What is good parenting? The perspectives of different professionals. *Family Court Review, 52*(1), 114–127.

Fairfax, A., Brehaut, J., Colman, I., Sikora, L., Kazakova, A., Chakraborty, P., & Potter, B. K. (2019). A systematic review of the association between coping strategies and quality of life among caregivers of children with chronic illness and/or disability. *BMC Pediatrics, 19*(1), 1–16.

Freidman, D. B. (1957). Parent development. *California Medicine, 86*(1), 25–28.

Galinksy, E. (1987). *The six stages of parenthood*. Perseus Books.

Ganjiwale, D., Ganjiwale, J., Sharma, B., & Mishra, B. (2016). Quality of life and coping strategies of caregivers of children with physical and mental disabilities. *Journal of Family Medicine and Primary Care, 5*(2), 343–348.

Hastings, R. P., & Taunt, H. (2002). Positive perceptions in families of children with developmental disabilities. *American Journal on Mental Retardation, 107*(2), 116–127.

Hutchison, L., Feder, M., Abar, B., & Winsler, A. (2016). Relations between parenting stress, parenting style, and child executive functioning for children with ADHD or autism. *Journal of Child and Family Studies, 25*(12), 3644–3656.

Individuals with Disabilities Education Act, 20 U.S.C. § 1400. (2004). https://sites.ed.gov/idea/

Kaufman, R. K. (2021). Autism Treatment Center of America Website: Autism and the myth of false hope. https://autismtreatmentcenter.org/knowledge-base/autism-and-the-myth-of-false-hope/

Kauser, S., Jevne, R. F., & Sobsey, D. (2003). Hope in families of children with developmental disabilities. *Journal on Developmental Disabilities, 10*(1), 35–46. [Google Scholar].

Küçük, E. E., & Alemdar, D. K. (2018). Life satisfaction and psychological status of mothers with disabled children: A descriptive study. *Community Mental Health Journal, 54*(1), 102–106.

Lee, M. H., Park, C., Matthews, A. K., & Hsieh, K. (2017). Differences in physical health, and health behaviors between family caregivers of children with and without disabilities. *Disability and Health Journal, 10*(4), 565–570.

Minnes, P., Perry, A., & Weiss, J. A. (2015). Predictors of distress and well-being in parents of young children with developmental delays and disabilities: The importance of parent perceptions. *Journal of Intellectual Disability Research, 59*(6), 551–560.

Mrazek, D. A., Mrazek, P., & Klinnert, M. (1995). Clinical assessment of parenting. *Journal of the American Academy of Child and Adolescent Psychiatry, 34*(3), 272–282.

Namkung, E. H., Song, J., Greenberg, J. S., Mailick, M. R., & Floyd, F. J. (2015). The relative risk of divorce in parents of children with developmental disabilities: Impacts of lifelong parenting. *American Journal on Intellectual and Developmental Disabilities, 120*(6), 514–526.

National Alliance on Mental Health Public Policy Platform. (2016, December). Revised twelfth edition.

National Center on Disability and Journalism. NCDJ style guide revised 2018. https://ncdj.org/style-guide/

National Down Syndrome Society Website. (2021). https://www.ndss.org/about-down-syndrome/down-syndrome/

Normile, B., & VanLandeghem, K. (2018, June). *State strategies to advance medical homes for children and youth with special health care needs*. National Academy for State Health Policy.

Olshansky, S. (1962). Chronic sorrow: A response to having a mentally defective child. *Social Casework, 43*(4), 190–193.

Pierron, A., Fond-Harmant, L., Laurent, A., & Alla, F. (2018). Supporting parenting to address social inequalities in health: A synthesis of systematic reviews. *BMC Public Health, 18*(1), 1–27.

Philo, J. (2011). *Different dream parenting: A practical guide to raising a child with special needs*. Discovery House Publishers.

Puka, K., Tavares, T. P., Anderson, K. K., Ferro, M. A., & Speechley, K. N. (2018). A systematic review of quality of life in parents of children with epilepsy. *Epilepsy and Behavior, 82*, 38–45.

Rule, A. R. L. (2018). I am that parent. *Journal of the American Medical Association, 319*(5), 445.

Rutter, M. (1998). Developmental catch-up, and deficit, following adoption after severe global early privation. *The Journal of Child Psychology and Psychiatry and Allied Disciplines, 39*(4), 465–476.

Shawler, P. M., & Sullivan, M. A. (2017). Parental stress, discipline strategies, and child behavior problems in families with young children with autism spectrum disorders. *Focus on Autism and Other Developmental Disabilities, 32*(2), 142–151.

Szatmari, P., Cost, K. T., Duku, E., Bennett, T., Elsabbagh, M., Georgiades, S., … Zwaigenbaum, L. (2021). Association of child and family attributes with outcomes in children with autism. *JAMA Network Open, 4*(3), e212530.

The Arc Alliance Website. (2017). Autism and ASD. https://thearcalliance.org/autism-asd/

The WHOQOL Group. (1994). The development of the World Health Organization quality of life assessment instrument (the WHOQOL). In J. Orley & W. Kuyken (Eds.), *Quality of life assessment: International perspectives* (pp. 41–57). Springer-Verlag.

Todorow, C., Connell, J., & Turchi, R. M. (2018). The medical home for children with autism spectrum disorder: An essential element whose time has come. *Current Opinion in Pediatrics, 30*(2), 311–317. https://doi.org/10.1097/MOP.0000000000000605

University of Michigan Medicine, C. S. Mott Children's Hospital Website. (2021). Autism resources for parents. https://www.mottchildren.org/conditions-treatments/autism-spectrum-disorders/resources-parents

Van Orne, J., Branson, K., & Cazzell, M. (2018). Boot camp for caregivers of children with medically complex conditions. *AACN Advanced Critical Care, 29*(4), 382–392.

Ventola, P., Lei, J., Paisley, C., Lebowitz, E., & Silverman, W. (2017). Parenting a child with ASD: Comparison of parenting style between ASD, anxiety, and typical development. *Journal of Autism and Developmental Disorders, 47*(9), 2873–2884.

Volkmar, F., Siegel, M., Woodbury-Smith, M., McCracken, J., State, M., & the American Academy of Child and Adolescent Psychiatry (AACAP) Committee on Quality Issues (CQI). (2014). Practice parameter for the assessment and treatment of children and adolescents with autism spectrum disorder. *Journal of the American Academy of Child and Adolescent Psychiatry, 53*(2), 237–257.

Wallander, J. L., & Koot, H. M. (2016). Quality of life in children: A critical examination of concepts, approaches, issues, and future directions. *Clinical Psychology Review, 45*, 131–143.

Williams, K., Jacoby, P., Whitehouse, A., Kim, R., Epstein, A., Murphy, N., … Downs, J. (2020). Functioning, participation, and quality of life in children with intellectual disability: An observational study. *Developmental Medicine and Child Neurology, 63*(1), 89–96.

Winnicott, D. W. (2016). *The collected works of DW Winnicott* (Vol. 12). Oxford University Press.

Wirth, B., & Kuznetsov, A. (2016). *Shared plan of care: A tool to support children and youth with special health care needs and their families*. National Center for Medical Home Implementation.

Woodhouse, S., Miah, A., & Rutter, M. (2018). A new look at the supposed risks of early institutional rearing. *Psychological Medicine, 48*(1), 1–10.

Systems of Care for Children and Youth with Autism Spectrum Disorders

4

Andres J. Pumariega

Abstract

Over recent decades, significant strides have been made in evidence-based care for children with autism spectrum disorders, yet we lack a comprehensive system of care approach to address the needs of children/youth/families with ASD, including clinical, support, educational, and social needs. We have fragmented components of a system that are poorly knit together so that navigating them are too challenging for families, especially for those with less education, poorer economic resources, and families of minority children/youth who are now the numerical minority as of 2020. This chapter explores the history of systems of care for children with ASD, the current status of that system, and envisions a system that meets the diverse needs of children and families.

Keywords

Autism spectrum disorders · Autism · Systems of care · Level of care

A. J. Pumariega (✉)
Division of Child and Adolescent Psychiatry, Department of Psychiatry, University of Florida College of Medicine, Gainesville, FL, USA
e-mail: pumariegaa@ufl.edu

Introduction

The prevalence of autism spectrum disorders (ASD) has been increasingly adjusted upward over the last one to two decades, and it is unclear whether this is a result of increased incidence or better identification. Over the past four decades, there have been significant advancements in the assessment, treatments, and services for children and youth with ASD. These have included early systematic screening, comprehensive assessments, applied behavioral analysis, social skills training, speech and OT interventions, and adjunctive pharmacotherapy such as atypical antipsychotics for irritability and aggression and the more effective identification and management of comorbid disorders such as ADHD and anxiety disorders.

In spite of these advances, there are still major challenges in accessing care for children with ASD and their families. These services have not grown and scaled up to the level at which we now identify children, youth, and adults with ASD and emotional/behavioral needs. They are also associated with a confusing array of portals of entry and agencies for different age groups, dispersed location of services, disparate criteria for qualification for services, and divergent funding streams. These result in significant fragmentation of these services across service systems and providers. While we have experienced recent

P. McPherson (ed.), *Handbook of Treatment Planning for Children with Autism and Other Neurodevelopmental Disorders*, Autism and Child Psychopathology Series,
https://doi.org/10.1007/978-3-031-06120-2_4

improvements in insurance coverage for services for ASD and recent inclusion in Medicaid and third-party private plans, at the same time, there has been relative defunding of special education services in public schools, which has adversely affected recently instituted and hard-won educational services for ASD. There is also unevenness of standards of care in skill and expertise in the care of ASD that diminish service effectiveness.

A growing concern is the potentially rising cost of care for people with ASD. Blaxill et al. (2021) estimated the future cost of ASD in the United States using a forecast model that accounts for the true historical increase in ASD, with inputs including ASD prevalence, census population projections, six cost categories, ten age brackets, inflation projections, and three future prevalence scenarios. They estimate that future ASD costs will increase dramatically: from total base-case costs of $223 (175–271) billion/year in 2020 to $589 billion/year in 2030, $1.36 trillion/year in 2040, and $5.54 (4.29–6.78) trillion/year by 2060. Rising prevalence, the shift from child- to adult-dominated costs, the transfer of costs from parents onto government, and the soaring total costs raise pressing policy questions and demand an urgent focus on prevention strategies, which present opportunities for savings.

This chapter outlines the trajectory of recognition of the special clinical and support needs of people with ASD and the development of services and technologies to address these needs before moving to the challenge of a fragmented service system for this population, and a possible model that might promote coherence, effectiveness, and address cost concerns.

Brief Historical Review of the Evolution of ASD Services

It is now believed that ASD have been present among humans for millennia, but it was not until relatively recently that it has been recognized as a distinct behavioral entity with specific neurodevelopmental characteristics. People with ASD were likely cared for at home, with naturalistic accommodations made by family and communities. The first documented effort at specific identification an intervention occurred around 1800 by Dr. Jean-Marc-Gaspard Itard, a French physician, who studied a boy found naked and abandoned in a forest. The boy did not speak, did not respond to language and seemed deaf, and spent his time rocking and shelling beans. Itard disagreed with the prevailing medical opinion that dismissed the boy as being an "imbecile," and he cared for the boy and experimented with compassionate behavioral approaches to help him, with his notes later published in the book titled the *Wild Boy of Aveyron* (Wolf, 2004).

For the next one- and one-half centuries, people with ASD continued to be confused and comingled with people with intellectual disabilities and served primarily in large congregate residential programs. These were either funded by state governments or charitable nonprofit entities, though largely serving well-to-do families. One example was the Massachusetts School for the Feeble-Minded, subsequently renamed the Fernald State School, founded in 1848 by Dr. Samuel Gridley Howe, a physician and anti-abolition activist who had previously had success founding a school for the blind in that state (married to Julia Ward Howe, author of the Civil War anthem, The Battle Hymn of the Republic). Interestingly, this program came about after Dr. Howe conducted a systematic survey of "idiocy" that included data from 574 people thoroughly examined by him and colleagues in 63 towns in Massachusetts. The survey, which was intended as an advocacy tool to support the need for state funded residential services, included numerous people who exhibited clear documented signs and symptoms of ASD. He also documented the deplorable living conditions of these individuals, largely adults, in almshouses, kept in cages, and left to wonder unclean and uncared for, making the ethical case for improved care and living conditions (Donvan & Zucker, 2016).

Helena Trafford Devereaux, a young Philadelphia teacher, took another route to the development of residential services originating from her experiences with disabled children underserved by the public education system. She began teaching some of these children in her own home, believing that children, regardless of the level of their disabilities, could learn and achieve personal growth in an environment tailored to their needs. In January 1918, Helena Devereux used $94 of saved and borrowed funds to rent a house in Devon, Pa. On May 1, 1918, she and her students took residence in the house that became known as "Devereux Stone." Soon after, her students numbered 12, and, in 1919, Devereux was able to purchase the rental property, as well as the neighboring estate. From that point, the school and subsequent organization saw continued growth, comprising different schools and facilities united under one name – Devereux Schools. In 1938, the Commonwealth of Pennsylvania granted Devereux a nonprofit charter, and it then started a Foundation that grew the programs beyond Devon, PA, to California in 1943. By that time, it also employed a director of Psychology and Education, Edward L. French, Ph.D., who subsequently became the director of the Devereux Foundation upon Helena Devereux's retirement in 1957. Today, Devereaux has campuses in Texas, Massachusetts, Connecticut, and Arizona, developed academic affiliations, and became established as a national organization with expertise in the care and treatment of people with ASD as well as other behavioral health needs (Devereaux Foundation, 2021).

In parallel with these developments, the care of people with ASD underwent major advances and transformations. These occurred largely within the academic sector, both in psychiatry, psychology, pedagogy/education, and disciplines such as occupational therapy and speech/language therapy. At the same time, the academic sector both grew its involvement in care provision as well as translation of early investigational findings to the fledgling clinical sector. These included the first formal diagnostic definitions for disorders that now come under ASD (autism by Dr. Leo Kenner in the 1930s to 1943, Dr. Hans Asperger in 1944), the early use of pharmacotherapy for aggression and agitation associated with autism (Dr. Lauretta Bender M.D. and others in the 1960s and 1970s with the advent of first generation antipsychotics), the first behavioral trials using aversives in the 1970s, and auditory integration training (Dr. Guy Berard in the 1970s). Research from academic centers also contributed to major shifts in the understanding of the phenomenology and mechanism of disease in autism and related disorders. Whereas psychoanalysis had influenced this understanding through some of the early theories around early childhood development, leading to such misinterpretations as the "refrigerator mothers or parents," "symbiotic psychosis," and "parentectomy" concepts, later academics such as Folstein and Rutter in 1977 conducted the first twin studies establishing the genetic heritability of autism, later replicated for many related childhood genetic disorders. The work of Kanner and Asperger was later refined and integrated into the diagnostic classifications for autism, Asperger's, and ASD in the *Diagnostic and Statistical Manual of the American Psychiatric Association* starting in 1980 as well as the development of systematic diagnostic tools such as the Autism Diagnostic Interview (ADI) and the Autism Diagnostic Observational Scale (ADOS). Ivar Lovaas PhD published his initial work on discrete trial learning and applied behavioral analysis in 1987, establishing a highly evidence-based methodology for treatment integrating behavioral psychology, occupational, and speech/language therapy which became the basis of subsequent early intervention services with young children and management of problematic behaviors for older children. Later, clinically impactful discoveries included the studies establishing the evidence of risperidone and aripiprazole in the management of aggression, irritabilities, and stereotypies in the early 2000s, the use of intensive behavioral, multisensory, and communication interventions for intervention with younger children, and the identification and treatment of comorbidities such as ADHD in children and youth with ASD, and toddler level screening and identification for ASD using evidence-based tools such as the M- CHAT (Wolf, 2004; Kanner, 1943; Asperger, 1944; Lovaas, 1987; Dawson et al., 2010; Robins et al., 2001).

Though the academic sector provided some services as part of the research and development work it engendered, the scope and level of these services were initially too limited to make a significant contribution to access to care. This was especially true given the rapid increase in identification of children, adolescent, and even adults with ASD as improved diagnostic criteria and phenomenological concepts were adopted. The broader access to services for people with ASD began through advocacy by organizations such as the Autism Society (founded in 1988), advocating for special education services, including behavioral services, to be delivered by public schools. This advocacy work led to the 1997 amendment for the inclusion of autism within the purview of the Individuals with Disabilities Act, originally passed by Congress in 1975 to ensure all children and youth ages 3–22 receive a free and appropriate public education regardless of any disability. For the first time, this allowed children with autism spectrum disorders access to the same level of education as other children.

In the 2000s, continued advocacy at the state and national level contributed to the start of insurance coverage for medical and applied behavioral analysis (ABA) services, which previously were denied by both Medicaid and private insurance. Most of these benefits were initially instituted state to state and capped separately from other medical and behavioral health services. It was not until the advent of the Affordable Care Act (ACA) and associated Mental Health Parity Federal legislation in 2010 that benefits and copays were set at equal levels as that of other medical and behavioral health conditions (Centers for Disease Control, 2021a, 2021b; U.S. Department of Health and Human Services, 2021).

The neurodiversity movement more recently has engaged people with ASD in their self-advocacy for recognition of rights, services, and supports while at the same time looking to de-pathologize ASD. They have focused on greater recognition of needs and rights of young adults with ASD, including an access of services and supports for transition age and adults. These include community and 4-year college campuses (also under IDEA and the Americans for Disabilities Act), as well as vocational rehabilitation services that provide employment preparation and support services, distinguishing them from those usually delivered for people with intellectual disabilities and including linguistic support and sensory accommodations (Hendricks & Wehman, 2009; Roux et al., 2013).

Service System Roles in the Care of People with ASD

Early Intervention Services

Subsequent to the inclusion of children with ASD in special education services, all 50 states instituted early intervention services for children suspected to be on the autism spectrum in the 0- to 3-year period. These services are perhaps some of the most valuable, as some studies have pointed to the reversal of signs and symptoms of autism in children without intellectual disabilities with early intensive intervention including ABA and occupational and speech therapy, usually over 20 hours weekly. These early intervention services are the closest to primary to secondary prevention level services available for ASD. Systematic screening by pediatricians with tools such as the M-CHAT for toddler age screening for ASD dovetails nicely to improve access to early intervention services. Education of parents is facilitating greater access to such services but often requires much support to families to persist through lengthy application processes (Robins et al., 2001; O'Malley, 2003; Reichow et al., 2018).

Educational System

Children with autism require comprehensive and intensive services, often combining special education, speech and language pathologists, occupational and physical therapy, behavioral services, and others, all to work together to plan, problem-solve, and administer a child's individualized educational program (IEP). These special education programs have been shown to greatly improve quality of life and allow children to succeed. An important aspect of this mandate was the inclusion of special education services for children ages 3 and above, even if not actively attending school, to enable early educational intervention in preparation to entry to preschool and kindergarten (Public Law 99-157; Education of the Handicapped Act, 1986). As more youth with ASD progress academically and attend college, extension of such services has been developed by colleges and universities, often including social supports and campus neurodiversity advocacy groups.

Specialty Services

Medical Specialties

The needs of children, youth, and adults with ASD are typically identified and addressed by the fields of Pediatrics, Developmental Pediatrics, Psychiatry/Child and Adolescent Psychiatry, and Neurology/Pediatric Neurology with the support of a broader array of specialists to address medical complexities. Parents often approach pediatricians with developmental concerns which may lead to referral to specialists with the expertise in clinical diagnostic evaluation and pharmacotherapy of interfering symptoms of ASD (mood, aggression, stereotypies) as well as comorbidities such as ADHD, tic disorders, mood disorders, and others. The training within each of these specialties covers the assessment and treatment of people with ASD from different perspectives, each with its own level of time and focus on people with ASD. There is no formal certification of specialization for ASD within or across these specialties, though some of these specialists may pursue either sub-subspecialty fellowship training in ASD or develop such expertise by virtue of their practice focus (Volkmar et al., 2014; Hyman et al., 2020). There is a growing recognition that the Medical Home Model of well-coordinated, patient-centered care is a good fit for the needs of children with ASD, especially those with medical complexity; however, a medical home may be less suited to coordinate mental health services (Todorow et al., 2018).

Psychological Specialties

Neuropsychology and Developmental Psychology are key in the systematic assessment of people with ASD, including establishing diagnoses and assessment of cognitive and intellectual function. Clinical psychologists may provide varying levels of diagnostic and therapeutic services including cognitive behavioral and social skills training. Behavioral psychologists specialize in behavioral intervention based on behavioral analysis and discrete trial learning, most often collaborating with masters' level and bachelor's level intervention specialists. Doctoral and masters' level Behavioral Psychologists can achieve Behavioral Analysis Certification when they meet criteria for specialization in ABA (Volkmar et al., 2014; Hyman et al., 2020).

Occupational Therapists, Speech/Language Therapists, and Physical Therapists

These specialties are essential in addressing sensory deficits, communication/language deficits, and motor deficits, respectively. They will often integrate cognitive behavioral approaches within their interventions or collaborate with behavioral psychologists and medical specialists (Volkmar et al., 2014; Hyman et al., 2020).

Private Providers

Many of the specialties identified above will practice independently or in small groups within their specialty or in multi-specialty groups. Some of these are oriented to serving children and youth with ASD or also with other developmental challenges (such as learning disabilities or even other behavioral health disorders). Private providers rarely have home- or community-based service options or case management to coordinate multi-agency services, even across other private providers, relying on families themselves in performing case management and coordination services. While care coordination through a pediatric medical home can improve outcomes and decrease parental stress, the cost of implementing a medical home can be burdensome for small and medium practices; clearly, systems at all levels have not finished the work of designing a seamless system of coordinated care for children (Simpser & Hudak, 2017).

Mental Health and Developmental Disorder/Intellectual Disability Agencies

The public sector was originally comprised of statewide agencies that provided services for mental health and developmental disorders. The original paradigm was laid out by the Community Mental Health Act of 1963, which established nonprofit and public community mental health centers in communities across the United States, with the goal of increasing access to ambulatory level services and reducing institutional care. Some states organized differently and had separate Developmental Disorder/Intellectual Disabilities agencies and as such either operate specialty centers or residential programs. Many of these agencies have been dismantled over time or integrated into broader mental health/behavioral health agencies. Additionally, community mental health centers have both become more generic (focusing more on mental health and addiction services) and also shifting from public funding and control to not for profit independent organizations. Many of these agencies specialize in case management and coordination services, which assist in the coordination of services by providers from different sectors, as well as in intensive home- and community-based services (Centers for Disease Control, 2021b).

The needs of children and youth on the spectrum who are in state custody (either in child welfare or juvenile justice custody or involvement) are primarily addressed through the respective state agencies. The prevalence of children with autism in foster care was 7.5% in 2001, increased to 10.5% in 2005, and then declined to 9.1% in 2007. When statistically controlled for children's age, race, sex, and state of residence, children with autism were 2.4 times more likely, and children with intellectual disability 1.9 times more likely, to enter foster care than typically developing children (Cidav et al., 2017). Currently, these agencies primarily provide case management services and use Medicaid or other state funding supports to arrange for services through community mental health or contracted private service providers, often segmented off from general services.

Some states have developed resource, coordination, and support programs for individuals with ASD rather than fund direct clinical services. These can often collaborate with established providers and academic centers. An example of these is the Centers for Autism and Related Disorders (CARDs) in the state of Florida. The Florida CARDs are a series of centers associated with different academic institutions, staffed by specialists in education, behavioral psychology, and case management. They provide families and individuals with ASD services such as referral support, case management, advocacy with school districts around educational services and rights, and socialization support services for youth with ASD, and adult transition services involving college campuses and vocational rehab agencies.

The adult developmental disabilities sector has traditionally served people with more severe intellectual disabilities, including many comorbid with ASD. This sector has been moving rapidly to deinstitutionalize individuals currently in state residential programs or facilities to community care largely involving small group home residences or home care, including the few youths in such programs. However, the behavioral support resources and staffing needed to successfully move such individuals to community care are very limited, placing them at risk of functioning at lower levels than their potential, which contributes to the frequent utilization of psychiatric emergency services, hospitalization, and overmedication. Following the US Supreme Court decision in *Olmstead v L.C* (1999), many states have been subject to class action lawsuits over the access and quality of care for its developmentally disabled citizens (O'Malley, 2003).

We are not only identifying a greater number of children with ASD but also witnessing greater numbers demonstrating the potential to achieve higher levels of education and independent function. This, combined with greater advocacy by persons with ASD and their families, has led to a growing interest in transition planning and services for adolescents moving into adulthood. For most young people, including those with ASD, adolescence and young adulthood are filled with new challenges, responsibilities, and opportunities. However, research suggests fewer young people with ASD have the same opportunities as their peers without ASD. These findings include high rates of unemployment or underemployment, low participation in education beyond high school, the majority continuing to live with family members or relatives, and limited opportunity for community or social activities, with nearly 40% spending little or no time with friends. In addition, individuals with ASD may experience changes in their ASD symptoms, behaviors, and co-occurring health conditions during adolescence and young adulthood. These changes can affect their ability to function and participate in the community. The Centers for Disease Control (CDC) has initiated various programs aimed at tracking adult outcomes and planning for future service needs at the state and local level CDC's most recent funding cycle for the Autism and Developmental Disabilities Monitoring (ADDM). Network includes support for five sites to follow up on 16-year-olds who had been identified with ASD by 8 years of age. It will provide valuable information on transition planning in special education services and potential service needs after high school. CDC's Study to Explore Early Development (SEED) began identifying children with ASD in the mid-2000s, and these children are now beginning the transition from adolescence to adulthood. Through SEED Teen, CDC is tracking the changes that occur during this transition period to learn about factors that may promote more successful transitions and better outcomes in young adults with ASD (Hendricks & Wehman, 2009; Roux et al., 2013; Dudley et al., 2019; Centers for Disease Control, 2021b). The healthcare transition of emerging adults is addressed in the final chapter of this volume.

Academic Sector

The academic sector has more recently grown its clinical services in response to both reductions in state funding and research support, leveraging its unique status in housing specialty and subspecialty services and training programs and combining translational/clinical research and service. These academic centers are often multidisciplinary, including many if not most of the above listed specialties, offering services ranging from diagnostic assessments to treatment services and supports. Some include both specialty inpatient services and community-based services. Programs may be located within medical schools/centers or main campus academic institutions and can range from specialty and multispecialty clinics to multidisciplinary centers. Those with larger scope with multiple levels of care include the Kennedy Krieger Institute at John Hopkins and Marcus Autism Center in Atlanta. Academic centers are currently growing in many parts of the United States providing added alternatives for evidence-based services. Such services are highly subscribed and rarely have ready access due to long wait times and insurance requirements.

The National Institute of Child Health and Human Development (NICHHD) supports an initiative to grow such academic centers, the Autism Centers of Excellence (ACE) Program. The ACE program is a trans-NIH initiative that supports large-scale multidisciplinary studies on autism spectrum disorders, focused on determining the disorders' etiologies and potential treatment interventions. The program includes ACE research *centers*, which foster collaboration between teams of specialists who share the same facility to address a particular research problem in depth, as well as ACE research *networks*, which consist of researchers at many facilities throughout the country, all of whom work together on a single research question.

The ACE program is funded various grant mechanisms out of the NIH, with support coming from NICHD's Intellectual and Developmental Disorders (IDD) Branch, the National Institute of Mental Health, the National Institute for Deafness and Other Communication Disorders, the National Institute of Neurological Disorders and Stroke, and the National Institute of Environmental Health Sciences. These institutes are also active members of the NIH Autism Coordinating Committee, which was created in 1997 in response to a request from Congress to enhance the quality, pace, and coordination of NIH autism research. These institutes are also members of the federal Interagency Autism Coordinating Committee, which includes representatives from various agencies within the US Department of Health and Human Services and other governmental agencies, including the US Department of Education (National Institute for Child Health and Human Development, 2021).

Residential Treatment Centers

Residential treatment centers serving people with ASD continue to operate across the United States. Where access formerly depended on primarily on private funding or charitable or state support, private third-party insurance and even public insurance have provided support for this sector/level of care. Typically, youth who are served in these programs have multiple problems and complex needs. Whereas some poorly run nonaccredited institutions may operate very basic near-custodial services with few specialist and evidence-based interventions, the great majority have adopted interdisciplinary specialty approaches with an intensive treatment model often based on ABA technology. Certification by The Joint Commission (TJC) or the Commission on Accreditation of Rehabilitation Facilities (CARF International) is essential and often required by states for qualification for Medicaid and private insurance funding of services.

Role of Funding in Transforming and Shaping ASD Services

As with most health and human services, technological advances in care and funding of care are the prime drivers in shaping the service system. The services provided to people with ASD in the last century and before were primarily residential and custodial and received meager funding by states or charitable organizations. Some charitable and state funding went toward some of early specialty services around the middle of the twentieth century. Access to those services were limited, and qualification was haphazard, depending on subjective criteria. Healthcare inequity was the norm as wealthy individuals could access private residential programs due to their direct payments or even their charitable donations.

Funding by school districts for special services under IDEA were the first services for ASD that were universally accessible to the overall population, racial disparities, and disparities of access resulting from regional funding disparities notwithstanding. Though these services were crucial in providing a level of support for some independent function, they were often restricted by eligibility criteria, often leaning toward less functional children with more obvious problems and needs that impacted their ability to make educational progress. Unfortunately, educational and school-based services have been curtailed more recently due to their high level of demand as there is higher identification of (and possibly higher prevalence) of ASD.

More recently, expanded insurance funding under Medicaid and under third-party insurance, both due to the ACA, have served to expand access to clinical services (ambulatory specialty, behavioral, and even residential services) and some educational support services (such as OT and speech therapy) outside of public schools. Unfortunately, there are arbitrary barriers that have been set up around access to such services. For example, though there is fairly ready access to medical specialty services, at times, the reimbursement level is low, and many medical providers (such as child and adolescent psychiatrists) will not accept direct insurance reimbursement at all, and some will not accept lower paying insurances. Insurance funding for many other services such as applied behavioral analysis and residential services require systematic psychological assessment using the ADOS or ADI, and as a result, there are long queues waiting for such assessment services. Additionally, with the defunding of special education, access to behavioral services within schools has decreased, while insurance companies do not allow for funding for ABA services to cross into or collaborate with schools. Such restrictions render many behavioral interventions ineffective since they cannot be developed and applied across all the child's functional settings (Centers for Disease Control, 2021a; U.S. Department of Health and Human Services, 2021).

The most challenging aspect of all is the multiple portals of entry and multiple types of providers and agencies that need to be coordinated to effectively address the needs of many if not most people with ASD. Such complexity, without highly coordinated care, often leaves parents bewildered, confused, and overwhelmed, even those with high levels of education and health literacy. Sometimes, those at the lowest socioeconomic levels who qualify for public case management services have better care coordination, though it also can falter at critical transitions, such as the transition to adult services. Adult services for ASD are significantly less developed and often are tied into public services for those with serious persistent mental illness (Centers for Disease Control, 2021a).

Racial/Ethnic Disparities

The challenges encountered in accessing and coordinating services for ASD are compounded by the structural racism within our provider agencies and systems, resulting in significant racial/ethnic disparities in identification, diagnosis, and service access and quality. A number of recent studies have begun to document such disparities. Mandell et al. (2009) studied 2568 children aged 8 years identified as meeting surveillance criteria for ASD through record abstraction from multiple sources. A total of 58% of children had a documented autism spectrum disorder. However, in adjusted analyses, Black (odds ratio [OR] = 0.79; 95% confidence interval [CI] = 0.64, 0.96), Hispanic (OR = 0.76; CI = 0.56, 0.99), or other race/ethnicity (OR = 0.65; CI = 0.43, 0.97) children were less likely than White children to have documented ASD. This disparity persisted for Black children, regardless of IQ, and was concentrated for children of other ethnicities when IQ lower than 70. Lim et al. (2020) reviewed eight studies and found that children of immigrants with ASD are diagnosed at a later age, those with limited English proficiency receive fewer service hours, and barriers include long wait times, language barriers, and limited health literacy. Magana et al. (2015), examining data from the 2005/2006 and 2009/2010 National Survey of Children with Special Health Care Needs, found racial/ethnic disparities in the quality of provider interactions were substantial in both 2005/2006 and 2009/2010. Black and Latino parents were significantly less likely than White parents to report that their provider spent enough time with their child and was sensitive to the family's values. Quebles et al. (2020) examined data for 2576 children with ASD 6–18 years old from the Autism Treatment Network (ATN) dataset. Multivariable logistic regression for age, gender, DSM-IV-TR ASD diagnosis (Autistic Disorder, PDD-NOS, Asperger's Disorder), and parents' education did not show any racial or ethnic differences in behavioral challenges, conduct problems, or sleep disturbances. Black children had lower odds of total problem behaviors; Asian children had lower odds of hyperactivity vs Whites. Racial and ethnic minority children had lower odds of total problem behaviors and conduct problems compared to Whites. Diverse children had lower odds of medication use across range of different problems except for sleep disturbance.

Obviously, culturally competent services that address access, cultural values/beliefs, and acceptability of services and provide modifications that address cultural differences are critical (Pumariega et al., 2013). At the same time, funding for services for underserved populations is also key. For example, LaClair and colleagues (2019) used quasi-difference-in-difference models to determine the effect of Medicaid waiver generosity on racial/ethnic disparities in ASD. Unmet needs among Black vs White children with ASD were roughly cut in half (13% decrease), with implementation of an average generosity waiver. No significant differences were seen for Hispanic ethnicity.

Framing ASD Services Within a Community-Based Systems of Care Model

Though we have made significant strides in evidence-based care for ASD, yet we lack a comprehensive system of care approach to address the needs of children/youth/families with ASD, including clinical, support, educational, and social needs. We have fragmented components of a system that are poorly knit together so that navigating them are too challenging for families, especially for those with less education, poorer economic resources, and families of minority children/youth who are now the numerical minority as of 2020.

Over the last 35 years, the community-based system of care model and principles have been developed in response to similar problems with fragmentation of care and lack of appropriate access in child mental health, particularly for children and families with multiple problems and needs. These principles are based on a flexible and individualized approach to service delivery for the child and family within the home and community as an alternative to treatment in out-of-home settings while attending to family and systems issues that impact such care. The key principles include access to a comprehensive array of services, treatment individualized to the child's needs, treatment in the least restrictive environment possible, full utilization of family and community resources, full participation of families and youth as partners in service planning and delivery, interagency coordination, the use of case management for service coordination, no ejection or rejection from services due to lack of "treatability" or "cooperation" with interventions, early identification and intervention, smooth transition of youth into the adult service system, effective advocacy efforts, and non-discriminating, culturally sensitive services (Winters et al., 2007) (see Table 4.1).

Table 4.1 The Community-based Systems of Care Model shares key principles with the Pediatric Medical Home Model

The Community-Based Systems of Care Model	The Pediatric Medical Home Model
1. Full participation of families and youth as partners in services planning and delivery of culturally sensitive, non-discriminating services	1. Developmentally appropriate and culturally competent family-centered care with shared decision-making guiding care
2. Provision of a comprehensive array of services	2. Provision of comprehensive primary care
3. Treatment guided by an individualized treatment plan developed collaboratively with the family	3. Care guided by a shared plan of care developed in collaboration with the family
4. Early identification and intervention	4. Coordination with early intervention programs
5. The use of case management for individualized service coordination	5. Care coordination with pediatric medical subspecialists and surgical specialists
6. Full utilization of family and community resources	6. Continuity of care between ambulatory and inpatient settings
7. Smooth transition of youth into the adult service system, effective advocacy efforts	7. Organized, well-planned transitions, including transition to adult-oriented health care, work, and independence
8. Interagency coordination at a systems level	8. Providing clear and unbiased information
9. Treatment in the least restrictive environment possible	9. Providing family access to a comprehensive, central record
10. No ejection or rejection from services due to lack of "treatability" or "cooperation" with interventions	10. All insurance accepted, with changes accommodated

Adapted from Winters et al. (2007); Medical Home Initiatives for Children With Special Needs Project Advisory Committee (2002)

Family-driven care is a cornerstone of the system-of-care model and has had a significant influence on national policy for both child and adult mental health (33–35). The child and family drive the clinical planning process through determining the goals and desired outcomes of services, selecting the composition of the interagency service planning team, evaluating the effectiveness of services, and having a meaningful role in all decisions, including those that impact funding of services. The interagency planning team typically has representatives from all the agencies and sectors involved with the child, and the team process facilitates interagency and interdisciplinary collaboration. The complementary contributions of various team members function synergistically in identifying system and community resources to promote better outcomes (Winters et al., 2007).

For children with complex problems involved in multiple child-serving agencies, assessment and treatment planning are primarily accomplished through interdisciplinary clinical teams. These teams bring together different clinical and support resources to address the child's needs to supporting him/her and their family in their community environment. Teams use the wraparound process, a specific model of a child- and family-driven team planning process that has been empirically tested within systems of care. Wraparound is a definable, integrated planning process that results in a unique set of community services and natural supports that are individualized for a child and family to achieve a set of positive outcomes. The wraparound process builds on the strengths of the child and family, is community-based (using a balance of formal and informal supports), is outcome-driven, and provides unconditional care. The use of a strength-based orientation and discussion of needs rather than problems promote more active engagement by families in service planning activities. Interventions designed to reinforce strengths of the child and family may include nontraditional therapies such as specific skills training or mentored work experiences that remediate or offset deficits. These interventions generally are not included in traditional categorical funding and may require flexible funds that are not assigned to specific service types. Care management is key for the wraparound process so that different services and different interventions can be well-coordinated and integrated for greatest effectiveness, and not duplicated (Winters et al., 2007).

Family participation is also facilitated through the parallel development of child and family teams (CFTs). CFTs are composed primarily of nonprofessional members led by the consumer family, usually a parent. In cases of older youth as consumers, the youth may serve as team leader. Empowering youth and families to assume a central role in outlining treatment goals and planning requires the involvement of specially trained individuals who can guide families to develop such goals. CFTs collaborate with interdisciplinary teams and professionals in agencies providing services. The CFT creates an overall care plan, including a crisis plan. The clinical team then negotiates their role in the crisis and care plans. This negotiation further educates families about how their child's needs could be addressed through treatment and enables professionals to learn about the realities faced by the family (Winters et al., 2007).

More recently, service quality, cost-effectiveness, and outcomes and integration of evidence-based practices have received greater emphasis within community-based systems of care programs. An example of such emphasis has been the multisite national evaluation of the Comprehensive Mental Health Services Program for Children and Their Families. This program, which has funded over 100 local and regional systems of care programs, has evaluations both over baseline ratings at the start of the programs and matched control evaluations. These have demonstrated significant improvement in child and family function using objective measures, reduced suicidality, reduced racial/ethnic disparities, increased stability of living situation, reduced hospitalizations, and reduced cost of care in other service sectors such as education, juvenile justice, child welfare, and general health, with correlations of outcomes to fidelity to system of care principles (U.S. Department of Health and Human Services, 2015).

To date, there has been limited application of the community-based system of care principles and model in the realm of services for people with ASD. However, the parallels between the challenges in caring for children with serious emotional disturbances and people on the autism spectrum are significant, and these populations often overlap. The ideal application of the system of care model for people with ASD would have basic medical, developmental, and behavioral health services coordinated through the Pediatric Medical Home, which is increasingly recognized at the "hub" for ASD services and supports. The Pediatric Medical Home shares many principles and elements with the community-based mental health systems model (Asarnow et al., 2017) (see Table 4.1). An adjunctive entity that has recently surfaced nationally and is closely tied to pediatric medical homes is Pediatric Behavioral Health Collaborative Programs. These provide the supports that are often necessary for pediatric primary care providers to access entry level behavioral health (and often developmental) services, including consultation (telephonic or televideo) with child and adolescent psychiatrists, psychologists, and licensed social workers, assistance with care coordination services to assist families in accessing community resources, ongoing training and skill building support, and technical assistance for practices to integrate behavioral health services within the practice's care processes. This model has been effective in facilitating primary management of entry level to moderate behavioral health needs in primary care. This collaboration can also identify children and youth with ASD with complex needs who require a more interdisciplinary coordinated care approach (Asarnow et al., 2015; Pumariega et al., 2016).

For children, youth, and adults with ASD who have such complex needs, more intensive care could be coordinated at the local level using interdisciplinary teams and child and family teams. These teams could pursue comprehensive assessment and treatment planning, bring in the necessary medical and behavioral specialists/disciplines from their respective sectors (private, public, academic, nonprofit, etc.) to negotiate service/treatment plans with families and affected persons, implement such plans, address arbitrary barriers to access to care and care coordination (e.g., between private/academic and public providers, across schools and applied behavioral analysis providers), and blend funding sources for different types and levels of services. Such interdisciplinary teams would be significantly more clinically effective and cost-effective. State entities that currently provide limited support and referral services could provide the oversight, structure, and case management support and serve a convening function for such teams, with incentives from enhanced funding for team participation (as opposed to solo treatment in silos) and empower such teams by streamlining eligibility and access procedures. More uniform standards around the qualifications of service providers and application of evidence-based interventions to fidelity by such care coordinating entities could also greatly enhance such approaches. The primary care provider within the Pediatric Medical Home and the pediatric behavioral health collaborative program would continue to be central and engaged within these teams.

A core technology for the operation of such teams would need to be behavioral analysis and discrete trial learning with multiple baseline designs, with all interventions the youth or adult receive being coordinated and measured in terms of reduction of agreed-upon target behavioral outcomes. Multiple baseline designs lend themselves nicely to objective measurement of outcomes for individual sand could be used in the aggregate to learn more about treatment and service effectiveness.

The orientation to least restrictive environment of care for system of care programs could bring greater emphasis on home-, school-, and community-based services for people with ASD, leaving residential services for highly unmanageable youth and adults. Objective level of care tools based on systems of care principles could be used to determine level of care intensity need. One such tool, the CALOCUS/CASII (Fallon et al., 2006) is not only directly applicable to this goal but is designed to be used within developmental disorders settings. There are positive outcomes that could result from this greater emphasis, with identification of key services. For example, Mandell et al. (2012) examined the use of respite care and therapeutic services, based on procedure codes, versus hospitalizations associated with a diagnosis of ASD. They found that each $1000 increase in spending on respite care during the preceding 60 days resulted in an 8% decrease in the odds of hospitalization in adjusted analysis. The use of therapeutic services was not associated with reduced risk of hospitalization.

A system of case model with a public health orientation would also enhance data collection to assist in advocacy for improved service funding. We continue to find inadequate funding, especially from the private insurance sector, for community-based services that often result in higher institutionalization and lower community functionality for people with ASD. For example, Wang et al. (2013) compared healthcare costs and service use for autism spectrum disorder (ASD) between Medicaid and private insurance, using 2003 insurance claims data in 24 states. In terms of costs and service use per child with ASD, Medicaid had higher total healthcare costs ($22,653 vs. $5254), higher ASD-specific costs ($7438 vs. $928), higher psychotropic medication costs ($1468 vs. $875), more speech therapy visits (13.0 vs. 3.6 visits), more occupational/physical therapy visits (6.4 vs. 0.9 visits), and more behavior modification/social skills visits (3.8 vs. 1.1 visits) than private insurance (all $p < 0.0001$). In multivariate analysis, being enrolled in Medicaid had the largest effect on costs, after controlling for other variables. These findings emphasize the need for continued efforts to improve private insurance coverage of autism. More recently, Liu et al. (2021) used 2008–2013 Medicaid data to conduct a retrospective cohort study to evaluate the effect of Medicaid home- and community-based services (HCBS) waiver programs on emergency department (ED) utilizations among youth with autism spectrum disorder (ASD). They showed that annual ED utilization rates were 13.5% and 18.8% for individuals on autism-specific and intellectual and developmental disabilities (IDD) waivers, respectively, vs. 28.5% for those without a waiver. Compared to no waivers, autism-specific waivers (adjusted odds ratio, 0.62; 95% confidence interval, [0.58–0.66]) and IDD waivers (0.65; [0.64–0.66]) were strongly associated with reduced ED utilization. These findings suggest that HCBS waivers are effective in reducing the incidence of ED visits among youth with ASD.

Conclusion

It is clear that services for people with ASD will continue to grow in scope and demand over the coming decades as we better identify and diagnose affected individuals. Scientific advances, particularly in prevention science and neurobiology, might mitigate the societal and family costs and burdens that are associated with this growth. However, equal savings and increase in effectiveness can result from service system reform that can right-size services (enhance community-based and reduce higher cost hospital and residential) and enhance service coordination/integration and family and youth engagement.

References

Asarnow, J. R., Rozenman, M., Wiblin, J., & Zeltzer, L. (2015). Integrated medical-behavioral care compared with usual primary care for child and adolescent behavioral health: A meta-analysis. *JAMA Pediatrics, 169*(10), 929–937.

Asarnow, J. R., Kolko, D. J., Miranda, J., & Kazak, A. E. (2017). The pediatric patient-centered medical home: Innovative models for improving behavioral health. *American Psychologist, 72*(1), 13.

Asperger, H. (1944). Die "autistichen Psychopathen" im Kindersalter. Archive fur psychiatrie und Nervenkrankheiten. (U. Frith, Trans.) (Ed.) (1991). In *Autism and Asperger's syndrome* (Vol. 117, pp. 76–136). Cambridge University Press.

Blaxill, M., Rogers, T., & Nevison, C. (2021). Autism Tsunami: The impact of rising prevalence on the societal cost of autism in the United States. *Journal of Autism and Developmental Disorders*. https://doi.org/10.1007/s10803-021-05120. Accessed on August 1, 2021.

Centers for Disease Control. (2021a). *Accessing services for autism spectrum disorder*. https://www.cdc.gov/ncbddd/autism/accessing-services-for-autism-spectrum-disorder.html. Last accessed on August 2, 2021.

Centers for Disease Control. (2021b). *Treatment and intervention services for autism spectrum disorder*. https://www.cdc.gov/ncbddd/autism/treatment.html. Last accessed on August 1, 2021.

Cidav, Z., Munson, J., Estes, A., Dawson, G., Rogers, S., & Mandell, D. (2017). Cost offset associated with early start Denver model for children with autism. *Journal of the American Academy of Child and Adolescent Psychiatry, 9*, 777–783. https://doi.org/10.1016/j.jaac.2017.06.007. PMID: 28838582.

Dawson, G., Rogers, S., Munson, J., Smith, M., Winter, J., Greenson, J., Donaldson, A., & Varley, J. (2010). Randomized, controlled trial of an intervention for toddlers with autism: The early start Denver model. *Pediatrics, 125*(1), e17–e23.

Devereaux Foundation. (2021). *The history of Devereaux advanced behavioral health*. https://www.devereux.org/site/SPageServer/?pagename=history. Last accessed on August 2, 2021.

Donvan, J., & Zucker, C. (2016, January/February). The early history of autism in America. *Smithsonian Magazine*. https://www.smithsonianmag.com/science-nature/early-history-autism-america-180957684/. Last accessed on August 2, 2021.

Dudley, K. M., Klinger, M., Meyer, A., Powell, P., & Klinger, L. (2019). Understanding service usage and needs for adults with ASD: The importance of living situation. *Journal of Autism and Developmental Disorders, 49*(2), 556–568.

Education of the Handicapped Act Amendments of 1986, 20 U.S.C. §140 et seq. (1986). https://www.govinfo.gov/content/pkg/STATUTE-100/pdf/STATUTE-100-Pg1145.pdf. Last accessed on August 14, 2021.

Fallon, T., Pumariega, A. J., Sowers, W., Klaehn, R., Huffine, C., Vaughan, T., Winters, N., Chenven, M., Marx, L., Zachik, A., Heffron, W., & Grimes, K. (2006). A level of care instrument for children's systems of care: Construction, reliability, and validity. *Journal of Child and Family Studies, 15*, 140–152.

Hendricks, D. R., & Wehman, P. (2009). Transition from school to adulthood for youth with autism spectrum disorders. *Focus on Autism and Other Developmental Disabilities, 24*(2), 77–88.

Hyman, S. L., Levey, S. E., & Myers, S. M. (2020). Council on Children with Disabilities, Section on Developmental and Behavioral Pediatrics. Identification, evaluation, and management of children with autism spectrum disorder. *Pediatrics, 145*(1), e20193447.

Kanner, L. (1943). Autistic disturbances of affective contact. *Nervous Child, 2*, 217–250.

LaClair, M., Mandell, D., Dick, A., Iskandarani, K., Stein, B., & Leslie, D. (2019). The effect of Medicaid waivers on ameliorating racial/ethnic disparities among children with autism. *Health Services Research, 54*, 912–919.

Lim, N., O'Reilly, M., Sigafoos, J., Lancioni, G., & Sanchez, N. (2020). A review of barriers experienced by immigrant parents of children with autism when accessing service. *Review Journal of Autism and Developmental Disorders, 8*, 366–372. Published online. https://doi.org/10.1007/s40489-020-00216-9. Accessed on August 1, 2021.

Liu, G., Velott, D., Kong, L., Dick, A., Mandell, D., Stein, B., Murray, M., Ba, D., Cidav, Z., & Leslie, D. (2021). The association of the Medicaid 1915(c) home and community-based services waivers with emergency department utilization among youth with autism spectrum disorder. *Journal of Autism and Developmental Disorders, 52*(4), 1587–1597. https://doi.org/10.1007/s10803-021-05060-2. Last accessed on August 1, 2021.

Lovaas, O. I. (1987). Behavioral treatment and normal educational and intellectual functioning in young autistic children. *Journal of Consulting and Clinical Psychology, 55*(1), 3–9. https://doi.org/10.1037/0022-006x.55.1.3. ISSN 1939-117. PMID 3571656.

Magana, S., Parish, S., & Son, E. (2015). Have racial and ethnic disparities in the quality of health care relationships changed for children with developmental disabilities and ASD? *American Journal on Intellectual and Developmental Disabilities, 120*(6), 504–513.

Mandell, D., Wiggins, L., Carpenter, L., Daniels, J., DiGuiseppi, C., Durkin, M., Giarelli, E., Morrier, M., Nicholas, J., Pinto-Martin, J., Shattuck, P., Thomas, K., Yeargin-Allsop, M., & Kirby, R. (2009). Racial/ethnic disparities in the identification of children with autism spectrum disorders. *American Journal of Public Health, 99*, 493–498.

Mandell, D., Xie, M., Morales, K., Lawer, L., McCarthy, M., & Marcus, S. (2012). The interplay of outpatient services and psychiatric hospitalization among Medicaid-enrolled children with autism spectrum disorders. *Archives of Pediatrics and Adolescent Medicine, 166*(1), 68–73. https://doi.org/10.1001/archpediatrics.2011.714

Medical Home Initiatives for Children With Special Needs Project Advisory Committee. (2002). Reaffirmed 2008. The medical home. *Pediatrics, 110*(1), 184–186.

National Institute for Child Health and Human Development, National Institutes of Health, U.S. Department of Health and Human Services. (2021). *Autism Centers of Excellence (ACE) Program*. https://www.nichd.nih.gov/research/supported/ace. Last accessed on August 1, 2021.

O'Malley, K. (2003). Youth with comorbid disorders. In A. J. Pumariega & N. C. Winters (Eds.), *Handbook of community systems of care; the new child & adolescent community psychiatry* (pp. 276–315). Jossey Bass Publishers.

Olmstead v L.C 527 U.S. 581 (1999).

Pumariega, A. J., Rothe, E., Mian, A., Carlisle, L., Toppelberg, C., Harris, T., Gogineni, R., Webb, S., Smith, J., & American Academy of Child and Adolescent Psychiatry (AACAP) Committee on Quality Issues (CQI). (2013). Practice parameter for cultural competence in child and adolescent psychiatric practice. *Journal of the American Academy of Child and Adolescent Psychiatry, 52*(10), 1101–1115.

Pumariega, A. J., Roberts, M., Naydock, G., Solkhah, R., Hanbury, R., Kairys, S., & Kelley, P. (2016). Pediatric Psychiatry Collaborative: New Jersey model program and cases. *New Jersey Pediatrics, Fall 2016*, 6–11. ISSN 2375-477X

Quebles, I., Solomon, O., Smith, K., Rao, S., Lu, F., Azen, C., Anaya, G., & Yin, L. (2020). Racial and ethnic differences in behavioral problems and medication use among children with autism spectrum disorders. *American Journal on Intellectual and Developmental Disabilities, 125*(5), 369–388. https://doi.org/10.1352/1944-7558-125.5.369

Reichow, B., et al. (2018). Early intensive behavioral intervention (EIBI) for young children with autism spectrum disorders (ASD). *Cochrane Database of Systematic Reviews, 5*, CD009260.

Robins, D., Fein, D., Barton, M., & Green, J. (2001). The modified checklist for autism in toddlers: An initial study investigating the early detection of autism and pervasive developmental disorders. *Journal of Autism and Developmental Disorders, 31*(2), 131–144.

Roux, A. M., et al. (2013). Postsecondary employment experiences among young adults with an autism spectrum disorder. *Journal of the American Academy of Child and Adolescent Psychiatry, 52*(9), 931–939.

Simpser, E., & Hudak, M. L. (2017). Financing of pediatric home health care. *Pediatrics, 139*(3), e20164202.

Todorow, C., Connell, J., & Turchi, R. M. (2018). The medical home for children with autism spectrum disorder: An essential element whose time has come. *Current Opinion in Pediatrics, 30*(2), 311–317.

U.S. Department of Health and Human Services, Substance Abuse and Mental Health Services Administration Center for Mental Health Services, Child, Adolescent and Family Branch. (2015). *The comprehensive community mental health services for children with serious emotional disturbances program, report to Congress, 2015*. Publication No. PEP16-CMHI2015.

U.S. Department of Health and Human Services. (2021). *The affordable care act and autism and related conditions*. https://www.hhs.gov/programs/topic-sites/autism/aca-and-autism/index.html. Last accessed on July 31, 2021.

Volkmar, F., Siegel, M., Woodbury-Smith, M., King, B., McCracken, J., State, M., & American Academy of Child and Adolescent Psychiatry (AACAP) Committee on Quality Issues (CQI). (2014). Practice parameter for the assessment and treatment of children and adolescents with autism spectrum disorder. *Journal of the American Academy of Child and Adolescent Psychiatry, 53*(2), 237–257.

Wang, L., Mandell, D., Lawer, L., Cidav, Z., & Leslie, D. (2013). Healthcare service use and costs for autism spectrum disorder: A comparison between Medicaid and private insurance. *Journal of Autism and Developmental Disorders, 43*, 1057–1064.

Winters, N. C., Pumariega, A. J., & Work Group on Community-Based Systems of Care, Committee on Community Psychiatry, & Work Group on Quality Issues Practice Parameters for Child Mental Health in Systems of Care. (2007). Practice parameter on child and adolescent mental health care in community systems of care. *Journal of the American Academy of Child and Adolescent Psychiatry, 46*(2), 284–299.

Wolf, S. (2004). The history of autism. *European Child and Adolescent Psychiatry, 13*, 201–208. https://doi.org/10.1007/s00787-004-0363-5

5 Progress Monitoring During the Treatment of Autism and Developmental Disorders

Celeste Tevis, Megan Callahan, and Johnny L. Matson

Abstract

Autism spectrum disorder (ASD) is characterized by deficits in social interaction and communication, as well as repetitive and restricted behaviors. As the prevalence rates of ASD have steadily increased since the 1990s, there is a growing need to understand the optimal treatment options available for ASD and their efficacy. Monitoring the progress of individuals' response to interventions is crucial to manage any developmental challenges those with ASD may encounter throughout the life span. However, autism and developmental disorders are multifaceted, often presenting with a wide range of symptoms, impacting several developmental domains. Additionally, there are currently no standardized measures that can be utilized for the purpose of progress monitoring, further adding complications. The aims, challenges, limitations, and measures utilized for progress monitored among individuals with ASD are discussed.

Keywords

Progress monitoring · ASD · Developmental disorders · Comorbidities · Treatment outcomes · Response to treatment · Autism · Medical home

C. Tevis (✉) · M. Callahan · J. L. Matson
Department of Psychology, Louisiana State University, Baton Rouge, LA, USA
e-mail: ctevis1@lsu.edu

Autism spectrum disorder (ASD) is a neurodevelopmental disorder characterized by deficits in social interaction and communication, as well as the presence of repetitive and restricted behaviors (American Psychiatric Association, 2013). Since the 1990s, prevalence rates of ASD have steadily risen (Matson & Kozlowski, 2011; Bolte & Diehl, 2013) and are most currently estimated to impact 1 in 54 children (CDC, 2020). As a result, there is an increased need to identify optimal treatments for ASD. Previously, the focal point of research regarding ASD measurement tools was on their use in early diagnosis, as opposed to their use in assessing progress throughout interventions (Bolte & Diehl, 2013). While accurate identification of ASD serves as a crucial initial step, monitoring progress throughout interventions is necessary for managing developmental challenges that arise throughout the life span (Eapen et al., 2016).

ASD is multifaceted and commonly presents with a wide array of heterogeneous symptoms (Eapen et al., 2016; Bolte & Diehl, 2013). The

P. McPherson (ed.), *Handbook of Treatment Planning for Children with Autism and Other Neurodevelopmental Disorders*, Autism and Child Psychopathology Series,
https://doi.org/10.1007/978-3-031-06120-2_5

level of impairment across social, cognitive, behavioral, and communication skills differs among individuals with ASD. In addition to the core symptoms, comorbid conditions, like attention deficit hyperactivity disorder (ADHD), oppositional defiant disorder, and intellectual disability (ID), commonly co-occur in those with ASD (Frazier et al., 2011; Simonoff et al., 2008). Consequently, the central features of ASD may not be the exclusive targets of intervention for children. For example, challenging behaviors, like aggression and self-injury, commonly require management. More detailed progress monitoring is essential for children with multiple co-occurring conditions or behaviors compared to those with fewer comorbidities, making progress monitoring a highly individualized and complex process (Eapen et al., 2016). For children receiving care in a pediatric medical home, progress monitoring should be addressed in the shared plan of care (Todorow et al., 2018). Chapters 1 and 2 can be referenced for a detailed review of the medical home model of care.

Additionally, there are a variety of ways to measure progress among those with ASD (Hanratty et al., 2015). While several measures exist to assess symptoms of ASD, there are currently no standardized measures expressly designed for progress monitoring purposes (Bolte & Diehl, 2013; Copeland & Buch, 2013). As a result, diagnostic measures are often utilized to meet monitoring requirements (Copeland & Buch, 2013). The importance of progress monitoring, existing challenges, and future directions will be discussed. The literature surrounding current tools used to assess treatment response targeting ASD and related comorbid conditions is discussed below. The psychometric properties of these measurement tools in individuals with ASD are provided in the later part of the chapter.

Progress Monitoring Comorbid Conditions

Anxiety Disorders

According to a recent meta-analysis, as many as 40% of children with ASD meet criteria for one or more comorbid anxiety disorders (Van Steensel et al., 2011; White et al., 2009). Specific phobia (30–44%) is the most common anxiety disorder in those with ASD, followed by obsessive compulsive disorder (OCD) (17–37%), generalized anxiety disorder (15–35%), and separation anxiety (9–38%) (Zaboski & Storch, 2018). Comorbid anxiety among those with ASD can be associated with an increased likelihood of sleep problems, self-injurious behaviors, gastrointestinal illness, parental stress, and symptoms of depression (Mazurek & Petroski, 2015; Kerns et al., 2015; Mazurek et al., 2013; Kerns et al., 2017). However, the relationship between anxiety disorders and ASD is multifaceted, as the symptoms of anxiety can be difficult to extricate from the core symptoms of ASD. There is some intersection between symptoms of anxiety and ASD, such as social avoidance and compulsive behavior (Kerns et al., 2017; Lecavalier et al., 2014). Additionally, a high prevalence of children with ASD, estimated between 50% and 80%, also have accompanying intellectual disabilities (Mpaka et al., 2016). Impairments in expressive language skills make self- and parent-report measures less reliable among children with ASD and lower cognitive skills (Lecavalier et al., 2014). Comorbid anxiety can impact the severity of ASD symptoms; thus, it is crucial to monitor the treatment of anxiety symptoms among those with ASD (Kerns et al., 2017).

Despite the high prevalence of comorbid anxiety disorders among those with ASD, there is a lack of consensus surrounding how anxiety symptoms should be measured. Approximately 36 parent-, child-, and teacher-report measures have been used to measure anxiety symptoms in children with ASD, though very few measures have been validated on children with ASD (Lecavalier et al., 2014). In a recent review, Lecavalier et al. (2014) reviewed 10 measures detailed in over 38 research studies. Of the measures, a select few were considered to be appropriate for monitoring anxiety among children with ASD. The first is the Child and Adolescent Symptom Inventory—5th Edition (CASI-5, Gadow & Sprafkin, 2002) which is a behavior rating scale for DSM-5 behavioral and emotional disorders utilized for children between 5 and

18 years of age. As part of the CASI-5, there are individual items that assess across several anxiety disorders including generalized anxiety disorder, separation anxiety, and social anxiety. In a randomized controlled trial, White et al. (2013) used the Child and Adolescent Symptom Inventory—4th Edition Revised (CASI-4R; Gadow & Sprafkin, 2002; Hallett et al., 2013) anxiety scale as the primary outcome measure of anxiety symptoms among children with ASD and at least one comorbid anxiety disorder. The Multidimensional Anxiety Scale for Children (MASC; March et al., 1997) was also found to be useful for measuring anxiety symptoms in those with ASD. The MASC can be completed by caregivers or as a self-report measure for children between 8 and 19 years of age. The items assess several subscales including the physical symptoms scale, social anxiety scale, harm avoidance scale, separation/panic scale, and total anxiety scale. The MASC, especially the self-report form, is dependent on language abilities; however, the total score on the parent report form of the MASC has shown to be an effective treatment outcome measure among children with ASD (Wood et al., 2009; Storch et al., 2013). Lastly, the Pediatric Anxiety Rating Scale (PARS; Ginsburg et al., 2011) is a measure that can be utilized to assess severity of anxiety symptoms in children between 6 and 17 years of age. The PARS is a semi-structured interview for children and caregivers. Similarly to the MASC, the child-interview section of the PARS requires fluent expressive language skills, which may limit its utility among children with ASD (Lecavalier et al., 2014). While the PARS was found to have acceptable validity when used to measure symptoms of anxiety among children with ASD, its internal consistency reliability was low (Lecavalier et al., 2014). However, a study completed by Maddox et al. (2020) utilized a modified PARS for children with ASD aged 6–17 years of age. This modified PARS is a child and caregiver interview that clinicians can complete with both individuals together. This modified version of the PARS is not as dependent on the child's verbal skills and expression of anxiety symptoms but rather emphasizes observable signs of anxiety that caregivers may witness.

Sleep Problems

Children with ASD are at a higher risk for experiencing sleep problems than their typically developing peer counterparts. As many as 50–80% of children with ASD experience sleep problems, such as frequent awakenings during the night, early wake times, or fragmented sleep (Abel et al., 2017; Hodge et al., 2012). Prior studies have suggested that sleep difficulties among children with ASD may worsen the core symptoms of ASD and contribute to challenging behaviors, like aggression and self-injury (Fadini et al., 2015; Abel et al., 2017). Thus, sleep problems are essential for clinicians to identify and monitor, so that they do not impact the progress of treatment (Abel et al., 2017).

Sleep questionnaires can be used to both assess and monitor sleep problems in children with ASD. One common measure, the Children's Sleep Habits Questionnaire (CSHQ; Owens et al., 2000), is a caregiver-report questionnaire that assesses bedtime resistance, sleep-onset delay, sleep anxiety, nighttime waking, parasomnias, sleep-disordered breathing, and daytime sleepiness (Owens et al., 2000). On the CSHQ, parents rate how frequently their children experienced sleep problems within the past week. While the CSHQ was originally developed to identify sleep problems in typically developing children aged 4–10 years old, it is also used in clinical populations (Moore et al., 2017). Currently, the CSHQ is the most widely used standardized measure designed to identify and monitor sleep problems in children with ASD (Hodge et al., 2012; Moore et al., 2017). According to a study completed by Giannotti et al. (2008), consisting of 104 children with ASD and 162 typically developing children, the CSHQ was found to demonstrate adequate internal consistency among participants with ASD.

The Modified Simonds and Parraga Sleep Questionnaire (MSPSQ; Simonds & Parraga,

1982) is another measure that identifies sleep disturbances in children between 5 and 18 years of age. The MSPSQ has been modified and used to both assess and monitor treatment outcomes for children with ASD and other developmental disabilities (Moore et al., 2017; Wiggs & Stores, 2004). The MSPSQ includes 51 items organized across two parts, one used to assess sleep quality and quantity, and another used to identify specific sleep difficulties. While it is similar to the CSHQ in that it assesses a range of sleep problems, it also obtains information advantageous to treatment planning (Moore et al., 2017). Additionally, the Family Inventory of Sleep Habits (FISH; Malow et al., 2009) is a measure designed to assess sleep hygiene of individuals with ASD aged 3–10. It probes for information related to children's habits before bedtime, bedtime routines, sleep environment, and caregiver behaviors surrounding bedtime. A study completed by Malow et al. (2015) examined whether caregiver sleep education provided in either an individual or group format would be most effective for caregivers of children with ASD. Caregiver questionnaires, including the FISH and CSHQ, were utilized to identify the initial sleep difficulties and posttreatment outcomes of the children with ASD. The FISH can be particularly useful in treatment planning and monitoring treatment outcomes for children with ASD, as it was designed to be utilized in children with ASD (Moore et al., 2017).

Additionally, sleep diaries, also known as sleep logs, can be used to record and monitor sleep problems in children with ASD. Sleep diaries require caregivers to report on their child's sleep quality over the previous night for a period of 2 weeks or longer to ensure adequate validity. Typically, sleep diaries will include information regarding the child's bed time and waking times, daytime sleeping, the onset time of sleep, and any sleep-waking activity (Hodge et al., 2012). Several studies using samples of children with ASD have supported the use of sleep diaries to assess and monitor variables impacting sleep. Prior studies that examined children with ASD found strong correlations between parent sleep diary reports and data from actigraphy (i.e., an objective measure that monitors an individual's nighttime movement) (Hodge et al., 2012; Goodlin-Jones et al., 2008; Allik et al., 2006). The findings from the aforementioned studies suggest that sleep diaries can be particularly useful in assessing and monitoring sleep and wake times, though they may not provide as accurate information regarding sleep latency compared to objective measures, like actigraphy (Hodge et al., 2012).

Challenging Behaviors

Challenging behaviors, also referred to as problem behaviors, are behaviors that impact an individual's ability to function and often can cause harm, such as aggression, property destruction, and self-injurious behavior (Minshawi et al., 2014). While challenging behaviors are not a core feature of ASD, they are prevalent in an estimated 64.3–94.3% of children with ASD (Murphy et al., 2009). The rate at which individuals engage in challenging behaviors is associated with increased ASD severity, impairments in adaptive skills, and comorbid intellectual disability (ID) (Baeza-Velasco et al., 2014; Baghdadli et al., 2008). A range of negative effects of challenging behaviors have been identified. Some of these effects include bodily injury, limited social relationships, impairments in adaptive behavior, poor academic skills, and decreased quality of life (Baghdadli et al., 2008; Herzinger & Campbell, 2007). As a result, frequent progress monitoring of challenging behaviors throughout evidence-based treatment is essential to ensure decreased frequency and intensity of behaviors.

Considering the high prevalence of challenging behaviors in those with ASD, there is a need for validated measures that assess and progress monitor challenging behaviors in children with ASD (Minshawi et al., 2014). Currently, there are several caregiver-rated measures that are used to identify challenging behaviors in those with ASD and developmental delays, as well as typically developing individuals. The Aberrant Behavior Checklist (ABC; Aman et al., 1985a) can be used to both assess the core symptoms of ASD and various comorbid emotional and behavioral problems across several domains including irritability, agitation and crying, lethargy/social withdrawal, stereotypic behavior, hyperactivity/noncompliance, and inappropriate speech (Aman et al., 1985a). It was originally developed to be utilized as a measure of treatment outcome in individuals with ID but has since been used and normed on individuals with a range of developmental delays associated with behavioral problems such as ASD, Down syndrome, and Fragile X syndrome (Kat et al., 2020). It has been utilized extensively in both pediatric and adult populations due to the high reliability and validity of the measure (Aman et al., 1985b; Schmidt et al., 2013). The ABC is especially advantageous for monitoring treatment outcomes, as it is more comprehensive than other measures used to identify behavior problems in individuals with ASD like the Autism Behavior Checklist or Social Response Scale (Kat et al., 2020).

While the ABC was originally developed for use among individuals in residential facilities, revisions to the measure have been made to allow for more appropriate usage among individuals in home and school settings; this measure is referred to as the Aberrant Behavior Checklist—Community (ABC-C) (Aman & Singh, 1994; Schmidt et al., 2013). The ABC has been used to monitor the outcomes of behavioral interventions in both research (Aman et al., 2009) and in practice. For example, the ABC has been a useful tool in identifying participants for forms of research interventions, as well as monitoring comorbid behavioral difficulties among those with genetic syndromes or neurodevelopmental disorders. While the ABC has mainly been used to assess children, adolescents, and adults, some studies have evaluated behavioral and pharmacological treatment effects with the ABC in children under 5 years of age; this is most likely due to the lack of other available measures for young children (Schmidt et al., 2013; Brown et al., 2002; Chadwick et al., 2000). However, in a study completed by Schmidt et al. (2013), researchers found that the original 5-factor structure of the ABC-C was not supported in children under 5 years of age, and it may either underestimate or overestimate behavior problems in this population. Thus, the utility of the ABC-C for young children under 5 years of age is not fully supported and established currently (Schmidt et al., 2013).

Additionally, the Behavior Problems Inventory-01 (BPI-01; Rojahn et al., 2001) is a narrowband measure that assesses behavioral problems in individuals with ID and developmental disabilities. It contains items that measure self-injurious behavior, stereotypic behavior, and aggressive/destructive behavior on both a five-point frequency scale (never = 0, monthly = 1, weekly = 2, daily = 3, hourly = 4) and a four-point severity scale (no problem = 0, slight problem = 1, moderate problem = 2, severe problem = 3). It was designed to assess behavioral problems in individuals, measure treatment outcomes, and assist with surveys used for administrative decision-making (Rojahn et al., 2001). While the BPI-01 was found to have adequate reliability in adults with intellectual disabilities (González et al., 2009), it was not specifically developed or validated on infants and toddlers (Rojahn et al., 2013; Minshawi et al., 2014).

There are limited measures that identify and monitor behavioral difficulties in children with ASD specifically, especially infants and toddlers with ASD. While two common measures, the Behavior Assessment System for Children Third Edition (BASC-3; Reynolds & Kamphaus, 2015) and the Child Behavior Checklist (CBCL; Achenbach & Rescorla, 2000) both assess behavior problems in young children under 5 years of age, they were both designed for typically developing populations; there is also a lack of research on contact validity for use of the these measures among children with ASD (Hanratty et al., 2015). However, infants and toddlers with ASD are at a greater risk for developing challenging behaviors than young children with other developmental disabilities (Hartley et al., 2008). The Baby and Infant Screen for Children with Autism Traits, Part 3 (BISCUIT-Part 3; Matson et al., 2009) is a caregiver-rated measure that was designed to be used in children between 17 and 37 months of age with this purpose in mind. The BISCUIT-Part 3 includes 17 different items that assess a range of behavioral topographies including self-injurious behaviors, stereotypic behaviors, disruptive behaviors, and aggressive behaviors. Clinical cutoff scores are presented on the measure for no to minimal impairment, moderate impairment, and severe impairment. Additionally, the Autism Spectrum Disorder—Behavior Problems for Children (ASD-BPC; Matson et al., 2008) and Autism Spectrum Disorder—Behavior Problems for Adults (ASD-BPA; Matson & Rivet, 2007) are measures that identify behavioral problems in individuals with ASD specifically. The ASD-BPC is a measure completed by caregivers of children between 2 and 16 years of age made up of 18 items that are rated on a three-point Likert scale (0 = not a problem or impairment, 1 = mild problem or impairment, 2 = severe problem or impairment). The ASD-BPC is made up of two domains including internalizing and externalizing behaviors. The ASD-BPA is a caregiver-rated screening measure that can be used to determine the frequency, intensity, function, and duration of challenging behaviors in adults with ASD. It is made up of 19 items across 3 subscales including aggression/destruction, disruptive behavior, and self-injurious behavior, and a total score. The items are rated on a binary scale (0 = not a problem, no impairment, 1 = problem, impairment). The ASD-BPA can be followed up by the BP-01, direct measures, or functional assessment depending upon the necessity of the individual. The ASD-BPA identifies a range of challenging behaviors in adults with ASD and can be administered in a short amount of time. As a result, it can serve as a method of monitoring treatment outcomes in adults with ASD (Matson & Rivet, 2007).

Depression

Depression is a common diagnosis in children and adolescents, with a prevalence rate of nearly 12% of children in the population (DeFilippis, 2018). Depression is associated with increased risk of medical illness, suicidality, and poor functioning in children and adolescents alike (Chandrasekhar & Sikich, 2015). Prior researchers have suggested that individuals with ASD are nearly four times more likely to develop depression than their typically developing peers, suggesting the importance of regular screening, monitoring, and treatment of depression (Hudson et al., 2019). However, comorbid depression diagnoses can be difficult to identify in those with ASD, as depression symptoms can present differently and overlap with ASD. For example, depression can be accompanied by social withdrawal and interpersonal difficulties, which is also characteristic of ASD (DeFilippis, 2018). Additionally, there is a lack of validated diagnostic tools that can be used among individuals with ASD (Chandrasekhar & Sikich, 2015; DeFilippis, 2018).

Several questionnaires have been utilized to identify and monitor depression symptoms in those with ASD, though there is a lack of standardized progress monitoring tools that can be utilized in this population (DeFilippis, 2018). The Children's Depression Inventory—Second Edition (CDI-2; Kovacs, 2010) is utilized commonly among children with ASD. It is a brief self-, parent-, or teacher-report measure for children between 7 and 17 years of age that can be used to identify the cognitive, affective, and behavioral symptoms of depression. The self-report version of the CDI-2 is a clinically validated measure of depression with adequate sensitivity (80%) and specificity (84%) in typically developing children (Kovacs, 2010; Kovacs & Staff, 2003). While the specificity and sensitivity of the measure have been found to be lower when utilized among children between 10 and 17 with ASD, it has been used to identify and monitor depression symptoms among them (Mazefsky & Oswald, 2011). The parent-report version of the CDI-2 has also been utilized to measure and monitor depressive symptoms among children and adolescents with ASD and developmental delays (Gotham et al., 2015). However, Mazefsky and Oswald (2011) suggest that self-report measures, like the CDI-2, should be interpreted with caution, even though they do provide useful information regarding depressive symptoms. As a result, clinicians may choose to use both the self-report and parent-report version of the CDI-2 when evaluating and monitoring depressive symptoms.

The BASC-3 is also commonly used to identify symptoms of depression and anxiety, though it was not created to be a diagnostic tool (Reynolds & Kamphaus, 2015). Prior studies have compared the use of the BASC in individuals with ASD and typically developing children on the internalizing composite and its related subscales. For example, Mahan and Matson (2011) examined the differences on the BASC-2 parent-rating scales between a typically developing sample and children with ASD between 6 and 16 years of age. The children with ASD scored significantly higher on the depression subscale; however, there were not significant differences between the scores on the anxiety subscales among those with ASD and those typically developing individuals. This study supports the valid use of the BASC among those with ASD when screening and monitoring symptoms of depression, as it replicated the findings of higher internalizing symptoms in children with ASD.

Attention Deficit/Hyperactivity Disorder

While approximately 7% of the population has a diagnosis of attention deficit/hyperactivity disorder (ADHD), between 30% and 80% of individuals with ASD have a comorbid diagnosis of ADHD (Hollingdale et al., 2020; Young et al., 2020). ADHD is characterized by deficits in attention, as well as symptoms of hyperactivity and impulsivity (APA, 2013). While the conditions independently have established standards on their identification and treatment, the understanding of ASD and ADHD as comorbid conditions is less founded (Young et al., 2020). ASD and ADHD were not formally recognized as co-occurring conditions until the *Diagnostic and Statistical Manual of Mental Disorders* (5th ed.; DSM-5; APA, 2013). Prior researchers suggested that lower cognitive functioning, increased social impairments, and delays in adaptive skills are more prevalent among individuals with comorbid ASD and ADHD compared to individuals with ASD only (Rao & Landa, 2013). Thus, there is an increased need for an understanding of the identification and monitoring of the conditions together (Young et al., 2020).

Several narrowband rating scales can be used to both screen for symptoms of ADHD and monitor response to interventions (Young et al., 2020). One such rating scale is called the ADHD Rating Scale—5 for Children and Adolescents (ADHD-RS-5; Dupaul et al., 2016). The ADHD-RS-5 contains both parent- and teacher-rating scales that can be used with children between the ages of 5 and 17 years of age (Dupaul et al., 2016). The items on the ADHD-RS-5 are closely linked to the symptom domains specified in the DSM-5, including both inattention and hyperactivity symptoms (Dupaul et al., 2016; Epstein & Weiss, 2012). One study completed by Yerys et al. (2017) examined the use of the ADHD-RS-IV among children with ASD, as it is a rating scale that is regularly used to screen for ADHD symptoms in children with ASD. However, limitations existed with factorial validity when used in individuals with ASD. Thus, caution is advised when using narrowband rating scales, particularly the ADHD-RS-IV/ADHD-RS-5, to measure response to treatment (Yerys et al., 2017). The Conners' Rating Scales—Revised (CRS-R; Conners, 1997) contains parent, teacher, and adolescent self-report scales for individuals between 6 and 18 years of age. This measure contains several content, index, and symptom scales that can be used to identify specific difficulties related to ADHD symptomology and related comorbidities. The CRS-R provides large normative samples and t-scores based on age and sex. The CRS-R is a valuable option for clinicians to consider as symptom-based outcome measures, as it has adequate psychometric properties (Epstein & Weiss, 2012). The Vanderbilt ADHD Diagnostic Rating Scale (VADRS; Collett et al., 2003) is another rating scale for children between 6 and 12 years of age that assesses both symptoms of ADHD and their impact on behavior and academic performance. In addition to the ADHD symptom scales, there are performance scales included in the measure to assess impairment in math, reading, writing, and interpersonal relationships (Epstein & Weiss, 2012). Two versions of the VADRS are available, including a parent- and teacher-rated scale. Clinicians be inclined to use the VADRS to monitor changes in the frequency of ADHD symptoms and any medication side effects, as the measure is available in the public domain (Epstein & Weiss, 2012).

Progress Monitoring Methods

Progress monitoring can be defined as "the simple repeated measurement of student performance toward a long-range goal" (Deno, 1985). While children with ASD could benefit from ongoing evaluation of academic skills, like math and reading, impairment in social communication is a defining characteristic of ASD (APA, 2013; Witmer et al., 2015). Interventions aimed at the core features of ASD, as well as any accompanying challenging behaviors, are most likely to be beneficial. As described by Salvia et al. (2013), several methods are commonly utilized to monitor ongoing intervention. These methods include interview techniques, rating scales, and systematic observations, all of which are accompanied by both advantages and disadvantages (Witmer et al., 2015). While interview techniques can provide qualitative information, they risk susceptibility to bias. Rating scales completed by informants, such as parents, teachers, or children, can provide a rating of a particular behavior based on the informant's observations over a period of time. While they can be completed relatively quickly and assist in the detection of low base rate behaviors, they may not specifically target the behavior of interest and may be susceptible to response bias. Systematic observation techniques allow for a target behavior to be measured at the time and place of its occurrence. These techniques are a central approach to data collection in applied behavior analysis (ABA) to treat children with ASD, as they are less vulnerable to bias; however, it is debatable whether lay people, like parents and teachers, can use these techniques reliably to monitor progress regularly across settings (Witmer et al., 2015; Fisher et al., 2011).

Progress Monitoring Challenges

While ABA treatment techniques are well supported by research, children with ASD do not all experience uniform success; this further highlights the need for effective measures to assess response to treatment among children. Measuring progress in areas of social communication can be difficult as the changes often vary across children and interventions (Grzadzinski et al., 2020). For example, changes in social communication behaviors may be related to either an increased quantity of the behavior or its overall quality (Grzadzinski et al., 2020; Kasari et al., 2015). Consider two social communication goals that are often targeted in ABA therapy: prosody and manding. For treatments targeting language prosody (e.g., rhythm and intonation of speech), clinicians may measure the quality of a child's spoken phrases. Contrarily, changes in communicative behaviors that are expected to increase in frequency, such as mands (e.g., requests), would most adequately be examined in terms of the behavior's quantity. Additionally, interventions vary in their focal points, with some emphasizing behaviors across symptoms of ASD and others focusing on more general social communication skills, like joint attention. Thus, it is crucial for measures used to assess progress to account for "subtle, though clinically meaningful" changes across many different areas of social communication, including the quality and quantity of behaviors (Grzadzinski et al., 2020, p. 2; Anagnostou et al., 2015).

As discussed by Bolte and Diehl (2013), there is very limited consistency in the measures used to assess changes in treatment response for individuals with ASD. After the review of 195 intervention trials, researchers found that 289 different measurement tools were utilized in an attempt to assess behavior changes related to treatment. Only three measures, including the Aberrant Behavior Checklist (ABC), the Clinical Global Impression Scale (CGI), and the Vineland Adaptive Behavior Scales (VABS), were consistently included in more than 2% of the studies.

Progress Monitoring Limitations

Limited Validity

One of the primary challenges associated with progress monitoring is the lack of valid and reliable tools that can assess change across time. As a result, clinicians often utilize tools designed for diagnostic purposes which are not typically sensitive to changes that occur as a result of intervention. Also, many of these measures were developed for use with typically developing children, often making delayed developmental trajectories difficult to monitor across time (Eapen et al., 2016). Several measures have been criticized for their lack of validity as tools used to assess treatment progress (Grzadzinski et al., 2020; Eapen et al., 2016). For example, McConachie et al. (2015) completed a review of tools to measure outcome for children with ASD, finding that three of the most commonly utilized measures for progress monitoring, the ABC, CGI, and VABS, along with several others, were not recommended. These tools were found to be limited in their ability to detect change in response to intervention, as they were not developed for the purpose of progress monitoring and often required extensive training (Grzadzinski et al., 2020).

Contrarily, after reviewing 38 measures, Anagnostou et al. (2015) found 6, including the ABC and VABS, to be sufficient at measuring treatment response among children with ASD, though some limitations were noted (Grzadzinski et al., 2020). While the ABC was found to adequately measure treatment effects in children with ASD, Fragile X, and related developmental delays, it was not able to measure some deficits in social communication (e.g., impairments in nonverbal communication and quality of social initiations). Additionally, while the VABS was found to adequately characterize social and communication skills of those with ASD, it has not been found to be consistently sensitive to change over time; this is particularly a challenge for interventions lasting shorter than 6 months (Anagnostou et al., 2015; Grzadzinski et al., 2020).

Potential Bias

Several current measures utilized for progress monitoring purposes rely on either caregiver or clinician report. For example, the CGI is commonly used as an outcome measure, particularly regarding medication therapy. Similarly, measures designed to assess adaptive functioning, such as the Adaptive Behavior Assessment System (ABAS) and the VABS, require a caregiver report of the child's current skills. These measures, as well as other similar measures, can emphasize placebo effects, exceeding any small changes that may be related to treatment (Guastella et al., 2015; Grzadzinski et al., 2020). Because clinicians and caregivers often take an active role in treatment, the threat of biased results can be high. For instance, when completing a caregiver-rated measure, any changes observed may be the result of caregiver's understanding of whether or not the child was receiving treatment, as opposed to the actual treatment effects (Guastella et al., 2015; Grzadzinski et al., 2020).

Considering this limitation, one structured observation measure, the Early Social-Communication Scales (ESCS), has been designed to adequately assess changes in verbal and nonverbal communication skills among typically developing or developmentally delayed children (Mundy et al., 2013; Anagnostou et al., 2015). It has been found to be an adequate measure for treatment outcome. However, it requires considerable training for reliable administration, and it may only be utilized for typically developing children between 8 and 30 months or developmentally delayed children with a verbal age that falls in the aforementioned range (Grzadzinski et al., 2020; Mundy et al., 2013).

Limited Measures that Evaluate ASD Symptoms

While most interventions for children with ASD target ASD-specific symptoms, many of the commonly utilized treatment outcome measures were not designed to assess ASD-specific symptoms (Grzadzinski et al., 2020; Kasari et al., 2002; Green et al., 2010). For example, the VABS measures global adaptive functioning in individuals across several domains, including communication, socialization, daily living, and motor skills (Sparrow et al., 2016). Additionally, the ABC assesses a range of challenging behaviors across various settings, such as irritability, social withdrawal, stereotypic behavior, noncompliance, and inappropriate speech (Aman et al., 1985a). Children with ASD can have associated adaptive functioning difficulties or challenging behaviors, so measuring any improvements in these deficits as a result of treatment can be clinically informative. However, measures that directly assess ASD-specific symptoms, including deficits in social communication and repetitive behaviors, are required to evaluate whether there are improvements of the core features of ASD as a result of treatment (Grzadzinski et al., 2020).

In order to assess changes in the core symptoms of ASD, previous studies have utilized the ADOS (Lord et al., 2000; Grzadzinski et al., 2020). However, the ADOS was designed to be utilized as a diagnostic tool, rather than a measure of treatment outcomes. Prior researchers have used both the raw scores from the subscales and the Calibrated Severity Scores (CSS), which are considered to be more sensitive to changes over time. The raw scores were generally not able to meaningfully identify changes associated with treatment, as changes in raw scores were observed even among children in control groups (Grzadzinski et al., 2020; Green et al., 2010; Gutstein et al., 2007). The ADOS CSS was found to effectively evaluate change over the course of several years; however, it does not adequately measure changes associated with brief interventions (Anagnostou et al., 2015; Grzadzinski et al., 2020). Considering these limitations, it is recommended that diagnostic measures, like the ADOS, not be utilized to measure treatment outcomes (Grzadzinski et al., 2020).

Individualized Progress Monitoring

According to the National Institute for Health and Care Excellence (2013), intervention and treatment plans for ASD should be individualized, research based, continually monitored for progress, and frequently revised. As children with ASD develop into adolescents and young adults, it becomes increasingly difficult to deliver efficacious treatments as their particular goals and needs evolve (Eapen et al., 2016). As early as the elementary school years, children with ASD may be educated in ABA therapy clinics or exclusively special education schools, while others may be integrated into the general education curriculum. However, as children age, they may be exposed to an increasing variety of environments including advanced educational placements, work settings, or supported employment settings. Clinicians should aim to include input from individuals with ASD into treatment planning, along with the goals of any caregivers that may be involved. While elements of an individualized treatment plan are expected to change as the individual progresses, there are crucial core elements. These elements include long-term goals for the individual with ASD including a transition plan, assessment of current performance of adaptive skills, measurable goals over a specific interval of time, a method in which to monitor progress, and a review and revision of the mental health and medical provider treatment plans which are captured in the overall shared plan of care (Eapen et al., 2016).

Progress Monitoring Throughout the Life Span

There are four primary stages throughout the life span that are often accompanied with different goals, as individuals' abilities and environments change. Generally, these developmental stages include the preschool years, primary years, high school years, and adult years. During the preschool years, treatment goals typically focus on impairments in receptive and expressive language, social skills, behavioral difficulties, motor skills, and increasing adaptive independence. However, as individuals with ASD age, treatment goals typically "shift from assessing specific developmental domains and abilities to assessing participation in education, employment, or civic life" (Eapen et al., 2016, p. 90). Changes in progress monitoring should also be considered during periods of assessment and reassessment of individuals with ASD. For example, there are several significant transitions throughout the life span that may be stressful for individuals with ASD and their families; these include the time immediately following a diagnosis, the start of school or educational programs, the transition from one educational placement to another, and post high school as students move into vocational placements. During these time periods in particular, goal-specific assessment and progress monitoring are crucial. Topics related to independent living, daily living skills, personal and sexual relationships, and driving may also be considered. Contact information to any relevant resources or services that may assist individuals with ASD and their families during major life transitions should be incorporated into the progress monitoring framework (Eapen et al., 2016).

Developments in Progress Monitoring

When choosing appropriate treatment outcome measurements, there are several factors that should be considered. Clinicians should recognize that not all children will respond to treatments in the same way (Vivanti et al., 2014; Grzadzinski et al., 2020). There are currently no measurement tools that are comprehensive enough to assess all changes. Additionally, measurement tools may overidentify children that did not change as a result of intervention, or, conversely, underidentify children that underwent changes as a result of the intervention. Below are several highlighted measures that can be especially useful in assessing response to intervention among individuals with ASD (Grzadzinski et al., 2020).

Treatment Outcome Measures for ASD

Change across the core symptoms of ASD is among the most crucial measure of change for interventions aimed to improve ASD (Matson, 2007). However, according to Rogers (1998), the vast majority of research investigating treatment outcomes for ASD did not utilize one of the main ASD outcome measures. A review conducted by Bolte and Diehl (2013) investigated specific measurement tools used to identify response to treatment in ASD interventions between 2001 and 2010. In the review, 195 articles investigating clinical trials involved in the treatment of ASD were included. Over 289 unique measurement tools were utilized to measure treatment outcomes in the included articles. From these measurement tools, approximately 62% were only used in one study over the 10-year period of time. The most utilized measurement tools were the Aberrant Behavior Checklist, Clinical Global Inventory, and Vineland Adaptive Behavior Scales, present in nearly 3% of the included studies. Within the articles, over 600 various target skills were also identified to monitor; these skills were divided into different categories, such as autism severity, adaptive functioning, communication/language, and behavioral difficulties. While core features of ASD, like social skills and communication/language, were frequently monitored in the studies, there was very little consistency in the measures used.

In another review completed by McConachie et al. (2015), 184 journal articles published between 1992 and 2013 were identified for review. The articles included all investigated measures used to assess treatment outcomes for early intervention among children with ASD up to 6 years of age. Throughout the review of the articles, over 130 measurement tools were identified for use in assessing response to treatment. While there were no studies that met inclusion criteria for 75 of the measures, psychometric properties for 57 of the measurement tools were examined. Measures assessing several target skills, including autism severity, global outcome, social skills, and cognitive ability, were included.

In a recent review by Brugha et al. (2015), researchers examined 31 articles discussing treatment outcome measures in treatment trials for adolescents and adults with ASD. In the review, the researchers considered outcome measures that identified the core symptoms of ASD, commonly comorbid symptoms, like anxiety, challenging behaviors, and cognitive impairments, and overall quality of life. A range of different studies were included, such as retrospective assessment studies, case series, and randomized and placebo-controlled trials. Researchers found that there was a lack of focus on the assessment of outcomes related to the core symptoms of ASD overall. The Ritvo-Freeman Real-life Rating Scale (RF-RLRS; Freeman et al., 1986) was used most commonly in the studies to measure treatment outcomes across the core symptoms of ASD, with the social relatedness subscale showing the most change. Overall, the RF-RLRS was found to have low interrater reliability, though it may have utility when it is used in pharmacological studies and for direct observation and monitoring. The Social Responsiveness Scale (SRS; Constantino & Gruber, 2005) was also used to measure outcomes in the core symptoms of ASD and two studies noted improvements in social skills across time. None of the included studies used the ADOS or other related observational tools to assess change in ASD severity, as these instruments were designed for diagnostic purposes. However, most of the outcome measures included in the review were not specific to ASD symptoms, like the CGI and the VABS.

Additionally, Howard (2019) evaluated tools that were intended to measure treatment outcomes for parent-mediated interventions for ASD. Parent-mediated interventions are particularly crucial for early intervention and often involve trained professionals teaching parents intervention techniques that be used in home and community settings. In prior studies, researchers have found improved communication and social skills among children involved in parent-mediated interventions (McConachie & Diggle, 2007). In the studies reviewed, several different measurement tools were used to identify response to treatment, including measures that emphasize child outcomes, parent outcomes, and parent-child outcomes. In general, the results of the review were variable, as no one measure showed treatment effects that were consistent. However, within the categories of tools reviewed, particular measures performed better than others. For example, with respect to parent-mediated interventions, measures assessing social skills and adaptive behavior were more likely to show a treatment effect, though it was not consistent. Additionally, the parent-child interaction variables, like child initiations, showed significant increases in response to treatment in various studies. As demonstrated by the review, the utilization of various measurement tools can make it increasingly difficult to assess treatment outcomes. An overview of the aforementioned reviews can be found in Table 5.1.

ASD-Specific Measures

Autism Treatment Evaluation Checklist

The Autism Treatment Evaluation Checklist (ATEC; Rimland & Edelson, 1999) is a checklist to monitor improvements due to interventions in children with ASD. There are a total of 72 items and 4 subtests: Speech/Language Communication (14 items), Sociability (20 items), Sensory/Cognitive Awareness (18 items), and Health/Physical/Behavior (25 items). The measure is one page to be completed by parents, caregivers, and/or teachers.

A study by Magiati et al. (2011) investigated the ATEC utility in monitoring progress with promising findings. The ATEC was found to have good content validity, as the total and subscale scores were significantly correlated with age

Table 5.1 Review of treatment outcome measures for ASD

Authors	Intervention type	Number of studies included	Measures used
Bolte and Diehl (2013)	Global functioning Core ASD symptoms Behavior problems	8 3 9 5 10	Vineland Adaptive Behavior Scales (VABS) Bayley Scales of Infant Development (Bayley) Clinical Global Impressions (CGI) Childhood Autism Rating Scale (CARS) Aberrant Behavior Checklist (ABC)
McConachie et al. (2015)	Core ASD symptoms Behavior problems Global functioning	13; 3 intervention evaluation studies 8; 4 intervention evaluation studies 1 1 2; 1 intervention evaluation study 1 3 1 1; 1 intervention evaluation study 67; 24 intervention evaluation studies 9; 4 intervention evaluation studies	Childhood Autism Rating Scale (CARS) Gilliam Autism Rating Scale (GARS) BISCUIT-Part 1 Autism Impact Measure (AIM) Social Communication Questionnaire (SCQ) Autism Treatment Evaluation Checklist (ATEC) Aberrant Behavior Checklist (ABC) BISCUIT-Part 3 Home Situations Questionnaire-Pervasive Developmental Disorders Version Vineland Adaptive Behavior Scales (VABS) Psychoeducational Profile—Revised/ Psychoeducational Profile—Third Edition
Brugha et al. (2015)	Core ASD symptoms Global functioning	3; used primarily for pharmacological research 1 2 1 6 1 9	Ritvo-Freeman Real-life Rating Scale (RF-RLRS) Autism Behavior Checklist Social Responsiveness Scale Childhood Autism Rating Scale (CARS) Yale-Brown Obsessive-Compulsive Scale (Y-BOCS) The Repetitive Behavior Scale Clinical Global Impression (CGI)
	Behaviors/Mood	2 3 2	Aberrant Behavior Checklist (ABC) Maladaptive Subscale of Vineland Adaptive Behavior Scale (VABS) Positive and Negative Affect Schedule (PANAS)
Howard (2019)	Child outcome Communication Adaptive behavior Behaviors Parent outcome Parental stress Family functioning	1 1 6 2 1 2 4 1 1	Early Social Communication Scales Vineland Adaptive Behavior Scales (VABS) (Communication Domain) Vineland Adaptive Behavior Scales (VABS) Child Behavior Checklist (CBCL) Preschool Behavior Checklist Developmental Behavior Checklist Parenting Stress Index (PSI) Stress-Arousal Checklist Mc Master Family Assessment Device

equivalences and raw scores from standardized measures. It was also found to have good predictive validity, as the initial score predicted the progress made over time and the overall outcome at the follow-up administration.

Childhood Autism Rating Scale

The Childhood Autism Rating Scale, Second Edition (CARS2; Schopler et al., 2010) is a diagnostic measure for children with ASD, aged 2 years and older. It consists of three forms: CARS-2 Standard Version (CARS2-ST), CARS-2 High Functioning Version (CARS-2-HF), and CARS-2 Questionnaire for Parents or Caregivers (CARS2-QPC). The Standard Version is for individuals below the age of 6 or individuals with an IQ score below 79. The High Functioning Version is for children 6 years and older with higher intellectual functioning (IQ above 80) and verbal abilities. Unlike the first two, which are scored by a clinician, the Questionnaire for Parents or Caregivers is not scored but rather is used to inform the CARS2-ST and CARS2-HF. The Standard and High Functioning Versions have 15 categories, rated between 1 and 4, with .5 increments. The categories for the CARS-ST include relating to people, imitation, emotional response, body use, object use, adaptation to change, visual response, listening response, taste, smell, and touch response and use, fear or nervousness, verbal communication, nonverbal communication, activity level, and level and consistency of intellectual response. The CARS-HF categories include social-emotional understanding, emotional expression and regulation of emotions, relating to people, body use, object use in play, adaptation to change/restricted interests, visual response, listening response, taste, smell, and touch response and use, fear or anxiety, verbal communication, nonverbal communication, thinking/cognitive integration skills, and level and consistency of intellectual response. The raw score is calculated by adding the domains, which is then converted to a T-score and percentile. It is a level C measure that takes approximately 5–10 minutes to score.

The CARS2-ST was found to have an internal consistency of .93 and item-to-total correlations between .43 and .81. The CARS2-HF had item-to-total correlations between .53 and .88. The CARS2-HF was found to have an interrater reliability of .95, while interrater reliability for the CARS2-ST was not provided. A factor analysis supported a two-factor solution for the CARS-ST: (1) communication and sensory issues and (2) emotional issues, while a three-factor solution was found for the CARS2-HF: (1) social and emotional issues, (2) cognitive functioning and verbal abilities, and (3) sensory issues. Both the CARS2-ST and CARS2-HF were found to have convergent validity with the ADOS (Lord et al., 1999; .79 and .77 respectively). However, correlations were .38 and .47 for the standard and High Functioning version, respectively, when compared to the Social Responsiveness Scale (Constantino & Gruber, 2005).

Psychoeducational Profile

The Psychoeducational Profile, Third Edition (PEP-3; Schopler et al., 2005) is a measure for individuals 2–7 years of age with autism spectrum disorder and other communication disorders to assist with individualized educational programs (IEP), confirm diagnoses, and assess results of educational interventions. The measure has a performance portion which assesses ten areas: cognitive verbal/preverbal, expressive language, receptive language, fine motor, gross motor, visual-motor imitation, affective expression, social reciprocity, characteristic motor behaviors, and characteristic verbal behaviors. This section consists of 172 total items, which yield composite scores for each of the 3 factors: communication, motor, and maladaptive behaviors (Schopler et al., 2005). Additionally, the PEP-3 has a parent/caregiver report, which consists of three subtests: problem behaviors, personal self-care, and adaptive behavior. The PEP-3 is a level B measure, for individuals with a master's degree in psychology, education, or related field, and takes 45–90 minutes to administer.

The PEP-3 has demonstrated a high reliability, ranging between .92 and .97 for developmental subtests, .90 and .93 for maladaptive subtests, and .84 and .90 for caregiver report subtests (Conrod & Marcus, 2013). The parent/caregiver report also had high interrater reliability, ranging from .70 to 91 for problem behavior, .65 to 1.00 for personal self-care, and .52 to .90 for adaptive behavior (Conrod & Marcus, 2013). Convergent validity was supported with the Vineland subtests related to developmental skills, self-care skills, and behaviors related to ASD. On the other hand, it showed divergent validity with lower correlations with VABS motor and daily living skills scores. The PEP-3 was negatively correlated with the CARS and Autism Behavior Checklist-Second Edition (Conrod & Marcus, 2013).

Social Responsiveness Scale

The Social Responsiveness Scale, Second Edition (SRS-2; Constantino & Gruber, 2012) measures ASD symptom severity in children and adolescents 4–18 years of age. There are four rating forms, based on age of the individual: the Preschool Form (2 years 6 months to 4 years 6 months), School-Age Form (4–18 years), and Adult Form (19–89 years). The Adult Form has two versions for self-report and for parents, spouses, friends, and relatives to complete. The SRS-2 consists of 65 items and 5 subscales: Social Awareness (8 items), Social Cognition (12 items), Social Communication (22 items), Social Motivation (11 items), and Restricted Interests and Repetitive Behavior (12 items). The items are rated on a Likert-type scale ranging from 1 (not true) to 4 (always true). It takes approximately 15–30 minutes to complete.

The SRS-2 was found to have an internal consistency which ranged between .94 and .96 across all age groups, with high internal consistency across age, gender, and clinical subgroups (Bruni, 2014). The Preschool Form was found to have an interrater reliability of .61, while the School-Age Form had higher interrater reliability at .77. The interrater reliability for the Adult Self-Report Form compared with various familial raters ranged between .61 and .92 (Bruni, 2014). A confirmatory factor analysis found a good fit for the two-factor structure: (1) social communication and interaction and (2) restricted interests and repetitive behaviors. The School-Age Form was found to have concurrent validity with Social Communication Questionnaire (Rutter et al., 2001) and CARS (Schopler et al., 1980), among other rating scales of social communication (Bruni, 2014).

Developmental Measures

Bayley Scales of Infant and Toddler Development

The Bayley Scales of Infant and Toddler Development, Third Edition (Bayley-III; Bayley, 2006) is a standardized test, which consists of five scales for children between 1 and 42 months of age. The scale consists of five scales to assess and measure developmental functioning: cognitive, language, motor, social-emotional, and adaptive behavior. The first three scales are direct assessments, while the latter two are parent-informed measures. The Bayley-III provides scaled scores, composite scores, percentile ranks, confidence intervals, and developmental age equivalences for the scales and subtests. In addition, the test kit contains most of the needed stimulus materials. The Bayley-III is a B-level test and takes approximately 50–90 minutes to administer, based on the child's age.

The cognitive scale contains a total of 91 items. The language scale contains a total of 97 items, with a Receptive Communication subtest with 49 items and an Expressive Communication subtest with 48 items. The Motor scale contains 138 items, also with 2 subtests. The Fine Motor subtest contains 66 items, and the Gross Motor subtest contains 72 items. The Social-Emotional scale consists of a questionnaire of 35 items, rated from 0 (can't tell) to 5 (all of the time). Lastly, the Adaptive Behavior scale consists of a questionnaire of 241 items, rated from 0 (is not able) to 3 (always when needed).

The Bayley-III has demonstrated good internal consistency with reliability ranging from .91 to .93 for the Cognitive, Language, and Motor scales and .86 to .91 for the subtests (Albers & Grieve, 2007). The internal consistency ranged from .76 to .94 for the Social-Emotional scale and ranged from .79 to .86 for the Adaptive Behavior scale. Test-retest reliability of the Cognitive, Language, and Motor scales was .67

to .80 for children 2–4 months of age and .83 to .94 for children 33–42 months of age and had an average stability higher than .80 across all age groups (Albers & Grieve, 2007). The Adaptive Behavior scale had a test-retest reliability of .80 and higher for the domains, with stability increasing with age.

A confirmatory factor analysis supports a three-factor model of the Cognitive, Language, and Motor scales, except in the 0–6-month age group in which a two-factor model was also supported (Albers & Grieve, 2007). The Bayley-III had a correlation of .60 between the Cognitive composite and the Mental Index Score on previous edition of the Bayley, the Bayley Scales of Infant and Toddler Development, Second Edition (BSID-II; Bayley, 1993). It also had a correlation of .60 between the motor composite scores on the Bayley-III and BSID-II (Albers & Grieve, 2007). The Bayley-III also showed high correlations between the Cognitive and Language composite scores and the Verbal, Performance, and Full-Scale scores of the Wechsler Preschool and Primary Scale of Intelligence-Third Edition (Wechsler, 2002).

The fourth edition of the Bayley Scales of Infant and Toddler Development (Bayley-4; Bayley & Aylward, 2019) was released in the United States in 2019. The Bayley-4 adopts polytomous scoring, on a scale of 0–2, rather than 0 or 1 in the Bayley-III. The new version has kept the same 5 scales (i.e., Cognitive, Language, Motor, Social-Emotional, and Adaptive Behavior). The adaptive Behavior scale reduced the items to 120 and the total administration time for the Bayley was reduced from 30–90 minutes to 30–70 minutes. Lastly, the Social-Emotional and Adaptive Behavior scales have remote options for caregivers.

For the Cognitive, Language, and Motor scales, Pearson (2019a) reports the internal consistency ranges between .93 and .95 for the subtests and between .95 and .96 for the composite scores. The internal consistency ranges between .85 and .91 for the Social-Emotional scale and between .91 and .98 for the Adaptive Behavior scale (Pearson, 2019a). They also reported test-retest reliability ranged between .81 and .85 for the Cognitive, Language and Motor subtests and composites and between .72 and .87 for the Adaptive Behavior scale (Pearson, 2019a). Pearson also reports a classification accuracy of .82 for developmental delay and .89 for language delay (Pearson, 2019b).

Adaptive Measure

Vineland Adaptive Behavior Scales

The Vineland Adaptive Behavior Scales, Third Edition (Vineland-3; Sparrow et al., 2016) is a measure of adaptive functioning for individuals birth to 90 years of age. There are three forms: Interview, Parent/Caregiver, and Teacher, which each has a Comprehensive and Domain-Level versions. The two versions have three domains, which include Communication, Daily Living Skills, and Socialization. The three domains comprise the Adaptive Behavior Composite (ABC). The Comprehensive version includes an additional nine core and five optional subdomains. Adaptive raw scores of the subdomains are converted to v-scale scores and determine standard scores for the subdomain. The Parent Caregiver has 502 and 180 items for the Comprehensive and Domain-level versions, respectively; the Teacher Form has 333 items. Each item is scored using a Likert-type scale from 0 (never) to 2 (usually or often); however, some questions require only yes (2) or no (0). The Vineland-3 is a level B measure and takes 10–40 minutes to administer, depending on the form.

The Vineland-3 has demonstrated excellent reliability, with the internal consistency between .94 and. 99 for the Comprehensive Form adaptive domains and ABCs and .86 and .99 for the Domain-Level adaptive domains and ABCs, across all age groups (Pepperdine & McCrimmon, 2018). The test-retest reliability ranged between .64 and .94 for the Comprehensive Form adaptive domains and ABCs and between .63 and .92 for the Domain-Level adaptive domains and ABCs (Pepperdine & McCrimmon, 2018). Further, interrater reliability ranged from .61 to .87 for the Comprehensive Form adaptive domains and

ABCs, with the exception of the Socialization adaptive domain the in the Teacher Form which had an interrater reliability of .46 for individuals 3–5 years of age. The interrater reliability ranged from .58 to .93 for the Domain-Level adaptive domains and ABCs.

The Parent/Caregiver Comprehensive form has demonstrated concurrent validity with the Bayley-III, with moderate to high correlations, ranging from .67 to .81 (Pepperdine & McCrimmon, 2018). The Parent/Caregiver and Teacher Forms also demonstrated concurrent validity with the Adaptive Behavior Assessment System-Third Edition (ABAS-3; Harrison & Oakland, 2015), with correlations between .75 and .88 for the Teacher Forms and .41 and .98 for the Parent/Caregiver forms.

Behavioral Measures

Aberrant Behavior Checklist

The Aberrant Behavior Checklist – Community (Aman & Singh, 1994) is a behavior rating scale to evaluate treatment effects in individuals with developmental and intellectual disabilities. The ABC consists of 58 items rated from 0, indicating not at all a problem to 3, indicating a severe problem. There are five factors: Irritability (15 items), Lethargy/Social Withdrawal (16 items), Stereotypy (7 items), Hyperactivity/Noncompliance (16 items), and Inappropriate Speech (4 items). Scores are generated for each factor, but a total score is not recommended (Aman & Singh, 2017).

Internal consistency of the ABC was found to be very good, ranging from .86 to .95, with test-retest reliability ranging from .96 to .99 (Gaddis & Grill, 1995). Interrater reliability is found to be lower, ranging from .55 to .69 (Gaddis & Grill, 1995).

A study by Norris et al. (2019) investigated the structural validity in a sample of individuals 2–14 years of age with ASD. A confirmatory factor analysis supported the original five-factor solution. The factors, excluding Inappropriate Speech, were all negatively correlated with the Socialization, Daily Living Skills, and Communication domains of the Vineland Adaptive Behavior Scales, Second Edition (VABS-II; Sparrow et al., 2005).

Behavior Assessment System for Children

The Behavior Assessment System for Children, Third Edition (BASC-3; Reynolds & Kamphaus, 2015) is a behavioral assessment for individuals between 2 and 21 years of age. There are three rating scales: Parent Rating Scale, Teacher Rating Scales, and Self-Report of Personality. The Parent and Teacher scales are for individuals between 2 and 21 years of age, while the Self-Report scale is for ages 6 through 25 years. The BASC-3 includes validity indicators, composite scales, and four indexes. In addition, there is a Parenting Relationship Questionnaire (BASC-3 PRQ), Behavioral and Emotional Screening System (BASC-3 BESS), and Flex Monitor. The Flex Monitor (BASC-3 Flex Monitor) is a method to monitor behavioral and emotional functioning changes in response to treatment and intervention for individuals 2–18 years of age. The Flex Monitor also has three forms for self, teacher, and parent. It takes 5 minutes to complete each standard form, with custom forms varying in completion time.

Reliability coefficients for the BASC-3 PRS ranged from .76 to .97 in the general sample and between .71 and .98 for the clinical sample. Similarly, reliability coefficients for the TRS ranged from .77 to .98 in the general sample and .78 to .98 in the clinical sample. Lastly, the coefficients for the SRP ranged from .71 to .97 for the general sample and .57 to .96 in the clinical sample. Test-retest reliability was found to be greater than .80 for the PRS and TRS, and lower for the SRP, which ranged from .59 to .87 in the youngest group of respondents. Additionally, the BASC-3 had a wide range for internal reliability, between .32 and .84 for the TRS and between .47 and .87 for the PRS.

Children's Sleep Habits Questionnaire

The Children's Sleep Habits Questionnaire (CSHQ; Owens et al., 2000) is a parent questionnaire for children between 4 and 10 years of age experiencing sleep difficulties. There are 35 items which cover 8 sleep domains such as bedtime behavior and sleep onset, sleep duration, anxiety around sleep, behavior occurring during sleep and night waking, sleep-disordered breathing, parasomnias, and morning waking/daytime sleepiness. Each item is rated as "rarely" (occurring 0–1 time per week), "sometimes" (occurring 2–4 times per week), or "usually" (5–7 times per week).

Owens et al. (2000) found the internal consistency to be .68 in a community sample, with the internal consistency for the subdomains ranging between .36 and .70. The internal consistency for the clinical sample was found to be .78, ranging from .44 to .83 for the subdomains. Test-retest reliability was acceptable, ranging from .62 to .79 in a community sample. The CBCL was found to have a sensitivity of .80 and specificity of .72 with a cutoff score of 41 (Owens et al., 2000).

Anxiety and Depression Measures

Child Depression Inventory

The Child Depression Inventory, Second Edition (CDI 2; Kovacs & Staff, 2011) is a self-report measure to assess depression symptoms in children and adolescents 7–17 years of age. There is a full version (28 items) as well as a short version, primarily used for screening. In addition, there is a Teacher Form (12 items) and a Parent Form (17 items). All forms except for the short form have two subscales: Emotional Problems, with Negative Mood/Physical Symptoms and Negative Self-esteem as subscales, and Functional Problems, with Ineffectiveness and Interpersonal Problems as subscales.

The internal consistency was found to range between .67 and .91 for total and subscales across all age and sex groups (Bae, 2012). The CDI 2 was found to have convergent validity with the Beck Depression Inventory – Youth version (BDI-Y; Beck et al., 2001) and Conners Comprehensive Behavior Rating Scales (Conners CBRS; Conners, 2008).

Multidimensional Anxiety Scale for Children

The Multidimensional Anxiety Scale for Children, Second Edition (MASC; March, 2013) is a self-report and parent-report measure to identify and treat anxiety in children between 8 and 19 years of age. It consists of six scales and four subscales for a total of 50 items. The scales include Separation Anxiety/Phobias, Generalized Anxiety Disorder (GAD) Index, Social Anxiety: Total, Obsessions and Compulsions, Physical Symptoms: total, and Harm Avoidance. The subscales are Humiliation/Rejection and Performance Fears under Social Anxiety: Total and Panic and Tense/Restless under Physical Symptoms. The scores yield a total score, Anxiety Probability scale, scale scores, and an Inconsistency index.

Exploratory factor analyses yielded a four-factor solution: Physical Symptoms, Social Anxiety, Separation/Panic, and Harm Avoidance (Fraccaro et al., 2015). The MASC-2 was found to have acceptable internal consistency, with a coefficient alpha for the Total Score of .92 for the self-report and .89 for the parent report. Test-retest reliability was found to be between .80 and .94 for both forms. The MASC-2 demonstrates discriminative validity, as individuals with separation anxiety disorder, GAD, and social phobia scored highest on the respective scales (Fraccaro et al., 2015). It also demonstrated convergent validity with the Beck Youth Inventory – Anxiety (BYI-A; Beck et al., 2001), the GAD Scale of the Conners Comprehensive Behaviour Rating Scales – Self Report (Conners CBRS-SR; Conners, 2008) and parent report.

Pediatric Anxiety Rating Scales

The Pediatric Anxiety Rating Scales (PARS; RUPP, Research Units on Pediatric Psychopharmacology Anxiety Study Group,

2002) is a clinician-rated measure of the severity of anxiety symptoms in children between 6 and 17 years of age. Clinicians rate a total of 50 items, informed by separate child and parent interviews. Each item is rated on a 6-point scale ranging from 0 (none) to 1–5 (minimal to extreme) on seven dimensions: number of symptoms, frequency, severity of distress associated with anxiety symptoms, severity of physical symptoms, avoidance, interference at home, and interference out-of-home. A 5-item total is also available, which excludes number of symptoms and physical symptoms, which may be used for medication trials.

A study by Storch and colleagues (2012) investigated the reliability and validity of the PARS in children 7–17 years of age with autism. They found the internal consistency to be .59. Test-retest reliability was .83 and .86 for the interrater reliability. The PARS demonstrated convergent validity with the CGI-Severity, the MASC-P, and the internalizing and anxiety scales of the CBCL. Further, it demonstrated divergent validity with the ADOS, and the externalizing, attention, delinquent, and aggressive scales of the CBCL.

Attention Deficit/Hyperactivity Disorder Measures

ADHD Symptoms Rating Scale

The ADHD Symptoms Rating Scale (ADHD-SRS; Holland et al., 1998) is a measure to assess ADHD in school-aged individuals (K – 12). There is a total of 56 items on both the parent and Teacher Forms. An Exploratory Factor Analysis revealed a two-factor solution: Hyperactive-Impulsive and Inattention.

The ADHD-SRS was found to have excellent reliability with an internal consistency of .98 for the Parent Form and .99 for the Teacher Form (Holland et al., 1998). The test-retest reliability ranged between .95 and .97 for the total and subscale scores. Further, the ADHD-SRS was found to have convergent validity with other measures of ADHD (e.g., Attention Deficit Disorders Evaluation Scale (ADDES; McCarney, 1995a, b) and the Conners' Teacher Rating Scale (CTRS)) (Table 5.2).

Summary

With the prevalence of ASD rising since the 1990s, there is an increasing need for effective treatment options. In order to maximize treatment benefits and track response to treatments, ongoing progress monitoring is essential. ASD is multidimensional and can present with a wide range of symptoms (Eapen et al., 2016). Core symptoms of ASD, like impairments in social communication, are often targeted in interventions (Bolte & Diehl, 2013). However, deficits across various domains, including communication, behavioral and social skills, and cognitive abilities, vary among individuals with ASD. In addition to the core symptoms, comorbid conditions like challenging behaviors, sleep problems, anxiety, depression, and ADHD can be present in those with ASD (Frazier et al., 2011; Simonoff et al., 2008). As a result, symptoms of co-occurring conditions are often targets of intervention for children with ASD. More detailed progress monitoring is often necessary for those with multiple comorbid conditions compared to those with few (Eapen et al., 2016).

Since additional symptoms can present in conjunction with core ASD symptoms, interventions and treatment plans for ASD should be highly individualized, research based, progress monitored, and revised often (National Institute for Health and Care Excellence, 2013). Clinicians should aim to include input from caregivers in treatment planning goals and progress monitoring. While there are several different methods and tools that are used to monitor response to interventions among individuals with ASD, there are currently no standardized measures expressly designed for progress monitoring purposes (Bolte & Diehl, 2013). In spite of the challenges and limitations associated with progress monitoring, several evidence- based tools can be used to assess response to treatment in individuals from infancy to late adulthood.

Table 5.2 Progress monitoring measures

Domain	Measure	Age	Informant	Target behavior(s)
ASD	*Autism Treatment Evaluation Checklist* (Rimland & Edelson, 1999)	Children, unspecified	Parents/caregiver and/or teachers	Speech/language communication, sociability, sensory/cognitive awareness, health/physical/behavior
	Childhood Autism Rating Scale (Schopler et al., 2010)	2+ years	Clinician	Relating to people, imitation, emotional response, object use, body use, adaptation to change, visual response, listening response, taste, smell, and touch response and use, fear or nervousness, communication, activity level, intellectual response
	Psychoeducational Profile, Third Edition (Schopler et al., 2005)	2–7 years	Parent/caregiver, clinician	Language, motor, imitation, affect
	Social Responsiveness Scale, Second Edition (Constantino & Gruber, 2012)	4–18 years	Self-report, parents, spouse, friends, relatives, teachers	Social awareness, social cognition, social communication, social motivation, restricted interests and repetitive behavior
Developmental	*Bayley Scales of Infant and Toddler Development, Third Edition* (Bayley, 2006	1–42 months	Parent/caregiver, clinician	Cognitive, language, motor, social-emotional, and adaptive behavior
Adaptive	*Vineland Adaptive Behavior Scales, Third Edition* (Sparrow et al., 2016)	Birth–90 years	Parent/caregiver, teacher, clinician	Communication, daily living skills, socialization, and motor
Challenging behaviors	*Aberrant Behavior Checklist* (Aman et al., 1985a)	Individuals with developmental and intellectual disabilities	Parent/caregiver	Irritability, lethargy/social withdrawal, stereotypy, hyperactivity/noncompliance, and inappropriate speech
	Behavior Assessment for Children, Third Edition (Reynolds & Kamphaus, 2015)	2–21 years	Self-report, parent/caregiver, teacher	Externalizing problems, internalizing problems, adaptive skills
Sleep difficulties	*Children's Sleep Habits Questionniare* (Owens et al., 2000)	4–10 years	Parent/caregiver	Bedtime behavior, sleep onset, duration, anxiety around sleep, parasomnias, sleepiness
Depression	*Child Depression Inventory* (Kovacs & Staff, 2011)	7–17 years	Self-report, parent/caregiver, teacher	Emotional problems and functional problems
Anxiety	*Multidimensional Anxiety Scale for Children, Second Edition* (March, 2013)	8–19 years	Self-report, parent/caregiver	Anxiety/phobias, generalized anxiety disorder, social anxiety, obsessions and compulsions, and physical symptoms
	Pediatric Anxiety Rating Scales (RUPP, 2002)	6–17 years	Clinician	Number of symptoms, frequency, severity of distress, severity of physical symptoms, avoidance, interference
ADHD	*ADHD Symptoms Rating Scale* (Holland et al., 1998)	K – 12	Parent/caregiver, teacher	Hyperactive-impulsive and inattention

References

Abel, E. A., Schwichtenberg, A. J., & Brodhead, M. T. (2017). Assessing sleep problems within the context of early behavioral intervention for autism spectrum disorder. *Sleep, 40*, 340–341.

Achenbach, T. M., & Rescorla, L. A. (2000). *Manual for the ASEBA Preschool forms & profiles*. University of Vermont, Research Center for Children, Youth, & Families.

Albers, C. A., & Grieve, A. J. (2007). Test review: Harcourt assessment. *Journal of Psychoeducational Assessment*, *25*(2), 180–190. https://doi-org.libezp.lib.lsu.edu/10.1177/0734282906297199. Bayley, N. (2006). Bayley scales of infant and toddler development–third edition. San Antonio, TX.

Allik, H., Larsson, J., & Smedje, H. (2006). Sleep patterns of school-age children with Asperger syndrome or high-functioning autism. *Journal of Autism and Developmental Disorders, 36*, 585–595.

Aman, M., & Singh, N. (1994). *The aberrant behavior checklist-community*. Slosson Education Publications, Inc..

Aman, M. G., & Singh, N. N. (2017). *Aberrant behavior checklist manual* (2nd ed.). Slosson Educational Publications, Inc..

Aman, M. G., Singh, N. N., Stewart, A. W., & Field, C. J. (1985a). The aberrant behavior checklist: A behavior rating scale for the assessment of treatment effects. *American Journal of Mental Deficiency, 89*(5), 485–491.

Aman, M. G., Singh, N. N., Stewart, A. W., & Field, C. J. (1985b). Psychometric characteristics of the aberrant behavior checklist. *American Journal of Mental Deficiency, 89*(5), 492–502.

Aman, M. G., McDougle, C. J., Scahill, L., Handen, B., Arnold, L. E., Johnson, C., Stigler, K. A., Bearss, K., Butter, E., Swiezy, N. B., Sukhodolsky, D. D., Ramadan, Y., Pozdol, S. L., Nikolov, R., Lecavalier, L., Kohn, A. E., Koenig, K., Hollway, J. A., Korzekwa, P., … Wagner, A. (2009). Medication and parent training in children with pervasive developmental disorders and serious behavior problems: Results from a randomized clinical trial. *Journal of the American Academy of Child and Adolescent Psychiatry, 48*, 1143–1154.

American Psychiatric Association. (2013). *Diagnostic and statistical manual of mental disorders* (5th ed., p. 473). American Psychiatric Association.

Anagnostou, E., Jones, N., Huerta, M., Halladay, A. K., Wang, P., Scahill, L., Horrigan, J. P., Kasari, C., Lord, C., Choi, D., Sullivan, K., & Dawson, G. (2015). Measuring social communication behaviors as a treatment endpoint in individuals with autism spectrum disorder. *Autism, 19*, 622–636.

Bae, Y. (2012). Test review: Children's depression inventory 2 (CDI 2). *Journal of Psychoeducational Assessment, 30*(3), 304–308. Kovacs, M. "Children's depression inventory 2 (CDI 2)" (2nd ed.). Multi-Health Systems Inc.; 2011.

Baeza-Velasco, C., Michelon, C., Rattaz, C., & Baghdadli, A. (2014). Are aberrant behavioral patterns associated with the adaptive behavior trajectories of teenagers with autism spectrum disorders? *Research in Autism Spectrum Disorders, 8*(3), 304–311.

Baghdadli, A., Picot, M. C., Pry, R., Michelon, C., Burzstein, C., Lazartigues, C., & Aussiloux, A. (2008). What factors are related to a negative outcome of self-injurious behavior during childhood in pervasive developmental disorders? *Journal of Applied Research in Intellectual Disabilities, 21*(2), 142–149.

Bayley, N. (1993). *Bayley scales of infant development* (2nd ed.). The Psychological Corporation.

Bayley, N. (2006). *Bayley scales of infant and toddler development* (3rd ed.). The Psychological Corporation.

Bayley, N., & Aylward, G. P. (2019). *Bayley scales of infant and toddler development* (4th ed.). The Psychological Corporation.

Beck, J. S., Beck, A. T., & Jolly, J. B. (2001). *Manual for the Beck youth inventories of emotional and social impairment*. The Psychological Corporation.

Bolte, E. E., & Diehl, J. J. (2013). Measurement tools and target symptoms/skills used to assess treatment response for individuals with autism spectrum disorder. *Journal of Autism and Developmental Disorders, 43*, 2491–2501.

Brown, E. C., Aman, M. G., & Havercamp, S. M. (2002). Factor analysis and norms for parent ratings on the aberrant behavior checklist-community for young people in special education. *Research in Developmental Disabilities, 23*, 45–60.

Brugha, T. S., Doos, L., Tempier, A., Einfeld, S., & Howlin, P. (2015). Outcome measures in intervention trials for adults with autism spectrum disorders; A systematic review of assessments of core autism features and associated emotional and behavioral problems. *International Journal of Methods in Psychiatric Research, 24*(2), 99–115.

Bruni, T. P. (2014). Test review: Social responsiveness scale–Second edition (SRS-2). *Journal of Psychoeducational Assessment, 32*(4), 365–369.

Center for Disease Control and Prevention. (2020). Autism spectrum disorder (ASD): Data & statistics. Retrieved 1 May 2021. https://www.cdc.gov/ncbddd/autism/data.html

Chadwick, O., Piroth, N., Walker, J., Bernard, S., & Taylor, E. (2000). Factors affecting the risk of behaviour problems in children with severe intellectual disability. *Journal of Intellectual Disability Research, 44*(2), 108–123.

Chandrasekhar, T., & Sikich, L. (2015). Challenges in the diagnosis and treatment of depression in autism spectrum disorders across the lifespan. *Dialogues in Clinical Neuroscience, 17*(2), 219–227.

Collett, B. R., Ohan, J. L., & Myers, K. M. (2003). Ten-year review of rating scales. V: Scales assessing attention-deficit/hyperactivity disorder. *Journal*

of the American Academy of Child and Adolescent Psychiatry, 42, 1015–1037.
Conners, C. K. (1997). *Conners' rating scales-Revised technical manual*. Multi-Health Systems.
Conners, C. K. (2008). *Conners comprehensive behavior rating scales (Conners CBRS)*. MultiHealth Systems, Inc.
Conrod, E., & Marcus, L. (2013). Psychoeducational profile – Revised (PEP-3). In F. R. Volkmar (Ed.), *Encyclopedia of autism spectrum disorders*. Springer. https://doi.org/10.1007/978-1-4419-1698-3_936
Constantino, J. N., & Gruber, C. P. (2005). *Social responsiveness scale (SRS)*. Western Psychological Services.
Constantino, J. N., & Gruber, C. P. (2012). *Social responsiveness scale–Second edition (SRS-2)*. Western Psychological Services.
Copeland, L., & Buch, G. (2013). Early intervention issues in autism spectrum disorders. *Autism, 3*(1), 1–7.
DeFilippis, M. (2018). Depression in children and adolescents with autism spectrum disorder. *Children, 5*(9), 112.
Deno, S.L. (1985). Curriculum-based measurement: The emerging alternative. *Exceptional Children, 52*, 219–232.
DuPaul, G. J., Power, T. J., Anastopoulos, A. D., & Reid, R. (2016). *ADHD rating scale-5 for children and adolescents: Checklists, norms, and clinical interpretation*. Guilford Press.
Eapen, V., Williams, K., Roberts, J., Rinehart, N., & McGillivray, J. (2016). Monitoring progress in autism spectrum disorder. In J. L. Matson (Ed.), *Handbook of assessment and diagnosis of autism spectrum disorder* (pp. 87–115). Springer International Publishing.
Epstein, J. N., & Weiss, M. D. (2012). Assessing treatment outcomes in attention-deficit/hyperactivity disorder: A narrative review. *The Primary Care Companion for CNS Disorders, 14*(6), 1–30.
Fadini, C. C., Lamônica, D. A., Fett-Conte, A. C., Osório, E., Zuculo, G. M., Giacheti, C. M., & Pinato, L. (2015). Influence of sleep disorders on the behavior of individuals with autism spectrum disorder. *Frontiers in Human Neuroscience, 9*, 347.
Fisher, W. W., Piazza, C. C., & Roane, H. S. (2011). *Handbook of applied behavior analysis*. Guilford Press.
Fraccaro, R. L., Stelnicki, A. M., & Nordstokke, D. W. (2015). Test review: Multidimensional anxiety scale for children by JS March. *Canadian Journal of School Psychology, 30*(1), 70–77.
Frazier, T. W., Shattuck, P. T., Narendorf, S. C., Cooper, B. P., Wagner, M., & Spitznagel, E. L. (2011). Prevalence and correlates of psychotropic medication use in adolescents with an autism spectrum disorder with and without care-giver-reported attention-deficit/hyperactivity disorder. *Journal of Child and Adolescent Psychopharmacology, 21*(6), 571–579.
Freeman, B. J., Ritvo, E. R., Yokota, A., & Ritvo, A. (1986). A scale for rating symptoms of patients with the syndrome of autism in real life settings. *Journal of the American Academy of Child and Adolescent Psychiatry, 25*(1), 130–136.
Gaddis, L., & Grill, J. J. (1995). Aberrant behavior checklist. *The twelfth mental measurements yearbook*, p. 12.
Gadow, K. D., & Sprafkin, J. (2002). *Child symptom inventory-4 screening and norms manual*. Checkmate Plus.
Giannotti, F., Cortesi, F., Cerquiglini, A., Miraglia, D., Vagnoni, C., Sebastiani, T., & Bernabei, P. (2008). An investigation of sleep characteristics, EEG, abnormalities and epilepsy in developmentally regressed and non-regressed children with autism. *Journal of Autism and Developmental Disorders, 38*(10), 1888–1897.
Ginsburg, G. S., Keeton, C. P., Drazdowski, T. K., & Riddle, M. A. (2011). The utility of clinicians ratings of anxiety using the pediatric anxiety rating scale (PARS). *Child and Youth Care Forum, 2*(40), 93–110.
González, M., Dixon, D. R., Rojahn, J., Esbensen, A. J., Matson, J. L., Terlonge, C., & Smith, K. R. (2009). The behavior problems inventory: Reliability and factor validity in institutionalized adults with intellectual disabilities. *Journal of Applied Research in Intellectual Disabilities, 22*(3), 223–235.
Goodlin-Jones, B. L., Tang, K., Liu, J., & Anders, T. F. (2008). Sleep patterns in preschool-age children with autism, developmental delay, and typical development. *Journal of the American Academy of Child Adolescent Psychiatry, 47*(8), 930–938.
Gotham, K., Brunwasser, S. M., & Lord, C. (2015). Depressive and anxiety symptom trajectories from school age through young adulthood in samples with autism spectrum disorder and developmental delay. *Journal of American Academy of Child and Adolescent Psychiatry, 54*(5), 369–376.
Green, J., Charman, T., McConachie, H., Aldred, C., Slonims, V., Howlin, P., Le Couteur, A., Leadbitter, K., Hudry, K., Byford, S., Barrett, B., Temple, K., Macdonald, W., & Pickles, A. (2010). Parent-mediated communication-focused treatment in children with autism (PACT): A randomized controlled trial. *Lancet, 375*(9752), 2152–2160.
Grzadzinski, R., Janvier, D., & Kim, H. S. (2020). Recent developments in treatment outcome measures for young children with autism spectrum disorder (ASD). *Seminars in Pediatric Neurology, 34*, 1–13.
Guastella, A. J., Gray, K. M., Rinehart, N. J., Alvares, G. A., Tongue, B. J., Hickie, I. B., Keating, C. M., Caccioti-Saija, C., & Einfeld, S. L. (2015). The effects of a course of intranasal oxytocin on social behaviors in youth diagnosed with autism spectrum disorders: A randomized controlled trial. *Journal of Child Psychology and Psychiatry, 56*(4), 444–452.
Gutstein, S., Burgess, A. F., & Montfort, K. (2007). Evaluation of the relationship development intervention program. *Autism, 11*(5), 397–411.
Hallett, V., Lecavalier, L., Sukhodolsky, D. G., Cipriano, N., Aman, M. G., McCracken, J. T., McDougle, C. J., Tierney, E., King, B. H., Hollander, E., Sikich, L., Bregman, J., Anagnostou, E., Donnelly, C., Katsovich, L., Dukes, K., Vitiello, B., Gadow, K., & Scahill, L.

(2013). Exploring the manifestations of anxiety in children with autism spectrum disorders. *Journal of Autism and Developmental Disorders, 43*, 2341–2352.

Hanratty, J., Livingstone, N., Robalino, S., Terwee, C. B., Glod, M., Oono, I. P., Rodgers, J., Macdonald, G., & McConachie, H. (2015). Systematic review of the measurement properties of tools used to measure behavior problems in young children with autism. *PLoS One, 10*(12), 1–21.

Harrison, P., & Oakland, T. (2015). *Adaptive behavior assessment system, third edition (ABAS-3)*. Pearson.

Hartley, S. L., Sikora, D. M., & McCoy, R. (2008). Prevalence and risk factors of maladaptive behaviour in young children with autistic disorder. *Journal of Intellectual Disability Research, 52*(10), 819–829.

Herzinger, C. V., & Campbell, J. M. (2007). Comparing functional assessment methodologies: A quantitative synthesis. *Journal of Autism and Developmental Disorders, 37*(8), 1430–1445.

Hodge, D., Parnell, A. M. N., Hoffman, C. D., & Sweeney, D. P. (2012). Methods for assessing sleep in children with autism spectrum disorders: A review. *Research in Autism Spectrum Disorders, 6*, 1337–1344.

Holland, M. L., Gimpel, G. A., & Merrell, K. W. (1998). Innovations in assessing ADHD: Development, psychometric properties, and factor structure of the ADHD Symptoms Rating Scale (ADHD-SRS). *Journal of Psychopathology and Behavioral Assessment, 20*(4), 307–332.

Hollingdale, J., Woodhouse, E., Young, S., Fridman, A., & Mandy, W. (2020). Autistic spectrum disorder symptoms in children and adolescents with attention-deficit/hyperactivity disorder: A meta-analytical review. *Psychological Medicine, 50*, 2254.

Howard, J. (2019). A review on measuring treatment response in parent-mediated autism early interventions. *Pediatric Medicine, 2*, 1–16.

Hudson, C. C., Hall, L., & Harkness, K. L. (2019). Prevalence of depressive disorders in individuals with autism spectrum disorder: A meta-analysis. *Journal of Abnormal Child Psychology, 47*(1), 165–175.

Kasari, C. (2002). Assessing change in early intervention programs for children with autism. *Journal of autism and developmental disorders, 32*(5), 447–461. https://doi.org/10.1023/a:1020546006971.

Kasari, C., Gulsrud, A., Paparella, T., Hellemann, G., & Berry, K. (2015). Randomized comparative efficacy study of parent-mediated interventions for toddlers with autism. *Journal of Consulting and Clinical Psychology, 83*, 554–563.

Kat, S., Xu, L., Guo, Y., Ma, J., Ma, Z., Yang, Y., Wang, H., Li, X., & Liu, J. (2020). Reliability and validity of the simplified Chinese version of the aberrant behavior checklist in Chinese autism population. *Frontiers in Psychology, 11*, 1–8.

Kerns, C. M., Kendall, P. C., Zickgraf, H., Franklin, M. E., Miller, J., & Herrington, J. (2015). Not to be overshadowed or overlooked: Functional impairments associated with comorbid anxiety disorders in youth with ASD. *Behavior Therapy, 46*, 29–39.

Kerns, C. M., Wood, J. J., Kendall, P. C., Renno, P., Crawford, E. A., Mercado, R. J., Fujii, C., Collier, A., Hoff, A., Kagan, E. R., Small, B. J., Lewin, A. B., & Storch, E. A. (2017). The treatment of anxiety in autism spectrum disorder (TAASD) study: Rationale, design, and methods. *Journal of Child and Family Studies, 25*(6), 1889–1902.

Kovacs, M. (2010). *Children's depression inventory 2nd edition (CDI2)*. Multi-Health Systems Inc.

Kovacs, M. & Staff, M.H.S. (2003). Children's depression inventory (CDI): Technical.

Kovacs, M., & Staff, M. (2011). *Children's depression inventory 2nd edition (CDI 2): Technical manual*. Multi-Health Systems.

Lecavalier, L., Wood, J. J., Halladay, A. K., Jones, N. E., Aman, M. G., Cook, E. H., Handen, B. L., King, B. H., Pearson, D. A., Hallett, V., Sullivan, K. A., Grondhuis, S., Bishop, S. L., Horrigan, J. P., Dawson, G., & Scahill, L. (2014). Measuring anxiety as a treatment endpoint in youth with autism spectrum disorder. *Journal of Autism and Developmental Disorders, 44*(5), 1128–1143.

Lord, C., Rutter, M., DiLavore, P., & Risi, S. (1999). *Autism diagnostic observation schedule: Manual*. Western Psychological Services.

Lord, C., Risi, S., Lambrecht, L., Cook, E. H., Jr., Leventhal, B. L., DiLavore, P. C., Pickles, A., & Rutter, M. (2000). The autism diagnostic observation schedule--generic: A standard measure of social and communication deficits associated with the spectrum of autism. *Journal of Autism and Developmental Disorders, 30*(3), 205–223.

Maddox, B. B., Lecavalier, L., Miller, J. S., Pritchett, J., Hollway, J., White, S. W., Gillespie, S., Evans, A. N., Schultz, R. T., Herrington, J. D., Bearss, K., & Scahill, L. (2020). Reliability and validity of the pediatric anxiety rating scale modified for autism spectrum disorder. *Autism, 24*(7), 1773–1782.

Magiati, I., Moss, J., Yates, R., Charman, T., & Howlin, P. (2011). Is the autism treatment evaluation checklist a useful tool for monitoring progress in children with autism spectrum disorders? *Journal of Intellectual Disability Research, 55*(3), 302–312.

Mahan, S., & Matson, J. L. (2011). Children and adolescents with autism spectrum disorders compared to typically developing controls on the behavioral assessment system for children, second edition (BASC-2). *Research in Autism Spectrum Disorders, 5*(1), 119–125.

Malow, B. A., Crowe, C., Henderson, L., McGrew, S. G., Wang, L., Song, Y., & Stone, W. L. (2009). A sleep habits questionnaire for children with autism spectrum disorder. *Journal of Child Neurology, 24*, 19–24.

Malow, B. A., Adkins, K. W., Reynolds, A., Weiss, S. K., Loh, A., Fawkes, D., Katz, T., Goldman, S. E., Madduri, N., Hundley, R., & Clemons, T. (2015). Parent-based sleep education for children with autism spectrum disorders. *Journal for Autism and Developmental Disorders, 44*(1), 1–22.

March, J. S. (2013). *Multidimensional anxiety scale for children (MASC 2): Technical manual*. Multi-Health Systems Incorporated.

March, J. S., Parker, J. D., Sullivan, K., Stallings, P., & Conners, C. K. (1997). The multidimensional anxiety scale for children (MASC): Factor structure, reliability, and validity. *Journal of the American Academy of Child and Adolescent Psychiatry, 36*(4), 554–565.

Matson, J. L. (2007). Determining treatment outcome in early intervention programs for autism spectrum disorders: A critical analysis of measurement issues in learning based interventions. *Research in Developmental Disabilities, 28*(2), 207–218.

Matson, J. L., & Kozlowski, A. M. (2011). The increasing prevalence of autism spectrum disorders. *Research in Autism Spectrum Disorders, 5*(1), 418–425.

Matson, J. L., & Rivet, T. T. (2007). A validity study of the autism spectrum disorders—Behavior problems for adults (ASD-BPA) scale. *Journal of Developmental and Physical Disabilities, 19*, 557–564.

Matson, J. L., Gonzalez, M. L., & Rivet, T. T. (2008). Reliability of the autism spectrum disorder-behavior problems for children (ASD-BPC). *Research in Autism Spectrum Disorders, 2*(4), 696–706.

Matson, J. L., Boisjoli, J., Rojahn, J., & Hess, J. (2009). A factor analysis of challenging behaviors assessed with the Baby and Infant Screen for Children with aUtism Traits (BISCUIT-Part 3). *Research in Autism Spectrum Disorders, 3*, 714–722.

Mazefsky, C. A., & Oswald, D. P. (2011). Preliminary evidence suggesting caution in the use of psychiatric self-report measures with adolescents with high-functioning autism spectrum disorders. *Research in Autism Spectrum Disorders, 5*(1), 164–174.

Mazurek, M. O., & Petroski, G. F. (2015). Sleep problems in children with autism spectrum disorder: Examining the contributions of sensory over-responsivity and anxiety. *Sleep Medicine, 16*(2), 270–279.

Mazurek, M. O., Vasa, R. A., Kalb, L. G., Kanne, S. M., Rosenberg, D., Keefer, A., Murray, D. S., Freedman, B., & Lowery, L. (2013). Anxiety, sensory over-responsivity, and gastrointestinal problems in children with autism spectrum disorders. *Journal of Abnormal Child Psychology, 41*, 165–176.

McCarney, S. B. (1995a). *Attention deficit disorders evaluation scale – Home version technical manual* (2nd ed.). Hawthorne Educational Services, Inc.

McCarney, S. B. (1995b). *Attention deficit disorders evaluation scale-school version technical manual* (2nd ed.). Hawthorne Educational Services, Inc.

McConachie, H., & Diggle, T. (2007). Parent implemented early intervention for young children with autism spectrum disorder: A systematic review. *Journal of Evaluation in Clinical Practice, 13*(1), 120–129.

McConachie, H., Parr, J. R., Glod, M., Hanratty, J., Livingstone, N., Oono, I. P., Robalino, S., Baird, G., Beresford, B., Charman, T., Garland, D., Green, J., Gringras, P., Jones, G., Law, J., Le Couteur, A. S., Macdonald, G., McColl, E. M., Morris, C., … Williams, K. (2015). Systematic review of tools to measure outcomes for young children with autism spectrum disorder. *Health Technology Assessment, 19*(1), 1–506.

Minshawi, N. F., Hurwitz, S., Fodstad, J. C., Biebl, S., Morriss, D. H., & McDougle, C. J. (2014). The association between self-injurious behaviors and autism spectrum disorders. *Psychology Research and Behavior Management, 7*, 125–136.

Moore, M., Evans, V., Hanvey, G., & Johnson, C. (2017). Assessment of sleep in children with autism spectrum disorder. *Children, 4*(8), 72.

Mpaka, D. M., Okitundu, D. L., Ndjukendi, A. O., N'situ, A. M., Kinsala, S. Y., Mukau, J. E., Ngoma, V. M., Kashala-Abotnes, E., Ma-Miezi-Mampunza, S., Vogels, A., & Steyaert, J. (2016). Prevalence and comorbidities of autism among children referred to the outpatient clinics for neurodevelopmental disorders. *The Pan African Medical Journal, 25*, 82.

Mundy, P., Delgado, C., Block, J., Venezia, M., Hogan, A., & Seibert, J. (2013). Early social communication scales (ESCS).

Murphy, O., Healy, O., & Leader, G. (2009). Risk factors for challenging behaviors among 157 children with autism spectrum disorder in Ireland. *Research in Autism Spectrum Disorders, 3*(2), 474–482.

National Institute for Health and Care Excellence. (2013). *Autism: The management and support of children and young people on the autism spectrum (NICE clinical guidelines)*. NICE.

Norris, M., Aman, M. G., Mazurek, M. O., Scherr, J. F., & Butter, E. M. (2019). Psychometric characteristics of the aberrant behavior checklist in a well-defined sample of youth with autism spectrum disorder. *Research in Autism Spectrum Disorders, 62*, 1–9.

Owens, J. A., Spirito, A., & McGuinn, M. (2000). The children's sleep habits questionnaire (CSHQ): Psychometric properties of a survey instrument for school-aged children. *Sleep, 23*(8), 1043–1052.

Pearson Assessments. (2019a). Bayley-III vs. Bayley-4: What's changed? https://www.pearsonassessments.com/content/dam/school/global/clinical/us/assets/bayley-4/bayley-3-to-4-comparison.pdf

Pearson Assessments. (2019b). By the numbers. https://www.pearsonassessments.com/content/dam/school/global/clinical/us/assets/bayley-4/bayley4-infographic.pdf

Pepperdine, C. R., & McCrimmon, A. W. (2018). Test review: Vineland Adaptive Behavior Scales, (Vineland-3) by Sparrow, S. S, Cicchetti, D. V, & Saulnier, C. A. *Canadian Journal of School Psychology, 33*(2), 157–163.

Rao, P. A., & Landa, R. J. (2013). Association between severity of behavioral phenotype and comorbid attention deficit hyperactivity disorder symptoms in children with autism spectrum disorders. *Autism, 18*, 272–280.

Research Units on Pediatric Psychopharmacology Anxiety Study Group. (2002). The pediatric anxiety rating scale (PARS): Development and psychometric properties. *Journal of the American Academy of Child and Adolescent Psychiatry, 41*(9), 1061–1069.

Reynolds, C. R., & Kamphaus, R. W. (2015). *Behavior assessment system for children–Third edition (BASC-3)*. Pearson.

Rimland, B., & Edelson, S. M. (1999). Autism treatment checklist (ATEC). Accessed at https://www.autism.org/autism-treatment-evaluation-checklist/

Rogers, S. J. (1998). Empirically supported comprehensive treatments for young children with autism. *Journal of Clinical Child Psychology, 27*(2), 168–179.

Rojahn, J., Matson, J. L., Lott, D., Esbensen, A. J., & Smalls, Y. (2001). The behavior problems inventory: An instrument for the assessment of self-injury, stereotyped behavior, and aggression/destruction in individuals with developmental disabilities. *Journal of Autism and Developmental Disorders, 31*(6), 577–588.

Rojahn, J., Schroeder, S. R., Mayo-Ortega, L., Oyama-Ganiko, R., LeBlanc, J., Marquis, J., & Berke, E. (2013). Validity and reliability of the behavior problems inventory, the aberrant behavior checklist, and the repetitive behavior scale-revised among infants and toddlers at risk for intellectual or developmental disabilities: A multi-method assessment approach. *Research in Developmental Disabilities, 34*, 1804–1814.

Rutter, M., Bailey, A., & Lord, C. (2001). *Social Communication Questionnaire (SCQ)*. Western Psychological Services.

Salvia, J., Ysseldyke, J., & Bolt, S. (2013). *Assessment in special and inclusive education* (12th ed.). Cengage Learning.

Schmidt, J. D., Huete, J. M., Fodstad, J. C., Chin, M. D., & Kurtz, P. F. (2013). An evaluation of the aberrant behavior checklist for children under age 5. *Research in Developmental Disabilities, 34*(4), 1190–1197.

Schopler, E., Reichler, R., Devellis, R., & Daly, K. (1980). Toward objective classification of childhood autism: Childhood Autism Rating Scale (CARS). *Journal of Autism and Developmental Disorders, 10*(1), 91–103.

Schopler, E., Lansing, M., Reichler, R., & Marcus, L. (2005). *Psychoeducational profile: Third edition (PEP-3)*. Pro-Ed.

Schopler, E., Van Bourgondien, M. E., Wellman, G. J., & Love, S. R. (2010). *Childhood autism rating scale* (2nd ed.). Western psychology services.

Simonds, J., & Parraga, H. (1982). Prevalence of sleep disorders and sleep behaviors in children and adolescents. *Journal of American Academy of Child and Adolescent Psychiatry, 21*, 383–388.

Simonoff, E., Pickles, A., Charman, T., Chandler, S., Loucas, T., & Baird, G. (2008). Psychiatric disorders in children with autism spectrum disorders: Prevalence, comorbidity and associated factors in a population-driven sample. *Journal of the American Academy Child and Adolescent Psychiatry, 47*(8), 921–929.

Sparrow, S. S., Cicchetti, D. V., & Balla, D. A. (2005). *Vineland adaptive behavior scales* (2nd ed.). American Guidance Service.

Sparrow, S. S., Cicchetti, D. V., & Saulnier, C. A. (2016). *Vineland adaptive behavior scales, third edition (Vineland-3)*. Pearson.

Storch, E. A., Arnold, E. B., Lewin, A. B., Nadeau, J. M., Jones, A. M., De Nadai, A. S., & Murphy, T. K. (2013). The effect of cognitive-behavioral therapy versus treatment as usual for anxiety in children with autism spectrum disorders: A randomized, controlled trial. *Journal of the American Academy of Child and Adolescent Psychiatry, 52*(2), 132–142.

Storch, E. A., Wood, J. J., Ehrenreich-May, J., Jones, A. M., Park, J. M., Lewin, A. B., & Murphy, T. K. (2012). Convergent and discriminant validity and reliability of the pediatric anxiety rating scale in youth with autism spectrum disorders. *Journal of Autism and Developmental Disorders, 42*, 2374–2382.

Todorow, C., James, C., & Turchi, R. M. (2018). The medical home for children with autism spectrum disorder: An essential element whose time has come. *Current Opinion in Pediatrics, 30*(2), 311–317.

Van Steensel, F. J. A., Bögels, S. M., & Perrin, S. (2011). Anxiety disorders in children and adolescents with autism spectrum disorders: A meta-analysis. *Clinical Child and Family Psychology Review, 14*, 302–317.

Vivanti, G., Prior, M., Williams, K., & Dissanayake, C. (2014). Predictors of outcomes in autism early intervention: Why don't we know more? *Frontiers in Pediatrics, 2*(58), 1–10.

Wechsler, D. (2002). *The Wechsler Preschool and Primary Scale of Intelligence-Third Edition*. San Antonio, TX: The Psychological Corporation.

White, S. W., Oswald, D., Ollendick, T., & Scahill, L. (2009). Anxiety in children and adolescents with autism spectrum disorders. *Clinical Psychology Review, 29*, 216–229.

White, S. W., Ollendick, T., Albano, A. M., Oswald, D., Johnson, C., Southam-Gerow, M. A., Kim, I., & Scahill, L. (2013). Randomized controlled trial: Multimodal anxiety and social skill intervention for adolescents with autism spectrum disorder. *Journal of Autism and Developmental Disorders, 43*(2), 382–394.

Wiggs, L., & Stores, G. (2004). Sleep patterns and sleep disorders in children with autistic spectrum disorders: Insights using parent report and actigraphy. *Developmental Medicine and Child Neurology, 46*, 372–380.

Witmer, S. E., Nasamran, A., Parikh, P. J., Schmitt, H. A., & Clinton, M. C. (2015). Using parents and teachers to monitor progress among children with ASD: A review of intervention research. *Focus on Autism and Other Developmental Disabilities, 30*(2), 67–85.

Wood, J. J., Drahota, A., Sze, K. M., Har, K., Chiu, A., & Langer, D. (2009). Cognitive behavioral therapy for anxiety in children with autism spectrum disorders: A randomized, controlled trial. *Journal of Child Psychology and Psychiatry, 50*, 224–234.

Yerys, B. E., Nissley-Tsiopinis, J., de Marchena, A., Watkins, M. W., Antezana, L., Power, T. J., & Schultz, R. T. (2017). Evaluation of the ADHD rating scale in youth with autism. *Journal of Autism and Developmental Disorders, 47*(1), 90–100.

Young, S., Hollingdale, J., Absound, M., Bolton, P., Branney, P., Colley, W., Craze, E., Dave, M., Deeley, Q., Farrag, E., Gudjonsson, G., Hill, P., Liang, H., Murphy, C., Mackintosh, P., Murin, M., O'Reagan, F., Ourgin, D., Rios, P., … Woodhouse, E. (2020). Guidance for identification and treatment of individuals with attention deficit/hyperactivity disorder and autism spectrum disorder based upon expert consensus. *BMC Medicine, 18*(146), 1–29.

Zaboski, B. A., & Storch, E. A. (2018). Comorbid autism spectrum disorder and anxiety disorders: A brief review. *Future Neurology, 13*(1), 31–37.

Part II

Professional Perspectives on Treatment

6 Allergy and Immunology in Autism Spectrum Disorder and Other Neurodevelopmental Disorders

Dexter A. Matrana, Elizabeth R. Smith, and Luke A. Wall

Abstract

Many unique caveats exist regarding the allergic/immunologic (A/I) evaluation of patients with autism spectrum disorder (ASD) or neurodevelopmental disorders (NDD). This chapter includes a review of common allergic conditions and the unique approach to testing in this patient population. An overview of primary immunodeficiency and discussion of which patients with ASD/NDD should undergo immunologic workup is also provided. Guidance is offered regarding the indications for referral to an A/I specialist, preparing the patient for the visit, and the types of testing which might occur in the A/I office including strengths and limitations of testing. An overview of treatments of A/I disorders is included, as well as a discussion of dietary considerations in patients with ASD/NDD and concerns for food allergy.

Keywords

Autism · Neurodevelopmental disorders · Primary immunodeficiency · Allergy · Asthma · Atopic dermatitis

Introduction

Allergy/immunology (A/I) is a field that focuses specifically on the diagnosis and management of allergic diseases, asthma, and immune disorders. Most providers treat both adult and pediatric patients, typically in the outpatient setting. However, some hospital systems, especially at academic centers, will offer inpatient consultations. Allergist/immunologists first complete residency training in either internal medicine, pediatrics, or combined internal medicine and pediatrics. They then spend an additional 2–3 years in a fellowship program focused solely on allergy/immunology.

Allergic conditions can affect many different organ systems including the skin, respiratory system, and digestive tract. Some disorders include rhinitis, conjunctivitis, eczema, eosinophilic esophagitis, angioedema, anaphylaxis, asthma, urticaria, and adverse reactions to foods, drugs, and insect stings. Allergists determine if a reaction seems to be IgE-mediated, prompting concern for a rapid onset potentially life-threatening

D. A. Matrana · E. R. Smith · L. A. Wall (✉)
Division of Allergy/Immunology, Department of Pediatrics, Louisiana State University Health Sciences Center, Children's Hospital New Orleans, New Orleans, LA, USA
e-mail: lwall@lsuhsc.edu

P. McPherson (ed.), *Handbook of Treatment Planning for Children with Autism and Other Neurodevelopmental Disorders*, Autism and Child Psychopathology Series,
https://doi.org/10.1007/978-3-031-06120-2_6

allergic reaction. These reactions are typically specific for a certain trigger and resolve quickly with appropriate treatment.

While the immune system is very complex and has multiple associated disorders involving both overaction and underproduction, allergist/immunologists typically focus on primary immune deficiencies. These patients often present with recurrent infections and may have a predisposition to infections with unusual organisms. This chapter will review some of the more common immunodeficiencies. Discussion will also include which patients should be referred to a specialist and various treatment modalities.

Brief Review of Basic Immunology

The immune system can be divided into an innate arm and an adaptive arm. The innate arm consists of the mucosal barriers, skin, phagocytes (neutrophils, macrophages), natural killer cells, complement, and several types of specialized receptors. This is the initial defense against invading microorganisms. While this arm of the immune system works to defend the body, the specialized adaptive immune system is undergoing "training" to attack the offending agents. T cells and B cells make up the adaptive arm of the immune system. T cells can be divided into many subtypes but for the sake of this brief review, T cells broadly consist of cytotoxic T cells (CD8+) and helper T cells (CD4+). Cytotoxic T cells recognize infected cells (from intracellular bacteria, fungi, or viruses) and kill them, forming cellular immunity. Helper T cells initiate activation of cytotoxic T cells and B cells. B cells, after activation, divide into either memory or plasma cells, the latter produces antibodies against extracellular organisms, forming humoral immunity (Abbas et al., 2018).

Referring to an Allergist/Immunologist

When to refer or consider the possibility of referral may be difficult as primary immunodeficiency disorders (PID) can have an elusive clinical presentation. PID is much more common than once thought and has been noted to occur with a frequency of 1 per 2000 of live births (Bonilla et al., 2015). Despite this, patients may go undiagnosed until later in life or even postmortem (Rezaei & Bonilla, 2016). Frequently, concern for diagnosis of PID arises when a patient experiences an excess of infections or infection with an opportunistic pathogen; however, there are physical exam findings and even other system involvement in certain PIDs that can be utilized for diagnosis. In regard to age of onset, one should not rely on this factor exclusively. For example, some patients are born prematurely without their mothers' serum immunoglobulins while others are protected for the first 3 months of their life because of their mothers' serum immunoglobulins. Infants having T cell or innate defects can develop signs soon after birth. In some primary immunodeficiencies, manifestation of symptoms occurs in adulthood. Here we will discuss some criteria, signs, and symptoms that should raise concern for need of referral to an allergist/immunologist (Chinen et al., 2019).

Naturally, there are multiple signs and symptoms that the provider should be aware of in order to consider PID. A frequently used constellation of signs prompting consideration of PID consists of diarrhea, multiple episodes of sinopulmonary infections, and failure to thrive. Simply having frequent episodes of sinusitis, bronchitis, or pneumonia is enough to consider PID. Having infections that cover multiple areas of the body and are recurrent is another concerning sign for PID. Patients who present with infections from opportunistic and/or uncommon microorganisms should be evaluated for PID, particularly when these infections are recurrent in nature (Chinen et al., 2019). Another often overlooked consideration is an adverse effect of immunization, in which a patient becomes ill with live-attenuated vaccines (Chinen et al., 2019; Lederman, 2016). While evidence-based guidelines are somewhat limited regarding the specific frequency of infections which might indicate PID, for the sake of clarity we recommend utilizing Table 6.1.

It should be noted that many of the warning signs in the table are late findings which may suggest that significant morbidity or organ damage due to infections could already be present.

Table 6.1 Jeffrey Modell Foundation's 10 warning signs of primary immunodeficiency (Modell et al., 2018)

Four or more new ear infections in 1 year	Two or more serious sinus infections in 1 year
Two or more pneumonias in 1 year	Recurrent, deep skin or organ abscess
Two or more deep-seated infections including septicemia	Persistent thrush in mouth or fungal infection on skin
Two or more months on antibiotics with little effect	Need for IV antibiotics to clear infection
Failure of an infant to gain weight or grow normally	Family history of PID

If a patient has 2 or more of these 10 signs, consider referral to A/I specialist

Clinicians should maintain a high index of suspicion regarding PID and should refer any patient who experiences recurrent infections, difficult to treat infections, or an infection with an unusual organism for evaluation of PID even if the patient does not specifically fit the warning signs criteria.

Primary Immunodeficiencies

The signs and symptoms as well as Table 6.1 help to screen for PID; briefly we will review the most common primary immunodeficiencies with their associated presentations. A simplified overview of these disorders can be found in Table 6.2.

Humoral Immunity

Humoral immunity predominantly defends against extracellular organisms. Sinopulmonary infections with encapsulated bacteria is the most common clinical presentation of a humoral immune defect. Some common deficiencies in humoral immunity are described below.

Common Variable Immunodeficiency (CVID)

CVID is the most common symptomatic primary immunodeficiency and is defined as a heterogeneous group that has a deficiency in production of more than one antibody isotype. Patients have low IgG plus at least one other low immunoglobulin isotype (IgA or IgM) and weak polysaccharide vaccine response. This diagnosis does require the patient to be 4 years of age or older; however, frequently patients are not diagnosed until later in life. Therefore, we recommend to continue a high index of suspicion despite the patient being an adult. The general associated symptoms include recurrent otitis media, sinusitis, and bronchitis, and the pattern seen is an increase in the frequency and severity of upper respiratory tract infections (URI) with an associated increase in lower respiratory infections. Many patients experience pneumonia, although this is not always the case. In addition, patients will frequently complain of diarrhea that is either chronic or recurrent in nature. While viral infections are not as common as they are in some other PIDs, patients with humoral defects are at increased risk for, hepatitis B and C, gastrointestinal viral illnesses, and viral meningitis/encephalitis (Hwangpo & Schroeder, 2019). It is important to recall that these patients frequently go underdiagnosed for some time. Therefore, clinicians should have a low threshold for referring the patient for evaluation by an A/I specialist.

IgA Deficiency

Selective IgA deficiency is the most common asymptomatic immune deficiency. Rarely, some may present with recurrent sinopulmonary infections. These patients have an increased risk for the development of autoimmune conditions. In addition, nearly 20% of patients with IgA deficiency may also exhibit allergic symptoms such as allergic rhinitis, allergic conjunctivitis, eczema, and urticaria. It is pertinent to note that the infant to toddler with frequent sinusitis, otitis media, and pulmonary infections should be considered for IgA deficiency; particularly those with associated allergic symptoms (Hwangpo & Schroeder, 2019). Based on clinical presentation alone, it is not possible to distinguish a child with symptomatic Selective IgA Deficiency from a patient with a more substantial humoral defect.

Table 6.2 Selected primary immunodeficiencies

Disease	Age of Dx/Onset	Signs/symptoms	Microorganisms	Labs/findings
CVID	Diagnosed at age 4 or older	Sinusitis, otitis media, bronchitis, pneumonia, diarrhea	Most bacterial, such as: *S. pneumonia, H, influenzae and the protozoan G. lamblia*	Decrease in IgG and IgA or IgM. Failed vaccine response
IgA deficiency	Frequently diagnosed by age 1 but can be diagnosed at any age	Often silent. Some patients experience sinusitis, otitis media, bronchitis, pneumonia, and gastrointestinal infections	Encapsulated bacteria: Such as: *S. pneumonia, H. influenzae*	Decrease in serum IgA. Normal IgG, IgM, IgE
X-linked agammaglobulinemia/ autosomal recessive agammaglobulinemia	Symptoms can start 4–6 months after birth, often diagnosed by age 1 but can remain undiagnosed until childhood	Sinusitis, otitis media, bronchitis, diarrhea, and skin infections	Encapsulated bacteria such as *S. pneumonia, H. influenzae* or *G. lamblia. Enterovirus* caused meningitis	Decrease in serum IgG, IgA, IgM, IgE
Hyper-IgM syndrome	Many patients present by first year of life; however, nearly all will present by age 4	Sinusitis, otitis media, bronchitis, or diarrhea. Invasive opportunistic infections in X-linked form	Encapsulated bacteria as above as well as *Mycobacterium tuberculosis, Pneumocystis jiroveci, cytomegalovirus,* or *Cryptococcus neoformans*	Low IgG (as well as IgA, IgE) with normal, possibly elevated IgM
IgG subclass deficiency	Varies with each deficiency	Recurrent sinopulmonary infections	Encapsulated bacterial and viral infections	Normal IgG, IgM, IgA with low IgG subclass
CID	Varies with mutation	Recurrent infections; some with autoimmune disease, some with neurological diseases	Depending on mutation: encapsulated bacteria, fungal, or viral	CBC; if concerned immediate referral to A/I specialist
SCID	Typically prior to 6 months old	Recurrent bacterial, viral, or fungal infections that are severe. Frequently failure to thrive	Opportunistic bacteria, fungi such as candida, and even common viral infections	CBC; if concerned immediate referral to A/I specialist
CGD	Generally, infancy however can be diagnosed in adulthood	Abscess of the skin, lungs, and liver	Catalase + bacteria such as *S. aureus, Salmonella, and Klebsiella,* as well as the fungus *Aspergillus*	Phagocyte oxidase activity: DHR oxidation test

(continued)

Table 6.2 (continued)

Disease	Age of Dx/Onset	Signs/symptoms	Microorganisms	Labs/findings
LAD	Variable; generally, in first year of life	Cellulitis, ulcerations, bacterial, and fungal respiratory infection. Oral pathologies may present as well	Variable: *S. aureus*, *P. aeruginosa*, and many others	CBC: Neutrophils increased
Complement	Variable	Recurrent sinopulmonary infections, some present with autoimmune symptoms. Some with invasive meningitis or sepsis	*Neisseria* infections in particular deficiencies	CH50 is preliminary screening lab

Note: signs and symptoms are not entirely inclusive and all are not required to make the diagnosis. Further the infectious agents listed above include the most common culprits and is not an exhaustive list of all culprits
Abbreviations: *CVID* common variable immunodeficiency, *CID* combined immunodeficiency, *SCID* severe combined immunodeficiency, *CGD* chronic granulomatous disease, *LAD* leukocyte adhesion deficiency, *Dx* diagnosis

Agammaglobulinemia

Agammaglobulinemia, the inability to make immunoglobulins, is divided into two distinct genetic entities: X-linked and autosomal recessive. In X-linked agammaglobulinemia, the patient's B cells do not mature, resulting in almost complete absence of B cells and minimal production of all immunoglobulins (Hwangpo & Schroeder, 2019). These patients are generally symptomatic in the first 2 years of life (Bonilla et al., 2015; Hwangpo & Schroeder, 2019). Naturally, due to the genetic nature of this disease, boys are the affected children and females may be silent carriers of the mutation. Sinopulmonary infections, occurring with increased frequency and severity, are the most common presenting features. In addition, these patients can experience skin infections such as cellulitis, impetigo, and boils at an increased rate than the average child. These patients may also experience diarrhea. Autosomal recessive agammaglobulinemia presents in the same manner as the X-linked form. However, both males and females are affected. Patients will experience similar symptoms as described above (Bonilla et al., 2015; Hwangpo & Schroeder, 2019).

Hyper-IgM Syndromes

Hyper-IgM syndromes, also known as Class-Switch Recombination Defects, are described as patients with decreased levels of IgG, IgA, and IgE but with normal to elevated levels of IgM. Furthermore, the quantity of B cells will be within normal limits. This disease is due to the inability of B lymphocytes to class-switch from production of IgM to the other immunoglobulins. As with the previously described humoral immunodeficiencies, these patients present with an increased risk of infection with encapsulated bacteria; however, based on the many genetic defects that can give rise to this disease, one patient with hyper-IgM syndrome will not always present as another with hyper-IgM syndrome. Over 50% of these patients will experience chronic diarrhea. Additionally, they can present with abscesses, oral ulcers with associated neutropenia, enlarged lymph nodes, as well as splenomegaly. Another subset of patients with hyper-IgM syndrome may also present with autoimmune hepatitis, thrombocytopenia, or even hemolytic anemia. Patients with the X-linked form have a defect in T cell communication and are susceptible to life-threatening opportunistic infections (Bonilla et al., 2015; Hwangpo & Schroeder, 2019).

IgG Subclass Deficiency

The IgG immunoglobulin possesses four subgroups: IgG1, IgG2, IgG3, and IgG4. There are patients who are deficient in one or more of these classes which could be associated with a poor response to vaccination in some patients. Patients who are deficient in IgG2 may find themselves unable to mount an appropriate response to the pneumococcal polysaccharide infection; these patients will often, but not always, present with increased cases of *Streptococcus pneumoniae* pneumonia. It is important to note that there are patients who have decreased IgG subclasses that remain asymptomatic and are able to appropriately respond to immunizations (Bonilla et al., 2015; Holland & Uzel, 2019).

T Cells and Cellular Immunity

As discussed previously, T cells are the main cells involved in cellular immunity thus a deficiency increases the risk for viral and intracellular infections. However, due to the role of the T-helper cells, many patients with T cell deficiency will present with combined immunodeficiency, i.e., defects in the humoral and cellular immunity. Briefly we will review cellular immunodeficiency presentations then focus on the even more serious severe combined immunodeficiency. It is important to note that newborn screening will help to identify certain PIDs, most notably, severe combined immunodeficiency, depending on local government regulations (Bonilla et al., 2015; Roifman, 2019). Cellular immunodeficiencies are comprised of a host of genetic defects resulting in susceptibility to different infections; however, the common theme is increased risk for intracellular bacteria, certain fungal infections, and viruses. These patients will frequently present with severe illness in the setting of common viral infections such as RSV and rotavirus and can also present with diarrhea and frequent and/or severe candida infections. In addition, they are further susceptible to infections by mycobacteria and salmonella (McCusker et al., 2018).

Combined Immunodeficiency (CID) and Severe Combined Immunodeficiency (SCID)

CID and SCID may present similarly despite the designation of "severe." The main difference between these classifications is the quantity of T cells, in which <300 results in a diagnosis of severe. Patients who are affected by SCID are susceptible to a host of microorganisms and usually present very early in life, younger than 1 year of age and frequently prior to 6 months of age. These patients will present with chronic diarrhea, recurrent infections, and infections with opportunistic microorganisms such as *Candida* species and *Pneumocystis jiroveci*. In addition, these patients will exhibit failure to thrive and some will have a severe rash. SCID is considered to be a life-threatening condition, necessitating rapid diagnosis and management. Frequently, due to screening, these newborns will be diagnosed at birth. However, due to the nature of the many genetic defects, not all SCID diseases will be caught by newborn screening (McCusker et al., 2018).

Patients with CID may present with similar symptoms as SCID or its own set of symptoms. There are a multitude of diseases that fall under the CID class, with wide variability in clinical presentation. Most patients will present with recurrent infections (such as viral and opportunistic), gastrointestinal diseases, autoimmune diseases, and even neurological symptoms (Bonilla et al., 2015; McCusker et al., 2018; Roifman, 2019). DiGeorge syndrome, caused by 22q11.2 microdeletion, may be classified as a combined immune deficiency. This syndrome is discussed in detail in the section on Neurodevelopmental Disorders with Associated Immune Defects.

Innate Deficiencies

Phagocytic deficiencies can be characterized by decrease number of phagocytic cells, inability of phagocytes to move to the correct site such as lymphocyte adhesion deficiency, inability to generate the appropriate intracellular chemicals such as chronic granulomatous disease, and so on. There are many phagocytic defects that can arise. These patients will experience severe pyogenic bacterial infections and fungal infections such as cellulitis, abscesses, respiratory infections with bacterial and fungal microorganisms, and many other symptoms. Briefly, we will review some of the phagocytic deficiencies (Bonilla et al., 2015; Holland & Uzel, 2019).

Chronic Granulomatous Disease (CGD)

CGD is due to phagocytes being unable to generate superoxide after ingesting a microorganism. Initial onset of disease can span from infancy to adulthood based on which mutation the patient has acquired. These patients will experience infections from catalase positive bacteria such *as Staphylococcus aureus, Burkholderia cepacia, S. marcescens, Nocardia* species, and the fungus *Aspergillus*. Liver abscess with *S. aureus* and pulmonary abscess with *Aspergillus* are typical of CGD. As the name suggests, these patients also form granulomas, which can be located on the skin, in the respiratory tract, and even in the gastrointestinal tract. The granulomas can result in diarrhea, abdominal pain, and malabsorption (Bonilla et al., 2015; Holland & Uzel, 2019).

Leukocyte Adhesion Deficiency (LAD)

LAD arises from mutations that result in leukocytes being unable to move from the bloodstream into the extracellular matrix. As there are three distinct entities of LAD, some patients may present differently; however, the most common symptoms include recurrent bacterial infections, decrease in appropriate wound healing, and delayed separation of umbilical cord. These patients, depending on which mutation is present, will experience cellulitis, periodontitis, gingivitis, oral ulcers, and pulmonary infections. Some associated features can include dysfunctional platelets, stunted growth, and cognitive delay in some forms of this disease. As one could surmise, due to white blood cells being unable to navigate to the extracellular matrix, nearly all patients will have elevated neutrophil counts when peripheral blood counts are measured (Bonilla et al., 2015; Holland & Uzel, 2019).

Complement and Its Deficiencies

The complement system's role in the host defense involves direct injury to microorganisms and optimization of microorganism ingestion by phagocytes, a process referred to as opsonization. Complement's opsonization of bacteria is most important for host defense against extracellular encapsulated bacteria. With this knowledge, one can ascertain that a deficiency in the complement will result in increased risk of infections with encapsulated bacteria, most notably *Neisseria spp., S. pneumoniae, H. influenzae*, and other encapsulated bacteria. These patients may present with bacterial sinopulmonary infections, some even with associated autoimmune diseases. It is paramount that patients presenting with recurrent episodes of *Neisseria* infections be assessed for complement deficiency (Atkinson et al., 2019; Bonilla et al., 2015).

Selected Neurodevelopmental Disorders with Associated Immune Defects

Now that we have reviewed some of the more common immunodeficiencies, we will discuss the most common neurodevelopmental disorders with known associated immune defects. This discussion is not exhaustive, as many other such disorders exist as compiled by Ming and Stiehm (2008, 2016). A summary of syndromes with known or associated immune defects can be found in Table 6.3.

Cornelia de Lange Syndrome

Cornelia de Lange syndrome (CdLS) is a neurodevelopmental disorder which presents with short stature, intellectual disability, and bone abnormalities in the arms, hands, and fingers. These patients are described as having facial features that are distinguished by eyebrows that meet in medial aspect of the face, low set ears, and a small, upturned nose. Some CdLS patients have been known to exhibit behavior similar to that of certain ASD behaviors (Kline et al., 2018). Patients with CdLS are described as experiencing increased respiratory infections and middle ear infections which prompted a study in 2013 by Children's Hospital of Philadelphia that assessed immunodeficiency in patients with CdLS (Jyonouchi et al., 2013). Jyonouchi et al.'s (2013) study of those patients found an increase in frequency of respiratory infections, pneumonia, ear infections, as well as some other infections. Of the 45 patients assessed, nine patients were found to have a deficiency in some aspect of their antibodies. Our review of the literature did not reveal another study like this performed. However, Björkman et al. (2018) noted that patients with CdLS can have gene deletions resulting in decreased immunoglobulin gene diversity. With this information, one can extrapolate that patients with CdLS and recurrent sinopulmonary infections, recurrent ear infections, and recurrent skin infections should raise concern for a possible antibody deficiency.

Fetal Alcohol Syndrome

A diagnosis of fetal alcohol syndrome (FAS) is made when the appropriate prominent facial features, growth deficits, and CNS abnormalities are present. This diagnosis can only be made after all other possible diagnoses associated with each individual feature has been ruled out. These patients have the stereotypical smooth philtrum, thin vermillion border, and small palpebral fissures. Patients with fetal alcohol syndrome frequently present with developmental and or cognitive delay (Parker et al., 2010). This diagnosis does not have a strong association with immunodeficiency; however, Johnson et al. (1981) stated that the majority of the 13 FAS patients assessed had increased episodes of infection. In addition, all 13 patients had some decrease in the quantity or quality of their immune system (Johnson et al., 1981). To date, our authors remain unable to find a similar study produced. However, Gautheir (2015) suggests that alcohol may disrupt the immune system in the developing fetus. This statement relies on studies that show neonates with mothers who drank while pregnant were more likely to develop an infec-

Table 6.3 Selected syndromes with neurodevelopmental delay and immune defects

Disease	Infections	Immune defects	Tests to order
Cornelia de Lange syndrome	Increased sinusitis, ear infections, skin infections, and oral candidiasis	Some patients exhibit antibody deficiencies	CBC, IgG, IgA, IgM, IgE
Fetal alcohol syndrome	Possible increased rate of infection	Not well described	CBC, IgG, IgA, IgM, IgE
Rubinstein-Taybi	Increased sinusitis, increased pneumonia, and increase in ear infections	Some patients exhibit antibody deficiencies: humoral defects, failure to respond to pneumococcal immunization as well as diphtheria and tetanus immunizations	CBC, IgG, IgA, IgM, IgE, IgG subclasses
Smith-Magenis syndrome	Increased sinusitis, increased pneumonia, and increase in ear infections	Some patients exhibit antibody deficiencies: humoral defects, failure to respond to pneumococcal immunization	IgG subclasses, IgG, IgA, IgM, IgE
Trisomy 21	Variable	Variable	Dependent on patient; referral to AI recommended
DiGeorge	Dependent on severity	Highly variable	Dependent on severity; referral to AI recommended

Chinen et al. (2011), Perkins et al. (2017), Herriot and Miedzybrodzka (2016), Jyonouchi et al. (2013), Lougaris et al. (2016), Pasic (2015), Saettini et al. (2020), Sullivan (2008), Villella et al. (2000)

tion. The paper also states that alcohol use in the pregnant mother led to increased risk of infection in the premature infant as well as other aspects of the immune system being affected (Gauthier, 2015; Johnson et al., 1981). With the limited data present, it is not feasible to develop a group of symptoms that these patients may present with; however, it would be beneficial to have the frequently ill FAS patient assessed by an AI specialist.

Rubinstein-Taybi Syndrome (RTS)

Rubinstein-Taybi syndrome is characterized by intellectual disability, short stature, broad thumbs, and halluces with distinctive facial features: downward slanted eyes, low-hanging nasal septum, high palate, and high arched eyebrows (Kumar et al., 2012). Patients with RTS are known to experience increased infections particularly those involving the respiratory tract (Lougaris et al., 2016; Villella et al., 2000). Several case reports and case studies have asserted that there's a connection between RTS and immune deficits. Case reports have shown phagocytic defects, CVID, and severe B cell deficiency with hypogammaglobulinemia (Herriot & Miedzybrodzka 2016; Lougaris et al., 2016; Pasic, 2015; Saettini et al., 2020; Villella et al., 2000). It is evident that while there is no strong association, there are certainly cases to assert that some patients with RTS may also have an immunodeficiency, most likely humoral in nature. Should a patient present with recurrent sinopulmonary infection, diarrhea, otitis media, or with failure to respond to vaccinations, then this patient should be referred to an AI specialist.

Smith-Magenis Syndrome

Smith-Magenis syndrome is characterized by intellectual disability, delayed speech and language skills, and craniofacial features that include a prominent lower jaw (Gropman et al., 2006). This syndrome has a known association with otitis media, but frank immunodeficiency was not frequently described. Chinen et al. (2011) suggested that some Smith-Magenis patients possess a particular mutation which not only results in the syndrome but also have decreased response to pneumococcal vaccination. Furthermore, Perkins et al. (2017) assessed patients with the most commonly deleted causative gene in Smith-Magenis, 17p.11.2, for associated immunodeficiency. After 76 patients with Smith-Magenis syndrome and the associated 17p.11.2 deletion were studied, it was found that these patients had an increase frequency of otitis media, upper respiratory infections, sinusitis, pneumonia, and even some with gastroenteritis and cellulitis. These patients displayed different immunodeficiencies that consisted of decreased antibodies to pathogen-specific antigens, decreased response to pneumococcal vaccinations, as well as decreased class-switched memory B cells (Chinen et al., 2011; Perkins et al., 2017). Smith-Magenis patients presenting with increased sinopulmonary infections, skin infections, or gastrointestinal infections as well as those who fail to respond to pneumococcal vaccination should be evaluated by an AI specialist.

Trisomy 21 (Down Syndrome)

Trisomy of the 21st chromosome (also known as Down syndrome) is the most common chromosomal aneuploidy, affecting 1:700 live births in the United States with a variable phenotype (Parker et al., 2010). Nearly every organ system has the potential to be involved in this syndrome, the scope of which is far beyond the limits of this chapter. It is important to recognize that these patients have numerous anatomical and functional abnormalities requiring a multidisciplinary approach to their care, best met by a well-coordinated medical home. Developmental delay and cognitive impairment are typical of Down syndrome. Important and relative to the field of AI specifically, these patients experience common infections such as otitis media, sinusitis, pneumonia, bronchitis, and croup with increased frequency, duration, and severity compared to the general population (Ram & Chinen, 2011; Mitchell et al., 2003). Patients with Trisomy 21 have a higher incidence of acute lung injury following pulmonary infections, and if mechanical ventilation is required, they suffer a higher incidence of acute respiratory distress syndrome (ARDS) (Ram & Chinen, 2011). Patients are also at a higher risk of death due to sepsis (Ram & Chinen, 2011). Reasons for increased susceptibility to infections are numerous and can be considered in two main categories: immunologic and non-immunologic.

Non-immunologic abnormalities which may predispose to infections are numerous. Anatomical craniofacial abnormalities occur in nearly all patients and often contribute to otitis media and sinusitis. Lower airway abnormalities could be severe enough to warrant tracheostomy in some patients. However, more typically, lower airway abnormalities are relative mild and may be an occult factor contributing to lower airway infections. Functional abnormalities, such as gastroesophageal reflux/aspiration and hypotonia, may contribute to increased frequency, severity, and duration of infections (Wall & Zambrano, 2019). In addition, behavioral and cognitive variables factor into the increased risk for infections. For example, many patients with Trisomy 21 are very socially outgoing and may have physical contact with other people through hugs and handshakes very frequently. Yet, in the setting of cognitive impairment, it may be difficult for these individuals to realize the importance of infection-reducing measures such as avoidance of touching soiled surfaces, avoidance of embracing someone who is ill, avoidance of placing hands or objects in the mouth, and the importance of effective handwashing.

Immunologic abnormalities in patients with Trisomy 21 are numerous and occur on a wide spectrum of severity. Broad categories of clinically relevant immune abnormalities include T cell and B cell defects. Patients could have a small thymus, low T cell counts (sometimes low enough to cause low T cell receptor excision circles (TRECs) on the newborn screen), as well as reduced T cell in vitro proliferation (Ram & Chinen, 2011; Joshi et al., 2011). Autoreactive T cells may also occur, contributing to autoimmunity (Giménez-Barcons et al., 2014). Regarding B cell abnormalities, patients may have reduced total B cell numbers, reduced percentage of memory B cells, and impaired antibody response to polysaccharide vaccines (Ram & Chinen, 2011; Joshi et al., 2011). Immunoglobulin levels are normal or elevated in most patients. However, low immunoglobulins have been documented (Ram & Chinen, 2011; Joshi et al., 2011). Regarding neutrophil abnormalities, function is overall intact enough so that patients do not typically manifest "neutrophil defect" type of infections. Neutrophil respiratory burst capacity is typically normal. However, neutrophil chemotaxis may be suboptimal and could be a contributing factor to infections (Ram & Chinen, 2011). Abnormalities in compartments of innate immunity, including natural killer (NK) cells and Toll-like receptors (TLRs) have been the subject of research. The clinical impact of innate abnormalities in these patients is not entirely clear. The exact reason why patients with Trisomy 21 are prone to such a wide array of immune defects remains unclear. The increased gene dose of four interferon receptors encoded on chromosome 21 is proposed to play a major role (Sullivan et al., 2017).

22q11.2 Deletion Syndrome

22q11.2 deletion syndrome (DiGeorge syndrome) has historically been characterized by a constellation of cardiac anomalies, hypocalcemia, and impaired T cell production (Sullivan, 2008). The underlying cause is attributed to 22q11.2 microdeletion in 90% of patients (Sullivan, 2008). Deletions of this region often include more than 35 genes (Sullivan, 2008). The *TBX1* gene, located in this region, is mostly responsible for the phenotype (Sullivan, 2008). It encodes for a transcription factor important to embryogenesis and is involved in development of the pharyngeal arches and pouches. Disruption carries the potential for impact on the thymus and parathyroid glands (Sullivan, 2008). However, mutations in numerous other genes have been known to lead to this disorder (Sullivan, 2008). The nosology surrounding DiGeorge syndrome is somewhat confusing and may be used differently by various medical specialties. In general, complete DiGeorge syndrome refers to a patient with absent thymic tissue. Partial DiGeorge syndrome refers to a patient who has some degree of thymic impairment. Velocardiofacial syndrome (VCFS) is used for a patient with 22q11.2 microdeletion but with no thymic/immune impairment. In order to minimize confusion, currently it is recommended that the term 22q11.2 deletion syndrome should be used to refer to patients who carry a microdeletion in this chromosomal region, regardless of the clinical manifestations (Sullivan, 2008). While most patients have a de novo mutation, once the deletion occurs, it is inherited in an autosomal dominant fashion.

Some degree of immune impairment is present in 75% of patients secondary to thymic hypoplasia (McDonald-McGinn & Sullivan, 2011). Immune dysfunction presents with a wide spectrum of severity. The rare infant with absent thymic tissue (<1% of patients) requires a thymic transplant to sustain life (McDonald-McGinn & Sullivan, 2011). An alternative approach, if a fully matched donor is available, is to transplant peripheral blood T cells (McDonald-McGinn & Sullivan, 2011). Most patients have mild to moderate immune disorders. Antibody dysfunction such as selective IgA deficiency, hypogammaglobulinemia, and impaired vaccine response is seen in a minority of patients (Sullivan, 2008). Autoimmunity is seen in approximately 10% of patients (McDonald-McGinn & Sullivan, 2011).

Infants with 22q11.2 deletion syndrome are typically born with abnormal facial features, which may be subtle or profound. Some common features include hooded eyelids, bulbous nasal tip, small chin, posteriorly rotated ears, and crumpled ear helix (McDonald-McGinn & Sullivan, 2011). Overt cleft palate may be seen in 11%, with submucous cleft in 16% (McDonald-McGinn & Sullivan, 2011). Velopharyngeal insufficiency due to an abnormally short soft palate (velum) which does not seal appropriately is present in over 40% of patients and may contribute to difficulties with feeding and phonation (McDonald-McGinn & Sullivan, 2011).

Microcephaly is a relatively common, although not consistent, feature. In addition, gray matter has been demonstrated to be reduced in the frontal cortices, cingulate gyrus, and cerebellum by MRI (McDonald-McGinn & Sullivan, 2011). Cognitive impairment is very common in 22q11.2 deletion syndrome. In fact, one study demonstrated that only 18% had average IQ and up to 30% were deemed mentally retarded based on the Wechsler IQ Scale (McDonald-McGinn & Sullivan, 2011). Behavioral and psychiatric issues are common. Attention deficit hyperactivity disorder may be seen in over 50% of patients (McDonald-McGinn & Sullivan, 2011). Autism spectrum disorders may be found in over 40% of children with VCFS (Antshel et al., 2007). Frank psychiatric disorders are relatively common in older patients, with diagnoses including bipolar disorder and schizophrenia/schizoaffective disorder found in 10–30% of older adults (McDonald-McGinn & Sullivan, 2011).

Autism Spectrum Disorder and the Immune System

While a comprehensive review of the topic of nervous system development would be outside of the scope of this handbook, we will review some of the currently available knowledge on the immune system and its effect on the nervous system (NS), autism spectrum disorder (ASD), and neurodevelopmental delay (NDD). The immune system's many cells and chemical mediators have been implicated in the development of the NS, indicated mostly by animal studies. These include but are not limited to the microglia, the macrophages of the brain, whose role consists of angiogenesis, formation of synapses, pruning of synapses, and many other important functions (Morimoto & Nakajima, 2019). In addition, T cells and B cells have been shown to assist in the development of the NS and cytokines have been shown to assist in proliferation, differentiation, and survival of neurons (Bauer et al., 2007; Morimoto & Nakajima, 2019; Tanabe & Yamashita, 2018).

The development of ASD possesses known genetic, environmental, and other risk factors, although the exact mechanism of this diagnosis remains largely unknown (Hsiao, 2013). Despite this, there have been significant implications of the immune system's role in the development of ASD. One very popular hypothesis is maternal immune activation, in which it is postulated that secretion of maternal cytokines can result in development of ASD (Hsiao, 2013; Minakova & Warner, 2018). Secretion of cytokines can of course be from an illness that the mother is experiencing. However, infection is not always the causative mechanism. Indeed, this hypothesis does not stand alone considering how frequently a mother becomes sick while pregnant yet does not give birth to a child with ASD. The thought behind this process is a "two-hit" model in which the fetus is predisposed either by genetics or by epigenetic changes. Epigenetic changes are described as alterations in gene expression, without alteration of the genetic code, often due to environmental influences or exposures. After the initial "insult," the fetus has an increased susceptibility to the development of ASD with exposure to particular maternal cytokines (Minakova & Warner, 2018). In addition, autoantibodies have been implicated in the development of ASD (Meltzer & Van de Water, 2016; Edmiston et al., 2017). However, it should be noted that research in this arena remains fairly new. Certainly, much remains to be learned.

Regarding decreased immunity in the person with ASD, there have been case reports which suggest patients with ASD prior to the age of 2 years old have an increased risk of infection. However, despite an extensive literature search, there is very little data linking ASD with associated immunodeficiencies and as such most patients with ASD seem to possess a normal functioning immune system for host defense. The diagnosis of ASD surely does not preclude these patients from also having an immunodeficiency; therefore, always having a high index of suspicion in an ASD patient with recurrent infections will serve the provider well (Atladóttir et al., 2010; Sabourin et al., 2019). It must also be considered that many syndromic conditions could have ASD as one feature. This includes 22q11.2 microdeletion, as mentioned above.

Vaccinology

Vaccines and their indications, exclusions, schedule, and efficacy are most commonly covered via infectious disease. Here we will briefly review autism and the misconception surrounding its association with vaccinations as this is a common topic for the parent of an autistic child. In 1998, Wakefield et al. published an article in Lancet which suggested the MMR vaccination had an association with autism spectrum disorder. Since that time, immense fear, anxiety, and distrust of the medical society have flourished regarding immunizations (Gerber, 2009). Wakefield's paper has since been retracted and its suggestion has been disproven time and time again (Centers for Disease Control and Prevention's [CDC], 2020). A multitude of studies have assessed childhood vaccines and autism, all of which lead to the unwavering stance that vaccines do not cause autism. The practitioner should dissuade guardians from believing such to prevent deviation from vaccine schedules. For further information, please see the Centers for Disease Control and Prevention's (CDC) section on vaccines and autism at https://www.cdc.gov/vaccinesafety/concerns/autism.html.

Allergy

Allergic Disease in ASD

Atopy, or allergic conditions, appears to be relatively common in patients with ASD. Altarac (2008) showed an increased prevalence of allergies in ASD patients compared to the average person; respiratory allergies were experienced in 26.4% of ASD patients vs 14.9% in the general population. In addition, the prevalence of skin allergy was increased in ASD (14.9% vs 9.8%) and food allergy was also increased in ASD (14% vs 3.5%). However, this study showed that while there was an increased prevalence of respiratory and skin allergies in patients with ASD, the only statistically associated allergy with ASD was food allergy (Altarac, 2008). Another study showed a slight increase in prevalence of allergies in patients with ASD, no increase in prevalence of asthma, and again marked increase in prevalence of food allergies in comparison to the general population (Lyall et al., 2015). It is important to note that the aforementioned studies were produced via surveys to mothers and Lyall assessed patients aged 2–5 only. National Health Interview Survey (NHIS) data from 1997 to 2016 found an association between food allergies and ASD in 1868 children with ASD aged 3–17 years at a prevalence over twice that of neurotypical peers. Respiratory and skin allergies were also shown to have an association with ASD, albeit not nearly as large as food allergy (Xu et al., 2018). When interpreting the aforementioned studies, one must consider that all involved surveys rather than allergy testing or blinded food challenges. It is known that even in the general population, subjective report of food allergy may overestimate the true prevalence of food allergy. With this information, we believe it is imperative that this chapter also have a dedicated allergy section including signs, symptoms, concerns, and when to refer for specific allergies. A summary of common allergic disorders can be found in Table 6.4.

Food Allergy

Food allergy is estimated to affect up to 1–2% of the general population (Bock & Sampson, 2016). The most common food allergens consist of cow's milk, eggs, peanut, tree nuts, soybean, wheat, fish, and shellfish. These are referred to as the "Big 8." It is important to note that food allergy is the most common cause of outpatient anaphylaxis and therefore must be treated with the utmost concern (Kim & Burks, 2018). Patients with food allergies can present with cutaneous, gastrointestinal, or even respiratory signs and symptoms. Cutaneous signs and symptoms are the most common presentation of food allergies and include urticaria, angioedema, or pruritic dermatitis. Gastrointestinal signs and symptoms include nausea, vomiting, diarrhea, cramping, and abdominal pain. Patients can also experience pruritus of the mouth, eyes, and ears, sneezing, and rhinorrhea. Respiratory signs and symptoms can be severe, including bronchospasm or edema of the larynx. Despite the division of these symptoms into their respective anatomical categories, patients may present with any combination of the above signs and symptoms (Kim & Burks, 2018; Bock & Sampson, 2016).

The most feared complication of food allergy is anaphylaxis. The clinical diagnosis of anaphylaxis is made when a patient presents with any two of the following systems: skin, gastrointestinal, respiratory, or cardiovascular after recently ingesting food. The associated symptoms of each system are as such urticaria, angioedema, pruritus, or rash (skin); nausea, vomiting, diarrhea, or cramping (gastrointestinal); wheezing, cough, or shortness of breath (respiratory); and tachycardia or hypotension (cardiovascular). In addition to the above criteria, a diagnosis of anaphylaxis can also be made when pulmonary and/or cardiovascular collapse occurs in a patient after ingestion of a known food allergen. Patients presenting with anaphylaxis must be treated immediately with intramuscular epinephrine, prescribed an epinephrine auto-injector and educated on its use, and should be referred to A/I specialist (Kim & Burks, 2018; Bock & Sampson, 2016; Sampson et al., 2014). For children at risk of anaphylaxis, the medical home coordinator should assist the parent in ensuring that the school and other care settings are prepared to prevent, recognize, and respond to anaphylaxis.

A very important point to consider in the care of a patient with autism is that many are nonverbal and therefore are unable to articulate the reasoning behind their dislike for a food. Patients may only experience isolated symptoms such as abdominal pain or cramping, oral pruritus, or otic pruritus. These symptoms may be all that the patient experiences when they encounter a food to which they are allergic and yet may not have the ability to express the reason they are avoiding a particular food or display behavioral challenges. The possibility of food allergies should be considered in patients who are unable to communi-

Table 6.4 Summary of common allergic disorders

Disease	Signs and symptoms	Treatment	When to refer
Food allergy	Oral and otic pruritus. Urticaria, angioedema. Abdominal pain, nausea vomiting, diarrhea. Laryngeal edema. Bronchospasm	Avoidance plan; prescribe epinephrine auto-injector for accidental exposure	Uncertain of causative agent; assess for more than one food allergy
Allergic rhinitis	Rhinorrhea, sneezing, congestion of the nasal mucosa, postnatal drip, nasal pruritus. Nasal crease, allergic shiners, Dennie-Morgan lines	Intranasal corticosteroids Oral antihistamine Intranasal antihistamines	Patient desires immunotherapy, adverse effects from standard treatment, identification of causative agents, and/or failure of initial management
Rhinosinusitis	Nasal congestion, rhinorrhea, postnasal drip, cough, headache, facial pain. Tenderness over sinuses, periorbital edema. Boggy and erythematous nasal mucosa	Dependent on acute vs chronic; viral vs bacterial vs fungal. See society guidelines	Recurrent or chronic sinusitis. Occurring with allergic symptoms. Sinusitis with polyps
Asthma	Wheezing, shortness of breath, chest tightness, and cough	Many agents available. Refer to clinical guidelines	Evaluation for particular phenotypes such as allergic, occupational, or exercise induced asthma. Difficult to control asthma. Consideration of advanced therapies such as biologics
Atopic dermatitis	Intense pruritus, dry skin, dermatitis, flexural involvement (adults), facial and extensor involvement (infants)	Skin hydration, avoidance of fragrant lotions/ointments, topical steroids, other topical agents, antihistamines	Difficult to control AD, those who do not respond in an appropriate manner, patients with concern for other allergic disorders, or possible need/desire for immunotherapy and/or biologics

cate why they strongly dislike specific foods. This must be weighed with the often-observed pattern in which children with autism are inherently very selective with foods, which occurs independent of allergy. In addition, the food allergy testing pitfall whereby clinically irrelevant positive test results are very common makes it impractical to cast a wide net for food testing and expect useful results. This is discussed in greater detail in the "allergy testing" section of this chapter.

Allergic Rhinitis

Allergic rhinitis generally develops in childhood; however, it can develop at any age. Symptoms of allergic rhinitis consist of rhinorrhea, sneezing, congestion of the nasal mucosa, postnasal drip, and nasal pruritus (Ricketti & Ricketti, 2018). As noted in the aforementioned studies, there appears to be an increased prevalence in respiratory allergies in persons with ASD, and Xu et al. (2018) even suggests an association between ASD and allergic rhinitis (AR). With this information, it is pertinent to remember that the patient with ASD is equally at risk, if not more, than the neurotypical patient to develop AR. Allergic rhinitis is more than a host of uncomfortable symptoms. Patients with untreated allergic rhinitis can also experience fatigue, daytime somnolence, and decreased work productivity (Dykewicz et al., 2017). Patients with allergic rhinitis will frequently have dark circles under their eyes which are often referred to as allergic shiners and "lines" under their lower eyelids which are called Dennie-Morgan lines. In addition, one may note a crease across the nose referred to as the allergic crease which is secondary to the patient's repetitive motion of wiping the nose. Patients with allergic rhinitis, particularly at a young age, push up on their nose to wipe away

Table 6.5 Information regarding allergy/immunology clinic: what to expect and how to prepare

Condition	What to expect	How to prepare	Why is the visit important?
Allergic conditions in a child with ASD/NDD	Expect to be in clinic for at least 1 hour. A blood sample might be taken after the visit. Skin testing may be needed (sometimes this is scheduled at a different time). Your doctor will have to review your child's history and consider many details prior to deciding on the need for testing	Some AI clinics want patients to stop taking oral antihistamines (allergy medications taken by mouth) 5 days prior to the visit in case skin testing is needed. Contact the clinic in advance to ask if any medications should be stopped. At the visit, let your doctor know which mediations your child is taking. Bring any prior test results (allergies or immune system) with you. Be prepared to give a detailed history about possible triggers of allergic symptoms you have noticed in the past. Bring games to occupy your child and if your child has a security object (such as a special toy) consider bringing it. Some clinics allow snacks, but do not bring foods containing peanut or tree nuts as many patients in the clinic are allergic	Much misinformation exists regarding allergies and ASD. It is important to see a board-certified AI who is experienced with children. If your AI doctor wants to focus on history alone (without testing) or very focused testing (as opposed to large panels), this is because the history you provide is more valuable that testing. Testing can yield + results in patients who tolerate foods, leading to unnecessary dietary restrictions. If your child has difficulty verbalizing symptoms, clearly explain this to your doctor
Infections in a child with ASD/NDD	Expect to be in clinic for at least 1 hour. A blood sample might be taken after the visit	Bring any prior test results (allergies or immune system) with you. Include any previous imaging or cultures of infections (sinus infections, pneumonia, abscess, etc.). Be prepared to provide family history regarding weak immune system, autoimmunity, or severe childhood illness. Bring games to occupy your child and if your child has a security object (such as a special toy) consider bringing it. Some clinics allow snacks, but do not bring foods containing peanut or tree nuts as many patients in the clinic are allergic	If your child is having an unusual pattern of infections, or staying sick, it is very important for the immune system to be tested. Some children have anatomical abnormalities which increase infections. However, children with anatomical abnormalities can also have weaknesses of the immune system. An abnormal immune system can occur in anyone, but for ASD/NDD due to a genetic abnormality, it is possible that the same gene plays a role in the immune system for some children

Boguniewicz and Leung (2016, 2020), Novak and Leung (2016), Ong and Leung (2018)

mucus, a motion called the allergic salute which may often be observed in clinic. Allergic rhinitis can be divided into seasonal, perennial, and episodic. Episodic, for example, is when a patient without a cat comes across a cat and begins to suffer from the symptoms described above. The diagnosis of allergic rhinitis can be often performed with a good history and physical examination. Treatment includes attempted avoidance of triggers when applicable (cats, dogs, dust mites) and oral antihistamines as well as intranasal sprays. Intranasal corticosteroids remain the mainstay treatment for allergic rhinitis (Dykewicz et al., 2017; Gentile et al., 2016). We recommend starting with intranasal corticosteroids and adding oral antihistamine when intranasal corticosteroid is not enough. It is imperative to note that the side effect profile for intranasal corticosteroids is quite favorable but some side effects include pruritus, burning, and/or sneezing with use; a slightly more common side effect is epistaxis. Epistaxis should result in immediate cessation of use. One should refer to an AI specialist if the patient desires immunotherapy, identification of causative agents, has experienced adverse effects from standard treatment, or if the symptoms reduce the quality of life of the patient (Gentile et al., 2016).

Rhinosinusitis

As discussed in the immunology section, certain patients with immunodeficiency will present with recurrent sinusitis. For this reason, it is pertinent to review sinusitis presentation and treatment; while recurrent sinusitis may be a presentation for those with immunodeficiency, the non-immunocompromised patient can also experience sinusitis. Patients with sinusitis frequently present similarly to allergic rhinitis with symptoms such as nasal congestion, rhinorrhea, and postnasal drip. In addition to those related symptoms, these patients can also present with cough, headache, and facial pain. Causes of sinusitis include viral, bacterial, fungal, and noninfectious etiologies. The category of noninfectious is comprised of allergic causes and environmental factors that can act as an irritant such tobacco or inhaled chemicals. It is imperative to note that sinusitis can be a sequelae of allergic rhinitis hence the name rhinosinusitis (Epstein & Bernstein, 2018; Peters et al., 2014).

For the noncomplicated patient with acute sinusitis (<12 weeks), the most common cause is viral and therefore does not require antibiotic treatment but rather supportive treatment with options such as nasal irrigation, intranasal corticosteroids, or even the combination of both (Epstein & Bernstein, 2018). However, should the patient begin to improve and then worsen, present with a fever or facial pain, or fail to have some improvement after 10 days, then the diagnosis of acute bacterial sinusitis should be considered and treatment with antibiotics considered (Chow et al., 2012; Epstein & Bernstein, 2018; Peters et al., 2014). If a patient is presenting with frequent allergic symptoms and sinusitis or recurrent sinusitis (three or more episodes in a year), we recommend referral to an AI specialist.

Chronic rhinosinusitis (CRS) is defined as signs and symptoms of sinusitis lasting 12 weeks or longer with objective data which can consist of radiographic imaging exhibiting sinusitis or purulent mucus as well as other objective findings (Chow et al., 2012; Peters et al., 2014). Should the patient receive a diagnosis of CRS, it is our recommendation that the patient be referred to an AI specialist and an otolaryngologist.

Sinusitis can be further divided into fungal rhinosinusitis, chronic rhinosinusitis with and without polyps, and so forth; discussion of these topics would be out of the scope of this text but for the curious provider we recommend the American Academy of Allergy, Asthma, and Immunology and the Infectious Disease Society of America as excellent sources for review (Chow et al., 2012; Peters et al., 2014).

Asthma

The Global Initiative for Asthma (GINA) defines asthma as "a heterogeneous disease, usually characterized by chronic airway inflammation. It is defined by the history of respiratory symptoms such as wheeze, shortness of breath, chest tightness, and cough that vary over time and in intensity, together with variable expiratory airflow limitation" (Reddel et al., 2015, p. 624). Asthma can be divided into many types such as allergic asthma, nonallergic asthma, adult-onset asthma, occupational asthma, exercise-induced asthma, and many more (Greenberger & Stevens, 2018; Reddel et al., 2015, 2019). Briefly, we will discuss allergic asthma which often coincides with eczema and other allergies such as food or allergic rhinitis. This disease process is characterized by the allergen being inhaled by the patient resulting in activation of bronchial mast cells (Greenberger & Stevens, 2018; Reddel et al., 2019). Naturally, this disease process is driven by exposure to the offending agent; some examples of which are pollen, cockroach allergens, dust mites, cat and dog dander, and mold spores (Greenberger & Stevens, 2018). Due to the many phenotypes of asthma, the diagnosis and management/treatment of asthma varies with each phenotype. The diagnosis of asthma requires a clinical confirmation of symptoms consistent with asthma and, ideally, supportive spirometry findings. A very important limitation of spirometry, which must be noted for infants, toddlers, and those with ASD and NDD, is that the procedure requires the ability to follow commands and change actions rapidly throughout the test. Therefore, for the aforementioned patients, the diagnosis of asthma is often based solely on a clinical approach using their signs, symptoms, and clinical history, without using spirometry. Treatment is based on the frequency of patients' symptoms with the long-term goal of minimizing episodes, controlling symptoms, and decreasing the risk of exacerbations and serious sequelae such as death due to asthma. Treatment should be personalized for each patient according to the most up-to-date GINA guidelines. Referral to an AI specialist and/or pulmonologist is appropriate, especially for difficult to control asthma (Greenberger & Stevens, 2018; Reddel et al., 2015, 2019). In addition, asthma patients who also have allergy-related signs and symptoms should be referred to an AI specialist.

Atopic Dermatitis (AD)

AD may be mild and easily managed by primary providers or refractory, requiring specialized immunotherapy. Reports of the prevalence of AD in children with ASD range widely. Billeci et al. (2015) suggested an association between ASD and AD, whereas the CHARGE study did not find an association (Lyall et al., 2015). In the NHIS study, parents reported skin allergies nearly twice as often in children with ASD compared to neurotypical peers (Xu et al., 2018). Atopic dermatitis typically presents in childhood and can have significant effects on the patient's quality of life leading to poorer school/work performance in comparison to peers. In addition, these patients can suffer from decreased sleep as well as psychiatric disorders such as anxiety and depression (Ong & Leung, 2018). The diagnosis of AD is based on physical findings such as pruritic, dry skin, and eczematous rashes, typically located on the flexor folds or extensor surfaces of the extremities. The eczematous rash is more commonly noted on the face and extensor surfaces in infants and toddlers (Ong & Leung, 2018; Schneider et al., 2013). Some diseases that may be confused for AD include scabies, cutaneous T cell lymphoma, zinc deficiency, contact dermatitis, or even SCID (Boguniewicz & Leung, 2016; Boguniewicz & Leung, 2020; Novak & Leung, 2016; Ong & Leung, 2018). It is important to note that skin infections frequently occur in the patient with AD and increase the difficulty of treatment (Boguniewicz & Leung, 2020). Indeed, patients with AD can also possess a food allergy (Ong & Leung, 2018) and even further in some cases, food allergies may exacerbate AD (Boguniewicz & Leung, 2016, 2020). In addition, aeroallergens have been shown to induce and even worsen AD presentations (Boguniewicz & Leung, 2016, 2020; Ong & Leung, 2018). Treatment varies depending on the presentation of the disease. Many patients can be simply managed with appropriate skin hydration and skin moisturizers. In those who fail to respond to hydration and moisturization, topical steroids are indicated (Boguniewicz & Leung, 2016, 2020; Novak & Leung, 2016; Ong & Leung, 2018). Patients with difficult to control AD, comorbid allergic disorders, or possible need/desire for immunotherapy, referral to an AI specialist should be considered.

Allergy Testing

The AI specialist tailors the evaluation for potential allergens to the individual patient's history. Two main categories of testing modalities are used for IgE-mediated allergies: skin testing and serum (blood) testing. Serum testing utilizes IgE to assess for a causative allergen by measuring the level of IgE to the suspected allergen with a particular cutoff indicating positive or negative. Skin testing involves introduction of an allergen extract into the skin via skin prick to assess for a localized reaction which is graded by the diameter of the wheal and erythema as positive or negative result.

Serum Allergy Testing

Serum allergy testing detects circulating specific-IgE and is typically used when skin testing is not possible or not practical for the patient. Importantly, allergy medications do not interfere with serum allergy testing. An older modality, radioallergosorbent assay (RAST), may have the potential to yield an increased number of false-positive results (Bernstein et al., 2008). RAST has widely been replaced by newer, more reliable modalities. Serum allergy testing is typically resulted within a few days and is available through many commercial labs.

Skin Testing Process

Skin testing most often involves use of a commercially available allergen extract and a prick device. The device is dipped into the allergen extract, and then the skin is "pricked" using aseptic technique either on the back or the arms. Any positive result develops within 15–20 minutes and is manifest as a wheal (area of induration) and flare (area of erythema). Multiple devices are available. A common feature is a sharp point which is intended to only prick through the epidermis, approximately 1 mm deep (Bernstein et al., 2008). The technique typically does not cause bleeding and causes very little discomfort. Devices used may be single prick or a multi-prick device. The figures represent only some examples of multiple products on the market and the authors do not endorse a specific device. Multi-prick devices have the advantage that several allergens can be tested with one quick maneuver. This is especially advantageous when testing for environmental allergens because there is typically the need to test larger panels, not a single allergen. In young children or patients with ASD or NDD, a multi-prick-type device allows testing to be done quickly with one brief maneuver. All skin testing should include a positive control (histamine) and a negative control (Bernstein et al., 2008). The histamine prick should induce a wheal and flare which demonstrates that the test is working appropriately. This, of course, induces itching. In some patients, numerous allergens demonstrate positive reactions, and this reactivity to multiple allergens may lead to intense itching. The patient cannot scratch the area until the test results are interpreted and recorded. Distraction methods such as games may be useful to distract the patient's focus from the itching. The test site may be treated with a topical steroid cream as soon as the results are recorded, and the positive test sites resolve relatively quickly. The risk for any systemic allergic reaction due to skin prick testing is very low (Bernstein et al., 2008). However, skin testing should only be performed by a trained allergist in an appropriate medical setting.

Oral Food Challenge

Food allergy testing also utilizes oral food challenges (OFC). Generally, OFCs are reserved for ambiguous results and/or history such as a positive IgE in a patient with minimal to no symptoms upon ingestion of the food. In addition, OFC can also be used to assess if a patient developed tolerance to a certain allergen over time, as some children "grow out" of their allergy. A subset of an OFC is called a single-blinded OFC in which the patient is unaware of the foods that they are ingesting but the provider is aware. This subset of OFC is mainly used when there is uncertainty of the results of a prior OFC (Sampson et al., 2014).

Limitations and Potential Pitfalls

As noted with nearly all diagnostic studies, allergy testing can be both falsely positive and falsely negative; in addition to this, the results of these studies can be affected by the use of particular medications. Serum and skin testing should only be utilized when suspicious for a particular allergic entity; otherwise, a positive result in the setting of a patient without any clinical manifestations may be clinically irrelevant. This "positive" test suddenly raises many questions and often results in further unnecessary testing. Regardless of the allergy testing method used, clinically irrelevant positive results are common, even in individuals without any allergy symptoms (Bartuzi et al., 2017). Therefore, extraneous testing (especially large food allergen panels) is discouraged (Bernstein et al., 2008; Sampson et al., 2014).

Food Allergy Testing

A workup for a food allergy should only be performed when a patient presents with the previously reviewed symptoms; however, a patient with ASD or NDD may not be able to communicate these symptoms but rather will strongly avoid particular foods. In this setting, one should first rule out signs of anaphylaxis and if negative then start a shared-decision discussion about allergy testing. This shared-decision discussion should review the following: the utility of food allergen testing in that particular patient; risk of testing, particularly in regards to false results; and the next step that will need to be taken for both a negative and positive result. This abundance of caution to testing arises from clinically irrelevant sensitizations which may lead to unnecessary further testing or unnecessary treatment. One such treatment for food allergy is a food elimination diet which involves cutting out all aspects of a particular food/allergen to assess resolution of symptoms. Another treatment, similar in design, yet preferred, is avoidance of the food/allergen which causes the symptoms and/or avoidance by the patient. The difference is that a patient that is allergic to a particular allergen may not be allergic to it in all its forms. For example, a patient that is allergic to eggs may not be allergic to baked egg therefore an elimination diet would exclude a tolerated food, the baked egg, despite the patient having no reaction to it. This treatment, while necessary for a truly allergic patient, is not appropriate for a patient that is asymptomatic with ingestion of the allergen of concern; in addition, it is even more inappropriate for a patient with ASD or NDD to be forced to unnecessarily eliminate or avoid foods without reason (Mehta et al., 2014; Sampson et al., 2014).

Patients with ASD or NDD may already have a very limited diet due to multiple factors including oral aversion, inability to perform the mechanics of consuming certain foods, or very specific preference toward select foods. An additional food exclusion adds unnecessary complexity to the patient's daily life. Increased anxiety regarding fear of allergic reaction and increased social isolation due to being excluded from meals shared at social gathering or in the classroom may significantly impair quality of life. There is also the possibility of contributing to delayed growth and vitamin/dietary deficiencies (Mehta et al., 2014; Sampson et al., 2014). In addition, inappropriate food avoidance may theoretically result in an increased risk of immediate IgE-mediated reaction to the food should the patient desire to try the food later on in life (Chang et al., 2016). This occurs because immunologic tolerance to a specific food may be lost if the immune system is not exposed to the food (Chang et al., 2016).

A patient who has a detectable specific IgE (positive allergy test) to a given food, yet consumes the food regularly without any immediate reaction, under most circumstances should continue to consume the food on a regular basis in order to maintain immunologic tolerance. If clinical tolerance is uncertain, consultation should occur with an allergist to determine if an oral food challenge is appropriate (Sampson et al., 2014).

Medications Which Interfere with Skin Testing

Antihistamines interfere with skin prick testing and must be withheld prior to testing. First- and second-generation H1 antihistamines must with withheld a minimum of 2–3 days prior to testing (Bernstein et al., 2008)). However, in many scenarios, these medications could interfere with skin testing over a longer period of time (5–7 days) (Bernstein et al., 2008). Therefore, the most practical advice is to ask the patient to stop taking antihistamines 1 week prior to the allergy appointment if this temporary discontinuation can be tolerated based on the patient's potential resurgence of symptoms. H2 antihistamines may have a mild effect and should be withheld for 24 hours prior to testing (Bernstein et al., 2008). Many centrally acting medications such as anticholinergics, antidepressants, and other behavior-altering medications may interfere with skin test results. Tricyclic antidepressants have potent effects on skin testing. For example, imipramine may suppress skin testing for a mean of 10 days and doxepin a mean of 6 days (Bernstein et al., 2008). Only in unique scenarios is it essential to withdraw such antidepressants for the purpose of skin testing, as there is typically the option to perform serum testing (which is not inhibited by medications). Therefore, antidepressants and psychiatric medications should not be withheld routinely in advance of the initial allergy appointment.

Future Treatment Options

AI is a rapidly advancing field with the AI specialist contributing critical knowledge and skills to medical home team's care of children with ASD/NDD. As diagnoses become more specific, in part due to advancements in genetic testing promoting a greater understanding of the immune mechanism driving some diseases, treatment options have become more targeted and individualized.

Many new therapeutic approaches have been introduced over the past few years and additional treatment options are on the horizon. One such category of therapeutics is "biologics." The biologics that A/I specialist speak of when treating allergic/immunologic disorders are monoclonal antibodies. These are antibodies made in a lab to target one particular antigen. Monoclonal antibodies are used throughout many subspecialties with the general intention of inhibiting some aspect of a cell or its signaling pathway. The biologics that are used in the field of A/I function to reduce the overactivity of the immune system in particular allergic disorders. Biologics have led to major advancements in treatment as these therapeutics not only contribute to symptomatic improvement but have the potential to alter the immune system and lead to remission. Currently, biologics are available for the treatment of severe asthma, atopic dermatitis, urticaria, and nasal polyposis and are often started by a specialist after failure of traditional treatment modalities.

Another realm currently undergoing extensive research is the field of food allergy. Oral immunotherapy (OIT) for the treatment of food allergy has gained a lot of attention in recent years. OIT involves feeding increasing amounts of an allergen to an individual with the goal of providing protection from accidental ingestion, but it is not currently considered a cure. Currently, in the United States, this option is only approved for the treatment of peanut allergy. Furthermore, OIT may carry significant risks including anaphylaxis and only a subset of allergy patients may find that this is the right treatment for them (Dunlop, 2020).

With these advancements in research and new treatments becoming more readily available, ideally, patients will continue to see significant improvement in their symptoms which may not have been possible a few years ago.

Conclusion

The field of A/I has changed immensely from the past and continues to evolve exponentially. It is imperative that the provider remains abreast of the clinical manifestations and treatments discussed particularly for those who may not be able to communicate for themselves. Patients with particular NDD are at an increased risk for certain immunodeficiencies and as such should be screened appropriately. In addition to this, any patient can possess an immunodeficiency and therefore a high index of suspicion is required. Persons with ASD appear to have an increased prevalence of allergic disorders but most notably exhibit a high frequency of food allergy. Testing should only be ordered by an A/I specialist, as the limitations of such testing must be considered and discussed with the caregiver in advance. In conclusion, it is pertinent to refer patients with ADS or NDD to an A/I specialist when allergic or immunologic questions arise or when the caregiver or primary clinician is uncertain on how to proceed.

References

Abbas, A. K., Hitchman, H. L., & Pillai, S. (2018). Properties and overview of immune responses. In *Cellular and molecular immunology* (9th ed., pp. 1–11). Elsevier.

Altarac, M. (2008). Prevalence of allergies in us children with autism. *Annals of Epidemiology, 18*(9), 725. https://doi.org/10.1016/j.annepidem.2008.08.056

Antshel, K. M., Aneja, A., Strunge, L., Peebles, J., Fremont, W. P., Stallone, K., AbdulSabur, N., Higgins, A. M., Shprintzen, R. J., & Kates, W. R. (2007). Autistic spectrum disorders in velo-cardio facial syndrome (22q11. 2 deletion). *Journal of Autism and Developmental Disorders, 37*(9), 1776–1786.

Atkinson, J. P., Du Clos, T. W., Mold, C., Kulkarni, H., Hourcade, D., & Wu, X. (2019). 21 – The human complement system: Basic concepts and clinical relevance. In R. R. Rich, T. A. Fleisher, W. T. Shearer, H. W. Schroeder, A. J. Frew, & C. M. Weyand (Eds.), *Clinical immunology* (5th ed., pp. 299–317.e291). Elsevier. https://doi.org/10.1016/B978-0-7020-6896-6.00021-1

Atladóttir, H. O., Thorsen, P., Østergaard, L., Schendel, D. E., Lemcke, S., Abdallah, M., & Parner, E. T. (2010). Maternal infection requiring hospitalization during pregnancy and autism spectrum disorders. *Journal of Autism and Developmental Disorders, 40*(12), 1423–1430. https://doi.org/10.1007/s10803-010-1006-y

Bartuzi, Z., Kaczmarski, M., Czerwionka-Szaflarska, M., Małaczyńska, T., & Krogulska, A. (2017). The diagnosis and management of food allergies. Position paper of the Food Allergy Section the Polish Society of Allergology. *Advances in Dermatology and Allergology/Postępy Dermatologii i Alergologii, 34*(5), 391.

Bauer, S., Kerr, B. J., & Patterson, P. H. (2007). The neuropoietic cytokine family in development, plasticity, disease and injury. *Nature Reviews Neuroscience, 8*(3), 221–232. https://doi.org/10.1038/nrn2054

Bernstein, I. L., et al. (2008). Allergy diagnostic testing: an updated practice parameter. *Annals of Allergy, Asthma & Immunology, 100*(3), S1–S148.

Billeci, L., Tonacci, A., Tartarisco, G., Ruta, L., Pioggia, G., & Gangemi, S. (2015). Association between atopic dermatitis and autism spectrum disorders: A systematic review. *American Journal of Clinical Dermatology, 16*(5), 371–388. https://doi.org/10.1007/s40257-015-0145-5

Björkman, A., Du, L., van der Burg, M., Cormier-Daire, V., Borck, G., Pié, J., Anderlid, B.-M., Hammarström, L., Ström, L., de Villartay, J.-P., Kipling, D., Dunn Walters, D., & Pan-Hammarström, Q. (2018). Reduced immunoglobulin gene diversity in patients with Cornelia de Lange syndrome. *The Journal of Allergy and Clinical Immunology, 141*(1), 408–411. e408. https://doi.org/10.1016/j.jaci.2017.06.043

Bock, S. A., & Sampson, H. A. (2016). 41 – Evaluation of food allergy. In D. Y. M. Leung, S. J. Szefler, F. A. Bonilla, C. A. Akdis, & H. A. Sampson (Eds.), *Pediatric allergy: Principles and practice* (3rd ed., pp. 371–376.e372). Elsevier.

Boguniewicz, M., & Leung, D. Y. M. (2016). 51 – Management of atopic dermatitis. In D. Y. M. Leung, S. J. Szefler, F. A. Bonilla, C. A. Akdis, & H. A. Sampson (Eds.), *Pediatric allergy: Principles and practice* (3rd ed., pp. 448–457.e442). Elsevier.

Boguniewicz, M., & Leung, D. Y. M. (2020). 33 - atopic dermatitis. In A. W. Burks, S. T. Holgate, R. E. O'Hehir, L. B. Bacharier, D. H. Broide, G. K. Hershey, & J. R. Stokes Peebles (Eds.), *Middleton's allergy: Principles and practice* (9th ed., pp. 530–552). Elsevier Health Sciences.

Bonilla, F. A., Khan, D. A., Ballas, Z. K., Chinen, J., Frank, M. M., Hsu, J. T., Keller, M., Kobrynski, L. J., Komarow, H. D., Mazer, B., Nelson, R. P., Jr, Orange, J. S., Routes, J. M., Shearer, W. T., Sorensen, R. U., Verbsky, J. W., Bernstein, D. I., Blessing-Moore, J., Lang, D., Nicklas, R. A., … Joint Task Force on Practice Parameters, representing the American Academy of Allergy, Asthma & Immunology; the American College of Allergy, Asthma & Immunology; and the Joint Council of Allergy, Asthma & Immunology. (2015). Practice parameter for the diagnosis and management of primary immunodeficiency *The Journal of Allergy and Clinical Immunology, 136*(5), 1186–205. e2078. https://doi.org/10.1016/j.jaci.2015.04.049.

Centers for Disease Control and Prevention. (2020). *Autism and vaccines*. https://www.cdc.gov/vaccinesafety/concerns/autism.html

Chang, A., Robison, R., Cai, M., & Singh, A. M. (2016). Natural history of food-triggered atopic dermatitis and development of immediate reactions in children. *The Journal of Allergy and Clinical Immunology: In Practice, 4*(2), 229–236.

Chinen, J., Paul, M. E., & Shearer, W. T. (2019). 32 – Approach to the evaluation of the patient with suspected immunodeficiency. In R. R. Rich, T. A. Fleisher, W. T. Shearer, H. W. Schroeder, A. J. Frew, & C. M. Weyand (Eds.), *Clinical immunology* (5th ed., pp. 451–461.e451).

Chinen, J., Martinez-Gallo, M., Gu, W., Cols, M., Cerutti, A., Radigan, L., Zhang, L., Potocki, L., Withers, M., Lupski, J. R., & Cunningham-Rundles, C. (2011). Transmembrane activator and CAML interactor (TACI) haploinsufficiency results in B-cell dysfunction in patients with Smith-Magenis syndrome. *The Journal of Allergy and Clinical Immunology, 127*(6), 1579–1586. https://doi.org/10.1016/j.jaci.2011.02.046

Chow, A. W., Benninger, M. S., Brook, I., Brozek, J. L., Goldstein, E. J. C., Hicks, L. A., Pankey, G. A., Seleznick, M., Volturo, G., Wald, E. R., & File, T. M.,

Jr. (2012). IDSA clinical practice guideline for acute bacterial rhinosinusitis in children and adults. *Clinical Infectious Diseases, 54*(8), e72–e112. https://doi.org/10.1093/cid/cis370

Dunlop, J. H. (2020). Oral immunotherapy for treatment of peanut allergy. *Journal of Investigative Medicine, 68*(6), 1152–1155.

Dykewicz, M. S., Wallace, D. V., Baroody, F., Bernstein, J., Craig, T., Finegold, I., Huang, F., Larenas-Linnemann, D., Meltzer, E., Steven, G., Bernstein, D. I., Blessing-Moore, J., Dinakar, C., Greenhawt, M., Horner, C. C., Khan, D. A., Lang, D., Oppenheimer, J., Portnoy, J. M., … Wallace, D. V. (2017). Treatment of seasonal allergic rhinitis: An evidence-based focused 2017 guideline update. *Annals of Allergy, Asthma & Immunology, 119*(6), 489–511.e441. https://doi.org/10.1016/j.anai.2017.08.012

Edmiston, E., Ashwood, P., & Van de Water, J. (2017). Autoimmunity, autoantibodies, and autism Spectrum disorder. *Biological Psychiatry, 81*(5), 383–390. https://doi.org/10.1016/j.biopsych.2016.08.031

Epstein, T. G., & Bernstein, D. I. (2018). 27 – Nasal polyposis, rhinosinusitis, and nonallergic rhinitis. In L. C. Grammer & P. A. Greenberger (Eds.), *Patterson's allergic diseases* (8th ed., pp. 624–637). Lippincott Williams & Wilkins.

Gauthier, T. W. (2015). Prenatal alcohol exposure and the developing immune system. *Alcohol Research: Current Reviews, 37*(2), 279–285.

Gentile, D. A., Pleskovic, N., Bartholow, A., & Skoner, D. P. (2016). 24 – Allergic rhinitis. In D. Y. M. Leung, S. J. Szefler, F. A. Bonilla, C. A. Akdis, & H. A. Sampson (Eds.), *Pediatric allergy: Principles and practice* (3rd ed., pp. 210–218.e212). Elsevier. https://doi.org/10.1016/B978-0-323-29875-9.00024-0

Giménez-Barcons, M., Casteràs, A., del Pilar Armengol, M., Porta, E., Correa, P. A., Marín, A., ... & Colobran, R. (2014). Autoimmune predisposition in Down syndrome may result from a partial central tolerance failure due to insufficient intrathymic expression of AIRE and peripheral antigens. *The Journal of Immunology, 193*(8), 3872–3879.

Greenberger, P. A., & Stevens, W. S. (2018). 19 – Asthma. In L. C. Grammer & P. A. Greenberger (Eds.), *Patterson's allergic diseases* (8th ed., pp. 423–493). Lippincott Williams & Wilkins.

Gropman, A. L., Duncan, W. C., & Smith, A. C. (2006). Neurologic and developmental features of the Smith-Magenis syndrome (del 17p11.2). *Pediatric Neurology, 34*(5), 337–350. https://doi.org/10.1016/j.pediatrneurol.2005.08.018

Herriot, R., & Miedzybrodzka, Z. (2016). Antibody deficiency in Rubinstein-Taybi syndrome. *Clinical Genetics, 89*(3), 355–358. https://doi.org/10.1111/cge.12671

Holland, S. M., & Uzel, G. (2019). 22 – Phagocyte deficiencies. In R. R. Rich, T. A. Fleisher, W. T. Shearer, H. W. Schroeder, A. J. Frew, & C. M. Weyand (Eds.), *Clinical immunology* (5th ed., pp. 319–333.e311).

Hsiao, E. Y. (2013). Chapter nine – Immune dysregulation in autism spectrum disorder. In G. Konopka (Ed.), *International review of neurobiology* (Vol. 113, pp. 269–302). Academic Press. https://doi.org/10.1016/B978-0-12-418700-9.00009-5

Hwangpo, T., & Schroeder, H. W. (2019). 34 – Primary antibody deficiencies. In R. R. Rich, T. A. Fleisher, W. T. Shearer, H. W. Schroeder, A. J. Frew, & C. M. Weyand (Eds.), *Clinical immunology* (5th ed., pp. 471–487.e471).

Johnson, S., Knight, R., Marmer, D. J., & Steele, R. W. (1981). Immune deficiency in fetal alcohol syndrome. *Pediatric Research, 15*(6), 908–911. https://doi.org/10.1203/00006450-198106000-00005

Joshi, A. Y., Abraham, R. S., Snyder, M. R., & Boyce, T. G. (2011). Immune evaluation and vaccine responses in down syndrome: Evidence of immunodeficiency? *Vaccine, 29*(31), 5040–5046.

Jyonouchi, S., Orange, J., Sullivan, K. E., Krantz, I., & Deardorff, M. (2013). Immunologic features of Cornelia de Lange syndrome. *Pediatrics, 132*(2), e484–e489. https://doi.org/10.1542/peds.2012-3815

Kim, E., & Burks, W. (2018). 18 - food allergy. In L. C. Grammer & P. A. Greenberger (Eds.), *Patterson's allergic diseases* (8th ed., pp. 401–421). Lippincott Williams & Wilkins.

Kline, A. D., Moss, J. F., Selicorni, A., Bisgaard, A. M., Deardorff, M. A., Gillett, P. M., Ishman, S. L., Kerr, L. M., Levin, A. V., Mulder, P. A., Ramos, F. J., Wierzba, J., Ajmone, P. F., Axtell, D., Blagowidow, N., Cereda, A., Costantino, A., Cormier-Daire, V., FitzPatrick, D., … Hennekam, R. C. (2018). Diagnosis and management of Cornelia de Lange syndrome: First international consensus statement. *Nature Reviews Genetics, 19*(10), 649–666. https://doi.org/10.1038/s41576-018-0031-0

Kumar, S., Suthar, R., Panigrahi, I., & Marwaha, R. K. (2012). Rubinstein-Taybi syndrome: Clinical profile of 11 patients and review of literature. *Indian Journal of Human Genetics, 18*(2), 161–166. https://doi.org/10.4103/0971-6866.100751

Lederman, H. M. (2016). 7 – Approach to the child with recurrent infections including molecular diagnostics. In D. Y. M. Leung, S. J. Szefler, F. A. Bonilla, C. A. Akdis, & H. A. Sampson (Eds.), *Pediatric allergy: Principles and practice* (3rd ed., pp. 63–70.e61). Elsevier.

Lougaris, V., Facchini, E., Baronio, M., Lorenzini, T., Moratto, D., Specchia, F., & Plebani, A. (2016). Progressive severe B cell deficiency in pediatric Rubinstein-Taybi syndrome. *Clinical Immunology, 173*, 181–183. https://doi.org/10.1016/j.clim.2016.10.019

Lyall, K., Van de Water, J., Ashwood, P., & Hertz-Picciotto, I. (2015). Asthma and allergies in children with autism Spectrum disorders: Results from the CHARGE study. *Autism Research, 8*(5), 567–574. https://doi.org/10.1002/aur.1471

Mehta, H., Ramesh, M., Feuille, E., Groetch, M., & Wang, J. (2014). Growth comparison in children with and without food allergies in 2 different demographic populations. *The Journal of pediatrics, 165*(4), 842–848. https://doi.org/10.1016/j.jpeds.2014.06.003

Meltzer, A., & Van de Water, J. (2016). The role of the immune system in autism spectrum disorder. *Neuropsychopharmacology, 42*(1), 284–298. https://doi.org/10.1038/npp.2016.158

McCusker, C., Upton, J., & Warrington, R. (2018). Primary immunodeficiency. *Allergy, Asthma, and Clinical Immunology, 14*(Suppl 2), 61. https://doi.org/10.1186/s13223-018-0290-5

McDonald-McGinn, D. M., & Sullivan, K. E. (2011). Chromosome 22q11. 2 deletion syndrome (DiGeorge syndrome/velocardiofacial syndrome). *Medicine, 90*(1), 1–8.

Minakova, E., & Warner, B. B. (2018). Maternal immune activation, central nervous system development and behavioral phenotypes. *Birth Defects Research, 110*(20), 1539–1550. https://doi.org/10.1002/bdr2.1416

Ming, J. E., & Stiehm, E. R. (2008). Genetic syndromic immunodeficiencies with antibody defects. *Immunology and Allergy Clinics of North America, 28*(4), 715–736. https://doi.org/10.1016/j.iac.2008.06.007

Ming, J. E., & Stiehm, E. R. (2016). *Primary immunodeficiency diseases: Definition, diagnosis, and management* (2nd ed.). N. Rezaei, A. Aghamohammadi, & L. D. Notarangelo (Eds.). Springer.

Mitchell, R. B., Call, E., & Kelly, J. (2003). Ear, nose and throat disorders in children with down syndrome. *The Laryngoscope, 113*(2), 259–263.

Modell, V., Orange, J. S., Quinn, J., & Modell, F. (2018). Global report on primary immunodeficiencies: 2018 update from the Jeffrey Modell centers network on disease classification, regional trends, treatment modalities, and physician reported outcomes. *Immunologic Research, 66*(3), 367–380. https://doi.org/10.1007/s12026-018-8996-5

Morimoto, K., & Nakajima, K. (2019). Role of the immune system in the development of the central nervous system. *Frontiers in Neuroscience, 13*. https://doi.org/10.3389/fnins.2019.00916

Novak, N., & Leung, D. Y. M. (2016). 50 - role of barrier dysfunction and immune response in atopic dermatitis. In D. Y. M. Leung, S. J. Szefler, F. A. Bonilla, C. A. Akdis, & H. A. Sampson (Eds.), *Pediatric allergy: Principles and practice* (3rd ed., pp. 438–447.e433). Elsevier. https://doi.org/10.1016/B978-0-323-29875-9.00050-1

Ong, P. Y., & Leung, D. Y. M. (2018). 29 – Atopic dermatitis. In L. C. Grammer & P. A. Greenberger (Eds.), *Patterson's allergic diseases* (8th ed., pp. 665–673). Lippincott Williams & Wilkins.

Parker, S. E., Mai, C. T., Canfield, M. A., Rickard, R., Wang, Y., Meyer, R. E., Anderson, P., Mason, C. A., Collins, J. S., Kirby, R. S., Correa, A., & National Birth Defects Prevention Network. Updated National Birth. (2010). Prevalence estimates for selected birth defects in the United States, 2004–2006. *Birth Defects Research. Part A, Clinical and Molecular Teratology, 88*(12), 1008–1016. https://doi.org/10.1002/bdra.20735. Epub 2010 Sep 28.

Pasic, S. (2015). Rubinstein-Taybi syndrome associated with humoral immunodeficiency. *Journal of Investigational Allergology & Clinical Immunology, 25*(2), 137–138.

Perkins, T., Rosenberg, J. M., Le Coz, C., Alaimo, J. T., Trofa, M., Mullegama, S. V., Antaya, R. J., Jyonouchi, S., Elsea, S. H., Utz, P. J., Meffre, E., & Romberg, N. (2017). Smith-Magenis syndrome patients often display antibody deficiency but not other immune pathologies. *Journal of Allergy and Clinical Immunological Practice, 5*(5), 1344–1350. e1343. https://doi.org/10.1016/j.jaip.2017.01.028. Epub 2017 Mar 9.

Peters, A. T., Spector, S., Hsu, J., Hamilos, D. L., Baroody, F. M., Chandra, R. K., Grammer, L. C., Kennedy, D. W., Cohen, N. A., Kaliner, M. A., Wald, E. R., Karagianis, A., Slavin, R. G., & Joint Task Force on Practice Parameters, representing the American Academy of Allergy, Asthma and Immunology, the American College of Allergy, Asthma and Immunology, and the Joint Council of Allergy, Asthma and Immunology. (2014). Diagnosis and management of rhinosinusitis: A practice parameter update. *Annals of Allergy, Asthma & Immunology, 113*(4), 347–385. https://doi.org/10.1016/j.anai.2014.07.025

Ram, G., & Chinen, J. (2011). Infections and immunodeficiency in down syndrome. *Clinical and Experimental Immunology, 164*(1), 9–16.

Reddel, H. K., Bateman, E. D., Becker, A., Boulet, L.-P., Cruz, A. A., Drazen, J. M., Haahtela, T., Hurd, S. S., Inoue, H., de Jongste, J. C., Lemanske, R. F., Levy, M. L., Byrne, P. M., Paggiaro, P., Pedersen, S. E., Pizzichini, E., Soto-Quiroz, M., Szefler, S. J., Wong, G. W. K., & FitzGerald, J. M. (2015). A summary of the new GINA strategy: A roadmap to asthma control. *European Respiratory Journal, 46*(3), 622. https://doi.org/10.1183/13993003.00853-2015

Reddel, H. K., FitzGerald, J. M., Bateman, E. D., Bacharier, L. B., Becker, A., Brusselle, G., Buhl, R., Cruz, A. A., Fleming, L., Inoue, H., Ko, F. W., Krishnan, J. A., Levy, M. L., Lin, J., Pedersen, S. E., Sheikh, A., Yorgancioglu, A., & Boulet, L. P. (2019). GINA 2019: A fundamental change in asthma management: Treatment of asthma with short-acting bronchodilators alone is no longer recommended for adults and adolescents. *The European Respiratory Journal, 53*(6), 1901046. https://doi.org/10.1183/13993003.01046-2019

Rezaei, N., & Bonilla, F. (2016). Introduction on primary immunodeficiency diseases. In N. Rezaei, A. Aghamohammadi, & L. D. Notarangelo (Eds.), *Primary immunodeficiency diseases: Definition, diagnosis, and management* (p. 2). Springer.

Ricketti, P. A., & Ricketti, A. J. (2018). 26- allergic rhinitis. In L. C. Grammer & P. A. Greenberger (Eds.), *Patterson's allergic diseases* (8th ed., pp. 595–623). Lippincott Williams & Wilkins.

Roifman, C. M. (2019). 35 – Primary T-cell immunodeficiencies. In R. R. Rich, T. A. Fleisher, W. T. Shearer, H. W. Schroeder, A. J. Frew, & C. M. Weyand (Eds.), *Clinical immunology* (5th ed., pp. 489–508.e481).

Sabourin, K. R., Reynolds, A., Schendel, D., Rosenberg, S., Croen, L. A., Pinto-Martin, J. A., Schieve, L. A., Newschaffer, C., Lee, L.-C., & DiGuiseppi, C. (2019). Infections in children with autism spectrum disorder: Study to Explore Early Development (SEED). *Autism Research, 12*(1), 136–146. https://doi.org/10.1002/aur.2012

Saettini, F., Herriot, R., Prada, E., Nizon, M., Zama, D., Marzollo, A., Romaniouk, I., Lougaris, V., Cortesi, M., Morreale, A., Kosaki, R., Cardinale, F., Ricci, S., Domínguez-Garrido, E., Montin, D., Vincent, M., Milani, D., Biondi, A., Gervasini, C., & Badolato, R. (2020). Prevalence of immunological defects in a cohort of 97 Rubinstein-Taybi syndrome patients. *Journal of Clinical Immunology, 40*(6), 851–860. https://doi.org/10.1007/s10875-020-00808-4

Sampson, H. A., Aceves, S., Bock, S. A., James, J., Jones, S., Lang, D., Nadeau, K., Nowak-Wegrzyn, A., Oppenheimer, J., Perry, T. T., Randolph, C., Sicherer, S. H., Simon, R. A., Vickery, B. P., Wood, R., Joint Task Force on Practice Parameters, Bernstein, D., Blessing-Moore, J., Khan, D., Lang, D., … Wood, R. (2014). Food allergy: A Practice parameter update-2014. *The Journal of Allergy and Clinical Immunology, 134*(5), 1016–25.e43. https://doi.org/10.1016/j.jaci.2014.05.013.

Schneider, L., Tilles, S., Lio, P., Boguniewicz, M., Beck, L., LeBovidge, J., Novak, N., Bernstein, D., Blessing-Moore, J., Khan, D., Lang, D., Nicklas, R., Oppenheimer, J., Portnoy, J., Randolph, C., Schuller, D., Spector, S., Tilles, S., & Wallace, D. (2013). Atopic dermatitis: A practice parameter update 2012. *The Journal of Allergy and Clinical Immunology, 131*(2), 295–9.e27. https://doi.org/10.1016/j.jaci.2012.12.672

Sullivan, K. E. (2008 May 1). Chromosome 22q11. 2 deletion syndrome: DiGeorge syndrome/velocardiofacial syndrome. *Immunology and Allergy Clinics of North America., 28*(2), 353–366.

Sullivan, K. D., Evans, D., Pandey, A., Hraha, T. H., Smith, K. P., Markham, N., Rachubinski, A. L., Wolter-Warmerdam, K., Hickey, F., Espinosa, J. M., & Blumenthal, T. (2017 Nov 1). Trisomy 21 causes changes in the circulating proteome indicative of chronic autoinflammation. *Scientific Reports, 7*(1), 1–1.

Tanabe, S., & Yamashita, T. (2018). The role of immune cells in brain development and neurodevelopmental diseases. *International Immunology, 30*(10), 437–444. https://doi.org/10.1093/intimm/dxy041

Villella, A., Bialostocky, D., Lori, E., Meyerson, H., & Hostoffer, R. W. (2000). Rubinstein-Taybi syndrome with humoral and cellular defects: A case report. *Archives of Disease in Childhood, 83*(4), 360–361. https://doi.org/10.1136/adc.83.4.360

Wall, L. A., & Zambrano, R. M. (2019). Down syndrome. In I. MacKay & N. Rose (Eds.), *Encyclopedia of medical immunology*. Springer.

Xu, G., Snetselaar, L. G., Jing, J., Liu, B., Strathearn, L., & Bao, W. (2018). Association of food allergy and other allergic conditions with autism spectrum disorder in children. *JAMA Network Open, 1*(2), e180279–e180279. https://doi.org/10.1001/jamanetworkopen.2018.0279

7 The Pediatric Cardiologist's Role in the Care of Children with Autism and Other Intellectual and Developmental Disabilities

Ryan Jones, Kristen Sandoz, and Kelvin Lee

Abstract

Pediatric cardiology has made tremendous advances over the last 50 years resulting in increased life expectancy for children with congenital heart disease. This chapter introduces the general role of the cardiologist on the pediatric medical home team and the significant contributions made by the pediatric cardiologist to the care of children with autism spectrum disorders and other intellectual and developmental disabilities. The focus is on expectations for assessment and treatment of common cardiac conditions, assisting families in preparation of various appointments, and the new findings of the association between congenital heart disease and autism spectrum disorders.

Keywords

Pediatric cardiology · Autism · ASD · IDD · Developmental disability · Congenital heart disease · Intellectual disability

R. Jones (✉) · K. Sandoz
Louisiana State University Health Sciences Center, New Orleans, LA, USA
e-mail: ryan.jones@lsuhs.edu

K. Lee
AdventHealth for Children, AdventHealth Medical Group, Orlando, FL, USA

Pediatric cardiologists treat congenital and acquired heart disease in children. Acquired heart disease is not present at birth and includes conditions such as infection or inflammation of the heart and hypertension. Congenital heart disease (CHD) is present at birth, occurring in approximately nine infants per 1000 live births (Liu et al., 2019). A genetic syndrome is present in about 25% of all children with CHD (Pierpont et al., 2018). The most common genetic syndromes associated with CHD include Down, Noonan, CHARGE, Turner, and Williams syndromes and 22q11.2 microdeletion (Ko, 2015). Pierpont et al. (2018) have detailed over 30 syndromes associated with heart disease. These syndromes are associated with autism spectrum disorders (AuSD[1]) and other intellectual and developmental disabilities (IDD) (Calcagni et al., 2017). Conversely, CHDs are at increased risk of developmental delay (Mussatto et al., 2014). While the association between CHD and AuSD is generally a new topic of study, many advances in research have been made to delineate the link between the two (Sigmon et al., 2019).

This chapter introduces the general role of the cardiologist on the pediatric medical home team and the significant contributions made by the

[1] Because pediatric cardiologists refer to atrial septal defects as ASDs, this chapter will use "AuSD" to refer to autism spectrum disorders

P. McPherson (ed.), *Handbook of Treatment Planning for Children with Autism and Other Neurodevelopmental Disorders*, Autism and Child Psychopathology Series,
https://doi.org/10.1007/978-3-031-06120-2_7

pediatric cardiologist to the care of children with AuSD/IDD. The focus is on expectations for assessment and treatment of common cardiac conditions, assisting families in preparation of various appointments, and the new findings of the association between congenital heart disease and autism spectrum disorders.

The Role of Cardiology in the Pediatric Medical Home

The role of cardiology in pediatrics is essential to the health and well-being of the pediatric population, namely, those with CHD or acquired heart disease. Gilboa et al. (2016) estimated that over two million infants, children, adolescents, and adults are living with CHD in the United States. Many cardiac defects are diagnosed prenatally via ultrasound, and cardiologists are fortunately able to become involved in the care of these patients early on. Being prepared and anticipating future obstacles associated with the diagnosed defect(s) early on have led to improved survival of cardiac patients with over 97% of infants with CHD living to adulthood (Mandalenakis et al., 2020). Other defects are diagnosed shortly after birth when an infant presents with cyanosis, respiratory failure, or poor feeding, and the cardiologist is consulted as an essential part of the diagnostic team. In addition, the primary care pediatrician may identify potential cardiac issues with the presentation of heart-related symptoms or as part of routine screenings as the child develops. At this point, the shared plan of care commonly includes referral to a pediatric cardiologist for further diagnosis and management. Many patients with CHD have comorbidities; thus practicing a transdisciplinary approach, such as the medical home model, is vital for the well-being of these patients. Collaborating with primary care pediatricians, cardiovascular surgeons, and physicians of other subspecialties is therefore an important aspect of a pediatric cardiologist's practice. Pediatric cardiologists are vital members of the medical home team, contributing to the care of children with heart conditions before for birth and across childhood and adolescence.

Scope of Congenital Heart Disease

The epidemiology of individual occurrences and outcomes of CHD allows one to accurately evaluate the true impact of congenital cardiac defects on both patients and society as a whole (Allen et al., 2013). The burden of CHD on children, families, and society is significant, with a greater burden on children with AuSD or IDD as they are more likely than peers to experience CHD with crude odds ratios of 4.6 and 9.1, respectively (Razzaghi et al., 2015). According to the Baltimore-Washington Infant Study conducted in the North Eastern United States, the birth prevalence of congenital heart defects is estimated at 48 per 10,000 births (Ferencz et al., 1985; Ferencz et al., 1989). This landmark study reported septal defects as most common with ventricular septal defects most prevalent with a rate of 11.2 per 10,000 births. The least common CHD is tricuspid atresia with a rate of 0.4 per 10,000 births. The Centers for Disease Control and Prevention's (CDC) Metropolitan Atlanta Congenital Defects Program (MACDP) study was conducted from 1998 to 2005 in four counties in the South Eastern United States estimated a CHD prevalence rate of 81.4 per 10,000 births (Reller et al., 2008). This study included stillbirths and pregnancy terminations. Mai et al. (2019) reported a CHD prevalence of 19.3 per 10,000 births, noting a recent increase in the prevalence of tetralogy of Fallot and atrioventricular septal defect, with the increase in the latter possibly due to an increase in trisomy 18 and Down syndrome. The difference in prevalence rates between the two studies is most likely due to methodology rather than biology; for example, the MACDP examined 20 CHD while Mai included the 12 critical CHDs (Allen et al., 2013).

Mortality of Congenital Heart Defects

The mortality of CHD is of critical concern to the field of pediatrics due to the considerable risk for mortality and morbidity. Congenital heart defects are the leading cause of infant deaths due to congenital anomalies, accounting for 40% of deaths in children with birth defects (Lopez et al., 2020). Infancy is a particularly high risk period with almost half of the deaths due to CHD occurring in infants under 1 year of age (Gilboa et al., 2016). According to the CDC, cardiac defects are the cause of 1 in 24 neonatal deaths overall (4.2%) as well as 1 in 4 neonatal deaths due to birth defects (24.5%) (Chakraborty et al., 2018). Experiencing greater socioeconomic inequity has been associated with increased mortality due to CHD (Best et al., 2019). Mandatory screening policies, implemented in some states, have resulted in a significant decrease in infant mortality related to critical CHD (Abouk et al., 2017).

Recommendations for Making Referrals

Access to pediatric subspecialists, including cardiologists, is limited in some areas with long wait times for appointments (Greydanus & Joav Merrick, 2020; Macy et al., 2020). Because appointments may be difficult to secure in some regions, it is vital that primary care providers communicate the level of urgency of the referral. Clearly communicating the clinical state of the individual patient and the suspected extent of underlying disease that is present is vital to ensuring timely care. First and foremost, performing a comprehensive history and physical exam is primary and necessary to evaluating the likelihood of heart disease in children. In the majority of infants and children with severe cardiac disease, clinical manifestations may present as a combination of findings, whereas other children with lesser disease may present with an isolated finding or even asymptomatically. Evaluation of patients with suspected cardiac disease is most easily divided into the two categories of hemodynamically stable and hemodynamically unstable patients (Kane, 2019). Making the distinction between the two is of utmost importance to achieve the most favorable outcome. Children who hemodynamically unstable may have inadequate blood flow to the body and require emergent care (Potes et al., 2017). Only after the hemodynamic status is evaluated can an appropriate cardiology referral be made at the proper time in the diagnostic process.

For those patients with hemodynamic instability such as cardiogenic shock, emergent referral to a pediatric cardiologist should be made (Kane, 2019). Cardiogenic shock may manifest as impaired perfusion, respiratory difficulty, and/or hepatomegaly. Emergent care is required in either an emergency department or intensive care unit. For those patients who are hemodynamically stable but have the potential to progress to hemodynamic instability, cardiology referral must occur within days to 1 week. For those who are hemodynamically stable with potential heart defects that are unlikely to progress to hemodynamic instability, cardiology referral may occur nonurgently and may occur within weeks to even months. In hemodynamically stable patients, further diagnostic evaluation with ECG, chest x-ray, pulse oximetry, and/or echocardiogram may be beneficial. Presentations, by age group, that suggestive of underlying cardiac disease and require further evaluation are presented here, followed by details of the cardiovascular evaluation.

Infants

Congenital heart disease most often presents in infants and is evident in the behavior or appearance of the child, with abnormalities often observed by parents and/or health-care providers (Allen et al., 2013). The primary daily energy expenditure put forth by infants is feeding. Therefore, changes or difficulty in feeding should clue providers to possible underlying cardiac disease. Most infants feed 2–3 ounces every 2–3 hours and should be able to complete a feeding in under 30 minutes. A significant decrease in volume or increase in amount of time to finish a feed should prompt further history and evaluation. Pausing during feeds due to rapid respiratory rate, fatigue or by falling asleep, diaphoresis, and/or labored breathing while feeding are possible signs of heart failure (Marino et al., 2019). Cyanosis, whether central or peripheral, may be considered a normal variant or may indicate a significant cardiac abnormality, and it is imperative to distinguish the two. Acrocyanosis as evidenced by blueness of the distal extremities, while distressing to caregivers, is normal. On the other hand, central cyanosis as noted by blueness of the tongue and oral mucosa is most likely secondary to cardiac or respiratory abnormality.

Toddlers and Preschoolers

As with infants, breathing and feeding abnormalities should alert the physician to a possible cardiac etiology. As toddlers with CHD develop and begin to participate in more vigorous physical activity, they may fatigue more quickly than counterparts of similar age. Therefore, it is common practice to ask parents about fatigability and exercise intolerance in these patients in comparison to other children of similar age (Broeders, 2020). Developmental and growth history is also vital at this age. Children with CHD will often display delay in developmental milestones as well as weight and/or height. Therefore, children with breathing or feeding abnormalities, exercise intolerance, developmental delay, or failure to thrive require further evaluation to distinguish between a cardiac versus noncardiac cause of symptoms. If there are any abnormalities on cardiac physical exam or further diagnostic studies such as ECG or chest x-ray, a pediatric cardiologist consultation is appropriate.

Older Children and Adolescents

As children grow older and are able to become the primary historian, they may be able to voice symptoms that can alert the physician to underlying cardiac etiology. It is still always appropriate to question parents of the pediatric population, especially in those patients with AuSd/IDD who may have social and communication barriers.

Exercise intolerance should prompt further history and examination. While exercise intolerance is a nonspecific symptom, cardiac etiology should be higher on the differential if there is an abnormal cardiac examination, if there is a family history of sudden cardiac death, and most importantly, if syncope is associated with exertional activity (Kane, 2019; Miyake et al., 2016). In these patients, an ECG and chest x-ray should be performed along with referral to a pediatric cardiologist.

Although respiratory symptoms may be due to a number of different organ systems, respiratory issues are commonly seen in pediatric patients with cardiovascular anomalies. As with heart rate and blood pressure, respiratory rates vary with different age groups. Patients with left-to-right shunts may present with tachypnea due to the pulmonary overcirculation resulting from increased blood flow to the lungs. Retractions may accompany the tachypnea as well. If there is cardiac failure or pulmonary overcirculation, rales may be auscultated on respiratory examination. Although wheezing is often seen with respiratory disorders such as asthma or bronchiolitis, it may also be seen with pulmonary hypertension or increased left ventricular end diastolic pressure. Cardiac etiology must always be kept in the differential if there is no identifiable respiratory cause for the wheezing (Kane, 2019).

The Cardiovascular Examination

Cardiovascular examination should include thorough evaluation of vital signs and peripheral pulses, as well as inspection, palpation, and auscultation of the chest. In regard to cardiac murmurs, there are certain distinguishing factors that require a cardiology referral given the increased likelihood of a cardiac abnormality. Murmurs of grade 3 intensity, defined as a moderately loud murmur not associated with a palpable precordial thrill, or greater, warrant referral to a pediatric cardiologist. Murmurs that are holosystolic or diastolic and harsh or blowing in quality are more likely to be pathologic and require referral. Abnormal heart sounds including an aberrant S2 (single S2, loud P2, fixed or wide splitting), the presence of an S3 or S4, a systolic click, or a friction rub also require referral to a cardiac specialist. A difference of greater than 10 mmHg in systolic blood pressure between the right arm and leg or weaker or absent pulses in the lower extremities may allude to coarctation of the aorta and without a doubt necessitate referral (Etoom & Ratnapalan, 2014; Naik & Shah, 2013).

The American Academy of Pediatrics recommends that all newborns be screened for critical CHD with pulse oximetry (Diller et al., 2018). Newborns with abnormal pulse oximetry screening at birth require emergent referral to a pediatric cardiologist given the high probability of a congenital cardiac anomaly. Chest radiographs with cardiomegaly or pulmonary edema without a noncardiac cause and abnormal ECGs are additional criteria for referring to a pediatric cardiologist.

Association of Congenital Heart Defects and Autism Spectrum Disorders

Children born with CHD today have greater enhanced survival than earlier surgical eras due to the remarkable innovations and advances in surgical technique (Best et al., 2019). This improved survival has been associated with significant comorbidities, including neurodevelopmental and social challenges. There has been a long-term, well-known association between congenital cardiac disease and general neurodevelopmental and social delays (Morton et al., 2017). Pierpont et al. (2018) has reviewed the association of syndromic CHD in children with IDD in great detail, which will not be reviewed here. The association between CHD and AuSD is, however, less well understood and studied (Sigmon et al., 2019). There has been ongoing research to discover a link between these disorders, identify subgroups of patients with CHD who are at highest risk of AuSD, and most notably, what physicians can do about it (Calderon et al., 2019). This research is providing physicians with the tools to be able to appropriately counsel parents and families of children with CHD as well as utilize proper screening tools for AuSD.

In a 1:3 nested case-control study of the Military Health System (MHS) administrative database performed by Sigmon et al. (2019), it was established that CHD is associated with an increased odds of developing AuSD. Among the subgroups of CHDs, atrial septal defects (ASDs) and ventricular septal defects (VSDs) were found to have the highest association with AuSD. Left heart obstructive lesions (including hypoplastic left heart syndrome, aortic stenosis, and coarctation of the aorta) were found to have a significant association only in the initial analysis. After adjustment for covariates, there were increased odds of AuSD in children with CHD. For any CHD, the odds ratio (OR) was 1.32 (confidence interval 1.10–1.59). Defects associated with the most significant risk of AuSD were ASDs (OR 1.72; confidence interval 1.07–2.74) and VSDs (OR 1.65; confidence interval 1.21–2.25). Left heart obstructive lesions were found to have increased odds in the sensitivity analysis; however, the association was no longer significant (1.25, confidence interval 0.88–1.77). Physicians diagnosing an ASD or VSD in children should be aware of the increased odds of developing an AuSD. The causal relationship between CHD and AuSD is an area requiring further study. There is ongoing research that has suggested that a genetic mechanism may be the causal factor, as discussed later in this chapter. On the other hand, there may be environmental or clinical factors that act as the causal association between the two (Calderon et al., 2019).

Surgery for CHD has been associated with AuSD symptoms (Bean Jaworski et al., 2017; Calderon et al., 2019). A large-scale longitudinal study by Bean Jaworski et al. (2017) found an association between infantile surgery for CHD and screening positive for AuSD on neurodevelopmental evaluation. When utilizing conservative cutoff values, parent-reported developmental issues at 4 years of age and delayed sternal closure following cardiac surgery were significantly associated with positive AuSD screening results. While patients with CHD necessitating infantile surgery were at greater risk for screening positive on an AuSD screener when compared with the general population, it is essential to recognize that this risk is related to an assortment of genetic, operative management, and developmental elements. These findings highlight the fact that although cardiac surgical intervention is crucial for these children, it also presents a myriad of risk factors that may impede neurobehavioral development. The medical home team should be alerted to the importance of screening for AuSD in children who required surgery for a CHD.

A history of transposition of the great arteries necessitating the neonatal arterial switch operation was associated with less success with false belief tasks as compared to controls (Calderon et al., 2010). False belief tasks are a type of task used in theory of mind studies that demonstrate the ability to infer that another person does not possess knowledge that oneself possesses. Theory of mind is known to develop later, if at all, in those with AuSD as compared to children without AuSD as stated by Thompson (2017). A later study by Calderon et al. (2012) revealed that patients with prenatal diagnoses of transposition of the great arteries achieved superior theory of mind scores compared to patients with postnatal diagnoses. While these findings are not specifically diagnostic of AuSD, it elucidates that children with CHDs need regular screening for neurodevelopmental complications, including but not limited to AuSD. It further emphasizes that children with CHD, even without a concomitant diagnosis of AuSD/IDD, may benefit from services to evaluate and, if necessary, assist with neurological and cognitive development.

In several samples of children with CHD, repetitive behaviors and compromised social communication have been recognized (Nattel et al., 2017). It was also noted that many children with CHD have an increased rate of alexithymia (difficulty identifying, describing, and expressing one's own emotions), which is common in persons with AuSD. Like Calderon et al. (2012), Nattel et al. (2017) emphasize the difficulty with theory of mind in children with CHD.

Tsao et al. (2017) found that the risk for both AuSD and attention deficit hyperactivity disorder (ADHD) was significantly greater in children with CHD than in those without CHD. In this retrospective case-control study, children under 18 years of age with CHD diagnosis were identified from the National Health Insurance Research Database in Taiwan and were compared to controls without CHD after being matched for age and gender. The incidence rates of both ADHD and autism were greater in the case group with CHD than in the control group ($p < 0.001$ for both ADHD and AuSD).

The preceding research studies accentuate the important association between CHD and AuSD. This knowledge informs physicians of the necessity of early screening for AuSD so that appropriate counsel can be provided to parents and families of those with CHD. Screening also promotes comprehensive evaluation and referral to critical resources and treatment to minimize the impact of the core symptoms of AuSD.

Cardiac Considerations in Select Mental Disorders

In addition to congenital heart disease, children with AuSD frequently have comorbid medical or mental health conditions (Hossain et al., 2020; Jang & Matson, 2015). The medical home team should be alert for conditions that may indicate a need for cardiac evaluation or testing. These issues can be either of a primary cardiac etiology or due to side effects of treatment for noncardiac problems. Screening for potential cardiac complications due to side effects of medications is frequently performed in the outpatient setting; however, abnormal screening will require referral for a cardiology evaluation.

Attention Deficit Hyperactivity Disorder (ADHD)

ADHD is a common mental disorder experienced in individuals with AuSD/IDD, with a pooled prevalence of 28% in persons with AuSD (Lai et al., 2019; Siegel et al., 2020). Stimulant medications, the first-line treatment for ADHD, are known to have significant side effects that can affect the cardiac status. Stimulants can increase the basal heart rate as well as contribute to pathologic arrhythmias. As many of these medications have adrenergic properties, they may also increase the systolic blood pressure causing hypertension (Barry et al., 2012). Due to these potential side effects, the practitioner should monitor for cardiac complications in patients taking stimulant medications.

Currently, a black box warning is present for all stimulant medication for use in individuals with cardiac disorders (Barry et al., 2012). This warning echoes the 2008 statement by the American Heart Association and American Academy of Pediatrics on the cardiac evaluation of individuals with ADHD (Vetter et al., 2008). Initially, the statement recommended an EKG prior to starting stimulant medication, but this was later retracted. The recommendation was due to the known risk of sudden cardiac death (SCD) with stimulant medications in certain predisposed populations (Gould et al., 2009). Subsequent to the initial recommendation, clarifications were published which state that ECG is not mandatory for all patients with ADHD receiving a stimulant medication but rather for those with risk factors present. Currently, it remains a class II indication to obtain an ECG in individuals with ADHD. Thus, it is reasonable to obtain an ECG prior to initiation of stimulant medication but that is up to the physician's discretion. Currently, treatment of those without significant cardiovascular risks should not be withheld solely due to lack of an ECG (Hammerness et al., 2011).

Anxiety Disorders

Anxiety is a frequent issue in individuals with AuSD/IDD (Lai et al., 2019; Stewart et al., 2020). Anxiety may trigger an autonomic response which, in turn, can lead to cardiovascular symptoms. This nervous system response may cause sinus tachycardia, which is not a pathologic dysrhythmia; still, a rapid pounding heart beat can cause significant patient concern and worsen the anxiety. Anxiety as a cause of palpitations or tachycardia is a diagnosis of exclusion and while it should be considered for any patient with these symptoms, the necessity of an evaluation to rule out a pathologic dysrhythmia should not be minimized due to the patient having anxiety. Beta-blocker medications are sometimes used to help with tachycardia and anxiety as they can minimize these symptoms after evaluating for pathologic dysrhythmias (Altamura et al., 2013). In addition, selective serotonin receptor inhibitors have been utilized for anxiety. This class of medications has been described to potentially prolong the QT interval and can lead to an increased risk of dysrhythmia and will be discussed below.

Depression

The prevalence of depression in persons with AuSD ranges from 2.5% to 47.1% (Hossain et al., 2020). Depression is more common in individuals with AuSD without ID (Hofvander et al., 2009). Adolescents with IDD experience depression twice as often as peers without IDD (Eyre et al., 2019). While depression is a common comorbid diagnosis in persons with AuSD, little research is available on the treatment of depression in AuSD (DeFilippis, 2018). Pharmacologic therapy used for the treatment of depression may include SSRIs, tricyclic antidepressants, mood stabilizers, antipsychotics, or SNRIs. Many of these medications can potentially prolong the QT interval on ECG which can result in life-threatening arrhythmias (Hasnain & Vieweg, 2014; McNally et al., 2007). While the instances of these life-threatening arrhythmias are low, caution should be used with individuals predisposed to this complication or when these medications are used in combination. Limited data is available in pediatrics on QT prolongation due to these medications. Uchida et al. (2015) reviewed the records of 297 children receiving antidepressants, with 2% having a prolonged QT intervals on ECG after starting SSRI medications; however, there was no baseline comparison ECG. After a thorough history is obtained, including family history of sudden death, the need for an EKG should be considered on a case-by-case basis. Obtaining an ECG is reasonable in patients who have any increased risk factors including children with underlying heart disease or a family history of heart disease. Emergent treatment should not be withheld but caution may be required in individuals with significant cardiac risk. Caution should include a discussion of risks and alerting the family to symptoms that may indicate cardiac distress such as dizziness, felling faint or fainting, palpitations, or an irregular heartbeat.

Responding to Symptoms

Patients commonly present with a symptom that could indicate a cardiac issue and requires further assessment. These symptoms include chest pain, syncope, and palpitations. While the differential diagnosis includes cardiac conditions, other diagnoses should be considered.

Chest Pain

A common complaint in adolescents is chest pain, which is most often of noncardiac etiology. The examiner should inquire about the character, location, duration, and onset of pain, as well as any associated alleviating and exacerbating factors. Chest pain with exercise is more suspicious of cardiac origin and should be evaluated further. Chest pain associated with exercise may manifest in patients with congenital coronary artery abnormalities, acquired coronary artery abnormalities due to Kawasaki syndrome, significant left ventricular hypertrophy, or may be caused by noncardiac conditions such as exercise-induced bronchospasm (Selbst, 2010). Patients with chest pain who have findings on history, physical examination, or ECG that imply a cardiac etiology (such as recurrent chest pain with exercise) necessitate urgent consultation with a pediatric cardiologist (Kane, 2019). Acute chest pain is most commonly secondary to inflammatory conditions such as pericarditis and myocarditis, especially if associated with fever and viral symptoms.

Syncope

As noted above, syncope may warrant referral to a pediatric cardiologist (Miyake et al., 2016). Typically, syncope is of benign origin in children; however, it may be life threatening if it is associated with an underlying cardiac abnormality. It is vital to inquire about presyncopal symptoms. If syncope is associated with presyncopal symptoms such as change in position (e.g., standing from a sitting position), warm environment, lightheadedness, visual changes, a sense of warmth, or nausea, the cause is most often vasovagal in nature. If the child has completely recovered from the event and there are no concerning cardiac symptoms or findings on physical exam, these patients may usually be followed by their primary care provider. On the contrary, any syncope that is associated with exercise/exertion, chest pain, a history of congenital or acquired heart disease, or a family history of syncope or sudden cardiac death warrants urgent referral to a pediatric cardiologist, as a cardiac cause of the syncope is more likely.

Cardiac diseases that cause syncope include congenital heart diseases that cause left ventricular outflow tract obstruction such as aortic stenosis or that can generate ventricular arrhythmias like congenital coronary artery abnormalities. Electrophysiologic abnormalities may also cause ventricular arrhythmias with consequent syncope or cardiac arrest. Examples include Wolff-Parkinson-White (WPW), long QT syndrome, and Brugada syndrome. Other causes of ventricular arrhythmias may be due to cardiac muscle disease such as hypertrophic cardiomyopathy, dilated cardiomyopathy, and arrythmogenic right ventricular dysplasia. Reduced ventricular systolic function and pulmonary hypertension may also cause syncope secondary to reduced cardiac output (Zavala et al., 2020).

Palpitations

Palpitations are another common complaint of children and adolescents. According to Kane (2019), abnormal heart rate or rhythm may signify a cardiac abnormality. It is imperative to note the frequency and duration of the palpitations, whether the symptoms occur at rest versus exertion, and the presence or absence of associated symptoms including chest pain, shortness of breath, or fatigue. It is also important to query the parents of the child about additional symptoms such as rapid breathing, pallor, or diaphoresis (Allen et al., 2013). After performing a thorough history of symptoms and family history, a physical examination is the next step in differentiating between a cardiac versus noncardiac etiology. Overall appearance of the patient, vital signs, and cardiovascular and respiratory examination are the focus. It is important to note that normal heart rate differs with age. Tachycardia and bradycardia may be the result of cardiac and noncardiac causes, and ECG is warranted to assess the etiology of an abnormal heart rate. Patients with tachycardia who are symptomatic require urgent referral to a pediatric cardiologist.

While reviewing vital signs, special attention should be given to the patient's blood pressure. Hypertension may clue the provider in to a possible cardiac cause. As with heart rate, blood pressure differs with age as well. Blood pressure should be taken in both upper and lower extremities in order to evaluate for coarctation of the aorta. For primary care providers, it is vital to determine if the elevated blood pressure is due to primary hypertension or if it is secondary hypertension due to another disorder. The evaluation to determine between the two is outside the scope of this discussion, but either primary or secondary hypertension may warrant referral to a pediatric cardiologist.

The Assessment and Treatment of Common Cardiac Conditions

Clinical Examination

Cardiac clinical examinations are generally considered noninvasive, but that determination depends on the perspective of the patient. Initially at a cardiology clinic visit, vital signs will need to be taken including height, weight, respiratory rate (counted by observation), heart rate and blood pressure (checked with a blood pressure cuff), and oxygen saturation (checked with a finger probe). The cardiac physical examination includes a typical physical examination with the addition of the possibility of needing to evaluate the patient in different positions (lying, sitting, and standing). For children with sensory issues, these procedures could be unnerving. When referring children to the pediatric cardiologist, it is helpful to note any accommodations that might improve the experience for the child and family.

Electrocardiogram

The electrocardiogram is a commonly used diagnostic test that analyzes the electrical activity of the heart. Ten electrodes are placed on specific areas on the limbs and chest in order to give data on 12 "leads" that each give information on a specific area of the heart. Four electrodes are placed on the limbs (one for each of the arms and legs) and six are placed on the chest (these are known as the "precordial leads"). Once the leads are placed and the patient is lying still, the test is completed in about 10 seconds. The test is usually done with the patient in a recumbent position, though information can be obtained with the patient in other positions.

An electrocardiogram is used as a screening tool to give information on many different aspects of the hearts condition and function. It gives information on the heart rate (how fast the heart is beating), heart rhythm (the coordination and timing of the electrical signal moving to different areas of the heart), axis (the vector of electrical conduction), the size of the heart chambers, as well as information about irritation or injury the heart might have sustained.

Heart rate tells us whether the patient's heart is beating at a normal speed, beating slower than normal (bradycardia), or faster than normal (tachycardia). The normal range for heart rates for pediatric patients varies by age, with infants having a higher expected heart rate (100–150 beats per minute) and adolescents having a lower expected heart rate (60–100 beats per minute). Toddlers and younger children fall somewhere in between (around 80–120 beats per minute). A given patient of any age can increase the heart rate above the normal ranges at times of activation of their sympathetic nervous system (fight-or-flight response) or at times of illness or infection. Anxiety and stress can thus result in higher heart rates as can excitement or physical exertion.

Heart rhythm describes the origin and pathway of the electrical activity in the heart. Normal heart rhythm is described as "sinus rhythm" in which the electrical signal originates at the sinus node, which is located in the right atrium. The sinus node receives innervation from the vagus nerve, which allows the nervous system to control the rate at which the heart beats. In sinus rhythm, after the sinus node fires off its electrical signal (depolarizes), the signal conducts to the atrioventricular node in the center of the heart from which the signal is then distributed out to the ventricles of the heart that do the work of pumping the blood to the lungs or the body. Abnormalities in heart rhythm include arrhythmias and heart block. In heart block, the connection between the sinus node and the ventricles does not allow for one-to-one conduction of the electrical signal. In arrhythmias, the origin of the electrical signal maybe from somewhere in the heart other than the sinus node or the conduction may be through pathways outside the normal electrical pathways. These include, but are not limited to, premature beats (either atrial or ventricular), reentrant tachycardia, or ectopic tachycardia (again either atrial or ventricular).

An electrocardiogram can give information about the size of the heart's chambers based on the voltages received by the electrodes and the vectors of electrical conduction. Higher voltages can signify larger heart chambers and differences in the axis of the electrical conduction can signify differences in relative sizes of heart chambers. However, as opposed to the diagnostic accuracy that an electrocardiogram provides in heart rhythm diagnoses, it would be more appropriate to characterize the electrocardiogram as a preliminary screening tool with regard to heart chamber dilation or hypertrophy. A study of 5000 patients found a false-positive rate between 77 and 82 percent when screening for left ventricular hypertrophy but only a 6–7 percent false-negative rate for the same diagnosis. This means that while an electrocardiogram positive for left ventricular hypertrophy has a high chance for the patient actually having normal left ventricular size, very few patients who actually have left ventricular hypertrophy will undergo electrocardiogram and falsely be led to believe that their hearts are normal.

An electrocardiogram is a painless test that takes less than 10 minutes but does require that the child be still for about a minute. Social stories and videos are widely available to explain the procedure to children and families. As the test requires cleaning small areas of the skin with an alcohol wipe and placing stickers on the skin, alerting the technician performing the test about sensory issues is helpful.

Echocardiogram

An echocardiogram is a diagnostic ultrasound of the heart. It is the primary tool utilized by pediatric cardiologists to diagnose structural and functional defects of the heart in patients of all ages. Because it uses only high frequency sound waves (inaudible to the human ear), it is a noninvasive test with little-to-no associated morbidity or side effects. There is zero radiation exposure to the patient, and there is no risk of injury in a typical transthoracic (over the chest) echocardiogram. A typical complete echocardiogram takes between 20 and 45 minutes, though more limited studies focusing on a particular area of interest of cardiac anatomy might take as little as 5–15 minutes. However, most cardiologists will prefer that the initial echocardiogram is complete so as to rule out most major structural heart defects with a normal study. The test requires the patient to be lying back in a recumbent position, lying still for the majority of the study. A gel is placed on the echocardiogram probe in order to transmit the ultrasonic waves, and then the probe and gel are put onto to the chest in order to acquire pictures. Areas that are in contact with the probe and gel include below the breast about halfway between the nipple and mid-axillary line, as well as the area just to the left of the sternum, and up near the suprasternal notch just below the neck. The gel can be warmed (it is in most cardiology offices) and no vibrations are felt from the ultrasound wave. If the patient is unable to lie still or tolerate the physical touch of the echocardiogram probe, doing the echocardiogram under conscious sedation is an option. This would require weighing the risks and benefits of the information gained by the study with the risk of sedation for a given individual patient.

The echocardiogram provides direct, real-time information about the structure and function of the heart. Abnormalities in heart structure, such as abnormal holes, blocked or leaking heart valves, or blocked or abnormal blood vessels, can be seen and diagnosed. The echocardiogram also provides information about how the heart is functioning (squeezing to pump blood) and some information about structures surrounding the heart such as the pericardial sac surrounding the heart, the liver, and the lungs. Oftentimes, an echocardiogram can be a major source of information from which cardiologist can rule in or rule out structural heart disease.

Cardiac Computerized Tomography Imaging (Cardiac CT)

A cardiac CT scan is a type of cardiac imaging consisting of multiple x-ray images reconstructed by computer into a two- or three-dimensional image. CT scans can provide information about the size and shape of cardiac structures and blood vessels and provide detailed information about the relationships between those structures and blood vessels. Sometimes intravenous (IV) contrast is injected during the scan to enhance the detail of the imaging. The CT scanner uses ionizing radiation similar to an x-ray (also similar to the radiation in sunlight). As CT technology has improved, the dose of radiation needed to acquire images has decreased. The scan requires the patient to lie down onto a bed that slides through the doughnut-shaped scanner. The patient needs to stay still throughout the scan, but the scan can usually be done quickly, taking from between 5 seconds to a few minutes. Patients with claustrophobia may be uncomfortable in the scanner and anxiolytic medication might need to be taken in order to tolerate the scan. In most cases, general anesthesia is not necessary for the short duration of the test. Social stories and picture cards can be used to prepare the child for the test.

Cardiac Magnetic Resonance Imaging (Cardiac MRI)

Cardiac MRI is a type of cardiac imaging that uses magnetic fields to create images of the heart and blood vessels. These images are functional, meaning information about the moving heart and flow of blood in the heart can be obtained which is an advantage over CT scan imaging. These images are generated without the use of any radiation which is another advantage over CT scan imaging. However, the amount of time needed to acquire the images is much longer, taking between 1 and 3 hours. In cardiac MRI imaging, the patient is required to lie down on a bed that is then put into the tunnel-like MRI scanner. Like the CT scanner, claustrophobia can be an issue in the MRI scanner. IV contrast may also be necessary to obtain the images. In addition, the patient will often need to hold his or her breath at certain times throughout the image acquisition. Loud noises including banging and beeping sounds might be distressing for some patients. Some MRI scanners are able to provide headphones that can play music or otherwise neutralize those sounds. However, because of these factors, general anesthesia is much more commonly used to obtain cardiac MRI images, especially in pediatrics. The choice between cardiac CT scan imaging and cardiac MRI imaging will depend on the type of information that needs to be obtained. As with other tests, social stories, picture cards, or videos of the procedure can be used to prepare the child and improve the experience.

Managing Side Effects of Cardiac Medications

Depending on their indication, cardiac medications might be necessary to prevent morbidity and mortality in cardiac patients Table 7.1.

Arrhythmias can be life threatening as can hemodynamic abnormalities, and so medications might be necessary to manage these symptoms. Nevertheless, side effects such as dizziness or fatigue can often be managed by adjusting dosages or by changing to different medications in the same class. Additionally, if a medication causes a specific side effect upon initiation of a medication, some side effects (like dizziness) can regress as the patient's body becomes used to the changes in hemodynamics.

Cardiac Catheterization

Cardiac catheterization is a procedure through which small tubes or "catheters" are guided through the large blood vessels in the upper leg or the neck and into the heart. These catheters can measure pressures and oxygen saturation and can inject x-ray dye in order to image and map out blood flow through the heart and blood vessels. The catheterization may be diagnostic or therapeutic or both. In a therapeutic (or "interventional") catheterization, devices can be guided through the catheter in order to treat structural heart defects. Balloons or metal stents can be used to enlarge heart valves or blood vessels, and other devices can be used to close off abnormal holes or vessels. An electrophysiology study is a version of a catheterization where electrodes are guided into the heart that can map out the cardiac electrical system and sometimes be used to eliminate causes of arrhythmia. Cardiac catheterization and electrophysiology procedures usually take more than an hour and sometimes many hours. Because of the length of this procedure, most pediatric catheterization procedures are done under general anesthesia. The patient is usually brought into the catheterization lab awake, and then anesthesia is administered. Oftentimes a family member can accompany the patient back into the lab up until the point of going to sleep. Comfort items like stuffed animals, toys, or special blankets can be helpful up until that point as well.

Cardiac Surgery

Cardiac surgery is one of the scariest and most intimidating events in a patient's life and the lives of their parents. Open-heart surgery requires the patient be put under general anesthesia, intubated (have a breathing tube placed), the sternum to be opened exposing the heart, and then, in most cases, requires the heart to be stopped to allow the surgeon to operate. A specialized machine called a cardiopulmonary bypass machine does the work of the heart and lungs while the heart is stopped. Cardiac surgeries usually take several hours after which the chest is closed and the patient returned to the cardiac intensive care unit. Sometimes the patient can be extubated (have the breathing tube removed) right after surgery. Sometimes the patient will need to remain sedated with the breathing tube in place for several days after surgery. All types of heart surgery will require the patient to remain in bed, relatively still for the first day after surgery so the sternum can heal up. Multiple IV lines might be needed as well as lines in the large blood vessels of the body. Chest tubes which allow bleeding and other fluids to drain from around the heart and lungs are often left in place for several days after the surgery. After the breathing tube is removed, oxygen may still need to be delivered via a nasal cannula. Patients with challenges regarding touch, lights, and sounds will be tested by the intensive care unit setting and additional anxiolytic medication might be needed to keep them from endangering themselves. Depending on the progress after surgery, in most cases, the lines and tubes will start to be able to be removed 48–72 hours after surgery, though in some cases this may take longer to happen. Patients will be typically ready for discharge once several goals have been achieved: the patient will usually need to be breathing room air (without supplemental oxygen), not requiring IV medications, and tak-

Table 7.1 Characteristics of common cardiac medications

Medication type	Example medications	Mechanism of action	Indications	Side effects	Notes
Diuretics	Furosemide, chlorothiazide, aldactone	Encourages urination which can be used to pull fluid from around lungs and other "third-space" areas	Can be used before heart surgery to manage uncorrected conditions or after surgery to manage postoperative effusions	Electrolyte imbalances, dehydration	At high doses of diuretic medications, electrolytes might need to be monitored which could require blood draw or finger prick
Beta-blockers	Propranolol, atenolol, metoprolol	Inhibit beta-adrenergic receptor sites for the endogenous catecholamines epinephrine (adrenaline) and norepinephrine thereby inhibiting the sympathetic nervous system which controls the "fight-or-flight" response	Can be used as anti-arrhythmic or as antihypertensive medications. They can also be used as long term treatment for poor heart function	Bradycardia, hypotension, dizziness, fatigue, and (rarely) depressed mood	Cause a decrease in heart rate and blood pressure, which can be either intended effects or side effects of these medications, depending on their indication
Angiotensin-converting enzyme inhibitors (ACE inhibitors)	Captopril, enalapril, lisinopril	Decrease the formation of the hormone angiotensin II (which is a vasoconstrictor), thereby lowering blood pressure	Antihypertensive medications. Can also be used for the treatment of poor heart function	Hypotension, dizziness, fatigue	Do not affect the heart rate directly, though if a patient taking them were to become hypotensive, a reflexive tachycardia might occur
Anticoagulant and antiplatelet medications	Warfarin, heparin, low-molecular-weight heparin, aspirin, clopidogrel	Interfere with coagulation cascade, thereby inhibiting the formation of blood clots	In patients with recent heart surgery (especially with artificial heart valves), these medications are used to prevent the formation of clots that could damage a heart valve or blood vessel or break off and go to the lungs or brain (causing a stroke)	Bleeding, bruising, (rarely) stroke	Patients on these medications will be prone to easy bruising and should avoid activities where the chance of violent or heavy impact is a possibility

(continued)

Table 7.1 (continued)

Medication type	Example medications	Mechanism of action	Indications	Side effects	Notes
Cardiac glycosides	Digoxin	Increases utilization of intracellular calcium in cardiac myocytes thereby increasing heart contractility. Also stimulates parasympathetic nervous system, slowing conduction at the atrioventricular node and thereby decreasing rate of cardiac conduction	Can be used as an antiarrhythmic medication (especially for atrial fibrillation or atrial flutter) or to augment heart function	Nausea, vomiting, diarrhea, blurred vision, confusion, or dizziness	Digoxin was first isolated in 1930 from the foxglove plant

ing adequate nutrition (usually meaning, eating all food by mouth). Strategies to help prepare patients for surgery such as preoperative surveys of behavioral triggers and comfort strategies can help ensure proper resources are available to ensure comfort and a feeling of safety during the postoperative hospital stay.

Additional Guidance for the Primary Care Pediatrician

The American Heart Association (AHA) and the American Academy of Pediatrics (AAP) collaboratively published a statement in 2012 specifying recommendations for routine neurodevelopmental assessment and management of children with CHD. Children of all ages should be screened for autism spectrum disorders. However, enhanced surveillance and screening for AuSD in children with CHD is judicious since recent studies suggest increased risk (Marino et al., 2012). According to the AAP current guidelines, autism screening should be performed at 18 and 24 months with a standardized tool at the primary care pediatrician's office during well-child appointments. There are several screening tools available. Older children and adolescents should be screened for social and behavioral concerns at yearly well-child visits as well. Further screening should be provided at any time if the pediatrician or primary care provider is concerned for signs and/or symptoms of autism spectrum disorders. Ultimately, children who do not pass AuSD screening tools should be referred for a specialized diagnostic evaluation for AuSD. Proper referrals can then be made for early intervention services to provide adequate support for the child's social and neurodevelopmental well-being (Hyman et al., 2020).

Assisting the Child and Family in Preparing for Appointments

Once a child has been referred to a pediatric cardiologist, it is important to educate patients and their parents on what to expect when visiting a subspecialist. This is especially important for children with AuSD and developmental delays, as many of these patients have social and cognitive difficulties, and there may be some degree of uncertainty and anxiety when visiting a subspecialist. Much like visiting a primary care provider, if visiting a cardiologist outpatient in a clinic setting, nursing staff will obtain vital signs and a brief medical history. The cardiologist will then perform a thorough general medical history followed by a detailed cardiologic history. The physician will then perform a comprehensive physical examination, focusing most on the cardiovascular system. This includes, but is not limited to, reviewing vital signs, assessing the general appearance of the patient, evaluating the overall oxygenation of the patient by examining the general color, auscultating the heart and lungs, palpating for the point of maximal impulse of the heart, assessing strength of peripheral and central pulses, palpating the liver, and assessing for any edema. The physical exam, in most circumstances, will be followed by performing an ECG and an echocardiogram if clinically indicated and available. It may be beneficial to have appropriate nursing and child life staff, if available, to assist in these tests if needed. From there, the cardiologist may order laboratory studies and further imaging if necessary. If clinically indicated, the child may be referred for additional studies such as heart catheterizations, ablations, or cardiac surgery. All of these studies will be explained in detail to the patient and his/her guardian, and appropriate follow-up will be scheduled.

If being evaluated by a cardiologist in the hospital setting, the history and physical examination will be performed just as above. Further studies such as cardiac procedures or surgeries may need to be performed more quickly if clinically indicated, depending on the emergent status of the patient and his/her condition. Again, all studies will be thoroughly explained to the patient and his/her guardian if time allows. Lifesaving, emergent procedures will be performed as rapidly as possible to achieve the best possible outcome.

The pediatric cardiologist will remain in contact with the patient's primary care pediatrician to streamline the patient's care. It is essential to practice a transdisciplinary approach to cardiac patients to ensure efficient, ongoing communication between all providers. When referring a patient to a pediatric cardiologist, it is especially helpful to note any accommodations that would assist the child and family to have a more pleasant and productive experience.

Future Directions and Upcoming Research

Cardiology has made tremendous advances over the last half century with 97% of children born with CHD expected to reach adulthood (Mandalenakis et al., 2020). The association between CHD and AuSD/IDD is an up-and-coming field of research. With improved diagnostic practices and interventions widely accessible, increasing attention is being turned to an exciting area of study: identifying the etiology of CHD at the genetic level. With continued research and investigation into these associations, there is hope for new interventions to improve patient quality of life.

Many IDDs, including Down syndrome, 22q11.2 deletion syndrome, Williams syndrome, CHARGE syndrome, and others, have increased rates of AuSD and CHD (Calcagni et al., 2017; Richards et al., 2015). While the association is clear, the correlation is not (Sigmon et al., 2019). Insights into the molecular mechanisms underlying syndromic CHD is beginning to identify the genes responsible for heart defects, including chromosomal aneuploidy in Down syndrome and chromosomal microdeletion in 22q11.2 deletion syndrome (Calcagni et al., 2017). The ultimate goal is that these advances in understanding the etiology of CHD will enhance efforts to prevent CHD, ushering in a new era in the practice of pediatric cardiology.

References

Abouk, R., Grosse, S. D., Ailes, E. C., & Oster, M. E. (2017). Association of US state implementation of newborn screening policies for critical congenital heart disease with early infant cardiac deaths. *Journal of the American Medical Association, 318*(21), 2111–2118.

Allen, H. D., Driscoll, D. J., Shaddy, R. E., & Feltes, T. F. (2013). *Moss & Adams' heart disease in infants, children, and adolescents: Including the fetus and young adult*. Lippincott Williams & Wilkins.

Altamura, A. C., Moliterno, D., Paletta, S., Maffini, M., Mauri, M. C., & Bareggi, S. (2013). Understanding the pharmacokinetics of anxiolytic drugs. *Expert Opinion on Drug Metabolism & Toxicology, 9*(4), 423–440.

Barry, C. L., Martin, A., & Busch, S. H. (2012). ADHD medication use following FDA risk warnings. *The*

Journal of Mental Health Policy and Economics, 15(3), 119.

Bean Jaworski, J. L., Flynn, T., Burnham, N., Chittams, J. L., Sammarco, T., Gerdes, M., … Zackai, E. H. (2017). Rates of autism and potential risk factors in children with congenital heart defects. *Congenital Heart Disease, 12*(4), 421–429. https://doi.org/10.1111/chd.12461

Best, K. E., Vieira, R., Glinianaia, S. V., & Rankin, J. (2019). Socio-economic inequalities in mortality in children with congenital heart disease: A systematic review and meta-analysis. *Paediatric and Perinatal Epidemiology, 33*(4), 291–309.

Broeders, W. H. M. (2020). Early motor development and exercise capacity association in children with a congenital heart disease: An explorative study.

Calcagni, G., Unolt, M., Digilio, M. C., Baban, A., Versacci, P., Tartaglia, M., … Marino, B. (2017). Congenital heart disease and genetic syndromes: New insights into molecular mechanisms. *Expert Review of Molecular Diagnostics, 17*(9), 861–870.

Calderon, J., Angeard, N., Moutier, S., Plumet, M.-H., Jambaqué, I., & Bonnet, D. (2012). Impact of prenatal diagnosis on neurocognitive outcomes in children with transposition of the great arteries. *The Journal of Pediatrics, 161*(1), 94–98.

Calderon, J., Bellinger, D. C., & Newburger, J. W. (2019). Autism and congenital heart disease: Evidence and unresolved questions. *Pediatrics, 144*(5), e20192752.

Calderon, J., Bonnet, D., Courtin, C., Concordet, S., Plumet, M. H., & Angeard, N. (2010). Executive function and theory of mind in school-aged children after neonatal corrective cardiac surgery for transposition of the great arteries. *Developmental Medicine and Child Neurology, 52*(12), 1139–1144. https://doi.org/10.1111/j.1469-8749.2010.03735.x

Chakraborty, A., Gorla, S. R., & Swaminathan, S. (2018). Impact of prenatal diagnosis of complex congenital heart disease on neonatal and infant morbidity and mortality. *Prenatal Diagnosis, 38*(12), 958–963.

DeFilippis, M. (2018). Depression in children and adolescents with autism spectrum disorder. *Children, 5*(9), 112.

Diller, C. L., Kelleman, M. S., Kupke, K. G., Quary, S. C., Kochilas, L. K., & Oster, M. E. (2018). A modified algorithm for critical congenital heart disease screening using pulse oximetry. *Pediatrics, 141*(5), e20174065.

Etoom, Y., & Ratnapalan, S. (2014). Evaluation of children with heart murmurs. *Clinical Pediatrics, 53*(2), 111–117.

Eyre, O., Hughes, R. A., Thapar, A. K., Leibenluft, E., Stringaris, A., Davey Smith, G., … Thapar, A. (2019). Childhood neurodevelopmental difficulties and risk of adolescent depression: The role of irritability. *Journal of Child Psychology and Psychiatry, 60*(8), 866–874.

Ferencz, C., Neill, C. A., Boughman, J. A., Rubin, J. D., Brenner, J. I., & Perry, L. W. (1989). Congenital cardiovascular malformations associated with chromosome abnormalities: An epidemiologic study. *The Journal of Pediatrics, 114*(1), 79–86.

Ferencz, C., Rubin, J. D., McCarter, R. J., Brenner, J. I., Neill, C. A., Perry, L. W., … Downing, J. W. (1985). Congenital heart disease: Prevalence at live birth: The Baltimore-Washington infant study. *American Journal of Epidemiology, 121*(1), 31–36.

Gilboa, S. M., Devine, O. J., Kucik, J. E., Oster, M. E., Riehle-Colarusso, T., Nembhard, W. N., … Marelli, A. J. (2016). Congenital heart defects in the United States: Estimating the magnitude of the affected population in 2010. *Circulation, 134*(2), 101–109.

Gould, M. S., Walsh, B. T., Munfakh, J. L., Kleinman, M., Duan, N., Olfson, M., … Cooper, T. (2009). Sudden death and use of stimulant medications in youths. *American Journal of Psychiatry, 166*(9), 992–1001.

Greydanus, D. E., & Joav Merrick, M. D. (2020). Pediatric cardiology. *Journal of Alternative Medicine Research, 12*(4), 275–277.

Hammerness, P. G., Perrin, J. M., Shelley-Abrahamson, R., & Wilens, T. E. (2011). Cardiovascular risk of stimulant treatment in pediatric attention-deficit/hyperactivity disorder: Update and clinical recommendations. *Journal of the American Academy of Child & Adolescent Psychiatry, 50*(10), 978–990.

Hasnain, M., & Vieweg, W. V. R. (2014). QTc interval prolongation and torsade de pointes associated with second-generation antipsychotics and antidepressants: A comprehensive review. *CNS Drugs, 28*(10), 887–920.

Hofvander, B., Delorme, R., Chaste, P., Nydén, A., Wentz, E., Ståhlberg, O., … Gillberg, C. (2009). Psychiatric and psychosocial problems in adults with normal-intelligence autism spectrum disorders. *BMC Psychiatry, 9*(1), 1–9.

Hossain, M. M., Khan, N., Sultana, A., Ma, P., McKyer, E. L. J., Ahmed, H. U., & Purohit, N. (2020). Prevalence of comorbid psychiatric disorders among people with autism spectrum disorder: An umbrella review of systematic reviews and meta-analyses. *Psychiatry Research, 287*, 112922.

Hyman, S. L., Levy, S. E., & Myers, S. M. (2020). Identification, evaluation, and management of children with autism spectrum disorder. *Pediatrics, 145*(1), e20193447.

Jang, J., & Matson, J. L. (2015). Autism severity as a predictor of comorbid conditions. *Journal of Developmental and Physical Disabilities, 27*(3), 405–415.

Kane, D. A. (2019). *Suspected heart disease in infants and children: Criteria for referral*. Retrieved February 17, 2020, from https://www.uptodate.com/contents/suspected-heart-disease-in-infants-and-children-criteria-for-referral?search=recommendationsforcardiologyreferral&source=search_result&selectedTitle=1~150&usage_type=default&display_rank=1#H19

Ko, J. M. (2015). Genetic syndromes associated with congenital heart disease. *Korean Circulation Journal, 45*(5), 357–361. https://doi.org/10.4070/kcj.2015.45.5.357

Lai, M.-C., Kassee, C., Besney, R., Bonato, S., Hull, L., Mandy, W., ... Ameis, S. H. (2019). Prevalence of co-occurring mental health diagnoses in the autism population: A systematic review and meta-analysis. *The Lancet Psychiatry, 6*(10), 819–829.

Liu, Y., Chen, S., Zühlke, L., Black, G. C., Choy, M.-K., Li, N., & Keavney, B. D. (2019). Global birth prevalence of congenital heart defects 1970–2017: Updated systematic review and meta-analysis of 260 studies. *International Journal of Epidemiology, 48*(2), 455–463.

Lopez, K. N., Morris, S. A., Tejtel, K. S., Espaillat, A., & Salemi, J. (2020). US mortality due to congenital heart disease across the lifespan from 1999–2017 exposes persistent racial/ethnic disparities. *medRxiv, 142*, 1132.

Macy, M. L., Leslie, L. K., Turner, A., & Freed, G. L. (2020). Growth and changes in the pediatric medical subspecialty workforce pipeline. *Pediatric Research, 89*, 1297–1303.

Mai, C. T., Isenburg, J. L., Canfield, M. A., Meyer, R. E., Correa, A., Alverson, C. J., Lupo, P. J., Riehle-Colarusso, T., Cho, S. J., Aggarwal, D., Kirby, R. S., National Birth Defects Prevention Network. (2019). National population-based estimates for major birth defects, 2010–2014. *Birth Defects Research, 111*(18), 1420–1435. https://doi.org/10.1002/bdr2.1589. Epub 2019 Oct 3. PMID: 31580536.

Mandalenakis, Z., Giang, K. W., Eriksson, P., Liden, H., Synnergren, M., Wåhlander, H., ... Dellborg, M. (2020). Survival in children with congenital heart disease: Have we reached a peak at 97%? *Journal of the American Heart Association, 9*(22), e017704.

Marino, B. S., Lipkin, P. H., Newburger, J. W., Peacock, G., Gerdes, M., Gaynor, J. W., ... Johnson, W. H., Jr. (2012). Neurodevelopmental outcomes in children with congenital heart disease: Evaluation and management: A scientific statement from the American Heart Association. *Circulation, 126*(9), 1143–1172.

Marino, L. V., Johnson, M. J., Davies, N. J., Kidd, C., Richens, T., Bharucha, T., & Darlington, A.-S. (2019). Development of feeding information for infants with CHD. *Cardiology in the Young, 29*(9), 1165–1171.

McNally, P., McNicholas, F., & Oslizlok, P. (2007). The QT interval and psychotropic medications in children. *European Child & Adolescent Psychiatry, 16*(1), 33–47.

Miyake, C. Y., Motonaga, K. S., Fischer-Colbrie, M. E., Chen, L., Hanisch, D. G., Balise, R. R., ... Dubin, A. M. (2016). Risk of cardiac disease and observations on lack of potential predictors by clinical history among children presenting for cardiac evaluation of mid-exertional syncope. *Cardiology in the Young, 26*(5), 894.

Morton, P. D., Ishibashi, N., & Jonas, R. A. (2017). Neurodevelopmental abnormalities and congenital heart disease: Insights into altered brain maturation. *Circulation Research, 120*(6), 960–977.

Mussatto, K. A., Hoffmann, R. G., Hoffman, G. M., Tweddell, J. S., Bear, L., Cao, Y., & Brosig, C. (2014). Risk and prevalence of developmental delay in young children with congenital heart disease. *Pediatrics, 133*(3), e570–e577.

Naik, R. J., & Shah, N. C. (2013). Teenage heart murmurs. *Pediatric Clinics of North America, 61*(1), 1–16.

Nattel, S. N., Adrianzen, L., Kessler, E. C., Andelfinger, G., Dehaes, M., Côté-Corriveau, G., & Trelles, M. P. (2017). Congenital heart disease and neurodevelopment: Clinical manifestations, genetics, mechanisms, and implications. *Canadian Journal of Cardiology, 33*(12), 1543–1555.

Pierpont, M. E., Brueckner, M., Chung, W. K., Garg, V., Lacro, R. V., McGuire, A. L., ... Roberts, A. (2018). Genetic basis for congenital heart disease: Revisited: A scientific statement from the American Heart Association. *Circulation, 138*(21), e653–e711.

Potes, C., Conroy, B., Xu-Wilson, M., Newth, C., Inwald, D., & Frassica, J. (2017). A clinical prediction model to identify patients at high risk of hemodynamic instability in the pediatric intensive care unit. *Critical Care, 21*(1), 1–8.

Razzaghi, H., Oster, M., & Reefhuis, J. (2015). Long-term outcomes in children with congenital heart disease: National Health Interview Survey. *The Journal of Pediatrics, 166*(1), 119–124.

Reller, M. D., Strickland, M. J., Riehle-Colarusso, T., Mahle, W. T., & Correa, A. (2008). Prevalence of congenital heart defects in metropolitan Atlanta, 1998–2005. *The Journal of Pediatrics, 153*(6), 807–813.

Richards, C., Jones, C., Groves, L., Moss, J., & Oliver, C. (2015). Prevalence of autism spectrum disorder phenomenology in genetic disorders: A systematic review and meta-analysis. *The Lancet Psychiatry, 2*(10), 909–916.

Selbst, S. M. (2010). Approach to the child with chest pain. *Pediatric Clinics of North America, 57*(6), 1221–1234.

Siegel, M., McGuire, K., Veenstra-VanderWeele, J., Stratigos, K., King, B., Bellonci, C., ... Walter, H. J. (2020). Practice parameter for the assessment and treatment of psychiatric disorders in children and adolescents with intellectual disability (intellectual developmental disorder). *Journal of the American Academy of Child & Adolescent Psychiatry, 59*(4), 468–496. https://doi.org/10.1016/j.jaac.2019.11.018

Sigmon, E. R., Kelleman, M., Susi, A., Nylund, C. M., & Oster, M. E. (2019). Congenital heart disease and autism: A case-control study. *Pediatrics, 144*(5), e20184114.

Stewart, S. L., Babcock, S. E., Li, Y., & Dave, H. P. (2020). A psychometric evaluation of the interRAI Child and Youth Mental Health instruments (ChYMH) anxiety scale in children with and without developmental disabilities. *BMC Psychiatry, 20*(1), 1–14.

Thompson, B. N. (2017). *Theory of Mind: Understanding Others in a Social World*. Retrieved February 16, 2020, from https://www.psychologytoday.com/

Tsao, P.-C., Lee, Y.-S., Jeng, M.-J., Hsu, J.-W., Huang, K.-L., Tsai, S.-J., ... Kou, Y. R. (2017). Additive effect of congenital heart disease and early developmental disorders on attention-deficit/hyperactivity disorder and autism spectrum disorder: A nationwide population-based longitudinal study. *European Child & Adolescent Psychiatry, 26*(11), 1351–1359.

Uchida, M., Spencer, A. E., Biederman, J., Castro, V. M., Kenworthy, T., Chan, J., … Perlis, R. H. (2015). A systematic evaluation of the QTc interval and antidepressants in youth: An electronic health record study. *Journal of Developmental & Behavioral Pediatrics, 36*(6), 434–439.

Vetter, V. L., Elia, J., Erickson, C., Berger, C., Berger, S., & Blum, N. (2008). American Heart Association Council on Cardiovascular Disease in the Young Congenital Cardiac Defects Committee American Heart Association Council on Cardiovascular Nursing. Cardiovascular monitoring of children and adolescents with heart disease receiving medications for attention deficit/hyperactivity disorder [corrected]: a scientific statement from the American Heart Association Council on Cardiovascular Disease in the Young Congenital Cardiac Defects Committee and the Council on Cardiovascular Nursing. *Circulation, 117*(18), 2407–2423.

Zavala, R., Metais, B., Tuckfield, L., DelVecchio, M., & Aronoff, S. (2020). Pediatric syncope: A systematic review. *Pediatric Emergency Care, 36*(9), 442.

8 Oral Health Treatment Planning: Dental Disease Prevention and Oral Health Promotion for Children with Autism Spectrum Disorder and Developmental Disabilities

Donald L. Chi and Leah I. Stein Duker

Abstract

Children with autism spectrum disorders (ASDs) and other intellectual and developmental disabilities (IDDs) are at increased risk for poor oral health. Optimal oral health helps children achieve their maximum potential by preventing systemic health complications and giving them opportunities to eat, sleep, play, socialize, and learn throughout the life course without having to experience the health, economic, and social sequelae of untreated dental diseases. In this chapter, we describe the barriers to optimal oral health. We adopt a behavioral lens that lends itself to immediate strategies that can address the known risk factors for common oral diseases. Our goal is to empower nondental health professionals who treat children with ASD to play a more active role in oral health promotion and dental disease prevention.

Keywords

Oral health · Autism spectrum disorders · Intellectual and developmental disabilities · Behaviors · Toothbrushing · Added sugars · Access to dental care

Introduction

The goal of pediatric dentistry is to ensure optimal oral health for all children. Oral health is recognized as a critical component of overall health. Poor oral health - which for children is mainly defined as tooth decay and gingivitis - is associated with adverse outcomes including nutritional deficiencies and systemic health conditions like heart disease, respiratory infections, and diabetes (Bui et al., 2019). It is well-established that children with autism spectrum disorders (ASDs) and intellectual and developmental disabilities (IDDs) are at greater risk of poor oral health (Ferrazzano et al., 2020; Wilson et al., 2019). In recognition of this fact, the American Academy of Pediatrics has highlighted the particular importance of ensuring optimal oral health in children

D. L. Chi (✉)
School of Dentistry and School of Public Health, University of Washington, Seattle, WA, USA
e-mail: dchi@uw.edu

L. I. Stein Duker
Chan Division of Occupational Science & Occupational Therapy, Ostrow School of Dentistry, University of Southern California, Los Angeles, CA, USA

P. McPherson (ed.), *Handbook of Treatment Planning for Children with Autism and Other Neurodevelopmental Disorders*, Autism and Child Psychopathology Series,
https://doi.org/10.1007/978-3-031-06120-2_8

with developmental disabilities by coordinating efforts between the medical and dental homes (Norwood & Slayton, 2013, reaffirmed 2019) Fig. 8.1.

To date, clinical and public health efforts to support the oral health of special needs populations have focused almost exclusively on access to dental care (Burgette & Chi, 2021). Not surprisingly, unmet dental care needs remain high for most children with ASD and IDD. One explanation is that other well-documented determinants of oral health, particularly sugar intake and fluoride exposure, have been largely overlooked or ignored by clinicians, researchers, and policymakers.

This chapter will introduce readers to the epidemiology and behavioral determinants of dental diseases, with a focus on the determinants of tooth decay that are amenable to change for children with ASD and other IDD. The primary goal is to empower nondental health professionals who treat children with ASD to have a more active role in oral health promotion and dental disease prevention. Health-care professionals can play an important role in keeping oral health at the forefront when providing care to children with ASD and other IDD. Communication between providers in the medical home and the dental home is critical to ensuring that all children with ASD and other IDD can benefit from patient-centered, humanistic care, and optimal oral health. We also hope that clinicians from outside of dentistry will share their knowledge with dental health professionals to ensure that all children with ASD and other IDD can benefit from patient-centered, humanistic care, and optimal oral health.

Common Oral Conditions

The first lesson for the medical home team is to understand conditions commonly treated by a pediatric dentist. These include gingivitis, malocclusion, and tooth decay Table 8.1.

Gingivitis

Gingivitis is a reversible inflammatory condition of the gingiva (gums). It occurs when food debris accumulates on tooth surfaces along the gums and manifests as spontaneous bleeding during brushing or flossing. Numerous genetic disorders are associated with gingivitis, including Ehlers-Danlos syndrome, Down syndrome, and glycogen storage disease (Pizzo et al., 2009). Gingivitis is common in individuals with IDD as well as children with ASD who present with poor hygiene (Petrovic et al., 2016, Pi et al., 2020). It is a self-limited disease and resolves in 7–10 days with daily brushing and flossing of the gums and teeth.

Malocclusion

Malocclusions occur when there is insufficient space in the mouth for teeth to erupt and typically present as crowded teeth that erupt outside of the

Fig. 8.1 Definition of dental home. American Academy of Pediatric Dentistry (2020b)

The dental home is the ongoing relationship between the dentist and the patient, inclusive of all aspects of oral health care delivered in a comprehensive, continuously accessible, coordinated, and family-centered way.

The dental home should be established no later than 12 months of age to help children and their families institute a lifetime of good oral health.

A dental home addresses anticipatory guidance and preventive, acute, and comprehensive oral health care and includes referral to dental specialists when appropriate.

American Academy of Pediatric Dentistry. (2020) Definition of dental home. The Reference Manual of Pediatric Dentistry. Chicago, Ill.: American Academy of Pediatric Dentistry; p.15

Table 8.1 Common oral conditions

Gingivitis	A reversible inflammatory condition of the gingiva (gums)
Malocclusion	Misalignment of the teeth
Tooth decay	Damage to a tooth's enamel that may lead to cavitations requiring treatment

normal U-shaped dental arches of the upper and lower jaws. Teeth can be rotated, displaced, or impacted under the gums and therefore may not be clinically visible in the mouth. Other types of malocclusions include excess overjet (flared upper incisors) and underbites in which the mandible is protruded. Malocclusions are common in IDD including Down syndrome and cerebral palsy (Ziegler & Spivack, 2018). At least one study reported that malocclusions are more prevalent in children with ASD (Fontaine-Sylvestre et al., 2017). Orthodontic appliances or braces are used to correct malocclusions. Severe malocclusions may also require surgical interventions to reposition the jaw.

Tooth Decay (Dental Caries or Cavities)

Tooth decay is the most prevalent childhood disease. It is a chronic disease with acute manifestations in which dietary sugars are metabolized by naturally occurring intraoral bacteria, creating acid byproducts that progressively demineralize tooth enamel. Fluoride helps to reverse demineralization caused by acids. When there is either too much sugar in the diet or insufficient fluoride to remineralize damaged enamel, visible white spots appear on the teeth. Over time, acids from sugary foods and beverages turn these white spots into dark cavitations that may require treatment. As the cavitations get larger, more invasive treatments frequently become necessary. Decayed teeth, which initially might have been repaired with small fillings, may require crowns, root canals, or even extractions if left untreated. Large cavities can cause pain, interfere with eating and sleeping, and even result in infections that may require hospitalization.

All tooth surfaces in the mouth are susceptible to decay, but tooth decay most commonly occurs on the smooth surfaces of incisors (front teeth), chewing surfaces of molars that are grooved (back teeth), and interproximal surfaces (between the teeth). At least one meta-analysis reported that tooth decay rates are similar for children with ASD and those without (Zhang et al., 2020), but another publication reported higher caries prevalence in children with ASD (Burgette & Rezaie, 2020). While further study is needed regarding the prevalence of caries in children with ASD, children with IDD tend to have a higher tooth decay prevalence than peers without IDD (Davis, 2009), mainly because of the inability to cooperate associated with lower IQ.

Nearly all clinical care in pediatric dentistry is aimed at preventing and managing the sequelae of tooth decay. While good oral health is important for all children, there are unique challenges encountered by children with ASD. Up to 40% of children with ASD carry a diagnosis of intellectual or developmental disability (Wickstrom et al., 2021). Therefore, the remainder of the chapter will focus on tooth decay risk factors and management considerations for children with ASD.

Addressing the Risk Factors for Tooth Decay

Pediatric dentists have identified numerous risk factors for tooth decay including diet, oral hygiene, the oral microbiome, and enamel defects (Kirthiga et al., 2019). The oral microbiome of children with ASD or other IDD may hold clues to oral health, gastrointestinal disturbances, and gut-brain axis (Hicks et al., 2018). Enamel defects have been associated with prenatal factors and genetic syndromes including Down, Usher, and 22q11 deletion (Ziegler & Spivack, 2018). It is critical that the medical home encourage early referral to a pediatric dentist so that risk factors can be identified and preventive interventions implemented.

Characteristics of ASD Associated with Tooth Decay

There are two characteristics specific to ASD that increase a child's risk for tooth decay. The first is the *degree of intellectual and/or developmental disability*, which can lead to varying ability to cooperate with oral health behaviors. Toothbrushing may be a daily struggle for children with ASD. Caregivers may use sweet treats to gain cooperation in daily life or as behavioral rewards while unwittingly increasing risk of tooth decay (Anderson et al., 2012; Marshall et al., 2010). Children with ASD may pouch foods in particular areas of the mouth (also referred to as packing or pocketing food; Sarcia, 2021; Levin et al., 2014), which make tooth surfaces in those areas susceptible to decay. Families may find that dental offices may not have adequately trained staff to manage uncooperative patient behaviors. An added complexity is that as children with ASD get older, they also get physically stronger, which can exacerbate these challenges for caregivers and dentists. Alerting the dental team to expected challenges and necessary accommodations prior to the appointment can promote a successful consultation.

The second characteristic common to children with ASD is *sensory sensitivities*, which have the potential to negatively impact oral care in the home and dental office environments. For example, exposure to oral hygiene-related stimuli in the home may produce challenges of not being able to tolerate flavors, textures, and tactile sensations, leading to the rejection of oral hygiene products like commercially available toothbrushes and toothpaste. Bright lights, disinfectant smells, and sounds of rotary equipment in dental offices can also trigger sensitivities and subsequent fight, flight, or fright reactions. Desensitization or comfort items from home may assist with calming in the dental office but in some cases dental care may require moderate sedation or treatment under general anesthesia.

Addressing Challenges to Good Oral Hygiene Behaviors

Toothbrushing twice daily with fluoridated toothpaste helps prevent tooth decay by promoting remineralization of damaged tooth enamel. Ideally, each toothbrushing episode should last about 2 minutes. Children need to be supervised until they can brush all tooth surfaces in the mouth. Some children with ASD may not be accustomed to the sensation of toothbrush bristles or find them overstimulating or painful. Gradually introducing pressure to the head, face, mouth, and surfaces of the teeth may help some children. Parents can seek the guidance of an occupational therapist or applied behavioral analysis therapist if additional practice to accommodate the toothbrush is needed. Once the toothbrush is accepted, children may refuse to accept toothpaste because of the strong flavor or gritty texture. Again, desensitization may be necessary using gel toothpastes or beginning with the tiniest amount of toothpaste then proceeding with small increases. Timers and apps can help caregivers of children with ASD develop toothbrushing habits, though most commercially available apps are not specifically designed to meet the needs of children with ASD. Autism Speaks (n.d.) has published a dental toolkit for families with instructional videos on brushing, guidance on creating an information sheet to assist dental professionals with child-specific accommodations, and tips to prepare children for dental appointments.

Limiting Dietary Sugars

Excess added sugar intake is a strong risk factor for tooth decay. Sugars are found in sugar-sweetened beverages (including juice and soda), in carbohydrate-rich foods (e.g., breads, crackers, cookies, candies), and in medications sweetened with sucrose. Almost all processed foods contain added sugar. Natural and all-organic sweeteners, including honey, brown sugar, cane sugar, and agave, also lead to tooth decay.

Medication and Tooth Decay

Chronic medication use is potentially problematic in three ways. The first is that many are commonly sweetened with sucrose, which exposes the teeth to sugar. Pharmacies may add artificially sweetened flavorings to medications. Second, medication use may lead to xerostomia (dry mouth). Regular intake of water offsets the deleterious effects of reduced salivary flow and buffering capacity that typically accompany medication-induced xerostomia. There are also commercially available saliva substitute products, though these require compliance. Finally, some medications may cause gingival hyperplasia leading to bleeding or painful gums. Many of the anticonvulsant medications commonly prescribed for seizure management in children with ASD or IDD note gingival hyperplasia as a side effect (Hatahira et al., 2017) Fig. 8.2.

Access to Dental Care

Dental checkups are recommended every 6 months within a dental home (AAPD, 2020b definition of dental home), though a more frequent recall intervention may be necessary for children at increased risk of tooth decay. The goal of dental care is to allow a dental health professional to monitor an individual's oral health at a frequency which allows for early restorative intervention when necessary and provides preventive care at regular intervals. Preventive dental care comes in the form of cleanings, pit-and-fissure sealants (to block naturally occurring grooves with polymerized resin), and fluoride varnish treatments. Few dentists provide effective behavior modification therapy to address suboptimal hygiene or dietary behaviors. This is the result of inadequate training, lack of reimbursement, limited availability of evidence-based strategies, and beliefs that behavior change is beyond the scope of dentistry (Chi & Kateeb, 2021). The Autism Speaks Autism Treatment Network (n.d.) has developed an *Autism Toolkit for Dental Professionals* to address issues commonly encountered by dentists. Children with ASD may have difficulties cooperating during dental visits, especially if they are new to the practice.

Once a dental home is established, some caregivers of children with ASD may be concerned about products containing fluoride, which could lead to irregular or skipped fluoride treatments (Rada, 2010). Although fluorides from all sources (e.g., community water, toothpaste, varnish) are considered safe for children, increasing numbers of caregivers are hesitant about fluorides (Chi, 2017; Chi & Basson, 2018), which mirrors current challenges with vaccine hesitancy (Chi, 2014a, 2014b). Medical providers are likely to get questions from caregivers about the safety, necessity, and frequency of fluoride their child

Aripiprazole	Lithium
Clobazam	Loxapine
Clonidine	Methylphenidate
Cyclobenzaprine	Mirtazapine
Diphenhydramine	Olanzapine
Fluoxetine	Quetiapine
Gabapentin	Risperidone
Guanfacine	Sertraline
Lisdexamfetamine	Valproic Acid
	Ziprasidone

Adapted from Wolff, A., Joshi, R. K., Ekström, J., Aframian, D., Pedersen, A. M. L., Proctor, G., ... & Dawes, C. (2017). A guide to medications inducing salivary gland dysfunction, xerostomia, and subjective sialorrhea: a systematic review sponsored by the world workshop on oral medicine VI. *Drugs in R&d*, *17*(1), 1-28.

Fig. 8.2 Partial list of medications associated with tooth decay. (Adapted from Wolff et al. (2017))

needs. It is important to emphasize that fluorides are safe and effective at preventing cavities in all children. Non-fluoridated toothpastes do not have any therapeutic or preventive benefit. Therefore, children should brush with fluoridated toothpaste. An adult should administer toothpaste for young children (smear for children under age 3 and a pea-sized amount for children 3 and older) and supervise toothbrushing to make sure children do not eat toothpaste (American Academy of Pediatric Dentistry, 2020d).

Some caregivers who are concerned about fluoride may also be concerned about silver-colored fillings, but most dental practices have discontinued use of this material not because of safety concerns but because of the environmental impact of mercury (DeRouen et al., 2006).

Older children with ASD may not have a dental home to which they can transition (Cruz et al., 2015). Many families of children with ASD report difficulties finding a dentist willing to treat their child or a dentist with whom the family is comfortable (Stein Duker et al., 2019; Taneja & Litt, 2020). Helping families of children with ASD know what to expect during a dental visit can help make visits more comfortable for the child, caregiver, and dental office staff.

Additional Risk Factor Modification Strategies

Strategies for Home

Many caregivers of children with ASD report difficulty managing their child's daily oral hygiene and that their child requires assistance with toothbrushing, regardless of age (Hage et al., 2020; Qiao et al., 2020; Stein et al., 2012). Toothbrushing time can also be contentious if the child resists. Therefore, it is not surprising that children with ASD brush their teeth significantly fewer times per week compared to other typically developing children (Hage et al., 2020; Qiao et al., 2020; Stein et al., 2012).

In addition, children with ASD frequently exhibit a more limited food repertoire and refuse more foods offered, including fruits and vegetables, in comparison to typically developing children (Bandini et al., 2010). These issues often persist as the children grow older (Bandini et al., 2017). Food selectivity, in conjunction with the use of food-based behavioral rewards, may lead to additional intake of sweets and starchy foods throughout the day that increases tooth decay risk (Anderson et al., 2012; Leiva-García et al., 2019; Margari et al., 2020; Marshall et al., 2010; Page et al., 2021).

Most oral care interventions for the home environment have centered on improving toothbrushing skills, frequently utilizing visual supports to depict toothbrushing methods and techniques. Use of simple flip charts, picture series, or posters with photographs or drawings of the steps involved in toothbrushing have consistently yielded improvements in brushing skills, brushing cooperation, plaque levels, and gingival inflammation (Al Batayneh et al., 2020; Du et al., 2021; Pilebro & Backman, 2005; Smutkeeree et al., 2020). It is unclear whether these approaches lead to significantly less dental disease. Similar improvements have also been reported with the use of toothbrushing visual schedules depicted using an iPad app (Lopez Cazaux et al., 2019) and social stories (Zhou et al., 2017) as well as with video-modeling interventions (Popple et al., 2016). Current research by the authors focus on the potential to harness mobile apps tailored to the needs of children with ASD based on user-center design approaches to improve hygiene frequency and quality (Jacobson et al., 2019).

Sensory sensitivities have also been linked to difficulties with toothbrushing in the home (Khrautieo et al., 2020; Stein et al., 2013). Therefore, in addition to interventions focused on use of visual pedagogies to improve toothbrushing skills and behaviors, utilizing sensory-related strategies may also be beneficial for children with ASD. For example, for children with gustatory or olfactory sensitivities, identifying mild tasting and smelling toothpastes may be beneficial. Some caregivers have even reported making their own odorless and mild tasting toothpaste using baking soda. The concern with homemade toothpaste is that it does not contain an optimal concentration of fluoride, which then makes toothbrushing ineffective at preventing tooth decay. For children who experience intraoral tactile sensitivities, gel toothpaste has a smoother texture and is less gritty. Toothpastes that do not produce foam while brushing may also be helpful. In addition, completing oral "warm-up" activities prior to toothbrushing may decrease sensitivities. Warm-up activities may include an oral massage or vibration to the face and mouth area. Lastly, for those with visual or auditory sensitivities, oral care activities can take place in locations away from the bathroom, like at the dinner table, on the sofa, in bed, or any other location the child will tolerate.

Sensory sensitivities have likewise been linked to feeding difficulties in children with ASD (Page et al., 2021), specifically greater food refusal, smaller food repertoires, and consumption of decreased variety of fruits and vegetables (Chistol et al., 2018). The two most common reasons for food refusals in children with ASD are food-related smell and taste sensitivities and food consistency or texture (Hubbard et al., 2014). However, little research has focused on modifying dietary behaviors in children with ASD to reduce caries risk (Floríndez et al., 2021). Efforts should be made to limit not only the total quantity of added sugars but to also reduce intake frequency. The American Heart Association recommends that no added sugar be given to children under age 2 and that added sugar intake be limited to ≤6 teaspoons per day for children ages 2 years and older (Vos et al., 2017). The American Academy of Pediatrics recommends no fruit juice for children under age 12 months, ≤4 ounces per day for children ages 1–3 years, ≤6 ounces per day for children ages 4–6 years, and ≤ 8 ounces per day for children ages 7 years and older (Heyman & Abrams, 2017).

Plain water is the ideal beverage, especially if it contains fluoride. However, many children with ASD are accustomed to sweet drinks. No spill training cups and bottles should be avoided to reduce the length of time sugary beverages are in physical contact with the teeth. A potential alternative to sugar-sweetened beverages includes those that contain artificial sweeteners. While there is ongoing debate in the scientific community about the safety of artificial sugars, the evidence suggests that these are generally considered safe for children in moderation, especially when weighed against the known adverse consequences associated with extreme added sugar intake (Magnuson et al., 2016, 2017; Sylvetsky & Rother, 2018).

Foods that are low in simple carbohydrates, including fruits, vegetables, dairy products, and nuts are the healthiest. Grazing throughout the day should be avoided when possible, which limits the amount of time that teeth are exposed to dietary sugars. Children should be monitored to ensure that food is not being pouched in the mouth.

Strategies for the Dentist's Office

Children must be able to successfully tolerate both the dental environment and dental procedures in order to undergo dental checkups and treatment. The dental environment will likely be unfamiliar for many children with ASD and filled with possibly overwhelming sensory stimuli, including bright lights, loud noises (e.g., other patients, voices, music, dental equipment), smells (e.g., disinfectant, perfumes, dental materials), tastes and textures (e.g., gritty toothpaste, fluoride varnish), touch (e.g., dentist touching in and around the mouth, plastic X-ray holders), and movement (e.g., reclining in the dental chair) experiences. Dental treatment and procedures require children to follow directions, lay still in the dental chair for long periods of time, and keep their mouth open. Treatment may also require the ability to tolerate administration of local anesthetic, smells of dental medicaments and dental materials, pressure from the clamp and rubber dam, sound of the handpiece, and the temperature and color of the curing light. Because of challenges tolerating these experiences, some children may require sedation or general anesthesia to complete care.

The use of traditional basic and advanced behavior guidance techniques, as described by the American Academy of Pediatric Dentistry (2020a), should be utilized as appropriate to facilitate dental care for children with ASD. However, because of the unique characteristics inherent in this population (e.g., challenges with communication, changes in routine, sensory sensitivities), it is essential to consider adoption of additional techniques and strategies when treatment planning. When attempting to facilitate a positive oral health care encounter for children with ASD, care accommodations should be considered, often with utilization of multiple strategies. It is important to note that every child with ASD is different, so modifications should be tailored to each child's unique strengths and challenges. Finding the right strategy or strategies will often take time and patience and may require a process of trial and error that occurs over multiple dental visits.

Research suggests that certain provider characteristics may be beneficial for providing successful care to children with ASD, with best practices including clear communication, a willingness to collaborate with caregivers and other care providers, flexibility in implementation of dental care procedures, and utilizing the same team for visits whenever possible to promote consistency. For example, beyond the traditional tell-show-do behavior guidance techniques commonly utilized in dentistry (AAPD, 2020a), descriptions of dental activities and instructions should be provided in unambiguous, clear communication appropriate to the receptive language level of the child and with enough time for the child to adequately process information without the child feeling rushed (Dailey & Brooks, 2019; Parry & Shepherd, 2018; Thomas et al., 2017). Collaborating with nondental colleagues (e.g., occupational therapists, behavioral therapists) to identify and implement strategies to support success in treating children with ASD has been reported as beneficial by both caregivers and dental professionals (Lewis et al., 2015; Stein Duker et al., 2019; Thomas et al., 2017). Flexibility from the dental team – often described as use of trial and error to identify and provide individualized strategies to enhance care – is commonly reported by caregivers and dental professionals as integral to improving the odds of a positive dental experience. Even simple modifications such as allowing checkups to occur in nontraditional locations (e.g., floor, standing, another chair) have been noted to improve experiences for children with ASD (Stein Duker et al., 2019; Thomas et al., 2017). Lastly, as many children with ASD exhibit rigid adherence to routines and distress resulting from changes that deviate from those routines, it may be helpful to utilize the same dental team and care processes during visits (Dailey & Brooks, 2019; Stein Duker, 2019; Stein Duker et al., 2019).

Strategies for Preparation Preparation for dental visits can include a variety of activities, with the goal of familiarizing the child with the procedures that will be encountered, providing important child-specific information to the dental team, and facilitating child comfort during the visit.

Social Stories™ Social Stories™ are commonly utilized for children with ASD in a number of settings to reduce negative behavior; increase positive behavior; manage transitions, anxiety, or novel situations; and teach new skills (Kokina & Kern, 2010). These easily implemented structured short stories, often illustrated and personalized, provide developmentally appropriate descriptive information about an upcoming event, enhancing predictability of the encounter as well as describing suitable behavioral responses during that activity (Gray, 2010; Kokina & Kern, 2010). Depending on the skills and preferences of the child, these stories can be delivered verbally by peers or adults, read independently, or delivered digitally (Reynhout & Carter, 2006; Smith et al., 2021).

To accurately reflect a child's dental encounters, social stories should be developed collaboratively by dental providers, caregivers, and other appropriate clinicians treating the child (e.g., occupational therapists). This will allow dentists to guide the steps of the story based on their preferred treatment approaches as well as include caregiver feedback regarding which aspects of care should be included in the social story (e.g., all steps of a dental visit vs. a social story focused solely on cooperating during dental x-rays). Social stories and other books describing dental encounters have been utilized with success to improve the dental visit experience for children with ASD (Anderson et al., 2017; Dailey & Brooks, 2019; Marion et al., 2016; Murshid, 2017; Nelson et al., 2015).

Practice Practice is also integral to increase the chance of a successful dental visit. Caregiver-led preparation activities may include asking the child to practice opening their mouth for periods of time, allowing a caregiver to count the child's teeth as the dentist does during the oral examination, and lying down on a bed or reclining chair during toothbrushing activities to simulate the positioning of the dental chair (Stein Duker et al., 2019). In-home preparation may also include reviewing videos of successful dental encounters (Cuvo et al., 2010; Nelson et al., 2015) or pictorial representations of the dental office and the activities that will take place during the dental visit.

Dentist-led preparation may include provision of practice exam kits to allow for visit-related rehearsals at home. These kits, offered by some dentists, are inexpensive and include items such as a disposable dental mirror, exam gloves, mask, as well as instructions for use (Nelson, 2019). Other items may also be included, such as an air and water syringe, impression trays, or bitewing radiograph tabs (Dailey & Brooks, 2019; Nelson, 2019; Stein Duker et al., 2019).

Obtaining Child-Specific Information to Guide Care Providing child-specific information to the dental team beyond what is typically included in a medical record can facilitate communication and build rapport between caregivers and the dental team, as well as inform accommodation plans for dental visits. If available, the medical home shared plan of care should be sent to the dentist's office ahead of the visit. A pre-visit questionnaire should include information such as communication ability and preferences, sensory processing challenges, likes and interests, and current therapies (Nelson, 2019; Pfeiffer & Stein Duker, 2016; Stein Duker, 2019). If a pre-visit questionnaire is completed, it may help to identify certain items that caregivers can bring to the dental visit to help comfort their child. These may include favorite toys or comfort items to alleviate fear or anxiety, fidget toys for use while waiting or during treatment, or even items to help minimize sensory triggers during care (e.g., sunglasses, weighted lap pad, headphones) (Dailey & Brooks, 2019; Nelson, 2019; Stein Duker et al., 2019).

Although most of the strategies mentioned above are implemented in the home environment, it is also important to note that in-office strategies may support care for children with ASD. For example, decreasing time spent in waiting areas may be beneficial as waiting for appointments can be particularly challenging for children with ASD (Holt & Parry, 2019; Lewis et al., 2015; Parry & Shepherd, 2018; Stein Duker et al., 2019; Thomas et al., 2017). Utilizing strategic scheduling to obtain first or last appointments of the day, allowing families to exit the clinic while waiting for the appointment, and using real-time text messaging to inform families that the dentist is ready to provide care may reduce in-clinic waiting and support positive dental care encounters for children with ASD (Holt & Parry, 2019; Nelson, 2019; Stein Duker et al., 2019). Additionally, desensitization or practice visits may help prepare children with ASD for successful dental encounters.

Strategies During Dental Care

Visual Supports In addition to utilizing a consistent and familiar dental team as mentioned previously, working with a dental provider to establish a set routine for dental visits can facilitate the use of visually based supports to benefit children with ASD. These supports can vary, including visual picture schedules, visually based communication boards, or even online applications (apps). In lieu of relying on verbal or written communication to guide a child through the activities undertaken in a dental visit, these techniques should utilize visually based media (e.g., pictures, drawings, videos) to divide tasks into discrete step-by-step actions, enhancing the structure and predictability of dental encounters and providing simplified tasks to complete during a visit (Balian et al., 2021; Stein Duker, 2019;

AAPD, 2020a). These techniques can be utilized in the home to prepare for the dental encounter or at the dental office during care and have been shown to be effective in improving cooperation for children with ASD (Balian et al., 2021; Cagetti et al., 2015; Lefer et al., 2019; Gledys Zink et al., 2018; Mah & Tsang, 2016). In addition, a visual indicator of when the visit or task will be completed may also be helpful. For example, using a visual clock, visual timer, or even sand or liquid motion timer may be appropriate for children with ASD during dental visits (Dailey & Brooks, 2019; Miller-Kuhaneck & Chisolm, 2012; Stein Duker, 2019).

Behavior Management Supports Many children with ASD and their families are familiar with behavioral approaches to modify behavior. In the dental office, behaviorally based techniques to improve dental care for children with ASD most commonly include practice or desensitization visits. This technique utilizes gradual exposure to positive experiences of the components of the dental visit (e.g., entering dental room, sitting in chair, reclining in chair, intraoral exam, prophylaxis, scaling) (AAPD, 2020a; Nelson, 2019). Considering the affinity for routine and consistency children with ASD exhibit, repeated exposure to these types of practice visits may be especially helpful for this population, with research suggesting these interventions reduce negative behaviors and anxiety during care (Al Humaid et al., 2016; Nelson, 2019; Nelson et al., 2017; Nilchian et al., 2017; Orellana et al., 2014). One research team found that over 85% of children with ASD were able to tolerate an oral examination using an intraoral mirror while seated in a dental chair following desensitization practices (Nelson et al., 2017) and that these skills were maintained for over 90% of those children for at least 2 years (Yost et al., 2019).

Although some dentists have reported that families are often willing to pay a small out-of-pocket fee for these desensitization visits (Nelson, 2019), others have described challenges with the nonreimbursable time and financial outlay required (Stein Duker et al., 2019). Similar fiscal concerns are noted in the United Kingdom, with children with ASD sometimes referred to other providers who are salaried and therefore believed to have additional time available to provide practice and desensitization visits (Parry & Shepherd, 2018).

Distraction techniques, specifically viewing of videos using glasses-style eyewear, as a behavior management strategy have also yielded preliminary success in improving cooperation during routine dental treatment procedures (Fakhruddin & El Batawi, 2017; Isong et al., 2014; Suresh & George, 2019).

Sensory-Based Supports Sensory processing differences are common in children with ASD and have been linked to difficulties tolerating care in the dental office (e.g., greater behavioral and physiological distress, increased use of restraint and/or pharmacological methods) (Cermak et al., 2015; Stein et al., 2011, 2013, 2014). Therefore, it is essential to consider the impact – both negative and positive – of the sensory environment on children during their dental care encounters.

The waiting room, as stated earlier, can be problematic for many children with ASD. In addition to difficulties waiting for care, the sensory stimuli present in the waiting room may lead to additional challenges for those with sensory sensitivities. Waiting rooms are frequently large, busy environments that are a source of sensory stimuli across multiple modalities (e.g., visual, auditory, olfactory), and the waiting rooms of dental clinics are no exception. Implementation of sensory-based modifications in the waiting room have the potential to diminish the intensity and noxiousness of the stimuli present and provide calming sensory stimuli (Hong et al., 2018). To lessen possible sensory triggers in the waiting room, modifications may include replacing fluorescent lights with softer lighting, dimming overhead lights, or using natural sunlight (visual adaptations) or providing headphones or earplugs or soundproofing the doors leading to treatment areas (auditory adaptations). Modifications designed to provide calming sensory stimuli may include an aquarium, bubble tube, or projecting slow-moving images on the wall (visual); classical or nature-based music (auditory); bean bag chairs, fidget toys, or weighted items such as lap pads or stuffed animals (tactile); calming aromatherapy scents (olfactory); or rocking chairs (vestibular) (Hong et al., 2018).

To minimize potentially noxious sensory stimuli during dental care, the use of inexpensive and easily accessible items is often reported to be successful (Stein Duker et al., 2017, 2019; Dailey & Brooks, 2019; Parry & Shepherd, 2018). These items may be brought by caregivers or provided by the dental team. To diminish the brightness of the lights during care, children can wear sunglasses or baseball hats, or the dentist can dim lights (Dailey & Brooks, 2019; Stein Duker et al., 2019). Earplugs, earmuffs, headphones, or even knit hats can be used to dull loud noises in the dental environment and projecting calming music or white noise may facilitate relaxation, and providers should consider treating the child in a private room if possible (Stein Duker et al., 2019; Pfeiffer et al., 2019; Parry & Shepherd, 2018; Miller-Kuhaneck & Chisolm, 2012). Applying deep pressure sensations to the child's body can take place prior to care (e.g., bear hug applied by caregiver) or during care by draping a weighted blanket or lead apron across the child's body to induce a calming effect (Chen et al., 2012, 2016, 2019; Dailey & Brooks, 2019; Edelson et al., 1999; Grandin, 2006; Miller-Kuhaneck & Chisolm, 2012; Stein Duker et al., 2019). Difficulties with oral product-related tastes and textures can be addressed by allowing children to choose the flavor of prophylaxis paste or fluoride or utilizing an unflavored pumice. For those with olfactory sensitivities, using calming aromatherapy scents and unscented nasal hoods and gloves, instructing staff to not wear fragrant perfumes or soaps, and not using highly scented office cleansers or air fresheners may be helpful (Miller-Kuhaneck & Chisolm, 2012). Lastly, for children who are sensitive to movement, allowing them to climb into an already-reclined dental chair may be beneficial (Miller-Kuhaneck & Chisolm, 2012; Stein Duker, 2019). It is also important to note that every child with ASD is different. Sensory-based modifications should be tailored and individualized to each child's specific needs, as they have the potential to benefit some children while distressing others.

Combining use of multiple sensory strategies, the Sensory Adapted Dental Environment (SADE), has been shown to be efficacious in decreasing physiological stress and anxiety, behavioral distress, and child perception of pain during routine dental cleanings for children with ASD and other developmental disabilities (Cermak et al., 2015; Shapiro et al. 2009a, 2009b; Kim et al., 2019). The SADE includes modifications to the visual, auditory, tactile, and vestibular (movement) sensory experiences during dental care administered in a private room. Visual adaptations include blackout curtains placed over the windows, all overhead and dental fluorescent lights turned off, use of a head-mounted lamp to direct light into the child's mouth, and slow-moving Snoezelen images are projected onto the ceiling. Rhythmic nature images and piano music are projected quietly into the room, and the dental chair is pre-reclined before the child enters (auditory and vestibular adaptations, respectively). Tactile stimuli include laying a lead apron over the child's chest and using a special wrap with "butterfly" wings to wrap around the child to provide deep pressure hugging sensations. In the Shapiro et al. (2009a, 2009b) study, a bass vibrator was connected to the dental chair to also provide vibratory sensations. These adaptations were portable, only a short amount of time was required for set up, and the total cost for all adaptations was minimal. Despite the relatively small number of studies that have examined use of these adapted dental environments, many dental clinics in the United States and worldwide have implemented these or similar strategies. Likewise, they are now included in the American Academy of Pediatric Dentistry's list of best practices for behavioral guidance for children with special health-care needs (AAPD, 2020a).

Specifics of Dental Treatment

Noninvasive Dental Treatment

One benefit of early diagnosis of dental diseases like tooth decay is the possibility of intervening with less invasive treatments. Silver diamine fluoride (SDF) is a medicament that can be placed on pre-cavitated or cavitated tooth decay. SDF helps to "freeze" decay (Horst, 2018). It is colorless and odorless and easy to apply even onto the teeth of pre-cooperative or uncooperative children. SDF leaves permanent gray or black stains on the teeth, but this is usually not a significant concern when placed on primary teeth that eventually exfoliate or permanent teeth in the back part of the mouth (Crystal et al., 2017, 2019). Conversations with images should be used to ensure that caregivers understand the staining that will result from SDF. SDF may allow dentists to observe cavities until the teeth exfoliate naturally or to postpone dental treatment until the patient is able to better cooperate. There is near universal adoption of SDF within the US pediatric dentistry residency programs (Crystal et al., 2020), which means that this modality of treatment is likely available from community dentists who are current on evidence-based care.

Severely decayed teeth may need to be extracted, which requires the administration of local anesthesia to numb the teeth and supporting areas of the mouth. When decay is less severe, restorative care, like fillings and stainless-steel crowns, may be indicated, which also requires local anesthesia and the child to cooperate for treatment. There are two noninvasive restorative techniques that eliminate the need for local anesthesia. The first is atraumatic/alternative restorative technique (ART) (AAPD, 2020c). ART involves removal of decayed tooth structure with either a hand or rotary instrument and restoration with a tooth-colored filling material. The second is placement of stainless-steel crowns onto primary molars without removal of decay (also called the Hall Crown technique) (Innes et al., 2011). There are specific clinical indications for Hall Crowns, which include a tooth that is asymptomatic, with no abscess of parulis, and the ability of the child to follow directions when fitting and cementing the crown. Previous research indicates that caregivers and children find the Hall Crown technique to be acceptable (Page et al., 2014).

Protective Stabilization

In some cases, protective stabilization may be indicated to ensure safe delivery of dental treatment. This can be done with the assistance of a caregiver or caregiver to help hold the child's hands or body during treatment or with the use of immobilization devices like a papoose board (AAPD, 2020a). Protective stabilization is the second most common advanced behavior guidance technique used for children with ASD (Loo et al., 2009). Stabilization is utilized with children with ASD 20–33% of the time, significantly more than with typically developing children (3%) (Loo et al., 2009; Marshall et al., 2008; Stein et al., 2012). Caregivers of children with ASD have consistently reported that the use of stabilization, either by the dental provider or themselves, is a very stressful process, while simultaneously agreeing that it is often necessary, especially as their children gets older and bigger (Stein Duker et al., 2017). Anecdotal reports indicate that once a child is stabilized there is a calming effect of the stabilization device (Marshall et al., 2008). This may be because of the deep "hugging" pressure sensations provided by the papoose wrap which induces a calming effect (Edelson et al., 1999). As discussed previously under the SADE study, the deep pressure sensations offered by the papoose board, lead apron, or weighted blanket may promote relaxation when used alone (Chen et al., 2016, 2019) or in conjunction with auditory and visual sensory adaptations (Cermak et al., 2015; Shapiro et al. 2009a, 2009b; Kim et al., 2019).

Sedation and General Anesthesia

When patient behaviors are extremely uncooperative or there is an extensive treatment plan that cannot be delivered in a traditional dental setting, dentists may offer in-office sedation (during which the child is given sedative medications but remains awake during treatment) or general anesthesia (when the child is put to sleep). Past research has shown that significantly larger proportions of caregivers of children with ASD reported that dentists had utilized pharmacological methods to complete a routine prophylaxis as compared to caregivers of typical children (37% vs. 5%) (Stein et al., 2012). While adverse events related to sedation and general anesthesia are rare, they do occur and the decision to adopt this treatment route should be done carefully (Lee et al., 2013). Children undergoing general anesthesia for dental treatment are many times more likely to subsequently undergo treatment under general anesthesia (Chi et al., 2010). As a result, there should be an emphasis on immediately improving oral health behaviors to prevent the need for additional restorative treatment in the future, with a special focus on reducing added sugar intake and brushing with fluoridated toothpaste.

Referrals for Specialty Dental Treatment

General dentists are trained to manage the dental care needs of most children with ASD. In some cases, a referral to a specialist may be needed for extremely uncooperative patients, children with complex medical conditions, and those requiring extensive restorative treatment, usually provided under sedation or general anesthesia. Pediatric dentists undergo 2–3 years of additional training beyond dental school and become adept at managing uncooperative behaviors and providing dental care to medically complex children. One of the best ways to find a pediatric dentist is to ask family members, friends, and colleagues for referrals. It is important to find a pediatric dentist who truly enjoys working with children and is current on evidence-based dental procedures, including SDF, ART, and Hall Crowns. This ensures that the dentist will consider the range of treatment possibilities that can be adopted to prevent and manage tooth decay. Second and third opinions should be sought when there are concerns about the dentist's chairside manners or the potential for overtreatment.

Transition to Adult-Centered Dental Care

Children and teens with ASD who are seen by a pediatric dentist may need to eventually transition to a dentist who can provide adult dental care services (e.g., crowns for permanent teeth, bridge, implants, permanent molar root canals, periodontal treatment, dentures). Past research has shown that there are multiple barriers to dental care transitions, especially for emerging adults with functional limitations (Chi, 2014a, 2014b). As a result, families of children with ASD may continue to rely on pediatric dentists for care well into adulthood (Bayarsaikhan et al., 2015; Cruz et al., 2015). Part of the transition planning is working with the pediatric dentist to identify general dentists in the community who are comfortable treating adults with ASD. There may be a need for a transition period in which both the pediatric and general dentist are involved in patient management. Abrupt transitions can be difficult for families of children with ASD, especially if the child has developed a comfortable relationship with the dentist and dental visits have become routinized. The essential part of successful transitions is for caregivers to initiate these conversations early on, especially because not all pediatric dental offices have a formal process of communicating transition plans to families.

Future Directions

There are multiple future priorities for promoting the oral health of children with ASD. In terms of clinical research, there is a need for behavioral research aimed at helping families and clinicians improve toothbrushing and dietary behaviors. While there is a growing body of literature on modifications that can improve the dental care experience for children with ASD, there is a need for additional work to improve outcomes associated with dental care visits. Almost all research on dental care transitions is observational and few interventions have been evaluated to ensure that children with ASD are able to continue accessing dental care as adults. There is also a need for clinical research aimed at preventing dental diseases in children with ASD, which may need to combine both behavioral and therapeutic approaches like SDF. In terms of policies, there is need for state-level and federal efforts that incorporate meaningful strategies, especially within Medicaid dental programs. Efforts should go beyond solely increasing dentist reimbursement rates and emphasize the specific barriers to oral health that could be addressed to improve outcomes. Finally, community- and population-level policies like healthy school lunches and vending machines, sugar-sweetened beverage taxes, and water fluoridation should be promoted as upstream strategies to protect the oral health of all children including those with ASD and other DD.

References

Al Batayneh, O. B., Nazer, T. S., & Owals, A. I. (2020). Effectiveness of a tooth-brushing programme using the picture exchange communication system (PECS) on gingival health of children with autism spectrum disorders. *European Archives of Paediatric Dentistry, 21*, 277–283.

Al Humaid, J., Tesini, D., Finkelman, M., & Loo, C. Y. (2016). Effectiveness of the d-termined program of repetitive tasking for children with autism spectrum disorder. *Journal of Dentistry for Children, 83*(1), 16–21.

American Academy of Pediatric Dentistry. (2020a). Behavior guidance for the pediatric dental patient. In *The reference manual of pediatric dentistry* (pp. 292–310). American Academy of Pediatric Dentistry.

American Academy of Pediatric Dentistry. (2020b). Definition of dental home. In *The reference manual of pediatric dentistry* (p. 15). American Academy of Pediatric Dentistry.

American Academy of Pediatric Dentistry. (2020c). Pediatric restorative dentistry. In *The reference manual of pediatric dentistry* (pp. 371–383). American Academy of Pediatric Dentistry.

American Academy of Pediatric Dentistry. (2020d). Policy on use of fluoride. In *The Reference Manual of Pediatric Dentistry* (pp. 66–67). Chicago, Ill.: American Academy of Pediatric Dentistry.

Anderson, K. L., Self, T. L., & Carlson, B. N. (2017). Interprofessional collaboration of dental hygiene and communication sciences & disorders students to meet oral health needs of children with autism. *Journal of Allied Health, 46*(4), e97–e101.

Anderson, S. E., Must, A., Curtin, C., & Bandini, L. G. (2012). Meals in our household: Reliability and initial validation of a questionnaire to assess child mealtime behaviors and family mealtime environments. *Journal of the Academy of Nutrition and Dietetics, 112*(2), 276–284.

Autism Speaks. (n.d.). *Dental tool kit.* Retrieved May 15, 2021, from https://www.autismspeaks.org/tool-kit/dentist-for-kids-with-autism

Autism Speaks Autism Treatment Network. (n.d.). *Autism toolkit for dental professionals.* Retrieved May15, 2021 from https://www.autismspeaks.org/sites/default/files/2018-08/Dental%20Professionals%20Tool%20Kit.pdf

Balian, A., Cirio, S., Salerno, C., Gerhard Wolf, T., Campus, G., & Cagetti, M. G. (2021). Is visual pedagogy effective in improving cooperation towards oral hygiene and dental care in children with autism spectrum disorder? A systematic review and meta-analysis. *International Journal of Environmental Research and Public Health, 18*(2), 789.

Bandini, L. G., Anderson, S. E., Curtin, C., Cermak, S., Evans, E. W., Scampini, R., & Must, A. (2010). Food selectivity in children with autism spectrum disorders and typically developing children. *The Journal of Pediatrics, 157*, 259–264.

Bandini, L. G., Curtin, C., Phillips, S., Anderson, S. E., Maslin, M., & Must, A. (2017). Changes in food selectivity in children with autism spectrum disorder. *Journal of Autism and Developmental Disorders, 47*, 439–446.

Bayarsaikhan, Z., Cruz, S., Neff, J., & Chi, D. L. (2015). Transitioning from pediatric to adult dental care for adolescents with special health care needs: Dentist perspectives–Part two. *Pediatric Dentistry, 37*(5), 447–451.

Burgette, J., & Chi, D. L. (2021). Behavioral and social determinants of oral health in children with special health care needs. *Pediatrics, 148*(2), e2021050886.

Bui, F. Q., Almeida-da-Silva, C. L. C., Huynh, B., Trinh, A., Liu, J., Woodward, J., … Ojcius, D. M. (2019). Association between periodontal pathogens and systemic disease. *Biomedical Journal, 42*(1), 27–35.

Burgette, J. M., & Rezaie, A. (2020). Association between ASD spectrum disorder and caregiver-reported dental caries in children. *JDR Clinical and Translational Research, 5*(3), 254–261.

Cagetti, M. G., Mastroberardino, S., Campus, G., et al. (2015). Dental care protocol based on visual supports for children with autism spectrum disorders. *Medicina Oral, Patología Oral y Cirugía Bucal, 20*, e598–e604.

Cermak, S. A., Stein Duker, L. I., Williams, M. E., et al. (2015). Sensory adapted dental environments to enhance oral care for children with autism spectrum disorders: A randomized controlled pilot study. *Journal of Autism and Developmental Disorders, 45*, 2876–2888.

Chen, H.-Y., Yang, H., Chi, H.-J., & Chen, H.-M. (2012). Physiological effects of deep touch pressure on anxiety alleviation: The weighted blanket approach. *Journal of Medical and Biological Engineering, 33*(5), 463–470.

Chen, H.-Y., Yang, H., Chi, H.-J., & Chen, H.-M. (2019). Parasympathetic effect of deep pressure input on third molar extraction in adolescents. *Journal of the Formosan Medical Association, 118*(9), 1317–1324.

Chen, H.-Y., Yang, H., Meng, L.-F., Chan, P.-Y. S., Yang, C.-Y., & Chen, H.-M. (2016). Effect of deep pressure input on parasympathetic system in patients with wisdom tooth surgery. *Journal of the Formosan Medical Association, 115*(10), 853–859.

Chi, D. L. (2014a). Caregivers who refuse preventive care for their children: The relationship between immunization and topical fluoride refusal. *American Journal of Public Health, 104*(7), 1327–1333.

Chi, D. L. (2014b). Medical care transition planning and dental care use for youth with special health care needs during the transition from adolescence to young adulthood: A preliminary explanatory model. *Maternal and Child Health Journal, 18*(4), 778–788.

Chi, D. L. (2017). Parent refusal of topical fluoride for their children: Clinical strategies and future research priorities to improve evidence-based pediatric dental practice. *Dental Clinics of North America, 61*(3), 607–617.

Chi, D. L., & Basson, A. A. (2018). Surveying dentists' perceptions of caregiver refusal of topical fluoride. *JDR Clinical & Translational Research, 3*(3), 314–320.

Chi, D. L., & Kateeb, E. T. (2021). Factors influencing dentists' willingness to treat Medicaid-enrolled adolescents. *Journal of Public Health Dentistry, 81*(1), 42–49.

Chi, D. L., Momany, E. T., Neff, J., Jones, M. P., Warren, J. J., Slayton, R. L., Weber-Gasparoni, K., & Damiano, P. C. (2010). Impact of chronic condition status and severity on dental treatment under general anesthesia for Medicaid-enrolled children in Iowa state. *Paediatric Anaesthesia, 20*(9), 856–865.

Chistol, L. T., Bandini, L. G., Must, A., Phillips, S., Cermak, S. A., & Curtin, C. (2018). Sensory sensitivity and food selectivity in children with autism spectrum disorder. *Journal of Autism and Developmental Disorders, 48*, 583–591.

Cruz, S., Neff, J., & Chi, D. L. (2015). Transitioning from pediatric to adult dental care for adolescents with special health care needs: Adolescent and caregiver perspectives–Part one. *Pediatric Dentistry, 37*(5), 442–446.

Crystal, Y. O., Janal, M. N., Hamilton, D. S., & Niederman, R. (2017). Caregiveral perceptions and acceptance of silver diamine fluoride staining. *Journal of the American Dental Association, 148*(7), 510–518.e4.

Crystal, Y. O., Janal, M. N., Yim, S., & Nelson, T. (2020). Teaching and utilization of silver diamine fluoride and hall-style crowns in US pediatric dentistry residency programs. *Journal of the American Dental Association (1939), 151*(10), 755–763.

Crystal, Y. O., Kreider, B., & Raveis, V. H. (2019). Caregiveral expressed concerns about silver diamine fluoride (SDF) treatment. *The Journal of Clinical Pediatric Dentistry, 43*(3), 155–160.

Cuvo, A. J., Godard, A., Huckfeldt, R., & DeMattei, R. (2010). Training children with autism spectrum disorders to be compliant with an oral assessment. *Research in Autism Spectrum Disorder, 4*(4), 681–696.

Dailey, J. C., & Brooks, J. K. (2019). Autism Spectrum disorder: Techniques for dental radiographic examinations. *Journal of Dental Hygiene, 93*(6), 35–34.

Davis, M. J. (2009). Issues in access to oral health care for special care patients. *Dental Clinics of North America, 53*(2), 169–181.

DeRouen, T. A., Martin, M. D., Leroux, B. G., Townes, B. D., Woods, J. S., Leitão, J., ... Martins, I. P. (2006). Neurobehavioral effects of dental amalgam in children: A randomized clinical trial. *JAMA, 295*(15), 1784–1792.

Du, R. Y., Lam, P. P. Y., You, C. K. Y., & McGrath, C. P. (2021). Evaluation of visual pedagogy in improving plaque control and gingival inflammation among preschool children with autism spectrum disorder: An interventional study. *International Journal of Paediatric Dentistry, 31*, 89–105.

Edelson, S. M., Edelson, M. G., Kerr, D. C. R., & Grandin, T. (1999). Behavioral and physiological effects of deep pressure on children with ASD: A pilot study evaluating the efficacy of Grandin's hug machine. *Journal of Autism and Developmental Disorders, 53*(2), 145–152.

Fakhruddin, K. S., & El Batawi, H. Y. (2017). Effectiveness of audiovisual distraction in behavior modification during dental caries assessment and sealant placement in children with autism spectrum disorder. *Dental Research Journal (Isfahan), 14*(3), 177–182.

Ferrazzano, G. F., Salerno, C., Bravaccio, C., Ingenito, A., Sangianantoni, G., & Cantile, T. (2020). Autism spectrum disorders and oral health status: Review of the literature. *European Journal of Paediatric Dentistry, 21*(1), 9–12.

Floríndez, L. I., Floríndez, D. C., Price, M. E., Floríndez, F. M., Como, D. H., Polido, J. C., Baezconde-Garbanati, L., Pyatak, E., & Cermak, S. A. (2021). Exploring eating challenges and food selectivity for latinx children with and without autism spectrum disorder using qualitative visual methodology: Implications for oral health. *International Journal of Environmental Research and Public Health, 18*(7), 3751.

Fontaine-Sylvestre, C., Roy, A., Rizkallah, J., Dabbagh, B., & Ferraz Dos Santos, B. (2017). Prevalence of malocclusion in Canadian children with ASD spectrum disorder. *American Journal of Orthodontics and Dentofacial Orthopedics, 152*(1), 38–41.

Gledys Zink, A., Cassola Molina, E., Baffi Diniz, M., Santos, M. T. B. R., & Guaré, O. (2018). Communication application for use during the first dental visit for children and adolescents with autism spectrum disorders. *Pediatric Dentistry, 40*(1), 18–22.

Grandin, T. (2006). *Thinking in pictures: My life with autism*. Vintage Books.

Gray, C. (2010). *The new social story* (10th ed.). Future Horizons.

Hage, S. R. V., Lopes-Herrera, S. A., Santos, T. H. F., Defense-Netvral, D. A., Martins, A., Sawasaki, L. Y., & Fernandes, F. D. M. (2020). Oral hygiene and habits of children with autism spectrum disorders and their families. *Journal of Clinical and Experimental Dentistry, 12*(8), e719–e724.

Hatahira, H., Abe, J., Hane, Y., Matsui, T., Sasaoka, S., Motooka, Y., … Nakamura, M. (2017). Drug-induced gingival hyperplasia: A retrospective study using spontaneous reporting system databases. *Journal of Pharmaceutical Health Care and Sciences, 3*(1), 1–11.

Heyman, M. B., & Abrams, S. A. (2017). Section on gastroenterology, hepatology, and nutrition; committee on nutrition. Fruit juice in infants, children, and adolescents: Current recommendations. *Pediatrics, 139*(6), e20170967.

Hicks, S. D., Uhlig, R., Afshari, P., Williams, J., Chroneos, M., Tierney-Aves, C., … Middleton, F. A. (2018). Oral microbiome activity in children with autism spectrum disorder. *Autism Research, 11*(9), 1286–1299.

Holt, S., & Parry, J. A. (2019). Caregiver reported experience of using a real-time text messaging service for dental appointments for children and young people with autism spectrum conditions: A pilot study. *Special Care in Dentistry, 39*, 84–88.

Hong, E. Y., Cermak, S. A., & Stein Duker, L. I. (2018). Occupational therapists' distinct value in creating a sensory-friendly waiting room. *American Occupational Therapy Association SIS Quarterly Practice Connections, 3*(3), 11–13.

Horst, J. A. (2018). Silver fluoride as a treatment for dental caries. *Advances in Dental Research, 29*(1), 135–140.

Hubbard, K. L., Anderson, S. E., Curtin, C., Must, A., & Bandini, L. G. (2014). A comparison of food refusal related to characteristics of food in children with autism spectrum disorder and typically developing children. *Journal of the Academy of Nutrition and Dietetics, 114*, 1981–1987.

Innes, N. P., Evans, D. J., & Stirrups, D. R. (2011). Sealing caries in primary molars: Randomized control trial, 5-year results. *Journal of Dental Research, 90*(12), 1405–1410.

Isong, I. A., Rao, S. R., Holifield, C., Iannuzzi, D., Hanson, E., Ware, J., & Nelson, L. P. (2014). Addressing dental fear in children with autism spectrum disorders: A randomized controlled pilot study using electronic screen media. *La Clinica Pediatrica, 53*(3), 230–223.

Jacobson, D., Jacobson, J., Leong, T., Lourenco, S., Mancl, L., & Chi, D. L. (2019). Evaluating child toothbrushing behavior changes associated with a mobile game app: A single arm pre/post pilot study. *Pediatric Dentistry, 41*(4), 299–303.

Khrautieo, T., Srimaneekarn, N., Rirattanapong, P., & Smutkeeree, A. (2020). Association of sensory sensitivities and toothbrushing cooperation in autism spectrum disorder. *International Journal of Paediatric Dentistry, 30*(4), 505–513.

Kim, G., Carrico, C., Ivey, C., & Wunsch, P. B. (2019). Impact of sensory adapted dental environment on children with developmental disabilities. *Special Care in Dentistry, 39*, 180–187.

Kirthiga, M., Murugan, M., Saikia, A., & Kirubakaran, R. (2019). Risk factors for early childhood caries: A systematic review and meta-analysis of case control and cohort studies. *Pediatric Dentistry, 41*(2), 95–112.

Kokina, A., & Kern, L. (2010). Social story interventions for students with autism spectrum disorders: A meta-analysis. *Journal of Autism and Developmental Disorders, 40*, 812–826.

Lee, H. H., Milgrom, P., Starks, H., & Burke, W. (2013). Trends in death associated with pediatric dental sedation and general anesthesia. *Paediatric Anaesthesia, 23*(8), 741–746.

Lefer, G., Rouches, A., Bourdon, P., & Lopez Cazaux, S. (2019). Training children with autism spectrum disorder to undergo oral assessment using a digital iPad® application. *European Archives of Paediatric Dentistry, 20*(2), 113–121.

Leiva-García, B., El, P., Planells Del Pozo, P., & Molina-López, J. (2019). Association between feeding problems and oral health status in children with autism spectrum disorder. *Journal of Autism and Developmental Disorders, 49*(12), 4997–5008.

Levin, D. S., Volkert, V. M., & Piazza, C. C. (2014). A multi-component treatment to reduce packing in children with feeding and autism spectrum disorders. *Behavior Modification, 38*, 940–963.

Lewis, C., Vigo, L., Novak, L., & Klein, E. J. (2015). Listening to caregivers: A qualitative look at the dental and oral care experiences of children with autism spectrum disorder. *Pediatric Dentistry, 37*, e98–e104.

Loo, C. Y., Graham, R. M., & Hughes, C. V. (2009). Behavior guidance in dental treatment of patients with autism spectrum disorder. *International Journal of Paediatric Dentistry, 19*(6), 390–398.

Lopez Cazaux, S., Lefer, G., Rouches, A., & Bourdon, P. (2019). Toothbrushing training programme using an iPad® for children and adolescents with autism. *European Archives of Paediatric Dentistry, 20*(3), 277–284.

Magnuson, B. A., Carakostas, M. C., Moore, N. H., Poulos, S. P., & Renwick, A. G. (2016). Biological fate of low-calorie sweeteners. *Nutrition Reviews, 74*(11), 670–689.

Magnuson, B. A., Roberts, A., & Nestmann, E. R. (2017). Critical review of the current literature on the safety of sucralose. *Food and Chemical Toxicology, 106*(Pt A), 324–355.

Mah, J. W. T., & Tsang, P. (2016). Visual schedule system in dental care for patients with autism: A pilot study.

The Journal of Clinical Pediatric Dentistry, 40(5), 393–399.

Margari, L., Marzulli, L., Gabellone, A., & de Giambattista, C. (2020). Eating and mealtime behaviors in patients with autism spectrum disorder: Current perspectives. *Neuropsychiatric Disease and Treatment, 16*, 2083–2102.

Marion, I. W., Nelson, T. M., Sheller, B., McKinney, C. M., & Scott, J. M. (2016). Dental stories for children with autism. *Special Care in Dentistry, 36*, 181–186.

Marshall, J., Sheller, B., & Mancl, L. (2010). Caries-risk assessment and caries status of children with autism. *Pediatric Dentistry, 32*, 69–75.

Marshall, J., Sheller, B., Mancl, L., & Williams, B. J. (2008). Parental attitudes regarding behavior guidance of dental patients with autism. *Pediatric Dentistry, 30*(5), 400–407.

Miller-Kuhaneck, H., & Chisolm, E. C. (2012). Improving dental visits for individuals with autism spectrum disorders through an understanding of sensory processing. *Special Care in Dentistry, 32*, 229–233.

Murshid, E. Z. (2017). Effectiveness of a preparatory aid in facilitating oral assessment in a group of Saudi children with autism spectrum disorders in Central Saudi Arabia. *Saudi Medical Journal, 38*, 533–540.

Norwood, K.W., Jr, & Slayton, R.L. (2013). Council on children with disabilities; Section on oral health. Oral health care for children with developmental disabilities. *Pediatrics, 131*(3), 614–619. https://doi.org/10.1542/peds.2012-3650

Nelson, T., Chim, A., Sheller, B. L., McKinney, C. M., & Scott, J. M. (2017). Predicting successful dental examinations for children with autism spectrum disorder in the context of a dental desensitization program. *Journal of the American Dental Association (1939), 148*(7), 485–492.

Nelson, T. M. (2019). Desensitization and therapeutic behavioral approaches to dental care. In T. Nelson & J. R. De Bord (Eds.), *Dental care for children with special needs: A clinical guide* (pp. 99–122). Springer.

Nelson, T. M., Sheller, B., Friedman, C. S., & Bernier, R. (2015). Educational and therapeutic behavioral approaches to providing dental care for patients with autism spectrum disorder. *Special Care in Dentistry, 35*, 105–113.

Nilchian, F., Shakibai, F., & Jarah, Z. T. (2017). Evaluation of visual pedagogy in dental check-ups and preventive practices among 6-12-year-old children with autism. *Journal of Autism and Developmental Disorders, 47*, 858–864.

Orellana, L. M., Martínez-Sanchis, S., & Silvestre, S. J. (2014). Training adults and children with an autism spectrum disorder to be compliant with a clinical dental assessment using a TEACCH-based approach. *Journal of Autism and Developmental Disorders, 44*, 776–785.

Page, L. A., Boyd, D. H., Davidson, S. E., McKay, S. K., Thomson, W. M., & Innes, N. P. (2014). Acceptability of the hall technique to caregivers and children. *The New Zealand Dental Journal, 110*(1), 12–17.

Page, S. D., Souders, M. C., Kral, T. V. E., Chao, A. M., & Pinto-Martin, J. (2021). Correlates of feeding difficulties among children with autism spectrum disorder: A systematic review. *Journal of Autism and Developmental Disorders, 51*, 255.

Parry, J., & Shepherd, J. (2018). Understanding oral health challenges for children and young people with autistic spectrum conditions: Views of families and the dental team. *Journal of Dentistry & Oral Health, 19*(4), 16–20.

Petrovic, B. B., Peric, T. O., Markovic, D. L. J., Bajkin, B. B., Petrovic, D., Blagojevic, D. B., & Vujkov, S. (2016). Unmet oral health needs among persons with intellectual disability. *Research in Developmental Disabilities, 59*, 370–377.

Pfeiffer, B., & Stein Duker, L. I. (2016). The impact of the sensory environment on patient care for individuals with autism spectrum disorders. In E. Giarelli & K. Fisher (Eds.), *Integrated health care for people with autism spectrum disorder: Interdisciplinary planning and delivery of care* (pp. 69–92). Charles C. Thomas Publishers.

Pfeiffer, B., Stein Duker, L. I., & Shui, C. A. (2019). Effectiveness of noise-attenuating headphones on physiological responses in children with autism spectrum disorders. *Frontiers in Integrative Neuroscience, 13*, 65.

Pi, X., Liu, C., Li, Z., Guo, H., Jiang, H., & Du, M. (2020). A meta-analysis of oral health status of children with ASD. *The Journal of Clinical Pediatric Dentistry, 44*(1), 1–7.

Pilebro, C., & Backman, B. (2005). Teaching oral hygiene to children with autism. *International Journal of Paediatric Dentistry, 15*(1), 1–9.

Pizzo, G., Re, D. L., Piscopo, M. R., Pizzo, I., & Giuliana, G. (2009). Genetic disorders and periodontal health: A literature review. *Medical Science Monitor, 15*(8), RA167–RA178.

Popple, B., Wall, C., Flink, L., Powell, K., Discepolo, K., Keck, D., Mademtzi, M., Volkmar, F., & Shic, F. (2016). Brief report: Remotely delivered video modeling for improving oral hygiene in children with ASD: A pilot study. *Journal of Autism and Developmental Disorders, 46*(8), 2791–2796.

Qiao, Y., Shi, H., Wang, H., Wang, M., & Chen, F. (2020). Oral health status of Chinese children with ASD spectrum disorders. *Frontiers in Psychiatry, 11*, 398.

Rada, R. E. (2010). Controversial issues in treating the dental patient with autism. *Journal of the American Dental Association (1939), 141*(8), 947–953. https://doi.org/10.14219/jada.archive.2010.0308

Reynhout, G., & Carter, M. (2006). Social stories for children with disabilities. *Journal of Autism and Developmental Disorders, 36*(4), 445–469.

Sarcia, B. (2021). The impact of applied behavior analysis to address mealtime behaviors of concern among individuals with autism spectrum disorder. *The Psychiatric Clinics of North America, 44*(1), 83–93.

Shapiro, M., Sgan-Cohen, H. D., Parush, S., & Melmed, R. N. (2009a). Influence of adapted environment on the anxiety of medically treated children with developmental disability. *The Journal of Pediatrics, 154*, 546–550.

Shapiro, M., Melmed, R. N., Sgan-Cohen, H. D., & Parush, S. (2009b). Effect of sensory adaptation on anxiety of children with developmental disabilities: A new approach. *Pediatric Dentistry, 31*, 222–228.
Smith, E., Constantin, A., Johnson, H., & Brosnan, M. (2021). Digitally-mediated social stories support children on the autism spectrum adapting to a change in a 'real-world' context. *Journal of Autism and Developmental Disorders, 51*(2), 514–526.
Smutkeeree, A., Khrautieo, T., Thamseupsilp, S., Srimaneekarn, N., Rirattanapong, P., & Wanpen, W. (2020). The effectiveness of visual pedagogy for toothbrushing in children with autism spectrum disorder. *Journal of International Society of Preventive & Community Dentistry, 10*(4), 415–423.
Stein Duker, L. I. (2019). Adapting oral care protocols to support children with sensory sensitivities: Occupational therapy and dentistry. In T. Nelson & J. R. De Bord (Eds.), *Dental care for children with special needs: A clinical guide* (pp. 77–98). Springer.
Stein Duker, L. I., Florindez, L. I., Como, D. H., Tran, C. F., Henwood, B. F., Polido, J. C., & Cermak, S. A. (2019). Strategies for success: A qualitative study of caregiver and dentist approaches to improving oral care for children with autism. *Pediatric Dentistry, 41*(1), 4E–12E.
Stein Duker, L. I., Henwood, B. F., Bluthenthal, R. N., Juhlin, E., Policy, J. C., & Cermak, S. A. (2017). Caregivers' perceptions of dental care challenges in male children with autism spectrum disorder: An initial qualitative exploration. *Research in Autism Spectrum Disorder, 39*, 63–72.
Stein, L. I., Lane, C. J., Williams, M. E., Dawson, M. E., Polido, J. C., & Cermak, S. A. (2014). Physiological and behavioral stress and anxiety in children with autism spectrum disorders during routine oral care. *BioMed Research International, 2014*, 694876.
Stein, L. I., Polido, J. C., & Cermak, S. A. (2013). Oral care and sensory over-responsivity in children with autism spectrum disorders. *Pediatric Dentistry, 35*, 230–235.
Stein, L. I., Polido, J. C., Najera, S. O., & Cermak, S. A. (2012). Oral care experiences and challenges in children with autism spectrum disorders. *Pediatric Dentistry, 34*(5), 387–391.
Stein, L. I., Polido, J. C., Mailloux, Z., Coleman, G. G., & Cermak, S. A. (2011). Oral care and sensory sensitivities in children with autism spectrum disorders. *Special Care in Dentistry, 31*(3), 102–110.
Suresh, L. R., & George, C. (2019). Virtual reality distraction on dental anxiety and behavior in children with autism spectrum disorder. *Journal of International Dental and Medical Research, 12*(3), 1004–1010.
Sylvetsky, A. C., & Rother, K. I. (2018). Nonnutritive sweeteners in weight management and chronic disease: A review. *Obesity, 26*(4), 635–640.
Taneja, N., & Litt, M. D. (2020). Caregivers' barriers to dental care for children with ASD spectrum disorder. *Journal of Dentistry for Children, 87*(2), 98–102.
Thomas, N., Blake, S., Morris, C., & Moles, D. R. (2017). Autism and primary care dentistry: Caregivers' experiences of taking children with autism or working diagnosis of autism for dental examinations. *International Journal of Paediatric Dentistry, 28*(2), 226–238.
Vos, M. B., Kaar, J. L., Welsh, J. A., Van Horn, L. V., Feig, D. I., Anderson, C. A. M., Patel, M. J., Cruz Munos, J., Krebs, N. F., Xanthakos, S. A., Johnson, R. K., & American Heart Association Nutrition Committee of the Council on Lifestyle and Cardiometabolic Health; Council on Clinical Cardiology; Council on Cardiovascular Disease in the Young; Council on Cardiovascular and Stroke Nursing; Council on Epidemiology and Prevention; Council on Functional Genomics and Translational Biology; and Council on Hypertension. (2017). Added sugars and cardiovascular disease risk in children: A scientific statement from the American heart association. *Circulation, 135*(19), e1017–e1034.
Wickstrom, J., Dell'Armo, K., Salzman, E., Hooker, J. L., Delehanty, A., Bishop, S., Tassé, M. J., Wetherby, A. M., Piergies, A. M.H., Damiano, D., Rauch, A., & Thurm, A. (2021). Systematic review: Recommendations for rehabilitation in ASD and ID from clinical practice guidelines. *Archives of Rehabilitation Research and Clinical Translation, 3*(3), 100140. https://doi.org/10.1016/j.arrct.2021.100140
Wolff, A., Joshi, R. K., Ekström, J., Aframian, D., Pedersen, A. M. L., Proctor, G., … Dawes, C. (2017). A guide to medications inducing salivary gland dysfunction, xerostomia, and subjective sialorrhea: A systematic review sponsored by the world workshop on oral medicine VI. *Drugs in R&D, 17*(1), 1–28.
Wilson, N. J., Lin, Z., Villarosa, A., & George, A. (2019). Oral health status and reported oral health problems in people with intellectual disability: A literature review. *Journal of Intellectual & Developmental Disability, 44*(3), 292–304.
Yost, Q., Nelson, T., Sheller, B., McKinney, C. M., Tressel, W., & Chim, A. N. (2019). Children with autism spectrum disorder are able to maintain dental skills: a two-year case review of desensitization treatment. *Pediatric Dentistry., 41*(5), 397–403.
Zhang, Y., Lin, L., Liu, J., Shi, L., & Lu, J. (2020). Dental caries status in autistic children: A meta-analysis. *Journal of Autism and Developmental Disorders, 50*(4), 1249–1257.
Ziegler, J., & Spivack, E. (2018). Nutritional and dental issues in patients with intellectual and developmental disabilities. *The Journal of the American Dental Association, 149*(4), 317–321.
Zhou, N., Wong, H. M., Wen, Y. F., & McGrath, C. (2017). Oral health status of children and adolescents with intellectual disabilities: A systematic review and meta-analysis. *Developmental Medicine and Child Neurology, 59*(10), 1019–1026.

9 The Registered Dietitian Nutritionist's Treatment of Children with Developmental Disabilities or Autism Spectrum Disorders

Nicole A. Withrow

Abstract

The relationship that people have with food is complex and that complexity often begins in early childhood. For some, complexity begins in infancy when metabolic syndromes or genetic disorders requiring special diets are identified. For others, neurological or physical disorders impair feeding, perhaps requiring enteral feeding. Working collaboratively with a registered dietitian nutritionist is important in supporting the overall health of a child with nutritional challenges. This chapter details the care provided by a registered dietitian nutritionist from referral and assessment to the complex interventions delivered to maximize a child's nutritional well-being.

Keywords

Autism spectrum disorders · ASD · Developmental disability · Nutrition · Food selectivity · Problematic eating · Sensory processing · Oral motor · Registered dietitian nutritionist · Food chaining

N. A. Withrow (✉)
University of Northern Colorado, Nutrition and Dietetics, Greeley, CO, USA
e-mail: Nicole.Withrow@unco.edu

Introduction

Registered dietitian nutritionists (RDNs) are nutrition experts who translate nutritional science into evidence-based recommendations designed to assist children and adolescents to live healthier lives. Nutrition science has come a long way since 1826 when Jean Anthelme Brillat-Savarin advised *You are what you eat* (Barabási et al., 2020, p. 33), and RDNs are the key to interpreting complex science and assisting patients in applying evidence-based science to their everyday lives. The complexity of macronutrients (carbohydrates, fats, and proteins), vitamins, and the 65 nutrient and food components tracked by the US Department of Agriculture Food Surveys Research Group is the tip of the iceberg. Foods contain thousands of chemical compounds which are not routinely tracked or understood, so it should be no surprise that application of food chemistry to human physiology is more than the sum of these components. The practical application of this science is dietetics. The RDN brings foundations in evidence-based science and expertise in nutritional interventions to the clinical practice of helping patients understand that what they eat matters and can impact their overall health.

Everyone eats, and everyone has their own preferences and traditions when preparing and eating food. The RDN takes the time to under-

P. McPherson (ed.), *Handbook of Treatment Planning for Children with Autism and Other Neurodevelopmental Disorders*, Autism and Child Psychopathology Series,
https://doi.org/10.1007/978-3-031-06120-2_9

stand these practices and then applies an evidence-based assessment, the Nutrition Care Process and care principles in a patient-centered, family-friendly manner. RDNs understand the importance of providing individualized care based on biochemical markers, anthropometric measurements, nutrition focused physical exams, and medical history. Care is delivered through advanced interviewing techniques, patient and family education, and counseling strategies to further treatment goals and ultimately improvements in health and wellness. Based on data from a nutritional assessment, RDNs develop treatment plans to minimize nutritional risks while promoting the greatest benefits possible for the patient. As with all care, close coordination with patients, families, and medical colleagues is critical.

Today's health-care system involves a complex mix of health-care practitioners who each bring their own education and training to the team in service of promoting a patient's health. RDNs must have a minimum of a Bachelor of Science in Nutrition and Dietetics and complete a 1000-hour nonpaid supervised practice internship. Once the internship is completed, passing a comprehensive examination is required to achieve registered dietitian status. In 2024, all RDNs will have to have a minimum of a master's level degree plus completion of the 1000-hour supervised practice internship. RDNs may specialize in an area of interest by completing additional training, education, and certification programs. These specializations are offered through postgraduate fellowships, continuing education through the Academy of Nutrition and Dietetics, certificate programs through universities and the Academy of Nutrition and Dietetics, training and education grants, and additional academic degrees. For example, if an RDN is interested in neurodevelopment, they could apply and complete a fellowship through the Leadership Education in Neurodevelopment and Related Disabilities (LEND) program that provides long-term, postgraduate level interdisciplinary training. The purpose of LEND is to improve the health of infants through adolescents with disabilities. The training is rigorous and unique but once completed the RDN will have a specialty in neurodevelopmental disabilities. Additional specializations include Certified Dietary Manager, Certified Health Education Specialist, Certified Diabetes Educator, and many more. To maintain registration, an RDN must complete 75 continuing education credits within a 5-year period. The continuing education credits can include additional training in a specialty area, attending professional conferences, completing webinars, participating in grand rounds, and additional coursework at a university. The registration is mandatory to maintain the RDN designation.

Due to the complexities of health care, there are no professionals who work in isolation. Collaboration is critical to provide the best possible health care. Collaboration not only improves the quality of health care but it also saves lives. The Joint Commission on Accreditation of Healthcare Organization (JCAHO) reported that if medical errors appeared on the National Center for Health Statistic's they would outnumber accidents, diabetes, and Alzheimer's diagnoses (O'Daniel & Rosenstein, 2008). A study out of Johns Hopkins reported that medical error represents the third leading cause of death in US hospitals (Makary & Daniel, 2016). Clearly, teamwork and good communication between medical team members is critical to patient safety. Working collaboratively with RDNs is one important step to support the overall health of a child with nutritional challenges because the RDN brings an in-depth understanding of the complexities of disease, diet, metabolism, and medication interactions to patient care. Typically, medical doctors do not have an extensive education in nutrition and dietetics as some medical school curriculums do not require an Introduction to Nutrition course which would be necessary to ensure a basic understanding of metabolism, nutrition, and dietetics (Kiraly et al., 2014). Furthermore, the approach of the RDN is to translate the science of nutrition into solutions for healthy living while the focus of medicine is to treat disease. These dialectical approaches can complement each other when collaboration is fostered.

Brief Overview of Pediatric Nutrition

Adequate nutrition is typically achieved by consuming a varied diet of fruits, vegetables, whole grains, lean proteins, low fat dairy, and healthy fats. The act of eating is a complex developmental process that slowly progresses during infancy from rooting and sucking reflexes to chewing and swallowing a variety of foods by toddlerhood. The development of proficiency with eating requires neurosensory and motor coordination at baseline but environmental influences increasingly impact nutrition as children develop (Gahagan, 2012). As children develop, nutritional needs vary. Breastfeeding meets the nutritional requirements of most infants. The World Health Organization recommends that infants be breastfed with the gradual introduction of solid foods, or complementary feeding, beginning at 6 months as the need for calories and iron begins to exceed that provided by breast milk alone (World Health, 2014). Adequate nutrition during the first 2 years of life is critical for neurodevelopment since deficiencies during this period increase risk of lifelong challenges such as learning disabilities and poor growth (Corkins et al., 2016).

While young children are dependent on parents for food, many parents will tell you that feeding their child and making sure they consume an adequate diet can be extremely challenging. Some research has shown that children's taste preferences are programmed by maternal food choices in utero (Mennella et al., 2001). Density of taste papillae on the tongue and genetics also has a role in taste preferences. For example, the bitter receptor gene, TAS2R38, influences the ability to perceive bitterness and children with this gene have been found to eat fewer vegetables, but that does not mean they will never eat vegetables; they just may need additional support to be successful (Duffy et al., 2004; Gahagan, 2012). Toddlers who are food selective, generally have adequate caloric intake but may lack certain nutrients for optimal growth and development (Corkins et al., 2016). Referral to a RDN should be considered when a child is not consuming a variety of foods and avoidance of certain food groups is occurring. According the MyPlate guidelines, children between the ages of 2 and 3 should try to consume between 1000 and 1400 calories a day by eating 1–1½ cups from the fruit group, 1–1½ cups from the vegetable group, 3–5 ounces of grains, 2–4 ounces of protein, and 2–2½ cups from the dairy group. Ages 4–8, children are consuming up to 2000 calories a day and approximately 2 cups of fruit, 2½ cups of vegetables, 6 ounces of whole grains, 5½ ounces of protein, and 2½ cups of dairy every day (MyPlate.gov, 2020). By the age of four, eating habits typically mirror the preferences and patterns of the family (Gahagan, 2012).

During the school years, children gain increasing independence with eating and food choices. The National Health and Nutrition Examination Survey has found that the overall dietary quality among school age children did not meet the US dietary guidelines with quality dropping as children aged (Banfield et al., 2016). Poor dietary quality places children at risk for obesity, heart disease, diabetes, and other chronic health conditions. Across childhood and adolescence, obesity is a growing concern. Dietary quality can be linked to many factors including the availability and cost of nutritious foods, having regularly scheduled family mealtimes, and screen use during meals (Gu & Tucker, 2017; Hingle, 2013; Trofholz et al., 2019). As school-age children enter adolescence, the body's demands for protein, calcium, vitamin D, and iron increase in order to support the rapid growth during this phase of development (Corkins et al., 2016). Teens may be referred to an RDN for weight management, eating disorders, sports, recommendations for supplements and/or dietary supplements, and overall general health.

The Role of the RDN in Treating Children with Developmental Disabilities

The relationship that people have with food is complex and that complexity often begins in early childhood. For some, complexity begins in infancy when metabolic syndromes or genetic disorders requiring special diets are identified. For others, neurological or physical disorders impair feeding, perhaps requiring enteral feeding. Feeding challenges in infancy place children at increased risk of later feeding issues, with a fifth of NICU graduates experiencing difficulties (Hoogewerf et al., 2017). Feeding disorders, mealtime challenges, and risk of overweight and obesity are common concerns in children with developmental disabilities. An increased risk of obesity has been associated with many genetic disorders, including Cornelia de Lange, Prader-Willi, Down, and Angelman syndromes (Kaur et al., 2017; Ptomey et al., 2020). Children with ASD and ID are also at risk for overweight and obesity with over 50% experiencing obesity (Krause et al., 2016). Children living with disability, whether overweight, underweight, or at a healthy weight, are at risk for being malnourished due to food selectivity (Grumstrup & Demchak, 2019). Gastrointestinal issues may also contribute to malnutrition and additional comorbidities. As a group, children with IDD are at increased risk for gastrointestinal issues, with underweight most common in children with functional abdominal pain (Pawłowska et al., 2018).

Feeding problems are also more common in children with disability (Morris et al., 2017; Saini et al., 2019). By the toddler and preschool years, mealtime challenges are a common parental concern for many families, but for parents of children with ASD who do not consume well-rounded meals, mealtimes are reported to be stressful, chaotic, lack positive interactions, and guilt laden (Marquenie et al., 2011). Families of children with IDD report mealtime challenges related to issues with restricted food preferences, texture and temperature sensitivities, and health conditions (Grumstrup & Demchak, 2019). These challenges may be driven by sensory processing as well as motor and behavioral issues that can cause food selectivity. Food selectivity has been defined differently among health-care professionals, especially in regard to patients with ASD. Food selectivity in ASD is different from "picky eating" since food selectivity is chronic and most likely involves aberrant mealtime behaviors caused by undiagnosed medical issues, poor sleep quality, sensory processing issues, and/or motor impairments (Withrow & Alvidrez, 2019). Sharp and colleagues (2013) published a meta-analysis reporting that children with ASD between birth and 18 years had a fivefold increase in problematic feeding issues compared to typically developing children. In addition, they found that children with ASD preferred high carbohydrate diets, snacks/fat, and/or processed foods while rejecting fruits and vegetables (Sharp et al., 2013). In 2016, an operationalized definition of food selectivity was developed to assist health-care professionals in categorizing severity. See Table 9.1. The World Health Organization has highlighted the importance of a multidisciplinary team approach to treatment in order to address the four domains of feeding disorders: medical, nutritional, feeding skills, and psychosocial issues (Goday et al., 2019).While the nutritional assessment of children with developmental disability is informed by diagnoses or syndromes, the evaluation is approached with a patient- and family-centered focus. Nutritional issues can be complex and are often multifactorial in etiology. If not addressed early, it can impact a child's health, development, and general well-being. The pediatric RDN offers expertise across these domains to inform the health-care team and medical home team. The RDN assesses all areas that impact eating while working with other health-care providers to establish the best treatment plan for the child. RDNs are trained to assess, educate, and monitor their patients' nutritional risk and communicate with primary caregivers and health-care providers. The role of an RDN may be slightly different with each patient, but at a minimum

Table 9.1 Categorization of food selectivity (Sharp et al. (2017), Clinical Handbook of Complex and Atypical Eating Disorders

Category	Criteria	Rationale
Severe food selectivity	Complete rejection of one or more food groups	Increases the risk of micro- and/or macronutrient deficiency (e.g., scurvy; iron deficiency anemia; kwashiorkor)
	Accepts five or fewer total food items	Further narrowing of the diet would eliminate additional food groups
Moderate food selectivity	Consumes two or fewer items in one or more food groups	Reflects a diet that may lack diversity of nutrient-dense foods: Further restriction increases likelihood of nutrient deficiency
	Regularly (weekly) accepts at least one item across the five food groups	Decreases likelihood of being diagnosed with a nutrient deficiency; however, intake may be limited to a handful of preferred items or involve high intake of a single food group
Mild food selectivity	Diet involves at least three or more items from each good group (15 total foods); more than half of items fall into one food group	Suggests low probability of nutrient deficiency while recognizing child may show preference for a certain food group (e.g., grains)
	Consistently (daily) accepts foods from all five food groups	Indicates the child maintains a consistent degree of dietary diversity

they should be involved while a child is being diagnosed with ASD or other IDDs since problematic feeding issues are common in approximately 90% of children with special needs (Withrow & Alvidrez, 2019). The treatments provided by a RDN vary widely, and this chapter will explore the treatment of feeding problems in children with ASD.

Overview of Feeding Problems in Children with ASD

Feeding problems are common in childhood; they reportedly occur in 25–35% of typically developing children and up to 89% of children with ASD (Sharp et al., 2010). It appears that feeding problems in children with ASD may be more problematic and long-standing (Seiverling et al., 2018). Feeding problems in children with ASD include food refusal, food selectivity, rituals related to food and mealtime, sensory processing aversions, oral motor impairments, and aberrant mealtime behaviors (Williams et al., 2000). These unusual eating habits can interfere with the child's ability to consume an adequate amount and variety of foods necessary for healthy growth and development. It has been suggested that there may be nutrient insufficiency in children with ASD compared to typically developing children (Elizabeth Cornish, 1998; Herndon et al., 2009; Schreck et al., 2004). Some evidence of nutrient imbalances including increased homocysteine and decreased folate and vitamins B_6, B_{12}, and D has been reported (Ranjan & Nasser, 2015). Additional research is needed to fully understand these issues. What it is not well understood is how feeding problems and dietary intake impact healthy growth and development as both overweight and underweight have been identified as concerns for children with ASD.

The core symptoms of ASD include persistent impairment of social behavior and a pattern of restricted or repetitive behaviors. Manifestations of restricted behavior may include unique sensory reactivity including oral and taste sensitivity (American Psychiatric Association. & American Psychiatric Association. DSM-5 Task Force., 2013). In addition, motor development and motor impairments are increasingly being suggested as an early marker for ASD and a significant influence on the development of the core features (Wilson et al., 2018). For example, impaired imitation of motor actions and fine motor, gross motor, and postural delays (Bhat et al., 2011) may make self-feeding difficult and make mealtimes a challenge for older children (Smile et al., 2021).

A Sensory Processing Connection to ASD and Nutritional Risk

Researchers began investigating sensory processing theories in children with ASD during the 1970s (Ayres & Tickle, 1980; Baranek et al., 1997). Ayres studied children with ASD and found that they often suffered with differences in processing sensory input which sometimes led to undesirable behaviors (Ayres, 1979; Ayres & Tickle, 1980). Parents have often reported unusual sensory responses in their children with ASD, and research has suggested that 30–100% of children with ASD experience some type of sensory processing dysfunction (Bandini et al., 2010; Birch et al., 1987; Greenspan & Weider, 1997; Leekam et al., 2007). A recent systematic review found a significant association between feeding difficulties and impaired sensory processing in all 11 studies examined with atypical oral oversensitivity correlated with food refusal, especially refusing fruits and vegetables (Page et al., 2021).

Four hypotheses pertaining to children with ASD and sensory processing can be characterized as overarousal, under-arousal, perceptual inconstancy, and impaired cross-modal processing. Children who suffer with overarousal sensory dysfunction appear to be easily aroused and reactive to sensory stimuli. They may also appear to be slower to adapt to stimuli in the environment when compared to typically developing children (Rogers et al., 2003). Under-arousal theories have been discussed by Rimland, founder of the Defeat Autism Now! (DAN!) movement (Rimland, 1964). Rimland (1964) suggested that children with ASD have an impaired ability to make an association between past and present experiences, thus preventing learning and generalization that can contribute to a lack of typical responses to stimuli. The perceptual inconstancy theory was developed by Ornitz and Ritvo (1968). These researchers suggested that children experience abnormalities in perceptual integration and processing motility patterns. They suggested that children with ASD have abnormal states of arousal due to brainstem abnormalities, which can result in over-excitation and/or over-inhibition. The cross-modal theory suggests that children with ASD may have abnormalities in the hippocampus region of the brain. These abnormalities can cause failure to process incoming sensory information from the same event in an appropriate way (Brock et al., 2002).

Eating is an activity that may be negatively affected by sensory processing difficulties (Ayres & Tickle, 1980; Cermak et al., 2010). Numerous parent and professional reports of children with ASD state that sensory factors, such as smell, taste, texture, color, and temperature, can affect whether a child will consume food (Cermak et al., 2010; Marí-Bauset et al., 2014). Some researchers have speculated that sensory sensitivities cause an increase in food selectivity in children with ASD. Children with ASD often experience hypersensitivity in and around the mouth (Schwarz, 2003). This can lead to the experience of spitting, coughing, gagging, or vomiting when trying new foods. It has been postulated that over time those experiences can limit the amount of nutritional intake, restrict the variety of foods, and create a negative interaction between the child and caregiver (Bernard-Bonnin, 2006; Case-Smith & Bryan, 1999; Harris et al., 2000). Oral defensiveness, which can be a part of tactile defensiveness, can be defined as someone avoiding certain textures of foods and activities around the mouth (Cermak et al., 2010). As discussed, sensory sensitivities may contribute to an increase in food selectivity in children with ASD. The texture and type of foods have been consistently reported to affect food intake in children with ASD (Minshew & Hobson, 2008).

A Motor Connection to ASD and Nutritional Risk

A wide range of motor impairments have been associated with ASDs. Children with ASD may have difficulty feeding because of oral motor impairments (Field et al., 2003; Şahan et al., 2021). Up to 50–100% of children with an ASD suffer with motor skill impairments (Ghaziuddin et al., 1998; Green et al., 2002; Ming et al., 2007). The DSM-IV does not include motor impairments as a core deficit of ASDs, but research has demonstrated that children with ASD often suffer with motor impairments (Fulceri et al., 2019) thus potentially impacting their ability to eat. Oral motor impairments include tongue thrusting, having a weak suck, poor lip closure, inability to pucker lips, blow bubbles, etc. (Geraghty et al., 2010).

In 1976, DeMyer highlighted motor impairments as fundamental to the expression of autism along with the social and language problems and restricted movement that define ASDs (DeMyer, 1976). In the past decade, it has been recognized that motor impairments can assist in identifying an ASD. As Mostofsky and colleagues point out, motor impairments in ASD offer information about the neurological basis of the disorder (Mostofsky et al., 2007). Motor symptoms are some of the earliest observable behaviors that are considered reliable in helping diagnose an ASD. Rather than consider motor impairments as a comorbid condition, several professionals are viewing motor differences as integral to understanding, diagnosing, and treating the symptoms of an ASD (Amato & Slavin, 1998; Bhat, 2020; Bhat et al., 2011; Gernsbacher et al., 2008).

Assessment of oral-motor and fine motor skills provide helpful information to guide the approach of intervention, although the science is not yet at a point where clear answers can be given to what motor features may or may not be included within any of the disorders that make up the spectrum. Oral motor impairments may impact feeding and eating which can lead to difficulty in using a cup and/or a straw and/or managing foods and liquids in an individual's mouth. Assessment of oral motor and fine motor skills may provide helpful information to create and guide interventions for children suffering with ASD.

A Behavioral Connection to ASD and Nutritional Risk

According to research higher rates of aberrant mealtime behaviors exist among children with ASD (Withrow & Alvidrez, 2019). Problematic eating behaviors include difficulty sitting for meals, strong preference for brands and packaging, more grazing, self-injurious behaviors when novel foods are presented, and more rigidity and inflexibility around where to eat, when to eat, what utensils to use, etc. Problematic eating behaviors appear to be long-standing and can have implications for nutritional risk if not addressed in early years.

Dietary Intake in Children with ASD

Dietary intake in children with ASD appears to differ from that of typically developing children (Elizabeth Cornish, 1998; Eaves & Ho, 2008; Herndon et al., 2009; Schreck et al., 2004). Even though dietary intake appears to differ between children with ASD compared to typically developing children, the research has reported conflicting results (Zimmer et al., 2012). Some researchers have suggested that children with ASD consume fewer vegetables, salads, and fresh fruit when compared to same aged typically developing children (Emond et al., 2010). Withrow and Hsueh (2019) compared MyPlate food group intakes in children with ASD ($n = 10$) between 2 and 8 years of age compared to typically developing children ($n = 27$). No differences were found in total calorie intake, but children with ASD consumed less variety in vegetables and protein. Mean differences showed that children with ASD showed less variety in all food groups when compared to typically developing children. This study supported previous findings that children with ASD exhibit more food selectivity and consume a limited variety of foods within groups (Withrow, 2019).

A study conducted by Herndon et al. (2009) investigated the nutrient intake of 46 children with ASD and 31 children who were typically developing. Parents were trained and asked to complete a 3-day food record for their enrolled child. Results from 3-day food records illustrated that children with ASD consumed significantly more nondairy proteins and significantly less dairy food items ($p < 0.05$) when compared to the typically developing children (Herndon et al., 2009). The children with ASD also consumed less calcium but more vitamin B6 and vitamin E. These results supported other existing data (Cornish, 1998; Schreck et al., 2004) about nutrient intake in children with ASD (Herndon et al., 2009). A subgroup of children with ASD who were consuming the gluten- and casein-free diet were evaluated as well. Findings illustrated that children on the GFCF diet consumed more vitamin E, but there were no other statistically significant differences between children with ASD consuming a typical diet compared to the children with ASD who were consuming a GFCF diet (Herndon et al., 2009).

Referring Children to the RDN

Comprehensive Screening

There are minimal validated comprehensive nutritional risk screening inventories for children with ASD. To my knowledge there is only one that encompasses the primary domains that impact eating, the SAMIE, but none that encompass both medical and behavioral issues that impact eating in children with IDD and ASD. Because nutritional complexity is common for children with ASD, the routine medical surveillance by medical home team will screen for feeding issues. Should concerns be identified, the Medical/Nutrition, Oral-Motor, Behavior, Sensory, and Environmental (MOBS[E)] approach© is a recognized approach to

The Sensory Processing, Aberrant Mealtime Behaviors, Motor Inventory for Eating

For the following questions, please check the number that best represents typical characteristics during the **past 3 months.**

1 NEVER: Has not occurred	2 RARELY: 25% of the time	3 SOMETIMES: 50% of the time	4 USUALLY: 75% of the time	5 ALWAYS: 100% of time

	NEVER →				ALWAYS
SENSORY PROCESSING:	**1**	**2**	**3**	**4**	**5**
1. Gags easily after he/she sees and/or smells food	☐	☐	☐	☐	☐
⚠ **Note:** Score of 3 or greater indicates need for referral.					
2. Smells food that is new or different	☐	☐	☐	☐	☐
3. Avoids colorful foods	☐	☐	☐	☐	☐
SENSORY PROCESSING, CONT.:	**1**	**2**	**3**	**4**	**5**
4. Only eats specific brands of food	☐	☐	☐	☐	☐
• **Example:** McDonald's french fries, PF Goldfish crackers, Danimals yogurt, GM Cheerios, etc.					
5. Avoids certain foods due to smell	☐	☐	☐	☐	☐
6. Prefers specific certain textures	☐	☐	☐	☐	☐
• **Example:** Crunchy, chewy, juicy, soft textures, etc.					
7. Prefers foods at certain temperatures	☐	☐	☐	☐	☐
8. Tries new foods	☐	☐	☐	☐	☐
9. Picks off toppings/crust on bread	☐	☐	☐	☐	☐
• **Example:** Will not eat toppings/crust, etc.					
10. Refuses a familiar food item if not prepared the same way	☐	☐	☐	☐	☐
11. Dips most food into a sauce	☐	☐	☐	☐	☐
• **Example:** Ketchup, ranch, Tabasco, etc.					
12. Prefers that foods don't touch each other on the plate	☐	☐	☐	☐	☐

Fig. 9.1 Example of a domain from the SAMIE. Withrow and Alvidrez (2019)

guide the medical home team (Smile et al., 2021). The MOBS^E conceptual framework is used as a systematic evaluation of factors contributing to problematic eating and factors that should be prioritized as areas to treat. The MOBS^E approach may be augmented with the newly developed comprehensive eating screening inventory, the Sensory Processing, Aberrant Mealtime Behaviors, Motor Inventory for Eating (SAMIE) (Withrow & Alvidrez, 2019). See Fig. 9.1.

The SAMIE can be used by health-care providers and primary caregivers to screen nutritional risk and simplify referrals. The SAMIE screens for nutritional risk by identifying the four primary domains that affect eating in individuals with ASD – aberrant mealtime behavior, eating skills, dietary intake, and sensory processing. The SAMIE can be easily completed within 10 minutes and provide the medical home team and all health-care providers with comprehensive information regarding the primary areas of concern affecting eating. In addition to the domains identified by the SAMIE and the MOBS^E, the Brief Autism Mealtime Behavior Inventory (BAMBI) can be used to reinforce the SAMIE findings. The BAMBI was originally developed to identify behavioral issues identified during mealtimes (Lukens & Linscheid, 2008). The BAMBI is completed by primary caregivers and consists of 18 questions that can easily be completed in 10 minutes.

While sensory processing, motor, and behavioral issues impact eating in children with IDD/ASD, undiagnosed or untreated medical issues can be concerning and impact eating. Children with ASD have an increased prevalence of gastrointestinal (GI) issues when compared to typically developing children (Celia et al., 2016). Due to the complexities of feeding and GI issues in children with ASD, a standard of practice was necessary. Therefore, an expert panel developed a guidance in the form of an algorithm (see Fig. 9.2) for the nutritional management of GI symptoms in children with ASD (Berry et al., 2015). The algorithm describes 11 steps in the nutrition evaluation process and management of GI concerns in children with ASD to guide development and application of a prescriptive diet and treatment plan.

Preparing the Family for an RDN Appointment

To prepare for an outpatient nutritional assessment, primary caregivers should have their child's most recent blood work. There are several labs that an RDN can refer to during the nutrition assessment, but minimal labs are vitamin D, calcium, a total iron study (total iron-binding capacity, transferrin, ferritin, unsaturated iron-binding capacity levels), complete blood count, lipid profile if overweight, and zinc levels. The RDN may request that the primary caregiver records all food and drink intake for 3 days, also known as a 3-day food record. The 3-day food record is a commonly used measure to assess dietary intake in clinical practice and research (Lanigan et al., 2004). Ziegler et al. found that families of children with an ASD were more successful in completing the 3-day food record when compared to a 24-hour recall (Zeigler, 2006). The 3-day food record allows primary caregivers to use measuring tools and visual aids to improve accuracy. The 3-day food record does not rely on memory, and primary care givers can plan meals in advance that will improve accuracy in recording the amount of food that their child has consumed. The 3-day food record approach has been found to have fewer errors and to be more representative of actual dietary intake when compared to a 5-day food frequency and a 24-hour recall in adults (Crawford et al., 1994). The 3-day food record should consist of two weekdays and at least one weekend day (e.g., Thursday, Friday, and Saturday) and consecutive days. See the sample instructions for caregivers (Fig. 9.3).

If the child is extremely food selective, a 3-day food record may not be necessary and the primary care giver can write a list of commonly eaten foods with typical serving sizes, brand names, and frequency of consumption. This will allow the RDN to analyze dietary intake and provide an estimate of macro- (carbohydrates, fat, and protein) and micro-nutrient (vitamins and minerals) intake.

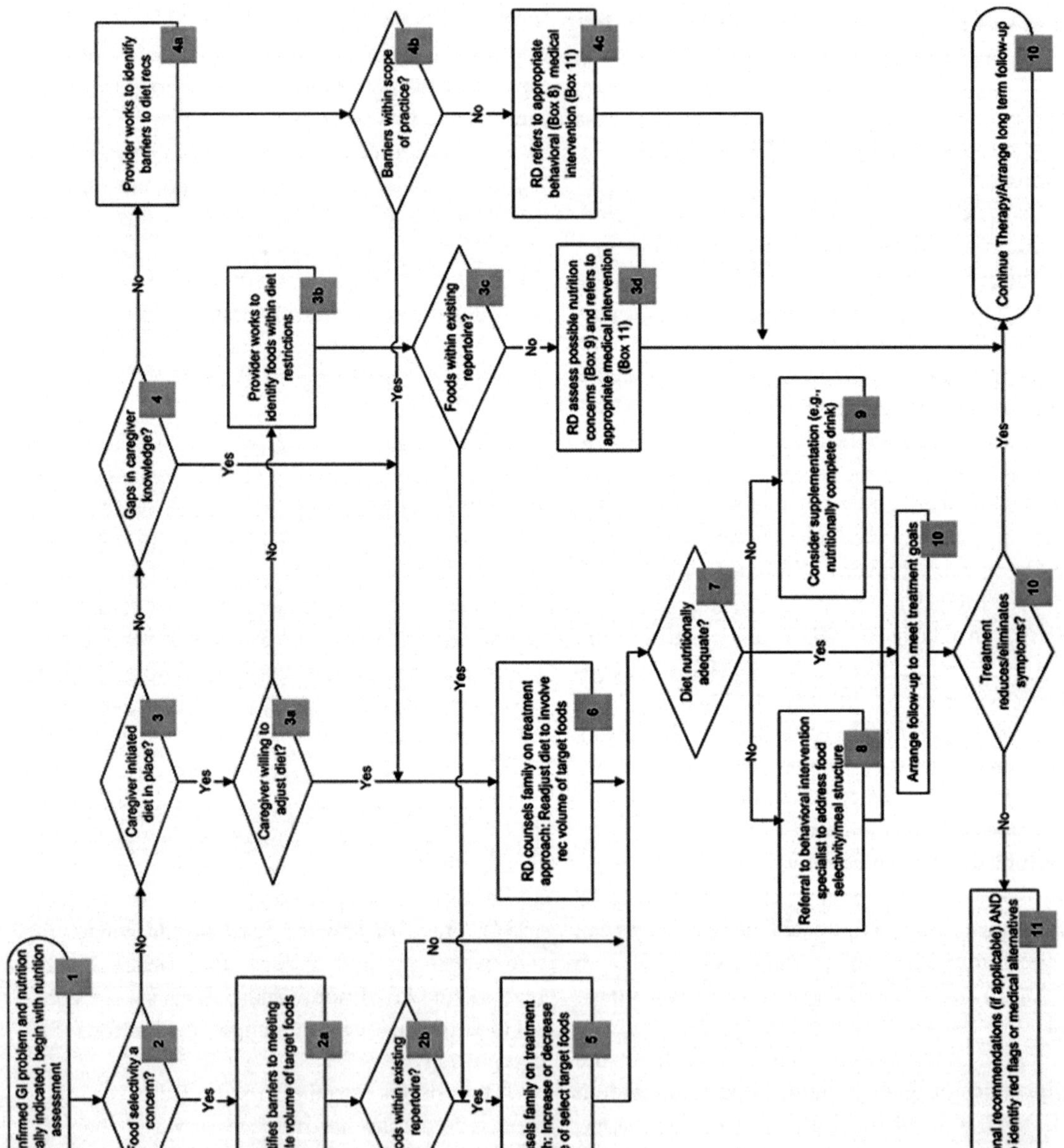

Fig. 9.2 Nutrition management of gastrointestinal symptoms in children with autism spectrum. Berry et al. (2015)

1. It is important to write down everything your child eats and drinks for 3 days. This includes water, all medications, and all supplements.
2. While completing this record, feed you child as you typically would.
3. Record on 2 weekdays and 1 one weekend day.
4. Record the date and time that the food, drink, medication, or supplement is taken.
5. For accuracy, it is best to record what your child has eaten immediately after each meal or snack.
6. Measure or weigh the amount of food eaten as accurately as possible.
7. Describe how the food was prepared (baked, fried, grilled, etc).
8. Be as specific as possible when describing the food (brand, fresh or frozen, etc.)
9. No amount is too small, list everything your child eats or drinks.

Fig. 9.3 Food record instructions for caregivers

What to Expect at the Appointment

A typical initial nutrition assessment typically lasts 60 minutes and will consist of obtaining a current weight and height so a body mass index (BMI) can be calculated along with a Nutrition Focused Physical Exam (NFPE) if the child permits. A NFPE is one component of the nutrition assessment that can provide necessary data to diagnose malnutrition and severity or identify risk of malnutrition. The RDN will review and discuss relevant nutritional labs with the child and primary caregiver. In addition, the RDN will most likely conduct a 24-hour food recall to assist in validating the 3-day food record and to capture potential changes in dietary intake. The RDN will analyze the 3-day food record and/or the 24-hour food recall to determine potential dietary risks. If food selectivity is a problem, the RDN can categorize the food selectivity as severe, moderate, or mild which will assist in providing recommendations for the child (Sharp et al., 2017). Severe food selectivity is defined as a complete rejection of one or more food groups and accepts five or fewer total food items. Moderate food selectivity is when a child consumes two or fewer items in one or more food groups and regularly (weekly) accepts at least one item across the five food groups and finally mild food selectivity involves at least three or more items from each food group (15 total foods) and consistently (daily) accepts foods from all five food groups. These categorizations have a prescriptive rationale for intervention. If a child is considered to be severely food selective, the RDN will recommend an intensive multidisciplinary intervention while a moderate food selectivity would be treated as outpatient/home intervention and a mild food selectivity would be less intense and the MyPlate or Ellyn Satter guidelines for introducing food can be used to educate the primary care givers and child. After discussing dietary intake and physical activity routines, the RDN will provide the child and primary caregiver recommendations to improve dietary intake and minimize nutritional risk. There will most likely be follow-up appointments similar to the initial consultation especially if a feeding intervention is necessary due to food selectivity.

The RDNs Treatment of Children with ASD

After the initial nutrition assessment, primary caregivers can expect the RDN to make recommendations to improve dietary intake and to reduce nutritional risk. Data collected during the nutrition assessment guides the RDN in selecting the appropriate nutrition diagnoses and recommendations. The RDN creates the nutrition intervention(s) that focus on remediating nutrition risk. Primary caregivers may be given nutrition handouts/informatics to help reinforce the information discussed during the appointment. Also, primary caregivers will be given short and long-term goals to improve their child's dietary intake. It is essential to educate parents that it may take a typically developing child up to 25 exposures of a new food before they consume it but double or triple that in a child with special needs. Due to food selectivity in children with ASD, the RDN needs to make recommendations based on the primary issue preventing a child from eating a variety of foods and to decrease nutritional risk. If a feeding intervention is needed, the RDN will discuss different methods and the primary caregiver will select the one that best fits their child. The two most popular feeding interventions are food chaining and the preferred and nonpreferred feeding method which are explained below. These two feeding interventions are implemented at home and can be easily modified by the RDN during subsequent appointments. Whichever intervention is selected, it should involve the primary caregiver and child. They should be able to decide on foods to target, how quickly to introduce new foods, how big a novel bite size is, and length of time to sit for meals. With more "buy in" from the child and parent, the more likely they will adhere to the feeding intervention. Parents are often supported separately from the child since they can be anxious about implementation; therefore checking in and coaching parents are critical.

Food chaining is a strategy that can help children consume new foods or foods they dislike by considering their food and taste preferences. The goal is to create "chains" between foods that are readily consumed while adding similar foods or changing food preparation so that the "targeted" food is similar in taste, texture, color, or brand. This method is successful when primary caregivers are consistent and do not make massive changes. For example, McDonald's French fries > purchase shoestring French fries and serve a couple when the child is consuming McDonald's French fries. Encourage the child to touch the different French fries and smell, lick, and bite while pairing with language as "same but different." It is okay if the child takes a bite and spits it out. Once the child starts consuming store-bought French fries, the McDonald's French fries are faded out, and a new food replaces it. Another example of food chaining is starting with Cheetos then introducing Cheetos puffs then try another brand of cheese puffs then introduce chickpea puffs then crunchy chickpeas and then chickpeas. There is no right or wrong way and most likely everyone will do things differently, but the end goal would be the "whole food" version of what the child prefers and move away from the "snack foods." Depending on the severity of food selectivity, the primary caregiver may want to "change" the shape instead of another type of potato, such as a steak fry or crinkle. If the child is receptive to different shapes and sizes of French fries, the primary caregiver can prepare French fries from a potato, and then the end goal is to have the child consume a fresh-baked potato. The preferred and nonpreferred method for feeding is to offer an accepted food, preferably a bite-size, and then offer a nonpreferred food item and have the child try it before offering a bite-size of a preferred food. This method requires a compliant child and is not always successful if the child experiences severe food selectivity. After the initial nutrition assessment, it is important that the RDN is monitoring and evaluating progress toward the specific goals. Often goals will need to be changed depending on the child.

Obesity

It is well known that obesity has dramatically increased over the past 20 years and is considered an "epidemic" and a "public health crisis" and can affect all children (Ogden et al., 2014). The prevalence of childhood obesity has tripled over the past two decades and has been estimated to affect approximately 8.4–20.5% of US children, ages 2–19 years, and the prevalence of overweight in children is approximately 34.5% (Ogden et al., 2014). According to Curtin et al. (2010), the prevalence of obesity in children with ASD was 30.4% compared to 24% in typically developing children. There is a small amount of evidence suggesting that children with ASD may be at a greater risk of obesity when compared to typically developing children (Curtin et al., 2005). It has been hypothesized that this may be due to the limited variety and consumption of healthy foods along with the lack of physical activity in children with ASD (Curtin et al., 2010).

Gastrointestinal Dysfunction

Interest in the connection between gastroenterology and ASDs is not new as it has been established that gastrointestinal disorders are more common in children with neurological disorders (Melmed et al., 2000). As early as the 1970s, ASDs and gastrointestinal (GI) symptoms were reported (Goodwin et al., 1971) with the goal of identifying a subtype of autism. Reports of the prevalence of GI symptoms in children with ASD vary from 9% to 91% (Buie et al., 2010). Possible explanations for gastrointestinal symptoms in children with ASD are multifactorial. GI symptoms in children with ASD could be due to genetics, immunologic factors, and embryologic and/or neurological factors. Gastrointestinal problems encompass a wide range of symptoms. The most common GI symptoms reported for children with ASD are reflux, chronic gastritis, abdominal pain, distention, food intolerance, food selectivity, constipation, and diarrhea (Erickson et al., 2005; Levy et al., 2007). It has been hypothesized that a subgroup of children with ASD do not produce and/or efficiently utilize the digestive enzymes necessary to break down certain proteins such as gluten and casein. This incomplete digestive process could potentially leave undigested peptides in the small intestine (Cass et al., 2008) and cause GI symptoms.

Another possible reason for GI dysfunction in children with ASD are food sensitivities (Elder et al., 2015). Food sensitivities can damage different tissues, and, as such, food sensitivities are thought to stem from an inability to digest food efficiently thus leading to inflammation and gut injury (Geir et al., 2020). Gluten and dairy are proposed to be the most common food sensitivities in children with ASD (Jyonouchi, 2009). A food sensitivity is suggested to have negative consequences in children with ASD and is associated with irritability, food refusal, sleep disturbances, constipation, and diarrhea in children with ASD (Leader et al., 2020; Levy et al., 2007).

Sleep Disturbances

Sleep disturbances are common in children with ASD with reports as high as 87% experiencing challenges (Cavalieri, 2016). Of particular interest to the RDN is restless leg syndrome. If the child has difficulties sleeping, it is important that an iron study is ordered. Restless leg syndrome has been correlated to low iron levels and if repleted sleep can be restored.

Dietary Treatment of ASD

There is much controversy and debate in the dietary treatment of ASDs since limited research is available and there is no cure at this point in time. Although no cure exists, there are numerous treatment options. Treatments have included educational, developmental, psychological, dietary, and pharmacological treatments, complementary and alternative medicine, occupational therapy, and speech and language therapy. According to the Green et al. (2004) survey of 552 parents, the mean number of interventions that families participated in at a given time for their child was seven (Green et al., 2004). Speech therapy was the most common type of intervention, followed by visual schedules, applied behavior analysis, and occupational therapy (Green et al., 2004). There has been general agreement within the professional community that the best type of treatment involves intensive developmental therapies, special education, and behavioral management (Volkmar et al., 2014).

Within the last decade dietary interventions have become increasingly popular among families of children with an ASD. Survey results indicate that 15% to 38% of families have tried and/or are currently using a dietary intervention to help treat symptoms of an ASD (Green et al., 2004). Most dietary interventions used to treat symptoms of an ASD involve eliminating at least one or more types of food from the child's diet such as wheat (gluten), milk (casein), soy, yeast, additives, sugar, eggs, and yeast (Cornish, 2002). The most common dietary intervention is a gluten- and casein-free (GFCF) diet. Families receive information regarding diet and nutritional therapies from other parents, Internet sites, unpublished sources, autism organizations, and complementary and alternative medical providers (Arnold et al., 2003). Many dietary interventions have shown little, if any, evidence supporting or refuting their efficacy and effectiveness; however, they continue to gain in popularity. Also, there is a small body of evidence linking the GFCF diet to suboptimal bone development, specifically reduced bone cortical thickness (Mulloy et al., 2010). Some dietary interventions also add additional stress for the family due the financial burden, time commitment, and possible increase in social isolation due to food restrictions.

Today, the use of complementary and alternative medicine (CAM) has steadily increased in western society, especially in the pediatric population. CAM interventions have been reported to occur at a rate as high as 85% in families who are treating a child with autism (Wong, 2009). Despite a lack of evidence-based trials, these treatments continue to gain in popularity among families and some providers. There are several types of CAM therapies being used to treat ASDs such as dietary interventions, biomedical, vitamins and minerals, chelation, and music therapies, cranial sacral, acupuncture, hyperbaric chambers, etc. Parents have discovered CAM interventions through the Internet, media, anecdotal reports, and autism support organizations (Levy & Hyman, 2002). The primary goal of CAM interventions is to prevent or treat illnesses "naturally," to promote a person's overall well-being, and to remediate symptoms of an ASD (Wong, 2009). Still, these approaches are not without risk. For example, chelation has been associated with renal impairment and death (James et al., 2015). The gluten-free, casein-free diet (GFCF) is the most common dietary CAM treatments by caregivers of children with ASD. Up-to-date information on CAM can be accessed at the National Center for Complementary and Integrative Health website (https://www.nccih.nih.gov/).

Conclusion

Few studies have been able to fully characterize the nutritional deficiencies and body mass index (BMI) status in children with ASD who are food selective. Food selectivity in children and adolescents with ASD and IDD can have serious outcomes if left untreated. Nutritional deficiencies can lead to growth impairments, immune dysfunction, sleep disturbances, increased risk of infections, and mortality (Allen, 2006). Conflicting results about nutritional status in children with ASD have been reported (Herndon et al., 2009; Emond et al., 2010; Levy et al., 2007; Schreck et al., 2004). Some research has stated dietary intakes of children with ASD do not differ from typically developing children (Emond et al., 2010; Levy et al., 2007), while other investigators have reported that children with ASD are not meeting dietary reference intakes (DRIs) for various nutrients (Herndon et al., 2009; Wei et al., 2010). Due to the high frequency of food selectivity in children with ASD and other developmental disabilities, it is important to monitor weight, height, and dietary intake. Prevention of nutritional risk is of the utmost importance since healthy growth and development needs to occur.

Further research is needed to look at factors that may contribute to food selectivity such as sensory processing impairments, oral motor difficulties, maladaptive behaviors, mealtime environment, and parental influence. Greater insight into why children with an ASD suffer with food selectivity is needed to develop appropriate nutritional interventions. RDNs should be involved at time of diagnosis and throughout the treatment of ASD and other developmental disabilities. RDNs value working closely with the child, primary caregiver, and support team so that nutritional deficiencies are avoided, and the best possible growth and development can occur.

If you are working with or raising children, it is important to understand that greater than 90% of children with ASD or IDD suffer with problematic eating for several reasons. It is vital that this is addressed so they can avoid nutritional risk and experience healthy growth and development. If someone does not know where to find a RDN, they can discuss with their MD and also contact eatright.org to local RDNs. Problematic eating can be treated, but it requires consistent and reasonable goals that can be developed by an RDN.

References

Allen, L. (2006). *Guidelines on food fortification with micronutrients*. World Health Organization.

Amato, J., & Slavin, D. (1998). A preliminary investigation of oromotor function in young verbal and nonverbal children with autism. *Infant-Toddler Intervention: The Transdisciplinary Journal, 8*, 175.

American Psychiatric Association., & American Psychiatric Association. DSM-5 Task Force. (2013). *Diagnostic and statistical manual of mental disorders: DSM-5* (5th ed.). American Psychiatric Association.

Arnold, G. L., Hyman, S. L., Mooney, R. A., & Kirby, R. S. (2003). Plasma amino acids profiles in children with autism: Potential risk of nutritional deficiencies. *Journal of Autism and Developmental Disorders, 33*(4), 449–454.

Ayres, A. J. (1979). Sensory integration and the child (Western Psychological Services, Los Angeles). *CPSC, 2000*(1), 1990–1994.

Ayres, A. J., & Tickle, L. S. (1980). Hyper-responsivity to touch and vestibular stimuli as a predictor of positive response to sensory integration procedures by autistic children. *American Journal of Occupational Therapy, 34*(6), 375–381.

Bandini, L. G., Anderson, S. E., Curtin, C., Cermak, S., Evans, E. W., Scampini, R., … Must, A. (2010). Food selectivity in children with autism spectrum disorders and typically developing children. *The Journal of Pediatrics, 157*(2), 259–264.

Banfield, E. C., Liu, Y., Davis, J. S., Chang, S., & Frazier-Wood, A. C. (2016). Poor adherence to US dietary guidelines for children and adolescents in the national health and nutrition examination survey population. *Journal of the Academy of Nutrition and Dietetics, 116*(1), 21–27. https://doi.org/10.1016/j.jand.2015.08.010

Barabási, A.-L., Menichetti, G., & Loscalzo, J. (2020). The unmapped chemical complexity of our diet. *Nature Food, 1*(1), 33–37.

Baranek, G. T., Foster, L. G., & Berkson, G. (1997). Tactile defensiveness and stereotyped behaviors. *American Journal of Occupational Therapy, 51*(2), 91–95.

Bernard-Bonnin, A.-C. (2006). Feeding problems of infants and toddlers. *Canadian Family Physician, 52*(10), 1247–1251.

Berry, R. C., Novak, P., Withrow, N., Schmidt, B., Rarback, S., Feucht, S., … Sharp, W. G. (2015). Nutrition management of gastrointestinal symptoms in children with autism spectrum disorder: Guideline from an expert panel. *Journal of the Academy of Nutrition and Dietetics, 115*(12), 1919–1927.

Bhat, A. N. (2020). Is motor impairment in autism spectrum disorder distinct from developmental coordination disorder? A report from the SPARK study. *Physical Therapy, 100*(4), 633–644.

Bhat, A. N., Landa, R. J., & Galloway, J. C. (2011). Current perspectives on motor functioning in infants, children, and adults with autism spectrum disorders. *Physical Therapy, 91*(7), 1116–1129.

Birch, L. L., McPheee, L., Shoba, B. C., Steinberg, L., & Krehbiel, R. (1987). "Clean up your plate": Effects of child feeding practices on the conditioning of meal size. *Learning and Motivation, 18*(3), 301–317.

Buie, T., Campbell, D. B., Fuchs, G. J., Furuta, G. T., Levy, J., VandeWater, J., … Beaudet, A. L. (2010). Evaluation, diagnosis, and treatment of gastrointestinal disorders in individuals with ASDs: A consensus report. *Pediatrics, 125*(Suppl. 1), S1–S18.

Case-Smith, J., & Bryan, T. (1999). The effects of occupational therapy with sensory integration emphasis on preschool-age children with autism. *American Journal of Occupational Therapy, 53*(5), 489–497.

Cavalieri, A. (2016). Sleep issues in children with autism spectrum disorder. *Pediatric Nursing, 42*(4), 169–174.

Celia, T., Freysteinson, W. W., & Frye, R. E. (2016). Concurrent medical conditions in autism Spectrum disorders. *Pediatric Nursing, 42*(5), 230–234.

Cermak, S. A., Curtin, C., & Bandini, L. G. (2010). Food selectivity and sensory sensitivity in children with autism spectrum disorders. *Journal of the American Dietetic Association, 110*(2), 238–246.

Corkins, M. R., Daniels, S. R., de Ferranti, S. D., Golden, N. H., Kim, J. H., Magge, S. N., & Schwarzenberg, S. J. (2016). Nutrition in children and adolescents. *Medical Clinics, 100*(6), 1217–1235.

Cornish, E. (1998). A balanced approach towards healthy eating in autism. *Journal of Human Nutrition and Dietetics, 11*(6), 501–509.

Cornish, E. (2002). Gluten and casein free diets in autism: A study of the effects on food choice and nutrition. *Journal of Human Nutrition and Dietetics, 15*(4), 261–269.

Curtin, C., Anderson, S. E., Must, A., & Bandini, L. (2010). The prevalence of obesity in children with autism: A secondary data analysis using nationally representative data from the National Survey of Children's Health. *BMC Pediatrics, 10*(1), 1–5.

Curtin, C., Bandini, L. G., Perrin, E. C., Tybor, D. J., & Must, A. (2005). Prevalence of overweight in children and adolescents with attention deficit hyperactivity disorder and autism spectrum disorders: A chart review. *BMC Pediatrics, 5*(1), 1–7.

Duffy, V. B., Davidson, A. C., Kidd, J. R., Kidd, K. K., Speed, W. C., Pakstis, A. J., … Bartoshuk, L. M. (2004). Bitter receptor gene (TAS2R38), 6-n-propylthiouracil (PROP) bitterness and alcohol intake. *Alcoholism, Clinical and Experimental Research, 28*(11), 1629–1637. https://doi.org/10.1097/01.alc.0000145789.55183.d4

Eaves, L. C., & Ho, H. H. (2008). Young adult outcome of autism spectrum disorders. *Journal of Autism and Developmental Disorders, 38*(4), 739–747.

Elder, J. H., Kreider, C. M., Schaefer, N. M., & de Laosa, M. B. (2015). A review of gluten-and casein-free diets for treatment of autism: 2005–2015. *Nutrition and Dietary Supplements, 7*, 87.

Erickson, C. A., Stigler, K. A., Corkins, M. R., Posey, D. J., Fitzgerald, J. F., & McDougle, C. J. (2005). Gastrointestinal factors in autistic disorder: A critical review. *Journal of Autism and Developmental Disorders, 35*(6), 713.

Field, D., Garland, M., & Williams, K. (2003). Correlates of specific childhood feeding problems. *Journal of Paediatrics and Child Health, 39*(4), 299–304.

Fulceri, F., Grossi, E., Contaldo, A., Narzisi, A., Apicella, F., Parrini, I., … Muratori, F. (2019). Motor skills as moderators of core symptoms in autism spectrum disorders: Preliminary data from an exploratory analysis with artificial neural networks. *Frontiers in Psychology, 9*, 2683–2683. https://doi.org/10.3389/fpsyg.2018.02683

Gahagan, S. (2012). The development of eating behavior-biology and context. *Journal of developmental and behavioral pediatrics: JDBP, 33*(3), 261.

Geir, B., Lyudmila, P., Maryam, D., Nagwa, A. M., Yuliya, S., Mona, A., & Salvatore, C. (2020). Gastrointestinal alterations in autism spectrum disorder: What do we know? *Neuroscience & Biobehavioral Reviews, 118*, 111.

Geraghty, M. E., Depasquale, G. M., & Lane, A. E. (2010). Nutritional intake and therapies in autism: A spectrum of what we know: Part 1. *ICAN: Infant, Child, & Adolescent Nutrition, 2*(1), 62–69. https://doi.org/10.1177/1941406409358437

Gernsbacher, M. A., Sauer, E. A., Geye, H. M., Schweigert, E. K., & Hill Goldsmith, H. (2008). Infant and toddler oral- and manual-motor skills predict later speech fluency in autism. *Journal of Child Psychology and Psychiatry, 49*(1), 43–50. https://doi.org/10.1111/j.1469-7610.2007.01820.x

Ghaziuddin, M., Weidmer-Mikhail, E., & Ghaziuddin, N. (1998). Comorbidity of Asperger syndrome: A preliminary report. *Journal of Intellectual Disability Research, 42*(4), 279–283.

Goday, P. S., Huh, S. Y., Silverman, A., Lukens, C. T., Dodrill, P., Cohen, S. S., … Gisel, E. (2019). Pediatric feeding disorder: Consensus definition and conceptual framework. *Journal of Pediatric Gastroenterology and Nutrition, 68*(1), 124.

Green, D., Baird, G., Barnett, A. L., Henderson, L., Huber, J., & Henderson, S. E. (2002). The severity and nature of motor impairment in Asperger's syndrome: A comparison with specific developmental disorder of motor function. *Journal of Child Psychology and Psychiatry, 43*(5), 655–668.

Greenspan, S., & Weider, S (1997). Developmental patterns and outcomes in infants and children with disorders in relating and communicating. A chart review of 200 cases of children with ASD. *Journal of developmental and learning disorders, 1*(1): 87–141.

Grumstrup, B. M., & Demchak, M. (2019). Parents of children with significant disabilities describe their Children's eating habits: A phenomenological study. *The Qualitative Report, 24*(1), 113–129.

Gu, X., & Tucker, K. L. (2017). Dietary quality of the US child and adolescent population: Trends from 1999 to 2012 and associations with the use of federal nutrition assistance programs. *The American Journal of Clinical Nutrition, 105*(1), 194–202. https://doi.org/10.3945/ajcn.116.135095

Harris, G., Blissett, J., & Johnson, R. (2000). Food refusal associated with illness. *Child Psychology and Psychiatry Review, 5*(4), 148–156.

Herndon, A. C., DiGuiseppi, C., Johnson, S. L., Leiferman, J., & Reynolds, A. (2009). Does nutritional intake differ between children with autism spectrum disorders and children with typical development? *Journal of Autism and Developmental Disorders, 39*(2), 212–222.

Hingle, M. (2013). Interventions to improve dietary quality in children. In *Diet quality* (pp. 119–134). Springer.

Hoogewerf, M., Ter Horst, H. J., Groen, H., Nieuwenhuis, T., Bos, A. F., & Van Dijk, M. W. G. (2017). The prevalence of feeding problems in children formerly treated in a neonatal intensive care unit. *Journal of Perinatology, 37*(5), 578–584.

James, S., Stevenson, S. W., Silove, N., & Williams, K. (2015). Chelation for autism spectrum disorder (ASD). *Cochrane Database of Systematic Reviews, 5*.

Jyonouchi, H. (2009). Food allergy and autism spectrum disorders: Is there a link? *Current Allergy and Asthma Reports, 9*(3), 194–201.

Kaur, Y., De Souza, R. J., Gibson, W. T., & Meyre, D. (2017). A systematic review of genetic syndromes with obesity. *Obesity Reviews, 18*(6), 603–634.

Kiraly, L. N., McClave, S. A., Neel, D., Evans, D. C., Martindale, R. G., & Hurt, R. T. (2014). Physician nutrition education. *Nutrition in Clinical Practice, 29*(3), 332–337.

Krause, S., Ware, R., McPherson, L., Lennox, N., & O'Callaghan, M. (2016). Obesity in adolescents with intellectual disability: Prevalence and associated characteristics. *Obesity Research & Clinical Practice, 10*(5), 520–530.

Lanigan, J. A., Wells, J. C. K., Lawson, M. S., Cole, T. J., & Lucas, A. (2004). Number of days needed to assess energy and nutrient intake in infants and young children between 6 months and 2 years of age. *European Journal of Clinical Nutrition, 58*(5), 745–750.

Leader, G., Tuohy, E., Chen, J. L., Mannion, A., & Gilroy, S. P. (2020). Feeding problems, gastrointestinal symptoms, challenging behavior and sensory issues in children and adolescents with autism spectrum disorder. *Journal of Autism and Developmental Disorders, 50*, 1401–1410.

Leekam, S. R., Nieto, C., Libby, S. J., Wing, L., & Gould, J. (2007). Describing the sensory abnormalities of children and adults with autism. *Journal of Autism and Developmental Disorders, 37*(5), 894–910.

Levy, S. E., & Hyman, S. L. (2002). Alternative/complementary approaches to treatment of children with autistic spectrum disorders. *Infants & Young Children, 14*(3), 33–42.

Levy, S. E., Souders, M. C., Ittenbach, R. F., Giarelli, E., Mulberg, A. E., & Pinto-Martin, J. A. (2007). Relationship of dietary intake to gastrointestinal symptoms in children with autistic spectrum disorders. *Biological Psychiatry, 61*(4), 492–497.

Makary, M. A., & Daniel, M. (2016). Medical error—The third leading cause of death in the US. *BMJ, 353*, i2139.

Marí-Bauset, S., Zazpe, I., Mari-Sanchis, A., Llopis-González, A., & Morales-Suárez-Varela, M. (2014). Food selectivity in autism spectrum disorders: A systematic review. *Journal of Child Neurology, 29*(11), 1554–1561.

Mennella, J. A., Jagnow, C. P., & Beauchamp, G. K. (2001). Prenatal and postnatal flavor learning by human infants. *Pediatrics, 107*(6), E88. https://doi.org/10.1542/peds.107.6.e88

Ming, X., Brimacombe, M., & Wagner, G. C. (2007). Prevalence of motor impairment in autism spectrum disorders. *Brain and Development, 29*(9), 565–570.

Minshew, N. J., & Hobson, J. A. (2008). Sensory sensitivities and performance on sensory perceptual tasks in high-functioning individuals with autism. *Journal of Autism and Developmental Disorders, 38*(8), 1485–1498.

Morris, N., Knight, R. M., Bruni, T., Sayers, L., & Drayton, A. (2017). Feeding disorders. *Child and Adolescent Psychiatric Clinics, 26*(3), 571–586.

Mostofsky, S. H., Burgess, M. P., & Gidley Larson, J. C. (2007). Increased motor cortex white matter volume predicts motor impairment in autism. *Brain, 130*(Pt 8), 2117–2122. https://doi.org/10.1093/brain/awm129

Mulloy, A., Lang, R., O'Reilly, M., Sigafoos, J., Lancioni, G., & Rispoli, M. (2010). Gluten-free and casein-free diets in the treatment of autism spectrum disorders: A systematic review. *Research in Autism Spectrum Disorders, 4*(3), 328–339.

O'Daniel, M., & Rosenstein, A. H. (2008). Professional communication and team collaboration. In Hughes, R.G. (Ed.), *Patient Safety and Quality: An Evidence-Based Handbook for Nurses*. Rockville, MD: AHRQ Publication.

Ogden, C. L., Carroll, M. D., Kit, B. K., & Flegal, K. M. (2014). Prevalence of childhood and adult obesity in the United States, 2011-2012. *JAMA, 311*(8), 806–814.

Ornitz, E. M., & Ritvo, E. R. (1968). Perceptual inconstancy in early infantile autism: The syndrome of early infant autism and its variants including certain cases of childhood schizophrenia. *Archives of General Psychiatry, 18*(1), 76–98.

Page, S. D., Souders, M. C., Kral, T. V. E., Chao, A. M., & Pinto-Martin, J. (2021). Correlates of feeding difficulties among children with autism spectrum disorder: A systematic review. *Journal of Autism and Developmental Disorders*.https://doi.org/10.1007/s10803-021-04947-4

Pawłowska, K., Umławska, W., & Iwańczak, B. (2018). A link between nutritional and growth states in pediatric patients with functional gastrointestinal disorders. *The Journal of Pediatrics, 199*, 171–177.

Ptomey, L. T., Walpitage, D. L., Mohseni, M., Dreyer Gillette, M. L., Davis, A. M., Forseth, B., … Waitman, L. R. (2020). Weight status and associated comorbidities in children and adults with down syndrome, autism spectrum disorder and intellectual and developmental disabilities. *Journal of Intellectual Disability Research, 64*(9), 725–737. https://doi.org/10.1111/jir.12767

Ranjan, S., & Nasser, J. A. (2015). Nutritional status of individuals with autism spectrum disorders: Do we know enough? *Advances in Nutrition, 6*(4), 397–407. https://doi.org/10.3945/an.114.007914

Rimland, B. (1964). *Infantile autism. New York: Appleton-Century-Crofts.*

Rogers, S. J., Hepburn, S., & Wehner, E. (2003). Parent reports of sensory symptoms in toddlers with autism and those with other developmental disorders. *Journal of Autism and Developmental Disorders, 33*(6), 631–642.

Şahan, A. K., Öztürk, N., Demir, N., Karaduman, A. A., & Arslan, S. S. (2021). A comparative analysis of chewing function and feeding behaviors in children with autism. *Dysphagia, 36*, 993–998.

Saini, V., Kadey, H. J., Paszek, K. J., & Roane, H. S. (2019). A systematic review of functional analysis in pediatric feeding disorders. *Journal of Applied Behavior Analysis, 52*(4), 1161–1175.

Schreck, K. A., Williams, K., & Smith, A. F. (2004). A comparison of eating behaviors between children with and without autism. *Journal of Autism and Developmental Disorders, 34*(4), 433–438.

Seiverling, L., Towle, P., Hendy, H. M., & Pantelides, J. (2018). Prevalence of feeding problems in young children with and without autism spectrum disorder: A chart review study. *Journal of Early Intervention, 40*(4), 335–346.

Sharp, W. G., Jaquess, D. L., Morton, J. F., & Herzinger, C. V. (2010). Pediatric feeding disorders: A quantitative synthesis of treatment outcomes. *Clinical Child and Family Psychology Review, 13*(4), 348–365.

Sharp, W. G., Berry, R. C., McCracken, C., Nuhu, N. N., Marvel, E., Saulnier, C. A., ... & Jaquess, D. L. (2013). Feeding problems and nutrient intake in children with autism spectrum disorders: a meta-analysis and comprehensive review of the literature. *Journal of autism and developmental disorders, 43*(9), 2159–2173.

Sharp, W. G., & Postorino, V. A. (2017). Food selectivity in autism spectrum disorder. *Clinical handbook of complex and atypical eating disorders*, 126–148.

Smile, S., Raffaele, C., & Perlin, R. (2021). Re-imagining the physicians' role in the assessment of feeding challenges in children with autism spectrum disorder. *Paediatrics & Child Health, 26*(2), e73–e77.

Trofholz, A. C., Tate, A., Loth, K., Neumark-Sztainer, D., & Berge, J. M. (2019). Watching television while eating: Associations with dietary intake and weight status among a diverse sample of young children. *Journal of the Academy of Nutrition and Dietetics, 119*(9), 1462–1469.

USDA-Dietary Guidelines for Americans. (2020–2025). *MyPlate*. Dietary Guidelines.gov.

Volkmar, F., Siegel, M., Woodbury-Smith, M., King, B., McCracken, J., & State, M. (2014). Practice parameter for the assessment and treatment of children and adolescents with autism spectrum disorder. *Journal of the American Academy of Child & Adolescent Psychiatry, 53*(2), 237–257.

Williams, P. G., Dalrymple, N., & Neal, J. (2000). Eating habits of children with autism. *Pediatric Nursing, 26*(3), 259.

Wilson, R. B., Enticott, P. G., & Rinehart, N. J. (2018). Motor development and delay: Advances in assessment of motor skills in autism spectrum disorders. *Current Opinion in Neurology, 31*(2), 134–139.

Withrow, N. A., & Alvidrez, L. (2019). The development of the Sensory Processing, Aberrant Mealtime Behavior, Motor Inventory for Eating (SAMIE) in children with an ASD. *Advances in Autism, 5*, 268.

Wong, V. C. N. (2009). Use of complementary and alternative medicine (CAM) in autism spectrum disorder (ASD): Comparison of Chinese and western culture (part a). *Journal of Autism and Developmental Disorders, 39*(3), 454–463.

World Health, O. (2014). *Global nutrition targets 2025: Breastfeeding policy brief.* Retrieved from http://www.who.int/teams/nutrition-and-food-safety/global-targets-2025

Zimmer, M. H., Hart, L. C., Manning-Courtney, P., Murray, D. S., Bing, N. M., & Summer, S. (2012). Food variety as a predictor of nutritional status among children with autism. *Journal of Autism and Developmental Disorders, 42*(4), 549–556.

10 Commonly Encountered Endocrine Problems in Children with Developmental Disabilities

Neslihan Gungor, Karen Johal, and Marcia Rankine

Abstract

Medical problems within the context of pediatric endocrinology are frequently encountered in children with developmental disorders. The pediatric endocrinologist plays a key role in maintaining the healthcare of children with developmental disabilities through the assessment and treatment growth disorders, diabetes, thyroid disorders, obesity, and other endocrine disorders. Once an endocrine disorder is identified, a medical home or other multifaceted approaches to the management of endocrine disorders in children with developmental disabilities promote the best clinical and mental health outcomes. Collaboration among the general pediatrician, pediatric subspecialist, and developmental/mental healthcare team is encouraged as the exchange of information is vital to our patients' health.

Keywords

ASD · Autism · Endocrinology · Developmental disorders · IDD · Disability

N. Gungor (✉) · K. Johal · M. Rankine
Department of Pediatrics, Endocrinology Division, Louisiana State University Health Sciences Center, Shreveport, LA, USA
e-mail: neslihan.gungor@lsuhs.edu

Introduction

Developmental disabilities are a group of conditions resulting from impairments that affect a child's physical, learning, or behavioral functioning (Olusanya et al., 2018; Centers for Disease Control and Prevention, 2020). Early identification of developmental disorders is critical to the well-being of children and their families. As recommended by the American Academy of Pediatrics, developmental surveillance for identification of developmental problems should occur at regular intervals as it promotes further developmental and medical evaluation, diagnosis, and treatment, including early developmental intervention. Children diagnosed with developmental disorders should be identified as children with special health-care needs and chronic condition management initiated. The identification of a developmental disorder and its underlying etiology drives a range of services. Whether delivered in medical settings, the home, or the community, coordination of services is critical. Coordination of care for children with special health-care needs is best accomplished in a medical home. The medical home model promotes comprehensive treatment planning ranging from the medical treatment of the child to family planning for his or her parents (Duby et al., 2006).

Medical problems within the context of pediatric endocrinology are frequently encountered

P. McPherson (ed.), *Handbook of Treatment Planning for Children with Autism and Other Neurodevelopmental Disorders*, Autism and Child Psychopathology Series,
https://doi.org/10.1007/978-3-031-06120-2_10

during the evaluation of childhood developmental disorders. A thorough evaluation should consider biological, environmental, and established risk factors for delayed development. Vision screening and objective hearing evaluation, review of newborn metabolic screening test results and growth charts, and an update of environmental, medical, family, and social history for additional risk factors are integral to this evaluation. Identifying an etiology will be expected to affect various aspects of treatment planning, including specific medical treatments for improved health and function of the child, and therapeutic intervention programming.

This chapter will review a selected group of pediatric developmental disorders which benefit evaluation and management from an endocrinology perspective. Due to the wide spectrum of such problems and limitations to chapter size, it is not all inclusive and the reader is advised to refer to other resources.

Common Disorders Treated by Endocrinologists

Trisomy 21

Brief Overview and Epidemiology

Trisomy 21 is a genetic condition caused by chromosomal aneuploidy of chromosome 21, where a person is born with three copies of this gene rather than two. Most commonly referred to as Down syndrome, Trisomy 21 is one of three genetic variants that comprise Down syndrome, along with mosaicism and translocation. Trisomy 21 is the most common chromosome abnormality among live-born infants. Approximately 6000 babies are born each year with Down syndrome, equaling roughly to one baby for every 700 live births (Parker et al., 2010). Prenatal screening tests for Trisomy 21 include maternal serum alpha-fetoprotein (MSAFP), triple screening, quad screening, chorionic villus sampling, amniocentesis, percutaneous umbilical blood sampling (PUBS), and fetal ultrasound (Iwarsson et al., 2017). Children born with Trisomy 21 are at increased risk for and suffer from a wide range of health complaints including intellectual disability, congenital malformations, dysmorphic features, and medical conditions such as congenital heart disease, hematologic disorders, and visual and auditory disorders. The prevalence of autism in persons with Down syndrome ranges from 5.1% to 42% (DiGuiseppi et al., 2010; Oxelgren et al., 2017). Trisomy 21 is the most frequent form of intellectual disability caused by a microscopically demonstrable chromosomal aberration (Ostermaier, 2020).

Referral to the Pediatric Endocrinologist

Children born with Trisomy 21 are often referred to pediatric endocrinologists in infancy to establish care due to the increased risk of thyroid disorders, which left untreated can significantly impact development. Trisomy 21 is associated with a wide spectrum of thyroid dysfunction including autoimmune conditions (such as Hashimoto's thyroiditis and Graves' disease), subclinical hypothyroidism, and congenital hypothyroidism. Roughly 4–18% of persons born with Trisomy 21 will exhibit some level of thyroid disorder (Bull, 2011). Initial thyroid screening is provided by newborn metabolic screening, which is mandatory for newborns in every US state, the District of Columbia, and many US territories including Guam, USVI, and Puerto Rico (McCandless & Wright, 2020). Due to increased risk of thyroid dysfunction, AAP recommends repeated screening for thyroid disorders (obtaining serum TSH) at 6 months of age, 12 months of age, and then annually or sooner if child has symptoms that could be related to thyroid dysfunction. Purdy et al. (2014) noted the TSH surge at birth as a confounding variable in interpreting newborn screening results and found that only 54 of 121 screening euthyroid at birth remained euthyroid during the first 3 months, with 32.5% hypothyroid and 17.5% requiring treatment for hypothyroidism. As hypothyroidism is the most common of thyroid disorders in Trisomy 21, many patients will begin taking levothyroxine in infancy or early childhood. To establish the efficacy of treatment, serum-free T4 and TSH are monitored on a monthly basis or as indicated for the child's condition, by an endocrinologist or PCP in good contact with the patient's pediatric endocrinologist. Once the patient reaches a steady state where serum-free T4 and TSH levels have normalized, thyroid function testing is obtained periodically, which varies for the age of the child (mostly every 3–6 months). In hypothyroid children, thyroid hormone supplementation is critical for intellectual development in addition to its role in nearly all metabolic functioning, which makes this screening and early referral to pediatric endocrinology so crucial.

A subset of patients with Trisomy 21 will develop hyperthyroidism. This includes autoimmune conditions such as Graves' disease and the initial stages of Hashimoto's thyroiditis in some patients (Hashitoxicosis). Patients may present with tachycardia, diarrhea, proptosis (bulging of the eye which is the ophthalmopathy component of Graves' disease), and menstrual irregularities during puberty. Medical management consists of blocking excessive thyroid hormone production and controlling cardiovascular hyperreactivity with pharmacological agents such as beta blockers. Further management is planned per course. The management algorithm for Graves' disease encompasses medical therapy, radioactive iodine ablation, and surgical intervention (thyroidectomy) and must be individualized (Okawa et al., 2015).

Children born with Trisomy 21 are also at an increased risk of developing *autoimmune disorders* when compared with the general population (Whooten et al., 2018). This constellation of autoimmune disorders includes type 1 diabetes mellitus (T1DM), celiac disease, and vitiligo; however, researchers have found that children born with Trisomy 21 have similar incidence of T1DM in comparison with the general population though higher rates of other autoimmune conditions such as celiac disease and vitiligo.

Linear growth is impaired for children with Down syndrome throughout the growing period, but most markedly in infancy and in adolescence (Cronk, 1978; Cronk et al., 1988). Thus specific growth charts have been developed for children with Down syndrome which have slower growth velocity and shorter final height compared to typical growth charts. Growth is most reduced in children with severe congenital heart disease (Cronk, 1978; Zemel et al., 2015). The cause of growth retardation in Trisomy 21 is not clear. Low circulating levels of insulin-like growth factor 1 (IGF-1) and diminished spontaneous and stimulated secretion of growth hormone have been reported in some patients. Suboptimal endogenous growth hormone production as a result of hypothalamic dysfunction has been demonstrated. However Trisomy 21 is not a condition of growth hormone deficiency in general (Zemel et al., 2015). Children with Trisomy 21 should be monitored for growth patterns, including BMI, and children and families should receive counseling regarding healthy diet and structured exercise program. The prevalence of obesity is greater in Trisomy 21 than in the general population, which is thought to result from the reduced resting metabolic rate in children and adults with Trisomy 21. Linear growth could be disturbed in association with other disorders, such as hypothyroidism or celiac disease. Therefore, a child with Trisomy 21 will need additional evaluation if his/her growth pattern deviates from the published Down syndrome-specific growth charts, and consultation by a pediatric endocrinologist may be required.

Turner Syndrome

Definition and Epidemiology

Turner syndrome (TS) is a chromosomal condition that affects females and is characterized by complete or partial loss of one X chromosome. It is one of the common chromosomal anomalies in humans, actually the most common sex chromosome abnormality in females, with a frequency of about 1 in 2500 live female births. Most mutations with monosomy X (45,X) do not survive the prenatal period. In fact, 10% of all spontaneous abortions involve the 45,X genotype. Mosaicism (45,X/46,XX) occurs much more commonly than the aneuploid state (Gravholt et al., 2019). Persons with TS have four times the risk for autism as a comorbidity (Björlin Avdic et al.,

2021). A wide spectrum of clinical manifestations are common to TS including short stature, delayed puberty, ovarian dysgenesis, hypergonadotropic hypogonadism, dysmorphic features (such as webbed neck, low posterior hairline, shield-like chest with widely spaced nipples), congenital malformations of the heart, endocrine disorders such as type 1 and type 2 diabetes mellitus, and autoimmune disorders including thyroid disease. TS represents an important cause of short stature and ovarian insufficiency in females (Culen et al., 2017).

Problems Within the Scope of Pediatric Endocrinology

Short Stature

Short stature is the most common finding in TS and is sometimes the only presenting sign. The short stature associated with TS is largely due to the absence of one copy of the SHOX (short stature homeobox containing) gene, which is normally located on both the X and Y chromosomes. Short stature is defined as height two standard deviations below the mean for gender and age (below the third percentile). If not treated, the adult height of a girl with TS can be up to 20 cm shorter than that the normal female population. Patients' heights should be plotted along growth curves specific to TS. Because short stature in TS is not due to growth hormone deficiency, a growth hormone stimulation test is not required. Justification for management is contingent upon growth trajectory: if the child has evidence of growth failure (e.g., below 50th percentile height velocity (HV) observed over 6 months in the absence of other treatable cause of poor growth), the child is already short or has a strong likelihood of short stature (e.g., short parents and short predicted adult height or already pubertal at the time of diagnosis). Growth hormone replacement therapy has been approved by the FDA for management of girls with TS. Treatment involves therapy with growth hormone and is usually started once the patient falls below the 5% for height. Growth hormone replacement therapy should be monitored by height measurements every 4–6 months in the first year of therapy and at least biannually thereafter. IGF-1 laboratory measurements are used to monitor effectiveness and safety of treatment. Starting growth hormone early (such as at 4–6 years of life) has been shown to result with improved height outcomes. In general, height gain of about 1 cm/year of growth hormone therapy is a reasonable expectation (Gravholt et al., 2017).

Pubertal Delay

Primary hypogonadism is one of the most common features of TS. Persons with TS commonly experience premature ovarian failure. In the complete or partial absence of one X chromosome in a female, normal oocyte depletion is accelerated. At puberty, most 46,XX females have between 400,000 and 500,000 oocytes. In contrast, in girls with 45,X karyotype, nearly all the oocytes are diminished by 2 years of age. The ovaries will also involute, becoming streaks of connective tissue also known as streak ovaries. Turner syndrome is often first suspected when breast development does not occur within pubertal age range. More specifically this corresponds to no breast development at 12–13 years of age for girls in the United States. However, spontaneous breast development may be seen in 10–20% of females with TS. Adrenarche usually occurs normally. Adrenarche is characterized by production of weak androgens by the zona reticularis of the adrenal cortex, which contribute to the development of pubic hair, the sebaceous gland, and the apocrine (sweat) gland. A small percentage of TS patients will have spontaneous menstrual periods. Patients with TS that do have periods are at a greater risk for miscarriage and premature menopause. Treatment of hypergonadotropic hypogonadism consists of replacement therapy with estrogens, with subsequent addendum of progesterone. Typically, estrogen therapy is started between 11 and 12 years of age via transdermal route and increased to adult dosing over 2–3 years. Estrogen therapy induces and maintains secondary sexual development (breast development). Cyclic progesterone is added once breakthrough bleeding occurs or after 2 years of estrogen treatment. This regimen is continued to induce cyclic uterine bleeding (menses).

The endocrinologist can assist the medical home team initiate age-appropriate discussions with older teenagers and young adults regarding fertility and difficulties that may be encountered with conception. For individuals containing Y chromosome material (as discovered by karyotyping), gonadectomy is recommended (Gravholt et al., 2017). Cryopreservation of ovarian tissue and/or oocytes from selected preadolescent or adolescent girls with TS is under investigation as a technique that might permit reproduction later in life (Oktay & Bedoschi, 2014).

Thyroid Dysfunction and Autoimmunity

Turner syndrome patients have an increased risk of autoimmunity which encompasses thyroid problems, celiac disease, diabetes (type 1), and inflammatory bowel disease. It is estimated that up to 50% of patients with Turner syndrome are affected by Hashimoto's thyroiditis, an autoimmune form of thyroid disease. Hashimoto's thyroiditis is caused by lymphocytic infiltration and fibrosis of the thyroid and circulating antithyroid peroxidase and/or antithyroglobulin antibodies, with the former being more specific. A small group of patients with this disorder will present with hyperthyroid symptoms such as tachycardia, nervousness, and palpitations in the early course of the illness known as hashitoxicosis. This is associated with elevations in free T4 and TSH suppression and usually transitional. It is expected to transition into a euthyroid and subsequently hypothyroid state in most patients. Signs and symptoms of hypothyroidism may include one or more of the following: dry skin, hair loss, thyroid enlargement (goiter), fatigue, constipation, myxedema, exercise intolerance, depression, menorrhagia, and cold intolerance. The course may also be subclinical. In the hypothyroid phase, laboratory evaluation will usually show a low free T4 and an elevated TSH. The detection of antithyroid peroxidase and/or antithyroglobulin antibodies is suggestive, but up to 10% of Hashimoto thyroiditis patients can have negative antibodies. Treatment includes thyroid hormone replacement with levothyroxine. For euthyroid patients with positive antithyroid antibodies, thyroid function testing (TSH/Free T4) should be performed every 6 months or when indicated by clinical evaluation (Leung & Leung, 2019; Mincer & Jialal, 2020).

Graves' disease is another autoimmune thyroid disorder which can be seen in patients with TS. Graves' disease is a hyperthyroid state, characterized by thyroid-stimulating immunoglobulin. Patients have elevated thyroid hormones (T3, T4) and suppressed TSH. Patients require careful clinical and biochemical monitorization and management with antithyroid medications (thionamides: Methimazole, and sometimes Iodide) and medications to control cardiovascular hyperreactivity (beta blockers), as needed. Full management algorithm also includes definitive treatment options, such as radioactive iodine ablation and thyroidectomy, and is followed on a case-by-case basis.

Metabolic Syndrome and Diabetes Mellitus

Patients with TS are at risk for overweight conditions, insulin resistance, dyslipidemia, metabolic syndrome, and type 2 diabetes mellitus (T2DM). The prevalence of T2DM in young adults with TS ranges from 5% to 25% in different reports, which is substantially increased over the general population (Backeljauw, 2021). The risk of both type 1 and type 2 diabetes mellitus is about tenfold and fourfold increased in patients with TS across all ages in epidemiological studies. Clinical management guidelines recommend annual measurement of HbA1c with or without fasting plasma glucose starting at the age of 10 years (Gravholt et al., 2017). Counseling on healthy nutrition and physical activity should be started in early childhood, with careful monitorization of growth charts. Prevention of overweight/obesity is instrumental in prevention of metabolic syndrome and associated comorbidities/complications in this patient group which is already at high risk for cardiovascular morbidity.

Prader-Willi Syndrome

Definition and Epidemiology

Prader-Willi syndrome (PWS) is a genetic disorder resulting from the absence of paternal chromosomal expression of 15q11.2-q13 in the majority of cases, maternal uniparental disomy in approximately 25% of cases, and imprinting defects in the abovementioned regions (Heksch et al., 2017). PWS shows no gender predominance and affects approximately 1 in 10,000–30,000 persons worldwide (Rare Disease Database, 2018). It is the most common syndromic form of obesity. Diagnosis is made with genetic testing after clinical evaluation. Testing is done with a sequence, starting with DNA methylation analysis (at 5′ SNRPN). A systematic review noted the prevalence of autism in persons with PWS as nearly 27% (Bennett et al., 2015). A multidisciplinary approach is required to address PWS patients' various needs (McCandless, 2011).

Problems Within the Scope of Pediatric Endocrinology

Pediatric patients with PWS have a multitude of clinical signs and symptoms including facial dysmorphia, hypotonia (with corresponding weak suck and feeding difficulties in the infancy period), intellectual disability, adrenal insufficiency, hypothalamic dysfunction, hypogonadism/delayed puberty, and classically hyperphagia that can result in obesity and metabolic syndrome as well as type 2 diabetes mellitus in later childhood and adulthood.

Hypothalamic dysfunction can cause endocrine dysfunction in various arenas including growth hormone deficiency, hypogonadism, hypothyroidism, central adrenal insufficiency, and bone mineralization deficiencies. A PWS patient with growth failure (low height percentile and/or decreased height velocity and/or decreased height in comparison to midparental height) will be a candidate for treatment with recombinant human growth hormone. Most individuals with PWS will have growth hormone deficiency if tested, though formal testing may not be necessary if growth data confirm growth failure. Growth hormone deficiency naturally leads to short stature and decreased height velocity, which once a patient is referred to the pediatric endocrinologist can be treated with human growth hormone. Recombinant human growth hormone (hGH) administration can improve metabolic function as well as motor and intellectual abilities. However, safety measures must be taken as recombinant hGH administration can also increase severity of obstructive sleep apnea and has been shown to cause severe respiratory problems and fatalities. Referrals to ENT/sleep medicine may be necessary. In practice, hGH is primarily used for benefits other than increased height, including improved body composition and motor function (Emerick & Vogt, 2013). Due to bone mineralization deficiency, patients often have atypical body composition which leads toward increased adipose tissue deposition (Emerick & Vogt, 2013).

Cryptorchidism is very prevalent among males, and HGH administration can improve phenotypic appearance. However, FSH, LH, and testosterone are low and, though responsive to puberty, remain lower than typical pubertal males (Heksch et al., 2017). There is not a consensus among providers on how to address male hypogonadism in PWS. There are similar clinical and physical manifestations for female patients with PWS, with patients typically displaying Tanner stage 3 secondary sex characteristics. Oral estrogens and combination progesterone-estrogen contraceptive pills are used for adolescent girls with PWS.

Though previous studies have not found difference between unaffected newborns and those with PWS regarding newborn metabolic screening thyroid function tests, researchers advocate for more in-depth thyroid function screening for neonates with PWS as the majority of children affected will demonstrate some measure of central hypothyroidism (Heksch et al., 2017). Levothyroxine should be started as soon as thyroid dysfunction is suspected or diagnosed (Medeiros et al., 2013).

Central adrenal insufficiency is common in individuals with PWS; some studies estimate that up to 60% of these individuals exhibit some measure of adrenal suppression throughout their lifetimes. HGH administration can suppress adrenal synthesis of steroids – requiring parents of children with PWS as well as medical professionals to closely monitor patients when ill, as vague symptoms could be manifestations of an adrenal crisis (Heksch et al., 2017).

Patients with PWS have lifelong struggles with disordered eating as manifested by hyperphagia. These patients can also have obesity that manifests before hyperphagia begins (Miller, 2012). Obesity carries the complications seen in the general population such as metabolic syndrome and diabetes mellitus. Feeding must be very structured for these patients, and some researchers recommend a diet consisting of 30% fat, 45% carbohydrates, and 25% protein to promote healthy weight gain (Miller, 2012). Mealtimes should be structured as children with PWS can have fraught relationships with food leading to increased anxiety and dysfunctional behavior (Emerick & Vogt, 2013).

Russell-Silver Syndrome

Definition and Epidemiology

Russell-Silver syndrome (RSS) is a genetic disorder characterized by severe failure to thrive (IUGR in fetal life and subsequent postnatal growth delay), decreased height velocity, recurrent hypoglycemia, feeding issues (including GERD, constipation, swallowing dysfunction), and altered phenotypic expression. Upon physical exam, the clinician may note a thin child with a head circumference at appropriate size with average weight peers, pointed chin, short and curved fifth upper extremity digits (clinodactyly) bilaterally, and micrognathia among other traits (Wakeling et al., 2017). Per Genetics Home Reference at the US National Library of Medicine, incidence may range from 1 in 30,000 to 1 in 100,000 live births, though exact numbers are unknown. The US Department of Health and Human Services Genetic and Rare Diseases information center states that RSS is mostly a sporadic disorder with no clear-cut inheritance pattern, though an infrequent number of cases are heritable through autosomal dominant or recessive patterns. On a genetic level, deletions and/or duplications of chromosomes 7 and 11 have been linked to RSS. An underlying molecular cause can currently be identified in approximately 60% of patients clinically diagnosed with RSS. The most common underlying mechanisms are loss of methylation on chromosome 11p15 (11p15 LOM; seen in 30–60% of patients) and maternal uniparental disomy for chromosome 7 (upd(7) mat; seen in ~5–10% of patients). However, the molecular etiology remains unknown in a substantial proportion of patients. Cognitive disorders, including autism, appear to be more common with upd(7)mat chromosomal abnormalities, while children with 11p15 chromosomal abnormalities often have lower birth weight and more physical asymmetry than those born with chromosome 7 abnormalities. Diagnosis is mainly based on clinical features and molecular testing as available at the patient's health center. A scoring guide and algorithm have been proposed to aid in diagnosis (Wakeling et al., 2017).

Problems Within the Scope of Pediatric Endocrinology

Children with RSS are often referred to pediatric endocrinologists due to recurrent, severe hypoglycemia and severe growth retardation. Primary care providers should obtain a comprehensive medical history including nutrition evaluation and pertinent biochemical tests (such as comprehensive metabolic panel, urinalysis, specifically for ketones, etc.) when making a referral to pediatric endocrinology for recurrent hypoglycemia. Careful monitoring of weight and height on growth charts is important to guide treatment. Regarding growth and feeding, it is important that children who can eat by mouth do not rely on nasogastric feeding. It is equally important that refeeding is carefully monitored as rapid weight gain can lead to metabolic derangements (insulin resistance) and cardiovascular compromise in older children with RSS.

Fasting hypoglycemia, with a tendency to ketotic hypoglycemia, is of particular concern especially in younger children and infants (under age 5 years). The medical home team should assist parents with nutrition counseling and instructions to check urinary ketones and point of care glucose, as needed. Cornstarch can be added to last meal of the day in order to prevent ketotic hypoglycemia. In children with RSS, hypoglycemia can also be exacerbated by febrile illness; during which, these children may require hospitalization for IV fluids containing 10% dextrose.

Russell Silver syndrome is associated with a significant reduction in adult height (around −3 SDS) and an indication for growth-promoting GH treatment under the SGA registered license (Wakeling et al., 2017). Additional potential benefits of GH treatment are increase in appetite, lean body mass, and muscle power, which can result in improved mobility. Growth hormone therapy should be deferred until caloric deficits are addressed. Growth hormone stimulation testing is not recommended as this can potentiate hypoglycemia due to fasting requirement. Treatment with GH should be started early, starting at age 2–4 years. Monitoring IGF-1 during GH therapy is recommended; however, the interpretation of IGF-1 levels is difficult in children with RS.

Patients with RSS may have early bone age delay which is followed by rapid advancement typically around 8–9 years and sometimes much younger especially in nonvolitionally overfed children. Onset of puberty is usually within the normal range (8–13 years in girls and 9–14 years in boys), however, at the younger end of the spectrum. Adrenarche can be early and aggressive in comparison with children born with non-SRS SGA, particularly in those with 11p15 LOM (Wakeling et al., 2017). These trends should be taken into consideration during management of children with RSS. It is important to note that in patients with RSS, pediatric endocrinology may be one of several subspecialties including pediatric gastroenterology who may be needed to provide cohesive and comprehensive care in coordination with the medical home team.

Optic Nerve Hypoplasia Spectrum and Septo-optic Dysplasia

Definition and Epidemiology

Optic nerve hypoplasia (ONH) is a common complex congenital disorder of unknown cause, involving a spectrum of anatomic malformations and clinical manifestations ranging from isolated hypoplasia of uni- or bilateral optic nerves, with a variable degree of visual impairment, to extensive brain malformations, hypothalamic-pituitary dysfunction, neurocognitive disability, and/or autism spectrum disorders (Ryabets-Lienhard et al., 2016). ONH is the second leading cause of congenital visual impairment, following cortical visual impairment. Septo-optic dysplasia (de Morsier syndrome) is the mildest form of lobar holoprosencephaly and involves optic nerve hypoplasia, the absence of the septum pellucidum, and pituitary dysfunction (Sarnat & Flores-Sarnat, 2016). A majority (2/3) of affected infants have hypothalamic-pituitary dysfunction which can range from simple growth hormone deficiency to panhypopituitarism. Affected patients can also have hydrocephalus. Vision impairment, developmental delays, and endocrine hormone deficiencies are common.

Due to early observations of co-occurrence with agenesis of the septum pellucidum and hypopituitarism, ONH has long been recognized as part of the septo-optic dysplasia (SOD) syndrome, which is now considered a clinically inaccurate term that attributes prognostic importance of the hypothalamic-pituitary dysfunction development to the absent septum pellucidum and/or other midline brain malformations. More recent, larger studies have demonstrated ONH to be an independent risk factor for hypothalamic-pituitary dysfunction, with abnormalities of the septum pellucidum having no prognostic value (Ryabets-Lienhard et al., 2016).

There are variable reports on the prevalence of ONH. The most recent estimates from the United Kingdom reported a prevalence of 10.9 per 100,000 in children younger than 16 years of age (Patel et al., 2006) and, from Stockholm, Sweden, 17.3 per 100,000 children younger than 18 years of age (Teär Fahnehjelm et al., 2014). A 2013 report by the Mayo Clinic College of Medicine cited an annual local incidence of 2.4 per 100,000 children younger than 19 years of age (1 in 2287 live births) in Olmsted County, Minnesota (Mohney et al., 2013).

Infants born to younger mothers are at increased risk. Associated genes include SOX2,

SOX1, HESX1, and OTX2, but most patients do not have an identifiable genetic cause (Huang & Doherty, 2018). Other predisposing factors include prenatal exposure to drugs and alcohol.

Problems Within the Scope of Pediatric Endocrinology

A variable spectrum of hypothalamic pituitary dysfunction is detected in these patients.

Hypothalamic-pituitary dysfunction is seen in 60–80% of these patients, with growth hormone-releasing hormone and/or growth hormone deficiency occurring in approximately 70% of patients.

Growth hormone deficiency can present as neonatal hypoglycemia. Among infant males, hypopituitarism should be suspected when neonatal hypoglycemia, jaundice, and micropenis are found. Untreated, GH deficiency can lead to short stature, reduced musculature, adiposity, high-pitched voice, and a cherubic appearance. Replacement therapy with growth hormone is very successful, and patients can achieve normal adult height. TRH-TSH deficiency is seen in 35–43% these patients. With respect to tertiary/secondary adrenal insufficiency, CRH-ACTH deficiency is reported in 17–27%. Central diabetes insipidus (ADH deficiency) is reported in 4–5%.

Untreated congenital hypothyroidism can lead to short stature and mental retardation if replacement therapy is not started early. Most cases are identified by newborn metabolic screening programs. Confirmation is by measuring free T4 and TSH versus age-related norms. Important to note that central hypothyroidism may be missed if the state screening program is solely using TSH for screening.

Other Disorders

Attention-Deficit Hyperactivity Disorder

Definition and Epidemiology

Attention-deficit hyperactivity disorder (ADHD) is a disorder that manifests in childhood with symptoms of hyperactivity, impulsivity, and/or inattention. ADHD is one of the most common neurobehavioral disorders of childhood and can profoundly affect children's academic achievement, well-being, and social interactions. The symptoms affect cognitive, academic, behavioral, emotional, and social functioning (American Psychiatric Association, 2013; Wolraich et al., 2019).

ADHD is a chronic disease and the recommended treatment strategies for children with ADHD vary according to age. Children undergoing treatment for ADHD should be monitored regularly for adherence to the treatment plan, adverse effects of therapy (for those on medications), and response to therapy (e.g., the achievement of target goals and the occurrence of core symptoms).

Problems Within the Scope of Pediatric Endocrinology

Several types of medications are available to treat ADHD in children and adolescents: stimulants, atomoxetine, alpha-2-adrenergic agonists, and antidepressants. In a school-aged child or adolescent (≥6 years), a stimulant medication is usually prescribed as the first-line agent.

Decreased appetite and poor growth are among the adverse effects of stimulants. This can reflect negatively on weight and height gain in children and adolescents (Krull, 2021). Pediatric endocrinologists are frequently consulted on growth disorders, in particular, short stature in children. A medical history of ADHD and pharmacological therapy must be taken into consideration during evaluation.

With respect to decreased appetite, it is advisable to administer the medication at or after a meal; encourage the child to eat nutrient-dense foods before those with "empty calories" and offer food that the child likes for the noon meal, which is often affected. With respect to the resultant poor growth, drug holidays may be beneficial for children in whom stimulant therapy is associated with a growth trajectory that crosses two major percentiles (i.e., the 5th, 10th, 25th, 50th, 75th, 90th, and 95th) (Pliszka, 2007). Drug holidays should only be undertaken if they can be tolerated without marked impairment in functioning. Poor growth that does not respond to drug holidays should be discussed with families to determine preferences for continued treatment. Nutritional consultation to facilitate growth while taking medications may also be warranted.

This is another example pointing to the crucial importance of careful monitorization of growth charts in children with developmental disorders.

Fetal Alcohol Spectrum Disorder

Definition and Epidemiology

Fetal alcohol spectrum disorders (FASDs) are collection of clinical features and CNS abnormalities in the setting of prenatal alcohol exposure. Each of the disorders, fetal alcohol syndrome (FAS), partial FAS, alcohol-related birth defects, and alcohol-related neurodevelopmental disorders, will present with some combination of the following: facial dysmorphism (flat philtrum, truncated palpebral fissures, microcephaly, thin upper lip), short stature, CNS abnormalities (which may include seizures, intellectual disability, ADHD, anger management issues, auditory and/or visual dysfunction) and known maternal consumption of alcohol during the prenatal period (Turchi & Smith, 2018). Autism is also more common in children with FASD than typically developing peers (Lange et al., 2018). FASD may still be under-recognized and reported even today, though recent journals estimate approximately 5% of live births in the United States may be infants born with FASD (Williams & Smith, 2015). Of note, specific populations of children, such as foster children, may have higher rates of FASD than the general public per Washington State Fetal Alcohol Spectrum Disorders Inter-Agency Work Group, the nation's leader in FASD research and prevention.

Problems Within the Scope of Pediatric Endocrinology

Patients are often referred to pediatric endocrinology for short stature. Fetal alcohol spectrum disorder is characterized by pre- and postnatal growth retardation. Prior to referral, the primary care physician should make a careful diagnosis of FASD and exclude other similarly presenting disorders such as fragile X syndrome. A chromosomal microarray may be considered in selected patients. The clinician may want to ascertain midparental height and obtain general biochemical evaluation tests including but not limited to complete blood count, urinalysis, comprehensive metabolic panel, phosphorus, and thyroid studies prior to referral. The clinician should also be mindful of psychoactive medications prescribed to the patient with FASD such as stimulant for ADHD, as use of stimulant medications can independently lead to decreased height velocity.

Endocrine-Related Adverse Effects of Medications Commonly Used in the Treatment of Developmental/Behavioral Disorders

Discussion on Meds Causing Weight Gain

Overweight and Obesity

Obesity, whether in children or adults, is defined as an abnormal accumulation of excess fat. It is measured by the body mass index (BMI), which accounts for a child's weight-to-height ratio and is age- and sex-specific in children. The US Centers for Disease Control and Prevention (CDC) has published age- and gender-specific charts for weight, height, and body mass index (BMI) in children aged 2–20 years (Centers for Disease Control and Prevention, 2018). Overweight is defined by a BMI at or above the 85th percentile and below the 95th percentile, whereas obesity is defined as a BMI at or above the 95th percentile. However, the problem with using BMI as an index for obesity is that it does not distinguish the difference between fat and muscle tissue; therefore an athlete (who has more muscle tissue) may have an overestimated BMI and a child who is more sedentary (reduced muscle mass) may have an underestimated BMI. In the United States, childhood obesity has been and is still a serious public health problem, putting children and adolescents at risk. Studies have shown that for children and adolescents aged 2–19 years of age, the prevalence of obesity was 18.5% and to further break it down into age groups, the obesity prevalence was 13.9% in 2–5 year olds, 18.4% in 6–11 year olds, and 20.6% in 12–19 year olds (Centers for Disease Control and Prevention, 2021). The increase in frequency however not only applies to the general pediatric patient group but also to those with hormonal, monogenic, and syndromic disorders.

The prevalence of overweight and obesity in children and adolescents with developmental disorders is likely to mimic the general pediatric population and these patients will also benefit periodic monitorization of growth charts. As discussed throughout this chapter, a deviation from the age- and gender-specific growth curve (in some cases, the condition-specific growth charts such as in Trisomy 21 or Turner syndrome) should alert an astute clinician to the need for further assessment. Channeling up or down in growth charts may be the first manifestation of an endocrine problem. Various reasons for a weight change should be considered as weight change could point to a medication adverse effect or metabolic syndrome. Addressing weight change early can help prevention of future obesity.

Obesity cannot be singled down to one cause; it involves social, environmental, behavioral, and biological determinants. Certain determinants such as genetics play a major role in how our bodies metabolize food. Overall, this is a complex interaction that requires a multidisciplinary team to assess and help manage the outcomes and consequences. While most cases of obesity are "exogenous," resulting from excessive energy consumption compared to energy expenditure, some cases are "endogenous" as a result of hormonal, genetic, or syndromic disorders. Some common endogenous cases of obesity in children include hypothyroidism, Leptin deficiency, Cushing's syndrome, Prader-Willi syndrome, Trisomy 21, and septo-optic dysplasia. It is therefore vital to recognize and have a suspicion that weight gain could be due to endogenous obesity.

The first step in evaluating overweight or obese children is a detailed history and physical examination which can help assess the severity and the associated comorbidities. Every visit needs to start with accurate measurements of weight, height, and vitals. Appropriate growth charts should be used in conjunction with the clinical evaluation. History should include information about the child's eating habits, identifying caretakers who feed the child, identifying foods in high calories that can be reduced or eliminated, eating patterns, and overall lifestyle history. A 3-day food record should be requested to further assess dietary patterns (Lanigan et al., 2010). History of physical activity is also equally important and should include evaluation of time spent doing physical activity but also assessment of screen time.

Medication history should also be reviewed because many children with developmental disorders require medications that can promote weight gain – medications such as psychoactive drugs, antiepileptic drugs, and glucocorticoids (Skelton & Klish, 2020). Other important components also include review of systems, family history, and psychosocial history. For example, if there is a history of hypotonia and failure to thrive during infancy with subsequent developmental delays and rapid weight gain one would be concerned for a diagnosis of Prader-Willi syndrome (Skelton, 2021). Physical examination is also important because it could help identify certain genetic syndromes that are highly associated with obesity. The distribution of excess adipose tissue can also help distinguish the etiology of obesity – as generalized distribution of increased adiposity is likely from overeating/overfeeding while centripetal distribution of adipose tissues can suggest glucocorticoid excess, e.g., Cushing syndrome (Skelton, 2021).

For general pediatricians and subspecialists, continuity of care is always crucial because it allows for close monitoring of growth charts, BMI trends, history, and physical exams. During routine medical surveillance, it important to keep a close eye on these basic tools because they are the key to diagnosing or preventing serious conditions in children. Accelerating changes in weight/BMI, for example, would be alarming and would warrant further investigation into any changes or addition of any medications. Growth charts are helpful because they too can help distinguish between exogenous obesity and endogenous obesity. In children that present with exogenous obesity, studies have shown that their height chart should also follow the linear rise, meaning, obese children are typically tall for their age. Whereas in children with short stature, obesity is more likely to be a manifestation of endocrine/metabolic/genetic causes (August et al., 2008). For example, children with Prader-Willi syndrome are often short for their genetic potential and/or fail to have a pubertal growth spurt.

Laboratory testing is also used in many clinics when it comes to assessing obesity in children (Gungor, 2017). Although there are no standardized laboratory tests, many clinics will start off with getting a comprehensive metabolic panel, thyroid screening (TSH, free T4), HgbA1c, and a lipid panel. Once again a complete history and physical exam will help path the way for which labs need to be drawn. For example, the American Diabetes Association (ADA) recommends using HgbA1c as a diagnostic marker for type 2 diabetes and recommends screening any child over the age of 10 years who is overweight or obese and have two or more additional risk factors such as family history of type 2 diabetes in a first- or second-degree relative, those of high-risk ethnicity (Asian, African-American, Hispanic, or Native American), and/or those with signs of insulin resistance (acanthosis, nigricans, hypertension, dyslipidemia, polycystic ovarian syndrome) (Xu & Verre, 2018). As mentioned earlier, since there is no standardized laboratory testing for obesity, it ultimately comes down to the clinician's discretion.

Medications

As previously stated, a detailed history especially in regard to medications is important, especially in children with developmental disorders. Many medications that are used in the management of developmental disorders can have a great impact on weight gain. Table 10.1 lists some of the commonly used medications in children that can lead to weight gain (Gungor, 2017). For children with autism, studies have shown that many are treated with antipsychotic medications to help with behavioral aspects of the disorder. One specific study showed that weight gain was much more significant with olanzapine, clozapine, and risperidone than with any other antipsychotics and that out of these risperidone seemed to be very commonly prescribed (Reekie et al., 2015). Even insulin, a very common drug in treating diabetes, can cause weight gain; however, the benefits of insulin outweigh the side effects and therefore it is extremely important to educate these patients and their families about healthy lifestyles and making healthier choices because they are predisposed to gaining weight. Children with seizure disorders are also often put on antiepileptic medications such as valproate and carbamazepine. In one specific study in children, it was seen that valproate caused an average increase in weight by 5 kg (Privitera et al., 2003). Iatrogenic weight gain is often seen in children and can deter children from taking their medications; therefore it is important for the medical home team to counsel parents/caregivers for prevention of acceleration of the overweight condition and involve dietitians and other weight management specialists early on.

Complications of childhood obesity can have a severe negative impact on many organs of the body leading to serious physical and psychological comorbidities. Several of these complications can persist into adulthood and have severe consequences. These complications include but are not limited to hypertension, coronary artery disease, obstructive sleep apnea, nonalcoholic fatty liver disease, pancreatitis, and depression (Skelton, 2021). Many of these complications can be pre-

Table 10.1 Commonly used medications in children with developmental disorders that lead to weight gain

Insulin and insulinotropic agents
Beta-blockers
Corticosteroids
Cyproheptadine
Antipsychotics
Lithium carbonate
Antiepileptic medications
Tricyclic antidepressants

vented if childhood obesity is managed appropriately from the beginning. Primary and secondary preventions are important concepts and comprehensive intervention should target both. Multispecialty approach involving the entire family is important. Appropriate referrals should be considered as needed: such as pediatric weight management centers for dietary, pharmacological, and/or surgical therapy in extreme cases.

The medical home can play an important role in assessing and evaluating patients based on their individual needs in order to diagnose the underlying cause of childhood obesity. Once the etiology of obesity is identified, the multidisciplinary team can begin to address the multiple determinants. This involves collaborating outside of the clinic and reaching out to the community, other health-systems, and school systems. Pediatric obesity continues to rise in the general population but even more so in children and young adolescents including those with developmental disabilities. It is crucial to start the process of educating parents and children about obesity and its overall effects on life quality but also how to manage and help improve these issues.

Closing Remarks

The pediatric endocrinologist plays a key role in maintaining the health of children with developmental disabilities through the assessment and treatment growth disorders, diabetes, thyroid disorders, obesity, and other endocrine disorders. A medical home or other multifaceted approaches to the management of endocrine disorders in children with developmental disabilities promote the best clinical and mental health outcomes. Collaboration among the general pediatrician, pediatric subspecialist, and developmental/mental health-care team is encouraged as the exchange of information is vital to our patients' health. Given the frequently encountered problems involving physical growth (such as short stature, overweight) and endocrine/hormonal disorders, it is advisable to perform careful clinical and biochemical screening to identify and manage any associated conditions and consider referral to pediatric endocrinology if indicated (Table 10.2).

Table 10.2 Checklist for referral to the pediatric endocrinologist

1. Monitor CDC growth charts and condition-specific growth charts (i.e., Trisomy 21, Turner syndrome) and consider referral.
2. Offer counseling early with channeling up or down noted to height, weight, or BMI percentiles and order additional workup as needed.
3. Monitor glucose and thyroid function as indicated by condition specific health surveillance guides.
4. Referrals to the pediatric endocrinologist should include. The reason for referral The diagnosis(es) of the patient Current medications Shared plan of care Growth charts Genetic test results, if available Any pertinent laboratory studies (recommend 8 am studies, fasting preferred) • If referral is for overweight condition: fasting CMP, fasting lipid panel, free T4, TSH, and hemoglobin A1C • If referral is for a thyroid-related concern: free T4 and TSH (if applicable thyroid peroxidase and thyroglobulin antibodies)

References

American Psychiatric Association. (2013). *Attention-deficit/hyperactivity disorder*. APA.

August, G. P., Caprio, S., Fennoy, I., Freemark, M., Kaufman, F. R., Lustig, R. H., Silverstein, J. H., Speiser, P. W., Styne, D. M., & Montori, V. M. (2008). Prevention and treatment of pediatric obesity: An endocrine society clinical practice guideline based on expert opinion. *Journal of Clinical Endocrinology and Metabolism, 93*(12), 4576–4599. https://doi.org/10.1210/jc.2007-2458

Backeljauw, P. (2021). *Clinical manifestations and diagnosis of Turner syndrome*. https://www.uptodate.com/contents/clinical-manifestations-and-diagnosis-of-turner-syndrome?search=Clinical%20manifestations%20and%20diagnosis%20of%20Turner%20syndrome&source=search_result&selectedTitle=1~150&usage_type=default&display_rank=1

Bennett, J. A., Germani, T., Haqq, A. M., & Zwaigenbaum, L. (2015). Autism spectrum disorder in Prader–Willi syndrome: a systematic review. *American journal of medical genetics Part A, 167*(12), 2936–2944.

Björlin Avdic, H., Butwicka, A., Nordenström, A., Almqvist, C., Nordenskjöld, A., Engberg, H., & Frisén, L. (2021). Neurodevelopmental and psychiatric disorders in females with Turner syndrome: a population-based study. *Journal of neurodevelopmental disorders, 13*(1), 1–9.

Bull, M. J. (2011). Health supervision for children with Down syndrome. *Pediatrics, 128*(2), 393–406. https://doi.org/10.1542/peds.2011-1605

Centers for Disease Control and Prevention. (2018). *Defining childhood obesity.*

Centers for Disease Control and Prevention. (2020). *Facts about developmental disabilities*. https://www.cdc.gov/ncbddd/developmentaldisabilities/facts.html

Centers for Disease Control and Prevention. (2021). *Childhood obesity facts. Prevalence of childhood obesity in the United States*. https://www.cdc.gov/obesity/data/childhood.html#:%7E:text=Prevalence%20of%20Childhood%20Obesity%20in%20the%20United%20States&text=The%20prevalence%20of%20obesity%20was,to%2019%2Dyear%2Dolds.

Cronk, C. E. (1978). Growth of children with Down's syndrome: Birth to age 3 years. *Pediatrics, 61*(4), 564–568.

Cronk, C., Crocker, A. C., Pueschel, S. M., Shea, A. M., Zackai, E., Pickens, G., & Reed, R. B. (1988). Growth charts for children with Down syndrome: 1 month to 18 years of age. *Pediatrics, 81*(1), 102–110.

Culen, C., Ertl, D. A., Schubert, K., Bartha-Doering, L., & Haeusler, G. (2017). Care of girls and women with Turner syndrome: beyond growth and hormones. *Endocrine connections, 6*(4), R39–R51.

DiGuiseppi, C., Hepburn, S., Davis, J. M., Fidler, D. J., Hartway, S., Lee, N. R., ... & Robinson, C. (2010). Screening for autism spectrum disorders in children with Down syndrome: population prevalence and screening test characteristics. *Journal of developmental and behavioral pediatrics: JDBP, 31*(3), 181–191.

Duby, J. C., Lipkin, P. H., Macias, M. M., Wegner, L. M., Duncan, P., Hagan, J. F., Cooley, W. C., Swigonski, N., Biondich, P. G., Lollar, D., Ackermann, J., Brin, A., Crane, M., Gibson, A., Skipper, S. M., Steinberg-Hastings, D., Capers, M. (2006). Identifying infants and young children with developmental disorders in the medical home: An algorithm for developmental surveillance and screening. *Pediatrics, 118*(1), 405–420. https://doi.org/10.1542/peds.2006-1231

Emerick, J. E., & Vogt, K. S. (2013). Endocrine manifestations and management of Prader-Willi syndrome. *International Journal of Pediatric Endocrinology, 2013*(1), 14. https://doi.org/10.1186/1687-9856-2013-14

Gravholt, C. H., Andersen, N. H., Conway, G. S., Dekkers, O. M., Geffner, M. E., Klein, K. O., Lin, A. E., Mauras, N., Quigley, C. A., Rubin, K., Sandberg, D. E., Sas, T. C. J., Silberbach, M., Söderström-Anttila, V., Stochholm, K., van Alfen-van derVelden, J. A., Woelfle, J., & Backeljauw, P. F. (2017). Clinical practice guidelines for the care of girls and women with Turner syndrome: Proceedings from the 2016 Cincinnati International Turner Syndrome Meeting. *European Journal of Endocrinology, 177*(3), G1–G70. https://doi.org/10.1530/eje-17-0430

Gravholt, C. H., Viuff, M. H., Brun, S., Stochholm, K., & Andersen, N. H. (2019). Turner syndrome: Mechanisms and management. *National Review of Endocrinology, 15*(10), 601–614. https://doi.org/10.1038/s41574-019-0224-4

Gungor, N. (2017). Practical guide for management of children with obesity. In *Synopsis of current general pediatrics practice. Recent advances in pediatric medicine* (Vol. 1, pp. 117–143). Bentham Science Publishers.

Heksch, R., Kamboj, M., Anglin, K., & Obrynba, K. (2017). Review of Prader-Willi syndrome: The endocrine approach. *Translational Pediatrics, 6*(4), 274–285. https://doi.org/10.21037/tp.2017.09.04

Huang, S. B., & Doherty, D. (2018). Congenital malformations of the central nervous system. In C. A. Gleason & S. E. Juul (Eds.), *Avery's diseases of the newborn* (10th ed.). Elsevier.

Iwarsson, E., Jacobsson, B., Dagerhamn, J., Davidson, T., Bernabé, E., & Heibert Arnlind, M. (2017). Analysis of cell-free fetal DNA in maternal blood for detection of trisomy 21, 18 and 13 in a general pregnant population and in a high risk population–a systematic review and meta-analysis. *Acta obstetricia et gynecologica Scandinavica, 96*(1), 7–18.

Krull, K. (2021). *Attention deficit hyperactivity disorder in children and adolescents: Treatment with medications.* https://www.uptodate.com/contents/attention-deficit-hyperactivity-disorder-in-children-and-adolescents-treatment-with-medications?search=.%20Attention%20deficit%20hyperactivity%20disorder%20in%20children%20and%20adolescents:%20Treatment%20with%20medications&source=search_result&selectedTitle=1~150&usage_type=default&display_rank=1.

Lange, S., Rehm, J., Anagnostou, E., & Popova, S. (2018). Prevalence of externalizing disorders and Autism Spectrum Disorders among children with Fetal Alcohol Spectrum Disorder: systematic review and meta-analysis. *Biochemistry and cell biology, 96*(2), 241–251.

Lanigan, J. A., Wells, J. C. K., Lawson, M. S., & Lucas, A. (2001). Validation of food diary method for assessment of dietary energy and macronutrient intake in infants and children aged 6–24 months. *European journal of clinical nutrition, 55*(2), 124–129.

Lanigan, J., Barber, S., & Singhal, A. (2010). Prevention of obesity in preschool children. *Proceedings of the Nutrition Society, 69*(2), 204–210.

Leung, A. K. C., & Leung, A. A. C. (2019). Evaluation and management of the child with hypothyroidism. *World Journal of Pediatrics, 15*(2), 124–134. https://doi.org/10.1007/s12519-019-00230-w

McCandless, S. E., & Committee on Genetics. (2011). Health supervision for children with Prader-Willi syndrome. *Pediatrics, 127*(1), 195–204.

McCandless, S. E., & Wright, E. J. (2020). Mandatory newborn screening in the United States: history, current status, and existential challenges. *Birth defects research, 112*(4), 350–366.

Medeiros, C., Bordallo, A. P., Souza, F. M., & Collet-Solberg. (2013). Endocrine management of children with Prader-Willi syndrome. *Pediatric Health, Medicine and Therapeutics, 4*, 117–125.

Miller, J. L. (2012). Approach to the child with Prader-Willi syndrome. *Journal of Clinical Endocrinology and Metabolism, 97*(11), 3837–3844. https://doi.org/10.1210/jc.2012-2543

Mincer, D. L., & Jialal, I. (2020). *Hashimoto thyroiditis*. StatPearls Publishing. https://www.ncbi.nlm.nih.gov/books/NBK459262

Mohney, B. G., Young, R. C., & Diehl, N. (2013). Incidence and associated endocrine and neurologic abnormalities of optic nerve hypoplasia. *JAMA Ophthalmology, 131*(7), 898–902. https://doi.org/10.1001/jamaophthalmol.2013.65

Okawa, E. R., Grant, F. D., & Smith, J. R. (2015). Pediatric Graves' disease: Decisions regarding therapy. *Current Opinion in Pediatrics, 27*(4), 442–447. https://doi.org/10.1097/mop.0000000000000241

Oktay, K., & Bedoschi, G. (2014). Oocyte cryopreservation for fertility preservation in postpubertal female children at risk for premature ovarian failure due to accelerated follicle loss in Turner syndrome or cancer treatments. *Journal of Pediatric Adolescent Gynecology, 27*(6), 342–346. https://doi.org/10.1016/j.jpag.2014.01.003. Epub 2014 Sept 10. PMID: 25214440; PMCID: PMC4252563.

Olusanya, B. O., Davis, A. C., Wertlieb, D., Boo, N.-Y., Nair, M. K. C., Halpern, R., Kuper, H., Breinbauer, C., de Vries, P. J., Gladstone, M., Halfon, N., Kancherla, V., Mulaudzi, M. C., Kakooza-Mwesige, A., Ogbo, F. A., Olusanya, J. O., Williams, A. N., Wright, S. M., Manguerra, H., … Kassebaum, N. J. (2018). Developmental disabilities among children younger than 5 years in 195 countries and territories, 1990–2016: A systematic analysis for the Global Burden of Disease Study 2016. *The Lancet Global Health, 6*(10), e1100–e1121. https://doi.org/10.1016/S2214-109X(18)30309-7

Ostermaier, K. (2020). *Down syndrome: Clinical features and diagnosis*. Retrieved June from https://www.uptodate.com/contents/down-syndrome-clinical-features-and-diagnosis?search=Down%20Syndrome:%20Clinical%20features%20and%20diagnosis&source=search_result&selectedTitle=1~150&usage_type=default&display_rank=1

Oxelgren, U. W., Myrelid, Å., Annerén, G., Ekstam, B., Göransson, C., Holmbom, A., ... & Fernell, E. (2017). Prevalence of autism and attention-deficit–hyperactivity disorder in Down syndrome: a population-based study. *Developmental Medicine & Child Neurology, 59*(3), 276–283.

Parker, S. E., Mai, C. T., Canfield, M. A., Rickard, R., Wang, Y., Meyer, R. E., Anderson, P., Mason, C. A., Collins, J. S., Kirby, R. S., & Correa, A. (2010). Updated National Birth Prevalence estimates for selected birth defects in the United States, 2004-2006. *Birth Defects Research Part A: Clinical and Molecular Teratology, 88*(12), 1008–1016. https://doi.org/10.1002/bdra.20735

Patel, L., McNally, R. J., Harrison, E., Lloyd, I. C., & Clayton, P. E. (2006). Geographical distribution of optic nerve hypoplasia and septo-optic dysplasia in Northwest England. *Journal of Pediatrics, 148*(1), 85–88. https://doi.org/10.1016/j.jpeds.2005.07.031

Pliszka, S. (2007). Practice parameter for the assessment and treatment of children and adolescents with attention-deficit/hyperactivity disorder. *Journal of the American Academy of Child and Adolescent Psychiatry, 46*(7), 894–921. https://doi.org/10.1097/chi.0b013e318054e724

Privitera, M. D., Brodie, M. J., Mattson, R. H., Chadwick, D. W., Neto, W., & Wang, S. (2003). Topiramate, carbamazepine and valproate monotherapy: Double-blind comparison in newly diagnosed epilepsy. *Acta Neurologica Scandinavica, 107*(3), 165–175. https://doi.org/10.1034/j.1600-0404.2003.00093.x

Purdy, I. B., Singh, N., Brown, W. L., Vangala, S., & Devaskar, U. P. (2014). Revisiting early hypothyroidism screening in infants with Down syndrome. *Journal of Perinatology, 34*(12), 936–940.

Rare Disease Database. (2018). *Prader-Willi syndrome*. https://rarediseases.org/rare-diseases/prader-willi-syndrome/

Reekie, J., Hosking, S. P., Prakash, C., Kao, K. T., Juonala, M., & Sabin, M. A. (2015). The effect of antidepressants and antipsychotics on weight gain in children and adolescents. *Obesity Reviews, 16*(7), 566–580. https://doi.org/10.1111/obr.12284

Ryabets-Lienhard, A., Stewart, C., Borchert, M., & Geffner, M. E. (2016). The optic nerve hypoplasia spectrum: Review of the literature and clinical guidelines. *Advances in Pediatrics, 63*(1), 127–146. https://doi.org/10.1016/j.yapd.2016.04.009

Sarnat, H. B., & Flores-Sarnat, L. (2016). *Developmental disorders of the nervous system* (Vol. 86). https://www.clinicalkey.com/#!/content/book/3-s2.0-B9780323287838000892?scrollTo=%23hl0001291Developmental%20Disorders%20of%20the%20Nervous%20System%20Harvey%20B.%20Sarnat%20and%20Laura%20Flores-Sarnat.%20Bradley's%20Neurology%20in%20Clinical%20Practice,%2089,%201279-1300.e6

Skelton, J. A. (2021). Childhood Obesity Can Be an "Island in the Stream". Childhood Obesity

Skelton, A., & Klish, W. (2020). *Clinical evaluation of the child or adolescent with obesity*. Up to Date. https://www.uptodate.com/contents/clinical-evaluation-of-the-child-or-adolescent-with-obesity?search=obesity%20in%20children&source=search_result&selectedTitle=1~150&usage_type=default&display_rank=1

Teär Fahnehjelm, K., Dahl, S., Martin, L., & Ek, U. (2014). Optic nerve hypoplasia in children and adolescents; prevalence, ocular characteristics and behavioural problems. *Acta Ophthalmologica, 92*(6), 563–570. https://doi.org/10.1111/aos.12270

Turchi, R. M., & Smith, V. C. (2018). The role of integrated care in a medical home for patients with a fetal alcohol spectrum disorder. *Pediatrics, 142*(4), e20182333. https://doi.org/10.1542/peds.2018-2333

Wakeling, E. L., Brioude, F., Lokulo-Sodipe, O., O'Connell, S. M., Salem, J., Bliek, J., Canton, A. P., Chrzanowska, K. H., Davies, J. H., Dias, R. P.,

Dubern, B., Elbracht, M., Giabicani, E., Grimberg, A., Grønskov, K., Hokken-Koelega, A. C., Jorge, A. A., Kagami, M., Linglart, A., … Netchine, I. (2017). Diagnosis and management of Silver-Russell syndrome: First international consensus statement. *Nature Reviews Endocrinology, 13*(2), 105–124. https://doi.org/10.1038/nrendo.2016.138

Whooten, R., Schmitt, J., & Schwartz, A. (2018). Endocrine manifestations of Down syndrome. *Current Opinion in Endocrinology, Diabetes, and Obesity, 25*(1), 61–66. https://doi.org/10.1097/med.0000000000000382

Williams, J. F., & Smith, V. C. (2015). Fetal alcohol spectrum disorders. *Pediatrics, 136*(5), e1395–e1406. https://doi.org/10.1542/peds.2015-3113

Wolraich, M. L., Hagan, J. F., Jr., Allan, C., Chan, E., Davison, D., Earls, M., Evans, S. W., Flinn, S. K., Froehlich, T., Frost, J., Holbrook, J. R., Lehmann, C. U., Lessin, H. R., Okechukwu, K., Pierce, K. L., Winner, J. D., & Zurhellen, W. (2019). Clinical practice guideline for the diagnosis, evaluation, and treatment of attention-deficit/hyperactivity disorder in children and adolescents. *Pediatrics, 144*(4), e20192528. https://doi.org/10.1542/peds.2019-2528

Xu, H., & Verre, M. C. (2018). Type 2 diabetes mellitus in children. *American Family Physician, 98*, 590–593. https://www.aafp.org/afp/2018/1101/afp20181101p590.pdf

Zemel, B. S., Pipan, M., Stallings, V. A., Hall, W., Schadt, K., Freedman, D. S., & Thorpe, P. (2015). Growth charts for children with Down syndrome in the United States. *Pediatrics, 136*(5), e1204–e1211. https://doi.org/10.1542/peds.2015-1652

11 The INSPIRED Hospital Care of Children with Developmental Disability

Shavonne Temple and Pamela McPherson

Abstract

The work that needs to be done to ensure diverse patient populations receive equitable healthcare has been widely acknowledged. Children with autism and other developmental disabilities have specifically been identified as populations in need of equitable care. This chapter introduces the concept of INSPIRED care for children requiring accommodations during hospital care. INSPIRED is a pneumonic that we developed to provide a framework of recommendations to improve the care that children living with ASD (autism spectrum disorder)/IDD (intellectual developmental disability) receive in an inpatient hospital setting. INSPIRED stands for inform, network, screen/assess, prepare, implement, review, evaluate, and dream. The domains in the pneumonic can serve as a care plan guideline for staff at hospitals and other inpatient facilities. Although resources will vary among facilities, the goals outlined in the domains can be executed by the care team to promote quality patient care and professional development. Working toward the implementation and the practice of INSPIRED hospital care is a critical step in ensuring that all children receive intentional and inclusive healthcare while hospitalized.

Keywords

Autism · Intellectual disability · Inspired · Hospital · Communication · Care plan · Treatment plan

Introduction

The hospital care of children living with developmental disability should be approached with thoughtful, patient-centered planning. Hospital staff are often underprepared to adequately accommodate children and adolescents with autism spectrum disorder (ASD) and/or an intellectual developmental disability (IDD) in an inpatient setting. There are numerous, yet avoidable, obstacles that can complicate a child's stay: the medical team may lack the training to treat youth with ASD/IDD, direct care teams may lack integration with the interdisciplinary team, some hospital settings do not support patient's unique sensory and behavioral needs, and there are no generally accepted hospital protocols for accommodating persons with ASD/IDD. Multiple

S. Temple (✉) · P. McPherson
Department of Psychiatry and Behavioral Medicine, Louisiana State University Health Sciences Center, Shreveport, LA, USA
e-mail: snt001@lsuhs.edu

P. McPherson (ed.), *Handbook of Treatment Planning for Children with Autism and Other Neurodevelopmental Disorders*, Autism and Child Psychopathology Series,
https://doi.org/10.1007/978-3-031-06120-2_11

studies have shown that children with ASD/IDD are hospitalized at higher rates in comparison to their neurotypically developing peers (Romito et al., 2021). There are currently no national guidelines of best practices to follow when treating children with ASD/IDD in inpatient settings, but it is imperative that hospitals develop and implement practices to improve the healthcare outcomes of children with ASD/IDD. Several institutions across the country have implemented guidelines, called autism-specific care plans (ACPs), to address some of the disparities experienced by children with ASD while in the hospital (Broder-Fingert et al., 2016). Designing and implementing these ACPs have the potential to improve the experience for patients with ASD/IDD in the hospital setting by addressing disparities and expanding resources (Broder-Fingert et al., 2016; *Management-Patients-Learning-Disability-Autism-during-Pandemic.Pdf*, n.d.).

INSPIRED is a pneumonic that we developed to provide a framework of recommendations to improve the care that children living with ASD/IDD receive in an inpatient hospital setting. INSPIRED stands for inform, network, screen/assess, prepare, implement, review, evaluate, and dream. The domains in the pneumonic can serve as a care plan guideline for staff at hospitals and other inpatient facilities. Although resources will vary among facilities, the goals outlined in the domains should be executed by the care team to promote quality patient care and professional development. Working toward the implementation and the practice of INSPIRED hospital care is a critical step in ensuring that all children receive intentional and inclusive healthcare while hospitalized (Table 11.1).

Table 11.1 INSPIRED Hospital care for children with developmental disability

Domains	Description of the domain
*I*nform	Ensuring staff and patient are informed about expectations of hospital stay
*N*etwork	Establishing interdisciplinary teamwork that will help in caring for the patient
*S*creen/assess	Screen and assess level of care that will be needed
*P*repare	Prepare patients for the day-to-day experience in the hospital
*I*mplement	Implement and execute the treatment plan to ensure quality patient care and professional communication
*R*eview	Review and distribute patient satisfaction surveys to receive feedback and investigate barriers to discharge
*E*valuate	Evaluate and address barriers to care
*D*ream	Dream big and look toward a more inclusive future

Call for Accommodations

INSPIRED is a call for accommodating children living with disability during hospitalization. It is founded on moral responsibility, professional ethics, and the law. Medical professionals have a moral duty to provide for the needs of children in our care. The responsibility of the medical professionals toward children is articulated by the mission statement of the American Academy of Pediatrics: *to attain optimal physical, mental, and social health and well-being for all infants, children, adolescents and young adults* ("Mission Statement," p. 1). Our professional ethics admonish us to respect human dignity and rights while acting in the best interests of our patients (American Medical Association, 2001). While ethics admonish us to do the right thing, the law demands compliance. From the Rehabilitation Act of 1973, the first federal civil right law protecting persons with disability from discrimination, to the reauthorization and expansion of the Americans with Disability Act in 2011, the law dictates the legal obligation of healthcare systems to accommodate patients living with disability (Iezzoni & Agaronnik, 2020).

INSPIRED's accommodations to support the optimal health of children are necessary in all hospital settings. The National Health Interview Survey data for 2015–2018 documented a 17.8% prevalence of developmental disability among children 3–17 years of age (Zablotsky & Black, 2020), and one in six children living with intellectual developmental disability (IDD) experience more frequent emergency department (ED) visits and 80% more inpatient admissions than the general population, with higher associated costs (Lindgren et al., 2020). As the healthcare needs of children with IDD are often medically complex, the responsibility to accommodate the needs of children with IDD falls on medical professionals of every discipline and specialty. Children with IDD most commonly present to the ED with respiratory infections, superficial injury, or otitis media, with inpatient admissions mostly common due to epilepsy/convulsions, mood disorders, or pneumonia (Lindgren et al., 2020). Accommodating children with IDD requires looking beyond these presenting health issues. Family and individual needs must be considered as well as the social determinants of health and adverse childhood experiences that contribute to overall wellness and may influence how a person interacts with healthcare staff. The healthcare inequities faced by persons with IDD have been widely recognized, with a disproportionate burden on Black and Latinx children (Hall & Kurth, 2019; Krahn, 2019). Equitable comprehensive care requires that the impact of these factors on the individual be understood and addressed.

While patient health is the driving force behind the delivery of INSPIRED care, the law mandates that the needs of children with ASD/IDD be accommodated. The Rehabilitation Act of 1973 and the Americans with Disabilities Act (ADA) mandate accommodations for persons with disability. In addition to ensuring that hospitals are physically accessible, the ADA requires that communication be effective and that policies, practices, or procedures be reasonably modified if necessary ("The Americans with Disabilities Act of 1990," 1990). Reasonable modifications abound. Common examples include a communication board for children with speech disorders, translation services, or side rails on beds for children with seizures. Modifications must accommodate children and caregivers. The US Department of Justice Civil Rights Division has issued guidance for the implementation of the ADA in medical settings addressing a range of issues including communication, mobility, accessible design, and hearing or vision impairment (US DOJ Guidance Documents, 2020). Healthcare systems should address issues of access and quality through policy, staff education, and individualized accommodations according to patient's need (McClintock et al., 2018). To further protect the rights of persons with developmental disabilities, the Congress passed the Developmental Disabilities and Assistance and Bill of Rights Act of 2000 to

> assure that individuals with developmental disabilities and their families participate in the design of and have access to needed community services, individualized supports, and other forms of assistance that promote self-determination, independence, productivity, and integration and inclusion in all facets of community life. (p. 4)

The act defines a developmental disability as a

> severe, chronic disability of an individual that (i) is attributable to a mental or physical impairment or combination of mental and physical impairments; (ii) is manifested before the individual attains age 22; (iii) is likely to continue indefinitely; (iv) results in substantial functional limitations in 3 or more of the following areas of major life activity: (I) Self-care. (II) Receptive and expressive language. (III) Learning. (IV) Mobility. (V) Self-direction. (VI) Capacity for independent living. (VII) Economic self-sufficiency; and (v) reflects the individual's need for a combination and sequence of special, interdisciplinary, or generic services, individualized supports, or other forms of assistance that are of lifelong or extended duration and are individually planned and coordinated. (Developmental Disabilities Assistance and Bill of Rights Act of 2000, 2000, pp. 7–8)

The moral, ethical, and legal responsibilities of healthcare professionals to provide quality medical care to children with ASD/IDD provide the foundation for this call to healthcare systems to provide *INSPIRED* care.

The INSPIRED Hospital Care Framework

I, INFORM: Preparing for Hospitalization

INFORM encompasses two goals, informing the patient about the hospital and informing the hospital team about the patient. Nosocomial- or hospital-related medical complications are well studied and provide a familiar construct for considering the impact of hospitalization on the individual and family. The stress of hospitalization can be significant. Chang (2019) has proposed the Nosocomial stress model to recognize the significant psychosocial stresses associated with hospitalization. In line with this model, perceived threat with associated feelings of helplessness and fear has been associated with the development of PSTD (post-traumatic stress disorder) symptoms 30 days after discharge from the emergency department (Moss et al., 2020). The stress of hospitalization on children and parents has been widely studied, with stress disproportionately experienced by parents of children with ASD/IDD (Karpur et al., 2019; Muskat et al., 2015; Stremler et al., 2017). Careful preparation and attention to patient's and family's specific needs can minimize the risk of children or their parents becoming overly stressed by hospitalizations.

For planned hospitalizations, advance preparation for a hospital stay and care coordination can decrease stress on the patient and family and improve outcomes (Rauch, 2018). A developmentally appropriate explanation of the reasons for hospitalization and planned procedures can be incorporated into the informed consent/assent process for parents and older children. To minimize stress on younger children, care should be given to support parents. Toddlers and infants as young as 5 months typically assess new situations by referencing parental responses; therefore, the parent should not be unduly stressed in front of the child, if at all possible. Children with ASD generally exhibit deficits in social referencing, but a child's response to parental emotion can vary among children and in the same child over time (Sivaraman et al., 2020). It is not uncommon to witness severe distress in children with ASD when a trusted caregiver is overwhelmed. Preschoolers learn through play, creating an opportunity to prepare children for medical procedures with age-appropriate simulated experiences. For example, play with medical equipment, when appropriate, or toys similar to the equipment can make later procedures less frightening. Similarly, familiar toys may provide comfort during hospitalization and, if possible, during procedures. Young children also benefit from appropriate reassurance about recovery, that is, *Your body is powerful and we will help your body to heal itself*. Explanations offered to school-aged children with ASD/IDD should match developmental skills with frequent queries to confirm understanding (Harris et al., 2013). Of course, parents are the experts about their children, and the medical team should look to parents for guidance regarding the child's understanding and needs.

To inform the hospital team, the American Academy of Pediatrics and the American College of Emergency Physicians (2010) recommend that the medical home's shared plan of care include a standardized emergency information form (EIF) as the best method of ensuring rapid access to critical information when children with special healthcare needs require emergency care or hospitalization. An EIF may include a description of baseline functioning, any needed antibiotic prophylaxis, procedures to avoid, accommodations that may be calming or necessary for certain procedures, and details regarding equipment or technology that is in use. In addition to a detailed medical summary, the EIF may include advance directives and instruction regarding medical decision-making capacity. The EIF is available as an interactive electronic template that is an open source. The use of a standardized EIF is recommended to optimize patient care (American College of Emergency Physicians, 2010). It may be accessed electronically or obtained from the family or primary care provider.

Finally, INFORM reminds us of our responsibility to remain informed regarding best practices for the care for children living with disability. There have been repeated calls to incorporate a competency-based curriculum on the care disability into medical education (Ankam et al., 2019; Bowen et al., 2020; Jette & Field, 2007; Office of the Surgeon General, & Centers for Disease Control and Prevention, 2002). In 2019, the Alliance for Disability in Health Care Education proposed the *Core Competencies on Disability for Health Care Education* (Alliance for Disability in Health Care, 2019). The

INSPIRED's call to action echoes this call to implement the *Core Competencies on Disability for Health Care Education.*

N, NETWORK: Interdisciplinary Care

Ideally, everyone at the hospital should be trained to properly care for patients with ASD/IDD, but unfortunately, that is not always the case. Physicians report having significantly lower overall self-perceived competency in caring for children with ASD compared with children with other neurodevelopmental conditions, according to a survey of 539 pediatricians and family physicians across the nation (Broder-Fingert et al., 2016). It is recommended that hospitals consider assembling an interdisciplinary treatment team or list of specialists who can be called upon to assist with the care of children with ASD/IDD during hospitalization. Child life specialists, occupational therapists, child psychiatrists, social workers, and case managers are potential members of a cohesive team who can exponentially improve patients' hospital experience and overall health outcomes.

Child life specialists (CLSs) play an integral part in the care of children and adolescents while they are in the hospital. A CLS's role is to collaborate closely with the healthcare team to address psychosocial issues in order to help alleviate stress and anxiety related to hospitalization (Romito et al., 2021). CLSs tailor their interventions to a child's specific needs, and children with ASD especially benefit from this service as they typically have a difficult time adjusting to the hospital environment which can complicate and lengthen their stay. The emotional support, therapeutic play, and patient and family education provided by CLS improve patient outcomes by minimizing circumstances that may otherwise overwhelm the child and impede treatment (Romito et al., 2021). Two examples of effective interventions that child life specialists implement are sensitization strategies to prepare patients for medical procedures (i.e., a playful activity that includes placing a band around the patients' stomach to help desensitize them before percutaneous endoscopy gastrotomy tube placement) and coping/distracting techniques for patients to practice during times of distress (i.e., introducing breathing techniques, comfort holds, and guided imagery) (*Shm_pediatric_boost_implementation_guide.Pdf*, n.d.).

Occupational therapists (OTs) play a distinctive role in the care of children with ASD in the inpatient setting. They can assess how a patient is functioning and develop a plan enhancing the quality of the patient's hospital stay. OTs can conduct a sensory profile assessment and provide recommendations on sensory strategies that are feasible in the hospital setting (Broder-Fingert et al., 2016). They can serve as a great point of contact to gather information about environmental modifications that can be arranged, so the patient can be optimally accommodated (*Autismspeakstoolkit.Pdf.Pdf*, n.d.). OTs can also help the patients adjust to the environment by suggesting and teaching sensory integration and desensitization techniques, assisting with motor development and skills to promote appropriate exercise, and teaching self-advocacy tips, so the patients can communicate when something is wrong in the absence of their caregiver (*Autism Fact Sheet.Pdf*, n.d.).

Speech and language pathologists (SLPs) can help devise simple communication systems to facilitate clear communication between the child and the care team (Thom et al., 2020). Across settings, communication is a common barrier that complicates care. SLPs assist in assessing the patient's communication strengths and preferences and recommending communication modalities that should be used during the patient's visit (*Speech Therapy*, n.d.). SLPs also assist patients in enhancing their social interactions with medical staff and addressing swallowing- or feeding-related challenges that may occur (*Speech Therapy*, n.d.).

Child psychiatry should be consulted if comorbid psychiatric disorders complicate the admission or if psychopharmacologic interventions may be helpful to manage acute distress (Thom et al., 2020). Child psychiatrists assess mental status changes and behavioral challenges and recommend pharmacological and behavioral management strategies to maximize overall care. Child psychiatry can also be consulted for less urgent situations such as providing family education on mental illness or developmental disorders or informing the medical team of any new psychiatric protocols that may be recommended for the patient's condition and treatment plan. Child psychiatrists can lead in devising a holistic behavioral treatment plan for the patient.

The early involvement of case management and social workers plays a vital role in identifying and addressing barriers and social transitions that patient may experience after discharge and can prevent or reduce the rate of readmission. The case managers and social workers should ensure that the appropriate referrals are in place and communicated with the caretakers before discharge is finalized. The treatment team should pay particularly close attention to patients aging out of pediatric services because studies have shown that these patients are rarely included in the transition plan from pediatric inpatient services to adult care facilities and often get lost to follow-up (Straus et al., 2019). Transitioning to adult services is addressed in detail in the final chapter of this book.

Networking is not only important for healthcare professionals, but it is also beneficial to patients and their families as well. It can be difficult for children and their families to navigate complications with their illness at baseline and that difficulty is exacerbated during hospital visits. Support groups – whether through community programs, social media, or medical facilities – have shown to substantially reduce loneliness, stress, and anxiety and to improve coping skills. Support groups also foster community and connection which can have a profound beneficial impact on the patients' and caregivers' overall mental health. The healthcare team should inform patients and their caregivers about support groups and encourage them to explore their options. Hospitals should curate a list of local support resources for families. The website, https://www.elemy.com/studio/autism-resources/support-groups, is a great resource that makes it easy for users to find a variety of support groups in their geographical area.

S, SCREEN/ASSESS: Tell Me More About You

Gathering baseline information on the patient's intellectual ability, preferred communication style, and environmental triggers prior to their admission to the hospital has been shown to improve the quality of care of patients with ASD (*Managing Patients With Autism to Effectively Deliver Care*, n.d.). In addition to the standardized emergency information form discussed previously, an effective tool that has gained popularity in hospitals is the healthcare passport. A healthcare passport helps provide important information and facilitates communication to the medical team about the patient's specific needs and preferences (*Managing Patients With Autism to Effectively Deliver Care*, n.d.). In some cases, hospitals have these passports available and ask the caregiver to fill the form out prior to admission. In the case that the institution does not provide these passports, there are many online resources that have healthcare passport templates. For children receiving care in a medical home, information from the shared plan of care can inform the healthcare passport. Broder-Finger et al. at the Arkansas Children's Hospital conducted a retrospective pilot study that demonstrated that the use a five-question screening assessment helped to improve the care for patients with ASD (Broder-Fingert et al., 2016; *Managing Patients With Autism to Effectively Deliver Care*, n.d.). The assessment is used to identify the presence or absence of an ASD diagnosis and the presence or absence of difficulties with four behavioral components: communication, environment, safety, and transition (*Management-Patients-Learning-Disability-Autism-during-Pandemic.Pdf*, n.d.). See Fig. 11.1 for a sample screening questionnaire.

Even with referral information gathered in preparation for hospitalization, it is imperative

Question	Yes or no?
Has your child been diagnosed with ASD or an IDD?	
Does your child have difficulties with communication?	
Is your child sensitive to sounds, textures, clothing, touch, odors, colors or other elements in the environment?	
Does your child have meltdowns or behavioral outbursts?	
Does your child wander from home or school?	
Does your child have difficulties transitioning from one task to another?	

Fig. 11.1 Sample screening questionnaire

that the medical team directly assess four behavioral domains (communication, environment, safety, and transition) to ensure they can accommodate each patient's unique needs. The assessment tests communication strategies – whether that is through verbal speech, sign language, a communication device, facial expressions, or gestures (Broder-Fingert et al., 2016; *Managing Patients With Autism to Effectively Deliver Care*, n.d.). The assessment includes observation regarding the patient's response to the environment, including how the patient responds to fluctuations with lighting, sounds, odors, or colors (*Managing Patients With Autism to Effectively Deliver Care*, n.d.). Regarding safety, the assessment consists of asking parents whether the patient is prone to wandering or expressing agitation or self-injury when placed in distressing situations (*Managing Patients With Autism to Effectively Deliver Care*, n.d.). For children prone to wandering, accommodations might include a door or bed alarm or a sitter. Parents can also be asked if the child wears any type of tracking device. Lastly, the treatment team will conduct an ongoing assessment of how the patient handles changes in the environment with notation of any temporal or situational factors that increase or ameliorate distress (*Managing Patients With Autism to Effectively Deliver Care*, n.d.; Straus et al., 2019).

A screening tool or healthcare passport can assist medical staff by providing information about the patient to inform a hospital stay. It is also important that members of the medical team possess the skills to care for these patients once they are aware of their condition and the necessary accommodations. Many physicians outside of the field of psychiatry admit that they are uncomfortable or feel underprepared when taking care of patients with ASD and IDDs (Malik-Soni et al., 2021). Physicians in training (and other healthcare professionals) typically do not receive dedicated instruction about ASD and the complexities of the condition. Training may not afford adequate exposure to treating patients with ASD/IDD. Unaddressed implicit biases compounded with feelings of anxiety and discomfort can lead to healthcare professionals to deliver suboptimal medical care. Although this may be inadvertent, this can lead to healthcare professionals being brief during patient-provider encounters, offending patients and their families through the use of harmful, outdated language when speaking about patients' condition and making increased medical errors when diagnosing and treating the patient. Hospitals, medical schools, residency programs, and other healthcare professional programs must do their part and expand their curricula to include more information about ASD, its etiology, and evidence-based practices of treatment. From an individual standpoint, healthcare professionals must self-reflect and uncover and address any implicit biases and modify their behavior. Above all, we must remain humble and respect the lived experience of families and children. Further discussion about biases will be discussed later in the chapter, under the *Evaluation* domain.

P, PREPARATION: Day-to-Day Experience in the Hospital

The hospital team has the dual roles of preparing the team and preparing the child and family for the hospital stay. Preparing children for their stay in the hospital can be tough, as the hospital is not a fun place to be. Preparing children for the ins and outs of their day-to-day experience in the hospital can help increase positive interactions and improve the overall health outcomes for the patients. In addition to the hospital having measures in place for prospective patients, the parents and/or caretakers can also play a vital role in ensuring that the patient's stay goes as smoothly as possible. An abrupt change in routine can have a huge impact on the child's emotional and mental well-being (*Management-Patients-Learning-Disability-Autism-during-Pandemic.Pdf*, n.d.). Family members and/or caretakers can be encouraged to obtain items from home that can facilitate communication (i.e., alternative communication devices), relieve distress (i.e., favorite toy or gadget), and foster feelings of familiarity (i.e., bedsheets from home, favorite stuffed animal, sentimental photographs) (Thom et al., 2020).

Once a medical team receives the admit orders, they should prepare to make any necessary accommodations for the patients and their caregiver. The Rady Children's Hospital in San Diego encourages their employees to use the pneumonic, "ASSUME," when thinking about the course of action they should take when taking care of patients with ASD (*Autismspeakstoolkit.Pdf.Pdf*, n.d.). "'A' is for ask the right questions, 'S' is for speak in simple short phrases, 'S' is for speak to an autism expert, if necessary, 'U' is for uncover invisible rules and set clear expectations, 'M' is for motivate the child with rewards for compliance, and 'E' to expect the child to have a hard time so that you are prepared" (*Autismspeakstoolkit.Pdf.Pdf*, n.d.). This pneumonic can be useful to prompt members on the inpatient team to remember simple adjustments they can make while treating children with ASD/IDD.

Optimizing the medical and interdisciplinary team's communication style can facilitate effective communication between the hospital staff and the patient. Strategies for good communication between shifts and among medical professionals should be maintained, such as a standardized transition approach such as I-PASS (Starmer et al., 2012). The I-PASS Institute promotes an evidence-based model for hand-off patient care at shift changes with a standardized curriculum. Hand-off procedures are increasingly automated within electronic health records (Potts et al., 2018).

In addition to communication among team members, the medical team must maintain good communication with families. There are numerous tactics the medical team, along with the interdisciplinary team, can use to engage in communication with the patients throughout their visit. While it is extremely important to involve the family members and/or caretakers in the discussion about the patient's medical assessment and plan, the children should also be informed about what will be happening with them during their stay. It is important that the child is aware and prepared for each encounter that they may have with the medical team, and it is helpful to use creative techniques including play to introduce things that may be frightening or anxiety provoking (*Autismspeakstoolkit.Pdf.Pdf*, n.d.). The hospital team can implement a daily check-in/checkout with the child to prepare for each day and review what went well and what may need accommodation.

The family members and/or caretakers should be encouraged to share any specific concerns with the medical team and discuss strategies that might help minimize distress. If necessary, the medical team can make certain adjustments like allocating a clinician by gender and placing patient in a private room to avoid excess noise and activity (*Management-Patients-Learning-Disability-Autism-during-Pandemic.Pdf*, n.d.). Other interventions that can be helpful are keeping low lighting, encouraging therapeutic play, keeping staff switching to a minimum, providing dietary accommodations, and teaching coping and pain management strategies (Romito et al., 2021). These interventions can make a significant difference in keeping the child comfortable and cooperative. Some hospitals and medical facilities provide goodie bags and technological devices for their pediatric patients. The goodie bags contain items that can help soothe the children or redirect their frustration, such as writing materials, coloring books, puzzles, bubbles, and simple games (*Autismspeakstoolkit.Pdf.Pdf*, n.d.; Straus et al., 2019). Squishy toys and chewing nontoxic silicon chews can be used to relieve stress and anxiety; they can also keep the child occupied and stimulated between activities (*Autismspeakstoolkit.Pdf.Pdf*, n.d.). Allowing the child to watch videos during physical exams and procedures, counting out loud to communicate the passage of time, and taking blood samples by thumb prick as opposed to a needle can greatly reduce discomfort while enduring these tasks (Straus et al., 2019).

During a typical hospital stay, there are many procedures and routines that children undergo that can be extremely distressing without proper preparation and support. Cyclical vital checks, bed sheet changes, morning and afternoon rounds, lab drawing, and undergoing noninvasive and invasive procedures can be overwhelming for the patient and increase uncooperative behaviors and distrust for the medical team. Developmentally appropriate communication can help minimize adverse effects of hospitalization (Romito et al., 2021). First/then cards and social stories are great ways to communicate, with visuals, what they are to expect. First/then cards are used to set small goals for patients who are very young, unable to follow a visual schedule, have behavioral challenges, or have limited communication skills (Romito et al., 2021). These cards consist of two pictures – the "first" card displays the task that you would like the child to do, and the "then" card displays the reward the child will receive after completing the task. These cards can help acutely motivate the child to do small tasks like taking his/her medicine or cooperating with taking vitals/imaging/lab work. This concept is used to increase the patient's cooperativity and compliance through positive reinforcement. Social stories are used to express complex information in a digestible, eye-catching format. Social stories can serve as a visual schedule of the sequential events that will take place while child is hospitalized (*Autismspeakstoolkit.Pdf.Pdf*, n.d.). The story should have numerous illustrations and simple descriptions and should be easy to follow. These stories serve as a very effective tool to express expectations and set clear boundaries to facilitate cooperation for the new and unfamiliar encounters the children will have while in the hospital (*Autismspeakstoolkit.Pdf.Pdf*, n.d.).

I, IMPLEMENT: The Plan in Action

Communication is an essential thread running through all domains of the *INSPIRED* framework guidelines. The treatment plan provides a guide to quality patient care and communication between medical professionals. Successful implementation of an individualized hospital treatment plan requires ongoing communication between members of the medical team and with the family and patient. Communication includes sharing information, education as needed, and anticipation of possible challenges delivered at the knowledge and communication level of each person. Even with expert communication, care challenges will arise. Families are key in identifying evolving challenges and problem-solving solutions as they often recognize evolving issues early and have the most experience soothing their child. Parents are more likely to voice concerns and solutions if their expertise has been acknowledged and valued.

Understanding how children with ASD/IDD may communicate with you is also extremely important. Some children may be nonverbal, not use eye contact, or not show their emotions through facial expressions (*Autismspeakstoolkit.Pdf.Pdf*, n.d.). Identifying the patient's unique communication styles can alleviate unnecessary stressors for the inpatient teams (Straus et al.,

2019). Frequent check-ins with family members and/or caretakers will be necessary to avoid misinterpretation of the patient's actions. It is also important to inquire about the child's behavioral response to pain and discomfort – laughing, crying, trying to hurt themselves, and becoming withdrawn or quiet might all be responses to pain (*Management-Patients-Learning-Disability-Autism-during-Pandemic.Pdf*, n.d.). Even children with good communication may not articulate their level of pain the same as their neurotypical peers – they may say the pain is less acute than you may anticipate, so do not hesitate to ask the caretaker to interpret the patient's behavior (*Management-Patients-Learning-Disability-Autism-during-Pandemic.Pdf*, n.d.).

Child psychiatry should be consulted if comorbid psychiatric disorders complicate the admission. In addition to psychiatric disorders, common challenges leading to consultation with psychiatry include self-harming behaviors, aggression, and emotional distress (Becker et al., 2020). The first step in addressing any challenge is to fully characterize the behavior, noting antecedents/triggers and consequences, before identifying the function of the behavior – escape, attention, self-stimulation, obtaining a desired object, or response to a medical/mental health issue (Matson et al., 2012). This process, called a functional behavior assessment, allows a targeted intervention or coping plan to be formulated to ameliorate the challenging behavior.

Self-harming behaviors include self-injurious behavior (SIB), non-suicidal self-injury (NSSI) behavior, and suicidal behavior. SIB typically refers to head banging, hitting, pinching, scratching, biting, or other forms of self-injury by persons with developmental disability. The prevalence of SIB in children with ASD/IDD is 4–53%, with an increased prevalence in some genetic disorders including Lesch-Nyhan, Prader-Willi, and Fragile X syndromes (Erturk et al., 2018). More frequent and varied repetitive behaviors, lower IQ, and more severe SIB at home predicted SIB in a psychiatric hospital setting (Handen et al., 2018). NSSI is a self-injury to regulate mood or resolve an interpersonal difficulty and increases the risk of suicidal behaviors. Youth with ASD are at increased risk of self-harm and suicidal behaviors, but additional research is needed to better understand the prevalence and implications for general hospital care (Oliphant et al., 2020). Hospital policy typically offers direction regarding suicide risk assessment. In response to a recommendation of the Joint Commission, an international work group formulated detailed guidance on the pediatric suicide risk assessment in hospital settings highlighting the importance of using standardized instruments, identifying risk and protective factors, and completing safety planning for those at risk (Brahmbhatt et al., 2019; Joint, 2016).

Hospitalization stresses children and commonly leads to agitation or aggression in emergency departments and on pediatric units (Gerson et al., 2019; Malas et al., 2017). The American Association of Emergency Psychiatrists has issued a consensus statement on best practices for addressing agitation in emergency departments (Gerson et al., 2019). This statement aligns with the standard of care for crisis intervention with actions beginning with the least restrictive measures. Prevention, the least restrictive and best option, is encompassed in the *I*, *N*, *S*, and *P* of *INSPIRED* care.

Even with reasonable attempts to make the child's visit as comfortable as possible and alleviate prospective insults, moments of distress may inevitably arise, and the hospital team should have an actionable plan in place. In times of distress, children regress. The child's ability to understand spoken language can become more limited, so it is important to stay calm and avoid reacting reflexively (*Autismspeakstoolkit.Pdf.Pdf*, n.d.). In moments where the patients are inconsolable, give them space, use minimal speech (one to three words), and dim the lights to try and get the children in a calmer state (*Autismspeakstoolkit.Pdf.Pdf*, n.d.). It can also be helpful to use visual aids to communicate any instructions you may want them to follow and try to redirect their focus to one of the items in their goodie bag. If the caretaker is nearby, reach out to them to ask for strategies they use at home that help calm the patient. When agitation occurs, a functional behavior assessment should be undertaken to identify the function of the behavior. As children with ASD/IDD may have unique responses to pain, a physical etiology should be ruled out as part of the assessment. Intervention should progress from treating medical issues to removing sources of distress and offering comfort before considering pharmacologic interventions. Restraints should be used only as a last resort and in accordance with the Center for Medicare and Medicaid Services and Joint Commission requirements (Nunno et al., 2021).

R, REVIEW: How Did We Do?

A critical component of the *INSPIRED* care guidelines is the *Review* domain. Feedback is crucial for ensuring fidelity following the implementation of any framework. Hospitals typically review patient's concerns in the form of satisfaction surveys. These surveys can offer valuable guidance for achieving excellence in patient care. Patient surveys are a great tool to review and assess if the implementation of INSPIRED guidelines is resulting in patients and caregivers reporting that they are receiving equitable, patient-centered, inclusive care. These surveys should gather information about the patients' and caregivers' demographics (i.e., age, race, diagnosis, gender), health literacy (i.e., patients' and caregivers' understanding of their reason for admission and their understanding of the diagnosis and treatment plan), medication management (i.e., availability of proper formulation and/or compounding, dosing syringes, palatability/adherence), and safety concerns (i.e., warning signs and symptoms to pay attention to that may warrant emergent follow-up) (*Shm_pediatric_boost_implementation_guide.Pdf*, n.d.). The patient surveys can also include topics that can help hospitals/medical facilities learn more about the needs of their patient population in general. Surveys would benefit from including information related to social determinants of health to improve equity and increase understanding of provider bias (Daniel et al., 2018). These topics include gathering information about the patient's environment (i.e., Is the patient in foster or convalescent care? Does patient receive home health or rehab care?), transportation (i.e., Does patient/caregiver rely on public transportation), and access to care (i.e., Does patient's insurance serve as a barrier to receiving care?) (*Shm_pediatric_boost_implementation_guide.Pdf*, n.d.). Another important component, which may arguably be the most important component of the survey, should be to get direct feedback about how the patients and their caregivers feel they were treated by the inpatient team. An example of some questions that can be asked are listed in the Fig. 11.2 below.

Did the treatment team address your concerns and answer your questions?
Did your treatment team spend enough time explaining why your child was receiving labs/imaging/procedures while in the hospital?
Did the treatment team explain medical information clearly and in a way that you could easily understand?
Did the treatment team respect you and your child?
Did the treatment team prepare you to care for your child's medical needs after discharge?
Did the treatment team provide follow-up appointments for your child?
Would you recommend this hospital to others?
Are there any suggestions or comments that you have for us?

Fig. 11.2 Sample patient survey questions

After gathering feedback about how the patients and their caregivers perceived their stay, it is important to investigate barriers to discharging the patients. Inefficient discharge processes often lead to lower patient satisfaction scores, higher readmission rates, and poor clinical outcomes (*Shm_pediatric_boost_implementation_guide.Pdf*, n.d.). Some common barriers to discharge are unreliable metric to assess patient's/caregiver's preparedness for discharge, lack of standardization in the discharge process, inefficient method to assess medication safety, and incohesive flow of allocating patient's follow-up care. Children with behavioral health or neurodivergent concerns are at a higher risk of experiencing issues with being discharged and subsequent rehospitalization. According to a study done at the University of California (UC), one of the main culprits in the increased number of unplanned readmissions is fragmentation of care following hospital discharge (Berry et al., 2011). It was also found that approximately one in five pediatric patients discharged from the hospital experience an adverse event (an injury related to medical management, in contrast to complications of disease) that could have been prevented – with miscommunications being the leading cause of the adverse events (Forster et al., 2014). Here are numerous methods for hospitals to collect data. The University of California developed a tool called GAPP, which stands for General Assessment for Pediatric Patients. GAPP was designed to gather pertinent information about the pediatric populations. This tool is used to gather information to assess socioeconomic, behavioral, cultural, linguistic, and other factors that may contribute to a patient's barrier to discharge (*Shm_pediatric_boost_implementation_guide.Pdf*, n.d.). The GAPP tool developed by the University of California Integrating Patient Care and Health Professions Education team can be accessed at https://www.hospitalmedicine.org/globalassets/clinical-topics/clinical-pdf/shm_pediatric_boost_implementation_guide.pdf.

E, EVALUATE: Evaluate and Address Barriers to Care

The *INSPIRED* care guidelines are intended to assist pediatricians and other healthcare professionals with providing exceptional care to children with ASD/IDDs in the hospital setting, but it must be acknowledged that there will still be barriers to executing these practices. Some obstacles that complicate the care and treatment of children with ASD/IDD, and plausible solutions, were mentioned in the beginning of the chapter: a deficit of specialized knowledge and training among physicians, lack of integration between the interdisciplinary teams, and the hospital's lack of resources to support the patient's unique sensory, behavioral, and communicative needs. However, there are other barriers such as physician wellness (or lack thereof) and stigma about autism (Malik-Soni et al., 2021).

Physician wellness has recently been a popular topic in medicine, but there has not yet been an impactful shift in the culture that has led to any significant changes. A huge component that negatively impacts physician wellness is burnout, which can be caused by chronic overwork, constant stressors, and a grueling work environment (West et al., 2018). Physician burnout can result in physicians making diagnostic and other medical errors, appearing detached and/or exhibiting negative feelings toward patients, and increased rates of anxiety, depression, and suicidal ideation. When burnout ensues, the physicians and their patients suffer consequences including – but not limited to – reduced productivity, increased physician turnover, decreased patient satisfaction, and poor patient outcomes (Han et al., 2019). This can result in devastatingly disproportionate treatment outcomes for vulnerable populations such as children with ASD/IDDs. Implementations that can help increase physician wellness and reduce burnout are reducing time pressures, hiring staff to perform tasks unrelated to direct patient care, decreasing work hours (including strict following of Accreditation Council for Graduate Medical Education's work restriction guidelines), and increasing vacation time and paid leave (Malik-Soni et al., 2021; Patel et al., 2019). Allotted time for professional coaching, stress reduction classes, and accessible mental health counseling should also be readily accessible and available for those who wish to use those resources (Patel et al., 2019). Improving physician wellness will result in having happier, healthier healthcare providers and better management and treatment outcomes for patients.

Stigma about ASD is not only prevalent in the general population but it also perseverates among healthcare professionals. It is important for healthcare providers to know that stigma surrounding ASD deeply affects many patients and their caregivers, requiring additional time and attention to fully assess patient's needs. Patients and their families often suffer from feelings of rejection, isolation, depression, and anxiety. The effects of stigma of ASD impact access to equitable care and treatment services (Malik-Soni et al., 2021). In addition to healthcare professionals addressing their biases on their own, there needs to be systemic change to successfully combat bias and stigma. Faculties of medicine and organizations that accredit medical school curricula should promulgate healthcare standards that ensure trainees are learning about the best practices of care for people with ASD/IDDs (Malik-Soni et al., 2021). There should be more learning opportunities that prepare physicians and other healthcare professionals to apply appropriate tools and technologies to maximize approaches to care (Malik-Soni et al., 2021). Healthcare professionals should also support the advocacy and research groups that launch initiatives that reduce stigma around ASD to continue increasing healthcare outcomes for children with ASD (Malik-Soni et al., 2021).

D, DREAM: Dream for a Better Tomorrow

Although there is a lot of more work that needs to be done to ensure diverse patient populations receive equitable care, steps are being made in the right direction. For example, recommendations have recently been made to inform the practice of applied behavioral analysis with Black families (Čolić et al., 2021), and lived experience of persons with ASD/IDD is increasingly being heard as healthcare systems develop services (Benevides et al., 2020). These steps and the steps you take to work hard for your patients are victories. You must relish in the moment and appreciate your successes. Celebrate all victories, even the small ones that occur each shift. Actively look for victories. Celebrating success is a well-known managerial principle linked to high-quality workplace performance and an important component of self-care (Juran et al., 1974; Stubbe, 2017). We should all dream big and keep pushing for a more progressive and inclusive future. Working together within our medical teams and with patients and families to achieve INSPIRED care will undoubtedly result in success!

References

Alliance for Disability in Health Care Education. (2019). *Core competencies on disability for health care education*. Alliance for Disability in Health Care Education.

American College of Emergency Physicians. (2010). Emergency information form for children with special health care needs. *Annals of Emergency Medicine, 56*(3), 315.

American Medical Association. (2001). *AMA principles of medical ethics. Code of medical ethics*.

Ankam, N. S., Bosques, G., Sauter, C., Stiens, S., Therattil, M., Williams, F. H., et al. (2019). Competency-based curriculum development to meet the needs of people with disabilities: A call to action. *Academic Medicine, 94*(6), 781–788.

Autism fact sheet.pdf. (n.d.). Retrieved August 18, 2021, from https://www.aota.org/-/media/corporate/files/aboutot/professionals/whatisot/cy/fact-sheets/autism%20fact%20sheet.pdf

Autismspeakstoolkit.pdf.pdf. (n.d.). Retrieved May 27, 2021, from https://www.rchsd.org/documents/2015/03/autismspeakstoolkit.pdf/

Becker, J. E., Smith, J. R., & Hazen, E. P. (2020). Pediatric consultation-liaison psychiatry: An update and review. *Psychosomatics, 61*(5), 467–480. https://doi.org/10.1016/j.psym.2020.04.015

Benevides, T. W., Shore, S. M., Palmer, K., Duncan, P., Plank, A., Andresen, M.-L., et al. (2020). Listening to the autistic voice: Mental health priorities to guide research and practice in autism from a stakeholder-driven project. *Autism, 24*(4), 822–833.

Berry, J. G., Agrawal, R., Kuo, D. Z., et al. (2011). Characteristics of hospitalizations for patients who use a structured clinical care program for children with medical complexity. *The Journal of Pediatrics, 59*, 284–290.

Bowen, C. N., Havercamp, S. M., Bowen, S. K., & Nye, G. (2020). A call to action: Preparing a disability-competent health care workforce. *Disability and Health Journal, 13*(4), 100941.

Brahmbhatt, K., Kurtz, B. P., Afzal, K. I., Giles, L. L., Kowal, E. D., Johnson, K. P., et al. (2019). Suicide risk screening in pediatric hospitals: Clinical pathways to address a global health crisis. *Psychosomatics, 60*(1), 1–9.

Broder-Fingert, S., Shui, A., Ferrone, C., Iannuzzi, D., Cheng, E. R., Giauque, A., Connors, S., McDougle, C. J., Donelan, K., Neumeyer, A., & Kuhlthau, K. (2016). A pilot study of autism-specific care plans during hospital admission. *Pediatrics, 137*(Supplement 2), S196–S204. https://doi.org/10.1542/peds.2015-2851R

Čolić, M., Araiba, S., Lovelace, T. S., & Dababnah, S. (2021). Black caregivers' perspectives on racism in ASD services: Toward culturally responsive ABA practice. *Behavior Analysis in Practice*, 1–10. https://doi.org/10.1007/s40617-021-00577-5

Chang, B. P. (2019). Can hospitalization be hazardous to your health? A nosocomial based stress model forhospitalization. *General Hospital Psychiatry, 60*, 83–89.

Daniel, H., Bornstein, S. S., & Kane, G. C. (2018). Addressing social determinants to improve patient care and promote health equity: An American College of Physicians position paper. *Annals of Internal Medicine, 168*(8), 577–578.

Erturk, B., Machalicek, W., & Drew, C. (2018). Self-injurious behavior in children with developmental disabilities: A systematic review of behavioral intervention literature. *Behavior Modification, 42*(4), 498–542.

Forster, A. J., Clark, H. D., Menard, A., et al. (2014). Adverse events among medical patients after discharge from hospital. *CMAJ, 170*, 345–349.

Gerson, R., Malas, N., Feuer, V., Silver, G. H., Prasad, R., & Mroczkowski, M. M. (2019). Best practices for evaluation and treatment of agitated children and adolescents (BETA) in the emergency department: Consensus statement of the American Association for Emergency Psychiatry. *The Western Journal of Emergency Medicine, 20*(2), 409–418. https://doi.org/10.5811/westjem.2019.1.41344

Hall, J. P., & Kurth, N. K. (2019). A comparison of health disparities among Americans with intellectual disability and/or autism spectrum disorder and Americans with other disabilities. *Inclusion, 7*(3), 160–168.

Han, S., Shanafelt, T. D., Sinsky, C. A., Awad, K. M., Dyrbye, L. N., Fiscus, L. C., et al. (2019). Estimating the attributable cost of physician burnout in the United States. *Annals of Internal Medicine, 170*(11), 784–790.

Handen, B. L., Mazefsky, C. A., Gabriels, R. L., Pedersen, K. A., Wallace, M., & Siegel, M. (2018). Risk factors for self-injurious behavior in an inpatient psychiatric sample of children with autism spectrum disorder: A naturalistic observation study. *Journal of Autism and Developmental Disorders, 48*(11), 3678–3688.

Harris, T. B., Sibley, A., Rodriguez, C., & Brandt, M. L. (2013). Teaching the psychosocial aspects of pediatric surgery. *Seminars in Pediatric Surgery, 22*(3), 161–166. https://doi.org/10.1053/j.sempedsurg.2013.05.005

Iezzoni, L. I., & Agaronnik, N. D. (2020). Healthcare disparities for individuals with disability: Informing the practice. In *Disability as diversity* (pp. 15–31). Springer.

Jette, A. M., & Field, M. J. (2007). *The future of disability in America*. National Academies Press.

Joint, C. (2016). *Sentinel Event Alert# 56: Detecting and treating suicide ideation in all settings*. Joint Commission. In.

Juran, J. M., Gryna, F. M., & Bingham, R. S. M. (1974). *Quality control handbook*. McGraw-Hill. Kaming, P. F., Olomolaiye, P. O., Holt, G. D., & Harris, F. C. (1997). Factors influencing construction time and cost overruns on high-rise projects in Indonesia. *Construction Management and Economics, 15*, 83–94.

Karpur, A., Lello, A., Frazier, T., Dixon, P. J., & Shih, A. J. (2019). Health disparities among children with autism spectrum disorders: Analysis of the national survey of children's health 2016. *Journal of Autism and Developmental Disorders, 49*(4), 1652–1664.

Krahn, G. L. (2019). A call for better data on prevalence and health surveillance of people with intellectual and developmental disabilities. *Intellectual and Developmental Disabilities, 57*(5), 357–375.

Lindgren, S., Lauer, E., Momany, E., Cope, T., Royer, J., Cogan, L., et al. (2020). Disability, hospital care, and cost: Utilization of emergency and inpatient care by a cohort of children with intellectual and developmental disabilities. *The Journal of Pediatrics, 229*, 259–266.

Malas, N., Spital, L., Fischer, J., Kawai, Y., Cruz, D., & Keefer, P. (2017). National survey on pediatric acute agitation and behavioral escalation in academic inpatient pediatric care settings. *Psychosomatics, 58*(3), 299–306.

Malik-Soni, N., Shaker, A., Luck, H., Mullin, A. E., Wiley, R. E., Lewis, M. E. S., Fuentes, J., & Frazier, T. W. (2021). Tackling healthcare access barriers for individuals with autism from diagnosis to adulthood. *Pediatric Research*, 1–8. https://doi.org/10.1038/s41390-021-01465-y

Management-Patients-Learning-Disability-Autism--during-Pandemic.Pdf. (n.d.). Retrieved May 27, 2021, from https://www.nice.org.uk/media/default/about/covid-19/specialty-guides/management-patients-learning-disability-autism-during-pandemic.pdf

Managing Patients With Autism to Effectively Deliver Care. (n.d.). Retrieved May 27, 2021, from https://www.childrenshospitals.org/Newsroom/Childrens-Hospitals-Today/Issue-Archive/Issues/Fall-2014/Articles/Managing-patients-with-autism-to-effectively-deliver-care

Matson, J. L., Tureck, K., & Rieske, R. (2012). The Questions About Behavioral Function (QABF): Current status as a method of functional assessment. *Research in Developmental Disabilities, 33*(2), 630–634. https://doi.org/10.1016/j.ridd.2011.11.006

McClintock, H. F., Kurichi, J. E., Barg, F. K., Krueger, A., Colletti, P. M., Wearing, K. A., & Bogner, H. R. (2018). Health care access and quality for persons with disability: Patient and provider recommendations. *Disability and Health Journal, 11*(3), 382–389.

Mission Statement. Retrieved from https://www.aap.org/en/about-the-aap/

Moss, J., Roberts, M. B., Shea, L., Jones, C. W., Kilgannon, H., Edmondson, D. E., et al. (2020). Association between perceived threat and the development of posttraumatic stress disorder symptoms in patients with life-threatening medical emergencies. *Academic Emergency Medicine, 27*(2), 109–116. https://doi.org/10.1111/acem.13877

Muskat, B., Burnham Riosa, P., Nicholas, D. B., Roberts, W., Stoddart, K. P., & Zwaigenbaum, L. (2015). Autism comes to the hospital: The experiences of patients with autism spectrum disorder, their parents and health-care providers at two Canadian paediatric hospitals. *Autism, 19*(4), 482–490.

Nunno, M. A., McCabe, L. A., Izzo, C. V., Smith, E. G., Sellers, D. E., & Holden, M. J. (2021). A 26-year study of restraint fatalities among children and adolescents in the United States: A failure of organizational structures and processes. In *Child & youth care forum* (pp. 1–20). Springer US.

Office of the Surgeon General, & Centers for Disease Control and Prevention. (2002). *Closing the gap: A national blueprint to improve the health of persons with mental retardation*. U.S. Department of Health and Human Services, Public Health Service, Office of the Surgeon General.

Oliphant, R. Y. K., Smith, E. M., & Grahame, V. (2020). What is the prevalence of self-harming and suicidal behaviour in under 18s with ASD, with or without an intellectual disability? *Journal of Autism and Developmental Disorders, 50*(10), 3510–3524.

Patel, R. S., Sekhri, S., Bhimanadham, N. N., Imran, S., & Hossain, S. (2019). A review on strategies to manage physician burnout. *Cureus, 11*(6), e4805. https://doi.org/10.7759/cureus.4805

Potts, L., Ryan, C., Diegel-Vacek, L., & Murchek, A. (2018). Improving patient flow from the emergency department utilizing a standardized electronic nursing handoff process. *JONA: The Journal of Nursing Administration, 48*(9), 432–436.

Rauch, D. A. (2018). Physician's role in coordinating care of hospitalized children. *Pediatrics, 142*(2), e20181503.

Romito, B., Jewell, J., Jackson, M., & AAP Committee on Hospital Care; Association of Child Life Professionals. (2021). Child life services. *Pediatrics, 147*(1), e2020040261. https://doi.org/10.1542/peds.2020-040261

S.1809 – 106th Congress (1999-2000): Developmental Disabilities Assistance and Bill of Rights Act of 2000. (2000, October 30). http://www.congress.gov/

Shm_pediatric_boost_implementation_guide.Pdf. (n.d.). Retrieved August 16, 2021, from https://www.hospitalmedicine.org/globalassets/clinical-topics/clinical-pdf/shm_pediatric_boost_implementation_guide.pdf

Sivaraman, M., Virues-Ortega, J., & Roeyers, H. (2020). Social referencing skills in children with autism spectrum disorder: A systematic review. *Research in Autism Spectrum Disorders, 72*, 101528.

Speech Therapy. (n.d.). *Autism speaks*. Retrieved August 18, 2021, from https://www.autismspeaks.org/speech-therapy-autism

Starmer, A. J., Spector, N. D., Srivastava, R., Allen, A. D., Landrigan, C. P., Sectish, T. C., & I-PASS Study Group. (2012). I-pass, a mnemonic to standardize verbal handoffs. *Pediatrics, 129*(2), 201–204.

Straus, J., Coburn, S., Maskell, S., Pappagianopoulos, J., & Cantrell, K. (2019). Medical encounters for youth with autism spectrum disorder: A comprehensive review of environmental considerations and interventions. *Clinical Medicine Insights. Pediatrics, 13*, 1179556519842816. https://doi.org/10.1177/1179556519842816

Stremler, R., Haddad, S., Pullenayegum, E., & Parshuram, C. (2017). Psychological outcomes in parents of critically ill hospitalized children. *Journal of Pediatric Nursing, 34*, 36–43.

Stubbe, D. E. (2017). Optimizing empathy: Physician self-care as a crucial component of trauma-informed treatment. *Focus, 15*(4), 432–434.

The Americans with Disabilities Act of 1990, Pub. L. No. Pub. L. No. 101-336, 104 Stat. 328 Stat. (1990).

Thom, R. P., Hazen, M. M., McDougle, C. J., & Hazen, E. P. (2020). Providing inpatient medical care to children with autism spectrum disorder. *Hospital Pediatrics, 10*(10), 918–924. https://doi.org/10.1542/hpeds.2020-0140

U.S. Department of Justice. (2020). *Guidance documents*. https://www.justice.gov/guidance?keys=ADA&page=1. Accessed 7/11/2021.

West, C. P., Dyrbye, L. N., & Shanafelt, T. D. (2018). Physician burnout: Contributors, consequences and solutions. *Journal of Internal Medicine, 283*(6), 516–529.

Zablotsky, B., & Black, L. I. (2020). Prevalence of children aged 3–17 years with developmental disabilities, by urbanicity: United States, 2015–2018. *National Health Statistics Reports*, (139), 1–7.

12 Planning for Special Education Services: A Primer for Health-Care Professionals

Barzanna A. White

Abstract

Coordinating care with schools is a significant task carried out by the medical home coordinator and involves providing information to the school to support evaluations for services and guide the administration of medication or delivery of medical-related services. For the 16% of children receiving accommodations at school, special needs are documented in numerous formats including comprehensive evaluations, functional assessments of behavior, response to intervention plans, individualized family service plans, individualized educational plans, Section 504 plans, individualized health plans, behavior intervention plans, and individualized transition plans. Federal law mandates these plans with broad guidelines, with the states creating policy and local school districts responsible for implementation (Lancaster Central School District, 2021). This means that health-care providers must understand how federal law is interpreted by their state and enacted by local schools. This chapter follows the implementation of federal law from the 1950s to the present day to elucidate educational services for children with special needs.

B. A. White (✉)
Caddo Parish Schools, Shreveport, LA, USA
e-mail: bwhite@caddoschools.org

Keywords

PL 94-142 · Individualized education plan · Individualized accommodation plan · Functional behavioral assessments · Behavior intervention plans · Exceptionalities · Autism

Introduction

The promotion of public education in the 1800s has given rise to international discourse on the provision of education for all children including the introduction of the first intellectual tests developed independently by Binet and DeSanctis in 1905 (Cicciola et al., 2014). As testing categorized the abilities of individual children, governments were faced with complex policy decisions regarding approaches for the education of children requiring accommodations. Research has culminated in the recognition of inclusive education as the ideal (Kurth et al., 2018). Inclusive education has been promoted internationally by the United Nations and European Agency for Special Needs and Inclusive Education. For example, the United Nations Educational, Scientific and Cultural Organization's educational goals for 2030 Incheon Declaration state the following:

> particular attention is needed to ensure access to and outcomes of quality education and learning for children, youth and adults with disabilities (Mundial & UNICEF, 2016, p. 33).

P. McPherson (ed.), *Handbook of Treatment Planning for Children with Autism and Other Neurodevelopmental Disorders*, Autism and Child Psychopathology Series,
https://doi.org/10.1007/978-3-031-06120-2_12

The challenges for the education of youth with special needs lie at every level from national policy to mobilizing the local resources. This chapter will explore the American experience in addressing the challenge for providing a free and appropriate education for all children.

The National Health Interview Survey found that approximately 17% of children of ages 3–17 in the United States have a developmental disability and often require a range of medical and support services (Zablotsky et al., 2019). It is incumbent on professionals to document carefully the service needs and delivery to ensure clear communication and monitor progress. In medical settings, care is documented in a treatment plan or, in the medical home, as a shared plan of care. For most of the year, children spend the majority of their waking hours at school. Coordinating care with schools is a significant task carried out by the medical home coordinator and involves providing information to the school to support evaluations for services and guide the administration of medication or delivery of medical-related services. For the 16% of children receiving accommodations at school, special needs are documented in numerous formats including comprehensive evaluations, functional behavioral assessments, response to intervention plans, individualized family service plans, individualized educational plans, Section 504 plans, individualized health plans, behavior intervention plans, and individualized transition plans (National Center for Education Statistics, 2021) (Fig. 12.1).

Federal law mandates these plans with broad guidelines, with the states creating policy and local school districts responsible for implementation (Lancaster Central School District, 2021). This means that health-care providers must understand how federal law is interpreted by their state and enacted by local schools. This chapter follows the implementation of federal law from the 1950s to present day to elucidate educational services for children with special needs (Fig. 12.2).

A Historical Perspective: A Rocky Start

When it comes to trends and initiatives, school systems are often microcosms of society. The initial schools in the United States were primarily built on teaching core American values, that is, assimilating the new immigrants, along with teaching basic academic skills such as reading, writing, and arithmetic. At that time, to have more than an eighth-grade education was a luxury and echoed the zeitgeist of the times – that children were small adults and should be treated as such. This trend continued throughout the industrial revolution, as many children were sequestered from schools and forced to work in factories where they often performed dangerous jobs. Those who were lucky enough to attend school did so in a changing school climate. During the industrial revolution, schools mirrored factories. Every operation was structured and standardized – from canned curricula, to standardized assessment, to top-down structures. Schools were run on bell schedules, all students were treated as commodities, teachers were expected to teach all students with little autonomy or deviation, and individual student differences were primarily ignored. Children with special needs were often excluded from school altogether or removed from their family and sent to live in special school settings. In fact, this shift lasted for almost 100 years until the paradigm changed to schools focusing efforts on how students should be taught to thrive in a global economy.

Although these paradigm shifts are evident regarding regular education in the United States, it was not until the 1950s that trends related to educating those with disabilities were even discussed. Advocacy groups, along with the civil rights movement, began to gain momentum. One such advocacy group in particular, the Association for Retarded Citizens (the ARC), "began to develop and validate practices for children with disabilities and their families. These practices, in turn, laid the foundation for implementing effective programs and services of early intervention and special edu-

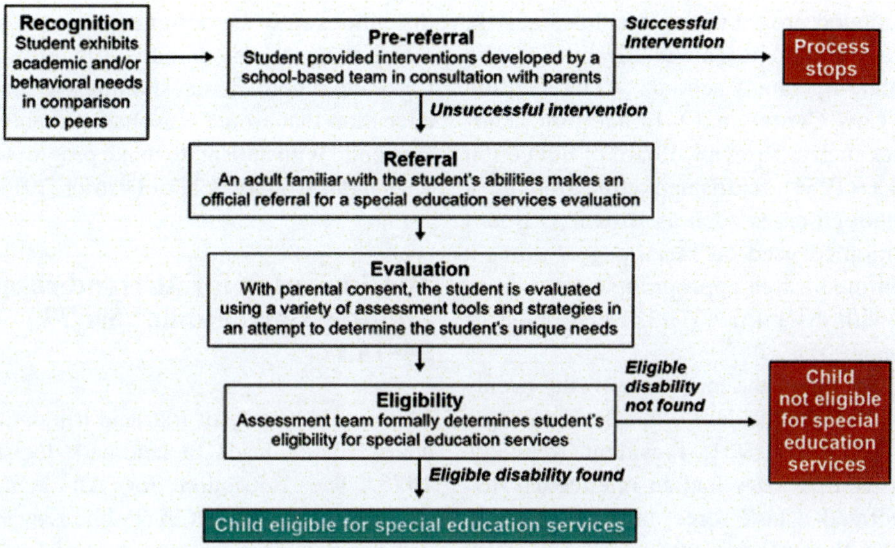

Fig. 12.1 Typical special education process for school-aged children and young adults. (Source: GAO analysis of information from selected state agencies and special education advocacy groups. | GAO-19-348)

1954 Brown v. Board of Education – reaffirmed a free appropriate education to students with disabilities.

1965 Elementary and Secondary Education Act (ESEA).

1973 Section 504 of the Rehabilitation Act – prohibits discrimination based on disability, 504 Plan(IAP).

1974 Family Educational Rights and Privacy Act (FERPA) – requires consent to release educational records.

1975 Education for All Handicapped Children Act (PL 94-142) – Free Appropriate Public Education (FAPE) - assures the rights of children with disabilities and their parents, assists states to provide appropriate education to children with disabilities documented in an IEP, and assesses and ensures effectiveness of these efforts.

1986 Education of the Handicapped Act (PL 99-457) – FAPE extended to 3–5-year-olds, Infant and Toddlers Program.

1990 EAHCA renamed IDEA (PL 101-476) – autism added as a disability, ITP, and expanded social work services.

1997 IDEA 1997 (PL 105-17) – ITP at 14years, changes to discipline policy, measurable goals emphasized and mediation of disputes.

2001 No Child Left Behind (NCLB) Act - Federal law designed to assess students academically which ushered in accountability for students, teachers, and schools, and included mandated yearly assessments based on A Nation at Risk.

2004 IDEA 2004 (PL 108-446) – required participation in state and local assessments, RTI, PBIS, and MTSS.

2015 Every student Succeeds Act (ESSA) replaced NCLB – which focuses on alternative assessments for 1% of students overall (10% of students in education), accountability goals, plans, and systems, interventions to reduce dropout rates, plans for low-performing schools, and consolidation of programs and block grants, and requires states to include data about bullying, school climate, and harassment in their annual state report card.

Fig. 12.2 Timeline of federal legislation on education

cation in states and localities across the country" (U.S. Department of Education, 2007, p. 3). Federal legislation from the 1950s to the late 1960s, which included the Training of Professional Personnel Act of 1959 (PL 86-158), the Elementary and Secondary Education Act (PL 89-10), the State Schools Act (PL 89-313), the Handicapped Children's Early Education Assistance Act of 1968 (PL 90-538), and many more statutes, provided professional development opportunities, film captioning for those who are deaf, grant assistance, early childhood programming for those at risk, and education for children with disabilities who attend state-operated schools (**The Public Interest Law Center**, n.d.). In addition, landmark cases such as Brown v. Board of Education of Topeka (1954) reaffirmed education as a right. Although cases such as Brown v. Board of Education focused on racial segregation, it also reaffirmed a free appropriate education to students with disabilities (U.S. Department of Education, 2007).

Perhaps some of the most noteworthy events occurred during President John F. Kennedy's time in office. In 1961, President Kennedy, whose sister Rosemary had an intellectual disability, formed a task force and subsequently created the National Institute on Child Health and Human Development. In addition, he advocated for mental health, increased funding for prevention, provided grants, and increased federal funding to train teachers on how to teach students with disabilities (John F. Kennedy Presidential Library and Museum, n.d.). Subsequent lawsuits filed by parents of children with disabilities, such as PARC (Pennsylvania Association for Retarded Children) v. Commonwealth of Pennsylvania (1971), began to shed light on discriminatory practices regarding those with intellectual disabilities (The Public Interest Law Center, n.d.). Many lawsuits were filed due to discriminatory practices because children were not able to access free and appropriate education due to their intellectual and other disabilities – a separate is not equal cry – that echoed the civil rights movement! Federal civil rights law addressing education includes Section 504 of the Rehabilitation Act of 1973, which bars discrimination against children with a physical or mental impairment. Section 504 is implemented in the schools as a Section 504 Plan which is discussed in detail later in this chapter.

Courts remain a venue for securing the educational rights of children. After due process options are exhausted in the school district, parents may seek legal action. This is just what happened in 2017 when the US Supreme Court held that an appropriate education for children with disabilities must be designed to promote benefit that is more than de minimis, suggesting that the Colorado school district should have provided an education that would reasonably enable Endrew F., a child with autism, to make progress (Endrew v. Douglas County School District, 2017).

The Education for All Handicapped Children's Act: Public Law (PL) 94-142

This sordid history of trial and tribulation in the courts finally leads to landmark legislation in 1975. The Education for All Handicapped Children's Act – EAHCA (Public Law 94-142) – was enacted. This law was designed to "support states and localities in protecting the rights of, meeting the individual needs of, and improving the results for infants, toddlers, children, and youth with disabilities and their families" (U.S. Department of Education, 2007, p. 1). With an estimated 1–2 million children in the United States with disabilities in the 1970s, PL 94-142 was designed with the following four overarching purposes:

1. "to assure that all children with disabilities have available to them…a free appropriate public education and related services to meet their unique needs"
2. "to assure the rights of children and their parents…are protected"
3. "to assist states and localities to provide for the education of all children with disabilities"
4. "to access and assure the effectiveness of efforts to educate all children with disabilities" (U.S. Department of Education, 2007, p. 3)

A free, appropriate public education (FAPE) is guaranteed to all children with disabilities, despite the severity of condition, under PL 94-142 (Dunn, 2013). This not only pertains to educational services but to additional services that may be necessary for children to benefit from special education. These are often referred to as related services and can include access to transportation and such "developmental, corrective, and other supportive services as are required to assist a student with an exceptionality" (Pastorek, 2009, p. 62). Some other examples of related services could include counseling, interpreting services, physical and occupational therapy, speech and language therapy, orientation and mobility instruction, and school health services.

Educationally Necessary Services

While medical professionals are familiar with medical necessity, one key concept that is often misunderstood by parents and professionals is related to the provision of services that are deemed "educationally necessary to benefit from special education" (Texas Council for Developmental Disabilities, 2013, p. 2). Whenever someone is paying for services, they always have the option to seek the best services available for the money. However, this is not the case under PL 94-142. Whatever service that will provide some benefit for the child to make academic, social, emotional, or other progress to meet their goals is deemed sufficient. For example, it is like purchasing a car that will provide you transportation but not one that has all of the amenities (heated and cooled seats, navigation, blind spot warning). Although no adequate explanation has ever been provided as to why topflight services are not required to ensure maximum benefit, most in the field assume that it is due to a lack of adequate funding. This disparity further highlights that equality is not the same as equity, especially due to the history of disproportionality of minority students and those from lower socioeconomic homes historically receiving special education services. Most states, especially after the School to Prison to Pipeline's report (Brownstein, 2015; Louisiana Center for Children's Rights, n.d.) and subsequent lawsuits filed that resulted in corrective action plans for school systems, now monitor disproportionality and discrepancy for their districts on a regular basis.

Implementing PL 94-142: Referral and Evaluation

Additional components of PL 94-142 include parental participation throughout the evaluation process, providing services in the least restrictive environment (LRE) based on the students' needs, due process that provides parents' safeguards for their children, nondiscriminatory assessment, and an individualized education program. While all states receiving federal funds for education must comply with PL 94-142, states develop their own guidelines to ensure compliance with the federal law. In Louisiana, the Louisiana Department of Education (LDOE) has developed the Louisiana's Educational Rights of Children with Disabilities: Special Education Processes and Procedural Safeguards (Pastorek, 2009). Most states have similar resources, and professionals are encouraged to seek guidance specifically pertinent to their state as terminology varies. For Louisiana, the process of a referral begins at the School Building Level Committee (SBLC). The SBLC is a regular education process, and the committee is usually composed of an administrator, counselor, regular education teacher, special education teacher, SBLC chair, and invited attendees (parents, mental health providers, advocates, etc.). This committee discusses children who may be at risk socially, emotionally, behaviorally, intellectually, and/or academically. Parents or professionals may ask to discuss their child in order to address concerns. This formal process is initiated by a letter from a parent, medical professional, school personnel, advocates, etc., with a response required within ten working days triggered by parent's request. This always includes a written notice with at least a 10-day period until the meeting, unless a parent waives his/her rights. In that case, they may simply sign a waiver, and the meeting may occur prior to the 10-day lapse. Progress monitoring, screenings, and interventions may be initiated as well as support services (e.g., counseling) for these students using a Multi-Tiered System of Support (MTSS) which will be discussed in greater detail along with response to intervention (RTI) (Pastorek, 2009).

The provisions of the EAHCA (PL 94-142) (United States Government Publishing Office, 1975) included the use of multidisciplinary teams. These teams are usually composed of school psychologists, social workers, educational diagnosticians, and teachers who work together during the evaluation or pupil appraisal process. Any tests that are utilized should have good psychometric properties (validity, reliability, norm groups) and should not be culturally, linguistically, or racially biased. Because parents have a unique perspective of their child, their input throughout the initial screening, referral, assessment, and placement processes is critical. Thus, parents are considered to be an important part of the evaluation, placement, and goal development process and are vital in the decision-making at all stages. Throughout the process, parents are ensured that written communication and due notice, confidentiality, the right to examine records, the ability to be a part of the process including classification and placement, and the right to obtain an external assessment from an independent evaluator at their expense are available. And whenever they disagree, they have the right to an advocate or legal representation.

Once the pupil appraisal special education evaluation begins, the team has 60 working days to complete the evaluation. Subsequent reevaluations occur every 3 years, or as may be needed, more frequently. Once the evaluation has been completed, "children with disabilities are to be educated, to the maximum extent appropriate, with students without disabilities" (Texas Council for Developmental Disabilities, 2013, p. 2). Taking students' needs into account, placements should be based on meeting the child's educational needs in the least restrictive environment (LRE). Teams should consider how the child could access the general education curriculum and provide a continuum of placements, from full inclusion to self-contained, based on the student's capability.

After the evaluation, an individualized educational plan (IEP) is developed. This process allows for up to 30 working days to develop the IEP which is a specifically designed plan that addresses the current level of academic functioning, includes the development of annual goals and instructional objectives, and addresses how these goals and objectives relate to the general curriculum. The IEP should also include the educational services that should be provided; any related services needed, including the type, initiation date, and length of service delivery; and an annual evaluation to determine whether the goals and objectives are being met. Testing accommodations are also a critical part of the IEP, as this determines what accommodations will be provided on federal, state, and locally mandated norm-referenced and criterion-referenced assessments. In most states, the majority of students are required to participate in statewide assessments unless they are incapable of engaging in the needed skills. Only 1% of a district's total student enrollment may be deemed exempt from standardized testing. Any student over this number who either takes no assessment or an alternative assessment will receive a score of "0." Therefore, it is better to have students take the state assessment and score a 2 rather than receive a "0" for anyone over the allotment (United States Publishing Office, 1975; Texas Council for Development Disabilities, 2013; Yell et al., 2018).

Amendments to PL 94-142

PL 94-142 was designed to provide services for children with disabilities from 3 to 21 years of age. In 1986, PL 99-457 created the Handicapped Infants and Toddlers Program to provide services to those with disabilities or developmental delays from birth through 2 years of age (Texas Council for Developmental Disabilities, 2013). Many of these students are identified via the Child Find and are usually assessed through multidisciplinary teams associated with local special education departments. While the Child Find requires school districts to identify individuals with needs from birth to the age of 21, pediatricians and the medical home team are important partners in referring children for services.

Individual Transition Plan (ITP)

A revision of PL 94-142, PL 101-476, occurred in 1990. After 15 years, the original law received a new name, Individuals with Disabilities Education Act (IDEA). This included expanding related services (social work and rehabilitation counseling) and adding autism and traumatic brain injury as categories of disabilities (Aleman, 1991). And most notable was the emphasis on the "people-first" language. In addition, a major provision of this law included the individual transition plan (ITP). A revision of PL 94-142, PL 105-17, occurred in 1997, and one component included a provision that transition plans begin as early as age 14, whereas the previous revision required students with IEPs to have an ITP prior to turning 16 years old. Any student whose IEP specifies programmatic or educational modifications, resource support, or adaptive strategies should have an ITP. These student-centered plans are designed to address a smooth transition from school to post-school functions and are developed to help ensure long-term success and ideally independent functioning, including vocational training and additional educational opportunities. Because many of these students may not be on a college track and some do not receive a high school diploma or equivalent, transitional plans were opportunities for school districts to provide vocational opportunities, to link families with outside agencies, and to assist parents to formulate realistic goals for their children's futures (Aleman, 1991; U.S. Department of Education, 2007).

At a minimum, all ITPs should address strengths, weaknesses, and interests, formulate measurable goals for high school, and list necessary services for students to achieve those goals. Because the process begins at age 14, goals are initially broad and become more focused as the student approaches graduation. Although ITPs would address different goals depending on a student's capability, the emphasis is to provide a realistic, attainable path to success no matter whether the student is college-bound or noncollege-bound. Depending on need, it is the IEP transition team's duty to ensure a smooth transition for ultimate success. What skills that are prioritized and taught vary greatly, depending on the individual, and could include personal care skills such as grooming or self-care, social skills, functional academic skills, and competencies needed for work. Opportunities specifically related to goals may also include teaching work skills via job shadowing and mentorships, establishing community connections, or navigating new programs, housing, or transportation services to ensure future success.

Students and their families are a valuable part of ITP process. Information gathered usually includes a student interview, a parent interview, a teacher observation, and other assessment measures. Many districts utilize school and community checklists, learning style inventories, interest inventories, motivation or self-determination inventories, and career interest surveys. Some districts assess the student's disposition to determine the "best fit" for their personality and aptitude. These ITPs are continuously evolving documents and are designed to be revised as needed. Generally, the ITP method follows a straightforward process. Step One (Information Gathering) would include interviewing the students, teachers, and parents as well as gathering and summarizing information such as the student's strengths, challenges, experiences, and interests. Step Two (Initial Transition Planning Meeting) would focus on reviewing pertinent student information and data; identifying needed assessments such as interests and communication, academic, social, and adaptive levels; and tentatively identifying long-term goals. Step Three (Second Transition Planning Meeting) is designed to complete any assessments, formalize short- and long-term goals, identify skills needed to be successful, and determine next steps. And finally, Step Four (IEP Review) allows the team to tweak goals and objectives to help ensure long-term success (Pastorek, 2009; Prince Edward Island Canada Department of Education, 2007).

Family interview questions are designed to be comprehensive and include questions related to the ability to successfully transition from high school to the adult world. Specific questions may address the type of living arrangements required based on the level of independence, whether full or partial financial independence is possible, what supports will be needed for life and leisure, the types of careers available, and strengths, challenges, and character traits that may align better with certain jobs. Engaged parents are better equipped to answer questions related to their child's functional skills, self-care skills, communication, problem-solving, responsibility, and overall disposition.

Student questions are very broad and should include favorite and least favorite academic classes, hobbies, interests for the future, work experiences and activities they enjoy, aspirations and dreams, and future career path (community college, 4-year university or college, mentorship/apprenticeship program, vocational training, or on-the-job learning). Many high school students are ill-equipped with a plan for their future, and if they have formulated a plan, most do not have specific guidance as to how to achieve it.

Teachers are trained keen observers and most special educators know their students with disabilities quite well. They frequently review their IEPs, provide accommodations, serve on various committees, attend multiple meetings, collaborate with general education personnel, counselors, and mental health professionals, and generally serve their students in smaller class sizes. Therefore, their input is vital for a well-rounded picture of the student. They are usually aware of how their students interact and communicate with each other, their preferences regarding classes and activities, how they deal with transitions, whether they may need additional time to process, their need for special accommodations (repeated directions, cooldown for sensory overload, or strategies for managing executive functioning issues), and how they respond to reinforcement, as well as school, personal, interpersonal, social, and community skills.

Assessments provide valuable tools in determining short-term and long-term ITP goals. Although a plethora of standardized and nonstandardized measures exist, IEP transition plan teams should select those that will provide them useful information. Examples of assessment tools used during the evaluation process are listed in Table 12.1.

Adaptive measures are designed to assess the social and emotional maturity, life skills, and other abilities in relation to one's peers. These include real-life skills such as cooking, cleaning, managing finances, working collaboratively with others, and following rules (American Addiction Centers, Inc., 2020).

As a general rule, aptitude tests focus on competencies learned usually outside of formal education, generally refer to a person's capacity to benefit from further education or training, and focus on the future (Anastasi, 1982), that is, the skills needed to determine a person's propensity to be successful.

Achievement tests measure the knowledge one has already attained, usually via formal education, measure attained ability, and focus on past learning. Because aptitude is correlated with achievement, colleges often require students to partake in and pass exams such as the American College Testing (ACT), Scholastic Assessment Test (SAT), or Graduate Record Examination (GRE). Other exams such as the Law School Admission Test (LSAT) and Medical College Admission Test (MCAT) are required for entry into law or medical school. For elementary and secondary students, many well-known achievement instruments may assist in making appropriate placement decisions and determining long-term goals.

Career choice and personal values measures are usually listed along with aptitude measures because measures often have aptitude properties, indicate personality styles, provide insight into decision-making abilities, and determine motivational characteristics, values, and even temperament. Some of the most popular measures include the following (VanDuzer, 2021):

Intelligence measures are designed to measure Spearman's original construct or general intelligence factor (g). Most tests now also measure a variety of other specific skill domains – specific intelligence (s) as well as fluid and crystallized intelligence. Fluid intelligence "refers to capacity

Table 12.1 Selected assessment tools

Adaptive measures	
Adaptive Behavior Assessment System – 3 (ABAS-3)	Assesses adaptive skills across the life span by comparing a series of ratings by parents, teachers, and individuals for ages birth–89. It was designed to assess those with learning disabilities, intellectual disabilities, developmental delays, sensory impairments, neurological disorders, physical impairments, and autism spectrum disorder (ABAS-3, 2015)
Adaptive Behavior Diagnostic Scale (ABDS)	Uses interview rating scales to assess conceptual, social, and practical domains (ages 2–21) that yield the overall Adaptive Behavior Index (ABDS, 2016)
Developmental Behavior Checklist 2 (DBC2)	Specifically designed rating scales used to assess those with developmental or intellectual disabilities in ages 4–18 (DBC2, 2018)
Goal-Oriented Assessment of Lifeskills (GOAL)	Examiner individually administers seven activities that assess functional motor skills in daily activity in ages 7–17 (GOAL, 2013).
Vineland Adaptive Behavior Scales, Third Edition (Vineland-3)	Considered the frontrunner in diagnosing developmental and intellectual disabilities in ages birth–90 (Vineland-3, 2016)
Achievement measures	
The Wechsler Individual Achievement Test 4 (WIAT-4), the Young Children's Achievement Test (YCAT), the TerraNova 3/the California Achievement Test (CAT 6), the Kaufman Test of Educational Achievement 3 (KTEA-3), the Woodcock-Johnson Tests of Achievement IV (WJ-A IV), and the Stanford Binet Intelligence Scales, Fifth Edition (SB-V) (Hood & Johnson, 2007)	Criterion- and norm-referenced tests are now part of the educational landscape due to NCLB and ESSA and their focus on accountability. Students in certain grades are assessed yearly to determine their academic progress, and schools are also assessed based on these results. In addition, some of the other individually administered assessment instruments are used to determine strengths and weaknesses during the formal evaluation process. The results from these instruments along with other indicators determine whether criteria for eligibility for special education or Section 504 services are met.
Career choice and personal values measures	
Myers-Briggs Type Indicator (MBTI) Career Aptitude Test	Identifies personality using readily observable behavior. Four dichotomous scales include extroversion vs. introversion, thinking vs. feeling, judging vs. perceiving, and sensing vs. intuiting (MBTI, 2021)
Holland Code Career Aptitude Test	Explores six interest areas: artistic, social, enterprising, conventional, investigative, and realistic (Holland Code Career Aptitude Test, 2017)
Motivational Appraisal of Personal Potential (MAPP) Career Aptitude Test	A comprehensive 71-question career assessment test (MAPP, 1995)
Keirsey Temperament Aptitude Sorter	Based on Dr. David Keirsey's model, which divides people into four temperaments: idealist, guardian, rational, and artisan Keirsey Assessments (2019)

(continued)

to learn new ways of problem solving and performing activities" whereas crystallized intelligence "refers to the accumulated knowledge of the world we have acquired through our lives" (Hood & Johnson, 2007, p. 60; Horn et al., 1981; Salthouse, 2004).

Interest measures are usually self-report measures designed to determine likes and dislikes of a variety of attitudes and activities in order to assist individuals with career choices, college majors, and work preferences. Additional information on more tests related to the aforementioned categories as well as those mental health measures, work, life skills, and personal values can be accessed via *The Twenty-First Mental Measurements Yearbook* (2021), which is a bibliography of all commercially available tests. Detailed information regarding the test's purpose, publisher, reviews, description, pricing, and reviews are provided.

Table 12.1 (continued)

The Princeton Review Career Quiz	Measures motivation, interests, interpersonal behavior, and stress management. Respondents are sorted into four categories – idea-centered, people-centered, production-centered, and procedure-centered (The Princeton Review Career Quiz, 2021).
CareerExplorer	Assesses interests and personality over 140 traits including determining the ideal workplace, whether your personality is a good match for your career path, what interests you, and whether where you live, your experiences, and your expected salary match your traits based on your personality (CareerExplorer, 2021)
Personal Values Assessment (PVA)	Designed for teens and allows them to focus on their personal values of who they are, what they value, how they make decisions, and how they respond to things that upset them (PVA, n.d.)
Intelligence measures	
Wechsler Intelligence Scale for Children – Fifth Edition (WISC-V)	Designed to assess a child's overall intellectual ability (ages 6-0–16-11) and five cognitive domains (verbal comprehension, visual-spatial, fluid reasoning, working memory, and processing speed) (WISC-V, 2014)
Stanford-Binet Intelligence Scales, Fifth Edition (SB-5)	Used to assess individuals between the ages of 2 and 85 years. This intelligence test provides a full-scale IQ, a brief IQ, verbal IQ, nonverbal IQ, and five factors (fluid, quantitative, knowledge, working memory, and visual-spatial) (SB-5, 2021).

(continued)

Table 12.1 (continued)

Woodcock-Johnson IV Tests of Cognitive Abilities (WJ IV)	Includes 18 tests that are used for measuring intellectual ability, academic domain-specific aptitudes, broad and narrow cognitive functioning, and related concepts in ages 2–90. There is also an academic assessment that is designed as a companion instrument to assess achievement (WJ IV, 2014).
Kaufman Assessment Battery for Children-II NU (KABC-II NU)	A clinical instrument for assessing cognitive development of children ages 3–18 of diverse backgrounds (KABC-II NU, 2018)
Interest measures	
Strong Interest Inventory	Assists individuals in exploring interests in six broad categories (investigative, artistic, realistic, enterprising, social, and conventional) to help them identify their work personality. It is useful in determining an appropriate college major, exploring career options, assessing career development by recognizing one's strengths and weaknesses, and helping employees engage as well as reintegrate into the workforce (Strong Interest Inventory, 2021).
Campbell Interest and Skill Survey (CISS)	Used to measure self-reported vocational interests and skills for ages 15 and above with a reading level of at least sixth grade. It features 7 orientation scales (influencing, organizing, helping, creating, analyzing, producing, and adventuring) and 29 basic scales (CISS, 1992).

Public Law 105-17 (1997 Amendments to PL 94-142)

The 1997 amendments brought about sweeping changes and significant reform to special education policies and practices. These included regulations regarding districtwide and statewide assessments, changes to the IEP, and how exceptional students who exhibit behavioral infractions could be addressed (Texas Council for Developmental Disabilities, 2013).

Prior to the latest revision, students who were identified and served under the special education realm were not required to participate in districtwide or statewide assessments uniformly across the states and nation. Therefore, assessment was often nonexistent or a hodgepodge in terms of measuring children with disabilities. As a result, many students often failed to thrive in classrooms where goals were either the same or similar for many years with no real consistency or design to reach minimum standards of these individuals' general education peers. Once PL 105-17 (U.S. Government Publishing Office, 1997) was enacted, students with disabilities were required to participate in districtwide and statewide assessments or be given alternative assessments based on the child's unique needs. This change not only impacted the ways most districts tested but ensured that states and districts that did not comply would be penalized significantly. Only 1% of a district's students (approximately 10% of the exceptional population) may opt out of assessment, and any number greater than this amount will result in a zero for all students over the cap who did not participate in the statewide assessment. Therefore, it benefitted a district for all students to take district and state assessments because even if a child scored five points, it was better than automatically receiving a zero. Prior to this, these scores were not a part of a school's or district's scores, and many districts only looked at disaggregated data to highlight their district's successes. Now, schools are evaluated on aggregated data, but major emphasis is placed upon disaggregated data to ensure that all populations and subpopulations are meeting benchmarks and standards and that disproportionality and discrepancy are kept in check. In addition, with exceptional students now being formally assessed, functional and developmental information collected through a variety of tools and strategies could be used to ensure that exceptional students were being measured against general education standards via their benchmarks and goals listed in their IEPs. Under these amendments, students were now required to be a part of the IEP, be involved with and progress in the general education curriculum, and have a regular educator be a part of the annual IEP meeting to address access to the general education curriculum (Texas Council for Developmental Disabilities, 2013; U.S. Department of Education, 2007).

One of the most controversial provisions of this revision was related to a student's behavior. The amendment was designed to make exceptional students accountable in regard to less serious discipline infractions. Exceptional students could be disciplined in similar ways to nonexceptional students for less severe infractions, as long as the misbehavior was not a manifestation of the student's disability. This even included a potential change in placement. This was good news for some and not for others, because it allowed schools to discipline more equitably, but also placed the responsibility of ensuring the implementation of the child's services as outlined in the IEP was done in accordance with the law. Therefore, any time a change of placement or a student is to receive up to 10 days of in-school suspension, out-of-school suspension, or expulsion, a manifestation determination review (MDR) must be conducted. The MDR basically addresses two questions: (1) Is the infraction related to the child's disability? (2) Did school personnel follow whatever is outlined in the child's IEP? If the infraction is not part of the child's disability and the school followed the IEP, then school personnel may discipline the exceptional student in the same manner as a nonexceptional student. But if the infraction is related to the student's disability, the student cannot be disciplined like a nonexceptional student. The same is true if the school did not follow the child's IEP, which is a legal document outlining services (U.S. Department of Education, 2007; Brownley, 2018).

And finally, states were now required to offer mediation to resolve disputes and information regarding the appeal process. Beginning at the district level, any disputes are to be addressed through due process determined by the state department of education in accordance with federal law. For an example, see the Louisiana Department of Education publication titled *Louisiana's Educational Rights of Children with Disabilities: Special Education Processes and Procedural Safeguards* (2020). Other states have similar booklets. If complaints cannot be resolved at the district or state level, parents may file an official complaint with the Office of Civil Rights (OCR) up to 365 days after the incident occurred. Fortunately, most disputes can be rectified at the local level through establishing rapport, by ensuring that parents are an integral part of the evaluation, assessment, and educational processes, by utilizing effective, open communication skills, and by allowing (within limits) advocates to be a part of the process.

Understanding the Individualized Educational Plans (IEPs) and the Individualized Accommodation Plans (IAPs)

Generally considered more comprehensive, the individualized educational plan (IEP) is a federal special education law for students with disabilities, whereas the individualized accommodation plan (IAP) is part of the Americans with Disabilities Act (Section 504 of the Rehabilitation Act of 1973), which "prohibits discrimination against people with disabilities who are in need of accommodations" (Colletti, 2018; U.S. Department of Education, 2020), and is part of the federal civil rights law. Provisions for both the IEP and the IAP are part of federal law which outline specific services to students and are individualized based on that student's needs. IEPs, because they are part of the Individuals with Disabilities Education Improvement Act (IDEIA), can be in place throughout the school years, as long as students still meet criteria when they are evaluated initially or reevaluated every 3 years. Renewed annually, IEPs tend to be more in-depth documents that outline needed services, accommodations, and specialized instruction and can include additional services, such as counseling, speech and language therapy, behavioral support, physical therapy, and occupational therapy. On the other hand, IAPs usually address how a student will learn at school but are usually reserved for more temporary accommodations, such as a student with a temporary injury (broken dominant hand), illness (during radiation treatment for cancer), or less severe condition such as mild asthma, voice defect, attention deficit hyperactivity disorder, dysgraphia, or dyslexia. These changes to the learning environment and provided services "enable all students to learn alongside their peers" (Understood Team, 2021, p. 2). Although these children may qualify for special education services as well, IAPs are preferred to accommodate those students who are capable of learning within the general education environment but simply need minor accommodations such as extended time, preferential seating, extra time on tests and assignments, an interpreter for American Sign Language (deaf or hard-of-hearing), a notetaker (while the person's dominate hand is in a cast), etc. And unlike IEPs, the equivalence of IAPs may be provided outside of the school realm. For example, college students who qualify under Section 504 may receive accommodations in their classes.

In order to be eligible for an IEP, there are two requirements. First, the child must have one of the following 13 disabilities listed below (Lee, 2021), and second, the disability must impact the child's educational performance (Pastorek, 2009; Louisiana Division of Administrators, n.d.). Table 12.2 lists the disabilities which are included in Bulletin 1508 (Pastorek, 2009).

To date, there is no guidance from the federal government that requires states to provide services for *gifted and talented* students. Many students may be considered for two or more exceptionalities such as ASD (autism spectrum disorder) and gifted, but unless the state recognizes gifted and talented, students would only receive services for the disability portion of the diagnosis (Tan et al., 2016). In fact, most states cannot even agree on the definition of gifted and talented. Yet of the "51 respondents, 44 report a state definition of giftedness" (National Association for Gifted Children and the Council of State Directors of Programs for the Gifted, 2019, p. 2). Despite the lack of uniformity, a few states provide total funding for their gifted and talented programs, some provide partial funding, and others do not fund anything related to gifted

Table 12.2 The disabilities are included in Bulletin 1508 (Pastorek, 2009)

Specific learning disability (SLD)	After intervention, the student does not achieve adequately for age in at least one of the following areas: mathematics calculation, mathematics problem-solving, written expression, basic reading skills, reading fluency skills, reading comprehension, listening comprehension, or oral expression. They may also have outside diagnoses such as auditory processing disorder, nonverbal learning disability, dyscalculia, dyslexia, or dysgraphia that, if severe enough, may meet criteria.
Other health impairments (OHI)	Chronic health problems resulting in limited strength, alertness, or vitality. Some conditions may include sickle cell anemia, epilepsy, chronic asthma, ADHD, hemophilia, lead poisoning (teratogen), cancer, rheumatic fever, HIV, Tourette's syndrome, nephritis, etc.
Autism spectrum disorder (ASD)	Autism, Asperger's disorder, pervasive developmental disorder not otherwise specified, Rett's disorder or childhood disintegrative disorder, and other coexisting conditions may be present, such as self-injurious behavior, enuresis, encopresis, Tourette's syndrome, fragile X, pica, depression, anxiety, etc.
Emotional disturbance (ED)	Does not include children who are socially maladjusted unless they have an emotional disturbance based on the following (Pastorek, 2009, p. 15): Inability to learn that cannot be explained by intellectual, sensory, or health factors: Inability to build or maintain satisfactory interpersonal relationships with peers and teachers Inappropriate types of behavior or feelings under normal circumstances General pervasive mood of unhappiness or depression A tendency to develop physical symptoms or fears associated with personal or school problems

(continued)

Table 12.2 (continued)

Speech/language impairment (SLI)	May include articulation, fluency, voice, or language impairments
Visual impairment (VI)	Includes blindness (20/200 or less distance in the better eye with correction), partial sight (20/70 or less distance with correction in the better eye), progressive loss of vision, blindness caused by a medical condition, or blindness in the peripheral field (no greater than 20°).
Hearing impairment (HI)	Includes deafness and hard of hearing (permanent or fluctuating hearing loss, unilateral hearing loss, or high frequency hearing loss).
Deaf-blindness (DB)	Concomitant hearing and visual impairments
Orthopedic impairment (OI)	Muscular or neuromuscular disabilities or skeletal deformities or abnormalities that impair functioning
Intellectual disability (ID)	Subaverage intelligence existing concurrently with deficits in adaptive behavior. Standardized IQ score ranges are identified below: *Mild ID* (55–70), *moderate ID* (40–54), *severe ID* (25–39), and *profound ID* (below 25)
Traumatic brain injury (TBI)	An acquired injury to the brain (after birth) that results in impairments in thinking, judgment, problem-solving, cognition, language, reasoning, attention, memory, speech, information processing, psychosocial behavior, or sensory, perceptual, or motor abilities
Multiple disabilities (MD)	Concomitant impairments

(continued)

or talented programs according to the 2018–2019 State of the States in Gifted Education Executive Summary (National Association for the Gifted and the Council of State Directors of Programs for the Gifted, 2019). State-by-state information can be found on the aforementioned websites. The Louisiana Department of Education (LDOE) does mandate identification of gifted and talented students, mandates services, and provides policy

Table 12.2 (continued)

Developmental delay (DD)	Developmental delays in at least one of the following areas: physical development, cognitive development, social or emotional development, adaptive development, or communication development. This exceptionality is used prior to the child's ninth birthday unless the condition is one that will not change such as TBI, autism, OHI, OI, deaf-blindness, HI, or VI. This allows the opportunity for students who may come from impoverished environments, those who fail to thrive, those not exposed to enrichments, and those exposed to adverse childhood experiences/ trauma possibly to catch up with their nondisabled peers.

guidelines on identifying students and providing services. And according to the National Association for Gifted Children's latest survey conducted in 2018–2019 (National Association for the Gifted and the Council of State Directors of Programs for the Gifted, 2019), Louisiana has point of contacts, credentialing guidelines, and a state's gifted education association and spends roughly $113,926,282.00 yearly to provide services to over 30,248 gifted students (National Association for the Gifted and the Council of State Directors of Programs for the Gifted, 2019).

To qualify for one of the 13 disabilities identified in Bulletin 1508, the disability must impact the child's educational performance or his/her ability to learn or benefit from the general education curriculum unless specialized instruction occurs (which would be provided via the IEP). The federal definition of disability states that it "must impact the educational performance" (Pastorek, 2009), and interpretation by some districts is vague. Therefore, some districts prefer to err on the side of caution and provide services when a child's grades decline from A's and B's to C's, while other districts take a more wait-and-see or wait-to-fail approach and only assess after a child's grades are D's and F's or when a child has failed one or more grades. The latter is unfortunate because being more than 2 years behind does not bode well for long-term success and often leads to those students dropping out of school prior to graduation.

Unlike IDEA, Section 504's definition is much broader. Under Section 504 of the Rehabilitation Act of 1973, a "disability must substantially limit one or more basic life activities [such as] learning, reading, communicating, and thinking" (Understood Team, 2021, pp. 3–4). There are two requirements to have an IAP. First, a child with a disability must struggle in school. And second, the disability must negatively impact the child's ability to learn (Understood Team, 2021).

What Is Included in an IEP Versus What Is Included in an IAP

IEPs are formalized plans created by a team of professionals (e.g., a regular education teacher, a special education teacher, a multidisciplinary staff member such as a school psychologist, social worker, or educational diagnostician), ancillary staff if those services are warranted, the parent/guardian, and often the child. IEPs should include annual goals, progress monitoring of the goals, present levels of academic and functional performance, specific services and when and how they will occur, information about participation in the general education curriculum, activities, and local and state assessments, accommodations, and modifications (**Louisiana Division of Education/Department of Education,** 2020). In addition, appropriate, measurable postsecondary goals must be based upon age-appropriate transition assessments related to training, education, employment, and, where appropriate, independent living skills, as well as the courses of study needed to assist the child in reaching those goals (Renner, 2018; Texas Project FIRST, n.d.).

Although many school districts have adopted standardized formats for IAPs, formalized written plans are not required under Section 504. But for clarity, ease of operations, consistency, and to reduce litigiousness, most school districts now provide some formalized plans based on best practices. Generally, these plans are direct and concise. Most address appropriate supports, services, and accommodations, who will provide the services, and who will monitor their effective implementation (Understood Team, 2021).

Public Law 108-446 (2004 Amendments to PL 94-142)

One of these revisions reversed one that was enacted in 1997 regarding benchmarks. According to this revision, short-term objectives were no longer required in an IEP for exceptional students with one exception – students who take alternative assessments (Texas Council for Developmental Disabilities, 2013) must continue to have benchmarks in their IEPs. In essence, this sent a clear message that the majority of exceptional students were to be compared to and assessed based on age-/grade-level peers and provided instruction based on the general education curriculum. In a nutshell, "Separate was not equal!" and parents of exceptional children were delighted with this sweeping change.

Somewhat controversial now, language from the No Child Left Behind was also included in this revision. Some of the tenets still remain that relate to limited English proficiency, academic subjects, standardized assessment, and highly qualified teachers. However, No Child Left Behind legislation did not yield the results intended and was replaced by the Every Student Succeeds Act (ESSA) in 2015 (The United States Department of Education, Every Student Succeeds Act (ESSA): A Comprehensive Guide). This federal legislation governs elementary and secondary education and still emphasizes accountability. However, districts are now allowed to set their own accountability goals as long as they address graduation rates, English language proficiency, and proficiency on tests (Klein, 2016). Low-performing schools must collect and assess data in order to identify and intervene where graduation rates are low (below 67%) and where subgroups are struggling (looking at disaggregated data) and identify students who are in the bottom 5% of performers (Klein, 2016). Additional changes included weighted student funding formulas, removing teacher evaluation systems that utilized student outcomes, improving access to early childhood education, utilizing consolidated block grant funding, using a portion of funds to address health and safety so students would be more well rounded, and providing evidence-based interventions for struggling students (The United States Department of Education, Every Student Succeeds Act (ESSA): A Comprehensive Guide, 2015).

One of the biggest changes was directly related to assessment. Rather than forcing states to use a discrepancy formula to determine whether a student has a learning disability, states could use alternative means to determine eligibility. The traditional discrepancy model utilized cognitive and achievement testing to determine if a learning disability was present. In essence, it was a mismatch between a students' cognitive ability and how they are progressing in school. Although it was considered an acceptable practice for years, for many, it was considered a "wait-and-see" or a "wait-to-fail" model. By utilizing a response to intervention (RTI) model, evidence-based practices are implemented, and progress monitoring takes place over weeks rather than years. If students fail to respond to interventions and/or instruction in the classroom setting, then the students may be considered eligible for services. In Louisiana, Bulletin 1508 is used and outlines everything including screening, assessment, multidisciplinary teams, RTI, exceptionalities, etc. The role of RTI was expanded further in the Individuals with Disabilities Education Improvement Act of 2004 and will be discussed shortly.

Individuals with Disabilities Education Improvement Act of 2004 (IDEIA or IDEA 2004)

Probably the most sweeping of all of the revisions of PL 94-142 happened in 2004 in order for the new law to be aligned with the No Child Left Behind (NCLB), which was replaced by ESSA in 2015 (Klein, 2016). Because one of the main tenants of NCLB was to ensure accountability for school systems, primarily measured through standardized tests, it was not surprising that measuring student progress was a key issue. Thus, progress of students with disabilities was to be aligned with the goals of their peers who were not identified as disabled. Progress was to be monitored regularly via measurable written goals, progress monitoring, and standardized testing or alternative assessment (for the 1% of the population who would not be assessed by standardized assessment measures).

Besides alignments, some of the optional requirements of past revisions were also solidified. First, transitional services were now to begin at age 16, thus allowing more time for students to explore potential career options, be further along in their coursework, and be closer to graduation. Second, the discrepancy model that was to be used at a district's discretion is no longer required for determining whether a student qualifies for a specific learning disability exceptionality. This squarely placed the focus on utilizing a response to intervention (RTI) model rather than a wait-to-fail model. Third, emphasis was placed on allowing school personnel and parents opportunities to work collaboratively via a dispute-resolution system in order to resolve their issues in a positive and constructive manner that ultimately benefits the child. Fourth, all interventions must be based on evidence-based practices and peer-reviewed literature. This includes all RTI interventions as well as services listed in IEPs (such as counseling, behavioral support, physical or occupational therapy, etc.). Fifth, the eligibility and evaluation process must be clarified and transparent so that states can ensure that all students can be identified early through coordinated systems including multidisciplinary, interagency agreements. Early identification is key to prevention and provides opportunities for some students to thrive (ESSA, 2015).

According to IDEIA (U.S. Government Printing Office, 2004, p. 3), "More minority children continue to be served in special education than would be expected from the percentage of minority students in the general school population. And African American children are identified as having intellectual disabilities and emotional disturbance at rates greater than their White counterparts" (U.S. Government Printing Office, 2004, p. 3). Although overidentification of minority students continues to be problematic in some districts, most state departments of education continue to monitor discrepancy and disproportionality on a regular basis. In fact, approximately 10 years ago, a number of school districts were placed in corrective action because of disproportionality and discrepancy issues. The ease of access and the transparency of data, although usually 1–2 years behind, make districts more accountable. And other child advocate groups, such as the Southern Poverty Law Center and professional organizations, continue to monitor the aforementioned issues.

Response to Intervention (RTI), Positive Behavioral Interventions and Supports (PBIS), and Multi-Tiered System of Supports (MTSS)

Without a doubt, one of the most sweeping changes to the landscape included in IDEIA 2004 outlined a multitiered approach model designed to identify and support all students. Not only does response to intervention (RTI) provide a framework for academic and behavioral supports, but it also emphasizes universal systematic screening of all children and focuses on high-quality instruction (**The RTI Action Network, n. d.**). Often schools utilize a Positive Behavioral Interventions and Supports (PBIS) or Multi-Tiered System of Support (MTSS), of which RTI is an integral part (Fig. 12.3).

At the universal level, Tier 1 level, or green level, all students receive high-quality instruction and positive behavior interventions and support. "Tier 1 supports serve as the foundation for behavior and academics. Schools provide these universal supports to all students. For most students, the core program gives them what they need to be successful and to prevent future problems" (OSEP Technical Assistance Center on Positive Behavioral Interventions and Supports, 2021, p. 1; National Center on Safe and Supportive Learning Environments, 2021).

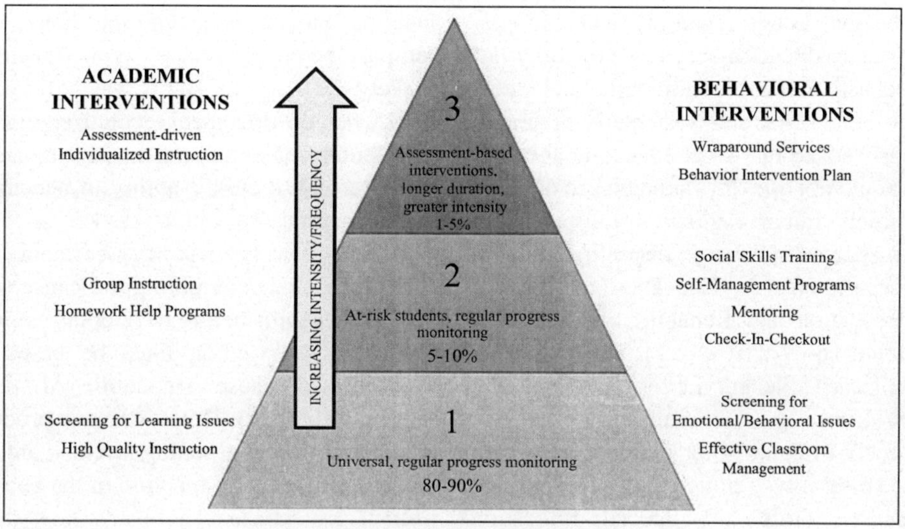

Fig. 12.3 Response to intervention (RTI) levels and examples of services

Depending on the school, success for all students without additional intervention ranges from 70% to 85%, with elementary schools in general being more successful than middle or high schools. Because some students will require additional interventions in order to be successful, academic small group instruction or behavioral support will be implemented. Known by most as the secondary level, Tier 2 level, or yellow level, these supports range on a continuum from minimal group support to extensive services. "This level of support focuses on skill deficits to groups of students with similar targeted needs" (OSEP Technical Assistance Center on Positive Behavioral Interventions and Supports, 2021, p. 1). Examples of such evidence-based topics could include behavioral supports such as check-in-checkout, social-emotional learning, conflict resolution, anger management, trauma-informed practices, and restorative practices. Even with these targeted interventions for academic and/or behavioral issues, a few students, approximately 1–5%, will also require tertiary level, Tier 3, or red supports. These are considered the most resource-intensive supports "due to the individualized approach of developing and carrying out interventions. At this level, schools typically rely on formal assessments to determine a student's need and to develop an individualized support plan" (OSEP Technical Assistance Center on Positive Behavioral Interventions and Supports, 2021, p. 1), which generally includes academic as well as behavioral goals. Pivotal points for successful implementation include universal screening, progress monitoring, staff working collaboratively rather than in silos, data-based problem-solving, systematic review of formative and summative data, adequate resources including human capital, and fidelity of program implementation (Sprague and Walker, 2004).

What Is PBIS?

RTI was developed to assist school staff in making educational decisions about individual student's needs including the intensity, frequency, and duration of evidence-based services. And although IDEIA 2004 expanded its use stating that RTI could be used as part of the special education evaluation, no formalized guidelines identifying the length or number of interventions were required before an evaluation could take place. Unfortunately, some school districts used this lack of specificity to delay initial evaluations. So, in 2016, Ruth Ryder, the acting director with the Office of Special Education Programs of the US Department of Education, disseminated a

Dear Colleague Letter regarding RTI and preschool special education services. Not only did the letter clarify RTI as a continuum of evidence-based comprehensive and systematic practices, but it clearly stated that local education agencies (LEAs) should not use "RTI strategies to delay or deny a timely initial evaluation for preschool children suspected of having a disability" (United States Department of Education, Office of Special Education and Rehabilitation Services, OSEP Memo 16-07, 2016, p. 1). This expanded her initial Dear Colleague Letter that stated that LEAs could not use "RTI strategies to delay or deny a timely evaluation for children suspected of having a disability" (United States Department of Education, Office of Special Education Services, OSEP Memo 11-07, 2011, p. 1).

And finally, and probably, one of the most controversial changes to the IDEIA 2004 was related to removal from school for a student with disabilities for certain offenses. These offenses are often what is referred to as "state-mandated offenses" such as illegal drugs, bodily harm, and behavior involving a weapon. For these offenses, no matter whether or not the infraction/behavior was related to or a manifestation of the child's disability, that student could be removed to an alternative interim setting for up to 45 school days. Students without disabilities may be removed for a full year according to most state statutes. Heated debates often ensue on both sides, as they attempt to justify the pros and cons of the aforementioned change.

Functional Behavioral Assessments (FBAs) and Behavior Intervention Plans (BIPs)

The tenants of behavioral assessment began in the field of psychology with theorists such as Pavlov, Thorndike, and Skinner. Application of the principles of applied behavior analysis was fast-tracked in the 1970s as token economy and reinforcement systems were applied to students with special needs. The field "originates from over 50 years of applied behavior analysis (ABA) research, which supports its practicality in understanding human behavior and helps simplify complex behavior change" (von Ravensberg & Blakely, 2014, p. 2). But it was not until IDEA 1997 when the document recognized that "physical, mental, and emotional disabilities can and do interfere with a child's ability to benefit from a free appropriate education" (Luker, 2021, p. 1).

A functional behavioral assessment (FBA) is "the process of determining the cause (or function) of behavior before developing an intervention. The intervention must be based on the hypothesized cause (function) of behavior" (Starin, 2011, p. 1). This systematic process aims at determining the setting events, antecedents that contribute to the function of the behavior, as well as any consequences and how they may increase or decrease the likelihood the behavior will continue or cease. Operationally defining the problem behavior, "identifying the environmental functions and setting events associated with the problem behavior" (U.S. Office of Special Education Programs, n.d., p. 1) are key components of the FBA. Thus, the FBA is not only an attempt to understand the student's behavior but is a primary tool that addresses four goals (Luker, 2021, p. 1): (1) to describe behavior, (2) to predict when and where the behavior will occur, (3) to identify possible reasons for why the child behaves the way he or she does, and (4) to develop intervention support strategies that conform to the IEP team's best understanding of why the behavior is occurring. Whether it is part of a manifestation review or standard practices outlined by IDEA 1997 and IDEA 2004, the law "requires every IEP team, at every IEP team meeting, to review the child's behavior and determine whether it significantly impedes the child's learning or the learning of others" (Luker, 2021, p. 1).

The FBA is a "systematic, reliable method of observing students in natural settings. This result-oriented process explicitly identifies problem behaviors, the specific actions that reliably predict the occurrence and nonoccurrence of problem behaviors, and how the behaviors may change across time" (von Ravensberg & Blakely, 2015, p. 1).

In order to effectively conduct a FBA, several sources of data are critical and should be triangulated for determining patterns as well as for accuracy. These include interviews; rating scales; observations within the school, home, and community settings; manipulating environmental events; data collection; and data analysis (Starin, 2011; Luker, 2021). Most effective FBAs not only describe the targeted behavior, but they also determine how to measure it appropriately, that is, frequency, intensity, or duration. Any prior interventions that were previously conducted are described along with their success or failure. Once these items are analyzed, functions for the areas are noted and hypothesized in Table 12.3 (Miller, 2019).

After a thorough FBA has been conducted, an individualized plan to support behavior is based on the functional assessment. A behavior intervention plan (BIP) is drafted "based on the FBA and guided by a reasonable understanding of why the behavior happens, [is] directed toward skill-building and environmental changes, is comprehensive: involving multiple intervention components, and is assessed off of its effectiveness-not just the change in the targeted behavior, but on the broader quality of life issues such as maintenance across time and generalization across settings" (Luker, 2021, p. 2). Behavior plans usually only address a couple of behaviors at a time, and the behaviors are selected based on need (safety concerns, such as for self-injurious behavior, always are prioritized before other behaviors that are targeted). Effective BIPs outline expected goals and objectives and identify specific interventions and how often they will occur, the person responsible for implementation, how the plan will be monitored and by whom, and when the plan will be reviewed, tweaked, updated, etc.

Table 12.3 Functional determinants of challenging behaviors

Affective regulation/ emotional reactivity	Identify emotional factors – anxiety, depression, anger, and poor self-concept – that play a role in organizing or directing problem behavior
Cognitive distortion	Identify distorted thoughts: inaccurate attributions, negative self-statements, and erroneous interpretations of events that play a role in organizing or directing problem behavior
Modeling	Identify the degree to which the behavior is copied, who they are copying the behavior from, and why they are copying the behavior
Family issues	Identify family issues that play a part in organizing and directing problem behavior
Physiological/ constitutional	Identify physiological and/or personality characteristics: developmental disabilities and temperament that play a part in organizing and directing problem behavior
Communicate need	Identify what the student is trying to say through the problem behavior
Curriculum/ instruction	Identify how instruction, curriculum, and educational environment play a part in organizing and directing problem behavior

A compendium of research is readily available on FBAs and BIPs, along with other related topics, and can be found on the Center on PBIS (OSEP Center on Positive Behavioral Interventions and Supports, 2021) website. One of the most comprehensive documents which includes an overview on FBAs and BIPs, behavioral principles, guidelines on conducting interviews, observing and summarizing behavior, features of BIP's implementation, and evaluation planning and teaming as well as forms is a recommended read (Loman et al., 2013). Two other noteworthy documents include a Dear Colleague Letter dated August 1, 2016 (United States Department of Education, Office of Special Education and Rehabilitation Services, OSEP Memo 16-11, 2016, p. 1), and a letter from the US Department of Education, Office of Special Education and Rehabilitation Services from Dr. M. Musgroves (2011) to a parent, G. Gallo, dated April 2, 2013 (United States Department of Education, Office of Special Education and Rehabilitation Services, OSEP Letter to Parent, 2013, p. 1). The first document outlines (1) procedural requirements regarding evaluation, eligibility determination, IEPs, and behavioral supports, (2) content requirements regarding behavioral support, (3) circumstances that may indicate potential denials of FAPE or of placement in the LRE, (4) implications for short-term disciplinary removals and other exclusionary disciplinary measures, and (5) additional information for parents and stakeholders. And the second letter addresses specific requirements of the FBA, assessment procedures within a RTI framework, and discipline questions and answers.

Like most federal laws, there is some guidance regarding FBAs and BIPs, but states have the leeway to adopt the provisions of the law as stated, restate them, or exceed the federal requirements. Only three states "exceeded IDEA's FBA provisions (California, Nevada, and Wisconsin)" (von Ravensberg & Blakely, 2015, p. 2). Unfortunately, many states continue to provide only minimum requirements which is far short of best practice. Therefore, it is vital for concerned parents, mental health professionals, and school staff to advocate for students as if they were their own children.

Individual Health Plans (IHPs)

Like so many adults, children have a variety of health needs that may require an Individual Health Plan (IHP), aka Emergency Care Plan (ECP). These plans are designed to address the student's medical concerns. Many students need to take medications at school, so an IHP is considered a formal document that "serves both the student and school" (Noble, 2015, p. 1). Designed with prevention in mind, IHPs allow students, families, and school personnel the opportunity to meet and discuss any health-care need that may impact the student medically or that interferes with school attendance as well as academic success (National Association of School Nurses, 2015).

Depending on the severity of medical conditions and how the conditions impact learning, IHPs may either be standalone plans, a part of the Section 504 IAP, or a part of the IDEA IEP. Standalone plans are less inclusive but still serve as a formal document that is created by school district personnel and usually managed by a school nurse whereas plans associated with IAPs and IEPs are generally more comprehensive and usually follow a standardized format.

Individual Health Plans/Emergency Care Plans

Usually based on state laws, IHPs are binding documents that provide students with a minimum standard of care for mild health conditions. Potential qualifying conditions include students who may have to take medication for health or mental health conditions during the day; those with mid conditions such as allergies, asthma, dysthymia, and self-harm; or those who require dietary needs, self-care needs, and rehabilitation, have treatment orders, need environmental modifications, etc. (www.blogs.svvsd.org). More substantial conditions should be considered for services within the Section 504 or IDEA process (School District 271, Coeur d'Alene, Idaho, 2016).

Standalone IHPs would be coordinated by a school nurse and would serve as the "accurate, centralized source of information about the student's medical needs and [provide] direction and authorization should a health need arise suddenly" (Noble, 2015, p. 1). At a minimum, plans should be reviewed at least yearly, as medications are often adjusted as health conditions change (Coeur d'Alene School District 271, 2016).

IHPs as Part of Section 504 of the Rehabilitation Act of 1973

This type of IHP is enacted when an individual has significant medical issues that "limit one or more major life activities, including school" (Noble, 2015, p. 2). These life activities "include, but are not limited to, caring for oneself, performing manual tasks, seeing, hearing, eating, sleeping, walking, standing, lifting, bending, speaking, breathing, learning, reading, concentrating, thinking, communicating, and working"

(U.S. Department of Education Office of Civil Rights, 2016, p. 4). And unlike IHPs that are associated with IEPs, there are no age limits on those connected to Section 504 plans. The plan is created by a team of individuals, including the student, parent/guardian, teacher, and principal/designee. Individualized accommodations (academic and nonacademic if needed) are provided to the student to provide a level playing field (equity in education). And any services or changes to the learning environment are documented within the IAP, which is updated at least annually (Coeur d'Alene School District 271, 2016; www.blogs.svvsd.org). Typically, these plans are managed by school counselors.

IHPs as Part of IDEA

As one might expect, qualifying for IDEA is more difficult than qualifying for Section 504 services. In order to receive special education services and have an IEP, "the student's disability must substantially interfere with their education and performance [which requires] specialized instruction" (Coeur d'Alene School District 271, 2016, p. 1). Designed for students ages 3–21, the IHP provides specialized and related services to students via an assessment process conducted by a multidisciplinary team which yields recommendations that translate into accommodations that are outlined in the IEP. And if the "IEP indicates that a child needs nursing services per the IHP, the child is also covered under Medicaid" U.S. Department of Education, n.d., p. 7). Under Bulletin 1508, a pupil appraisal coordinator (usually a school psychologist, social worker, or educational diagnostician) coordinates the evaluation process, and team members conduct assessments. Once a referral to a School Building Level Committee (SBLC) or equivalent school committee (depending on state) has been initiated and the prescreening process and interventions are complete, the multidisciplinary team has 60 days to complete the evaluation. Afterward, the IEP team has an additional 30 days to draft the IEP which includes any other documents such as the IHP. The IEP is renewed annually, but more often if needed, and a reevaluation is conducted every 3 years (Pastorek, 2009). The management of daily operations as listed in the IEP and IHP is overseen by a lead special education teacher who ensures all of the student's teachers are aware of accommodations and services listed in the aforementioned documents, whether the student is in self-contained classes, fully included, or somewhere in between.

As a general rule, good IHPs include background information, such as the health need/diagnosis, psychological concerns, transportation needs, dietary needs, equipment and supplies, medications, safety levels, activity level, whether self-care is needed, possible alerts, and emergency contact information. In addition, a list of medications, when they are administered and by whom, is included as well as an emergency plan, if necessary, which may include authorization for a nurse or other staff member to intervene. And effective IHPs also contain goals, action items, and interventions.

DSM-5 Versus ICD-11 Codes

After highlighting the many laws and policies related to operating with the school district, many outside professionals find it hard to grasp why schools used simplified exceptionality classifications rather than more comprehensive coding systems. In short, each organization and entity has reasons for preferring their system which aligns with their codes of practice. The Diagnostic and Statistical Manual – Fifth Edition (DSM-5, 2013) – is produced by the American Psychiatric Association, whereas the International Classification of Diseases (ICD-11, 2021) is published by the World Health Organization (2019). The DSM-5 primarily focuses on mental and behavioral disorders, whereas the ICD-11 is a categorization system for both physical and mental illnesses. Although some diagnoses meet criteria for exceptionalities under IDEA and Bulletin 1508, the issue must adversely impact the child's educational performance in order for the child to qualify for special education services. Therefore, many parents may need guidance and support after learning that outside diagnoses do not necessarily equate to school support via an IDEA evaluation and subsequent IEP or Section 504 assessment and subsequent IAP. And professionals who traverse between organizations must be well versed with all systems in order to assist their clients to navigate the agencies and systems in place.

Summary

Navigating the realm of the school system can often be a complex task for the medical home team, especially as federal laws, state laws, and local policies and procedures are constantly changing. Although keeping abreast of this ever-changing landscape can be quite daunting, it is important to remember that federal laws supersede state laws and state laws supersede local laws regarding the same area. Whenever federal laws are revised, updated, or changed, most of the time they are an attempt to keep up with the zeitgeist of the times as well as provide services that benefit students and ensure a reasonable standard for practice of care and implementation of services within the school system.

References

Adaptive Behavior Assessment System-3 (ABAS-3). (2015). Retrieved from https://www.wpspublish.com/abas-3-adaptive-behavior-assessment-system-third-edition

Adaptive Behavior Diagnostic Scale (ABDS). (2016). Retrieved from https://www.wpspublish.com/abds-adaptive-behavior-diagnostic-scale

Aleman, S. R. (1991). *Education of the handicapped act amendments of 1990, P.L. 101-476: A summary* (CRS Report No. ED343318). Retrieved from Congressional Research Service. https://files.eric.ed.gov/fulltext/ED343318.pdf

American Addiction Centers, Inc. (2020). Retrieved from https://americanaddictioncenters.org/

American Psychiatric Association. (2013). *Diagnostic and statistical manual of mental disorders-5* (5th ed.). American Psychiatric Association. Retrieved from https://doi.org/10.1176/appi.books.9780890425596

Anastasi, A. (1982). *Psychological testing* (5th ed.). MacMillan.

Brownley, B. (2018). *Handling a manifestation determination review (MDR).* Wrightslaw. Retrieved from https://www.wrightslaw.com/info/discipl.mdr.strategy.htm#:~:text=by%20Bill%20Brownley%2C%20Esq.,When%20a%20child&text=This%20hearing%2C%20a%20Manifestation%20Determination,child%20based%20on%20his%20disability

Brownstein, R. (2015). *Education law a tool for ending school-to-prison pipeline*. Retrieved from https://www.splcenter.org/news/2015/12/14/education-law-tool-ending-school-prison-pipeline

Campbell Interest and Skill Survey (CISS). (1992). Retrieved from https://www.pearsonassessments.com/store/usassessments/en/Store/Professional-Assessments/Career-Planning/Campbell-Interest-and-Skill-Survey/p/100000323.html

CareerExplorer. (2021). Retrieved from https://www.careerexplorer.com/career-test/

Center on Positive Behavioral Interventions and Supports. (2021). *IDEAs that work.* Retrieved from https://osepideasthatwork.org/center-positive-behavioral-interventions-and-supports

Cicciola, E., Foschi, R., & Lombardo, G. P. (2014). Making up intelligence scales: De Sanctis's and Binet's tests, 1905 and after. *History of Psychology, 17*(3), 223.

Coeur D'Alene Public Schools. (2016). *Assessments & system performance*. Retrieved from https://www.cdaschools.org/domain/282

Colletti, E. N. (2018). *IEP vs. 504 plan: What's the difference?* Retrieved from www.niche.com

Developmental Behavior Checklist 2 (DBC2). (2018). Retrieved from https://www.wpspublish.com/dbc2-developmental-behavior-checklist-2

Dunn, D. (2013). Public law 94-142. In F. R. Volkmar (Ed.), *Encyclopedia of autism spectrum disorders*. Springer. Retrieved from https://doi.org/10.1007/978-1-4419-1698-3_393

Endrew v. Douglas County School District 580 US _ (2017).

Goal-Oriented Assessment of Lifeskills (GOAL). (2013). Retrieved from https://www.wpspublish.com/goal-goal-oriented-assessment-of-lifeskills

Holland Career Code Aptitude Test. (2017). *RIASEC inventory online version*. Retrieved from https://www.hollandcodes.com/online.html

Hood, A. B., & Johnson, R. W. (2007). *Assessment in counseling: A guide to the use of psychological assessment procedures* (4th ed.). American Counseling Association.

Horn, J. L., Donaldson, G., & Engstrom, R. (1981). Apprehension, memory, and fluid intelligence decline in adulthood. *Research on Aging, 3*(1), 33–84. Retrieved from https://doi.org/10.1177/016402758131002

John F. Kennedy Presidential Library and Museum. (n.d.). *John F. Kennedy and people with intellectual disabilities*. Retrieved from https://www.jfklibrary.org/learn/about-jfk/jfk-in-history/john-f-kennedy-and-people-with-intellectual-disabilities

Kaufman Assessment Battery for Children-II Normative Update (KABC-II NU). (2018). Retrieved from https://www.pearsonassessments.com/store/usassessments/en/Store/Professional-Assessments/Cognition-%26-Neuro/Gifted-%26-Talented/Kaufman-Assessment-Battery-for-Children-%7C-Second-Edition-Normative-Update/p/100000088.html

Keirsey Assessments. (2019). Retrieved from https://keirsey.com

Klein, A. (2016, March 31). The Every Student Succeeds Act: An ESSA overview. *Education Week*. Retrieved from https://www.edweek.org/policy-politics/the-every-student-succeeds-act-an-ESSA-overview/2016/03

Kurth, J. A., Miller, A. L., Toews, S. G., Thompson, J. R., Cortés, M., Dahal, M. H., … Wangare, F. (2018). Inclusive education: Perspectives on implementation and practice from international experts. *Intellectual and Developmental Disabilities, 56*(6), 471–485.

Lancaster Central School District. (2021). *What is PBIS?* Retrieved from https://www.lancasterschools.org/

Lee, A. M. I. (2021). *The 13 disability categories under IDEA*. Retrieved from https://www.understood.org/en/school-learning/special-services/special-education-basics/conditions-covered-under-idea

Loman, S., Strickland-Cohen, M. K., Borgmeier, C., & Horner, R. (2013). *Basic FBA to BSP trainer's manual*. Retrieved from https://pdxscholar.library.pdx.edu/cgi/viewcontent.cgi?article=1000&context=sped_fac

Louisiana Center for Children's Rights. (n.d.). Retrieved from https://lakidsrights.org/we-advocate/

Louisiana Division of Education. (2020). The 13 disabilities defined by the Individuals with Disabilities Education Act (IDEA). Retrieved from https://www.esc1.net/cms/lib/TX21000366/Centricity/Domain/59/13ClassificationsUnderIDEA.pdf

Luker, T. (2021). *Bridges4Kids: Building partnerships between families, schools, and communities*, p. 1. Retrieved from https://www.bridges4kids.org/

MAPP (Motivational Appraisal Personal Potential) Career Assessment Test. (1995). Retrieved from www.assessment.com

Miller, J. (2019). *Multimodal functional behavioral assessment (MFBA)*. Retrieved from http://www.mfba.net/slides.html

Mundial, G. B., & UNICEF. (2016). *Education 2030: Incheon declaration and framework for action: Towards inclusive and equitable quality education and lifelong learning for all*. UNESCO Institute.

Musgrove, M. (2011). *RTI cannot be used to delay-deny an evaluation for eligibility under the Individuals with Disabilities Education Act (IDEA)* (OSEP Memo DTD 11-07). Retrieved from https://sites.ed.gov/idea/files/idea/policy/speced/guid/idea/memosdcltrs/osep11-07rtimemo.pdf

Musgrove, M. (2013, April 2). *Letter to Gallo, G.* Retrieved from https://sites.ed.gov/idea/files/idea/policy/speced/guid/idea/memosdcltrs/acc-12-017845r-ut-gallo-fba-4-2-13.pdf

Myers Briggs Type Indicator. (2021). Retrieved from https://www.myersbriggs.org/my-mbti-personality-type/mbti-basics/

National Association for Gifted Children and the Council of State Directors of Programs for the Gifted. (2018–2019). *2018–2019 State of the states in gifted education*. Retrieved from https://www.nagc.org/sites/default/files/key%20reports/NAGC-2018-2019-State%20of%20the%20States_Executive%20Summary.pdf

National Association of School Nurses. (2015). *School wellness: What the law looks like*. Retrieved from https://www.nasn.org/nasn/nasn-resources/practice-topics/school-wellness

National Center for Education Statistics. (2021). Retrieved from https://nces.ed.gov/programs/coe/indicator/cgg

National Center on Safe Supportive Learning Environments. (2021). Retrieved from https://safesupportivelearning.ed.gov/resources/office-special-education-programs-osep-technical-assistance-center-positive-behavioral

Noble, D. (2015). *IHPs, 504 plans, and IEPs: What's the difference?* The Oley Foundation. Retrieved from https://oley.org/page/IHP_IEP_Difference

OSEP Technical Assistance Center on Positive Behavior Interventions and Supports. (2021). *Positive Behavior Interventions and Supports* [website] www.pbis.org

Pastorek, P. G. (2009). *Bulletin 1508: Pupil appraisal handbook (Title 28: Part C1)*. Louisiana Department of Education. Retrieved from http://fhfofgno.org/wp-content/uploads/2015/02/louisiana-bulletin-1508.pdf

Personal Values Assessment. (n.d.). Retrieved from https://www.valuescentre.com/tools-assessments/pva/

Prince Edward Island Department of Education. (2007). *Secondary transition planning*. Charlottetown, PE CIA 7N8. Retrieved from https://transitioncoalition.org/

Princeton Review Career Quiz. (2021). Retrieved from https://www.princetonreview.com/quiz/career-quiz

Public Interest Law Center. (n.d.). www.pubinlaw.org

RTI Action Center. (n.d.). www.rtinetwork.org

Renner, R. (2018). *The similarities and differences between IDEA and IDEIA*. www.theclassroom.com

Ryder, R. E. (2016). *RTI cannot be used to delay-deny an evaluation for eligibility under the Individuals with Disabilities Education Act (IDEA)* (OSEP Memo DTD 11-07). Retrieved from https://sites.ed.gov/idea/files/idea/policy/speced/guid/idea/memosdcltrs/oseprtipreschoolmemo.pdf

Salthouse, T. A. (2004). What and when of cognitive aging. *Current Directions in Psychological Science, 13*(4), 140–144. Retrieved from https://doi.org/10.1111/j.0963-7214.2004.00293.x

Sprague, J. R., & Walker, H. M. (2004). *Safe and healthy schools*. Guilford Press. Retrieved from https://www.guilford.com/books/Safe-and-Healthy-Schools/Sprague-Walker/9781593851057/print

Starin, S. (2011). *Functional behavioral assessments: What, why, when, where, and who?* p. 1. Retrieved from https://www.wrightslaw.com/info/discipl.fab.starin.htm

Strong Interest Inventory. (2021). Retrieved from https://www.themyersbriggs.com/en-US/Products-and-Services/Strong

Tan, L., Hughes, C., & Foster, J. (2016). Abilities, disabilities and possibilities: A qualitative study exploring the academic and social experiences of gifted and talented students who have co-occurring learning disabilities. *Journal of Pedagogic Development, 6*.

Texas Council for Developmental Disabilities. (2013). *Project IDEAL: Special education public policy*. Retrieved from www.projectIDEALonline.org/v/special-education-public-policy/

Texas Project FIRST. (n.d.). *Post-secondary goals*. Retrieved from http://texasprojectfirst.org/node/226

The Public Interest Law Center. (n.d.). *Public education*. Retrieved from https://www.pubintlaw.org/practice-area/public-education/archive/news/

The RTI Action Network. (n.d.). Retrieved from http://www.rtinetwork.org/about-us

U.S. Dept. of Ed. (n.d.). St. Vrain Valley Schools website. www.blogs.svvsd.org

U.S. Department of Education. (2007). *Archived: A 25 year history of the IDEA*. Retrieved from https://www2.ed.gov/policy/speced/leg/idea/history.html

U.S. Department of Education. (2015). *Every student succeeds act (ESSA)*. Retrieved from https://www.ed.gov/essa?src=ft

U.S. Department of Education. (2020) www.ed.gov, Laws and Guidance Section, ESSA.

U.S. Department of Education, Office of Civil Rights. (2016). *Parent and educator resource guide to section 504 in public elementary and secondary schools*.

U.S. Department of Education, Office of Special Education and Rehabilitation Services. (2013). *OSEP letter to parent*.

U.S. Department of Education, Office of Special Education and Rehabilitation Services. (2016). *OSEP Memo 16-11*.

U.S. Department of Education, Office of Special Education Services. (2011). *OSEP Memo 11-07*.

U.S. Government Printing Office. (2004). *U.S. Government Printing Office 2004 annual report*. Retrieved from https://www.govinfo.gov/app/details/GOVPUB-GP-f3aa3907720859265f2a3d9bf2102f14

U.S. Government Publishing Office. (1975). Retrieved from https://www.govinfo.gov/

U.S. Government Publishing Office. (1997). *Title I – Amendments to the Individuals with Disabilities Education Act*. Public Law 105-17. Retrieved from https://www.congress.gov/105/plaws/publ17/PLAW-105publ17.pdf

U.S. Office of Special Education Programs. (n.d.). *Welcome to OSEP*, p. 1. Retrieved from https://www2.ed.gov/about/offices/list/osers/osep/index.html

Understood Team. (2021). *The difference between IEPs and 504 plans*. Retrieved from https://www.understood.org/en/school-learning/special-services/504-plan/the-difference-between-ieps-and-504-plans

VanDuzer, T. (2021). *StudentTUTOR – Top 11 career aptitude tests for high school students*. Retrieved from https://student-tutor.com/

Vineland Adaptive Behavior Scales-Third Edition (Vineland-3). (2016). Retrieved from https://www.pearsonassessments.com/store/usassessments/en/Store/Professional-Assessments/Behavior/Adaptive/Vineland-Adaptive-Behavior-Scales-%7C-Third-Edition/p/100001622.html

von Ravensberg, H., & Blakely, A. (2014). When to use functional behavioral assessment? *Best Practice vs. Legal Guidance*. Retrieved from https://www.semanticscholar.org/paper/When-to-Use-Functional-Behavioral-Assessment-Best-Ravensberg-Blakely/0d99abd11b045e7593eb0a9c0258b848257ce27e

von Ravensberg, H., & Blakely, A. (2015). *When to use functional behavioral assessment? A state-by-state analysis of the law*, p. 1. Retrieved from https://www.semanticscholar.org/paper/When-to-Use-Functional-Behavioral-Assessment-A-of-Ravensberg-Blakely/9c3c1ee5299ea94804b23d08b9af9f584fd4a92f

Wechsler Intelligence Scale for Children-V (WISC-V). (2014). Retrieved from https://www.pearsonassessments.com/store/usassessments/en/Store/Professional-Assessments/Cognition-%26-Neuro/Gifted-%26-Talented/Wechsler-Intelligence-Scale-for-Children-%7C-Fifth-Edition-/p/100000771.html

Woodcock-Johnson IV (WJ IV). (2014). Retrieved from https://riversideinsights.com/solutions/woodcock_johnson_iv

World Health Organization. (2019). *International statistical classification of diseases and related health problems* (11th ed.). WHO. Retrieved from https://icd.who.int/

Yell, M. L., Shriner, J. G., Thomas, S. S., & Katsiyannis, A. (2018). Special education law for leaders and administrators of special education. In J. B. Crockett, B. Billingsley, & M. L. Boscardin (Eds.), *Handbook of leadership and administration for special education* (2nd ed.). Routledge.

Zablotsky, B., Black, L. I., Maenner, M. J., Schieve, L. A., Danielson, M. L., Bitsko, R. H., … Boyle, C. A. (2019). Prevalence and trends of developmental disabilities among children in the United States: 2009–2017. *Pediatrics, 144*(4), e20190811.

13 Pediatric Gastroenterology and Children with Autism Spectrum Disorder and Developmental Disabilities

Rebecca Wallace, Amanda Glinky, Jessica Picone, Russell Zwiener, and Erin Reuther

Abstract

Children and adolescents with autism spectrum disorder (ASD) and developmental disabilities have a higher prevalence of gastrointestinal conditions compared to the general population due to structural and behavioral components. This chapter aims to provide a clinical guide in understanding the etiology of gastrointestinal processes and conditions, presentation in children with ASD and developmental disabilities, management in the medical home, utilization of referrals to gastroenterological specialists, and communication with families about evidence-based treatments.

Keywords

Autism spectrum disorder · Organic GI conditions · Functional GI conditions · Gastroenterology · Evidence-based treatments · Family communication

Introduction

Pediatric gastroenterology is the study and practice of treating diseases and disorders of the gastrointestinal (GI) system within pediatric populations including infants, children, and adolescents. Pediatric gastroenterologists specialize in caring for conditions that affect any part of the GI tract, liver, pancreas, and nutrition. Diseases and conditions managed by pediatric gastroenterologists include celiac disease, gastroesophageal reflux disease (GERD), Crohn's and ulcerative colitis, hepatitis, pancreatitis, Hirschsprung's disease, and functional gastrointestinal disorders such as chronic functional abdominal pain, functional constipation, and irritable bowel syndrome, among others. Common presenting symptoms for pediatric GI include abdominal pain, constipation, diarrhea with or without bleeding, weight loss, and weight gain.

Children and adolescents with autism spectrum disorder (ASD) and other developmental delays have higher incidence and occurrence of gastrointestinal diseases and functional gastrointestinal disorders than their typically developing peers. In a sample of children with ASD matched with typically developing siblings, more children with ASD had GI problems (42% had GI problems) than their typically developing siblings (12% had GI problems) (Wang et al., 2011). The most common GI problems in children with ASD

R. Wallace (✉) · A. Glinky · J. Picone · R. Zwiener
E. Reuther
Children's Hospital New Orleans,
New Orleans, LA, USA
e-mail: rebecca.wallace@lcmchealth.org

P. McPherson (ed.), *Handbook of Treatment Planning for Children with Autism and Other Neurodevelopmental Disorders*, Autism and Child Psychopathology Series,
https://doi.org/10.1007/978-3-031-06120-2_13

were constipation and chronic diarrhea. Likewise, in a study assessing children referred to a pediatric clinic for complex feeding problems, approximately 75% of children referred had developmental delay (Burklow et al., 1998).

In this chapter, we will discuss the most common gastroenterology problems one is likely to encounter when working with children and adolescents with ASD and developmental delay, how to manage children with ASD and GI problems within a medical home, and information about when it would be most helpful to refer to a pediatric gastroenterologist or other specialists working within pediatric GI, such as registered dieticians and pediatric health psychologists.

Universal Monitoring from GI Perspective

Gastrointestinal problems are common in children with ASD and developmental delay, in general. Conditions range from very common disorders like constipation, diarrhea, and gastroesophageal reflux disease (GERD) to much less common conditions like gastritis, esophagitis, allergic gastrointestinal conditions, and small intestinal bacterial overgrowth (SIBO). Typical presentations of GI conditions in children with ASD vary from presentations of other children. Symptoms may vary in range from asymptomatic to abnormal posturing, self-injurious behavior to the abdomen, developmental/behavioral regression, and sleep disturbances due to possible atypical neurosensory responses compared to children without ASD. It is important to monitor and screen for certain disorders, particularly those covered in this chapter, as children with ASD likely present with atypical signs and symptoms. Early identification of GI symptoms and conditions is crucial for the best possible outcomes for children with developmental disabilities and ASD.

Monitoring Physical Exam

Advancements in medical technology have proven to be remarkably beneficial in the diagnostic process; however, the physical exam remains one of the clinicians' most important tools (Ediger & Hill, 2014).

Growth Chart Monitoring: Identification of Failure to Thrive and Obesity

Normal predicted growth over time reflects overall health and nutritional status of any child. Identifying when growth falls outside the realm of normal is important. Typically, growth under age two is assessed using WHO (World Health Organization) growth charts and CDC (Center for Disease Control) growth charts over the age of two. WHO and CDC growth charts are readily available at http://www.cdc.gov/growthcharts/. Special growth charts have been developed for some developmental disabilities including Down syndrome, Williams syndrome, Turner syndrome, and prematurity (Center for Disease Control Growth Charts, 2002. The CDC growth charts for children with special health care needs.).

Failure to Thrive

Poor weight gain that results in severe malnutrition can cause persistent short stature, secondary immune deficiency, and permanent damage to the brain and central nervous system (Kistin & Frank, 2019). Failure to thrive usually falls into four categories: inadequate nutrient intake, inadequate appetite or inability to eat large amounts, inadequate nutrient absorption or increased losses, and lastly increased nutrient requirements or ineffective metabolic utilization. Medical, nutritional, developmental/behavioral, and psychosocial factors may all contribute (Krebs et al., 2003; Kuczmarski et al., 2000). Assessment of the growth trajectory on a growth chart and proportionality may provide clues to the etiology of diminished weight. Failure to thrive is covered further in the organic condition section of the chapter.

Overweight and Obesity

A child is defined as overweight if their BMI is greater than the 85th but less than the 95th percentile for age and sex (DiBaise et al., 2019; Cole

& Lanham, 2011). This condition escalates to obesity when their BMI is greater than the 95th percentile for age and sex. According to the 2015–2016 National Health and Nutrition Examination Survey, using measured heights and weights, an estimated 18.5% of US children and adolescents of ages 2–19 years have obesity. The rapid rise in the prevalence of obesity has enabled pediatricians and other pediatric health-care providers play an increasingly important role in the early identification and prevention of childhood obesity and its associated comorbid conditions. Early diagnosis and treatment of obesity in children are crucial for the successful management of pediatric health (Dennison & Boyer, 2004; DiBaise et al., 2019). Identify patients who may be at risk by looking at such factors as family history, birth weight, and socioeconomic, ethnic, cultural, or environmental factors. Monitor trends in growth over time, and help families to identify concerning patterns when they first emerge.

Abdominal Exam

The pediatric abdominal examination generally consists of four components: appearance, auscultation, percussion, and palpation. Looking at the shape and contour of the abdomen for symmetry, pulsations, and vascular irregularities can help us to determine when more attention is needed. For example, a distended abdomen may indicate a variety of pathologies including but not limited to fecal retention, ascites (fluid in the abdomen), or even a bowel obstruction. Listening to hyperactive bowel sounds, or lack thereof, may lead us down the road of gastroenteritis or intestinal obstruction, respectively. When percussing the abdomen, a fluid wave can be produced when the examiner strikes one flank area with the tips of the fingers of the one hand and detects gentle pressure with the other hand on the opposite flank likely representing ascites (Goday et al., 2019). Abdominal organs are relatively easy to palpate in children because, in general, the abdominal wall is thin. Palpation is helpful for determining liver, kidney, and spleen size and for detecting other abnormal abdominal masses.

Skin and Physical Exam Findings Consistent with Micro- and Macronutrient Deficiencies

The most common micronutrient deficiencies are essential fatty acids; the B vitamins including riboflavin, niacin, pyridoxine, biotin, and B12; and fat-soluble vitamins including vitamin A, vitamin D, vitamin E, vitamin K, iron, zinc, iodine, vitamin C, and folate. Cutaneous abnormalities involving the skin, hair, and nails are commonly seen in patients with micronutrient deficiencies (MND). The greatest risk factors for MND in the USA include the presence of chronic medical disorders including malabsorptive conditions and pancreatic insufficiency or those who may have restricted diets of decreased access to food (Navarro et al., 2016). DiBaise et al. (2019) offer an excellent review of the cutaneous manifestations of specific micronutrient deficiencies with notation of the common causes for each.

Overall limited dietary intake affects the primary macronutrients including carbohydrates, proteins, and fats. Macronutrients are the chemical compounds consumed in the largest quantities and provide bulk energy (Perrin et al., 2003). Inadequate dietary intake may relate to social, psychological, medical, and physiologic issues. The major forms of malnutrition are marasmus (wasting) and kwashiorkor (edematous malnutrition). Low protein intake can lead to edema in the abdominal cavity. Fat and carbohydrate malabsorption can lead to diarrhea and weight loss. See figure below for signs of malnutrition.

Taking Stool Histories: Frequency and Consistency Including Bristol Stool Chart

Many components go into taking a stool history. The Bristol stool chart categorizes stool shape and consistency from category one of small hard lumps to category seven of entirely liquid stool (Saad et al., 2010). The chart is a helpful assessment and instructional tool for both parents and providers. Walking a family visually through the chart to describe their child's stool can provide a completely different stool history than a history elicited without this valuable visual aid. Other components of a stool history include consistency, frequency, color, urgency, volume, and odor.

Screening Lab and Imaging Evaluation

Screening is defined as testing for disease in an individual or population who appears to be healthy. Clinical judgment based on the history and physical examination guides screening to identify children who are at increased risk of disease and warrant additional testing (Pulikkan et al., 2019). Screening may include universal or selective and may be targeted to a risk-based population. Universal screening includes complete blood count (CBC) monitoring for iron deficiency or other forms of anemia along with lead screening to assess for poisoning. In the newborn period, universal screening includes measurement of bilirubin for the evaluation of jaundice. Targeted screening may include the measurement of hemoglobin A1C and fasting lipids in patients with obesity to rule out diabetes or increased cholesterol or triglycerides. Imaging screening for the most part is targeted to assess for a particular physical exam finding or symptom. Standard abdominal X-ray scan is an invaluable tool for the assessment of stool burden and air-fluid levels when considering obstruction or in the evaluation of a suspected ingestion of a foreign body.

Treatment Planning for GI Care: Referrals

Concerning Findings That Warrant Immediate Referral for Hospital Admission/ED (Emergency Department) Evaluation

The diagnosis of many disorders is often challenging in patients with ASD and developmental delay who may be unable to give a reliable medical history. Therefore, it is important for clinicians and primary care takers to have a good idea of what may warrant immediate referral to the emergency department or hospital admission. Excessive irritability and lethargy, beyond normal presentation, are clues that something bigger may be going on. Worrisome physical exam findings include but not limited to a hard, distended abdomen, somnolence, or pale skin in the setting of severe anemia. Indications of new-onset or worsening self-injurious behavior (SIB) may lead a caretaker down the path that something more serious may be going on or that a child may be in pain. Do not ignore the worries of caretakers who know these children best. A thorough medical examination is needed to rule out other causes and identify any potential medical reasons that may be causing the new SIB such as a headache, toothache, stomach problems, dermatological issues, or digestive ailments (Ristic, 2005).

Referral to Outside Providers

For complex presentations of GI conditions, a pediatric gastroenterologist can assist primary care providers by guiding a family and child through appropriate tests to establish a diagnosis and recommend a treatment plan. When making a referral to a pediatric gastroenterologist, consider including the following information with your referral—a growth chart with several points over time, a recent clinical information that led to the referral, a 3-day food diary, and a 7-day stool diary. For failure to thrive referrals, providers should also consider completing a screening for calprotectin and celiac disease prior to submitting the referral and including this testing in the referral information.

The gastroenterologist coordinates care with a pediatric health psychologist specializing in the treatment of children with ASD and GI diagnoses. Pediatric psychologists often help families learn behavioral strategies to help children manage pain and can participate in behavioral treatment of many GI conditions including functional constipation, encopresis, and functional pain. Some centers provide multispecialty clinics in which pediatric gastroenterologists and pediatric psychologists work together to care for patients with complex gastrointestinal needs.

Overview of Specific GI Conditions

Pediatric GI conditions are best understood in two categories—organic and functional. Organic conditions are any medical disease process by which there is physical disease or abnormal function of tissues of the body. Organic conditions contrast functional conditions in which there is no physical damage or inflammation but there is altered sensory processing such as nerve overactivity or maladaptive socio-emotional response to pain. Despite the lack of physical findings of disease, functional conditions are real and valid reasons for pain and symptoms of gastrointestinal disorders. Many children who initially present with abdominal pain are eventually diagnosed with functional pain when no organic source is found. Organic conditions are covered in this section while functional conditions are addressed later in the chapter.

Organic Conditions

Failure to Thrive

Failure to thrive (FTT) is the condition of inadequate growth or the inability to maintain consistent growth over time (Cole & Lanham, 2011). Every child should be monitored, over time, on the appropriate growth chart for their age, gender, and medical condition. Note that aside from the standard growth charts published by the Centers for Disease Control and Prevention and World Health Organization, there are also specialized growth charts for certain medical conditions including trisomy 21 (CDC, 2020). The diagnosis of failure to thrive is made through review of the child's growth chart showing failure to meet growth expectations over time. The condition is typically a result of simple math: either not enough calories are going in or too many calories are going out. Excess calories are expended either through gastrointestinal loss via malabsorption of calories and loss through the gastrointestinal tract or metabolic loss where too many calories are expended by the body such as in cardiac or endocrine disorders.

The workup of failure to thrive in the primary care setting includes a detailed history discussing enteral intake including a dietary recall and feeding history. Feeding history should include any abrupt changes or any perceived physical issues with eating. It may be helpful to watch the child eat in your office. Physical examination should look at the mouth and dentition and for skin findings concerning for nutritional deficiencies. Screening laboratory evaluation should be considered, including blood counts, electrolytes, liver and kidney function, and inflammatory markers, as well as a 3-day calorie count looking for sources of insufficient calorie intake. Referral for a formal evaluation by a registered dietician-nutritionist can be very helpful. All persistent cases of failure to thrive that do not resolve with increased calorie intake and without red flags for cardiac or endocrine issues should be referred to gastroenterology for further evaluation.

Vomiting

Vomiting is a common complaint in pediatrics. It is a nonspecific symptom which can represent problems in many different organ systems including the gastrointestinal, renal, and central nervous systems. Rumination which is covered later in this chapter is easily and commonly confused with vomiting. When eliciting a history of vomiting, it is important to well characterize the vomiting episodes. Determine the duration of symptoms, frequency of the episodes, content of the vomitus, and if it has any temporal association with certain foods or times of the day. Concerning features of vomitus include vomit that is bloody or coffee ground in nature, indicative for gastrointestinal bleeding, as well as grass green colored emesis which is concerning for bile content and would direct you toward considering a bowel obstruction, ileus, or even malrotation of the bowels (Wyllie et al., 2021). Emesis upon standing from lying flat without nausea is concerning for increased intracranial pressure and should be worked up as such (Berkowitz, 2014). Emesis with only certain types of foods such as dairy or with other symptoms including rashes, hives, swelling, and respiratory difficulty would be concerning for an allergy or anaphylaxis depending on the severity and acuity of symptoms.

All patients with a significant history of vomiting should be weighed and plotted over time on their growth chart looking for changes in growth trajectory. Vomiting with growth failure or weight loss is highly concerning. Abdominal examination should be performed looking for guarding, tenderness, distension, and quality of bowel sounds. If any concerning features are found on examination, consider imaging with plain film x-rays or ultrasound vs CT scan if findings seem to be more acutely concerning in nature. Laboratory workup should include a complete metabolic panel to look at electrolytes, kidney function, and serum bicarbonate. Obtain a urinalysis to look for ketones, glucose, and signs of infection, which can be masqueraded as vomiting. Urine or serum pregnancy testing is indicated for females beyond menstrual age as all children with communication difficulties are at a higher risk for sexual abuse. Always consider hospitalization for persistent vomiting with electrolyte derangements or concerning physical examination findings. When there is growth failure, children may need to be hospitalized to monitor for refeeding syndrome. Always calculate an anion gap, and if elevated, consider persistent vomiting as a presentation of new-onset diabetes in the setting of diabetic ketoacidosis or another metabolic condition.

Celiac Disease

Celiac disease is an autoimmune small bowel enteropathy caused by a sensitivity to gluten-containing foods (Ediger & Hill, 2014). It is commonly seen in adults and children with other autoimmune conditions such as type I diabetes, but it is also known to be associated with Down, Williams, and Turner's syndromes (Ediger & Hill, 2014). Patients can present in varied ways including growth failure, pubertal delay, abdominal pain, diarrhea, constipation, and vomiting (Ediger & Hill, 2014).

All patients should be screened for family history of celiac disease. On physical examination, look for blistering pruritic rashes and edema, as severe cases can present with protein-losing enteropathy. If history and physical examination suggest celiac disease, send total serum immunoglobulin A level in addition to anti-tissue transglutaminase (anti-tTg) testing as you refer to pediatric gastroenterology. Anti-tTg IgA is 95% sensitive and 96% specific for celiac disease and is a good screening test (Ediger & Hill, 2014). Further testing for other antibodies and human leukocyte antigen genetic testing can be left to subspecialty care but is less commonly needed for diagnosis. This condition is best diagnosed through esophagogastroduodenoscopy performed by a trained pediatric gastroenterologist. Do not advise patients to avoid gluten-containing foods prior to referral and endoscopy as doing so could mask the diagnosis during endoscopy. Villous blunting in the duodenum, the endoscopic hallmark of celiac disease, is masked when gluten is eliminated from the diet, and there is healing of the mucosa.

All patients with a diagnosis of celiac disease will remain on a gluten-free diet avoiding wheat, rye, and barley, for life as the mainstay of treatment (Ediger & Hill, 2014). As this diagnosis carries lifelong implications, it is important that families feel supported in making this lifestyle change. They should be offered consultation and support with a registered dietician-nutritionist under medical supervision (Ediger & Hill, 2014). All patients on restrictive diets need to be followed, longitudinally, to ensure that restrictive diets do not enable the development of nutrient deficiencies.

Gas and Bloating

Gassiness, bloating, and abdominal pain are very common complaints for all children but can be particularly challenging to those with developmental delay. Gas is either swallowed, released from foods we eat, or created through the breakdown of foods. Air swallowing is common in children with developmental delay, especially in those who hold their mouths open. This air can be expelled from the stomach through burping or venting if a gastrostomy is present. Note that patients with a Nissen fundoplication may be unable to belch which makes their abdominal gas more difficult for them to self-regulate.

Gas is not naturally present in large amounts in the small bowel, but it may be seen in patients who swallow air or in those who have small intestinal bacterial overgrowth related to conditions such as carbohydrate malabsorption or postsurgical changes that involve an absence of an ileocecal valve or have intestinal dysmotility. The ileocecal valve is a natural stop that prevents colonic bacteria from migrating into the remainder of the small intestine. Even if patients have an ileocecal valve, bacteria can make their way into the small intestine and overpopulate in the presence of readily available carbohydrates, generating more gas, causing discomfort and bloating. This condition called small intestine bacterial overgrowth is managed through the cycling of antibiotics such as metronidazole or gentamicin (Wyllie et al., 2021). Diagnosis is made through elevated breath hydrogen testing or cultures of aspirates of duodenal fluid obtained through endoscopy (Wyllie et al., 2021). Empiric treatment may also be considered without invasive testing, but it is recommended that this be done under the supervision of a pediatric gastroenterologist.

In any patient with gas and bloating, it is important to illicit a good history and physical exam including abdominal percussion. Consider an x-ray to assess gas patterns and stool burden. Never hesitate to refer a patient with a concerning abdominal exam to the emergency room for evaluation. Send stool studies for malabsorption including stool pH and reducing substances if you are concerned about carbohydrate malabsorption contributing to gas production. Measuring stool magnesium and phosphorus levels can help determine if a child is taking too many laxatives, which can create abdominal discomfort.

Trials of dietary changes to decrease intake of nonabsorbable foods like sorbitol and fiber-rich vegetables and beans may improve symptoms. A trial of lactose elimination may help, as well. Simethicone is a popularly used agent but may not be effective. All refractory cases should be referred to an experienced pediatric gastroenterologist for further evaluation.

Inflammatory Bowel Disease

Inflammatory bowel disease (IBD) is a group of autoimmune gastrointestinal disease including Crohn's disease and ulcerative colitis. These conditions often present with one or more of the following symptoms: failure to thrive, abdominal pain, anorexia, diarrhea, blood in stools, urgency and nocturnal stooling (Berkowitz, 2014). While symptoms may wax and wane, suspicion of IBD may first arise in the primary care setting. The workup includes a review of the growth chart, vital signs, and physical examination. Lab work including complete metabolic panel looking for low albumin and electrolyte derangements, elevated sedimentation rate or c-reactive protein looking for inflammation, and complete blood count looking for anemia and elevated platelets should follow as indicated. These lab findings are not essential to make the diagnosis but can point toward IBD. Infectious stool studies should be sent to rule out acute infections including *Yersinia* and *Campylobacter* prior to referral to pediatric gastroenterology.

Any patient with hemodynamic instability or concern for an acute abdomen should be immediately referred for emergency room evaluation. Blood transfusion may be needed to stabilize patients with significant rectal bleeding. IBD is diagnosed through diagnostic upper and lower endoscopy (Berkowitz, 2014) which can be performed in clinically stable patients. Steroid-sparing treatment regimens are preferred; however, patients with severe disease at diagnosis may require steroids to control disease and bridge to induction of therapy with biologic agents, the mainstay of treatment.

Management of IBD is done primarily by pediatric gastroenterologists; however, primary care physicians play an important role. IBD patients on biologic therapy need annual flu shots, screening eye exams, screening for immunity to hepatitis B virus and inactive tuberculosis infection (Shapiro et al., 2016). They should also be monitored for psychological and behavioral consequences of this chronic, potentially debilitating medical condition, which requires joint responsibility between the subspecialty and primary care groups making management in a medical home ideal (Shapiro et al., 2016).

Disorders of the Liver

Pediatric liver disease outside of the neonatal and infant period is a very broad topic. This section covers the highlights of diagnoses that should not be missed in the primary care setting. When liver disease is suspected or established, it is important to discuss symptoms of bruising and bleeding, looking for synthetic liver dysfunction, as well as jaundice and pruritus. Note family history of liver disease and recent ill exposures. A screening laboratory evaluation of liver disease including a complete metabolic panel (CMP), fractionated bilirubin, gamma-glutamyl transferase (GGT), and an international normalized ratio (INR) is needed. Serum ammonia should be assessed to look for underlying, occult metabolic liver disease accompanied by hyperammonemia. Always consider an ultrasound of the right upper quadrant if biliary disease is suspected or established, especially if there is elevation of serum bilirubin.

Jaundice

Jaundice is the physical exam finding consistent with an elevated total serum bilirubin concentration. Levels as low as 2 mg/dL can cause appreciable jaundice in older children and adolescents (Berkowitz, 2014). In any child with appreciable jaundice or elevated bilirubin, the level should be fractionated.

Direct hyperbilirubinemia implies an issue with the removal of conjugated bilirubin from the liver into the GI tract. Direct bilirubin of greater than 1.4 mg/dL after 6 months of age is always abnormal and should be further evaluated (Wyllie et al., 2021). It is important to take the full clinical picture of the patient into account. In patients with fever or other systemic signs of infection, the hyperbilirubinemia may be attributed to the ongoing acute infection. Colicky right upper quadrant pain worse after eating fatty foods or in any child with sickle cell disease or blood disorder points toward cholelithiasis or choledocholithiasis. It is imperative that in any patient with a direct hyperbilirubinemia, you obtain liver enzymes including a GGT. Elevation of GGT and alkaline phosphatase is typically caused by cholestasis or obstruction in the path of excretion of bilirubin. If GGT is normal but AST (aspartate transaminase) and ALT (alanine aminotransferase) are elevated, this points to other systemic causes such as drugs, infection, and metabolic disease.

Indirect hyperbilirubinemia implies that there is an issue with the processing and conjugation of the bile within the hepatocytes. This can occur in one of two ways: either too much bilirubin is being recycled for the hepatocytes to keep up with the demand or there is a lack of adequate enzyme levels to keep up with the physiologically expected demand of processing the blood. Indirect bilirubin is conjugated into direct bilirubin through the activity of uridine diphosphoglucuronate glucuronosyltransferase (Wyllie et al., 2021). In type I Crigler-Najjar syndrome, this enzyme is absent leading to abrupt detection of the condition, typically detected in infancy. However, in Gilbert's syndrome and type II Crigler-Najjar syndrome, the enzyme is merely reduced in activity allowing these conditions to go undetected for much longer, well into childhood or even adulthood. While type I Crigler-Najjar syndrome is fatal, type II Crigler-Najjar can be medically managed by a trained hepatologist and Gilbert's syndrome often requires no treatment at all (Murray and Horslen, e.d, 2013). Other causes of indirect hyperbilirubinemia include hemolytic anemia, hypothyroidism, and drugs such as rifampin.

Acute Hepatitis and Liver Failure

Hepatitis is a broad term that refers to inflammation of the liver, represented by elevated serum transaminases. Causes of acute hepatitis include medication toxicity, such as acetaminophen and antiepileptics, infections, and metabolic liver disease. A thorough history including signs and symptoms of infection, including vomiting, diarrhea, abdominal pain, fevers, travel history, and home medications, is necessary to fully assess for hepatitis (Berkowitz, 2014). In a patient with a history of tremors and mood or behavioral disorders, Wilson's disease should be included in the differential diagnoses (Brewer, 2001). Sexual history may also be necessary in adolescents and patients at risk for abuse to consider sexually transmitted infections or fecal oral infections such as hepatitis A and E (Berkowitz, 2014). The physical exam should look for jaundice and bruising and assess the span of the liver and tenderness to palpation of the right upper quadrant.

All patients with acutely elevated liver enzymes should have an INR drawn to assess for synthetic liver dysfunction. In any patient with elevated INR, behavioral changes, or drug toxicity, a serum ammonia will be needed. If there is any history of behavioral issues, motor disorders, or a Kayser-Fleischer ring around the iris, consider serum ceruloplasmin to screen for Wilson's disease (Brewer, 2001). Elevated liver enzymes accompanied by an elevated INR should raise a high clinical suspicion for acute liver failure and be treated as a medical emergency with immediate referral to a center experienced in the treatment of acute liver failure in pediatrics. The cause of half of all cases of pediatric liver failure cases remains unknown; however, always suspect medication toxicity and send medication levels as this may require medical intervention. For example, questions about acetaminophen ingestion allow an accurate timeline to be established which is very important when giving N-acetylcysteine in acetaminophen toxicity. *Poison control can be reached at 1-800-222-1222 for assistance with any case of ingestion.*

Chronic Hepatitis

Chronic hepatitis is defined as elevated serum transaminases that last longer than 6 months. Causes are quite varied and include autoimmune hepatitis, chronic infections, medications, fatty infiltration of the liver, and unknown causes. When evaluating a patient for referral to gastroenterology for chronic hepatitis, consider screening tests in addition to liver enzymes including GGT, ceruloplasmin (looking for Wilson's disease), thyroid studies, lipid panel, and abdominal ultrasound to assess texture of the liver as well as span and status of the biliary tree. Send INR and glucose level to look at liver function. All cases of chronic hepatitis should be referred to pediatric gastroenterology for further evaluation and treatment. A liver biopsy may be considered should elevations persist with or without a clear cause of the hepatitis.

Functional GI Conditions

Functional GI disorders are extremely common among patients with GI symptoms. In a pediatric gastroenterology clinic, over half (52%) of patients met criteria for one or more functional gastrointestinal disorders (Rouster et al., 2016). Among children less than 4 years, 29% met criteria for functional constipation, 13% had infant regurgitation, and 10% had cyclic vomiting syndrome. While in children 4–18 years, 36% met criteria for irritable bowel syndrome, 19% had abdominal migraine, 17% had functional constipation, 7% had functional abdominal pain, 7% had aerophagia, and 7% had dyspepsia (Rouster et al., 2016). Treatment of functional gastrointestinal disorders works best when practitioners use a biopsychosocial model to understand and conceptualize patients' problems (Hyman, 1999). The biopsychosocial approach includes all aspects of a child's life to understand their symptoms including biological or organic factors, psychological factors, and social and environmental factors. Appropriate treatment of functional GI disorders includes interventions that address all areas across biological, psychological, and social domains that may be impacting a child's symptoms and often includes interventions from a multidisciplinary team (Hyman, 1999).

Selective Eating: Avoidant/Restrictive Food Intake Disorder

Feeding concerns are very common among children with developmental disabilities, occurring up to 14 times the rate seen in typically developing children (Mayes & Zickgraf, 2019). The most frequently reported concern by parents is selective eating behaviors (Mayes & Zickgraf, 2019). Selective eating may include limited food preferences based on appearance, color, taste, smell, or texture, as well as reduced interest in exploring new foods. For many children with ASD specifically, slight changes in food color or consistency can lead to significant emotional distress. Children may also cycle through food preferences. For instance, the child may only eat chicken nuggets for 3 weeks, and then the child's preference shifts to macaroni and cheese for several weeks. It is also common for some individuals with developmental disabilities to experience a lack of interest or desire to eat. If severe, these feeding challenges can result in weight loss or failure to make expected gains in weight, malnutrition, and failure to thrive. In these instances, eating behaviors may be consistent with an avoidant/restrictive food intake disorder (ARFID). ARFID is characterized by limited food intake due to lack of interest, avoidance based on sensory characteristics, concern about aversive consequences of eating that results in weight loss, poor growth, nutritional deficiencies, dependence on supplementation, and/or interference with psychosocial functioning (APA, 2013). When the eating disturbance is occurring in the context of another mental health disorder (such as ASD) and/or medical condition, it is important that the eating behaviors are significant enough to warrant specific clinical attention before making a diagnosis of ARFID (APA, 2013).

Among individuals with ASD, selective eating is often conceptualized as a component of the sensory sensitivities and/or rigid and repetitive behavior patterns associated with the condition. Thus, a child's preference to have the same meal for dinner every night, refusal to eat the same snack from a different package, or resistance to trying a new food is a function of the rigid thinking characteristic of ASD. Changes in routine or foods that challenge sensory sensitivities often cause significant emotional distress and/or behavioral dysregulation for the child, which can be challenging for caregivers to manage and may contribute to parenting behaviors that reinforce selective eating (e.g., decreased presentation of new foods to avoid an emotional outburst). It is important to note that feeding concerns can also be a function of motor skill deficits and may require further assessment with an occupational therapist.

For children presenting with feeding concerns, it is important to obtain medical history, growth trajectories, developmental history, and thorough history of eating behaviors to aid in differential diagnosis and treatment planning. Similarly, for children diagnosed with a developmental disability, it is important to monitor growth and screen for feeding problems. For many children, feeding concerns do not require formal intervention with a mental health professional. However, when feeding concerns warrant clinical attention, behavioral approaches have been shown to be most effective (Bryant-Waugh, 2019). Behavioral interventions may include positive reinforcement of target eating behaviors, structured practice with new foods to expand food preferences over time, and reduction of negativity or family conflict surrounding eating behaviors (Sarcia, 2020). For many children with developmental disabilities, feeding concerns may be addressed within the context of applied behavior analysis (ABA) therapy (Sarcia, 2020). For children who are experiencing health consequences such as weight loss, failure to meet expected gains, and/or malnutrition, pharmacological interventions can be useful as an adjunctive therapy to stimulate appetite and/or reduce anxiety including olanzapine and cyproheptadine (Bryant-Waugh, 2019). Nutritional supplements may also be recommended to support growth; in extreme circumstances, nutritional supplementation through an NG tube or G-tube is recommended. Referral to a gastroenterologist is appropriate when health consequences, such as those described above, become apparent. For some children with deficits in fine motor skills, feeding therapy with an occupational therapist is also warranted.

Rumination

Rumination syndrome is defined as the repeated effortless regurgitation of recently ingested food over at least 1 month if it is not caused by a medical condition (APA, 2013). Individuals may expel, rechew, and/or re-swallow the contents of their rumination. Rumination is differentiated from recurrent vomiting through the apparent effortless and voluntary nature of the behavior. Rumination is not typically associated with pain, nausea, or retching; however, some individuals do note symptoms of abdominal pain or nausea prior to rumination episodes (Srinath et al., 2014). Rumination is most likely to occur immediately after a meal and usually stops once the gastric contents become acidic (Srinath et al., 2014). Rumination disorder is mostly seen among infants and individuals with developmental disabilities, though it can be seen in typically developing children, adolescents, and adults (APA, 2013; Lang et al., 2011). Notably, ruminative behaviors can result in concern for malnutrition, weight loss, failure to make expected gains in growth, electrolyte abnormalities, and oral/dental problems (APA, 2013; Murray & Thomas, 2018). Rumination can also be associated with food avoidance or restriction for individuals who wish to avoid the potential social consequences of regurgitation (APA, 2013).

Several factors have been considered in the development and maintenance of rumination disorder. Among typically developing infants, rumination has been associated with neglectful and/or minimally stimulating environments in which the infant is attempting to self-soothe, self-stimulate, and/or elicit caregiving through this behavior (Berkowitz, 1999). Similarly, rumination often serves as a self-soothing or self-stimulatory behavior that is automatically reinforcing among individuals with developmental disabilities. Thus, the enjoyment of the behavior or the pleasant sensory experience occurring with rumination episodes serves as the reinforcer that maintains the behavior over time (APA, 2013; Lang et al., 2011). Additionally, rumination can serve as means to escape or avoid an unpleasant experience (e.g., avoid a disliked food to end the meal) or receive social interaction (e.g., attention during cleanup from the caregiver; Lang et al. (2011)). Thus, it is important to understand the function of the behavior for each patient to guide intervention efforts.

When considering a diagnosis of rumination disorder, it is important to obtain an adequate history which includes growth history, frequency and duration of the behavior, associated symptoms (e.g., pain, nausea), volume of food consumed, rate of consumption, parent-child relationship factors, mealtime environment and behaviors, and family response to rumination (Luiselli, 1989). Further, potential organic causes for regurgitation should be evaluated, and GERD should be evaluated if apnea, reactive airway disease, hematemesis, or food refusal is present (Berkowitz, 1999). Other considerations include gastroparesis and intestinal obstructions (Srinath et al., 2014). A referral to a gastroenterologist should be made if "red flag" symptoms are present, including significant weight loss and blood in stool or emesis. Several interventions may be considered once the diagnosis is made. Often, a referral to a licensed psychologist, behavior analyst, or mental health provider is warranted as behavioral strategies have the strongest empirical support for the management of rumination (Ellis et al., 1997; Lang et al., 2011). Efforts to improve the parent-child relationship have been shown to be effective, particularly for infantile rumination, including mealtimes in calm, soothing environment with increased attention from caregiver and reduced distractions (Ellis et al., 1997). Additionally, dietary changes (increasing food volume, restricting trigger foods, thickening formula/changing textures, meal pacing with slower rate of intake), alternative sensory stimulation/self-soothing strategies, and manipulation of social contingencies have been effective for individuals with developmental disabilities (Ellis et al., 1997; Lang et al., 2011). For high-functioning older children, adolescents, or adults, relaxation strategies (e.g., diaphragmatic breathing) have been taught as a competing response in conjunction with meal pacing and self-monitoring to address rumination (Murray & Thomas, 2018). Overall, it is important that the intervention be individually tailored to the needs of child and family.

Pica

Pica is another type of feeding disorder that is common among individuals with developmental disabilities and can lead to gastrointestinal complications (Call et al., 2015; Stiegler, 2005). According to DSM-5, pica is defined as the persistent ingestion of nonnutritive or nonfood items over at least 1 month (APA, 2013). Additionally, the behavior must be inappropriate for developmental level and cannot be a part of culturally sanctioned practice. Onset is most common in childhood, and the presence of pica in adulthood is often associated with developmental disability or another mental health condition (APA, 2013). Common examples of pica items for children and adolescents include plastic, cloth, dirt, paper, paint, rocks, soap, and feces (Call et al., 2015; Matson et al., 2011; Stiegler, 2005). Prevalence rates range from 4% to 25%, with higher prevalence rates among individuals with developmental disabilities relative to the general population (Matson et al., 2011; Mayes & Zickgraf, 2019). Notably, the strongest risk predictor of severe and potentially life-threatening presentations of pica is developmental disability, with more profound deficits being associated with increased risk for pica (Matson et al., 2011).

Environmental factors are the most well-supported factors contributing to the development and maintenance of pica behaviors, though iron and/or zinc nutritional deficiencies have also been investigated (Matson et al., 2011; Stiegler, 2005). More specifically, evidence suggests that pica among individuals with developmental disabilities is most frequently maintained by automatic reinforcement (Matson et al., 2011; Stiegler, 2005). Similar to rumination disorder, this means that the behavior itself and/or the positive consequences (e.g., sensory stimulation, enjoyment) reinforces the behavior more often than social contingencies (e.g., attention, escape, avoidance). As such, pica can be a challenging behavior to eliminate. Pica is associated with significant medical complications, including obstruction or perforation of the intestines, choking, infection, poisoning, malnutrition, enlarged colon, constipation, surgical intervention, and increased mortality rates (Stiegler, 2005). As such, it is important to identify, assess, and treat pica within a timely manner.

Williams and McAdam (2012) provided guidelines for the assessment, treatment, and prevention of pica. The authors recommended universal screening for pica, which may include interview with caregivers, direct observation in clinic, stool checks, and review of medical records. If there is concern for pica, it would also be important to rule out nutritional deficiencies that may be contributing to pica behaviors (Williams & McAdam, 2012). If an individual screens positive for pica behaviors, a referral for further assessment and treatment to a licensed psychologist and/or board-certified behavior analyst with experience in treating pica is warranted given the potential associated safety concerns (Williams & McAdam, 2012). If symptoms of pain, blood in stool, and other GI symptoms are present, referral to a gastroenterologist is appropriate. ABA and behavioral interventions are the most widely supported evidence-based treatments for pica (Call et al., 2015; Hagopian et al., 2011; Matson et al., 2013). Evidence for pharmacotherapy and nutritional supplements has been limited and would not be considered a first-line intervention (Matson et al., 2013; Stiegler, 2005). While working toward an effective intervention, efforts should be made to maintain the individual's safety (Williams & McAdam, 2012). This includes creating a pica-safe environment in which access to pica items and/or other potentially dangerous items should be removed and providing increased supervision in situations/areas in which items may be present and cannot be removed. Additionally, caregivers can make efforts to redirect the individual when pica behaviors are observed. For example, a parent or caregiver can prompt the child to put the crayon on the ground and then give the child a preferred item (e.g., toy, snack) followed by positive praise for engaging in more appropriate behavior.

Constipation and Encopresis

Functional constipation is a functional gastrointestinal disorder that is often seen in children and adolescents with developmental delay and ASD. The Rome IV criteria is the current standard for defining chronic constipation. See Fig. 13.1.

Often, chronic constipation begins with a painful bowel movement, and then the child develops subsequent withholding behavior to avoid another negative or painful experience when stooling. Over time, children can develop a pattern of fecal retention, which further perpetuates a cycle of large, painful, and difficult bowel movements due to constipation. Many children also develop overflow incontinence, a condition in which there is a large mass of stool in the rectum and children have leakage of stool into underclothes during the day or night. Encopresis is a psychological/psychiatric diagnosis which describes a pattern of bowel movements and passage of feces that occur into inappropriate places including clothing or the floor (American Psychiatric Association, 2013). In encopresis, these events must occur at least once per month for 3 months. Passage of feces in inappropriate places can be either intentional or unintentional, but it cannot be directly due to the effects of a substance or laxative. Encopresis can only be diagnosed in a child who is at least 4 years old or of an equivalent cognitive functioning level and can occur with or without chronic constipation (American Psychiatric Association, 2013). Functional constipation has peak onset between 0 and 4 years of age (Van den Berg et al., 2006) and often overlaps with toilet learning. Encopresis has a peak onset between 4 and 7 years (American Psychological Association, 2013).

Functional constipation also occurs more frequently in children with ASD and developmental delay than typically developing children. In fact, of all children referred to a pediatric gastroenterology clinic, over half (52%) met criteria for a functional disorder, with 29% of children less than 4 years having functional constipation and 17% of children 5 years or older having functional constipation (Rouster et al., 2016). Among toilet-trained children, those with ASD are more likely to have experienced constipation or report taking medication to treat constipation in the past year than their typically developing peers (Cuffman & Burkhart, 2021). Children with ASD and developmental delay are also likely to experience expected delays in meeting the developmental milestone of toilet training. It is important to not confuse a child's toilet training according to their own developmental trajectory with a disorder or disease. Children with ASD who are nonverbal or have limited verbal abilities may have trouble describing their experience if they have constipation. In children with ASD, constipation is likely to present as avoidance of sitting on the toilet, refusal to engage in toilet training, or appearance of pain or straining behavior when trying to have a bowel movement. As listed in the criteria above, infrequent bowel movements and those that are large in size are often present and can be observed by parents and caregivers.

Treatment of chronic functional constipation comes from a biopsychosocial perspective. When both constipation and encopresis are present, it is important to treat and manage the constipation first before addressing toileting concerns. Treatment of constipation usually includes the use of osmotic laxatives and/or stimulant laxatives depending on the severity of constipation. Pediatric gastroenterologists may also use hospital-based clean-outs for cases of fecal impaction in which a very large stool mass is in the rectum. In the medical or primary care setting, it is recommended that pediatricians treat constipation using a daily osmotic laxative such as polyethylene glycol to achieve regular bowel

Chronic constipation is defined as having two or more of the following symptoms within the past 3 months:

1) Straining during more than 25% of defecations.
2) Lumpy or hard stools more than 25% of defecations.
3) Sensation of incomplete evacuation more than 25% of defecations.
4) Sensation of anorectal obstruction/blockage more than 25% of defecations.
5) Manual maneuvers to facilitate more than 25% of defecations (e.g., digital evacuation, support of pelvic floor).
6) Fewer than three solid bowel movements per week.
7) Loose stools are rarely present without the use of laxatives, and insufficient criteria is met for irritable bowel syndrome.

Fig. 13.1 Rome IV criteria for chronic constipation. (Adapted from Drossman (2016)) movements that occur at least three times per week and are of soft consistency. This will help to assure painless bowel movements. Once soft and more frequent bowel movements are achieved, incontinence and soiling can be addressed by recommending scheduled toilet sits for 5 minutes that occur 20–30 minutes after each mealtime. The child should not be pressured to have a bowel movement every time they sit on the toilet. Instead, scheduled sits help provide opportunities for the child to practice sitting on the toilet, and those children who were previously toilet trained may start defecating in the toilet again. For children who have never been toilet trained, it is important to not attempt or move forward with toilet training until constipation is managed. Children with ASD often take longer than their same-aged peers to learn toileting. It is important to allow them to learn at their own developmental pace.

It is recommended that pediatricians refer to a pediatric gastroenterologist if constipation is severe (such as a very large mass of stool in the rectum) or persists despite the use of a daily osmotic laxative. For children who have fecal incontinence or are having difficulty learning toileting behavior, referral to a psychologist who specializes in developmental delay and ASD or referral to applied behavior analysis treatment is recommended. Often in pediatric GI clinics, pediatric gastroenterologists and pediatric psychologists work together to develop treatment plans for children with chronic constipation and encopresis. When making a referral, it is helpful to include a stool diary to assess the frequency and consistency of bowel movements in the past 2–3 weeks, as well as a list of medications that have been trialed and any reactions or results of past medications.

Functional Abdominal Pain

Functional abdominal pain (FAP) is one of the most common presenting problems within pediatric gastroenterology clinics and accounts for 2–4% of pediatrician visits (Hyams, 1999; Srinath et al., 2014). FAP is characterized by recurrent or chronic abdominal pain that is not attributable to organic disease or another condition. Symptoms can range from postprandial fullness, early satiety, nausea, and upper abdominal pain or burning to pain associated with defecation (with changes in frequency and stool form) to episodic periods of intense pain (Brusaferro et al., 2018; Hyams, 1999). FAP syndromes include irritable bowel syndrome (IBS), functional dyspepsia, abdominal migraine, FAP not otherwise specified, and aerophagia.

Diagnoses are often made through the use of the more recently formulated Rome IV criteria or Apley's criteria in addition to lack of red flag symptoms and normal testing results following any indicated clinical workup (Banez & Cunningham, 2009; Brusaferro et al., 2018; Srinath et al., 2014). Family history of functional disorders and the presence of psychosocial stressors and/or psychiatric symptoms/disorders (viz., anxiety and depression) can also provide valuable data when making a diagnosis given high comorbidity rates with functional GI disorders (Srinath et al., 2014). When considering a diagnosis of a functional abdominal pain syndrome, it is necessary to gain an understanding of the child's nutritional history, growth history, general psychiatric history, and social factors in addition to the physical examination. Organic causes of symptoms should be ruled out with consideration for gastroesophageal reflux disease (GERD), ulcers, inflammatory bowel disease, appendicitis, gallbladder disease, and oncologic conditions (Probst & Burgess, 2014). Notably, a clinician should carefully and mindfully select diagnostic tools, tests, and procedures as continuing to explore organic causes with repeated and/or unnecessary tests and procedures can exacerbate functional symptoms and increase family anxiety that a serious or life-threatening disease is being "missed" (Hyams, 1999).

Unique challenges may emerge when assessing abdominal pain for children diagnosed with ASD or a developmental delay. Depending on the child's level of functioning, the child may have difficulty communicating about or describing symptoms to caregivers and physicians. Moreover, limited understanding, sensory sensitivities, and/or selective eating habits may complicate the physical exam, participation in diagnostic tests and procedures, and ruling out other conditions. It can be beneficial to have caregivers use a chart or diary to log symptoms and potential factors that may be related to pain symptoms (e.g., food intake, bowel movements, activity level) in order to better characterize symptoms and develop a treatment plan.

FAP syndromes are often treated utilizing a multimodal approach (Banez & Cunningham, 2009; Brusaferro et al., 2018). For some children and families, providing general reassurance and managing symptoms through simple behavioral changes (e.g., minimizing the use of ibuprofen which may worsen dyspepsia, reducing intake of a specific food that exacerbates symptoms, increasing fiber in diet, adding a stool softener) are sufficient to manage symptoms (Brusaferro et al., 2018; Hyams, 1999). Additionally, addressing psychosocial stressors and psychiatric symptoms may help to alleviate symptoms including addressing difficulties at school or participating in counseling services to address underlying mood or anxiety symptoms. Proactive coping strategies such as relaxation (e.g., diaphragmatic breathing, guided imagery, progressive muscle relaxation) and active distraction (e.g., taking a walk, completing a craft, drawing) can be effective tools (Banez & Cunningham, 2009). It is important to use consistent language about the child's pain and convey an understanding that the child *is* experiencing pain even if the pain is not attributable to organic disease. Families benefit from having a clear conceptualization of the pain experience, and this allows for increased engagement and commitment to the treatment plan. For children with more persistent or severe symptoms who are exhibiting significant disability due to their symptoms or when a diagnosis and/or treatment plan cannot be made, a referral to specialists (e.g., pediatric gastroenterologist and psychologist) may be indicated.

Evidence-based care for chronic pain with associated disability utilizes a rehabilitation model with a focus on increasing day-to-day functioning (e.g., time spent at school, return to typical activities and hobbies) rather than a primary focus on reducing pain (Banez & Cunningham, 2009; Probst & Burgess, 2014). Specific psychotherapies to target chronic pain (e.g., cognitive behavioral therapy) and behavioral modifications within the environment (e.g., positive reinforcement for active management of pain and increasing function) are also effective for the management of FAP syndromes (Banez & Cunningham, 2009; Brusaferro et al., 2018; Probst & Burgess, 2014). It is important to consider level of cognitive functioning when considering psychological interventions for children with developmental delays as behavioral interventions are more likely to be effective for children with lower levels of cognitive functioning.

There is limited evidence on effectiveness of pharmacological interventions for FAP among children and adolescents (Brusaferro et al., 2018; Hyams, 1999; Probst & Burgess, 2014). There is initial evidence to support the use of some antispasmodics (peppermint oil, drotaverine, dicyclomine, hyoscine [scopolamine], mebeverine), antidepressants (amitriptyline, imipramine, and citalopram), and antihistamines (cyproheptadine). Probiotics and other medications to target symptomatic relief have also been used. Comprehensive reviews of pharmacologic treatment are available (see Brusaferro et al., 2018; Probst & Burgess, 2014).

Interventions and Working with Families

Talking to Families

The relationship between caregivers and medical providers can be a complex one; this is especially true when the patient has ASD. This complication stems from the fact that there is no clear etiology or cure for ASD. In addition, caregivers historically are integral in diagnostics, research, advocacy, and treatments, perhaps more so than any other childhood condition (Goin-Kochel et al., 2009). When working with patients diagnosed with ASD, it is important to be aware of the history of parental involvement. Initially, ASD was attributed to parenting styles referred to as "refrigerator parents," indicating the symptoms stemmed from neglectful and emotionally detached parents (Silverman & Brosco, 2007). While this was disproven early on, some caregivers continue to feel a sense of guilt and pressure from the treatment team. Caregivers also sometimes feel frustrated that "enough" is not being done to help their child. Early research on symptomatology and potential behavioral treatment was based on parents taking on the responsibility of learning more about the condition, its effects, and possible ways of managing symptoms (Goin-Kochel et al., 2009). While some caregivers have gained greater acceptance of the chronicity of ASD, some treatments continue to aim at a cure.

While behavioral interventions have been established to improve symptom management, parents continue to seek medical treatments for the chronic condition. Behavioral treatments are often intense in rigid structure and time consuming making them impractical and challenging for families to implement in the long term (Silverman & Brosco, 2007). This, combined with the Western culture's emphasis on medicinal-based treatments, motivates parents to seek out the latest, progressive treatment, often called complementary and alternative medicine (CAM)-based treatments. CAM treatments are not well established, scientific evidence-based treatments but rather word-of-mouth approaches that showed promising results for a small subset of patients with ASD, which make the larger population hopeful (Wong & Smith, 2006). These treatments tend to be more diet, lifestyle, and procedure based than psychopharmacologic. For most patients with ASD, CAM treatments may not be harmful per say; they also may not be helpful. They can lead to parents feeling defeated, frustrated, and further desperate for a solution and/or cure.

So how do you as a provider interact with caregivers in a professional, supportive way that allows them to feel part of the treatment team and heard, while also encouraging evidence-based care? At the outset, it is imperative to create a respectful, open communicative relationship with families. This encompasses patience, understanding, normalizing their perspective, open-ended conversations, and collaborating with them as a team. This approach may appear time consuming at first; however, it saves time and maintains trust in the long run. It is important to know where families are coming from. Many caregivers are initially nervous to ask questions about treatment, especially CAM treatments, due to fear of being dismissed, rejected, or beleaguered. Experience has also taught them that professionals tend to show automatic preference for scientific evidence-based treatments, yet the family has not found them successful or practical for their child (Wong & Smith, 2006). When caregivers bring up questions about their child's symptoms and treatment options, listen to their concern and questions. Explore their perspective, and explain the challenges of their suggested treatments, including potential harms and limited evidence for success. Also provide an alternative evidence-based approach that might be helpful. See Fig. 13.2 for example of communication with family.

Complementary and Alternative Medicine (CAM)-Based Approaches

As a clinician, it is important to not only educate yourself on the scientific evidence-based treatments; it is also important to be familiar with the trending CAM treatments. This knowledge will allow you to understand the online advocacy for the CAM therapy, prevalence of its use in the community, and potential gains and harms of the approach (Wong & Smith, 2006). The challenge with these approaches is that the evidence-based research evaluating them is limited and most information is anecdotal, at best. Even if a treatment does not present direct harms, if it has not been thoroughly evaluated from a scientific perspective, it should not be recommended. This section will focus on the current community-trending approaches to ASD that involve the gastrointestinal systems. There is no scientific evidence that specialized diets have any impact on the symptoms of ASD.

1) When caregivers broach the topic of concerning symptoms and exploring treatment options, engage active listening skills. This includes the following:
   a. Maintain open body language:
      i. Face family and engage in comfortable eye contact.
      ii. Remove barriers between you.
      iii. Maintain non-judgmental tone.
      iv. Ignore any non-urgent interruptions.
   b. Repeat or paraphrase what they discussed:
      i. "I hear you are really struggling with your child's ___."
      ii. "Let me make sure I understand this approach (repeat approach)."
   c. Use open-ended questions:
      i. "Tell me more about your concerns/ this treatment."
      ii. "What is it about this treatment that appeals to you?"
      iii. "How do you suppose this treatment might be for your family?"
      iv. "Who shared their experience with this approach with you?"
2) Normalize their concern and interest in exploring other options:
   a. Example: "Many families in similar circumstances investigate dietary changes to address behavioral concerns."
   b. "It is natural and being a good caregiver to want the best and latest treatment for your child."
3) Use team-oriented statements:
   a. "Help me understand more about this."
   b. "Let us both make efforts to investigate the best treatment options for your child."
4) Speak with honesty from a *fact, evidence-based perspective*:
   a. It is acceptable to acknowledge that you are not aware of the treatment option and will have to explore it further in order to make a statement about it.
   b. It can be helpful to explain to families the rationale for evidence-based treatments.
   c. Review evidence-based treatment options in a supportive manner.
   d. Explain how interventions should have robust effects.
5) Express your concerns using facts and speaking directly about that particular child.
   a. Focus on symptom reduction and improved quality of life.
   b. "I hear your concerns, my concerns with this treatment are that there is the treatment suggested has not been adequately evaluated by the medical community to show that it works, and your child already struggles to eat enough to grow."
   c. "We have to consider the potential benefits and side effects of this treatment and how they affect your child."
6) Make a plan to discuss the topic more in detail at next visit, or offer to arrange a visit that can be dedicated to the topic so that you have more time to gather information about the topic and can be fully attentive to the conversation rather than trying to squeeze it into current conversation.

Fig. 13.2 Guideline for talking to parents about thier child's condition

Diets

Red Dye Diet

The red dye diet is based on the premise that artificial food dyes, especially Red 40, are linked to behavioral issues in children, especially those with ASD and attention-deficit hyperactivity disorder (Bakthavachalu et al., 2020). Food dye is added to food to increase the visual appeal of food; therefore, it is prevalent in foods aimed at childhood consumption. Red 40 is the predominantly used dye of the nine FDA-approved artificial food colorings making it an easy focus. A meta-analysis of a few UK studies found a link between food dye and hyperactivity (Metz et al., 2005). However, these studies were observational and small in nature and do not account for other possible contributing factors (overall diet, environment, etc.). A medical test that can evaluate possible allergic reactions or sensitivities to food dyes does not exist; therefore, any link is observational and subject to interpretation. There are not any evidence-based studies that have found a causal link between ASD and food dyes (Hartman & Patel, 2020).

The most significant harm with the red dye diet is limited nutritional intake. Children with

ASD tend to be very selective eaters and this diet further limits their food options. Furthermore, food dyes are prevalent in most processed and some unprocessed foods, which can make it burdensome for families to avoid, especially when dining outside of the home. A good way to discuss this with families is to explain that no direct link for symptom reduction has been reputably found between ASD and food dyes and that any beneficial results that were found have thus far been on an individual level and very minimal. Before trying such an elimination diet, families need to consider their specific child's eating habits and if their child would be able to keep up an adequate level of food intake with taking away so many food choices.

Gluten-Free and Casein-Free Diet Without Evidence of Celiac Disease or Allergies

A more established dietary approach is the gluten-free and casein-free diet followed in patients who do not have serologic or tissue evidence of true celiac disease, which is covered earlier in the chapter. This approach is based on the theory that gluten and casein peptides, binding proteins in wheat and milk, respectively, can trigger an immune response resulting in gastrointestinal inflammation which in turn can lead to damage in the central nervous system through defects in the immunological pathways (Bakırhan & Sanlier, 2017; Mulloy et al., 2009). In addition, these theories suggest that individuals with ASD have an increase in intestinal permeabilities, naturally making them more prone to the negative effects of gluten and casein. This diet approach has been researched more than other ASD-related CAM treatments, although results continue to be inconclusive (Lee et al., 2018). A struggle with the studies is that they are small population samples and based on parent or professional observation rather than biochemical findings. These limited studies are also, thus far, nonreproducible as several reproductions of these studies have found no significant improvement in physiological functioning or behavior.

The most significant harms related to the gluten-free and casein-free diet are nutritional deficits and decreased cortical bone thickness (Bakırhan & Sanlier, 2017), both of which result in growth and developmental delays. In addition, limiting gluten and casein foods from diet highly restrict available foods beyond some patients' already highly selective food preferences. A good way to discuss this with families is to recognize that research on this type of diet is still being done and currently results are too inconsistent to encourage families to continue the diet. Stress the importance of the child's growth and development, which this diet might impact negatively.

Ketogenic Diet

The ketogenic diet is extremely low in carbohydrates, moderate in protein, and high in fat and is based on research evaluating the effectiveness in controlling drug-resistant seizures. It suggests that ASD is linked to metabolic conditions such as epilepsy and the goal is to increase serum ketones which in turn increases mitochondrial functioning. Its aim is behavioral symptom reduction through reducing metabolic disturbances (Bakırhan & Sanlier, 2017). Research is based primarily on mice and patients with epilepsy; some studies are just now being done in children with ASD although they are not randomized. While this research is still new and ongoing, major limitations include factors such as small sample size, education, age and development throughout study (done over 3 months to a year), and additional supplements to maintain health and maximize diet effects such as medium-chain triglyceride oil (Lee et al., 2018). Studies show improvements in social-affective symptoms which can be correlational, although factors discussed above likely have a large effect on these findings. Studies denied any improvements in restrictive or repetitive behaviors. Lastly, given the novelty of the diet, long-term effects are largely unknown at this time (Lee et al., 2018).

The primary concern of ketogenic diet is potential decrease in linear growth (Bakırhan & Sanlier, 2017). Also, common side effects of the diet include diarrhea, vomiting, fatigue, constipation, dehydration, weight loss, acidosis, and hypoglycemia, which are even harder to manage in children with ASD especially if the child has limited abilities to communicate or tolerate changes to their body (Mulloy et al., 2009). The diet itself is also difficult to maintain and further limits nutritional options. A good way to speak to families about this is to acknowledge it as a novel approach that is currently being studied and speak to the unknowns of long-term effects and challenges that a highly restricted diet can pose. When discussing dietary changes, take the opportunity to educate families about the importance of a balanced diet of proteins, carbohydrates, and fats for growth and development.

Microbiome Management

Microbiome-related theories focus on the role that microorganisms in the intestinal tract have on the overall human body's functioning, especially metabolic and immunologic, and how the levels of microbiomes in individuals with ASD may be less diverse yet those present more active. For instance, there appears to be an overgrowth of various bacteria and *Candida* yet less presence of carbohydrate-degrading and carbohydrate-fermenting bacteria of genera *Prevotella*, *Coprococcus*, and *Veillonellaceae* (Fowlie et al., 2018). These changes are speculated to originate from in vitro development, from early dietary choices, and from exposure to antibiotics, yet no definitive correlation has been found (Rosenfeld, 2015). This theory also focuses on the gut-brain axis (GBA) which is the physiological structure through which the central nervous system communicates between the gastrointestinal, immune, and endocrine pathways (Vendrik et al., 2020). This area of study is of particular interest given the high comorbidity between ASD- and GI-related conditions (Rosenfeld, 2015). While studies have focused on evaluating the relationship between microbiome and ASD, no definitive link, causation, or consequence has been found (Pulikkan et al., 2019). Below are the CAM treatment approaches currently linked to microbiome theory.

Fecal Transplant

Fecal transplant, also called microbiota transfer therapy (MTT), is based on the microbiota theory that individuals with ASD have altered microbiota and a treatment strategy is to transplant donor fecal microbiota (Fowlie et al., 2018; Kang et al., 2020). Research is evaluating its effectiveness in neurological conditions such as ASD, Alzheimer's disease, and epilepsy by reason of the GBA theory. Fecal transplant is an accepted treatment for recurrent *Clostridioides difficile* infections, an infection often associated with high-frequency antibiotic use (Kang et al., 2020). It involves administering a solution of healthy donor fecal matter into the intestinal tract of the recipient, typically through a gastroenterology tube inserted in the nose, although some studies mixed matter with a drink, frozen encapsulated feces or performed enema (Vendrik et al., 2020). Improvements were often gastrointestinal related, transient, and difficult to correlate due to additional changes to individual's treatment or environment and inconsistency. Common side effects are nausea, fever, and mild allergic reactions.

The most significant concern is that this CAM treatment is not approved by the US Food and Drug Administration and highly experimental. It also can be an aversive treatment to administer as often it involves invasive techniques. Furthermore, the psychological effect of having someone else's fecal matter inserted into their body can be distressing to the individual and caregivers. Currently, the potential improvements do not seem significant or lasting enough to outweigh the costs. A good way to speak to families about this approach is to highlight the experimental aspect of this approach, and the limited research proving it is effectiveness. Also, review the potential negative effects and stress that the treatment itself poses. If families mentioned that they have a provider willing to complete treatment, educate them on assessing the goals of the provider and the provider's background in ASD research and treatment.

Probiotics

The use of probiotics is based on the theory that individuals with ASD have altered microbiota, for which daily probiotic use will improve in turn improving gastroenterology and behavioral symptoms (Fowlie et al., 2018). Few cases have shown improvement in "core autism symptom" after prolonged use; however, these cases do not specify what these symptoms are or their degree of improvement (Fowlie et al., 2018). Navarro et al. (2016) acknowledge that the relationship between microbiota disturbances and ASD is not well understood; therefore, any evidence of improvement with probiotics are unclear correlations. Studies are typically small cohorts with results that show probiotics may improve symptoms; however, results are inconsistent (Vendrik et al., 2020). Also, there are usually several confounding factors including other dietary changes, developmental factors, and behavioral therapeutic techniques. Given the limited harm in probiotic use, several medical providers do recommend probiotic use despite unclear evidence or etiology of benefits found (Navarro et al., 2016).

The addition of probiotics poses little harm; however, it also poses little gains. Probiotics can cause worsening gastrointestinal issues, increase expense to the family, and be a challenge to get individuals with ASD to take. When discussing probiotics with caregivers, it is important to explain that evidence for symptom and behavior improvement is inconsistent and limited, noting that it might help gastroenterology symptoms and might aid in collaboration when families feel strongly in trying probiotics despite their limited evidence for beneficial use.

Conclusion

Caregivers are seeking treatment for ASD, a noncurable condition with a complex treatment history. More so, than with many other conditions, families are susceptible to trending, potential treatments due to the lack of concrete understanding of its cause and the pathophysiology of the condition as a whole. Parents have the best interest of their child when questioning and seeking CAM treatments yet do not always understand the science behind them and importance of evidence-based care. Caregivers benefit from education about what evidence-based care is and why it is important. When discussing treatment plans with families, always speak from a place of understanding, patience, and collaboration to promote the therapeutic alliance between parent, patient, and medical team.

Acknowledgment We would like to dedicate this chapter to the memory of Dr. Paul Hyman. Dr. Hyman was a friend and mentor and was a pioneer and leader in the field of pediatric gastroenterology. Dr. Hyman gained national and international recognition and in 2020 was honored with the Distinguished Service Award by the North American Society for Pediatric Gastroenterology, Hepatology, and Nutrition (NASPGHAN). Dr. Hyman was committed to improving the lives of his patients, contributing new research to the field to help the lives of others, and training students and fellows in the field of pediatric neurogastroenterology. Dr. Hyman specialized his research in the treatment of functional gastrointestinal disorders, including pain-associated disability syndrome (PADS) and pediatric motility disorders. He believed in practicing and researching within the biopsychosocial model and in the benefits of a multidisciplinary approach in helping his most complex patients.

References

American Psychiatric Association. (2013). *Diagnostic and statistical manual of mental disorders* (5th ed.). American Psychiatric Association.

Bakırhan, H., & Sanlier, N. (2017). Current nutritional approaches in managing autism spectrum disorder: A review. *Nutritional Neuroscience, 22*, 1–11. https://doi.org/10.1080/1028415X.2017.1358481

Bakthavachalu, P., Kannan, S. M., & Qoronfleh, M. W. (2020). Food color and autism: A meta-analysis. *Advances in Neurobiology, 24*, 481–504. https://doi.org/10.1007/978-3-030-30402-7_15

Banez, G. A., & Cunningham, C. L. (2009). Abdominal pain-related gastrointestinal disorders. In M. C. Roberts & R. G. Steele (Eds.), *Handbook of pediatric psychology* (4th ed., pp. 403–419). The Guillford Press.

Berkowitz, C. (1999). Nonorganic failure to thrive and infant rumination syndrome. In P. E. Hyman (Ed.), *Pediatric functional gastrointestinal disorders* (pp. 4.1–4.10). Academy Professional Information Services, Inc..

Berkowitz, C. D. (2014). *Berkowitz's pediatrics: A primary care approach*. American Academy of Pediatrics.

Brewer, G. J. (2001). *Wilson's disease: A clinician's guide to recognition, diagnosis, and management*. Kluwer Academic.

Brusaferro, A., Farinelli, E., Zenzeri, L., Cozzah, R., & Esposito, S. (2018). The management of paediatric functional abdominal pain disorders: Latest evidence. *Pediatric Drugs, 20*, 235–247. https://doi.org/10.1007/s40272-018-0287-z

Bryant-Waugh, R. (2019). Feeding and eating disorders in children. *Psychiatric Clinics of North America, 42*(1), 157–167. http://doi-org.lib-e2.lib.ttu.edu/10.1016/j.psc.2018.10.005

Burklow, K., Phelps, A. N., Schultz, J. R., McConnell, K., & Rudolph, C. (1998). Classifying complex pediatric feeding disorders. *Journal of Pediatric Gastroenterology & Nutrition, 27*(2), 143–147.

Call, N. A., Simmons, C. A., Mevers, J. E. L., & Alvarez, J. P. (2015). Clinical outcomes of behavioral treatments for pica in children with developmental disabilities. *Journal of Autism and Developmental Disorders, 45*(7), 2105–2114. https://doi-org.lib-e2.lib.ttu.edu/10.1007/s10803-015-2375-z

CDC. (2020). *Growth charts for children with down syndrome*. Retrieved December 22, 2020, from https://www.cdc.gov/ncbddd/birthdefects/downsyndrome/growth-charts.html

Center for Disease Control Growth Charts (2002). Washington, DC: U.S. Department of Health and Human Services, Office of Disease Prevention and Health Promotion.

Cole, S. Z., & Lanham, J. S. (2011). Failure to thrive: An update [Abstract]. *American Family Physician, 83*(7), 829–834.

Cuffman, C., & Burkhart, K. (2021). Constipation prevalence and perceptions: Comparison of children and adolescents with ASD and other developmental-behavioral disorders. *Research in Autism Spectrum Disorders, 80*, 101710. https://doi.org/10.1016/j.rasd.2020.101710

Dennison, B. A., & Boyer, P. S. (2004). Risk evaluation in pediatric practice aids in prevention of childhood overweight. *Pediatric Annals, 33*, 25–30.

DiBaise, M., et al. (2019). Hair, nails, and skin: Differentiating cutaneous manifestations of micronutrient deficiency. *Nutrition in Clinical Practice, 34*, 490–503.

Drossman, D. A. (2016). *Rome IV functional GI disorders – Disorders of gut-brain interaction*. Rome Foundation.

Ediger, T. R., & Hill, I. D. (2014). Celiac disease. *Pediatrics in Review, 35*(10), 409–416. https://doi.org/10.1542/pir.35-10-409

Ellis, C. R., Parr, T. S., Singh, N. N., & Wechsler, H. A. (1997). Rumination. In N. N. Singh (Ed.), *Prevention and treatment of severe behavior problems: Models and methods in developmental disabilities* (pp. 237–252). Thomson Brooks/Cole Publishing Co..

Fowlie, G., Cohen, N., & Ming, X. (2018). The perturbance of microbiome and gut-brain axis in autism spectrum disorders. *International Journal of Molecular Sciences, 19*, 2251. https://doi.org/10.3390/ijms19082251

Goday, P. S., Huh, S. Y., Silverman, A., Lukens, C. T., Dodrill, P., Cohen, S. S., Delaney, A. L., Feuling, M. B., Noel, R. J., Gisel, E., Kenzer, A., Kessler, D. B., Kraus de Camargo, O., Browne, J., & Phalen, J. A. (2019). Pediatric feeding disorder. *Journal of Pediatric Gastroenterology & Nutrition, 68*(1), 124–129. https://doi.org/10.1097/mpg.0000000000002188

Goin-Kochel, R., Mackintosh, V., & Myers, B. (2009). Parental reports on the efficacy of treatments and therapies for their children with autism spectrum disorders. *Research in Autism Spectrum Disorders, 3*, 528–537. https://doi.org/10.1016/j.rasd.2008.11.001

Hagopian, L. P., Rooker, G. W., & Rolider, N. U. (2011). Identifying empirically supported treatments for pica in individuals with intellectual disabilities. *Research in Developmental Disabilities, 32*(6), 2114–2120. https://doi-org.lib-e2.lib.ttu.edu/10.1016/j.ridd.2011.07.042

Hartman, R. E., & Patel, D. (2020). Dietary approaches to the management of autism spectrum disorders. In M. Essa & M. Qoronfleh (Eds.), *Personalized food intervention and therapy for autism spectrum disorder management. Advances in neurobiology* (Vol. 24). Springer. https://doi.org/10.1007/978-3-030-30402-7_19

Hyams, J. S. (1999). Chronic and recurrent abdominal pain. In P. E. Hyman (Ed.), *Pediatric functional gastrointestinal disorders* (pp. 7.1–7.22). Academy Professional Information Services, Inc..

Hyman, P. E. (1999). *Pediatric functional disorders*. Academy Professional Information Services, Inc..

Kang, D.-W., Adams, J. B., Vargason, T., Santiago, M., Hahn, J., & Krajmalnik-Brown, R. (2020). Distinct fecal and plasma metabolites in children with autism spectrum disorders and their modulation after microbiota transfer therapy. *mSphere, 5*, e00314-20. https://doi.org/10.1128/mSphere.00314-20

Kistin, C. J., & Frank, D. A. (2019). Failure to thrive. In M. Augustyn & B. Zuckerman (Eds.), *Zuckerman Parker handbook of developmental and behavioral pediatrics for primary care* (4th ed., p. 256). Wolters Kluwer.

Krebs, N., Baker, R., Greer, F., Heman, M., Jaksic, T., Lifshitz, F., et al. (2003). American Academy of Pediatrics Committee on Nutrition. Prevention of pediatric overweight and obesity. *Pediatrics, 112*, 424–430.

Kuczmarski, R. J., Ogden, C. L., Grummer-Strawn, L. M., Flegal, K. M., Guo, S. S., Wei, R., et al. (2000). CDC growth charts: United States. *Advance Data*, (314), 1–27.

Lang, R., Mulloy, A., Giesbers, S., Pfeiffer, B., Delaune, E., Didden, R., Sigafoos, J., Lancioni, G., & O'Reilly, M. (2011). Behavioral interventions for rumination and operant vomiting in individuals with intellectual disabilities: A systematic review. *Research in Developmental Disabilities, 32*(6), 2193–2205. https://doi-org.lib-e2.lib.ttu.edu/10.1016/j.ridd.2011.06.011

Lee, R., Corley, M. J., Pang, A., Arakaki, G., Abbott, L., Nishimoto, M., Miyamoto, R., Lee, E., Yamamoto, S., Maunakea, A. K., Lum-Jones, A., & Wong, M. (2018). A modified ketogenic gluten-free diet with MCT improves behavior in children with autism spectrum disorder. *Physiology & Behavior, 188*, 205–211. https://doi.org/10.1016/j.physbeh.2018.02.006

Luiselli, J. K. (1989). Behavioral assessment and treatment of pediatric feeding disorders in developmental disabilities. In M. Hersen, R. M. Eisler, & P. M. Miller (Eds.), *Progress in behavior modification* (Vol. 24, pp. 91–131). Sage Publications, Inc..

Matson, J. L., Belva, B., Hattier, M. A., & Matson, M. L. (2011). Pica in persons with developmental disabilities: Characteristics, diagnosis, and assessment. *Research in Autism Spectrum Disorders, 5*(4), 1459–1464. https://doi-org.lib-e2.lib.ttu.edu/10.1016/j.rasd.2011.02.006

Matson, J. L., Hattier, M. A., Belva, B., & Matson, M. L. (2013). Pica in persons with developmental disabilities: Approaches to treatment. *Research in Developmental Disabilities, 34*(9), 2564–2571. https://doi-org.lib-e2.lib.ttu.edu/10.1016/j.ridd.2013.05.018

Mayes, S. D., & Zickgraf, H. (2019). Atypical eating behaviors in children and adolescents with autism, ADHD, other disorders, and typical development. *Research in Autism Spectrum Disorders, 64*, 76–83. https://doi-org.lib-e2.lib.ttu.edu/10.1016/j.rasd.2019.04.002

Metz, B., Mulick, J. A., & Butter, E. M. (2005). Autism: A late-20th-century fad magnet. In J. W. Jacobson, R. M. Foxx, & J. A. Mulick (Eds.), *Controversial therapies for developmental disabilities: Fad, fashion and science in professional practice* (pp. 237–263). Lawrence Erlbaum Associates Publishers.

Mulloy, A., Lang, R., O'Reilly, M., Sigafoos, J., Lancioni, G., & Rispoli, M. (2009). *Gluten-free and casein-free diets in the treatment of autism spectrum disorders: A systematic review*. Retrieved November 23, 2020 from https://www.sciencedirect.com/science/article/abs/pii/S1750946709001111

Murray, H. B., & Thomas, J. J. (2018). Rumination disorder in adults: Cognitive-behavioral formulation and treatment. In L. K. Anderson, S. B. Murray, & W. H. Kaye (Eds.), *Clinical handbook of complex and atypical eating disorders* (pp. 253–269). Oxford University Press.

Navarro, F., Liu, Y., & Rhoads, J. M. (2016). Can probiotics benefit children with autism spectrum disorders? *World Journal of Gastroenterology, 22*(46), 10093–10102. https://doi.org/10.3748/wjg.v22.i46.10093

Perrin, E., Frank, D., Cole, C., et al. (2003). *Criteria for determining disability in infants and children: Failure to thrive. Evidence report/technology assessment no. 72. AHRQ publication no. 03-E026*. Agency for Healthcare Research and Quality.

Probst, L. A., & Burgess, J. L. (2014). Pharmacotherapy of functional disorders in children. In R. D. Anbar (Ed.), *Functional symptoms in pediatric disease: A clinical guide* (pp. 373–416). Springer Science + Business Media.

Pulikkan, J., Mazumder, A., & Grace, T. (2019). Role of the gut microbiome in autism spectrum disorders. *Advances in Experimental Medicine and Biology, 1118*, 253–269. https://doi.org/10.1007/978-3-030-05542-4_13

Ristic, R. F. (2005). *The journal of safe management of disruptive and assaultive behavior*. Crisis Prevention Institute, Inc.

Rosenfeld, C. (2015). Microbiome disturbances and autism spectrum disorders. *Drug Metabolism and Disposition, 43*(10), 1557–1571. https://doi.org/10.1124/dmd.115.063826

Rouster, A. S., Karpinski, A. C., Silver, D., Monagas, J., & Hyman, P. E. (2016). Functional gastrointestinal disorders dominate pediatric gastroenterology outpatient practice. *Journal of Pediatric Gastroenterology and Nutrition, 62*(6), 847–851.

Saad, R. J., Rao, S. S., Koch, K. L., Kuo, B., Parkman, H. P., McCallum, R. W., Sitrin, M. D., Wilding, G. E., Semler, J. R., & Chey, W. D. (2010). Do stool form and frequency correlate with whole-gut and colonic transit? Results from a multicenter study in constipated individuals and healthy controls. *The American Journal of Gastroenterology, 105*(2), 403–411. https://doi.org/10.1038/ajg.2009.612

Sarcia, B. (2020). The impact of applied behavior analysis to address mealtime behaviors of concern among individuals with autism spectrum disorder. *Child and Adolescent Psychiatric Clinics of North America, 29*(3), 515–525. https://doi-org.lib-e2.lib.ttu.edu/10.1016/j.chc.2020.03.004

Shapiro, J. M., Subedi, S., & Leleiko, N. S. (2016). Inflammatory bowel disease. *Pediatrics in Review, 37*(8), 337–347. https://doi.org/10.1542/pir.2015-0110

Silverman, C., & Brosco, J. (2007). Understanding autism: Parents and pediatricians in historical perspective. *Archives of Pediatrics & Adolescent Medicine, 161*, 392–398. https://doi.org/10.1001/archpedi.161.4.392

Srinath, A. I., Turner, S. A., & Szigethy, E. (2014). Functional symptoms in gastroenterology: A punch to the gut. In R. D. Anbar (Ed.), *Functional symptoms in pediatric disease: A clinical guide* (pp. 59–85). Springer Science + Business Media.

Stiegler, L. N. (2005). Understanding pica behavior: A review for clinical and education professionals. *Focus on Autism and Other Developmental Disabilities, 20*(1), 27–38. https://doi-org.lib-e2.lib.ttu.edu/10.1177/10883576050200010301

U.S. Department of Health and Human Services and U.S. Department of Agriculture. (2015, December). *2015–2020 Dietary guidelines for Americans* (8th ed.). Available at http://health.gov/dietaryguidelines/2015/guidelines/

Van den Berg, M. M., Benninga, M. A., & Lorenzo, C. (2006). Epidemiology of childhood constipation: A systematic review. *American Journal of Gastroenterology, 101*(10), 2401–2409.

Vendrik, K., Ooijevaar, R., de Jong, P., Laman, J., van Oosten, B., van Hilten, J., Ducarmon, Q., Keller, J., Kuijper, E., & Contarino, M. (2020). Fecal microbiota transplantation in neurological disorders. *Frontiers in Cellular and Infection Microbiology, 10*, 98. https://doi.org/10.3389/fcimb.2020.00098

Wang, L. W., Tancredi, D. J., & Thomas, D. W. (2011). The prevalence of gastrointestinal problems in children across the United States with autism spectrum disorders from families with multiple affected members. *Journal of Developmental and Behavioral Pediatrics, 32*(5), 351–360. https://doi.org/10.1097/DBP.0b013e31821bd06a

Williams, D. E., & McAdam, D. (2012). Assessment, behavioral treatment, and prevention of pica: Clinical guidelines and recommendations for practitioners. *Research in Developmental Disabilities, 33*(6), 2050–2057. https://doi-org.lib-e2.lib.ttu.edu/10.1016/j.ridd.2012.04.001

Wong, H., & Smith, R. (2006). Patterns of complementary and alternative medical therapy use in children diagnosed with autism spectrum disorders. *Journal of Autism and Developmental Disorders, 36*, 901–909. https://doi.org/10.1007/s10803-006-0131-0

Wyllie, R., Hyams, J., & Kay, M. (Eds.). (2021). *Pediatric gastrointestinal and liver disease*. Elsevier.

14 An Approach to the Genetic Evaluation of Children with Autism Spectrum Disorders

Michael Marble and Pamela McPherson

Abstract

Over the past 20 years, new genomic technologies have greatly enhanced our understanding of disease across every medical specialty. Clinical geneticists are consulted to assess for the possible contribution of genetics to clinical conditions, provide information about recurrence risk, and support a family with the information to better understand their child's treatment needs and prognosis. This chapter details the clinical geneticist's evaluation of children with autism spectrum disorders, reviews basic genetics and the array of available genomic tests, and provides information about genetic counseling that families might receive after the diagnosis of an autism spectrum disorder.

Keywords

Clinical geneticist · Genomic · Autism spectrum disorder · Genetics · Treatment planning

M. Marble (✉)
University of New Mexico Health Sciences Center, Department of Pediatrics, Division of Pediatric Genetics, Albuquerque, NM, USA
e-mail: MMarble@salud.unm.edu

P. McPherson
Department of Psychiatry and Behavioral Medicine, Louisiana State University Health Sciences Center, Shreveport, LA, USA

Introduction

Advances in technology have placed the clinical geneticist's perspective front and center in the evaluation of children with rare and complex conditions, including children with autism spectrum disorder (ASD) and intellectual and developmental disabilities (IDD) (Costain et al., 2020). Neonatologists, general pediatricians, and other providers may request a genetic consult when routine examination and developmental surveillance indicate a possible genetic condition. The inclusion of a clinical geneticist to the clinical home team facilitates the diagnosis of genetic disorders, elucidates the possible contribution of genetics to associated clinical problems, provides information about recurrence risk, and supports a family with information to better understand their child's treatment needs and prognosis. Parents often experience a tremendous sense of relief upon learning the underlying etiology of their child's condition. When families know *what* they are dealing with, they may feel more confident moving forward with the medical home shared plan of care.

Clinical geneticists complete specialized training in clinical genetics and genomics with many geneticists also training in pediatrics or another medical specialty. Clinical geneticists are in great demand, with limited access to their expertise in many areas of the country (Chou et al., 2021). Clinical geneticists commonly prac-

P. McPherson (ed.), *Handbook of Treatment Planning for Children with Autism and Other Neurodevelopmental Disorders*, Autism and Child Psychopathology Series,
https://doi.org/10.1007/978-3-031-06120-2_14

tice in consultation to primary care providers and other specialists. For patients with inborn errors of metabolism, biochemical geneticists typically manage metabolic care directly. Children with ASD have increasingly been referred for genetic consultation in recent years (Vorstman et al., 2017).

An Overview of the Genetics of ASD

Autism spectrum disorder is a neurodevelopmental condition affecting behavior, communication, and social interactions (Miles & McCathren, 2005). ASD can occur as an isolated, non-syndromic trait or as one phenotypic feature of well-defined syndromes, chromosomal disorders, or metabolic conditions. Many patients with ASD are at risk for other morbidities, such as psychiatric, gastrointestinal, and various congenital anomalies that require specialized clinical monitoring and interventions best provided by a clinical home (Casanova et al., 2020; Todorow et al., 2018). The roles of the geneticist in evaluation of ASD include seeking an underlying cause, counseling the family about inheritance pattern/recurrence risk, and in the case of syndromic ASD alerting the family and primary care physician about the need for surveillance and interventions for potential associated clinical conditions. For example, ASD is often associated with epilepsy requiring close neurological monitoring (Ziats et al., 2021). Down and Williams syndromes are associated with an increased incidence of ASD (Kirchner & Walton, 2021) and require special health supervision for associated anomalies (Bull, 2011; Morris & Braddock, 2020). The 2013 American College of Clinical Genetics guidelines for ASD recommended that all individuals/families with ASD be offered a genetic evaluation (Schaefer & Mendelsohn, 2013).

Over the past several decades, genetic testing options and approaches to evaluation have evolved as new technologies have been developed and become less expensive. Genetic evaluation of ASD includes obtaining detailed information including prenatal history, clinical history, three-generation family history, careful physical exam, and appropriate genetic testing to provide a unifying genetic diagnosis. For an increasing number of patients, the genetic diagnosis informs treatment and assists parents in planning for their child's future. The genetic evaluation may reveal any of a number of chromosomal or single gene variants contributing to ASD. Some patients have deletions or duplications of chromosome material (copy number variants) impacting multiple genes, while other patients harbor pathogenic variants in individual genes. Single gene defects may cause non-syndromic autism or recognizable syndromes. In some cases, a single gene cause or copy number variant may predispose to ASD but exhibit reduced penetrance (Ziats et al., 2021). Epigenetic mechanisms such as disturbances of methylation and imprinting are another factor to consider in the genetic evaluation of patients with ASD (Waye & Cheng, 2018).

Inborn errors of metabolism are thought to contribute to a relatively small proportion of ASD overall but should be considered in the differential diagnosis if suggestive signs and symptoms are present (Schaefer & Mendelsohn, 2013). Inborn errors of metabolism are genetic disorders resulting in the disruption of metabolic pathways. For example, a clinical geneticist may order testing to identify defects in creatine synthesis or transport that can disrupt critical intracellular energy transport mechanisms leading to autistic features and intellectual disability (ID) in affected children (Longo et al., 2011). Developmental regression in a patient with features of ASD may suggest a lysosomal storage disease such as Sanfilippo syndrome (Schaefer & Mendelsohn, 2013). Spilioti et al. (2013) screened for metabolic disorders in 187 patients with features of ASD. Two patients were found to have succinic semialdehyde dehydrogenase deficiency, two had Lesch Nyhan syndrome, and one had phenylketonuria. Several other patients had elevated 3-hydroxyisovaleric acid in urine and abnormal glucose loading tests. There is some debate about whether to perform metabolic evaluations for patients with apparently non-syndromic ASD. A retrospective study of 274 patients with non-syndromic ASD reportedly showed a very low diagnostic yield for metabolic screening (Schiff et al., 2011). This study and other data suggest that routine screening for inborn errors of metabolism in patients with non-syndromic ASD will have a low return for establishing diagnoses (Asato et al., 2015). However, patients with ASD who have signs and symptoms of metabolic disease should undergo appropriate testing. Further studies on the metabolic contribution to the occurrence of ASD may refine testing recommendations in the future.

The remainder of this chapter provides an overview of the genetic causes of autism and suggests an approach to genetic assessment and diagnosis.

A Review of Basic Concepts in Genetics and Genetic Testing

Genetics has come a long way since 1865, when Gregor Mendel's experiments with peas demonstrated the transmission of heredity in discrete units and since the Danish botanist and geneticist, Wilhelm Johannsen, coined the term "gene" in 1909 (Johannsen, 1909). The human genome has been sequenced and new technologies have revolutionized the field of clinical genetics and paved the way for providing more robust answers to the questions of scientists, physicians, and families. Before exploring the genetics of ASD, this section will briefly review basic genetics and the common genetic tests.

The Building Blocks of Inheritance

An individual's genome, along with environmental factors, plays an important role in health and susceptibility to disease. A genome is the complete set of DNA which is tightly wound into chromosomes. In the 1880s, Wilhelm Waldeyer-Hartz coined the term *chromosome* to describe the elements of chromatin previously described by Walther Flemming in his study of cell division or mitosis (Wall, 2015). Chromosomes are stick-like structures in the nucleus of most cells and harbor the units of heredity, known as genes. The human genome normally consists of 23 pairs of chromosomes in each cell for a total of 46 chromosomes. Twenty-two of the pairs are autosomes (designated 1–22) while the remaining pair consists of the sex chromosomes (X and Y). Females typically have two X chromosomes while males have one X chromosome and one Y chromosome.

Clinical geneticists use their knowledge of genes and chromosomes, along with clinical information, to generate diagnostic hypotheses and order appropriate testing which may reveal the etiology of a child's condition. In the nucleus, there are over 20,000 genes which are arranged on autosomes and sex chromosomes. As there are two copies of each autosome, one inherited from the mother and one from the father, there are likewise two copies of each of the autosomal genes. The sex chromosomes are different in that females inherit an X chromosome from both the mother and father whereas males inherit an X chromosome from their mother and a Y chromosome from their father. Since males have only one X chromosome, they are at increased risk for disease when there is a mutation affecting a gene on the X chromosome. For example, a child may present with ASD, intellectual disability, unstable mood, and attentional difficulties with the clinical geneticist noting characteristic facial features of fragile X syndrome, hypermobile joints, and poor muscle tone leading to genetic testing. Results showing an increased number of triplet repeats within the *FMR1* gene confirm a diagnosis of fragile X syndrome. While the child in this example could be male or female, males with fragile X syndrome are at three times the risk for ASD and usually experience greater intellectual impairment than females (Hagerman et al., 2018; Klusek et al., 2014). Although males are more likely to be affected by mutations involving genes on the X chromosome, it is important to note that females can also have abnormal phenotypes related to mutated genes on the X chromosome. The *Atlas of X-Linked Intellectual Disability Syndromes* describes 150 distinct disorders with intellectual disability, caused by abnormalities affecting genes on the X chromosome (Stevenson et al., 2012). The inheritance pattern for these disorders is known as X-linked.

Disorders caused by mutations on autosomes can be autosomal recessive or autosomal dominant. Autosomal recessive conditions (such as phenylketonuria) are caused by pathogenic variants on both copies of the gene while autosomal dominant disorders (such as neurofibromatosis type 1) are caused by pathogenic variants on one copy of the gene.

Mitochondria have their own genome which contains 37 genes and is maternally inherited. There is some evidence for mitochondrial dysfunction in a small percentage of children with ASD (Rose et al., 2018; Rossignol & Frye, 2012). Not all mitochondrial dysfunctions have a genetic etiology. In children with mitochondrial dysfunction, gastrointestinal issues, seizures, infection, or fatigue may trigger neurodevelopmental regression (Rose et al., 2018). Testing for mitochondrial disease should be pursued in patients with ASD with characteristic presentations for this group of conditions.

In the 1990s, the concept of imprinted genes heralded modern epigenetics (Deichmann, 2016). For the vast majority of genes, both copies are able to encode for specific proteins. However, for some genes, only one copy is active while the other copy is turned off. For these genes, the parent of origin impacts which copy is active and which copy is silent. This phenomenon is known as imprinting and is mediated by methylation. For example, the H19 gene on chromosome 11 encodes for an untranslated messenger RNA which acts as a negative regulator of growth. H19 is normally methylated on the paternal copy of the gene and not expressed. The gene is not methylated on the maternal copy and is expressed. Hypomethylation of H19 leads to increased expression leading to poor growth in Russell-Silver syndrome. In contrast, hypermethylation of H19 causes under-regulation of growth and is one of the causes of the overgrowth condition, Beckwith-Wiedemann syndrome, which is associated with developmental disabilities and increased risk for cancer (Õunap, 2016). Prader-Willi syndrome, Angelman syndrome, and many other genetic conditions are associated with abnormalities of genetic imprinting (Monk et al., 2019; Nicholls et al., 1998).

Review of Selected Genetic Tests

Genetic testing may be ordered by the medical home team or the clinical geneticist to aid in the diagnosis of a genetic condition, inform prognosis and treatment, and better counsel parents regarding reproductive decisions. With the rapid advancement of genetic knowledge, deciding which test to order can be daunting to those outside the field of genetics. Formal or informal consultation with a clinical geneticist or genetic counselor can help resolve issues around testing. Online resources such as GeneReviews (https://www.ncbi.nlm.nih.gov/books/NBK1116/) provides helpful information about approach to diagnosis and testing for a great many genetic disorders.

The Karyotype

The karyotype provides a photograph of an individual's chromosomes. It reflects the number of chromosomes in a cell and detects some structural and numerical abnormalities. In the laboratory, special staining allows the chromosomes to be identified and arranged in pairs, creating an image referred to as the karyotype (see Fig. 14.1).

Each chromosome contains two parts, the p arm and the longer q arm, separated by a narrow point known as the centromere. The karyotype further delineates chromosomes by light and dark bands. Many numerical and structural abnormalities can be detected by karyotype analysis. For example, an extra copy of chromosome 21 (trisomy 21) causes Down syndrome. Females with only one X chromosome (45, X) have Turner syndrome. A karyotype is commonly ordered when a patient has features classic for a trisomy or other aneuploidy disorders, including trisomies 13, 18, and 21, monosomy X (Turner syndrome), and Klinefelter syndrome (47, XXY) that are easily identified by this method.

Chromosome Microarray (CMA)

A CMA allows for higher-resolution analysis than a karyotype and uses techniques such as

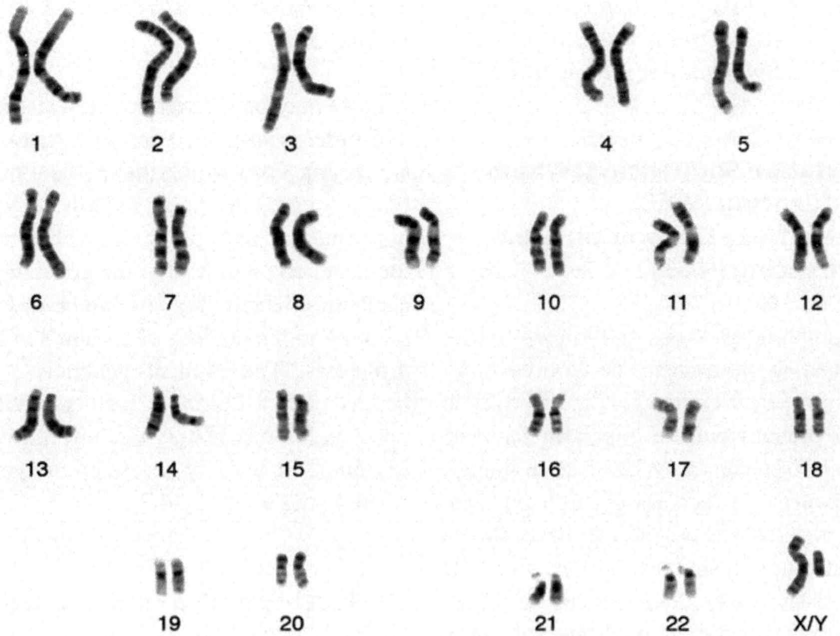

Fig. 14.1 A karyotype. (From the National Human Genome Research Institute https://www.genome.gov/genetics-glossary/Karyotype)

comparative genomic hybridization and single-nucleotide polymorphism (SNP) analysis to detect the presence of microdeletions and microduplications of chromosome material (copy number variants). A CMA/SNP array may reveal any of the following:

1. No deletions, duplications, or large regions of homozygosity are detected. A normal result, however, does not exclude a genetic cause of the patient's phenotype since CMA cannot identify single-nucleotide variants or small insertions or deletions (Malinowski et al., 2020).
2. Pathogenic deletion or duplication. The deletion or duplication causes abnormal phenotype.
3. Regions of homozygosity (ROH). Depending on the extent and chromosomal distribution, ROH may indicate evidence for a common ancestor, uniparental disomy, chromosome rearrangements, and increased risk for recessive disease (Wang et al., 2015).
4. Benign deletion or duplication. The deletion or duplication detected does not cause abnormal phenotype.
5. Deletion or duplication of uncertain significance. There is not enough information to classify the deletion or duplication as pathogenic or benign.

In 2010, the American Society for Human Genetics issued a consensus statement designating CMA as a first-tier clinical diagnostic test for unexplained IDD and ASD (Miller et al., 2010).

Fluorescence In Situ Hybridization (FISH)

FISH testing screens for well-defined specific deletions, duplications, and translocations of chromosome material using a fluorescently tagged piece of DNA that binds to the matching segment of patient's DNA highlighting the segment of concern. Clinical geneticists can use FISH to identify microdeletions, for example, FISH for 22q11.2 deletion is a test for velocardiofacial syndrome, also known as DiGeorge syndrome (Vickers & Gibson, 2019). Microdeletions and microduplications identifiable by FISH can also be detected by chromosome microarray. FISH is an appropriate test to confirm a microdeletion or microduplication syndrome in a patient with classic presentation.

Next-Generation Sequencing: Whole Exome Sequencing (WES) and Targeted Next-Generation Gene Panel Sequencing (NGS)

Genes are composed of exons and introns. Exons contain the coding sequences. The term "exome" refers to all of the exons in the genome. Introns do not code for proteins but are important for gene expression and regulation. WES and multigene sequencing panels are increasingly utilized diagnostic tests that as meta-analyses have shown increase the diagnostic yield for children with ASD and ID (Srivastava et al., 2019; Stefanski et al., 2021). WES provides the entire exome sequence, while multigene panels target candidate genes associated with the phenotype. The NIH (National Institutes of Health) Genetic Testing Registry describes a multitude of panels that have been developed for the identification of genes involved in ASD. A multidisciplinary consensus statement issued in 2019 called for exome sequencing as a first-tier clinical diagnostic test for children with neurodevelopmental disorders including ASD (Srivastava et al., 2019). The EuroGentest and the European Society of Human Genetics have published guidelines for NGS used in the diagnosis of genetic disorders (Matthijs et al., 2016).

Whole Genome Sequencing (WGS)

A WGS analyzes the entire genome sequence, including both exons and introns. Since the human genome was sequenced in 2003, the interest in and use of WGS for diagnostic testing has grown, and the cost of WGS has dropped.

Genomic technologies greatly increase the ability to determine the genetic etiology of ASD, ID, and other conditions. Finding the etiology may or may not impact management in specific cases. Still, it may provide an answer parents are looking for and the hope that future research may provide interventions.

The Genetic Evaluation of Children with ASD

A child may be referred to a clinical geneticist if a disorder is suspected or family history indicates a concern. Forwarding the medical home shared plan of care, perinatal records, developmental assessments, and psychological evaluations to the geneticist will further the goals of the genetic evaluation which may ultimately lead a family to a better understanding of a child's condition and prognosis. The clinical geneticist's evaluation begins with a thorough history, review of systems, and family history that informs the physical examination and any necessary imaging and laboratory studies (Fig. 14.2).

The History with a Focus on Genetics

The genetic consultation begins with detailed history. It is important to obtain a detailed history including prenatal history, history of present illness, review of systems, and family history. Families may be asked to complete questionnaires to facilitate the history taking process. The history can be taken by the clinical geneticist, genetic counselor, or combination thereof. The details of the prenatal history including overall maternal health and illness during pregnancy, teratogenic exposure, intrauterine growth restriction, congenital anomalies, pregnancy complications, and any special procedures or testing during pregnancy are important to ascertain. Circumstances of the delivery and perinatal period should be noted. Growth and developmental milestones should be assessed. The age of diagnosis, severity, results of previous genetic testing, and clinical course (such as developmental regression) may provide clues to the underlying cause.

The Review of Systems

A complete review of systems will provide information regarding involvement of organ systems which may direct testing toward a specific syndrome and will alert the primary care provider to

1. The referral should designate whether a genetic or a metabolic disorder is suspected.
2. When offering a referral to pediatric clinical geneticist to assess for a genetic etiology underlying the diagnosis of ASD, ask the family to be prepared to share a detailed three generation family history and detailed information regarding the child's perinatal history and development.
3. Inform the family that the comprehensive history will include queries regarding family history of ASD and other developmental, behavioral, learning, psychiatric, neurological, and general medical diagnoses.
4. The family should dress the child comfortably and be prepared for a thorough physical examination that will include measurements, a thorough examination of the skin for discoloration or other indications of a genetic disorder, and a careful inspection for dysmorphic features.
5. The family should be informed that laboratory studies may include a cheek swab, blood draw, urine sample and physical examination.
6. Ask the family to alert the geneticist of any accommodations that may be required during the appointment prior to the day of the appointment.
7. From the array of genomic tests, the clinical geneticist will choose the most appropriate. Additional testing may be required.
8. The test results will be reviewed with the family but may need to be revisited as new genomic knowledge becomes available.

Fig. 14.2 Preparing the family for an appointment with the clinical geneticist

the need for further evaluation and management of comorbidities.

Detailed Family History

A detailed three-generation family history is critical and may show a distinct inheritance pattern such as the X-linked pattern of fragile X syndrome or may reveal syndromic or other relevant phenotypic features in other family members.

Physical Examination

The physical exam may be unrevealing or may show features suggestive of a syndrome. For example, microcephaly is often present in Rett syndrome, other *MECP2*-related disorders, Angelman syndrome, and many other conditions. Macrocephaly is associated with *PTEN*-related disorders (PTEN hamartoma tumor syndrome), Sotos syndrome, and others. Postpubertal macroorchidism can be seen in fragile X syndrome. Cafe au lait spots and other skin findings may reveal neurofibromatosis or other syndromes with skin manifestations. There are a great many genetic conditions/syndromes associated with autism and other developmental disabilities; therefore, careful physical exam for dysmorphic features and congenital anomalies can be instrumental in achieving a diagnosis. The task of sorting through the vast number of syndromes and conditions which may exhibit autistic features and intellectual disability is aided by the use of online databases, books, and review articles, for example, the Online Mendelian Inheritance in Man (OMIM) database (omim.org), *Smith's Recognizable*

Patterns of Human Malformations (Jones et al., 2021), and the GeneReviews database (https://www.ncbi.nlm.nih.gov/books/NBK1116/).

Genetic Testing for Patients with ASD

Following medical guidelines for the genetic testing of children diagnosed with ASD, the medical home team may consult with a clinical geneticist for guidance regarding testing. The approach to genetic laboratory testing for patients with ASD is determined by the results of the patient's history, review of systems, family history, and physical exam findings. Many patients with ASD do not have an obvious cause or syndrome that would allow for a focused test specific to the diagnostic hypothesis.

Vignette 1: Marcus: The Evaluation of a Child with ASD Without an Obvious Syndrome or Other Causes

Marcus is a 4-year-old boy with developmental delay and a diagnosis of ASD. His medical home team has arranged Early Steps services, including applied behavioral analysis therapy, and referred him for genetic consultation. Marcus' mother reports that her pregnancy and delivery were unremarkable. A review of systems reveals that Marcus has no other health problems. A physical exam shows borderline large ears but is otherwise unremarkable. The family history is significant for a maternal uncle with intellectual disability.

In this scenario, the patient's findings are not definitive for a specific syndrome. The typical first-tier testing for this patient would be chromosome microarray (CMA) and fragile X syndrome DNA testing. CMA is appropriate in this situation as the patient does not exhibit features that would point to a definite and specific diagnosis; therefore, nontargeted screening for copy number variants via CMA is useful. Fragile X DNA testing is a targeted test which is appropriate in this case because fragile X syndrome is relatively common, and characteristic physical findings may not always be evident. Moreover, the family history shows evidence for X-linked inheritance.

CMA for ASD

In the report by Schaefer and Mendelsohn, (2013), cumulative data from various studies showed that approximately 10% of patients with ASD had clinically significant copy number variants identified by CMA. Miller et al. (2010) published a consensus statement in the *American Journal of Human Genetics* indicating that the diagnostic yield for CMA for ASD, intellectual disability, and multiple congenital anomalies was 15–20%. The authors pointed out that this is higher than the 3% diagnostic yield for karyotype (excluding Down syndrome and other recognizable chromosome disorders). Many copy number variants associated with autism spectrum disorder and intellectual disability have been described.

Fragile X Testing

Fragile X syndrome is a single gene disorder caused by expansion of a triplet repeat (CGG) in the FMR1 gene. Clinical features include intellectual disability and variable other features including ASD. Depending on the degree of CGG expansion, testing can show a normal result, a pre-mutation, or a full mutation. Those with premutations do not have fragile X syndrome; however, the mutation can expand across generations. In addition, pre-mutation carriers are at risk for fragile X-associated ataxia syndrome and premature ovarian failure. Patients with full mutations are at risk for intellectual disability, behavioral problems, and ASD. Since fragile X syndrome is relatively common, testing is often performed routinely along with CMA as a first-tier test for ASD of unknown cause. Fragile X syndrome is an X-linked disorder; therefore, the finding of an X-linked pattern of inheritance on pedigree analysis could be an important clue to this diagnosis (see Fig. 14.3). Estimates of diagnostic yield for fragile X testing in patients with ASD are in the range of 1–5% (Schaefer & Mendelsohn, 2013).

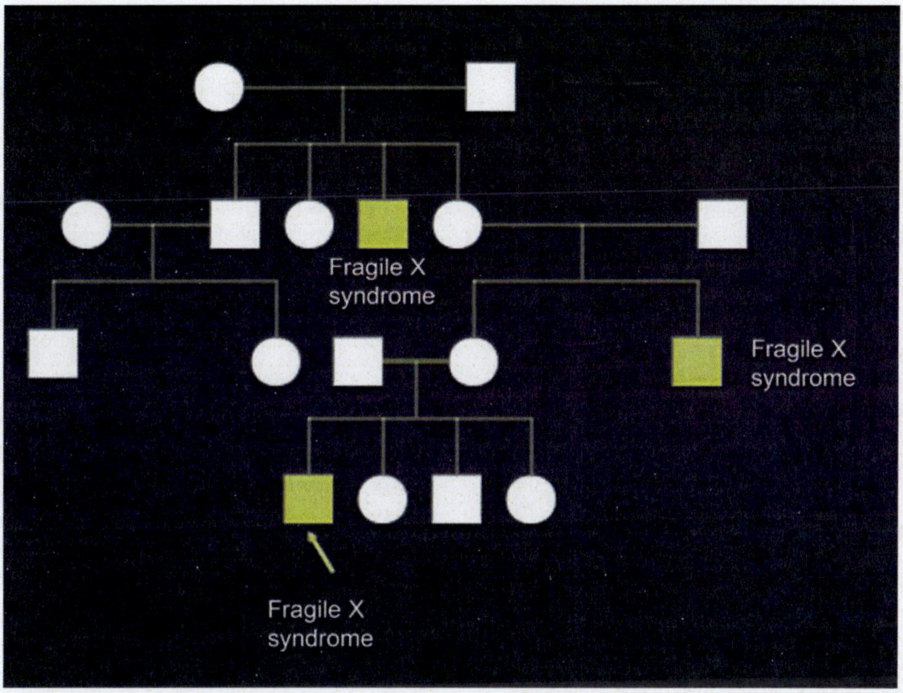

Fig. 14.3 Pedigree showing X-linked inhertance of fragile X syndrome

Vignette 1: Marcus' Outcome

Marcus' genetic testing reveals that he does have fragile X syndrome. The genetic counselor meets with his parents and explains Marcus' test results and the implications for other family members based on the *Consensus Guidelines* (Finucane et al., 2012). The medical home team implements the American Academy of Pediatrics Health Supervision Guidelines for fragile X syndrome (Hersh & Saul, 2011).

Vignette 2: Lucia: The Evaluation of a Child with ASD and Classic Features of a Microdeletion Syndrome

An eight-year-old girl Lucia is referred by her pediatrician for a genetic consultation after a psychological evaluation reveals ASD and intellectual disability. Lucia has a very outgoing personality and is noted to be very sensitive to loud or high-pitched sounds. The review of systems is significant for a history of neonatal hypercalcemia and supravalvular aortic stenosis which was repaired. Her facial features are consistent with Williams syndrome.

In this case, Lucia's history and physical findings are recognizable as Williams syndrome; therefore, it would be appropriate to order a targeted test to confirm the 7q11.23 microdeletion associated with this condition. This could be accomplished via fluorescence in situ hybridization (FISH) using a probe specific to Williams syndrome. The deletion would also be identified by CMA which could be ordered instead of FISH if the findings were only suggestive but not definitive for Williams syndrome. CMA would identify the Williams syndrome deletion but would also identify other copy number variants if Williams syndrome is not the correct diagnosis.

Florescence in situ hybridization (FISH) is a targeted test and is useful for diagnosis of well-defined microdeletion disorders such as DiGeorge syndrome, Williams syndrome, Smith Magennis syndrome, Angelman syndrome, Prader-Willi syndrome, and others.

Vignette 2: Lucia's Outcome

Lucia's family was relieved to receive the diagnosis of Williams syndrome. Genetic counseling is provided and the *Health Care Supervision for Children with Williams Syndrome* (Morris & Braddock, 2020) was initiated as well as educational accommodations and a health plan at her school.

Vignette 3: MaKayla: The Evaluation of a Child with ASD and Evidence for a Specific Single Gene Syndrome

When MaKayla's pediatrician noticed that her head circumference had plateaued, she was referred for a genetic evaluation. Her parents reported that MaKayla had normal development until 10 months of age when her head growth started to plateau and she began to regress developmentally. By 3 years of age, she exhibited constant hand-wringing movements and lost purposeful use of her hands. On exam, she had microcephaly.

MaKayla has typical features of Rett syndrome; therefore, a targeted DNA sequencing for the causative gene, *MECP2*, is indicated. Classic Rett syndrome is associated with normal development for the first 6–18 months of life followed by plateauing of development and loss of purposeful hand movements and the presence of microcephaly. These features should prompt *MECP2* sequence analysis.

Vignette 3: MaKayla: Outcome

MaKayla's *MECP2* sequence analysis reveals Rett syndrome. After genetic counseling was provided, her medical home team implemented the *Consensus Guidelines* to monitor MaKayla for health complications. Establishing a diagnosis of Rett syndrome is important, as it allows for monitoring for potential clinical issues such as prolonged QT, seizures, and apnea which are associated with this condition.

Vignette 4: Ali: The Evaluation of a Macrocephalic Child with Evidence for a Specific Single Gene Syndrome

Ali is a six-year-old boy who was recently diagnosed with ASD. On physical exam, he was noted to have macrocephaly. The family history was significant for his mother who has macrocephaly and thyroid cancer and has been diagnosed with Cowden syndrome, a phenotype within the spectrum of *PTEN* hamartoma tumor syndrome.

Cowden syndrome is an autosomal dominant condition caused by pathogenic variants in the *PTEN* gene. Studies suggest that patients with the combination of ASD and macrocephaly have a 10–20% chance of having a pathogenic variant in *PTEN* (Butler et al., 2005; Hansen-Kiss et al., 2017; Yehia et al., 2020). The diagnosis of *PTEN*-related disorders allows for screening for associated clinical issues and risk factors. *PTEN*-related disorders are associated with increased risk for malignancies such as breast and thyroid cancer. Starting at the age of diagnosis, children diagnosed with *PTEN*-related disorders are screened for thyroid cancer on an annual basis. Screening for other types of cancer is implemented at the appropriate time based on established guidelines for *PTEN* hamartoma tumor syndrome (https://www.ncbi.nlm.nih.gov/books/NBK1488/#phts.Management).

Vignette 4: Ali: Outcome

Targeted DNA testing showed that Ali is heterozygous for the pathogenic variant previously identified in his mother, confirming that Ali has a *PTEN*-related disorder. Therefore, arrangements were made for recommended screening for tumors associated with this condition.

Vignette 5: Juan: The Evaluation of a Child with ASD Who May Have a Single Gene Disorder but Whose Findings Are Not Indicative of a Specific Diagnosis

Juan is a five-year-old boy who was diagnosed with ASD when he started kindergarten. He is referred for a genetic evaluation which reveals core features of ASD and a nonspecific history and physical examination without dysmorphic features. Juan undergoes CMA and fragile X analysis which are negative.

Without history of physical findings to guide genetic testing, the geneticist considers the large number of other single gene disorders associated with ASD. These may be identified via single gene testing, multigene sequencing panels, and whole exome or whole genome sequencing. The geneticist orders whole exome sequencing for Juan.

Vignette 5: Juan: Outcome

The whole exome sequencing shows a heterozygous variant of uncertain significance in a gene associated with a form of autosomal dominant ASD and ID. It is noted in the report that the variant changes a highly conserved amino acid in the protein product of the gene, the variant is not found in population cohorts, and computer analysis predicts the variant to be disruptive to the protein. The geneticist tests both *unaffected* parents for the variant and they test negative. The geneticist reviews the testing results and tells the family that the variant is likely the cause of Juan's ASD since it occurred de novo in the affected family member and the predicted alteration to the protein is likely to be disruptive to protein function. At a one-year follow-up for the patient, the testing laboratory sends a revised report that the variant has been reclassified as pathogenic.

Diagnostic Yields and Genetic Testing Algorithms for ASD

Schaefer and Mendelsohn, (2013) summarized diagnostic yields for ASD utilizing CMA, fragile X testing, karyotype, *MECP2* and *PTEN* sequencing (when appropriate), and other testing. It was estimated that a diagnostic yield of 30–40% could be achieved using available technology and knowledge along with thorough evaluation. A two-tier diagnostic algorithm was proposed (Schaefer & Mendelsohn, 2013): In the first tier of the evaluation, if initial history (including three-generation pedigree) and physical exam do not lead to a specific syndrome diagnosis or evidence of metabolic/mitochondrial disorder, further evaluation would include chromosome microarray (oligonucleotide array-comparative genomic hybridization or SNP array) and fragile X syndrome DNA testing as appropriate. Second-tier testing included *MECP2* sequencing in all females with ASD, consideration of *MECP2* duplication testing in males, and *PTEN* testing only if the patient's head circumference is greater than two standard deviations above the mean. Brain imaging was recommended only if specific indicators were present. Hyman et al. (2020) proposed a modified but similar algorithm (adapted from Schaefer) along with consideration of referral to genetics and possible whole exome sequencing as part of the workup.

The advent of multigene panels, WES and WGS, has impacted the approach to genetic testing for patients with ASD. A meta-analysis by Srivastava et al. (2019) found WES to have a favorable diagnostic yield for neurodevelopmental disorders, including ASD and proposed WES as a *first-tier test*. Arteche-Lopez et al. (2021) reported that, as compared to CMA and fragile X syndrome testing, there is a higher diagnostic yield using WES for patients with ASD (for patients without clinical suspicion for fragile X syndrome). The authors proposed WES as a first-tier test for ASD. Recently, Manickam et al. (2021) published the ACMG Practice Guidelines for the use of whole exome sequencing and whole genome sequencing and recommended that these tests be considered as first- or second-tier tests for evaluation of patients with congenital anomalies, developmental delay, and intellectual disability. Although isolated autism was not a focus of the article, the authors indicated that the use of whole exome and whole genome sequencing is expected to have similar clinical utility for evaluation of isolated autism and a number of genetic conditions.

Future advances in genetic testing technologies and the cumulative experience of clinical geneticists, genetic counselors, and other providers will further shape and refine testing strategies/algorithms for patients with ASD. In a recent study by Harris et al. (2020), 72.2% of ASD patients with positive results on genetic testing received clinical recommendations based on the results, indicating the high clinical utility of genetic testing when a specific genetic etiology can be identified. For these children, the genetic testing expertise is important for appropriate clinical monitoring and interventions for associated conditions and risk factors.

ASD-Related Genetic Counseling

Genetic counseling and informed consent should be provided before genetic testing so that patients/families will understand the rationale for testing, potential for identifying variants of uncertain significance, limitations of genetic testing, possible detection of secondary findings, and other aspects. Posttest genetic counseling involves reviewing test results, discussing treatment and prognosis, discussing inheritance pattern and recurrence risk, informing family of available resources, and answering other questions the parents or individual may have. Genetic counseling and potential testing of other at-risk family members should also be discussed. For a patient with ASD of unknown cause, recurrence risk in future offspring of the parents is thought to be 3–10%. For patients with an identified genetic cause, the recurrence risk depends on the cause. Rett syndrome occurs as an isolated case in the family in about 99.5% of cases; therefore, recurrence risk is usually low. Fragile X syndrome recurrence risk for future sons is as high as 50% depending on whether the mother of an affected child has a full mutation or a pre-mutation and, in the case of pre-mutation, the degree of expansion of CGG repeats in *FMR1*. Many patients with ASD have de novo heterozygous variants in genes that are rare autosomal dominant causes of ASD. In these cases, recurrence risk is low although the possibility of mosaicism for the variant in a parent should be discussed. On the other hand, autosomal recessive disorders associated with ASD have a 25% recurrence risk. The identification of the cause of ASD in a child, along with appropriate genetic counseling, can give many parents a great sense of relief and assists them in future planning and management of their child's health care (Griesi-Oliveira & Sertié, 2017).

Future Directions

The explosion of genetic knowledge and new technologies over the past several decades has increased capabilities to establish molecular diagnoses in patients with ASD, intellectual disability, and many other conditions. Ordering multigene panels, whole exome sequencing, and whole genome sequencing is becoming more commonplace. Interpretation of sequence variants is a dynamic process due to the constant influx of new information. Patients who have undergone genomic testing can have their genetic data reanalyzed periodically in light of new scientific information. The American College of Medical Genetics and Genomics has issued guidance for the reevaluation and reanalysis of genomic test results (Deignan et al., 2019).

Ensuring access to the full range of genetic services, including testing and appropriate genetic counseling, is challenging due to shortages and unequal distribution of genetic personnel in some locations, as well as other barriers to care. Chou et al. (2021) identified serious gaps in the delivery of genetic services despite advances in knowledge and technology. The authors proposed increased use of telegenetics as a supplement to in-person clinic visits and as a tool for education and physician consultations. As primary care providers play an important role in caring for patients with genetic conditions, enhancing access to and exchange of knowledge with genetic providers is increasingly needed. Promotion of patient-centered care and the use of evidence-based guidelines are also important (Chou et al., 2021). A systematic review by

Brown et al. (2021) suggests that telemedicine enhances access, reduces costs, is efficacious, and has favorable patient satisfaction. The integration of genomic knowledge into electronic records, such as the issuing of alerts based on patient pharmacogenetics, will also likely have an increasing impact on genetic medicine.

References

Arteche-López, A., Gómez Rodríguez, M. J., Sánchez Calvin, M. T., Quesada-Espinosa, J. F., Lezana Rosales, J. M., Palma Milla, C., ... & Alvarez-Mora, M. I. (2021). Towards a change in the diagnostic algorithm of autism spectrum disorders: evidence supporting whole exome sequencing as a first-tier test. *Genes, 12*(4), 560.

Asato, M. R., Goldstein, A. C., & Schiff, M. (2015). Autism and inborn errors of metabolism: How much is enough? *Developmental Medicine & Child Neurology, 57*(9), 788–789.

Brown, E. G., Watts, I., Beales, E. R., Maudhoo, A., Hayward, J., Sheridan, E., & Rafi, I. (2021). Videoconferencing to deliver genetics services: A systematic review of telegenetics in light of the COVID-19 pandemic. *Genetics in Medicine, 23*(8), 1438–1449. https://doi.org/10.1038/s41436-021-01149-2

Bull, M. J. (2011). Health supervision for children with Down syndrome. *Pediatrics, 128*(2), 393–406. https://doi.org/10.1542/peds.2011-1605

Butler, M. G., Dasouki, M. J., Zhou, X.-P., Talebizadeh, Z., Brown, M., Takahashi, T. N., … Pilarski, R. (2005). Subset of individuals with autism spectrum disorders and extreme macrocephaly associated with germline PTEN tumour suppressor gene mutations. *Journal of Medical Genetics, 42*(4), 318–321.

Casanova, M. F., Frye, R. E., Gillberg, C., & Casanova, E. L. (2020). Comorbidity and autism spectrum disorder. *Frontiers in Psychiatry, 11*, 1273.

Chou, A. F., Duncan, A. R., Hallford, G., Kelley, D. M., & Dean, L. W. (2021). Barriers and strategies to integrate medical genetics and primary care in underserved populations: A scoping review. *Journal of Community Genetics, 12*(3), 291–309.

Costain, G., Walker, S., Marano, M., Veenma, D., Snell, M., Curtis, M., … Reuter, M. S. (2020). Genome sequencing as a diagnostic test in children with unexplained medical complexity. *JAMA Network Open, 3*(9), e2018109.

Deichmann, U. (2016). Epigenetics: The origins and evolution of a fashionable topic. *Developmental Biology, 416*(1), 249–254.

Deignan, J. L., Chung, W. K., Kearney, H. M., Monaghan, K. G., Rehder, C. W., Chao, E. C., & Committee, A. L. Q. A. (2019). Points to consider in the reevaluation and reanalysis of genomic test results: A statement of the American College of Medical Genetics and Genomics (ACMG). *Genetics in Medicine: Official Journal of the American College of Medical Genetics, 21*(6), 1267–1270. https://doi.org/10.1038/s41436-019-0478-1

Finucane, B., Abrams, L., Cronister, A., Archibald, A. D., Bennett, R. L., & McConkie-Rosell, A. (2012). Genetic counseling and testing for FMR1 gene mutations: Practice guidelines of the National Society of Genetic Counselors. *Journal of Genetic Counseling, 21*(6), 752–760. https://doi.org/10.1007/s10897-012-9524-8

Griesi-Oliveira, K., & Sertié, A. L. (2017). Autism spectrum disorders: An updated guide for genetic counseling. *Einstein (Sao Paulo, Brazil), 15*(2), 233–238. https://doi.org/10.1590/S1679-45082017RB4020

Hagerman, R. J., Protic, D., Rajaratnam, A., Salcedo-Arellano, M. J., Aydin, E. Y., & Schneider, A. (2018). Fragile X-associated neuropsychiatric disorders (FXAND). *Frontiers in Psychiatry, 9*, 564.

Hansen-Kiss, E., Beinkampen, S., Adler, B., Frazier, T., Prior, T., Erdman, S., … Herman, G. (2017). A retrospective chart review of the features of PTEN hamartoma tumour syndrome in children. *Journal of Medical Genetics, 54*(7), 471–478.

Harris, H. K., Sideridis, G. D., Barbaresi, W. J., & Harstad, E. (2020). Pathogenic yield of genetic testing in autism spectrum disorder. *Pediatrics, 146*(4), e20193211.

Hersh, J. H., & Saul, R. A. (2011). Health supervision for children with fragile X syndrome. *Pediatrics, 127*(5), 994–1006. https://doi.org/10.1542/peds.2010-3500

Hyman, S. L., Levy, S. E., Myers, S. M., Council on Children with disabilities, section on developmental and behavioral pediatrics (2020). Executive Summary: Identification, evaluation and management of children with autism spectrum disorder. *Pediatrics, 145*(1), e20193447. https://doi.org/10.1542/peds.2019-3447. Epub 2019 Dec 16. PMID: 31843864

Johannsen, W. (1909). *Elemente der exakten Erblichkeitslehre*. Fischer.

Jones, K. L., Jones, M. C., & Del Campo, M. (2021). *Smith's recognizable patterns of human malformation-E-book*. Elsevier Health Sciences.

Kirchner, R. M., & Walton, K. M. (2021). Symptoms of autism spectrum disorder in children with Down syndrome and Williams syndrome. *American Journal on Intellectual and Developmental Disabilities, 126*(1), 58–74.

Klusek, J., Martin, G. E., & Losh, M. (2014). Consistency between research and clinical diagnoses of autism among boys and girls with fragile X syndrome. *Journal of Intellectual Disability Research, 58*(10), 940–952.

Longo, N., Ardon, O., Vanzo, R., Schwartz, E., & Pasquali, M. (2011). Disorders of creatine transport and metabolism. *American Journal of Medical Genetics Part C, Seminars in Medical Genetics, 157c*(1), 72–78. https://doi.org/10.1002/ajmg.c.30292

Malinowski, J., Miller, D. T., Demmer, L., Gannon, J., Pereira, E. M., Schroeder, M. C., … Shen, J. (2020). Systematic evidence-based review: Outcomes from exome and genome sequencing for pediatric patients with congenital anomalies or intellectual disability. *Genetics in Medicine, 22*(6), 986–1004.

Manickam, K., McClain, M. R., Demmer, L. A., Biswas, S., Kearney, H. M., Malinowski, J., Massingham, L. J., Miller, D., Yu, T. W., Hisama, F. M. ACMG Board of Directors. (2021) Exome and genome sequencing for pediatric patients with congenital anomalies or intellectual disability: an evidence-based clinical guideline of the American College of Medical Genetics and Genomics (ACMG). *Genetics in Medicine, 23*(11), 2029–2037. https://doi.org/10.1038/s41436-021-01242-6. Epub 2021 Jul 1. PMID: 34211152.

Matthijs, G., Souche, E., Alders, M., Corveleyn, A., Eck, S., Feenstra, I., … Weiss, M. (2016). Guidelines for diagnostic next-generation sequencing. *European Journal of Human Genetics, 24*(1), 2–5.

Miles, J., & McCathren, R. (2005). Autism overview. *GeneReviews at GeneTests: Medical Genetics Information Resource*, 1–29.

Miller, D. T., Adam, M. P., Aradhya, S., Biesecker, L. G., Brothman, A. R., Carter, N. P., … Epstein, C. J. (2010). Consensus statement: Chromosomal microarray is a first-tier clinical diagnostic test for individuals with developmental disabilities or congenital anomalies. *The American Journal of Human Genetics, 86*(5), 749–764.

Monk, D., Mackay, D. J. G., Eggermann, T., Maher, E. R., & Riccio, A. (2019). Genomic imprinting disorders: Lessons on how genome, epigenome and environment interact. *Nature Reviews Genetics, 20*(4), 235–248.

Morris, C. A., & Braddock, S. R. (2020). Health care supervision for children with Williams syndrome. *Pediatrics, 145*(2), e20193761.

Nicholls, R. D., Saitoh, S., & Horsthemke, B. (1998). Imprinting in Prader–Willi and Angelman syndromes. *Trends in Genetics, 14*(5), 194–200.

Õunap, K. (2016). Silver-Russell syndrome and Beckwith-Wiedemann syndrome: Opposite phenotypes with heterogeneous molecular etiology. *Molecular Syndromology, 7*(3), 110–121. https://doi.org/10.1159/000447413

Rose, S., Niyazov, D. M., Rossignol, D. A., Goldenthal, M., Kahler, S. G., & Frye, R. E. (2018). Clinical and molecular characteristics of mitochondrial dysfunction in autism spectrum disorder. *Molecular Diagnosis & Therapy, 22*(5), 571–593.

Rossignol, D. A., & Frye, R. E. (2012). Mitochondrial dysfunction in autism spectrum disorders: A systematic review and meta-analysis. *Molecular Psychiatry, 17*(3), 290–314.

Schaefer, G. B., & Mendelsohn, N. J. (2013). Clinical genetics evaluation in identifying the etiology of autism spectrum disorders: 2013 guideline revisions. *Genetics in Medicine, 15*(5), 399–407.

Schiff, M., Benoist, J.-F., Aïssaoui, S., Boepsflug-Tanguy, O., Mouren, M.-C., De Baulny, H. O., & Delorme, R. (2011). Should metabolic diseases be systematically screened in nonsyndromic autism spectrum disorders? *PLoS One, 6*(7), e21932.

Spilioti, M., Evangeliou, A., Tramma, D., Theodoridou, Z., Metaxas, S., Michailidi, E., … Asprangathou, D. (2013). Evidence for treatable inborn errors of metabolism in a cohort of 187 Greek patients with autism spectrum disorder (ASD). *Frontiers in Human Neuroscience, 7*, 858.

Srivastava, S., Love-Nichols, J. A., Dies, K. A., Ledbetter, D. H., Martin, C. L., Chung, W. K., … Prock, L. (2019). Meta-analysis and multidisciplinary consensus statement: Exome sequencing is a first-tier clinical diagnostic test for individuals with neurodevelopmental disorders. *Genetics in Medicine, 21*(11), 2413–2421.

Stefanski, A., Calle-López, Y., Leu, C., Pérez-Palma, E., Pestana-Knight, E., & Lal, D. (2021). Clinical sequencing yield in epilepsy, autism spectrum disorder, and intellectual disability: A systematic review and meta-analysis. *Epilepsia, 62*(1), 143–151.

Stevenson, R. E., Schwartz, C. E., Rogers, R. C., & Rogers, R. C. (2012). *Atlas of X-linked intellectual disability syndromes*. Oxford University Press.

Todorow, C., Connell, J., & Turchi, R. M. (2018). The medical home for children with autism spectrum disorder: An essential element whose time has come. *Current Opinion in Pediatrics, 30*(2), 311–317.

Vickers, R. R., & Gibson, J. S. (2019). A review of the genomic analysis of children presenting with developmental delay/intellectual disability and associated dysmorphic features. *Cureus, 11*(1), e3873.

Vorstman, J. A. S., Parr, J. R., Moreno-De-Luca, D., Anney, R. J. L., Nurnberger, J. I., Jr., & Hallmayer, J. F. (2017). Autism genetics: Opportunities and challenges for clinical translation. *Nature Reviews Genetics, 18*(6), 362.

Wall, W. (2015). *The search for human chromosomes: A history of discovery*. Springer International Publishing.

Wang, J. C., et al. (2015). Regions of homozygosity identified by oligonucleotide SNP arrays: Evaluation the incidence and clinical utility. *European Journal of Human Genetics, 23*(5), 663–671.

Waye, M. M. Y., & Cheng, H. Y. (2018). Genetics and epigenetics of autism: A review. *Psychiatry and Clinical Neurosciences, 72*(4), 228–244.

Yehia, L., Keel, E., & Eng, C. (2020). The clinical spectrum of PTEN mutations. *Annual Review of Medicine, 71*, 103–116.

Ziats, C. A., Patterson, W. G., & Friez, M. (2021). Syndromic autism revisited: Review of the literature and lessons learned. *Pediatric Neurology, 114*, 21–25.

Prematurity and Autism Spectrum Disorder

15

Maria Gabriela Dominguez Garcia, Ramachandra Bhat, Shabih Manzar, Maroun J. Mhanna, and Nitin Walyat

Abstract

Preterm children are at increased risk for morbidity and mortality including long-term problems like behavioral and socioemotional difficulties because of the stress of interrupted physical and neural development. In 2020, approximately 360,000 infants required treatment in the NICU. Treatment planning for the care of these infants often begins with the obstetrician and neonatologist. There is evidence in the preterm population that altered brain development and injury may play an important role in the later development of ASD. The skilled interventions led by the neonatologist set the stage for an improved prognosis for preterm children. However, a multidisciplinary approach during the NICU stay and after discharge is very important in the diagnosis and management of the comorbidities and long-term sequelae of the premature infant.

Keywords

Neonatology · Autism spectrum disorder · Autism · ASD · Preterm · Prematurity

M. G. Dominguez Garcia (✉) · R. Bhat · S. Manzar · M. J. Mhanna · N. Walyat
Department of Pediatrics, Neonatal Perinatal Medicine, Louisiana State University, Shreveport, LA, USA
e-mail: maria.dominguezgarcia03@lsuhs.edu

Introduction

Treatment planning for the care of infants with medical complexity begins with neonatal and perinatal physicians. These specialists care for infants in neonatal intensive care units (NICU) and may be involved during the prenatal period. In 2020, approximately 360,000 infants required treatment in the NICU (Hamilton et al., 2021). Care delivered by the neonatologist is critical especially during the golden hour, the first hour after delivery, with quality care significantly improving lifelong outcomes (Sharma, 2017).

Specialized care is often required if infants are preterm or low birth weight. In the USA, 1 of every 10 infants is born before 37 weeks gestation (preterm), with 10.2% born preterm and 2.75% early preterm (less than 34 weeks) (Ferré et al., 2016). This percentage varies among different racial and ethnic groups with the higher rates of premature births by African American women compared to White women (Martin et al., 2019). With improving technology and new medical developments, the care of preterm infants has improved over the years with increased survival of all preterm infants, including the extremely preterm (<28 weeks) infant population. On graduation from the NICU, infants at risk are typically referred to medical homes or neonatal follow-up programs.

P. McPherson (ed.), *Handbook of Treatment Planning for Children with Autism and Other Neurodevelopmental Disorders*, Autism and Child Psychopathology Series,
https://doi.org/10.1007/978-3-031-06120-2_15

Infants may be admitted to the NICU to address a range of needs including prematurity, respiratory distress, sepsis, hyperbilirubinemia, feeding problems, congenital malformations, and hypoglycemia or other electrolyte abnormalities. Because of the stress of interrupted physical and neural development, all preterm infants are at higher risk for immediate morbidity and mortality and higher risk for long-term problems including developmental delay, vision problems, hearing problems, cerebral palsy, learning difficulties, and others (Agrawal et al., 2018). Kallioinen et al. (2017) have detailed the developmental follow-up of preterm children with useful graphics as a summary of the NICE (National Institute for Health and Healthcare Excellence) guidance. A clinical report from the American Academy of Pediatrics (AAP) notes that birth before 25 weeks gestation results in neurodevelopmental impairment for the majority of infants (Cummings, 2015). Infants born between 32 and 36 weeks gestation have twice the risk of neurodevelopmental disability at 2 years compared to full-term peers (Johnson et al., 2015). The presentation of preterm infants during the early school years is more predictive of adult functioning, with deficits during this time associated with ADHD (attention-deficit hyperactivity disorder), autism spectrum disorders (ASD), learning disorders, and addictive disorders (Synnes & Hicks, 2018). A full review of the sequelae of prematurity is beyond the scope of this chapter which focuses on current knowledge regarding the perinatal presentation of infants later diagnosed with ASD.

Overview of Autism Spectrum Disorders from the Neonatologist's Perspective

The prevalence of ASD has increased from 1 in 150 children in 2000 to 1 in 54 in 2020 per the Centers for Disease Control and Prevention (*Data & Statistics on Autism Spectrum Disorder*, 2020). ASD is more common among boys (1 in 42) than girls (1 in 189) in the general population. However, Allen et al. (2020) found a prevalence of ASD in preterm infants of 10.3% with an increased correlation between early birth week and ASD in female infants. They found that infants born <33 weeks gestation had a prevalence of 41% ASD for boys and 32% for girls, and after 33 weeks a prevalence of 20% ASD for males and 4% for females (Allen et al., 2020). For clinicians assessing children for ASD, queries about prematurity should include asking the gestational age. Fetal growth restriction is also a risk factor for ASD particularly in low birth weight infants and preterm infants with injury to the cerebellum (Agrawal et al., 2018). Abel et al. (2013) described that infants with fetal growth Z scores 1.5 standard deviation (SD) below and >2 SD above mean for gestational age also have greater risk for developing ASD (Abel et al., 2013). Other factors associated with fetal growth restriction and ASD include parental age, parental psychiatric history, socioeconomic status, congenital malformations, and maternal gestational diabetes (Abel et al., 2013).

There is evidence of an increased risk for autism is associated with advanced maternal age, smoking during pregnancy, multiparity, prematurity, low birth weight/growth restriction, low Apgar scores/perinatal asphyxia, pregnancies complicated by hypertension, bleeding, cesarean delivery, congenital malformations, chorioamnionitis, long NICU stays, and small-for-gestational-age and large-for-gestational-age infant birth weights (Abel et al., 2013; Hultman et al., 2002; Limperopoulos et al., 2008; Raghavan et al., 2019; Schieve et al., 2015). Some studies have shown that spontaneous preterm delivery and histological placental chorioamnionitis also increase the risk of ASD and other neurodevelopmental problems like ADHD (Raghavan et al., 2019).

The Brain of the Preterm Infant and ASD

Injury and Infection

ASD is a multifactorial disease in the preterm and full-term infant, with evidence that in the preterm population, brain development and injury may play an important role in the later development of ASD. The phenotype of preterm infants with ASD differs from those who were full-term and developed ASD. Preterm infants seem to have more impairment in areas like social interaction, communication, and an association with other comorbidities like global developmental delay and attention-deficient hyperactivity disorder (ADHD), but they have less stereotypic behaviors (Bokobza et al., 2019). Padilla et al. (2017) examined the neonatal records of 84 children born <27 weeks gestation comparing those diagnosed with ASD (27.4%) at 6.5 years of age to those without ASD, noting numerous insults experienced by the ASD group. The ASD group had lower mean birth weight; smaller head circumference at birth, term age, and 6.5 years; lower IQ scores; and higher frequency of neonatal complications like necrotizing enterocolitis, retinopathy of prematurity, surgical treatment of patent ductus arteriosus, and bronchopulmonary dysplasia.

The preterm infant is at high risk of brain injury (Johnson et al., 2015). The areas of the brain that commonly suffer damage include those known to be involved in producing ASD symptoms, including social deficits and sensory issues. Areas affected in preterm infants include the corpus callosum, amygdala, and cerebellum which play important roles in ASD by contributing to impairment in social competence, emotional responses, and cognitive impairment (Bokobza et al., 2019). Neonatal MRIs indicated smaller volumes of temporal, occipital, insular, and limbic cortices, and smaller mean volumes in gray matter and the cerebellum for infants who later developed ASD compared to those who did not (Padilla et al., 2017).

In addition to structural insults to the developing brain, infections are associated with increased risk of ASD. Infants born prematurely due to chorioamnionitis have a 16-fold increased risk of developing ASD. Infants with neonatal encephalopathy have a sixfold increased risk of developing ASD, with additional risk for epilepsy, cerebral palsy, congenital rubella, and hearing or visual impairments (Limperopoulos et al., 2008). Preterm delivery with maternal-fetal inflammation and/or infection is also correlated with the presence and severity of brain injury and later neurodevelopmental disturbance (Bokobza et al., 2019). A rise of pro-inflammatory cytokines in the blood of neonates may be a marker for the later development of ASD (Bokobza et al., 2019). Neuroinflammation involving the microglia and astrocytes in response to inflammation/infection has been shown to have increased activation in the cortex, cerebellum, and white matter of postmortem brains of children diagnosed with ASD (Bokobza et al., 2019). When screening children for ASD, query regarding perinatal infection or brain injury is important for predicting risk and prognosis.

Encephalopathy of Prematurity and Risk of ASD in Preterm Infants

Encephalopathy of prematurity (EOP) is a complicated condition in which a perinatal and postnatal insults inflict predominant white matter injury and less predominant gray matter brain injury with an impact on brain development, thus impairing long-term neurological function. Numerous insults can lead to EOP including anoxia, infection, inflammation, excitotoxicity, ischemia-reperfusion, and oxidative stress.

Neuroimaging for Identifying Neonates at Risk of ASD in Neonates Born Preterm

Structural and functional alteration of developing neonatal brain following preterm birth is predominantly determined by the perinatal and the ongoing postnatal destructive brain injuries. Furthermore, consequent disordered brain organization and maturational abnormalities also contribute to the structural and functional alteration leading to long-term neurodevelopment consequences, including the neuropsychological neurobehavioral abnormalities that define ASD. Histopathologically, the components of EOP include localized PVL (periventricular leukomalacia) (cystic and non-cystic PVL), diffuse white matter disease involving the subcortical white matter of the cerebrum (diffuse PVL), axonal/neuronal disease/injury with consequent cerebral and deep gray matter volume loss, and myelin sheath abnormalities. Review of predictive abilities of various neuroimaging markers of brain injury for early recognition of preterm infants at risk of ASD will facilitate aggressive developmental early interventions tailored against them (Volpe, 2009a, 2009b).

The odds of developing ASD in the presence of neurosonographic markers of brain injury have been evaluated in multicenter and single-center studies. A regional study (Movsas et al., 2013) involving a prospective cohort of 1105 low birth weight infants who were very meticulously followed and systematically evaluated for ASD showed a significant association of nonhemorrhagic ventriculomegaly, a marker of brain injury as validated by MRI studies (Inder et al., 2003), either in isolation or when occurred in association with PVL with the development of ASD later in life. A sevenfold higher risk of ASD was noted among the preterm infants with neurosonographic features of isolated ventriculomegaly as opposed to no abnormalities of neurosonography. Other markers of brain injury, occurring in isolation, were not associated with a later diagnosis of ASD. Conversely, another large study involving a cohort of extremely low gestational age preterm infants failed to demonstrate a significant association of either white matter disorders in isolation or occurring with IVH (intraventricular hemorrhage) (Campbell et al., 2021). In the Campbell et al.'s study, the white matter disorder was identified by the presence of periventricular echo-lucency and/or moderate to severe ventriculomegaly on a late cranial ultrasound scan that was performed after 2 weeks of postnatal age. The cohort was longitudinally followed for 10 years with standard diagnostic criteria using validated ASD evaluation tools used to define ASD. Similarly, a study based on NICHD NRN (National Institute of Child Health and Human Development Neonatal Research Network) generic database (Pappas et al., 2018) demonstrated no association between nonhemorrhagic ventriculomegaly and an increased risk of behavioral abnormalities defined using the Brief Infant-Toddler Social and Emotional Assessment tool. Even though inconstancies exist across the studies, the presence of ventriculomegaly on cranial ultrasound needs to be regarded as an important marker of preterm brain injury, necessitating a meticulous neurodevelopmental follow-up and an aggressive early developmental intervention. Cerebellar hemorrhage, a known cause of cerebellar injury, can be seen in association with other signs of brain injury or in isolation, with an average prevalence of rate of 2.3% in preterm infants. Cranial sonography performed through the mastoid window enables the detection of cerebellar hemorrhage with enhanced diagnostic accuracy. A recent meta-analysis (Pappas et al., 2018) of observational studies has proven a strong association of cerebellar hemorrhage with later development of non-motor adverse neurological consequences, including behavioral abnormalities encompassing the clinical spectrum of ASD.

Conventional MRI can also be complementary to cranial sonography for identifying brain lesions mentioned above that are associated with the later development of ASD. Furthermore, conventional MRI is more sensitive in identifying diffuse white matter injury, non-cystic PVL, corpus callosal abnormalities, and cortical gray matter abnormalities (Burkitt et al., 2019; Panigrahy et al., 2012). More advanced MRI evaluations, including volumetric, morphometric, functional, and DTI imaging modalities, have better sensitivities for the early recognition of preterm infants at risk for ASD (Padilla et al., 2017; Pandit et al., 2013; Panigrahy et al., 2012) but have limited practical application in routine setting owing to cost and limited availability.

Late Preterm Infants and ASD

As we have described, infants born prematurely and especially those born <28 weeks and with weight<1500 g are at increased risk of developing ASD. But infants born moderately preterm or late preterm have also increased risk of having ASD as well as speech delay requiring speech therapy, cognitive delay, motor delay, and cerebral palsy (Srinivas Jois, 2018). Guy et al. (2015) found that 14.5% of infants born between 32 and 36 weeks had a positive initial Modified Checklist for Autism in Toddlers (M-CHAT) screen when compared to controls (9.3%) and a 2.5% of moderate to late preterm infants had a true positive screen when followed up compared to 0.5% of controls. A review of neurodevelopmental outcomes of late preterm infants by Srinivas found that infants born as late preterm due to "medical indications" had higher risk of hyperactivity and high scores for ADHD, which is present in 65% of persons with ASD (Hossain et al., 2020; Srinivas Jois, 2018).

Monitoring for Early Signs of ASD

With the increased risk of ASD for preterm infants, it is vital that treatment planning includes parent education and referral for neurodevelopmental follow-up after discharge from the hospital. Ideally, this should include a medical home specializing in the care of preterm infants, referral to supports and services, and arranging for necessary durable medical equipment. Supports and services may include local, state, and federal programs that provide guidance, direct care, or financial support. While parent education is an ongoing process throughout the NICU admission, parents often feel overwhelmed at discharge and will require additional support to feel confident in caring for their medically complex child and to maintain their own self-care. Parents have emphasized the importance of maximizing the use of technology in providing education and contact to the medical home (Lakshmanan et al., 2019). Parents should be informed about the risk for ASD and assured that follow-up care will be including screening for a range of developmental issues including ASD (Lakshmanan et al., 2019). This includes following American Academy of Pediatrics clinical reports to administer the M-CHAT or other screens for ASD (Hyman et al., 2020). Premature infants fail M-CHAT screening up to 27% more often than full-term peers (Pineda et al., 2015). Interestingly, the study of 62 preterm infants by Pineda et al. (2015) found that social behavior, eye contact, social smile, attention, sensitivity to touch, hypotonia, head lag, and nystagmus in neonates were associated with later developmental issues but not with failing the M-CHAT screen at 2 years of age (Pineda et al., 2015). Newer data have described arm tone deficits and asymmetric visual tracking at month of age for infants later diagnosed with ASD, as well as decline in social interaction between 2 and 6 months of age (Pineda et al., 2015).

Klin et al. (2015) reviewed the research for the early diagnosis of ASD noting that at age two, toddlers with ASD showed less eye fixation to a caregiver than typically developing or developmentally delayed toddlers without ASD (Klin et al., 2015). In contrast to Pineda et al., the review described a decline in eye fixation from 2 months of age in infants later diagnosed with ASD and that children with the most rapid decline of eye fixation had the most social disabilities later in life (Klin et al., 2015).

Temperament has also been studied for associations with ASD. Vlaeminck et al. (2020) assessed temperament and sensory processing in preterm infants at 10, 18, 24, and 36 months and found that by the second year of life, both temperament and sensory processing had a good predictive value for ASD. In this study, the longer hospitalization length at birth was associated with poorer cognitive outcomes (Vlaeminck et al., 2020).

Other Risk Factors Associated with Prematurity and ASD

The nature-nurture debate of child development has raged for over a century with the nature citing the prominence of the infant as the genetic product of the biological father and mother. The field of epigenetics has called this view into question by noting the ability of environmental factors to alter genetic expression. Epigenetics focuses on the integrated model of genetic makeup, explaining how nature and nurture jointly influence phenotype. Thousands of research articles have contributed an increased understanding on the role of epigenetics in the development of ASD. In this section, we will focus on the five major parental categories of risk factors that have been associated with ASD.

Advanced Paternal Age

The genetic contribution of parents has been examined to better understand the maternal and paternal factors associated with ASD. Advanced maternal age has long been associated with increased risk for ASD (Sandin et al., 2012). More recently, advanced paternal age has been examined. With the changing demography, men are fathering children at a later age, and the prevalence of advanced paternal age (APA), defined variably as age over 35 or above 45 at first childbirth, is also increasing. Since 1980, proportion of infants born to fathers of APA increased from ~4% to 10% (Khandwala et al., 2018). With advanced paternal age, the rate of conception decreases mainly due to the decrease in fertility. Similarly, risks of complications such as preterm birth, gestational diabetes, intrauterine growth restriction, chromosomal and nonchromosomal birth defects, and childhood diseases like cancer and autism are associated with APA. Because of the maternal and fetal risks associated with APA, men delaying fatherhood past the age of 45 might consider sperm banking harmful to the mother and infant (Phillips et al., 2019).

Assisted Reproductive Technology

The Centers for Disease Control and Prevention (CDC) has collected assisted reproductive technology (ART) data since 1966 with number of ART-assisted births steadily increasing to nearly 60,000 in 2015 (Sunderam et al., 2018). It is therefore plausible to look for an association between ASD and assisted reproduction. A meta-analysis of 41 studies to assess the associations between assisted conception and ASD, developmental delay, and cerebral palsy was inconclusive (Hvidtjørn et al., 2009). Similarly, Diop et al. (2019) compared a large group of ART to a fertile group, noting that the odds of ASD were not statistically higher among children conceived with ART (OR 1.07; 95% CI 0.88–1.30). More research is needed to look further into the risk of ASD with ART.

Maternal Medication Use

Societal pressures and changing social dynamic have resulted in increase in the mental health issues including depression. Antidepressant use has increased among pregnant women. A Swedish retrospective cohort study examined the associations between first-trimester antidepressant exposure with birth complications and neurodevelopmental problems. They found first-trimester exposure to an antidepressant associated with all outcomes compared with unexposed offspring (preterm birth odds ratio [OR], 1.47 [95% CI, 1.40–1.55]; small for gestational age OR, 1.15 [95% CI, 1.06–1.25]; autism spectrum disorder hazard ratio [HR], 2.02 [95% CI, 1.80–2.26]; attention-deficit hyperactivity disorder HR, 2.21 [95% CI, 2.04–2.39]) (Sujan et al., 2017). In addition, antibiotics, metformin, calcium channel blockers, terbutaline, and antiseizure medications have been associated with ASD (Hisle-Gorman et al., 2018).

Maternal Obesity

Obesity is now prevalent in the USA with over a million women experiencing obesity during pregnancy in 2014 (Chen et al., 2018). Maternal obesity is a risk factor for many perinatal complications including hypertension, diabetes, low or high birth weight, and poor neurodevelopment outcome (Mitanchez & Chavatte-Palmer, 2018). Reynolds et al. (2014) reported an association between maternal obesity and ASD at 2 years of age in preterm infants (⩽30 weeks gestation). The 62 infants enrolled in the study underwent magnetic resonance imaging at term equivalent and then developmental testing including autism screen at 2 years of age with an odds ratio of 9.88

(P = 0.002) favoring a positive ASD screen (Reynolds et al., 2014).

Parental Psychiatric History

Parental severe mental illness (SMI) is associated with an increased risk of offspring autism spectrum disorder (ASD) and attention-deficit hyperactivity disorder (ADHD). McCoy et al. (2014) studied a cohort of offspring born to parents with SMI between 1992 and 2001 in Sweden. They used logistic and Cox regression to assess the associations between parental SMI. After controlling for measured covariates, maternal and paternal SMI were associated with an increased risk for preterm birth, low birth weight, and gestational age, and for offspring ASD and ADHD. A recent meta-analysis of nine studies found an association between parental mood disorders and ASD (Ayano et al., 2019). Parental ASD also places children at an increased risk for ASD (Leijdesdorff et al., 2017).

Inhospital Interventions to Mitigate the Risk of ASD in Preterm Infants

The Infant Brain Imaging Study, comparing infants at low and high familial risk for ASD, reported 96% accuracy in predicting a future diagnosis of ASD from scans of 6-month-old infants (Emerson et al., 2017). This study and others (Dickinson et al., 2021) highlight early variations in brain development among children who were diagnosed with ASD; however, preterm and low birth weight children were excluded from these studies. Although similar research regarding brain development among preterm is lacking, optimizing brain growth and development during the first year of life can mitigate the severity of neurodevelopmental outcomes. Neonatologists act to promote healthy brain development in preterm infants by monitoring every organ system and providing interventions to maximize health. Machine learning holds the promise of identifying which preterm infants will suffer poor neurological outcomes (Tataranno et al., 2021). The early identification of preterm infants at risk for ASD, and the early introduction of developmental interventions in the NICU among at-risk infants, is essential (Webb et al., 2014). Along with the initiation of all the validated early developmental interventions, the utilization of family nurture intervention (FNI) (Welch et al., 2020; Welch et al., 2015) in the NICU is of specific importance in the context of ASD prevention among the high-risk populations. The goal of FNI includes establishing an emotional bond and facilitating emotional communication between mother-infant dyad in order to decrease the separation-associated stress for both. A nurture specialist guiding FNI is responsible for calming and facilitating an emotional connection between mother and infant. The interventions combined with FNI involve kangaroo mother care (KMC), parents reading while providing the KMC, parents speaking to infants in their native language, and parents expressing their love and feelings to the infants. Touching while not providing the KMC is also part of FNI. Scent exchange by exchanging the cloths between the infant and mother is also a part of FNI. An RCT (randomized controlled trial) (Welch et al., 2020) conducted between 2008 and 2014 with longitudinal follow-up data has shown the beneficial impact of FNI not only in reducing the risk of ASD and other cognitive and behavioral impairments but also in facilitating better caregiving behaviors among the parents. Better parental caregiving behavior further improves social engagement, thus reducing the risk of ASD.

Fatty Acid Supplementation

Docosahexaenoic acid (DHA) and arachidonic acid (AA) dietary supplementations have been added to the diet of premature children to promote healthy brain development and possibly lower the risk of developing ASD; however, additional research is needed to confirm any benefits. DHA, an omega-3 long-chain polyunsaturated fatty acid, is a key component of the brain structure and helps the brain function better by reducing inflammation (Innis, 2005; Martínez & Mougan, 1998). DHA plays a major role in signaling pathways and neurotransmitter metabolism. Self-reporting maternal consumption of omega-3 fats, such as DHA, during the third trimester has been associated with a reduced risk of ASD by 40%. However, when serum levels of polyunsaturated fatty acids were measured during the third trimester, there was no correlation between serum levels and reduction in ASD (Huang et al., 2020). DHA crosses the placenta and accumulates in the brain during fetal development. Because brain growth accelerates during the third trimester and quadruples in size (Bouyssi-Kobar et al., 2016), preterm birth contributes to the disruption of transplacental transportation and decreases storage of DHA in preterm infants (Innis, 2005).

There is clinical evidence that premature infants, who are fed DHA- and AA-enriched formula, have an improvement in growth and development. Infants who are fed DHA- and AA-enriched formula have a higher Bayley mental and psychomotor developmental scores in comparison to controls (Clandinin et al., 2005). A small pilot study suggests that omega-3 and omega-6 fatty acid supplementation in premature infants (<29 weeks gestational age) is associated with a reduction in ASD symptoms, as assessed by the Brief Infant-Toddler Social and Emotional Assessment ASD scale (Keim et al., 2018b). This encouraging preliminary finding led the team to conduct a larger randomized controlled trial to determine if DHA supplementation improves neurodevelopmental outcomes in children born prematurely (the Omega Tots study). Three hundred seventy-seven children, who born prematurely, were enrolled and randomized to receive daily DHA plus AA or placebo for 6 months, and they were evaluated at 16–22 months of corrected age. The trial failed to show any significant improvement in cognitive development in children who received DHA plus AA supplementation. Bayley III cognitive scores were similar between children who did and did not receive DHA plus AA. However, in a subgroup of children who were extremely small at birth (birth weight less than a 1000 g), DHA plus AA supplementation was associated with lower scores on the Bayley III language testing. A similar association was seen on effortful control scores in children whose annual household income was greater than $35,000 (Keim et al., 2018a). Recently, a secondary analysis of the Omega Tots study was conducted showing that DHA plus AA supplementation decreased the risk of scoring at risk for ASD. Children who received DHA plus AA had an 11% lower risk on the Pervasive Developmental Disorders Screening Test-II, Stage 2, than controls (Boone et al., 2020).

Other studies have also failed to show any benefit of DHA supplementation on the developmental outcome of infants born prematurely. In one study, DHA supplementation in premature infants did not improve later language skills, when they were assessed by the MacArthur Communicative Development Inventory (MCDI) at 2 years of age, or behavior when assessed by the Strengths and Difficulties Questionnaire (SDQ), or temperament when assessed by the Short Temperament Scale for Children (STSC) at 3–5 years of age (Smithers et al., 2010). A study of high-dose DHA supplementation in preterm infants did not show an impact on IQ at 7 years of age as measured by the Wechsler Abbreviated Scale of Intelligence (WASI) (Collins et al., 2015).

One might speculate that the conflicting evidence of DHA and AA levels in children with ASD might be the reason behind the conflicting evidence to support DHA and AA dietary supplementation as a therapeutic intervention (Parletta et al., 2016; Yui et al., 2016). Future studies are needed to determine if there is a subgroup of children who were born prematurely and with ASD who would benefit from such early dietary intervention.

Neurodevelopmental Trajectory of Preterm Infants and Toddlers

Post-NICU Care

The transition from the NICU to home and follow-up care requires careful planning and coordination in order to promote best outcomes for medically complex children. The Federal Agency for Healthcare Research (*Transitioning Newborns from NICU to Home*, June 2021) and Quality (AHRQ) and the National Association of Neonatal Nurses (NANN) provide a toolkit that provides parents with critical information to facilitate the transition process (Agency for Healthcare Research and Quality, 2021; National Association of Neonatal Nurses, undated). The AHRQ toolkit includes information tailored to family and primary care provider needs as well as a health coach manual. NANN's *Baby Steps to Home* website (https://babystepstohome.com/) includes an extensive array of handouts to explain diagnoses, procedures, and specialized child care for families that can be customized for specific child needs. Even when provided with excellent resources, parents can feel overwhelmed by the responsibility of caring for a medically complex child. Astute medical teams will offer knowledge and social-emotional supports, including addressing social determinants of health, to meet families' needs more fully (Edwards & Horbar, 2021; Lakshmanan et al., 2019).

Like families, primary care providers may feel in need of additional guidance to address the needs of medically complex children (Hobbs et al., 2015). Pediatric home healthcare is commonly needed for medically complex infants, yet the process of identifying specific needs and prescribing and arranging services varies widely across locales and changes often, sometimes leaving families and professionals confused and frustrated (Fratantoni et al., 2019). Post-NICU care may be delivered in a program staffed by neonatologists, a medical home, or independent pediatricians. In all models care coordinators or health coaches can assist in establishing post-NICU care and routines. Voller (2018) has reviewed the full range of post-NICU care required by preterm infants (see Table 15.1), calling out the importance of screening for neurodevelopmental disorders.

Table 15.1 Follow-up Care for NICU graduates

Feeding, nutrition, and growth
Respiratory support
Vision screening
Hearing screening
Immunizations
Developmental sc10.1.reening
Motor screening
Neuroimaging/neurodevelopmental screening
Cognitive screening
Behavioral screening
Supports, services, and durable equipment

Adapted from Volmer (2018)

Ongoing Developmental Monitoring

Early diagnosis for early intervention is considered a crucial factor for good outcomes for children with ASD. As the prevalence of ASD has increased, it has raised concerns related to early detection through developmental surveillance in early life. Children born preterm carry an increased risk for developing ASD; however, their medical complexity may complicate assessment with ASD screening instruments. For example, the M-CHAT showed increased false positives and negatives for 2-year-old toddlers with a history of preterm birth (<28 weeks) (Joseph et al., 2017). The predictive validity of the M-CHAT and the Q-CHAT (Quantitative Modified Checklist for Autism in Toddlers) has not been established for children born preterm (Wong et al., 2014).

Zwaigenbaum et al. (2015) issued recommendations for the early recognition of ASD, highlighting developmental trajectory, motor delays, temperament, and atypical social-communication skills as potential markers for ASD. Chen et al. (2020) examined developmental trajectory as a marker for ASD in preterm infants (<32 weeks). At 5 years of age, children were classified as low-declining, high-declining, or high-stable based on serial neurodevelopmental assessments with the low-declining group 15 times more likely to develop ASD. The AAP recommends specific autism screening for all children at 18 months and 24 months of age in addition to global developmental screening (Johnson & Myers, 2007). The National Institute for Health and Healthcare Excellence (NICE) has published a guideline for monitoring the development of preterm infants and children (Kallioinen et al., 2017). The NICE guideline addresses a schedule of developmental screening with recommended screening and assessment instruments for late preterm (28–30 weeks) until the age of 2 and early preterm (<28 weeks) until the age of 4 (Kallioinen et al., 2017). While guidelines for preterm children do not specifically address screening of older children for ASD, Eryigit-Madzwamuse et al. (2015) described an increased rate of ASD in adults who were preterm (<32 weeks) or very low birth weight (<1500 g) (Eryigit-Madzwamuse et al., 2015).

Preterm children display greater social-communication difficulty and autistic behavior than the general population in early childhood as assessed by the Q-CHAT. The Quantitative Checklist for Autism in Toddlers (Q-CHAT) is a recent revision of the M-CHAT. It is a parent-completed questionnaire consisting of updated items, with each item having a five-point rating scale instead of a binary scoring system (Wong et al., 2014). The M-CHAT has been revised to the M-CHAT-R/F and now includes a follow-up interview that should be used together with the M-CHAT screen to improve the predictive value of the screening. It has been recommended that both parts of the screening should be completed (Luyster et al., 2011).

Recommendations

Evidence from the most rigorously conducted RCTs (French & Kennedy, 2018) indicates that skilled interventions led by the neonatologist and early interventions that focus on caregiver-child synchrony, joint attention, and engagement lead to positive changes in infant outcomes including increased social-communication behavior and decreased autistic characteristics. Providers caring for preterm children during their newborn period and after discharge should follow AAP recommendations for periodic assessments on social-emotional development and behavioral problems as well as psychosocial problems. Families would benefit from enrollment in specialized medical home or a high-risk follow-up clinic. Such clinics offer timely access to specialist evaluations when indicated by routine medical surveillance. Preterm children are at increased risk for behavioral and socioemotional difficulties due to brain immaturity and vulnerability during critical periods of development. It is important that clinicians are aware and monitor these difficulties to optimize the overall neurodevelopmental function of preterm children and their families.

References

Abel, K. M., Dalman, C., Svensson, A. C., Susser, E., Dal, H., Idring, S., … Magnusson, C. (2013). Deviance in fetal growth and risk of autism spectrum disorder. *The American Journal of Psychiatry, 170*(4), 391–398. https://doi.org/10.1176/appi.ajp.2012.12040543

Agency for Healthcare Research and Quality (2021). *Transitioning newborns form NICU to home: A resource toolkit*. Rockville, MD: Author. Retrieved from www.ahrq.gov/professionals/systems/hospital/nicu_toolkit/index.html. Accessed 3 July 2021.

Agrawal, S., Rao, S. C., Bulsara, M. K., & Patole, S. K. (2018). Prevalence of autism spectrum disorder in preterm infants: A meta-analysis. *Pediatrics, 142*(3), e20180134. https://doi.org/10.1542/peds.2018-0134

Allen, L., Leon-Attia, O., Shaham, M., Shefer, S., & Gabis, L. V. (2020). Autism risk linked to prematurity is more accentuated in girls. *PLoS One, 15*(8), e0236994. https://doi.org/10.1371/journal.pone.0236994

Ayano, G., Maravilla, J. C., & Alati, R. (2019). Risk of autistic spectrum disorder in offspring with parental mood disorders: A systematic review and meta-analysis. *Journal of Affective Disorders, 248*, 185–197. https://doi.org/10.1016/j.jad.2019.01.038

Bokobza, C., Van Steenwinckel, J., Mani, S., Mezger, V., Fleiss, B., & Gressens, P. (2019). Neuroinflammation in preterm babies and autism spectrum disorders. *Pediatric Research, 85*(2), 155–165. https://doi.org/10.1038/s41390-018-0208-4

Boone, K. M., Parrott, A., Rausch, J., Yeates, K. O., Klebanoff, M. A., Norris Turner, A., & Keim, S. A. (2020). Fatty acid supplementation and socioemotional outcomes: Secondary analysis of a randomized trial. *Pediatrics, 146*(4). https://doi.org/10.1542/peds.2020-0284

Bouyssi-Kobar, M., du Plessis, A. J., McCarter, R., Brossard-Racine, M., Murnick, J., Tinkleman, L., … Limperopoulos, C. (2016). Third trimester brain growth in preterm infants compared with in utero healthy fetuses. *Pediatrics, 138*(5). https://doi.org/10.1542/peds.2016-1640

Burkitt, K., Kang, O., Jyoti, R., Mohamed, A. L., & Chaudhari, T. (2019). Comparison of cranial ultrasound and MRI for detecting BRAIN injury in extremely preterm infants and correlation with neurological outcomes at 1 and 3 years. *European*

Journal of Pediatrics, 178(7), 1053–1061. https://doi.org/10.1007/s00431-019-03388-7

Campbell, H., Check, J., Kuban, K. C. K., Leviton, A., Joseph, R. M., Frazier, J. A., Douglass, L. M., Roell, K., Allred, E. N., Fordham, L. A., Hooper, S. R., Jara, H., Paneth, N., Mokrova, I., Ru, H., Santos, H. P., Jr., Fry, R. C., & O'Shea, T. M. (2021). Neonatal cranial ultrasound findings among infants born extremely preterm: Associations with neurodevelopmental outcomes at ten years of age. *The Journal of Pediatrics*. https://doi.org/10.1016/j.jpeds.2021.05.059

Chen, C., Xu, X., & Yan, Y. (2018). Estimated global overweight and obesity burden in pregnant women based on panel data model. *PLoS One, 13*(8), e0202183. https://doi.org/10.1371/journal.pone.0202183

Chen, L.-W., Wang, S.-T., Wang, L.-W., Kao, Y.-C., Chu, C.-L., Wu, C.-C., … Huang, C.-C. (2020). Early neurodevelopmental trajectories for autism spectrum disorder in children born very preterm. *Pediatrics, 146*(4), e20200297. https://doi.org/10.1542/peds.2020-0297

Clandinin, M. T., Van Aerde, J. E., Merkel, K. L., Harris, C. L., Springer, M. A., Hansen, J. W., & Diersen-Schade, D. A. (2005). Growth and development of preterm infants fed infant formulas containing docosahexaenoic acid and arachidonic acid. *The Journal of Pediatrics, 146*(4), 461–468. https://doi.org/10.1016/j.jpeds.2004.11.030

Collins, C. T., Gibson, R. A., Anderson, P. J., McPhee, A. J., Sullivan, T. R., Gould, J. F., … Makrides, M. (2015). Neurodevelopmental outcomes at 7 years' corrected age in preterm infants who were fed high-dose docosahexaenoic acid to term equivalent: A follow-up of a randomised controlled trial. *BMJ Open, 5*(3), e007314. https://doi.org/10.1136/bmjopen-2014-007314

Cummings, J. (2015). Antenatal counseling regarding resuscitation and intensive care before 25 weeks of gestation. *Pediatrics, 136*(3), 588–595. https://doi.org/10.1542/peds.2015-2336

Data & Statistics on Autism Spectrum Disorder. (2020). https://www.cdc.gov/ncbddd/autism/data.html

Dickinson, A., Daniel, M., Marin, A., Gaonkar, B., Dapretto, M., McDonald, N. M., & Jeste, S. (2021). Multivariate neural connectivity patterns in early infancy predict later autism symptoms. *Biological Psychiatry: Cognitive Neuroscience and Neuroimaging, 6*(1), 59–69. https://doi.org/10.1016/j.bpsc.2020.06.003

Diop, H., Cabral, H., Gopal, D., Cui, X., Stern, J. E., & Kotelchuck, M. (2019). Early autism spectrum disorders in children born to fertile, subfertile, and ART-treated women. *Maternal and Child Health Journal, 23*(11), 1489–1499. https://doi.org/10.1007/s10995-019-02770-z

Edwards, E. M., & Horbar, J. D. (2021). Following through: Interventions to improve long-term outcomes of preterm infants. *Seminars in Perinatology, 45*(4), 151414. https://doi.org/10.1016/j.semperi.2021.151414

Emerson, R. W., Adams, C., Nishino, T., Hazlett, H. C., Wolff, J. J., Zwaigenbaum, L., … Piven, J. (2017). Functional neuroimaging of high-risk 6-month-old infants predicts a diagnosis of autism at 24 months of age. *Science Translational Medicine, 9*(393). https://doi.org/10.1126/scitranslmed.aag2882

Eryigit-Madzwamuse, S., Strauss, V., Baumann, N., Bartmann, P., & Wolke, D. (2015). Personality of adults who were born very preterm. *Archives of Disease in Childhood: Fetal and Neonatal Edition, 100*(6), F524–F529. https://doi.org/10.1136/archdischild-2014-308007

Ferré, C., Callaghan, W., Olson, C., Sharma, A., & Barfield, W. (2016). Effects of maternal age and age-specific preterm birth rates on overall preterm birth rates – United States, 2007 and 2014. *MMWR. Morbidity and Mortality Weekly Report, 65*(43), 1181–1184. https://doi.org/10.15585/mmwr.mm6543a1

Fratantoni, K., Raisanen, J. C., Boss, R. D., Miller, J., Detwiler, K., Huff, S. M., … Donohue, P. K. (2019). The pediatric home health care process: Perspectives of prescribers, providers, and recipients. *Pediatrics, 144*(3), e20190897.

French, L., & Kennedy, E. M. M. (2018). Annual research review: Early intervention for infants and young children with, or at-risk of, autism spectrum disorder: A systematic review. *Journal of Child Psychology and Psychiatry, 59*(4), 444–456. https://doi.org/10.1111/jcpp.12828

Guy, A., Seaton, S. E., Boyle, E. M., Draper, E. S., Field, D. J., Manktelow, B. N., … Johnson, S. (2015). Infants born late/moderately preterm are at increased risk for a positive autism screen at 2 years of age. *The Journal of Pediatrics, 166*(2), 269–275.e263. https://doi.org/10.1016/j.jpeds.2014.10.053

Hamilton, B. E., Martin, J. A., & Osterman, M. J. (2021). Births: Provisional data for 2020.

Hisle-Gorman, E., Susi, A., Stokes, T., Gorman, G., Erdie-Lalena, C., & Nylund, C. M. (2018). Prenatal, perinatal, and neonatal risk factors of autism spectrum disorder. *Pediatric Research, 84*(2), 190–198. https://doi.org/10.1038/pr.2018.23

Hobbs, J. E., Hussey-Gardner, B. T., & Donahue, P. K. (2015). Pediatrician perspectives: Caring for NICU graduates in the community. *Contemporary Pediatrics, 32*(12), 18.

Hossain, M. M., Khan, N., Sultana, A., Ma, P., McKyer, E. L. J., Ahmed, H. U., & Purohit, N. (2020). Prevalence of comorbid psychiatric disorders among people with autism spectrum disorder: An umbrella review of systematic reviews and meta-analyses. *Psychiatry Research, 287*, 112922. https://doi.org/10.1016/j.psychres.2020.112922

Huang, Y., Iosif, A. M., Hansen, R. L., & Schmidt, R. J. (2020). Maternal polyunsaturated fatty acids and risk for autism spectrum disorder in the MARBLES high-risk study. *Autism, 24*(5), 1191–1200. https://doi.org/10.1177/1362361319877792. Epub 2020 Jan 21. PMID: 31958995

Hultman, C. M., Sparén, P., & Cnattingius, S. (2002). Perinatal risk factors for infantile autism. *Epidemiology, 13*(4), 417–423.

Hvidtjørn, D., Schieve, L., Schendel, D., Jacobsson, B., Svaerke, C., & Thorsen, P. (2009). Cerebral palsy, autism spectrum disorders, and developmental delay in children born after assisted conception: A systematic review and meta-analysis. *Archives of Pediatrics & Adolescent Medicine, 163*(1), 72–83. https://doi.org/10.1001/archpediatrics.2008.507

Hyman, S. L., Levy, S. E., & Myers, S. M. (2020). Identification, evaluation, and management of children with autism spectrum disorder. *Pediatrics, 145*(1), e20193447. https://doi.org/10.1542/peds.2019-3447

Inder, T. E., Wells, S. J., Mogridge, N. B., Spencer, C., & Volpe, J. J. (2003). Defining the nature of the cerebral abnormalities in the premature infant: a qualitative magnetic resonance imaging study. *The Journal of Pediatrics, 143*(2), 171–179. https://doi.org/10.1067/s0022-3476(03)00357-3

Innis, S. M. (2005). Essential fatty acid transfer and fetal development. *Placenta, 26*(Suppl A), S70–S75. https://doi.org/10.1016/j.placenta.2005.01.005

Johnson, C. P., & Myers, S. M. (2007). Identification and evaluation of children with autism spectrum disorders. *Pediatrics, 120*(5), 1183–1215. https://doi.org/10.1542/peds.2007-2361

Johnson, S., Evans, T. A., Draper, E. S., Field, D. J., Manktelow, B. N., Marlow, N., … Boyle, E. M. (2015). Neurodevelopmental outcomes following late and moderate prematurity: A population-based cohort study. *Archives of Disease in Childhood: Fetal and Neonatal Edition, 100*(4), F301–F308. https://doi.org/10.1136/archdischild-2014-307684

Joseph, R. M., Korzeniewski, S. J., Allred, E. N., O'Shea, T. M., Heeren, T., Frazier, J. A., … Kuban, K. (2017). Extremely low gestational age and very low birthweight for gestational age are risk factors for autism spectrum disorder in a large cohort study of 10-year-old children born at 23-27 weeks' gestation. *American Journal of Obstetrics and Gynecology, 216*(3), 304. e301–304. e316.

Kallioinen, M., Eadon, H., Murphy, M. S., & Baird, G. (2017). Developmental follow-up of children and young people born preterm: Summary of NICE guidance. *BMJ, 358*, j3514.

Keim, S. A., Boone, K. M., Klebanoff, M. A., Turner, A. N., Rausch, J., Nelin, M. A., … Sheppard, K. W. (2018a). Effect of docosahexaenoic acid supplementation vs placebo on developmental outcomes of toddlers born preterm: A randomized clinical trial. *JAMA Pediatrics, 172*(12), 1126–1134. https://doi.org/10.1001/jamapediatrics.2018.3082

Keim, S. A., Gracious, B., Boone, K. M., Klebanoff, M. A., Rogers, L. K., Rausch, J., … Rhoda, D. A. (2018b). ω-3 and ω-6 fatty acid supplementation may reduce autism symptoms based on parent report in preterm toddlers. *The Journal of Nutrition, 148*(2), 227–235. https://doi.org/10.1093/jn/nxx047

Khandwala, Y. S., Baker, V. L., Shaw, G. M., Stevenson, D. K., Lu, Y., & Eisenberg, M. L. (2018). Association of paternal age with perinatal outcomes between 2007 and 2016 in the United States: Population based cohort study. *BMJ, 363*, k4372.

Klin, A., Klaiman, C., & Jones, W. (2015). Reducing age of autism diagnosis: Developmental social neuroscience meets public health challenge. *Revista de Neurologia, 60*(Suppl 1(0)), S3–S11.

Lakshmanan, A., Kubicek, K., Williams, R., Robles, M., Vanderbilt, D. L., Mirzaian, C. B., … Kipke, M. (2019). Viewpoints from families for improving transition from NICU-to-home for infants with medical complexity at a safety net hospital: A qualitative study. *BMC Pediatrics, 19*(1), 223. https://doi.org/10.1186/s12887-019-1604-6

Leijdesdorff, S., van Doesum, K., Popma, A., Klaassen, R., & van Amelsvoort, T. (2017). Prevalence of psychopathology in children of parents with mental illness and/or addiction: An up to date narrative review. *Current Opinion in Psychiatry, 30*(4), 312–317. https://doi.org/10.1097/yco.0000000000000341

Limperopoulos, C., Bassan, H., Sullivan, N. R., Soul, J. S., Robertson, R. L., Moore, M., … du Plessis, A. J. (2008). Positive screening for autism in ex-preterm infants: Prevalence and risk factors. *Pediatrics, 121*(4), 758–765. https://doi.org/10.1542/peds.2007-2158

Luyster, R. J., Kuban, K. C., O'Shea, T. M., Paneth, N., Allred, E. N., & Leviton, A. (2011). The modified checklist for autism in toddlers in extremely low gestational age newborns: Individual items associated with motor, cognitive, vision and hearing limitations. *Paediatric and Perinatal Epidemiology, 25*(4), 366–376. https://doi.org/10.1111/j.1365-3016.2010.01187.x

Martin, J. A., Hamilton, B. E., Osterman, M. J. K., & Driscoll, A. K. (2019). Births: Final data for 2018. *National Vital Statistics Reports, 68*(13), 1–47.

Martínez, M., & Mougan, I. (1998). Fatty acid composition of human brain phospholipids during normal development. *Journal of Neurochemistry, 71*(6), 2528–2533. https://doi.org/10.1046/j.1471-4159.1998.71062528.x

McCoy, B. M., Rickert, M. E., Class, Q. A., Larsson, H., Lichtenstein, P., & D'Onofrio, B. M. (2014). Mediators of the association between parental severe mental illness and offspring neurodevelopmental problems. *Annals of Epidemiology, 24*(9), 629–634, 634.e621. https://doi.org/10.1016/j.annepidem.2014.05.010

Mitanchez, D., & Chavatte-Palmer, P. (2018). Review shows that maternal obesity induces serious adverse neonatal effects and is associated with childhood obesity in their offspring. *Acta Paediatrica, 107*(7), 1156–1165. https://doi.org/10.1111/apa.14269

Movsas, T. Z., Pinto-Martin, J. A., Whitaker, A. H., Feldman, J. F., Lorenz, J. M., Korzeniewski, S. J., Levy, S. E., & Paneth, N. (2013). Autism spectrum disorder is associated with ventricular enlargement in a low birth weight population. *The Journal of Pediatrics, 163*(1), 73–78. https://doi.org/10.1016/j.jpeds.2012.12.084

National Association of Neonatal Nurses. (undated). Baby steps to home. Retrieved from http://babystepstohome.com. Accessed 2 July 2021.

Padilla, N., Eklöf, E., Mårtensson, G. E., Bölte, S., Lagercrantz, H., & Ådén, U. (2017). Poor brain growth in extremely preterm neonates long before the onset of autism spectrum disorder symptoms. *Cerebral Cortex, 27*(2), 1245–1252. https://doi.org/10.1093/cercor/bhv300

Pandit, A. S., Ball, G., Edwards, A. D., & Counsell, S. J. (2013). Diffusion magnetic resonance imaging in preterm brain injury. *Neuroradiology, 55* Suppl 2, 65–95. https://doi.org/10.1007/s00234-013-1242-x

Panigrahy, A., Wisnowski, J. L., Furtado, A., Lepore, N., Paquette, L., & Bluml, S. (2012). Neuroimaging biomarkers of preterm brain injury: toward developing the preterm connectome. *Pediatric Radiology, 42* Suppl 1(0 1), S33–S61. https://doi.org/10.1007/s00247-011-2239-4

Pappas, A., Adams-Chapman, I., Shankaran, S., McDonald, S. A., Stoll, B. J., Laptook, A. R., Carlo, W. A., Van Meurs, K. P., Hintz, S. R., Carlson, M. D., Brumbaugh, J. E., Walsh, M. C., Wyckoff, M. H., Das, A., & Higgins, R. D. (2018). Neurodevelopmental and behavioral outcomes in extremely premature neonates with ventriculomegaly in the absence of periventricular-intraventricular hemorrhage. *JAMA Pediatrics, 172*(1), 32–42. https://doi.org/10.1001/jamapediatrics.2017.3545

Parletta, N., Niyonsenga, T., & Duff, J. (2016). Omega-3 and omega-6 polyunsaturated fatty acid levels and correlations with symptoms in children with attention deficit hyperactivity disorder, autistic spectrum disorder and typically developing controls. *PLoS One, 11*(5), e0156432. https://doi.org/10.1371/journal.pone.0156432

Phillips, N., Taylor, L., & Bachmann, G. (2019). Maternal, infant and childhood risks associated with advanced paternal age: The need for comprehensive counseling for men. *Maturitas, 125*, 81–84. https://doi.org/10.1016/j.maturitas.2019.03.020

Pineda, R., Melchior, K., Oberle, S., Inder, T., & Rogers, C. (2015). Assessment of autism symptoms during the neonatal period: Is there early evidence of autism risk? *The American Journal of Occupational Therapy: Official Publication of the American Occupational Therapy Association, 69*(4), 6904220010p6904220011–6904220010p6904220011. https://doi.org/10.5014/ajot.2015.015925

Raghavan, R., Helfrich, B. B., Cerda, S. R., Ji, Y., Burd, I., Wang, G., … Wang, X. (2019). Preterm birth subtypes, placental pathology findings, and risk of neurodevelopmental disabilities during childhood. *Placenta, 83*, 17–25. https://doi.org/10.1016/j.placenta.2019.06.374

Reynolds, L. C., Inder, T. E., Neil, J. J., Pineda, R. G., & Rogers, C. E. (2014). Maternal obesity and increased risk for autism and developmental delay among very preterm infants. *Journal of Perinatology, 34*(9), 688–692. https://doi.org/10.1038/jp.2014.80

Sandin, S., Hultman, C. M., Kolevzon, A., Gross, R., MacCabe, J. H., & Reichenberg, A. (2012). Advancing maternal age is associated with increasing risk for autism: A review and meta-analysis. *Journal of the American Academy of Child and Adolescent Psychiatry, 51*(5), 477–486.e471. https://doi.org/10.1016/j.jaac.2012.02.018

Schieve, L. A., Clayton, H. B., Durkin, M. S., Wingate, M. S., & Drews-Botsch, C. (2015). Comparison of perinatal risk factors associated with autism spectrum disorder (ASD), intellectual disability (ID), and co-occurring ASD and ID. *Journal of Autism and Developmental Disorders, 45*(8), 2361–2372. https://doi.org/10.1007/s10803-015-2402-0

Sharma, D. (2017). Golden hour of neonatal life: Need of the hour. *Maternal Health, Neonatology and Perinatology, 3*, 16. https://doi.org/10.1186/s40748-017-0057-x

Smithers, L. G., Collins, C. T., Simmonds, L. A., Gibson, R. A., McPhee, A., & Makrides, M. (2010). Feeding preterm infants milk with a higher dose of docosahexaenoic acid than that used in current practice does not influence language or behavior in early childhood: A follow-up study of a randomized controlled trial. *The American Journal of Clinical Nutrition, 91*(3), 628–634. https://doi.org/10.3945/ajcn.2009.28603

Srinivas Jois, R. (2018). Neurodevelopmental outcome of late-preterm infants: A pragmatic review. *Australian Journal of General Practice, 47*(11), 776–781.

Sujan, A. C., Rickert, M. E., Öberg, A. S., Quinn, P. D., Hernández-Díaz, S., Almqvist, C., . . . D'Onofrio, B. M. (2017). Associations of maternal antidepressant use during the first trimester of pregnancy with preterm birth, small for gestational age, autism spectrum disorder, and attention-deficit/hyperactivity disorder in offspring. *JAMA, 317*(15), 1553–1562. https://doi.org/10.1001/jama.2017.3413

Sunderam, S., Kissin, D. M., Crawford, S. B., Folger, S. G., Boulet, S. L., Warner, L., & Barfield, W. D. (2018). Assisted reproductive technology surveillance - United States, 2015. *MMWR Surveillance Summaries, 67*(3), 1–28. https://doi.org/10.15585/mmwr.ss6703a1

Synnes, A., & Hicks, M. (2018). Neurodevelopmental outcomes of preterm children at school age and beyond. *Clinics in Perinatology, 45*(3), 393–408. https://doi.org/10.1016/j.clp.2018.05.002

Tataranno, M. L., Vijlbrief, D. C., Dudink, J., & Benders, M. (2021). Precision medicine in neonates: A tailored approach to neonatal brain injury. *Frontiers in Pediatrics, 9*, 634092. https://doi.org/10.3389/fped.2021.634092

Transitioning Newborns from NICU to Home. (2021, June). Agency for Healthcare Research and Quality. https://www.ahrq.gov/patient-safety/settings/hospital/resource/nicu/index.html

Vlaeminck, F. P. P., Vermeirsch, J. P. P., Verhaeghe, L. U., Warreyn, P. P. P., & Roeyers, H. P. P. X. (2020). Predicting cognitive development and early symptoms of autism spectrum disorder in preterm children: The value of temperament and sensory processing. *Infant Behavior & Development, 59*, 101442.

Voller, S. M. (2018). Follow-up care for high-risk preterm infants. *Pediatric Annals, 47*(4), e142–e146.

Volpe, J. J. (2009a). Brain injury in premature infants: a complex amalgam of destructive and developmental disturbances. *The Lancet Neurology, 8*(1), 110–124. https://doi.org/10.1016/s1474-4422(08)70294-1

Volpe, J. J. (2009b). The encephalopathy of prematurity-brain injury and impaired brain development inextricably intertwined. *Seminars in Pediatric Neurology, 16*(4), 167–178. https://doi.org/10.1016/j.spen.2009.09.005

Webb, S. J., Jones, E. J. H., Kelly, J., & Dawson, G. (2014). The motivation for very early intervention for infants at high risk for autism spectrum disorders. *International Journal of Speech-Language Pathology, 16*(1), 36–42. https://doi.org/10.3109/17549507.2013.861018

Welch, M. G., Barone, J. L., Porges, S. W., Hane, A. A., Kwon, K. Y., Ludwig, R. J., … Myers, M. M. (2020). Family nurture intervention in the NICU increases autonomic regulation in mothers and children at 4-5 years of age: Follow-up results from a randomized controlled trial. *PLoS One, 15*(8), e0236930. https://doi.org/10.1371/journal.pone.0236930

Welch, M. G., Firestein, M. R., Austin, J., Hane, A. A., Stark, R. I., Hofer, M. A., … Myers, M. M. (2015). Family Nurture Intervention in the Neonatal Intensive Care Unit improves social-relatedness, attention, and neurodevelopment of preterm infants at 18 months in a randomized controlled trial. *Journal of Child Psychology and Psychiatry, 56*(11), 1202–1211. https://doi.org/10.1111/jcpp.12405

Wong, H. S., Huertas-Ceballos, A., Cowan, F. M., & Modi, N. (2014). Evaluation of early childhood social-communication difficulties in children born preterm using the Quantitative Checklist for Autism in Toddlers. *The Journal of Pediatrics, 164*(1), 26–33.e21. https://doi.org/10.1016/j.jpeds.2013.07.013

Yui, K., Imataka, G., Kawasak, Y., & Yamada, H. (2016). Increased ω-3 polyunsaturated fatty acid/arachidonic acid ratios and upregulation of signaling mediator in individuals with autism spectrum disorders. *Life Sciences, 145*, 205–212. https://doi.org/10.1016/j.lfs.2015.12.039

Zwaigenbaum, L., Bauman, M. L., Stone, W. L., Yirmiya, N., Estes, A., Hansen, R. L., … Fein, D. (2015). Early identification of autism spectrum disorder: Recommendations for practice and research. *Pediatrics, 136*(Supplement 1), S10–S40.

16 The Neurological Perspective: Autism Spectrum Disorders and Other Developmental Disabilities

Syed Ali Raza, Junaid Ansari, and Rosario M. Riel-Romero

Abstract

Neurologists provide valuable contributions to the care of children with autism spectrum disorders and other neurodevelopmental disorders, often addressing comorbid epilepsy and/or motor disorders. Children with ASD often have neurological comorbidities that are of utmost significance in addressing prognosis and treatment planning as well as informing research regarding the characterization of the autism spectrum disorder and the elucidation of ASD's underlying neural circuitry. The most common neurological abnormalities associated with ASD – stereotypies and praxis, motor deficits, epilepsy, gait and coordination, and sleeping abnormalities – are explored in this chapter. In addition the chapter explores recent neurological advances in unlocking the clues provided by these comorbidities to determine if they are specific to ASD and if they are causal in nature or an epiphenomena. The impact of recent advancements in the pathophysiology, epidemiology, and genetics of ASD spectrum on neurological interventions is explored, highlighting the hope for the better understanding and management of ASD through neurological research. Finally, this chapter details the manifestations of ASD along with evaluation and assessment from a neurological standpoint.

Keywords

Neurology · Autism · Epilepsy · Motor abnormalities · Genetics · Developmental disorders

Introduction

Neurodevelopmental disorders (NDD) are a heterogenous group of conditions featuring disturbances or delays in a variety of developmental domains including motor, cognition, language, and social faculties (Ismail & Shapiro, 2019; Jeste, 2015). The medical home team of children with NDD commonly includes a neurologist to lend expertise in nervous system development for assessment and treatment planning. This chapter will review the contribution of the neurologist to the care of children with neurodevelopmental disorders and comorbid epilepsy and/or motor disorders. After highlighting epilepsy and motor disorders generally, this chapter will discuss the care provided to children with autism spectrum disorder (ASD).

S. A. Raza (✉) · J. Ansari · R. M. Riel-Romero (✉)
Department of Neurology, Louisiana State University Health Sciences Center, Shreveport, LA, USA
e-mail: rrielr@lsuhsc.edu

P. McPherson (ed.), *Handbook of Treatment Planning for Children with Autism and Other Neurodevelopmental Disorders*, Autism and Child Psychopathology Series,
https://doi.org/10.1007/978-3-031-06120-2_16

ASD is a common, highly heritable neurodevelopmental disorder with a heterogenous presentation involving deficits in two major domains – social communication and reciprocity, and repetitive, restricted behaviors (Baumer & Spence, 2018). ASD prevalence has increased in the past few decades due to better recognition of the disease, broadening of the diagnostic criteria, as well as diagnostic substitution. There is a male preponderance with 4:1 male to female ratio (Christensen et al., 2016). The diagnosis of ASD is established with the contributions of medical home team members through taking a detailed developmental history from the parents and from observation of interaction between the child and the parents. Neurologists are commonly called upon to diagnose and appropriately manage neurological manifestations and comorbidities.

Epilepsy

The expertise of the pediatric neurologist is commonly sought for the identification and management of seizures in persons with developmental disorders. An increased incidence of epilepsy is found in many developmental disorders due to genetic, structural, and/or metabolic disturbances (Berg et al., 2017; Nickels et al., 2016). Genetic syndromes associated with epilepsy include Rett syndrome, Angelman syndrome, and fragile X syndrome (Berg et al., 2017). Tuberous sclerosis and neurofibromatosis are genetic disorders with a structural component contributing to epilepsy. Given the structural impact of cerebral palsy on the nervous system, the high prevalence of epilepsy in children with CP is not surprising nor is the fact that it can be particularly difficult to treat (Fiolita et al., 2020; Karatoprak et al., 2019). The prevalence estimates of epilepsy in children with ASD range from 4% to 38% (Lukmanji et al., 2019; Thomas et al., 2017). Epilepsy is diagnosed in persons with intellectual disability (ID) at a rate three to four times greater than the general population (Robertson et al., 2015).

Epilepsy co-occurs with ASDs with two peaks of onset, an initial peak that occurs in early childhood and a later peak in adolescence (Tuchman & Rapin, 2002). Conversely, ASD occurs in almost 46% of children with epilepsy (Matsuo et al., 2010). Coupled with the fact that children with intellectual and developmental disabilities (IDD) have an increased rate of comorbid ASD and epilepsy (Salpekar, 2018), patients with ASD may present to the neurologist for seizure evaluation at any age. Not surprising, with such medical complexity, seizure semiology is variable with complex partial, generalized, as well as mixed types being present – no single seizure semiology is most common (Jeste, 2011). A retrospective study involving 345 patients with ASD demonstrated 44% of paroxysmal abnormalities to be focal, 12% were generalized, and 42% were mixed (Parmeggiani et al., 2010). In this study, focal abnormalities were localized to temporal region in 31%, frontal in 18%, occipital in 13%, and parietal in 5% of the patients. A study of over a thousand Japanese children with ASD and epilepsy found more than 60% with frontal spikes, with significant involvement of the mirror neuron system which is known to contribute to symptoms of ASD (Yasuhara, 2010). A smaller retrospective study of 59 patients reported that seizures in the majority were found to be focal with or without secondary generalization (53.4%), generalized tonic-clonic in 19.2%, absence in 7.7%, polymorphic seizures in 4%, and preceding infantile spasms in 3% of the group (Pacheva et al., 2019). A British study of 150 children noted generalized tonic-clonic seizures to be the predominant seizure type (Bolton et al., 2011). In summary, the medical home team should be aware that seizure presentation is highly variable in children with ASD. See Table 16.1 for the seizure types associated with ASD in conformity with the International League Against Epilepsy (ILAE) classification scheme.

Risk factors for epilepsy in ASD include syndromic autism (Miles et al., 2005; Pavone et al., 2004), intellectual disability, and female sex (Amiet et al., 2008). Children with very early-onset seizures (e.g., infantile spasms) have an increased risk of ASD (Baumer & Spence, 2018). An important association exists between cognitive impairment and epilepsy in ASD (Ekinci et al., 2010) which is particularly robust in chil-

Table 16.1 Common seizure types in ASD/DD

Partial epilepsy/focal
Aware or impaired awareness
Motor onset
Non-motor onset
Generalized onset
Motor onset
Tonic-clonic
Other motor
Non-motor (absence)
Myoclonic epilepsy
Epileptic encephalopathies
Landau-Kleffner syndrome
Lennox-Gastaut syndrome
CSWS
West syndrome

dren with ASD in the context of tuberous sclerosis (Jeste et al., 2008). Even in the absence of epilepsy there is high prevalence of paroxysmal EEG abnormalities in ASD (Canitano, 2007; Jeste, 2011). A large retrospective study of children with ASD reported abnormal EEG discharges in 85% of the subjects, with the highest incidence of spikes in children with intellectual disability (Yasuhara, 2010). These paroxysmal EEG abnormalities could be either focal, generalized, or mixed with the focal abnormalities being mostly localized to the frontal lobe (Yasuhara, 2010). Given the prior finding of a significant incidence of spikes in children with intellectual disability, one could pose the question whether epileptic encephalopathy causes autism (Tharp, 2004). This is of relevance considering that pharmacotherapy aimed at spike suppression could potentially alter the developmental trajectory of patients with ASDs. The question being whether a pathophysiological association exists between epileptiform discharges and regression considering there is language regression seen in the setting of continuous slow wave spikes (Deonna & Roulet-Perez, 2010). More recent findings of the genetic overlap between ASD and epileptic encephalopathies add support to this possibility, but further research is necessary to fully understand this complex relationship (Srivastava & Sahin, 2017).

Pharmacological Management of Epilepsy

Management of epilepsy in ASD patients mostly follows the usual paradigm of epilepsy treatment. If a specific metabolic syndrome is found to be associated with the seizures in a patient with ASD, a specific treatment may be indicated. However, antiseizure mediations (ASMs) form the mainstay of treatment of epilepsy in ASD patients with the general treatment principles applied, although careful considerations should be made to minimize untoward behavioral side effects when choosing seizure medications due to the high rate of medical complexity in children with ASD and IDD. The common ASMs used for the management of epilepsy in ASD are reviewed in Table 16.2 including their most common side effects. Despite the frequent co-occurrence of seizures in ASD patients, there is relative dearth of studies evaluating effectiveness of ASMs (Frye et al., 2013). A survey conducted by the Autism Society of North Carolina with responders listing their medications as well as their satisfaction levels indicated that 15.2% of the individuals were receiving ASMs with the most common being carbamazepine, valproic acid, and phenytoin and the parents being satisfied in general with the ASM treatment (Aman et al., 1995). Of note, this study did not include newer ASMs including lamotrigine, lacosamide, and zonisamide, among others, that were not in common use or were approved after the study. Another survey study noted that valproate, lamotrigine, levetiracetam, and ethosuximide provided the best seizure control along with the least adverse effects among all the ASMs examined (Frye et al., 2011). As a general rule, newer ASMs such as lamotrigine, oxcarbazepine, topiramate, and levetiracetam have fewer side effects in comparison to older ASMs like phenobarbital, phenytoin, primidone, and carbamazepine. In our clinical experience, valproate, an older medication, has good efficacy for many patients with ASD. Potential side effects and spectrum of action are major considerations when prescribing an ASM. The broad-spectrum ASMs have effectiveness for a wide variety of seizures whereas narrow-spectrum ASMs are

Table 16.2 Antiseizure medications (ASMs) – spectrum and common adverse effects

AEDs	Spectrum of action	Adverse effects[a]
Valproate	Broad	Hepatotoxicity, hyperammonemia, weight gain, hair thinning, thrombocytopenia, and pancreatitis
Lamotrigine	Broad	Stevens-Johnson's syndrome (up-titrate slowly), abnormal liver function tests
Oxcarbazepine	Narrow	Hyponatremia
Levetiracetam (Keppra)	Broad	Behavioral side effects – agitation, aggression, and mood instability
Topiramate	Broad	Weight loss, psychomotor and cognitive slowing, metabolic acidosis, nephrolithiasis, and glaucoma
Lacosamide	Narrow	Dizziness, diplopia, and fatigue, precluded in abnormal heart rhythms
Carbamazepine	Narrow	Hyponatremia, dizziness, and ataxia
Zonisamide	Broad	Dizziness and loss of appetite
Ethosuximide	Narrow	Dizziness, nausea, vomiting, and sleep disturbance
Phenobarbital	Narrow	Drowsiness, lethargy, and hyperactivity
Phenytoin	Narrow	Abnormal liver function tests, gingival hyperplasia, ataxia, nystagmus, hirsutism, and coarsening of facial features
Clonazepam	Broad	Ataxia, cognitive dysfunction, and respiratory depression (rare)

[a]The US FDA (Food and Drug Administration) has issued a suicide risk warning on all ASMs. Note, most ASMs, if not all, are associated with detrimental effects on bone density/health aside from suicide risk

most applicable for focal, absence, or myoclonic seizures. While absence seizures – also called petit mal seizures – are generalized, focal onset seizure (previously known as partial seizure) means a seizure that occurs in a particular part of the brain. An ASM's side effect profile must be considered in relationship to the individual. For example, levetiracetam may exacerbate behavioral problems and is best avoided in children with behavioral abnormalities. Valproate is associated with weight gain and polycystic ovary disease limiting its use in girls and those with weight issues. Similarly, valproate is contraindicated in ASD patients with mitochondrial disorders as it may further impair mitochondrial functions. Valproate may increase abdominal distress in children with comorbid gastrointestinal disorders. This highlights the importance of communication between the patient, family, and medical team to fully understand each patient's unique challenges and needs.

Beyond the Pharmacological Management of Epilepsy

Non-epileptic treatments are typically used to treat seizures when ASMs are not effective. Multiple non-blinded and randomized controlled trials conducted in children showed effectiveness of ketogenic diet in epilepsy in children (Frye et al., 2013; Levy et al., 2012). In another retrospective case-control survey study, ketogenic diet was rated as the most favorable non-ASM treatment for improving seizures (Frye et al., 2011). In certain instances of patients with refractory epilepsy, particularly epileptic encephalopathies, intravenous immunoglobulin infusion and corticosteroids also have a role in treatment. In a retrospective review of 59 ASD patients with seizures who were treated with a vagus nerve stimulator (VNS), more than half experienced at least a 50% reduction in seizure frequency and significant improvement in the quality of life, suggesting a role of VNS in treatment of epilepsy in patients with ASD (Park, 2003). In cases where potential epileptogenic foci are visualized clearly with neuroimaging and identified as foci using various methodologies including magnetoencephalography (MEG), subdural intracranial recording grids, and/or intraoperative mapping, epilepsy surgery can also play a significant treatment role. This is particularly true of patients with syndromic autism, for instance, tuberous sclerosis complex and tumors. These interventions can be frightening for children and their parents, necessitating multiple discussions and careful preparation of the child for the procedure.

Motor Disorders

The early identification of motor deficits is imperative for various reasons. Firstly, motor deficits are objective and quantifiable, providing a framework for measurement and a proposed role to act as motor biomarkers following temporal characterization and specificity studies in relation to NDD/ASD (Swanson & Hazlett, 2019; Thurm et al., 2016; Varcin & Nelson Iii, 2016). Second, motor deficits and the underlying neural circuitry dysfunction can enable classification schemes or endophenotypes within the heterogenous spectrum of NDD. And finally, motor function is critical in development, language acquisition, social interaction, and learning. Therefore, better characterization of motor deficits, including developmental coordination, stereotypic movement, and tic disorders and their precursors, can foster early intervention which could improve functional and behavioral outcomes for children with NDD.

Motor impairments are far less common in ASD compared to social communication and interaction despite being one of the earliest signs of ASD. Motor concerns include vestibular control impairment, gross and fine motor abnormalities, and oculomotor issues. Others such as motor clumsiness, delayed development of hand dominance, and primitive reflexes are nonspecific neurological symptoms associated with ASD (Mosconi & Sweeney, 2015). Interestingly, it has been suggested that motor deficit patterns may be able to distinguish ASD subtypes. Asperger's syndrome is characterized with consistent motor "clumsiness." ASD patients without language delays often demonstrate saccade dysmetria or abnormal eye movements. Given the frequency of motor impairments in genetic syndromes of autism (Geschwind, 2009) and that they generally do not improve over early childhood (Van Waelvelde et al., 2010), an important question remains regarding their role in the diagnostic process and inclusion in the diagnostic criteria of ASD.

Early oral-motor skills and imitation predict language acquisition in infants who are later diagnosed with ASD (Gernsbacher et al., 2008; McDuffie et al., 2005). A study analyzing motor function and gait from home videos of children with ASD, developmental delay and typical controls, showed that between groups, ASD children showed delayed development of movements including lying supine, sitting, and walking (Ozonoff et al., 2008). Other studies in the first 2 years of life showed delays in motor development including postural asymmetry, developmental milestones, and the overall gross and fine motor movements (Esposito et al., 2011; Iverson & Wozniak, 2007; Provost et al., 2007). A major limitation of these studies is the use of retrospective home videos without standardization. A prospective study also demonstrated low performance in gross and fine motor skills in the ASD group of children when examined at 24 months of age in comparison to language-delayed group within the study (Landa & Garrett-Mayer, 2006).

An entire plethora of gait abnormalities have been reported in children with ASD. A meta-analysis of 41 studies investigating coordination, gait, arm movements, and postural instability in ASD compared to controls showed significant motor incoordination and postural instability in the ASD group (Fournier et al., 2010). This is despite the heterogeneity in methodology of the various studies included in the meta-analysis. In a subgroup analysis, attenuation of effects with increasing age was seen suggestive of improved motor function over time. There are no strict guidelines as to when these abnormalities should be brought to a neurologist's attention; however, referral should be considered when impairment limits activities of daily living. As with other ASD comorbidities, there may be tremendous heterogeneity in presentation, and in most of the cases, there is attenuation of motor abnormalities over time.

Motor delays are common in many NDD, including fragile X, Down, Angelman, Noonan, and deletion 22q11 syndromes, cerebral palsy, muscular dystrophies, ADHD (attention deficit hyperactivity disorder), and ASD. The American Academy of Pediatrics has established guidelines for the early identification and evaluation of motor delays that include periodic screening at well-child visits and recommendation for physical examination and neuroimaging (Noritz & Murphy, 2013). The American Academy of Neurology details the screening and neuroimaging recommendations for children with cerebral palsy in a practice parameter (Ashwal et al., 2004).

Motor Disorders: Developmental Coordination Disorder/Dyspraxias

In neurology, *praxis* concerns the neural networks supporting the ability to produce both meaningful and meaningless gestures. Deficits in praxis, or dyspraxias, result in the impaired performance of the complex gestures that correlate with social, communication, and behavioral deficits with the early onset of dyspraxias affecting learning and overall cognitive development. For children with ASD, dyspraxia has been attributed to impaired formation of spatial representation and poor motor execution (Dowell et al., 2009; Dziuk et al., 2007; Mostofsky & Ewen, 2011).

When a dyspraxia interferes with activities of living and motor skills are substantially below expectations, a developmental coordination disorder (DCD) should be considered. The deficits in DCD may involve *planning, sensorimotor coordination, motivation, goal directedness, regulation of motor activities, and skill learning* (Fletcher et al., 2020, p. 173). DCD may be diagnosed in children with ASD if criteria are met; however, when assigning the diagnosis of DCD to a child with ID, the level of motor impairment must exceed that predicted by the ID alone (American Psychiatric Association & American Psychiatric Association. DSM-5 Task Force, 2013). There is limited prevalence data for DCD, with general estimates ranging from 2% to 20% (Blank et al., 2019). Preterm and low-birthweight children are at increased risk for DCD as are children with ASD and ADHD (Edwards et al., 2011; Kopp et al., 2010). Neurological assessment to rule out neuromuscular diseases, neoplasm, metabolic disorders, or other conditions that may cause coordination difficulties is indicated. A physical or occupational therapist may administer the Movement Assessment Battery for Children and the Bruininks-Oseretsky Test of Motor Proficiency and provide indicated therapy (Bruininks & Bruininks, 2005; Henderson, 1992). To offer further guidance, the European Academy of Childhood Disability has published clinical practice recommendations for the assessment and management of DCD (Blank et al., 2019).

Motor Disorders: Stereotypies

Stereotypic movements *are repetitive, seemingly driven, and apparently purposeless motor behaviors* (American Psychiatric Association & American Psychiatric Association. DSM-5 Task Force, 2013, p. 77). The DSM-5 (2103) instructs that self-injurious behavior and a known medical condition, such as neurodevelopmental disorder, should be noted as specifiers. A systematic review of stereotypies in persons with developmental disabilities involving 44 studies and representing 11,331 participants found an average prevalence of stereotypy of 61%. Diagnoses represented were Down, fragile X, Prader-Willi syndromes, and ASD, with stereotypy most common in persons with ASD, at 88% (Chebli et al., 2016). Significant stereotypical behaviors have also been reported in children with Angelman, Cornelia de Lange, Cri du Chat, fragile X, Lowe, and Smith-Magenis syndromes (Moss et al., 2009). For children with Cornelia de Lange and Lesch-Nyhan syndromes, stereotypical behaviors are often associated with self-injurious behavior (Nyhan, 1976; Srivastava et al., 2020).

Stereotypies should be assessed, and intervention considered when the movement:

(a) Persists at similar levels past the age of two
(b) Is displayed with high intensity or frequency

(c) Appears atypical or unusual in its manifestation
(d) Interferes with an individual's functioning (Chebli et al., 2016, p. 107) from Didden et al. (2012)

The earliest descriptions of ASD reported stereotypical behaviors (Asperger & Frith, 1991; Kanner, 1943). Motor stereotypies or repetitive behaviors often are present early in children with ASD and are intimately linked to social communication and cognitive deficit (low functioning, lower IQ ASD groups) and can correlate with disease progression and severity (Jeste, 2011). In the current DSM-5 classification, "stereotypies" are the only neurological manifestation included in the diagnostic criteria for ASD (American Psychiatric Association & American Psychiatric Association. DSM-5 Task Force, 2013). These abnormalities include finger movements, body posturing, rocking, spinning, hand/arm flapping, full-body testing, toe walking, and repetitive jumping, among others. Additionally, vocal stereotypies including repetitive sounds or verbal echolalia are common in ASD. A systematic review and meta-analysis involving 8124 persons with ASD found younger age, lower intelligence, and greater severity of ASD, but not gender, to be correlated with a higher number of stereotypies (Melo et al., 2020).

Sleep Disorders

Sleep is necessary for healthy neurodevelopment, and neurodevelopment influences sleep in a complex relationship (McKenna & Reiss, 2018). Variation in neurophysiology and/or neuroanatomy contributes to the high prevalence of sleep disorders in persons with developmental disorders (Esbensen & Schwichtenberg, 2016; Fletcher et al., 2020). Short sleep duration, low sleep quality/efficiency, and circadian sleep desynchronization are common in children with ASD as are behavioral challenges related to sleep (Carmassi et al., 2019; Mazurek & Sohl, 2016). For persons with ID, poor sleep quality and shorter duration are reported (Surtees et al., 2018). Sleep disorders are common in many genetic disorders including fragile X syndrome, tuberous sclerosis, neurofibromatosis, and syndromes including Down, Williams, Smith-Magenis, Rett, Prader-Willi, Angelman, and Lesch-Nyhan (Stores, 2014). Epilepsy may be associated with hypersomnia, insomnia, obstructive sleep apnea, restless legs syndrome, and parasomnias (Latreille et al., 2018). It is critical that the medical home team inquire directly about sleep as caregivers may underestimate the contribution of sleep disorders to behavioral challenges (Hoffmire et al., 2014). The impact of medication on sleep must also be considered by the team.

Sleep impairments are prevalent in a significant proportion of individuals with ASD and NDD. The primary disorder in ASD is insomnia, and there have been both subjective and objective measures investigating sleep disturbances with the prior employing questionnaires and the latter actigraphy and polysomnography. Subjectively, multiple studies have reported difficulty initiating and maintaining sleep, restless sleep, co-sleeping, and early morning awakenings (Jeste, 2011). Biologically, sleep impairment may be explained by aberrant circadian rhythms. Genes controlling circadian rhythms (clock genes) may have a role in modulating melatonin for sleep regulation and in integrity of synaptic transmission in ASD (Bourgeron, 2007; Nicholas et al., 2007). Other studies have suggested melatonin dysregulation (Tordjman et al., 2005) which is further corroborated by the finding of exogenous melatonin therapy being effective in improving sleep in ASD. Further studies plotting the developmental trajectory of sleep impairment prospectively are warranted to see if there are clinical associations of sleep impairments such as periodic limb movements and/or restless legs syndrome in children with ASD. Focusing and managing behavioral issues associated with ASD may have significant impact with sleep. More discussion on the sleep abnormalities associated with ASD can be found in the sleep chapter of this edition, Chap. 22.

Neurological Considerations in the Diagnosis of ASD

Screening for ASD

Considering the heterogeneity of ASD, most recommendations suggest multidisciplinary evaluation, medical evaluation, and genetic testing. There has been controversy with regard to universal screening for ASD with the US Preventive Services Task Force stating that not enough data were present to advocate for universal screening of children for autism (Baumer & Spence, 2018). However, the American Association of Pediatrics endorses recommendations for developmental surveillance and specific ASD screening at ages 18 and 24 months and for all children who fail routine developmental surveillance (Hyman et al., 2020a, 2020b).

The early neurological abnormalities seen with ASD are often peculiar and if investigated can facilitate both screening and diagnosis at an early age resulting in prompt intervention and treatment to promote a better prognosis. As discussed above the motor deficits seen in ASD correlate with the severity and are one of the earliest abnormalities linked with social and cognitive deficits (Jeste, 2011). In the current era, there is quite a need for further clarification and elucidation of the clinical and scientific evidence with empirical data which may help in a definite association between motor and cognitive skills in ASD.

As of now there are no motor impairments or neuroimaging signatures that are part of the diagnostic criteria of ASD. Considering that motor deficits are one of the common changes seen in ASD, future research is likely to modify the diagnostic criteria. Similarly, given the quantifiable nature of motor impairments, it could also be used as a biomarker and for characterization of endophenotypes of ASD which would facilitate early clinical interventions aimed at improving outcomes.

Clinical Evaluation of ASD Symptoms

Neurologists have a special role in the evaluation of patients with suspected ASD considering the attention to neurological details that is requisite for making the diagnosis. Evaluation of patients with suspected ASD includes detailed history, physical examination, and appropriate investigations. The medical history should include birth history, age of parents at birth (older paternal age being a risk factor for ASD), perinatal risk factors, and pregnancy or delivery complications. Clinicians should assess for medical conditions commonly seen in patients with ASD including gastrointestinal concerns, sleep disturbances, seizures/epilepsy, and lead exposure. Family history is also of importance including history of epilepsy, genetic, metabolic, autoimmune, speech, intellectual, and ADHD/learning disorders. Metabolic disorders should also be considered in the differential diagnosis and ruled out. Finally, the DSM-5 ASD diagnostic criteria including specific inquiry for social communication deficits and repetitive behavior/restricted interests must be evaluated.

A detailed physical and neurological examination of the patient is a significant part of evaluation. ASD is often associated with various neurogenetic syndromes. Therefore, assessment for dysmorphic features (e.g., prominent ears in fragile X syndrome or facial features suggestive of metabolic anomalies), growth parameters (height, weight, and head circumference), and skin examination to assess for neurocutaneous features (e.g., hypopigmented spots or fibromas) are pertinent. Neurologic exams emphasizing the patient's attention, cranial nerves, muscle tone, motor coordination, reflexes, and gait are all significant aspects of evaluation of patients with ASD.

Neurogenetic syndromes commonly occur in 10–20% of patients with ASD (Bourgeron, 2016). Children with fragile X syndrome, tuberous sclerosis complex, 15q duplication, neurofibromatosis, Angelman syndrome, Prader-Willi syndrome, Rett syndrome, and Down syndrome have increased rates of autism in comparison to general population (Johnson et al., 2007). Chromosomal microarray testing is useful for detecting copy number variations, but in order to diagnose balanced translocations, whole exome sequencing (WES) or whole genome sequencing (WGS) may be needed. To identify genetic aberrations involving single nucleotide polymorphisms, WES/WGS is required (Sanchez Fernandez et al., 2019; Shendure et al., 2004).

Depending on the clinical phenotype, further investigations including brain MRI, EEG, and metabolic testing must be considered. The relevance and implications of neuroimaging and electroencephalography are detailed below. Neuroimaging is particularly useful in situations where there is microcephaly, hypertonia, and focal neurological deficit(s) or when concern for neurodegenerative disease process is a consideration. Often, neuroimaging may help in excluding cortical dysplasia, mesial temporal sclerosis, and cystic lesions which have been associated with ASD (Casanova et al., 2013). In cases where there are concerns for clinical or subclinical seizures, an EEG should be considered. EEG is also necessary in instances where regression is secondary to possible Landau-Kleffner syndrome or in cases of acquired epileptic aphasia. In situations where a metabolic defect is suspected, varied tests including plasma and urine amino acids, plasma acylcarnitine, and even checking lead levels should be considered. Audiologic testing is also important in children with language delay and diagnosis of ASD since it may be a treatable condition.

A framework for the evaluation of ASD from a neurological perspective with management options is presented in Table 16.4. The evaluation and management of ASD is best accomplished through a multidisciplinary approach including neurologists and other specialists as indicated. The medical home approach enables the needs of the child and family to be addressed in a comprehensive manner.

Neurobiology: Insights from MRI Findings and Electrophysiologic Studies

Neuroimaging

MRI is a useful imaging modality which can facilitate understanding of how the brain develops structurally and functionally in patients with ASD when compared to controls. Reproducibility of results is an issue to date, primarily due to motion artifacts (Power et al., 2019) and/or different scanning machines. However, structural studies employing diffusion tensor imaging (DTI) (Solso et al., 2016) and functional MRI (fMRI) (Clements et al., 2018) have enhanced our understanding of how altered neural circuits relate to the clinical syndrome of autism (Ecker et al., 2012; Langen et al., 2014). MRI studies support the heterogeneity of autism, demonstrating various subgroups with different neurobiological alterations to explain the symptomatology. Longitudinal studies with multiple brain MRIs of infants at high risk of developing autism during their first 2 years of life have detailed the structural changes associated with autism (Wolff et al., 2012). The current studies suggest the presence of disruption of neural pathways prior to the emergence of behavioral symptoms in autism which might help regarding the underlying mechanisms. The data from MRI studies has revealed differences in the neurobiology between young children diagnosed with autism and those without, specifically, differences in cortical thickness decrease in regions involving language, social cognition, and behavioral control (Smith et al., 2016). A major obstacle in using neuroimaging with MRI as a reliable biomarker in patients with autism is reproducibility (Uddin et al., 2017).

Language production and comprehension are the core features affected in autism. fMRI studies investigating circuits for language have demonstrated hyper-activation of the superior temporal gyrus and inferior frontal gyrus as well as hypo-activation of the bilateral middle temporal gyri (Emerson et al., 2017; Herringshaw et al., 2016).

In addition, these studies have identified correlates for challenges in processing emotions shown by faces and the "social brain" and deficits in attention (Herringshaw et al., 2016).

As research in this area progresses, MRI studies could be well suited to categorize subgroups of autism (Lombardo et al., 2015) as well as in differentiating from other neurodevelopmental anomalies (Carlisi et al., 2017). The use of MRI as a biomarker in response to treatments and for purposes of longitudinal tracking currently remains in its infancy.

Electrophysiological Studies

Traditionally, EEG has been used to diagnose comorbid epilepsy in patients with ASD (Levisohn, 2007) although it can also be used to study its mechanisms. EEGs have the benefit of being less expensive than MRIs, making the study of brain dynamics on a smaller timescale more economically feasible. Like EEG, magnetoencephalography (MEG) is noninvasive and records the activity of the brain surface; however, MEG provides higher spatial resolution than EEG. Both have been used to explore brain connectivity in persons with ASD, noting a more random connectivity pattern overall (O'Reilly et al., 2017).

EEG has revealed alterations in oscillatory activity in resting state in autism patients, with more slow waves, less alpha waves, and less intra- and interhemispheric asymmetry than in normal controls (Cantor et al., 1986). Parsing out endophenotypes of ASD using spatiospectral analyses to map out trajectories of EEG in infants is an application in development (Lefebvre et al., 2018; Tierney et al., 2012). Other studies focusing on mechanisms have used task-based modulation of cognitive faculties – examples being low-level perception and action observation – in people with ASD. A feature of ASD involves failure to mirror an observed action of another person (Chan & Han, 2020) which was based on altered mu wave suppression in autism (Oberman et al., 2005) previously but later was questioned empirically (Bernier et al., 2013). This implies a more complex picture of impaired executive functions and visual attention (Dumas et al., 2014). Additional studies have shown modulation of sensory processing in people with autism with observed changes in sensitivities and latency (Marco et al., 2011). It is also important to note here that interactive tasks that encompass real-time social interaction would allow study of brain activity in experimental contexts that would be more relevant to autistic symptoms than passive tasks which are used for most functional imaging studies.

Management

The management of patients with ASD involves multiple approaches including behavioral and educational therapies apart from addressing comorbidities (Table 16.3). For educational purposes, agencies in the United States, including the Centers for Disease Control and Prevention, the National Institutes of Health, and private foundations and groups (e.g., Autism Speaks, Autism Society of America, and American Academy of Pediatrics), have multiple resources

Table 16.3 Neurological management of ASD/DD

Neurological interventions
Neurological evaluation
Family education
Management of epilepsy
Antiseizure medications
Vagus nerve stimulator
Ketogenic diet
Other interventions
Steroids
IVIG (intravenous immune globulin)
Refer to medical home team for treatment
ASD – ABA, speech/language, OT, PT, etc.
Comorbid conditions – ID, mental health, etc.
Sleep hygiene
Future directions
Syndromic autism – Fragile X syndrome, Rett syndrome, Angelman syndrome, Prader-Willi syndrome
New avenues including gene editing
CRISPR/Cas9 modality
ASOs

Table 16.4 The neurologist's clinical approach to ASD

Detailed history	Detailed physical examination	Further testing
Birth history Age of parents at birth Perinatal risk factors Pregnancy complications Comorbid conditions Epilepsy/seizures Intermittent explosive disorder Motor impairments Developmental history Family history Metabolic disorders ADHD/learning disorders	Complete neurological exam (esp. checking for UMN (upper motor neuron) signs) Dysmorphic features Growth parameters Neurocutaneous examination Evaluate for social communication deficits and restrictive/repetitive domains	MRI (magnetic resonance imaging) Electroencephalography Video EEG (to evaluate seizure – like activities, that is, motor stereotypies and tics) Serum and urine testing for metabolic disorders Audiologic testing Genetic: Chromosomal microarray Fragile X WES – whole exome sequencing WGS – whole genome sequencing Metabolic disorder testing if warranted

to help the patients and their families affected with ASD. It is widely recognized that intensive early interventions for children with ASD lead to improved long-term outcomes and greater skill development (Dawson et al., 2010; Reichow, 2012). Applied behavior analysis (ABA) is currently considered the gold standard treatment for ASD based on learning skills and repetition and reinforcement to obviate maladaptive behaviors. ABA can be used to improve cognition, adaptive behaviors, communication skills, and socialization (Dawson et al., 2010; Reichow, 2012; Schreibman et al., 2015).

Neurologists assist with assessment and management of the motor impairments and epilepsy associated with ASD. Additionally, as discussed earlier, diagnosis and identification of neurogenetic syndromes associated with ASD can be managed with the assistance of a neurologist. For example, the continuous spikes and waves during sleep (CSWS) can masquerade as ASD (Tuchman, 2009). Treatment with high-dose nightly diazepam has been used for CSWS (Sanchez Fernandez et al., 2014). In addition, Landau-Kleffner syndrome (LKS), characterized by epileptic aphasia, clinical seizures, and abnormal EEG findings, can be misdiagnosed as autistic behavior. It is imperative to have clinical suspicion for LKS considering the imperative to prevent epileptic regression and start therapy including ASMs, nightly benzodiazepines, and corticosteroids (Frye et al., 2013; Tuchman, 2009). For epilepsy/seizures in patients with ASD, ASMs are prescribed by neurologists. Levetiracetam is often avoided in autistic patients with psychiatric comorbidities due to an association with depression and thoughts or self-harm. Depakote and Lamotrigine are more commonly used in persons with ASD (Frye et al., 2013).

Psychiatric and behavioral comorbidities are common with ASD. Cognitive behavioral therapy (CBT) has been shown to be beneficial for anxiety in high-functioning children with autism (Danial & Wood, 2013). Atypical antipsychotics, including risperidone and aripiprazole, have been FDA approved for treating aggression and irritability in children with ASD (Baumer & Spence, 2018). For ADHD, non-stimulant medications including norepinephrine reuptake inhibitors (atomoxetine) and α_2-adrenergic agonists (guanfacine and clonidine) are commonly used (Jain et al., 2011; Jeste, 2015; Kratochvil et al., 2002; Sallee et al., 2009). For anxiety or OCD (obsessive-compulsive disorder) coexisting with ASD, selective serotonin reuptake inhibitors (SSRIs) have been used with success (Baumer &

Spence, 2018). Note though, particular attention needs to be given in using tricyclic antidepressants (TCAs) and bupropion in patients with ASD, given their propensity to decrease the seizure threshold. For a more nuanced discussion of the role of the psychiatrist in caring for children with ASD and IDD, see Chap. 19 in this volume.

Various other treatments including dietary modifications, vitamin supplementation, acupuncture, chiropractic, chelation, hyperbaric oxygen, and immunologic agents have been used in ASD patients although inadequate scientific evidence exists to support these treatments (Levy & Hyman, 2015). Previously, there had been controversy regarding vaccinations and increased risk for ASD. Subsequent studies and CDC guidelines unequivocally negate any links between vaccination and ASD risk (Institute of Medicine, 2004).

Future Directions

With recent advancements in the genetic toolkit, neurologists are poised to develop new therapeutic strategies for ASD particularly for children with monogenic autism. Antisense oligonucleotides (ASOs) have recently been used for early-onset neurological disorders like spinal muscular atrophy (Mendell et al., 2017). For ASD resulting from high-confidence risk genes, ASOs may provide hope. Similarly, CRISPR/Cas9 as gene-editing technology may offer new approaches to treatment as our knowledge of ASD expands, providing more specific targets (Zhao et al., 2018). As science unravels these mysteries, neurologists will have new tools for the evaluation and treatment of ASD and other NDD, advancing the ability of the medical home team to enhance the quality of life for children.

References

Aman, M. G., Van Bourgondien, M. E., Wolford, P. L., & Sarphare, G. (1995). Psychotropic and anticonvulsant drugs in subjects with autism: Prevalence and patterns of use. *Journal of the American Academy of Child and Adolescent Psychiatry, 34*(12), 1672–1681. https://doi.org/10.1097/00004583-199512000-00018

American Psychiatric Association, & American Psychiatric Association. DSM-5 Task Force. (2013). *Diagnostic and statistical manual of mental disorders: DSM-5* (5th ed.). American Psychiatric Association. http://www.PsychiatryOnline.org

Amiet, C., Gourfinkel-An, I., Bouzamondo, A., Tordjman, S., Baulac, M., Lechat, P., Mottron, L., & Cohen, D. (2008). Epilepsy in autism is associated with intellectual disability and gender: Evidence from a meta-analysis. *Biological Psychiatry, 64*(7), 577–582. https://doi.org/10.1016/j.biopsych.2008.04.030

Ashwal, S., Russman, B. S., Blasco, P. A., Miller, G., Sandler, A., Shevell, M., & Stevenson, R. (2004). Practice parameter: Diagnostic assessment of the child with cerebral palsy: Report of the Quality Standards Subcommittee of the American Academy of Neurology and the Practice Committee of the Child Neurology Society. *Neurology, 62*(6), 851–863. https://doi.org/10.1212/01.wnl.0000117981.35364.1b

Asperger, H., & Frith, U. T. (1991). "Autistic psychopathy" in childhood. In U. Frith (Ed.), *Autism and Asperger syndrome* (p. 37). Cambridge University Press.

Baumer, N., & Spence, S. J. (2018). Evaluation and management of the child with autism spectrum disorder. *Continuum (Minneap Minn), 24*(1, Child Neurology), 248–275. https://doi.org/10.1212/CON.0000000000000578

Berg, A. T., Tarquinio, D., & Koh, S. (2017). Early life epilepsies are a comorbidity of developmental brain disorders. *Seminars in Pediatric Neurology, 24*(4), 251–263. https://doi.org/10.1016/j.spen.2017.10.008

Bernier, R., Aaronson, B., & McPartland, J. (2013). The role of imitation in the observed heterogeneity in EEG mu rhythm in autism and typical development. *Brain and Cognition, 82*(1), 69–75. https://doi.org/10.1016/j.bandc.2013.02.008

Blank, R., Barnett, A. L., Cairney, J., Green, D., Kirby, A., Polatajko, H., Rosenblum, S., Smits-Engelsman, B., Sugden, D., Wilson, P., & Vinçon, S. (2019). International clinical practice recommendations on the definition, diagnosis, assessment, intervention, and psychosocial aspects of developmental coordination disorder. *Developmental Medicine and Child Neurology, 61*(3), 242–285. https://doi.org/10.1111/dmcn.14132

Bolton, P. F., Carcani-Rathwell, I., Hutton, J., Goode, S., Howlin, P., & Rutter, M. (2011). Epilepsy in autism: Features and correlates. *The British Journal of Psychiatry, 198*(4), 289–294. https://doi.org/10.1192/bjp.bp.109.076877

Bourgeron, T. (2007). The possible interplay of synaptic and clock genes in autism spectrum disorders. *Cold Spring Harbor Symposia on Quantitative Biology, 72*, 645–654. https://doi.org/10.1101/sqb.2007.72.020

Bourgeron, T. (2016). Current knowledge on the genetics of autism and propositions for future research.

Comptes Rendus Biologies, 339(7–8), 300–307. https://doi.org/10.1016/j.crvi.2016.05.004

Bruininks, R. H., & Bruininks, B. D. (2005). *BOT2: Bruininks-Oseretsky test of motor proficiency*. AGS Publishing.

Canitano, R. (2007). Epilepsy in autism spectrum disorders. *European Child & Adolescent Psychiatry, 16*(1), 61–66. https://doi.org/10.1007/s00787-006-0563-2

Cantor, D. S., Thatcher, R. W., Hrybyk, M., & Kaye, H. (1986). Computerized EEG analyses of autistic children. *Journal of Autism and Developmental Disorders, 16*(2), 169–187. https://doi.org/10.1007/BF01531728

Carlisi, C. O., Norman, L., Murphy, C. M., Christakou, A., Chantiluke, K., Giampietro, V., Simmons, A., Brammer, M., Murphy, D. G., Consortium, M. A., Mataix-Cols, D., & Rubia, K. (2017). Disorder-specific and shared brain abnormalities during vigilance in autism and obsessive-compulsive disorder. *Biological Psychiatry: Cognitive Neuroscience and Neuroimaging, 2*(8), 644–654. https://doi.org/10.1016/j.bpsc.2016.12.005

Carmassi, C., Palagini, L., Caruso, D., Masci, I., Nobili, L., Vita, A., & Dell'Osso, L. (2019). Systematic review of sleep disturbances and circadian sleep desynchronization in autism spectrum disorder: Toward an integrative model of a self-reinforcing loop. *Frontiers in Psychiatry, 10*, 366.

Casanova, M. F., El-Baz, A. S., Kamat, S. S., Dombroski, B. A., Khalifa, F., Elnakib, A., Soliman, A., Allison-McNutt, A., & Switala, A. E. (2013). Focal cortical dysplasias in autism spectrum disorders. *Acta Neuropathologica Communications, 1*, 67. https://doi.org/10.1186/2051-5960-1-67

Chan, M. M. Y., & Han, Y. M. Y. (2020). Differential mirror neuron system (MNS) activation during action observation with and without social-emotional components in autism: A meta-analysis of neuroimaging studies. *Molecular Autism, 11*(1), 72. https://doi.org/10.1186/s13229-020-00374-x

Chebli, S. S., Martin, V., & Lanovaz, M. J. (2016). Prevalence of stereotypy in individuals with developmental disabilities: A systematic review. *Review Journal of Autism and Developmental Disorders, 3*(2), 107–118.

Christensen, D. L., Baio, J., Van Naarden Braun, K., Bilder, D., Charles, J., Constantino, J. N., Daniels, J., Durkin, M. S., Fitzgerald, R. T., Kurzius-Spencer, M., Lee, L. C., Pettygrove, S., Robinson, C., Schulz, E., Wells, C., Wingate, M. S., Zahorodny, W., Yeargin-Allsopp, M., & Centers for Disease Control and Prevention. (2016). Prevalence and characteristics of autism spectrum disorder among children aged 8 years--autism and developmental disabilities monitoring network, 11 sites, United States, 2012. *MMWR Surveillance Summaries, 65*(3), 1–23. https://doi.org/10.15585/mmwr.ss6503a1

Clements, C. C., Zoltowski, A. R., Yankowitz, L. D., Yerys, B. E., Schultz, R. T., & Herrington, J. D. (2018). Evaluation of the social motivation hypothesis of autism: A systematic review and meta-analysis. *JAMA Psychiatry, 75*(8), 797–808. https://doi.org/10.1001/jamapsychiatry.2018.1100

Danial, J. T., & Wood, J. J. (2013). Cognitive behavioral therapy for children with autism: Review and considerations for future research. *Journal of Developmental and Behavioral Pediatrics, 34*(9), 702–715. https://doi.org/10.1097/DBP.0b013e31829f676c

Dawson, G., Rogers, S., Munson, J., Smith, M., Winter, J., Greenson, J., Donaldson, A., & Varley, J. (2010). Randomized, controlled trial of an intervention for toddlers with autism: The early start Denver model. *Pediatrics, 125*(1), e17–e23. https://doi.org/10.1542/peds.2009-0958

Deonna, T., & Roulet-Perez, E. (2010). Early-onset acquired epileptic aphasia (Landau-Kleffner syndrome, LKS) and regressive autistic disorders with epileptic EEG abnormalities: The continuing debate. *Brain & Development, 32*(9), 746–752. https://doi.org/10.1016/j.braindev.2010.06.011

Didden, R., Sturmey, P., Sigafoos, J., Lang, R., O'Reilly, M. F., & Lancioni, G. E. (2012). Nature, prevalence, and characteristics of challenging behavior. In *Functional assessment for challenging behaviors* (pp. 25–44). Springer.

Dowell, L. R., Mahone, E. M., & Mostofsky, S. H. (2009). Associations of postural knowledge and basic motor skill with dyspraxia in autism: Implication for abnormalities in distributed connectivity and motor learning. *Neuropsychology, 23*(5), 563–570. https://doi.org/10.1037/a0015640

Dumas, G., Soussignan, R., Hugueville, L., Martinerie, J., & Nadel, J. (2014). Revisiting mu suppression in autism spectrum disorder. *Brain Research, 1585*, 108–119. https://doi.org/10.1016/j.brainres.2014.08.035

Dziuk, M. A., Gidley Larson, J. C., Apostu, A., Mahone, E. M., Denckla, M. B., & Mostofsky, S. H. (2007). Dyspraxia in autism: Association with motor, social, and communicative deficits. *Developmental Medicine and Child Neurology, 49*(10), 734–739. https://doi.org/10.1111/j.1469-8749.2007.00734.x

Ecker, C., Suckling, J., Deoni, S. C., Lombardo, M. V., Bullmore, E. T., Baron-Cohen, S., Catani, M., Jezzard, P., Barnes, A., Bailey, A. J., Williams, S. C., Murphy, D. G., & MRC AIMS Consortium. (2012). Brain anatomy and its relationship to behavior in adults with autism spectrum disorder: A multicenter magnetic resonance imaging study. *Archives of General Psychiatry, 69*(2), 195–209. https://doi.org/10.1001/archgenpsychiatry.2011.1251

Edwards, J., Berube, M., Erlandson, K., Haug, S., Johnstone, H., Meagher, M., Sarkodee-Adoo, S., & Zwicker, J. G. (2011). Developmental coordination disorder in school-aged children born very preterm and/or at very low birth weight: A systematic review. *Journal of Developmental & Behavioral Pediatrics, 32*(9), 678–687.

Ekinci, O., Arman, A. R., Isik, U., Bez, Y., & Berkem, M. (2010). EEG abnormalities and epilepsy in autistic spectrum disorders: Clinical and familial correlates.

Epilepsy & Behavior, 17(2), 178–182. https://doi.org/10.1016/j.yebeh.2009.11.014

Emerson, R. W., Adams, C., Nishino, T., Hazlett, H. C., Wolff, J. J., Zwaigenbaum, L., Constantino, J. N., Shen, M. D., Swanson, M. R., Elison, J. T., Kandala, S., Estes, A. M., Botteron, K. N., Collins, L., Dager, S. R., Evans, A. C., Gerig, G., Gu, H., McKinstry, R. C., … Piven, J. (2017). Functional neuroimaging of high-risk 6-month-old infants predicts a diagnosis of autism at 24 months of age. *Science Translational Medicine, 9*(393), eaag2882. https://doi.org/10.1126/scitranslmed.aag2882

Esbensen, A. J., & Schwichtenberg, A. J. (2016). Sleep in neurodevelopmental disorders. *International Review of Research in Developmental Disabilities, 51*, 153–191. https://doi.org/10.1016/bs.irrdd.2016.07.005

Esposito, G., Venuti, P., Apicella, F., & Muratori, F. (2011). Analysis of unsupported gait in toddlers with autism. *Brain & Development, 33*(5), 367–373. https://doi.org/10.1016/j.braindev.2010.07.006

Fiolita, D., Suwarba, I. G. N. M., & Mahalini, D. S. (2020). Characteristics of epilepsy in children with cerebral palsy. *Clinical Neurology and Neuroscience, 4*(2), 29.

Fletcher, F. E., Knowland, V., Walker, S., Gaskell, M. G., Norbury, C., & Henderson, L. M. (2020). Atypicalities in sleep and semantic consolidation in autism, *23*(3), e12906. https://doi.org/10.1111/desc.12906

Fournier, K. A., Hass, C. J., Naik, S. K., Lodha, N., & Cauraugh, J. H. (2010). Motor coordination in autism spectrum disorders: A synthesis and meta-analysis. *Journal of Autism and Developmental Disorders, 40*(10), 1227–1240. https://doi.org/10.1007/s10803-010-0981-3

Frye, R. E., Sreenivasula, S., & Adams, J. B. (2011). Traditional and non-traditional treatments for autism spectrum disorder with seizures: An on-line survey. *BMC Pediatrics, 11*, 37. https://doi.org/10.1186/1471-2431-11-37

Frye, R. E., Rossignol, D., Casanova, M. F., Brown, G. L., Martin, V., Edelson, S., Coben, R., Lewine, J., Slattery, J. C., Lau, C., Hardy, P., Fatemi, S. H., Folsom, T. D., Macfabe, D., & Adams, J. B. (2013). A review of traditional and novel treatments for seizures in autism spectrum disorder: Findings from a systematic review and expert panel. *Frontiers in Public Health, 1*, 31. https://doi.org/10.3389/fpubh.2013.00031

Gernsbacher, M. A., Sauer, E. A., Geye, H. M., Schweigert, E. K., & Hill Goldsmith, H. (2008). Infant and toddler oral- and manual-motor skills predict later speech fluency in autism. *Journal of Child Psychology and Psychiatry, 49*(1), 43–50. https://doi.org/10.1111/j.1469-7610.2007.01820.x

Geschwind, D. H. (2009). Advances in autism. *Annual Review of Medicine, 60*, 367–380. https://doi.org/10.1146/annurev.med.60.053107.121225

Henderson, S. E. (1992). *Movement assessment battery for children*. The Psychological Corporation.

Herringshaw, A. J., Ammons, C. J., DeRamus, T. P., & Kana, R. K. (2016). Hemispheric differences in language processing in autism spectrum disorders: A meta-analysis of neuroimaging studies. *Autism Research, 9*(10), 1046–1057. https://doi.org/10.1002/aur.1599

Hoffmire, C. A., Magyar, C. I., Connolly, H. V., Fernandez, I. D., & van Wijngaarden, E. (2014). High prevalence of sleep disorders and associated comorbidities in a community sample of children with Down syndrome. *Journal of Clinical Sleep Medicine, 10*(4), 411–419. https://doi.org/10.5664/jcsm.3618

Hyman, S. L., Levy, S. E., Myers, S. M., & Council on Children With Disabilities, Section on Developmental and Behavioral Pediatrics. (2020a). Executive summary: Identification, evaluation, and management of children with autism spectrum disorder. *Pediatrics, 145*(1), e20193448. https://doi.org/10.1542/peds.2019-3448

Hyman, S. L., Levy, S. E., Myers, S. M., & Council on Children With Disabilities, Section on Developmental and Behavioral Pediatrics. (2020b). Identification, evaluation, and management of children with autism spectrum disorder. *Pediatrics, 145*(1), e20193447. https://doi.org/10.1542/peds.2019-3447

Institute of Medicine (US) Immunization Safety Review Committee. (2004). *Immunization safety review: vaccines and autism*. The National Academies Press. https://doi.org/10.17226/10997

Ismail, F. Y., & Shapiro, B. K. (2019). What are neurodevelopmental disorders? *Current Opinion in Neurology, 32*(4), 611–616. https://doi.org/10.1097/WCO.0000000000000710

Iverson, J. M., & Wozniak, R. H. (2007). Variation in vocal-motor development in infant siblings of children with autism. *Journal of Autism and Developmental Disorders, 37*(1), 158–170. https://doi.org/10.1007/s10803-006-0339-z

Jain, R., Segal, S., Kollins, S. H., & Khayrallah, M. (2011). Clonidine extended-release tablets for pediatric patients with attention-deficit/hyperactivity disorder. *Journal of the American Academy of Child and Adolescent Psychiatry, 50*(2), 171–179. https://doi.org/10.1016/j.jaac.2010.11.005

Jeste, S. S. (2011). The neurology of autism spectrum disorders. *Current Opinion in Neurology, 24*(2), 132–139. https://doi.org/10.1097/WCO.0b013e3283446450

Jeste, S. S. (2015). Neurodevelopmental behavioral and cognitive disorders. *Continuum (Minneap Minn), 21*(3 Behavioral Neurology and Neuropsychiatry), 690–714. https://doi.org/10.1212/01.CON.0000466661.89908.3c

Jeste, S. S., Sahin, M., Bolton, P., Ploubidis, G. B., & Humphrey, A. (2008). Characterization of autism in young children with tuberous sclerosis complex. *Journal of Child Neurology, 23*(5), 520–525. https://doi.org/10.1177/0883073807309788

Johnson, C. P., Myers, S. M., & American Academy of Pediatrics Council on Children With Disabilities. (2007). Identification and evaluation of children with autism spectrum disorders. *Pediatrics, 120*(5), 1183–1215. https://doi.org/10.1542/peds.2007-2361

Kanner, L. (1943). Autistic disturbances of affective contact. *Nervous Child, 2*(3), 217–250.

Karatoprak, E., Sözen, G., & Saltık, S. (2019). Risk factors associated with epilepsy development in children with cerebral palsy. *Child's Nervous System, 35*(7), 1181–1187.

Kopp, S., Beckung, E., & Gillberg, C. (2010). Developmental coordination disorder and other motor control problems in girls with autism spectrum disorder and/or attention-deficit/hyperactivity disorder. *Research in Developmental Disabilities, 31*(2), 350–361.

Kratochvil, C. J., Heiligenstein, J. H., Dittmann, R., Spencer, T. J., Biederman, J., Wernicke, J., Newcorn, J. H., Casat, C., Milton, D., & Michelson, D. (2002). Atomoxetine and methylphenidate treatment in children with ADHD: A prospective, randomized, open-label trial. *Journal of the American Academy of Child and Adolescent Psychiatry, 41*(7), 776–784. https://doi.org/10.1097/00004583-200207000-00008

Landa, R., & Garrett-Mayer, E. (2006). Development in infants with autism spectrum disorders: A prospective study. *Journal of Child Psychology and Psychiatry, 47*(6), 629–638. https://doi.org/10.1111/j.1469-7610.2006.01531.x

Langen, M., Bos, D., Noordermeer, S. D., Nederveen, H., van Engeland, H., & Durston, S. (2014). Changes in the development of striatum are involved in repetitive behavior in autism. *Biological Psychiatry, 76*(5), 405–411. https://doi.org/10.1016/j.biopsych.2013.08.013

Latreille, V., Louis, E. K. S., & Pavlova, M. (2018). Co-morbid sleep disorders and epilepsy: A narrative review and case examples. *Epilepsy Research, 145*, 185–197.

Lefebvre, A., Delorme, R., Delanoe, C., Amsellem, F., Beggiato, A., Germanaud, D., Bourgeron, T., Toro, R., & Dumas, G. (2018). Alpha waves as a neuromarker of autism spectrum disorder: The challenge of reproducibility and heterogeneity. *Frontiers in Neuroscience, 12*, 662. https://doi.org/10.3389/fnins.2018.00662

Levisohn, P. M. (2007). The autism-epilepsy connection. *Epilepsia, 48*(Suppl 9), 33–35. https://doi.org/10.1111/j.1528-1167.2007.01399.x

Levy, S. E., & Hyman, S. L. (2015). Complementary and alternative medicine treatments for children with autism spectrum disorders. *Child and Adolescent Psychiatric Clinics of North America, 24*(1), 117–143. https://doi.org/10.1016/j.chc.2014.09.004

Levy, R. G., Cooper, P. N., & Giri, P. (2012). Ketogenic diet and other dietary treatments for epilepsy. *Cochrane Database of Systematic Reviews*, (3), CD001903. https://doi.org/10.1002/14651858.CD001903.pub2

Lombardo, M. V., Pierce, K., Eyler, L. T., Carter Barnes, C., Ahrens-Barbeau, C., Solso, S., Campbell, K., & Courchesne, E. (2015). Different functional neural substrates for good and poor language outcome in autism. *Neuron, 86*(2), 567–577. https://doi.org/10.1016/j.neuron.2015.03.023

Lukmanji, S., Manji, S. A., Kadhim, S., Sauro, K. M., Wirrell, E. C., Kwon, C. S., & Jette, N. (2019). The co-occurrence of epilepsy and autism: A systematic review. *Epilepsy and Behavior, 98*(Pt A), 238–248. https://doi.org/10.1016/j.yebeh.2019.07.037

Marco, E. J., Hinkley, L. B., Hill, S. S., & Nagarajan, S. S. (2011). Sensory processing in autism: A review of neurophysiologic findings. *Pediatric Research, 69*(5 Pt 2), 48R–54R. https://doi.org/10.1203/PDR.0b013e3182130c54

Matsuo, M., Maeda, T., Sasaki, K., Ishii, K., & Hamasaki, Y. (2010). Frequent association of autism spectrum disorder in patients with childhood onset epilepsy. *Brain & Development, 32*(9), 759–763. https://doi.org/10.1016/j.braindev.2010.05.005

Mazurek, M. O., & Sohl, K. (2016). Sleep and behavioral problems in children with autism spectrum disorder. *Journal of Autism and Developmental Disorders, 46*(6), 1906–1915. https://doi.org/10.1007/s10803-016-2723-7

McDuffie, A., Yoder, P., & Stone, W. (2005). Prelinguistic predictors of vocabulary in young children with autism spectrum disorders. *Journal of Speech, Language, and Hearing Research, 48*(5), 1080–1097. https://doi.org/10.1044/1092-4388(2005/075)

McKenna, H., & Reiss, I. K. M. (2018). The case for a chronobiological approach to neonatal care. *Early Human Development, 126*, 1–5. https://doi.org/10.1016/j.earlhumdev.2018.08.012

Melo, C., Ruano, L., Jorge, J., Pinto Ribeiro, T., Oliveira, G., Azevedo, L., & Temudo, T. (2020). Prevalence and determinants of motor stereotypies in autism spectrum disorder: A systematic review and meta-analysis. *Autism, 24*(3), 569–590.

Mendell, J. R., Al-Zaidy, S., Shell, R., Arnold, W. D., Rodino-Klapac, L. R., Prior, T. W., Lowes, L., Alfano, L., Berry, K., Church, K., Kissel, J. T., Nagendran, S., L'Italien, J., Sproule, D. M., Wells, C., Cardenas, J. A., Heitzer, M. D., Kaspar, A., Corcoran, S., … Kaspar, B. K. (2017). Single-dose gene-replacement therapy for spinal muscular atrophy. *The New England Journal of Medicine, 377*(18), 1713–1722. https://doi.org/10.1056/NEJMoa1706198

Miles, J. H., Takahashi, T. N., Bagby, S., Sahota, P. K., Vaslow, D. F., Wang, C. H., Hillman, R. E., & Farmer, J. E. (2005). Essential versus complex autism: Definition of fundamental prognostic subtypes. *American Journal of Medical Genetics. Part A, 135*(2), 171–180. https://doi.org/10.1002/ajmg.a.30590

Mosconi, M. W., & Sweeney, J. A. (2015). Sensorimotor dysfunctions as primary features of autism spectrum disorders. *Science China. Life Sciences, 58*(10), 1016–1023. https://doi.org/10.1007/s11427-015-4894-4

Moss, J., Oliver, C., Arron, K., Burbidge, C., & Berg, K. (2009). The prevalence and phenomenology of repetitive behavior in genetic syndromes. *Journal of Autism and Developmental Disorders, 39*(4), 572–588.

Mostofsky, S. H., & Ewen, J. B. (2011). Altered connectivity and action model formation in autism is autism. *The Neuroscientist, 17*(4), 437–448. https://doi.org/10.1177/1073858410392381

Nicholas, B., Rudrasingham, V., Nash, S., Kirov, G., Owen, M. J., & Wimpory, D. C. (2007). Association of Per1 and Npas2 with autistic disorder: Support for the clock genes/social timing hypothesis. *Molecular Psychiatry, 12*(6), 581–592. https://doi.org/10.1038/sj.mp.4001953

Nickels, K. C., Zaccariello, M. J., Hamiwka, L. D., & Wirrell, E. C. (2016). Cognitive and neurodevelopmental comorbidities in paediatric epilepsy. *Nature Reviews Neurology, 12*(8), 465.

Noritz, G. H., & Murphy, N. A. (2013). Motor delays: Early identification and evaluation. *Pediatrics, 131*(6), e2016. https://doi.org/10.1542/peds.2013-1056

Nyhan, W. L. (1976). Behavior in the Lesch-Nyhan syndrome. *Journal of Autism and Childhood Schizophrenia, 6*(3), 235–252.

O'Reilly, C., Lewis, J. D., & Elsabbagh, M. (2017). Is functional brain connectivity atypical in autism? A systematic review of EEG and MEG studies. *PLoS One, 12*(5), e0175870. https://doi.org/10.1371/journal.pone.0175870

Oberman, L. M., Hubbard, E. M., McCleery, J. P., Altschuler, E. L., Ramachandran, V. S., & Pineda, J. A. (2005). EEG evidence for mirror neuron dysfunction in autism spectrum disorders. *Brain Research. Cognitive Brain Research, 24*(2), 190–198. https://doi.org/10.1016/j.cogbrainres.2005.01.014

Ozonoff, S., Young, G. S., Goldring, S., Greiss-Hess, L., Herrera, A. M., Steele, J., Macari, S., Hepburn, S., & Rogers, S. J. (2008). Gross motor development, movement abnormalities, and early identification of autism. *Journal of Autism and Developmental Disorders, 38*(4), 644–656. https://doi.org/10.1007/s10803-007-0430-0

Pacheva, I., Ivanov, I., Yordanova, R., Gaberova, K., Galabova, F., Panova, M., Petkova, A., Timova, E., & Sotkova, I. (2019). Epilepsy in children with autistic spectrum disorder. *Children (Basel), 6*(2), 15. https://doi.org/10.3390/children6020015

Park, Y. D. (2003). The effects of vagus nerve stimulation therapy on patients with intractable seizures and either Landau-Kleffner syndrome or autism. *Epilepsy & Behavior, 4*(3), 286–290. https://doi.org/10.1016/s1525-5050(03)00080-5

Parmeggiani, A., Barcia, G., Posar, A., Raimondi, E., Santucci, M., & Scaduto, M. C. (2010). Epilepsy and EEG paroxysmal abnormalities in autism spectrum disorders. *Brain & Development, 32*(9), 783–789. https://doi.org/10.1016/j.braindev.2010.07.003

Pavone, P., Incorpora, G., Fiumara, A., Parano, E., Trifiletti, R. R., & Ruggieri, M. (2004). Epilepsy is not a prominent feature of primary autism. *Neuropediatrics, 35*(4), 207–210. https://doi.org/10.1055/s-2004-821079

Power, J. D., Silver, B. M., Silverman, M. R., Ajodan, E. L., Bos, D. J., & Jones, R. M. (2019). Customized head molds reduce motion during resting state fMRI scans. *NeuroImage, 189*, 141–149. https://doi.org/10.1016/j.neuroimage.2019.01.016

Provost, B., Lopez, B. R., & Heimerl, S. (2007). A comparison of motor delays in young children: Autism spectrum disorder, developmental delay, and developmental concerns. *Journal of Autism and Developmental Disorders, 37*(2), 321–328. https://doi.org/10.1007/s10803-006-0170-6

Reichow, B. (2012). Overview of meta-analyses on early intensive behavioral intervention for young children with autism spectrum disorders. *Journal of Autism and Developmental Disorders, 42*(4), 512–520. https://doi.org/10.1007/s10803-011-1218-9

Robertson, J., Hatton, C., Emerson, E., & Baines, S. (2015). Prevalence of epilepsy among people with intellectual disabilities: A systematic review. *Seizure, 29*, 46–62. https://doi.org/10.1016/j.seizure.2015.03.016

Sallee, F. R., McGough, J., Wigal, T., Donahue, J., Lyne, A., Biederman, J., & SPD503 Study Group. (2009). Guanfacine extended release in children and adolescents with attention-deficit/hyperactivity disorder: A placebo-controlled trial. *Journal of the American Academy of Child and Adolescent Psychiatry, 48*(2), 155–165. https://doi.org/10.1097/CHI.0b013e318191769e

Salpekar, J. (2018). Neuropsychiatric effects of epilepsy in developmental disorders. *Current Opinion in Psychiatry, 31*(2), 109–115. https://doi.org/10.1097/YCO.0000000000000392

Sanchez Fernandez, I., Chapman, K., Peters, J. M., Klehm, J., Jackson, M. C., Berg, A. T., & Loddenkemper, T. (2014). Treatment for continuous spikes and waves during sleep (CSWS): Survey on treatment choices in North America. *Epilepsia, 55*(7), 1099–1108. https://doi.org/10.1111/epi.12678

Sanchez Fernandez, I., Loddenkemper, T., Gainza-Lein, M., Sheidley, B. R., & Poduri, A. (2019). Diagnostic yield of genetic tests in epilepsy: A meta-analysis and cost-effectiveness study. *Neurology, 92*(5), e418–e428. https://doi.org/10.1212/WNL.0000000000006850

Schreibman, L., Dawson, G., Stahmer, A. C., Landa, R., Rogers, S. J., McGee, G. G., Kasari, C., Ingersoll, B., Kaiser, A. P., Bruinsma, Y., McNerney, E., Wetherby, A., & Halladay, A. (2015). Naturalistic developmental behavioral interventions: Empirically validated treatments for autism spectrum disorder. *Journal of Autism and Developmental Disorders, 45*(8), 2411–2428. https://doi.org/10.1007/s10803-015-2407-8

Shendure, J., Mitra, R. D., Varma, C., & Church, G. M. (2004). Advanced sequencing technologies: Methods and goals. *Nature Reviews. Genetics, 5*(5), 335–344. https://doi.org/10.1038/nrg1325

Smith, E., Thurm, A., Greenstein, D., Farmer, C., Swedo, S., Giedd, J., & Raznahan, A. (2016). Cortical thickness change in autism during early childhood. *Human Brain Mapping, 37*(7), 2616–2629. https://doi.org/10.1002/hbm.23195

Solso, S., Xu, R., Proudfoot, J., Hagler, D. J., Jr., Campbell, K., Venkatraman, V., Carter Barnes, C., Ahrens-Barbeau, C., Pierce, K., Dale, A., Eyler, L., & Courchesne, E. (2016). Diffusion tensor imaging provides evidence of possible axonal overconnectivity in frontal lobes in autism spectrum disorder toddlers.

Biological Psychiatry, 79(8), 676–684. https://doi.org/10.1016/j.biopsych.2015.06.029

Srivastava, S., Clark, B., Landy-Schmitt, C., Offermann, E. A., Kline, A. D., Wilkinson, S. T., & Grados, M. A. (2020). Repetitive and self-injurious behaviors in children with Cornelia de Lange syndrome. *Journal of Autism and Developmental Disorders, 51*(5), 1748–1758.

Srivastava, S., & Sahin, M. (2017). Autism spectrum disorder and epileptic encephalopathy: common causes, many questions. *Journal of Neurodevelopmental Disorders, 9*, 23–23. https://doi.org/10.1186/s11689-017-9202-0

Stores, G. (2014). *Sleep and its disorders in children and adolescents with a neurodevelopmental disorder: A review and clinical guide*. Cambridge University Press.

Surtees, A. D. R., Oliver, C., Jones, C. A., Evans, D. L., & Richards, C. (2018). Sleep duration and sleep quality in people with and without intellectual disability: A meta-analysis. *Sleep Medicine Reviews, 40*, 135–150. https://doi.org/10.1016/j.smrv.2017.11.003

Swanson, M. R., & Hazlett, H. C. (2019). White matter as a monitoring biomarker for neurodevelopmental disorder intervention studies. *Journal of Neurodevelopmental Disorders, 11*(1), 33.

Tharp, B. R. (2004). Epileptic encephalopathies and their relationship to developmental disorders: Do spikes cause autism? *Mental Retardation and Developmental Disabilities Research Reviews, 10*(2), 132–134. https://doi.org/10.1002/mrdd.20025

Thomas, S., Hovinga, M. E., Rai, D., & Lee, B. K. (2017). Brief report: Prevalence of co-occurring epilepsy and autism spectrum disorder: The US National Survey of Children's Health 2011–2012. *Journal of Autism and Developmental Disorders, 47*(1), 224–229.

Thurm, A., Tierney, E., Farmer, C., Albert, P., Joseph, L., Swedo, S., Bianconi, S., Bukelis, I., Wheeler, C., & Sarphare, G. (2016). Development, behavior, and biomarker characterization of Smith-Lemli-Opitz syndrome: An update. *Journal of Neurodevelopmental Disorders, 8*(1), 12.

Tierney, A. L., Gabard-Durnam, L., Vogel-Farley, V., Tager-Flusberg, H., & Nelson, C. A. (2012). Developmental trajectories of resting EEG power: An endophenotype of autism spectrum disorder. *PLoS One, 7*(6), e39127. https://doi.org/10.1371/journal.pone.0039127

Tordjman, S., Anderson, G. M., Pichard, N., Charbuy, H., & Touitou, Y. (2005). Nocturnal excretion of 6-sulphatoxymelatonin in children and adolescents with autistic disorder. *Biological Psychiatry, 57*(2), 134–138. https://doi.org/10.1016/j.biopsych.2004.11.003

Tuchman, R. (2009). CSWS-related autistic regression versus autistic regression without CSWS. *Epilepsia, 50*(Suppl 7), 18–20. https://doi.org/10.1111/j.1528-1167.2009.02212.x

Tuchman, R., & Rapin, I. (2002). Epilepsy in autism. *Lancet Neurology, 1*(6), 352–358. https://doi.org/10.1016/s1474-4422(02)00160-6

Uddin, L. Q., Dajani, D. R., Voorhies, W., Bednarz, H., & Kana, R. K. (2017). Progress and roadblocks in the search for brain-based biomarkers of autism and attention-deficit/hyperactivity disorder. *Translational Psychiatry, 7*(8), e1218. https://doi.org/10.1038/tp.2017.164

Van Waelvelde, H., Oostra, A., Dewitte, G., Van Den Broeck, C., & Jongmans, M. J. (2010). Stability of motor problems in young children with or at risk of autism spectrum disorders, ADHD, and or developmental coordination disorder. *Developmental Medicine and Child Neurology, 52*(8), e174–e178. https://doi.org/10.1111/j.1469-8749.2009.03606.x

Varcin, K. J., & Nelson Iii, C. A. (2016). A developmental neuroscience approach to the search for biomarkers in autism spectrum disorder. *Current Opinion in Neurology, 29*(2), 123.

Wolff, J. J., Gu, H., Gerig, G., Elison, J. T., Styner, M., Gouttard, S., Botteron, K. N., Dager, S. R., Dawson, G., Estes, A. M., Evans, A. C., Hazlett, H. C., Kostopoulos, P., McKinstry, R. C., Paterson, S. J., Schultz, R. T., Zwaigenbaum, L., Piven, J., & Network, I. (2012). Differences in white matter fiber tract development present from 6 to 24 months in infants with autism. *The American Journal of Psychiatry, 169*(6), 589–600. https://doi.org/10.1176/appi.ajp.2011.11091447

Yasuhara, A. (2010). Correlation between EEG abnormalities and symptoms of autism spectrum disorder (ASD). *Brain & Development, 32*(10), 791–798. https://doi.org/10.1016/j.braindev.2010.08.010

Zhao, H., Jiang, Y. H., & Zhang, Y. Q. (2018). Modeling autism in non-human primates: Opportunities and challenges. *Autism Research, 11*(5), 686–694. https://doi.org/10.1002/aur.1945

17 The Management of Chronic Pain in Children with Autism and Developmental Disability

Shawn E. McNeil, Jennifer Stanfill, Devin S. Reed, Chikezie N. Okeagu, Andrew P. Bourgeois, Elyse M. Cornett, and Alan D. Kaye

Abstract

Pain management for the general adult population can present its challenges. However, pain management for children with autism spectrum disorder (ASD) or developmental delay (DD) offers a unique set of obstacles. The first challenge comes in assessing the child's pain due to both communication difficulties and self-reports. The second difficulty presents in treating pain; children with ASD or DD have different coping skills and might benefit from techniques like hypno-analgesia and multisensory environments (MSE). Moreover, there are very few studies and limited research on effective pain management practices for children with ASD or DD. This chapter will focus on children with ASD/DD and discuss epidemiology, pain management, coordination of care, and challenges in treating this population.

Keywords

Autism · Treatment planning · Mental health · Psychologist · Physician · Therapist · Care-coordination · Neurodevelopmental disorder · Developmental disorder · Pain management

Introduction

Pain in children and teens is common and often underdiagnosed. A child's pain may present as headaches, abdominal pain, and chronic musculoskeletal and joint pain (Friedrichsdorf et al., 2016). While pain is prevalent in this age group, the focus of most pain management physicians and training programs involves adults and seniors (Mathews, 2011). This discrepancy may exist because there is very little hard evidence of how pain is experienced, perceived, and best managed

S. E. McNeil
Department of Psychiatry and Behavioral Medicine, LSU Health Shreveport - School of Medicine, Shreveport, LA, USA

J. Stanfill · D. S. Reed · C. N. Okeagu
Department of Anesthesiology, Louisiana State University Health Sciences Center, New Orleans, LA, USA

A. P. Bourgeois · E. M. Cornett (✉)
Department of Anesthesiology, Louisiana State University Health Sciences Center, Shreveport, Shreveport, LA, USA
e-mail: elyse.bradley@lsuhs.edu

A. D. Kaye
Department of Anesthesiology, Louisiana State University Health Sciences Center, New Orleans, LA, USA

Department of Anesthesiology, Louisiana State University Health Sciences Center, Shreveport, Shreveport, LA, USA

P. McPherson (ed.), *Handbook of Treatment Planning for Children with Autism and Other Neurodevelopmental Disorders*, Autism and Child Psychopathology Series,
https://doi.org/10.1007/978-3-031-06120-2_17

in children, particularly those with complex disorders like autism spectrum disorder (ASD) and developmental delay (DD). Children with intellectual or developmental disorders have higher rates of comorbidities, leading to higher rates of acute and chronic pain, and these children are frequently undertreated for pain when healthcare providers either don't know how to measure their pain or don't understand that some nonverbal actions may be manifestations of pain (Baldridge, 2010). Clearly, pain management specialists are vital medical home team members for children with ASD/DD who experience pain.

Pain is subjective and highly individualized, with assessment guided by patient report. Self-report is difficult for children who often lack the vocabulary, articulation, and/or understanding to describe their pain to caregivers. Complicating the assessment, most pain scales rely on self-report which is often difficult for children with ASD or DD. In addition, children may respond to pain in ways that are overlooked by caregivers, including being quiet or reserved, even disengaged, or resorting to violent behaviors. While children lack the vocabulary to describe their pain, children can still respond to pain with consistent, noticeable physiological signs such as tachycardia, sweating, wincing, weeping, jerking, and muscle tension (Mathews, 2011). Such signs allow the pain specialist to assess a child's pain and implement a treatment plan for pain management. The management of pain in children is critical because, if left unaddressed, chronic pain can significantly affect all domains of a child's development. If pain persists, it can cause emotional and psychological scars to develop that can affect future choices and overall health as they grow (Mathews, 2011). Since children of the same age group may differ widely in their perception and tolerance of pain, the assessment and management of pain requires that the treating clinician be skilled in assessing a child's level of development and understand the implications of ASD/DD for the assessment and management of pain (Mathews, 2011). In this chapter, we will discuss the assessment and management of pain in children with a focus on the challenges presented by the treatment of children with ASD/DD.

Among children in the United States, 14% have special healthcare needs that may complicate the assessment and treatment of their medical conditions. These children account for almost half of pediatric healthcare costs (McSpadden et al., 2012). These figures highlight the need for practitioners' appropriate training and a practical approach for delivering treatment to these children. Healthcare workers need the education and tools to appropriately assess children with special needs and access to a team that can support the treatment process. The importance of additional training for medical providers in pediatric pain is widely recognized (Bhandari et al., 2019). Training for all providers treating children is critical, as 40% of children and adolescents complain of pain at least once a week, and chronic pain affects at least 15–20% of children. Additionally, one and a half million children undergo surgery each year and in up to 20% of these cases the pain may progress and become chronic (Mathews, 2011). In this chapter we will discuss pain in specific developmental disorders, epidemiology of children with ASD/DD, characterization of pain in children, management of pain in children with ASD/DD, coordination of care, and challenges associated with pain treatment in children with ASD/DD.

Pain in Specific Developmental Disorders

Developmental disorders is a term used to categorize a diverse group of medical conditions that impair one or more developmental domains and begin early in development with the expectation that functional impairment will be lifelong. An average of 70% of children with developmental disorders experience chronic pain (Barney et al., 2020) with up to 90% of persons with cerebral palsy (CP) reporting chronic pain (Svedberg et al., 2008). The management of chronic pain in persons with developmental disorders is critical to maximizing participation in activities of daily living and promoting healthy development (Temple et al., 2012).

Autism Spectrum Disorder

ASD is a neurodevelopmental disorder that occurs at an early age. Under the DSM 5, ASD is characterized by a functional deficiency in social communication, minimal interests and repetitive behaviors, and sensory hypersensitivity (Taghizadeh et al., 2015). Children present early in life with deficits in information-processing skills, difficulties coping with stress, and have atypical cognitive profiles (Vlassakova & Emmanouil, 2016). Common symptoms included marked impairment in nonverbal behaviors such as eye-to-eye gaze, facial expression, and body postures and a loss of interest in social interaction (Sharma et al., 2018).

The prevalence of ASD has been steadily increasing over the past two decades, with current estimates reaching up to 1 in 36 children, with a four-to-five fold higher affinity for boys over girls, and this ratio is thought to be the same across all racial, ethnic, and socioeconomic backgrounds (Sharma et al., 2018). The prevalence of ASD in the United States was first reported in 2002, and at that time, it was estimated to be 66%. In 2012, the Centers for Disease Control and Prevention (CDC) estimated that approximately 1.5% of US children aged 8 had ASD. The DSM-5, released in 2013, addressed the variation in diagnostic approaches to ASD with the introduction of a "spectrum" of ASD. In the DSM-5, the DSM-IV diagnoses of pervasive developmental disorder (PDD) diagnoses, including autistic disorder, Asperger's disorder, childhood disintegrative disorder, and pervasive developmental disorder not otherwise specified (PDD-NOS), are now all classified as ASD (Hodges et al., 2020). Since the DSM-5 was released relatively recently, much of the epidemiologic data available comprises, at least partially, ASD and DD diagnoses that were made using older criteria. This further complicates attempts to estimate prevalence accurately. Much of the increase in incidence is attributed to improved identification owing to increased awareness and changing practices such as universal screening of children aged 18–24 months. As such, milder cases of ASD account for a sizeable portion of this increase, with far less dramatic changes in estimates of ASD with co-occurring intellectual disability (Lyall et al., 2017; Zablotsky et al., 2019).

The 2016–17 National Survey of Children's Health reported that parents of children with ASD were almost twice to report their child experienced frequent or chronic pain as compared to the reports of parents of typically developing (TD) peers. The Survey found frequent chronic pain in 15.6% of children with ASD compared to 8.2% for TD children, and if children with ASD had at least one developmental comorbidity, frequent or chronic pain was reported in 19.9% (Whitney & Shapiro, 2019). For children with ASD, the pain specialist must consider deficits in social interaction and atypical physical or social feedback in the assessment of pain and augment traditional assessment models with accommodations to improve reliability. Children with ASD may have trouble processing information reflected in the facial expressions of others. As such, this deficit may significantly interfere with the ability to use traditional pain assessment tools which rely on the recognition of facial expressions (Ely et al., 2016). Social-communicative deficits may explain why children with ASD seem to demonstrate a less outward reaction to pain or pleasure than other children. This finding correlates with several studies showing that children with ASD may have a reduced sensitivity to pain stimuli (Ely et al., 2016). Interestingly, other reports have cited children with ASD have heightened pain sensitivity as well (Clarke, 2015). Atypical pain thresholds are a recognized feature of the sensory hypo- and hyper-responsiveness associated with ASD (Clarke, 2015). For this reason, it has been documented that autistic individuals may deny pain but describe noxious stimuli as "discomfort" or simply "hurt." Pain specialists have highlighted the need for additional research on the management of pain for children with ASD (Whitney & Shapiro, 2019).

Cerebral Palsy

CP is a developmental disorder caused by abnormal fetal brain development or perinatal brain injury that are nonprogressive and results in impairment of movement, balance and/or posture. CP occurs in 1 in 345 children in the United States (CDC, 2020). Children with CP may experience pain for many reasons, including spasticity, muscle spasm, scoliosis, and hip dislocation (Massaro et al., 2013). In addition, pain often interferes with sleep, which in turn exacerbates pain (Stores, 2014). Children with CP most commonly report lower extremity pain that increases during adolescence and progresses with age (Mckinnon et al., 2019). Functional impairment may lead to a sedentary lifestyle that promotes obesity, which may further exacerbate joint pain (Wakefield et al., 2018). Studies have highlighted the negative impact of pain on overall quality of life for children with CP (Mckinnon et al., 2019).

Pain in Other Developmental Disorders

Children with intellectual disability experience an increased risk of pain with reduced access to treatment. The prevalence of pain may be increased for many reasons including increased risk of injury and co-morbid conditions (Doody & Bailey, 2019). Specific conditions associated with intellectual disability and pain include Down Syndrome (DS), fragile X spectrum disorder (FXSD), and neurofibromatosis Type 1 (NF1). In general, children with DS are more sensitive to pain than TD peers but slower to register pain (McGuire & Defrin, 2015). Fragile X spectrum disorder (FXSD) is associated abnormal pain processing that appears to be associated with the Fragile X mental retardation gene (Mei et al., 2020). While some individuals with FXSD may be less responsive to pain, the FXMR gene has been associated with an increased fibromyalgia and chronic muscle pain in females (Leehey et al., 2011). The prevalence of pain in persons with NF1 ranges from 29% to 70% and may be associated with neurofibromas, scoliosis, gastrointestinal tumors, or other conditions.

Overview of Pain in Children with ASD/DD

Since the dawn of medicine, pain causation and treatment has been at the forefront of medical research and widely discussed. While physician's through the millennia have sought to alleviate pain, the management of chronic pain as a medical specialty was only recognized in 1976 (Loeser & Schatman, 2017). For pain medicine board certification, a 1-year specialization fellowship is completed after general residency training, with most fellowships associated with an anesthesiology program. While pediatric pain management is a recognized need, it is not a recognized specialty (Bhandari et al., 2019). Physicians specializing in the treatment of pain recognize the biopsychosocial model and the value of an interdisciplinary approach to pain management. The International Association for the Study of Pain defines pain as *an unpleasant sensory and emotional experience associated with, or resembling that associated with, actual or potential tissue damage*, recognizing that elements the experience of pain can be unique to the individual (Raja et al., 2020, p. 1976). While great steps have been taken over the past centuries to better understand the mechanisms underlying pain and its perception, the mystery of pain lies in the unique experience of the individual. The mysteries of pain management continue to challenge the great medical minds of today, and pain remains one of the most misunderstood, underdiagnosed, and/or understood medical problems in children (Mathews, 2011).

Modern theories of pain date to 1965, when Melzack and Wall proposed the *gate control theory* of pain which introduced body–mind components of pain and revolutionized pain management (Katz & Rosenbloom, 2015). The fact that perception of pain may vary among individuals is further explained by the neuromatrix theory of pain. Melzack's neuromatrix theory of pain recognizes an individual specific pain neurosignature with multiple determinants, including genetics and stressors (Melzack & Katz, 2007). The biopsychosocial theory of pain expands on the neuromatrix theory, recognizing that the experience of pain extends beyond the biomedical model and includes the contributions of social determinants of health (Gatchel et al., 2007). Recently an enactive theory of pain, the 5E Process, broadens the biopsychosocial theory to encompass fully encompass the individual's experience of pain. The 5E Process hypothesizes that an individual's internal and external environments create the individual's experience of pain (Stilwell & Harman, 2019). See Table 17.1.

The classification of pain is complex and evolving. While no theory fully explains the experience of pain, Pain may be classified as nociceptive or neuropathic. Nociceptive pain may be somatic or visceral and occurs when receptors are stimulated and cause the transmission of an electrical signal along a nerve fiber to the central nervous system and ultimately to the brain, where each individual may uniquely perceive the stimuli (Yaksh, 2007). Visceral nociceptive pain in children presents most often as gastroenteritis or appendicitis (Leung & Sigalet, 2003). Sources of somatic nociceptive pain include developmental impairments or postoperative pain. Neuropathic pain is *pain caused by a lesion or disease of the somatosensory system* (Colloca et al., 2017, p. 1). Neuropathic pain is rare in children. While some syndromes closely mimic those seen in adults, the prevalence and course of the disorder can differ considerably in children, depending on development and external influences (Walco et al., 2010). Children have often been undertreated for both nociceptive and neuropathic pain due to the inability to verbalize pain and the challenges associated with assessing pediatric pain.

Table 17.1 Modern pain theories, the body–mind perspective and beyond

Gate control theory[a]	The transmission of a signal from stimulus to the brain is mediated by a gate-keeping function within the spinal cord which modulates nociceptive and nonnociceptive signals.
Neuromatrix theory[b]	The perception of pain, or neurosignature, is the active product of the subjective experience of the body-self neuromatrix of the brain.
Biopsychosocial theory[c]	A holistic approach acknowledging the roles of biological, psychological, and social factors in the perception and management of pain.
5E process[d]	Embodied (the physical body and the individual's experience of their body), embedded (in social and environmental contexts), enactive (dynamic process), emotive (affective framing and making sense), extended (making sense of the brain-body-environment connections).

[a]Melzack and Wall (1965), [b]Melzack (2001), [c]Loeser (1982), [d]Adapted from Stilwell and Harman (2019)

Children may be referred to a pain specialist for the management of acute or chronic pain. Acute pain is a common, adverse stimulus experienced by children. Acute pain may present as gastrointestinal distress, headache, or muscle strain and may be due to injury or illness. If not treated effectively, acute pain can have long-term physical or psychological sequelae such as anticipatory anxiety, lowered future pain thresholds, and reduced analgesic efficacy, translating to increased analgesic requirements (Wong et al., 2012a). Traditionally, pain that lasts less than 3 months is considered acute pain, but there is not a specific time frame that separates acute and chronic pain. According to The American Pain Society Task Force on Pediatrics, chronic pain has biopsychosocial underpinnings, may be episodic or ongoing, and may be the result of a specific medical condition or pain may be the disorder (Friedrichsdorf et al., 2016). While chronic pain is commonly encountered in pediatric practices and carries significant caregiver burden, the majority of children reporting chronic pain may experience impairment but are not disabled by it (Liossi & Howard, 2016). Of concern, are the significant number of children with repeated acute nociceptive pain episodes who develop chronic pain because of an underlying medical condition, also known as *chronic-on-acute pain* (Friedrichsdorf et al., 2016). Addressing pediatric chronic pain is critical to promote health development and to minimize the possibility of adult chronic pain (Hassett et al., 2013).

The Assessment and Management of Pain in Children with ASD/DD

The Assessment of Pain in Children with ASD/DD

Assessing pain is the first step in being able to manage pain in any population. Accurate pain assessment is essential and requires innovation and individualized considerations to facilitate the provision of quality care. Office visits for a patient with ASD/DD can be overwhelming – the lights, music, forms, and new interactions present many unique hurdles. Primary care providers and parents should alert the specialist of necessary accommodations prior to the consultation. Recommendations for the assessment and treatment of pain in children with significant impairment include scrutiny of the communication of pain with words or behaviors and broad consideration of the possible sources of pain (Hauer & Houtrow, 2017).

The assessment of pain typically requires questions to characterize the location, intensity, quality, duration, course, and exacerbating or ameliorating factors. Caregivers of these children may find it difficult to ascertain the level of pain the child is experiencing. Some individuals even believe that children with ASD are not as sensitive to pain, which may affect the caregiver's perception of that child's pain (Messmer et al., 2008). This highlights the role of educating parents, teachers, and other caregivers in recognizing when a child with special needs is experiencing pain. It also emphasizes the utility of a more comprehensive way to assess pain in children, such as the socio-communicative model by Dr. Kenneth Craig and his team, which helps evaluate the way a child communicates their pain (Messmer et al., 2008).

Several studies have shown that self-report is the most accurate way to determine acute pain intensity (Zabalia, 2013). Self-report requires the subject to use cognitive processes such as logic of classes and order relations (Zabalia, 2013). In other words, self-reporting requires being able to categorize, sort, and rank items. These skills are typically acquired in children over the age of seven with normal cognitive abilities. Children with ASD/DD often lack the skills to accurately self-report. In addition, children may struggle with temporal relationships. Children with ASD/DD often have difficulty with expressive and receptive communication; so even simple medical history questions such as "Where does it hurt?" can be confusing to the child. Patients with ASD/DD often perceive and communicate their pain differently. Their descriptive vocabulary may not align with common descriptors of pain quality. For example, a child may use "hot" to indicate any type of pain. Therefore, the clinician should employ simplistic language and use terms familiar to the patient. Pediatric patients with neurologic or developmental impairment present unique challenges in evaluating the presence or severity of pain, as they often cannot self-report their pain. Each child may display a unique cluster of responses and behaviors. This unique yet variable expression necessitates input from a consistent care provider with knowledge of the patient's baseline behavior patterns and response to painful events. Behaviors associated with pain include vocalizations such as crying or moaning, grimaced facial expressions, increased movement or stiffened tone, or the inability to be consoled.

The pain management specialist typically employs scales to augment the traditional self-report assessment. While self-reporting is the preferred method for pain assessment, practitioners may implement several clinically sensitive assessment tools to assist the diagnosis. Patients with ASD have been shown to reliably understand the seriation and graded responses necessary to complete such scales (Ely et al., 2016). Analog pain scales, or informant-report scales, completed with help from a caregiver, can aid in the assessment. If necessary, the tools may be adapted to accommodate the needs of the child. For instance, using a simplified version of the scales, such as limiting the face numbers to 4 options rather than 10 with Faces Pain Scale-Revised (FPS-R) and enlarged body parts on the Eland scale, improves the accuracy of self-reporting in these children (Zabalia, 2013). One tool, the Premature Infant Pain Profile (PIPP), consists of behavioral facial actions, physiologic indicators of heart rate and oxygen saturation, and contextual age variables (Witt et al., 2016). The FLACC (Face, Legs, Activity, Cry, Consolability) scale incorporates five behavioral components scored 0–2 for a maximal pain score of 10 (American Physical Therapy Association, 2020). While a daily caregiver's expertise cannot be replaced, assessment tools such as the FLACC scale, Non-communicating children's pain checklist–postoperative version (NCPC–PV), or Individualized Numeric Rating Scale (INRS) supplement pain assessment and are useful in monitoring pain throughout treatment (American Physical Therapy Association, 2020). The INRS offers the added benefit of being customizable to each child by adding specific behaviors they feature during atypical pain. The caveat in using these methods is that modifying a proven effective scale changes the effectiveness and makes it difficult to standardize the tools. With the limited data comparing tools, it is recommended for each institution to adopt its own standardized guidelines for routine assessment of pediatric pain.

Other than self-report, pain can be assessed in children using psychophysical assessment techniques like Quantitative Sensory Testing (QST). This psychophysical assessment requires cognitive skills like comprehension of abstract concepts, following instructions, and differentiating between stimuli (Barney et al., 2020). This type of assessment is based on reaction time, which might falsely elevate their pain threshold if slowed in a child without those cognitive skills. As an alternative assessment in children with ASD or DD, a modified QST (mQST) may be used. It compares a range of behavioral reactivity in a standardized way. However, while an mQST can establish pain sensitivity and sensory function, it will not establish a pain threshold (Barney et al., 2020).

Another option to assess pain in this population is for caregivers and/or healthcare providers, to gauge the subject's pain behaviors, called proxy-reporting. Pain behaviors are best evaluated by long-term caregivers, who can discern the individual's 'regular' behaviors versus 'pain' behaviors. Pain behaviors are facial and bodily responses to pain, including moaning, irritability, seeking comfort, a furrowed brow, and being uncooperative (Barney et al., 2020). Assessing these pain behaviors can then be used to create a personal profile for the individual to be used in the hospital and clinic setting to describe his/her common signs of pain. Caregivers and healthcare providers can also use the Visual Analog Scale (VAS). The issue is that the VAS is supposed to be used as a self-report method, but when used to judge someone else's pain, it merely becomes an estimate (Zabalia, 2013). Each of these pain assessment tools has its shortcomings. To overcome these flaws, the clinician should strive to use all of these tools, self-report with FPS-R, psychophysical methods like mQST, and proxy-reporting, thus providing a more complete and reliable assessment of pain in this unique patient population (Barney et al., 2020). While some caregivers may be astutely aware of their child's pain, other caregivers may find it difficult to ascertain the level of pain that the child is experiencing. Some individuals even believe that children with ASD are not as sensitive to pain, which may affect the caregiver's perception of that child's pain (Messmer et al., 2008). This highlights the role of educating parents, teachers, and other caregivers in recognizing when a child with special needs is experiencing pain. It also emphasizes the utility of a more comprehensive way to assess pain in children, such as the socio-communicative model by Dr. Kenneth Craig and his team, which helps evaluate the way a child communicates their pain (Messmer et al., 2008).

The assessment of pain includes a physical examination guided by the history and scales. While the physical examination of acute pain may be focal, chronic pain requires more thorough examination. Observation for subtle indications of pain as the child is positioned on the examination table or palpated should be noted as altered sensitivity is a common feature of ASD/DD. Heightened and reduced sensitivities to pain are common. Neuroscience studies show mammalian emotional feelings are organized within the primitive subcortical regions of the brain, making it unsurprising that patients with neurodevelopmental disorders may experience unusual pain types (*Autism Spectrum Disorder and Amplified Pain*, n.d.). Tactile sensitivity, or allodynia, may require clinicians to alter how they perform routine examinations, adapt to the patient's preferences, and explain each small step. Posture, trigger points, spasms, and physical limitation to range of motion should be noted.

The Management of Pain in Children with ASD/DD

The management goals of pediatric pain are to prevent, reduce, and control the pain and minimize suffering. This can be done by treating the underlying source or addressing related symptoms such as anxiety or distress (Birnie et al., 2018; Gai et al., 2020). The three P's of pain management: psychological, physical, and pharmacological methods of treatment can be used to treat children, teens, and adults (Sickkids Staff, 2020). Providing effective pain management to children with ASD and DD should involve a multidisciplinary approach that takes into account the physical and psychological needs of the child. Treatment may include pharmacological and nonpharmacological elements that are combined in an integrative approach. While the pharmacological options may seem straightforward, some nonpharmacological treatments may not be evident to practitioners outside of a multidisciplinary setting.

Regular pain assessment and cognitive therapies are helpful. Families and caregivers can assist in nonpharmacological treatments, which are particularly useful in reducing stress and anxiety. Distraction, relaxation, and massage have all been implemented to significantly reduce pain (Birnie et al., 2018). Pharmacologic therapy is often needed in conjunction with nonpharmacologic intervention to achieve desired pain relief or procedural success. Many developmentally disabled individuals also present with congenital anomalies or have underlying organ dysfunction. Developmentally disabled individuals have high incidences of gastroesophageal reflex, self-injury, and chronic pain, among many other comorbidities – however, exact prevalence is unestablished (May & Kennedy, 2010). Multimodal regimens are still necessary – focusing on treating underlying etiologies – however, children with intellectual disabilities often require less weight-based dosing than other children of similar size (Valkenburg et al., 2015).

There are additional concerns in children with ASD and DD due to the sensory challenges that these individuals face. The perception of sensory stimuli, including pain, among these children varies significantly from those without these conditions, often resulting in exaggerated sensory perception and a potentially magnified response to pain. This perception usually affects the child's mood as well as their behavior. Disturbances in behavior, including agitation, should be assessed and treated appropriately, including referral to an occupational therapist for assessment of sensory integration dysfunction. An additional consideration is that children with "medical complexity," including those with ASD and DD, frequently struggle to accurately localize the pain or describe details that may be medically relevant (Morse et al., 2020).

Pharmacological Management

Pharmacologic treatment management should incorporate multimodal analgesic therapy of both nonopioid and opioid analgesics in a stepwise fashion. Oral analgesics should be used when possible to avoid painful administration routes. The World Health Organization (WHO) developed a schematic ladder tool to help guide physicians in appropriate analgesic selection (Ventafridda et al., 1985). This was modified by the Hospital for Sick Children in Toronto,

Canada, with mild pain being treated with nonopioids and moderate to severe pain being treated with a combination of nonopioids and opioid agents (Wong et al., 2012b, p. 2). Nonopioid agents such as acetaminophen and nonsteroidal anti-inflammatory drugs are the most common drugs used in adults and children. Despite offering low analgesic potency and ceiling effect, nonopioids can exert an "opioid-sparing" effect by reducing opioid consumption and minimizing adverse side effects. The choice of opioid is based upon intensity and duration of pain, with oral morphine being the most commonly used in children due to its relative safety (Greco & Berde, 2005). It is recommended to avoid codeine and tramadol in children due to variability in metabolism resulting in fatal overdose (Greco & Berde, 2005). Common side effects of opioids include pruritis and constipation, which should proactively be treated to prevent additional child discomfort. It can be challenging to handle escalating pain in a pediatric patient. Patient-controlled analgesia, opioid equivalence dosing if altering opioids, and potentially considering pain consultation are all viable options.

An empirical trial of gabapentin can be initiated if no treatable cause of pain is readily identified, beneficially targeting visceral hyperalgesia and central neuropathic pain. Other agents, such as tricyclic antidepressants, clonidine, and opioids, can provide further comfort. Care must be taken that the analgesic regimen does not interact with a patient's home medication regimen. Developmentally disabled kids often are on antiepileptics that are potent inducers of the CYP 450 enzymes (Perucca, 2006). Comprehensive treatment of children with ASD or DD can be challenging, but regular assessment and input from caregivers can help guide a successful analgesic regimen. See Table 17.2.

Table 17.2 Pain assessment tools for children

Assessment tool	Description
Adolescent Pediatric Pain Tool (APPT)[a]	Self-report by children of age 8 and older assessing pain intensity, pattern, timing, and quality.
Face, Legs, Activity, Cry, Consolability scale (FLACC)[b]	FLACC scale is a measurement used to assess pain for children between the ages of 2 months and 7 years or individuals that are unable to communicate their pain. The scale is scored in a range of 0–10 with 0 representing no pain.
Faces Pain Scale – Revised (FPS-R)[c]	Pain intensity self-report by children 4 and older. Freely available and widely translated.
Non-communicating Children's Pain Checklist-Postoperative Version (NCPC-PV)[d]	Recommended for children and youth 18 years of age or younger. The scale involves a checklist of behaviors that a caregiver can assess that the child has experienced within the last 2 hours. A number is circled for each behavior with 0 = not at all, 1 = just a little 2 = fairly often, 3 = very often, NA-non applicable. Behaviors are categorized into vocal, social, facial, activity, physiological, and eating/sleeping.
Individualized Numeric Rating Scale (INRS)[e]	Is based solely on the child's individual pain indicators as described by parents and caregivers. It is an adaptation of the numeric rating scale thatincorporates the parents' (and/or caregivers) descriptions of their individual child's past and current responses to pain. Once described, the responses are then stratified on a scale from 0 to 10.
Wong-Baker FACES Pain Scale (WBFPS)[f]	Pain intensity self-report tool for children of age 3 and older. Freely available.

[a]Savedra et al. (1989), [b]American Physical Therapy Association (2020), [c]Bieri et al. (1990), [d]Breau et al. (2002), [e]Solodiuk et al. (2010), [f]Wong and Baker (1988)

Physical Management of Pain

In pain management practice, there are several physical techniques a clinician might use, including the use of multisensory environments (MSEs), physical therapy, thermal therapy, and exercise. While the concept varies, most MSEs have a similar theme: dimly lit spaces, relaxing music, aromatic oils, comfortable seating, and padded floors. The idea is to create a "gentle sensory stimulation that produces a calm and engrossing atmosphere (Breslin et al., 2020a)." The benefits of MSEs include relaxation of the patient, pain management, improved attention, and reduced maladaptive behaviors. In a study that examined the within-session effect of MSE on children with intellectual disabilities during a dental procedure, there was increased cooperation, decreased physiological distress, decreased duration and magnitude of discomfort, and decreased use of physical restraints (Breslin et al., 2020b). In another study, dentists reported cooperation in children with ASD using only a CD player, projector, and weighted body-hugging wrap (Cermak et al., 2015). The limitations of these studies include cost and the variety of sensory stimulation threshold among children with DD. The cost of care was shown to increase in these studies, but it is unclear whether the cost increased due to the longer duration of procedures or the MSE tools' cost.

Psychological Management of Pain

Nonpharmacological interventions may also seek to enhance a child's coping abilities and their overall emotional well-being. Generally, these strategies will target the patient's cognitions, emotions, behaviors, and sensory experiences. Aside from the psychological effect, many of these nonpharmacological interventions seek to change neural circuits that affect the cognitive pain response, thus providing long-term results (Wren et al., 2019). Collaboration with a child psychiatrist or psychologist may aid in assessing the family dynamics that underlie the perception of the child's pain. In addressing the mental health needs of pediatric patients in pain, various therapies may be employed. These range from traditional psychotherapies like cognitive behavioral therapy (CBT) and mindfulness techniques to other nontraditional modalities, such as music therapy, acupuncture, massage, and medical hypnosis. Integration of a child psychiatrist, psychologist or "child life specialist" as a member of the treatment team may help address the mental health concerns present (Wren et al., 2019).

The general population responds to pain using coping skills. Coping is an adaptive learned process to deal with stress and pain. The most effective coping skills in dealing with pain are active strategies like a distraction, as opposed to passive strategies like avoidance or seeking social support (Zabalia, 2013). Having ASD/DD may limit the diversity of coping skills in children. Multiple studies have shown children with ASD/DD tend to choose passive strategies like seeking social support. Children with CP are more likely to rely on social supports for pain management compared to typically developing (TD) peers (Chaleat-Valayer et al., 2019). Therefore, caregivers should be asked to identify current coping strategies, with successful strategies being reinforced and suggestions made on modifying strategies that are not successful.

Hypno-analgesia has proven to be a useful pain management tool. In a study of adolescents aged 13–20 with intellectual disability (mental age 4–10), patients employed a technique that consisted of thinking of a favorite place or hobby or listening to a simple but realistic story as a pain management strategy (Zabalia, 2013). These emotion-focused strategies are shown to reduce the anxiety caused by pain due to an uncontrollable event, such as iatrogenic pain or procedural pains. Children with ASD/DD typically are unable to utilize strategies like distracting themselves independently; however, if there is a caregiver present, the emotion-focused strategies are effective in children with ASD/DD with mental age greater than 4 years old. Virtual reality and mobile apps have shown promise for pain management (Ahmadpour et al., 2020; Thurnheer et al., 2018).

Research Challenges in the Study of Pain Management in Children with ASD or DD

While there is research addressing pain management in children with ASD or DD, it is clear that much more and complete research is needed. As previously mentioned, some studies have relied on proxy-reporting, which can be influenced by the reporter (Barney et al., 2020). Also, there is a need to identify valid outcome measures to use in treatment trials. The quality of evidence supporting observational pain scales differs. There is a need to recognize that results may vary according to verbal or nonverbal children with ASD or DD or between different degrees of intellectual disability. Adequately powered studies with sufficient sample sizes are another research challenge in studying pain management in children with ASD or DD. One solution could be to collaborate across countries and networks or create registers to systematically collect data on pain in these populations (Barney et al., 2020). Finally, further research of the pain mechanisms involved with children with ASD or DD is needed. One solution might be repurposing an existing assessment mechanism like mQST to understand better the physiological aspect of pain in children with ASD or DD.

Conclusion

Studying pain in children, particularly children with developmental disabilities, is challenging, as many of these children typically lack social and situational awareness, processing skills, and verbal capacity to reliably communicate and characterize their pain. Providing adequate pain relief to children with ASD/DD is best accomplished through a multidisciplinary approach that acknowledges the child's physical and psychological needs. Treatment can include pharmacological, physical, and psychological elements coordinated in an integrative approach. Primary care providers may refer the management of acute or chronic pain to the specialist when there are questions regarding pharmacological management, interventional pain management is required, or to assist to coordinate a pain management treatment team. Since many children who are special needs rely on caregivers, educating parents, teachers, and other caregivers in recognizing when a child with special needs is experiencing pain is of the utmost importance. Healthcare providers should work with the patient and the family or caregiver to help develop the best possible individualized treatment plan for each patient.

References

Ahmadpour, N., Keep, M., Janssen, A., Rouf, A. S., & Marthick, M. (2020). Design strategies for virtual reality interventions for managing pain and anxiety in children and adolescents: Scoping review. *JMIR Serious Games, 8*(1), e14565.

American Physical Therapy Association. (2020). *Face, legs, activity, cry, consolability (FLACC) behavioral pain scale*. APTA. https://www.apta.org/patient-care/evidence-based-practice-resources/test-measures/face-legs-activity-cry-consolability-flacc-behavioral-pain-scale

Autism Spectrum Disorder and Amplified Pain. (n.d.). Retrieved November 30, 2020, from https://www.hindawi.com/journals/crips/2015/930874/

Baldridge. (2010). *Figu 1. 110*(12).

Barney, C. C., Andersen, R. D., Defrin, R., Genik, L. M., McGuire, B. E., & Symons, F. J. (2020). Challenges in pain assessment and management among individuals with intellectual and developmental disabilities. *PAIN Reports, 5*(4), e821. https://doi.org/10.1097/pr9.0000000000000822

Bhandari, R. P., Goddard, J., Campbell, F., Sangster, M., & Stevens, B. (2019). Becoming a pediatric pain specialist: Training opportunities to advance the science and practice of pediatric pain treatment. *Pediatric Pain Letter, 21*(1), 10.

Bieri, D., Reeve, R. A., Champion, G. D., Addicoat, L., & Ziegler, J. B. (1990). The Faces Pain Scale for the self-assessment of the severity of pain experienced by children: development, initial validation, and preliminary investigation for ratio scale properties. *Pain, 41*(2), 139–150.

Birnie, K. A., Noel, M., Chambers, C. T., Uman, L. S., & Parker, J. A. (2018). Psychological interventions for needle-related procedural pain and distress in children and adolescents. *Cochrane Database of Systematic Reviews, 2018*(10), CD005179. John Wiley and Sons Ltd. https://doi.org/10.1002/14651858.CD005179.pub4

Breau, L. M., McGrath, P. J., Camfield, C. S., & Finley, G. A. (2002). Psychometric properties of the non-communicating children's pain checklist-revised. *Pain, 99*(1–2), 349–357.

Breslin, L., Guerra, N., Ganz, L., & Ervin, D. (2020a). Clinical utility of multisensory environments for people with intellectual and developmental disabilities: A scoping review. *The American Journal of Occupational Therapy, 74*(1), 1–12. https://doi.org/10.5014/ajot.2020.037267

Breslin, L., Guerra, N., Ganz, L., & Ervin, D. (2020b). Clinical utility of multisensory environments for people with intellectual and developmental disabilities: A scoping review. *The American Journal of Occupational Therapy, 74*(1), 7401205060p1–7401205060p12. https://doi.org/10.5014/ajot.2020.037267

CDC. (2020, March 2). *Learn more about cerebral palsy (CP) | CDC*. Centers for Disease Control and Prevention. https://www.cdc.gov/ncbddd/cp/index.html

Cermak, S. A., Stein Duker, L. I., Williams, M. E., Dawson, M. E., Lane, C. J., & Polido, J. C. (2015). Sensory adapted dental environments to enhance oral care for children with autism spectrum disorders: A randomized controlled pilot study. *Journal of Autism and Developmental Disorders, 45*(9), 2876–2888. https://doi.org/10.1007/s10803-015-2450-5

Chaleat-Valayer, E., Roumenoff, F., Bard-Pondarre, R., Ganne, C., Verdun, S., Lucet, A., & Bernard, J.-C. (2019). Pain coping strategies in children with cerebral palsy. *Developmental Medicine & Child Neurology, 61*(11), 1329–1335. https://doi.org/10.1111/dmcn.14204

Clarke, C. (2015). Autism spectrum disorder and amplified pain. *Case Reports in Psychiatry, 2015*, 1–4. https://doi.org/10.1155/2015/930874

Colloca, L., Ludman, T., Bouhassira, D., Baron, R., Dickenson, A. H., Yarnitsky, D., … Finnerup, N. B. (2017). Neuropathic pain. *Nature Reviews Disease Primers, 3*(1), 1–19.

Doody, O., & Bailey, M. E. (2019). Understanding pain physiology and its application to person with intellectual disability. *Journal of Intellectual Disabilities: JOID, 23*(1), 5–18. https://doi.org/10.1177/1744629517708680

Ely, E., Chen-Lim, M. L., Carpenter, K. M., Wallhauser, E., & Friedlaender, E. (2016). Pain assessment of children with autism spectrum disorders. *Journal of Developmental and Behavioral Pediatrics, 37*(1), 53–61. https://doi.org/10.1097/DBP.0000000000000240

Friedrichsdorf, S. J., Giordano, J., Desai Dakoji, K., Warmuth, A., Daughtry, C., & Schulz, C. A. (2016). Chronic pain in children and adolescents: Diagnosis and treatment of primary pain disorders in head, abdomen, muscles and joints. *Children, 3*(4), 42. https://doi.org/10.3390/children3040042

Gai, N., Naser, B., Hanley, J., Peliowski, A., Hayes, J., & Aoyama, K. (2020). A practical guide to acute pain management in children. *Journal of Anesthesia, 34*(3), 421–433. https://doi.org/10.1007/s00540-020-02767-x

Gatchel, R. J., Peng, Y. B., Peters, M. L., Fuchs, P. N., & Turk, D. C. (2007). The biopsychosocial approach to chronic pain: Scientific advances and future directions. *Psychological Bulletin, 133*(4), 581.

Greco, C., & Berde, C. (2005). Pain management for the hospitalized pediatric patient. *Pediatric Clinics of North America, 52*(4), 995–1027. https://doi.org/10.1016/j.pcl.2005.04.005

Hassett, A. L., Hilliard, P. E., Goesling, J., Clauw, D. J., Harte, S. E., & Brummett, C. M. (2013). Reports of chronic pain in childhood and adolescence among patients at a tertiary care pain clinic. *The Journal of Pain, 14*(11), 1390–1397.

Hauer, J., & Houtrow, A. J. (2017). Pain assessment and treatment in children with significant impairment of the central nervous system. *Pediatrics, 139*(6), e20171002. https://doi.org/10.1542/peds.2017-1002

Hodges, H., Fealko, C., & Soares, N. (2020). Autism spectrum disorder: Definition, epidemiology, causes, and clinical evaluation. *Translational Pediatrics, 9*(S1), S55–S65. https://doi.org/10.21037/tp.2019.09.09

Katz, J., & Rosenbloom, B. N. (2015). The golden anniversary of Melzack and Wall's gate control theory of pain: Celebrating 50 years of pain research and management. *Pain Research and Management, 20*(6), 285–286.

Leehey, M. A., Legg, W., Tassone, F., & Hagerman, R. (2011). Fibromyalgia in fragile X mental retardation 1 gene premutation carriers. *Rheumatology (Oxford, England), 50*(12), 2233–2236. https://doi.org/10.1093/rheumatology/ker273

Leung, A. K. C., & Sigalet, D. L. (2003). Acute abdominal pain in children. *American Family Physician, 67*(11), 2321–2326.

Liossi, C., & Howard, R. F. (2016). Pediatric chronic pain: Biopsychosocial assessment and formulation. *Pediatrics, 138*(5), e20160331. https://doi.org/10.1542/peds.2016-0331

Loeser, J. D. (1982). Concepts of pain. In J. Stanton-Hicks & R. Boaz (Eds.), *Chronic low back pain*. Raven Press.

Loeser, J. D., & Schatman, M. E. (2017). Chronic pain management in medical education: A disastrous omission. *Postgraduate Medicine, 129*(3), 332–335. Taylor & Francis.

Lyall, K., Croen, L., Daniels, J., Fallin, M. D., Ladd-Acosta, C., Lee, B. K., Park, B. Y., Snyder, N. W., Schendel, D., Volk, H., Windham, G. C., & Newschaffer, C. (2017). The changing epidemiology of autism spectrum disorders. *Annual Review of Public Health, 38*, 81–102. https://doi.org/10.1146/annurev-publhealth-031816-044318

Massaro, M., Pastore, S., Ventura, A., & Barbi, E. (2013). Pain in cognitively impaired children: A focus for general pediatricians. *European Journal of Pediatrics, 172*(1), 9–14. https://doi.org/10.1007/s00431-012-1720-x

Mathews, L. (2011). Pain in children: Neglected, unaddressed and mismanaged. *Indian Journal of Palliative Care, 17*(Suppl), S70–S73. https://doi.org/10.4103/0973-1075.76247

May, M. E., & Kennedy, C. H. (2010). Health and problem behavior among people with intellectual disabilities. *Behavior Analysis in Practice, 3*(2), 4–12. https://doi.org/10.1007/BF03391759

McGuire, B. E., & Defrin, R. (2015). Pain perception in people with Down syndrome: A synthesis of clinical and experimental research. *Frontiers in Behavioral Neuroscience, 9*, 194.

Mckinnon, C. T., Meehan, E. M., Harvey, A. R., Antolovich, G. C., & Morgan, P. E. (2019). Prevalence and characteristics of pain in children and young adults with cerebral palsy: A systematic review. *Developmental Medicine and Child Neurology, 61*(3), 305–314. https://doi.org/10.1111/dmcn.14111

McSpadden, C., Therrien, M., & McEwen, I. R. (2012). Care coordination for children with special health care needs and roles for physical therapists. *Pediatric Physical Therapy, 24*(1), 70–77. https://doi.org/10.1097/PEP.0b013e31823e0686

Mei, X., Yang, Y., Zhao, J., Wang, Y., Chen, Q., Qian, X., Li, X., & Feng, Z. (2020). Role of fragile X mental retardation protein in chronic pain. *Molecular Pain, 16*, 1744806920928619. https://doi.org/10.1177/1744806920928619

Melzack, R. (2001). Pain and the neuromatrix in the brain. *Journal of Dental Education, 65*(12), 1378–1382.

Melzack, R., & Katz, J. (2007). A conceptual framework for understanding pain in the human. In *Pain management* (pp. 3–10). Elsevier.

Melzack, R., & Wall, P. D. (1965). Pain mechanisms: A new theory. *Science, 150*(3699), 971–979.

Messmer, R. L., Nader, R., & Craig, K. D. (2008). Brief report: Judging pain intensity in children with autism undergoing venepuncture: The influence of facial activity. *Journal of Autism and Developmental Disorders, 38*(7), 1391–1394. https://doi.org/10.1007/s10803-007-0511-0

Morse, B. L., Solodiuk, J. C., Greco, C. D., Mauskar, S., & Hauer, J. (2020). Initial validation of GRASP: A differential diagnoses algorithm for children with medical complexity and an unknown source of pain. *Hospital Pediatrics, 10*(8), 633–640. https://doi.org/10.1542/hpeds.2019-0322

Perucca, E. (2006). Clinically relevant drug interactions with antiepileptic drugs. *British Journal of Clinical Pharmacology, 61*(3), 246–255. https://doi.org/10.1111/j.1365-2125.2005.02529.x

Raja, S. N., Carr, D. B., Cohen, M., Finnerup, N. B., Flor, H., Gibson, S., … Vader, K. (2020). The revised International Association for the Study of Pain definition of pain: Concepts, challenges, and compromises. *Pain, 161*(9), 1976–1982.

Savedra, M. C., Tesler, M. D., Holzemer, W. L., & Ward, J. A. (1989). *Adolescent pediatric pain tool (APPT): Preliminary user's manual*. University of California.

Sharma, S. R., Gonda, X., & Tarazi, F. I. (2018). Autism spectrum disorder: Classification, diagnosis and therapy. *Pharmacology and Therapeutics, 190*, 91–104. https://doi.org/10.1016/j.pharmthera.2018.05.007

Sickkids Staff. (2020). *Acute pain: How to treat and manage in young children*. https://www.aboutkidshealth.ca:443/article?contentid=3637&language=english

Solodiuk, J. C., Scott-Sutherland, J., Meyers, M., Myette, B., Shusterman, C., Karian, V. E., … Curley, M. A. (2010). Validation of the Individualized Numeric Rating Scale (INRS): A pain assessment tool for nonverbal children with intellectual disability. *Pain, 150*(2), 231–236.

Stilwell, P., & Harman, K. (2019). An enactive approach to pain: Beyond the biopsychosocial model. *Phenomenology and the Cognitive Sciences, 18*(4), 637–665.

Stores, G. (2014). *Sleep and its disorders in children and adolescents with a neurodevelopmental disorder: A review and clinical guide*. Cambridge University Press. https://doi.org/10.1017/CBO9780511972935

Svedberg, L. E., Englund, E., Malker, H., & Stener-Victorin, E. (2008). Parental perception of cold extremities and other accompanying symptoms in children with cerebral palsy. *European Journal of Paediatric Neurology, 12*(2), 89–96. https://doi.org/10.1016/j.ejpn.2007.06.004

Taghizadeh, N., Davidson, A., Williams, K., & Story, D. (2015). Autism spectrum disorder (ASD) and its perioperative management. *Paediatric Anaesthesia, 25*(11), 1076–1084. https://doi.org/10.1111/pan.12732

Temple, B., Dubé, C., McMillan, D., Secco, L., Kepron, E., Dittberner, K., Ediger, J., & Vipond, G. (2012). Pain in people with developmental disabilities: A scoping review. *Journal on Developmental Disabilities, 18*(1), 73–86.

Thurnheer, S. E., Gravestock, I., Pichierri, G., Steurer, J., & Burgstaller, J. M. (2018). Benefits of mobile apps in pain management: Systematic review. *JMIR mHealth and uHealth, 6*(10), e11231.

Valkenburg, A. J., de Leeuw, T. G., van Dijk, M., & Tibboel, D. (2015). Pain in intellectually disabled children: Towards evidence-based pharmacotherapy? *Paediatric Drugs, 17*, 339–348. https://doi.org/10.1007/s40272-015-0138-0

Ventafridda, V., Saita, L., Ripamonti, C., & De Conno, F. (1985). WHO guidelines for the use of analgesics in cancer pain. *International Journal of Tissue Reactions, 7*(1), 93–96.

Vlassakova, B. G., & Emmanouil, D. E. (2016). Perioperative considerations in children with autism spectrum disorder. *Current Opinion in Anaesthesiology, 29*(3), 359–366. https://doi.org/10.1097/ACO.0000000000000325

Wakefield, S. M., Sanderson, J., & McPherson, P. (2018). Assessment of obesity. In J. L. Matson (Ed.), *Handbook of childhood psychopathology and developmental disabilities assessment* (pp. 433–452). Springer International Publishing.

Walco, G. A., Dworkin, R. H., Krane, E. J., LeBel, A. A., & Treede, R.-D. (2010). Neuropathic pain in children: Special considerations. *Mayo Clinic Proceedings, 85*(3 Suppl), S33–S41. https://doi.org/10.4065/mcp.2009.0647

Whitney, D. G., & Shapiro, D. N. (2019). National prevalence of pain among children and adolescents with autism spectrum disorders. *JAMA Pediatrics, 173*(12), 1203–1205. https://doi.org/10.1001/jamapediatrics.2019.3826

Witt, N., Coynor, S., Edwards, C., & Bradshaw, H. (2016). A guide to pain assessment and management in the neonate. *Current Emergency and Hospital Medicine Reports, 4*, 1–10. https://doi.org/10.1007/s40138-016-0089-y

Wong, D. L., & Baker, C. M. (1988). Pain in children: comparison of assessment scales. *Pediatric Nursing, 14*(1), 9–17.

Wong, C., Lau, E., Palozzi, L., & Campbell, F. (2012a). Pain management in children: Part 1—Pain assessment tools and a brief review of nonpharmacological and pharmacological treatment options. *Canadian Pharmacists Journal, 145*(5), 222–225. Canadian Pharmacists Association. https://doi.org/10.3821/145.5.cpj222

Wong, C., Lau, E., Palozzi, L., & Campbell, F. (2012b). Pain management in children: Part 2—A transition from codeine to morphine for moderate to severe pain in children. *Canadian Pharmacists Journal, 145*(6), 276–279.e1. https://doi.org/10.3821/145.6.cpj276

Wren, A., Ross, A., D'Souza, G., Almgren, C., Feinstein, A., Marshall, A., & Golianu, B. (2019). Multidisciplinary pain management for pediatric patients with acute and chronic pain: A foundational treatment approach when prescribing opioids. *Children, 6*(2), 33. https://doi.org/10.3390/children6020033

Yaksh, T. L. (2007). Dynamics of the pain processing system. In *Pain management* (pp. 21–32). Elsevier.

Zabalia, M. (2013). Beyond misconceptions: Assessing pain in children with mild to moderate intellectual disability. *Frontiers in Public Health, 1*, 1–5. https://doi.org/10.3389/fpubh.2013.00023

Zablotsky, B., Black, L. I., Maenner, M. J., Schieve, L. A., Danielson, M. L., Bitsko, R. H., Blumberg, S. J., Kogan, M. D., & Boyle, C. A. (2019). Prevalence and trends of developmental disabilities among children in the United States: 2009–2017. *Pediatrics, 144*(4), e20190811. https://doi.org/10.1542/peds.2019-0811

Overview of Physical Therapy for Children with Autism and Other Intellectual and Developmental Disabilities

18

Sarah Murnane

Abstract

Physical therapists are critical members of the medical home team offering expert family-centered care for children with disorders or injury that impair musculoskeletal function, resulting in poor balance, a lack of coordination, gait abnormalities, weakness, pain, or other impairments. From infancy through adulthood, the role of the pediatric physical therapist is critical to the development and quality of life of children living with developmental disability. This chapter provides an overview of the general physical therapy assessment and interventions before focusing on the care of children with autism, cerebral palsy, and developmental delay.

Keywords

Physical therapy · Autism · Cerebral palsy · Developmental delay · IDD · ASD

S. Murnane (✉)
School of Allied Health Professions, LSU Health Shreveport, Shreveport, LA, USA
e-mail: sarah.murnane03@lsuhs.edu

Introduction

Physical therapists (PTs) are critical members of the medical home team offering expert family-centered care for children with disorders or injury that impair musculoskeletal function, resulting in poor balance, a lack of coordination, gait abnormalities, weakness, pain, or other impairments. PTs are independent healthcare professionals focusing on the diagnosis of impairment resulting from disorders, disease, or injury. While diagnosing autism spectrum disorders (ASD) or intellectual disabilities would be beyond the scope of PT practice, pediatric PTs assist in restoring or enhancing functionality for children through a variety of assessment and treatment techniques tailored to individual needs. PTs may also assist in assessing for and obtaining adaptive equipment such as mobility aids and orthotics. Children are typically referred to the PT because a delay in motor milestones has been identified by parents or providers. PT therapies may be delivered in the home, school, clinic, or hospital settings.

Initiating Physical Therapy Services

Physical therapists are movement specialists assisting children to participate fully in their homes, school, and communities. While direct

P. McPherson (ed.), *Handbook of Treatment Planning for Children with Autism and Other Neurodevelopmental Disorders*, Autism and Child Psychopathology Series,
https://doi.org/10.1007/978-3-031-06120-2_18

access laws in all states have empowered patients to seek PT care without a physician's referral, the scope of these laws and requirements of insurance providers vary. Additional information about direct access is available on the American Physical Therapy Association's website. State requirements are summarized at getpt.org. In the United States, about half of children with special healthcare needs receive health insurance through Medicaid or children's health insurance program (CHIP) (Musumeci & Chidambaram, 2018). Within federal guidelines, states determine how Medicaid is administered. Under Medicaid, PT is classified as a rehabilitation service, typically requiring a primary care provider's referral or a determination of eligibility for services through school-based programs.

The medical home team care coordinator can assist families in understanding access to local resources. In addition, when referring a child for PT services, the medical home should forward the shared plan of care (See Chap. 1 in this volume for more information on the SPoC) along with a prescription noting the diagnosis or condition to be addressed, therapy requested, frequency, therapy goals, and any patient-specific limitations or precautions (Houtrow et al., 2019). It is helpful to note any accommodations that will be necessary, including sensory issues, particularly sensitivity to touch or movement. In addition, the family will benefit from knowing what to expect during the appointment. It may be helpful to bring baby books to the appointment or for the parents to review family videos in order to be prepared to answer questions about motor milestones. Children should wear clothes that allow freedom of movement and bring any medical equipment they use. The American Academy of Pediatric Physical Therapy provides fact sheets and other resources for families on their website.

The Physical Therapy Assessment

When assessing the child at the first appointment, a physical therapist will complete an initial evaluation lasting 1–2 hours. This should include a thorough medical and social history, including medications the child is taking, any prior interventions the child has had, and any current services they are receiving. Documentation of major developmental milestones and any areas of current developmental concern should be reviewed. A review of systems is completed that may include gathering information on the gastrointestinal, integumentary, cardiovascular/ pulmonary, musculoskeletal, and neurologic systems, as well as sensory and communication issues to rule out any red flags as well as guide clinical decision-making. During the initial evaluation, the physical therapist may draw on a number of sources for information such as the caregiver's report, if school aged, the teacher's report, and if able to communicate, the child's report. Children's reports of motor strengths and challenges often differ from that of their parents, so understanding the child's perception is key to engaging and motivating the child in therapy (Jasmin et al., 2018).

A child's motor performance may be observed and measured through a variety of standardized assessments. See Table 18.1. Motor function is assessed in different positions such as sitting, standing, and walking. The quality of the motor function is also important such as speed, or time

Table 18.1 Sample of available standardized tests and measures used in pediatric assessments

Test	Areas assessed
Movement Assessment Battery for Children (MABC-2) (Henderson, 1992; Henderson et al., 2007)	Motor development age 3–16 years
Bruininks-Oseretsky Test of Motor Proficiency (BOT-2) (Bruininks & Bruininks, 2005)	Fine and gross motor development age 4–21 years
6-Minute Walk Test (6MWT) (Laboratories, 2002)	Endurance level
Functional Strength Measurement (FSM) (Aertssen et al., 2019)	Strength in arms and legs
School Function Assessment (SFA) (Coster, 1997)	Functional tasks that affect school
Gross Motor Performance Measure (GMPM) (Boyce et al., 1992)	Gross motor performance in children with CP
Peabody Developmental Motor Scales Second Edition (PDMS-2) (Folio, 1983)	Gross and fine motor development birth–age 5 years

to complete the movement, range, or available movement of the joint, symmetry, or evenness of movement, and control, or smoothness, coordination, and stability of the task. Deficits in muscular endurance, strength, cardiorespiratory fitness, and balance are also identified.

After the evaluation, the physical therapists will analyze the findings to form task-oriented interventions. Task-oriented interventions focus on functional tasks that the child and parent would like to master (Das & Ganesh, 2019). Goals that are relevant, measurable, and specific will be established. Additional research is needed to establish best practices for goal-setting with children (Pritchard-Wiart et al., 2019). Often times, physical therapists will provide education and a home exercise program to be completed by the parent or caregiver between therapy visits. For a home exercise program to be effective, parents must be able to envision the outcome of program adherence for their child, master the skills to implement the program, and be armed with coping strategies for challenges that arise (Gorgon, 2018).

Infants and Toddlers

Infants and toddlers are often identified for physical therapy services if they have abnormal tone; limited joint mobility; a diagnosed genetic, brain, spine, or neuromuscular disorder; or fail to meet major developmental milestones. The American Academy of Pediatrics recommends a developmental screening at a child's 9-, 18-, and 30-month well visit (Council on Children With et al., 2006). Some insurance companies do not cover a 30-month well child visit, so the screen can occur at 24 months, which may capture a larger percentage of children who could benefit from an Early Intervention Program (EIP) (Marks, 2020). A parent or healthcare provider can also request an evaluation of the child's skills and abilities to see if they qualify for physical therapy early intervention services. The EIP for Infants and Toddlers with Disabilities is for children from birth to age 3 to provide services and support for children with disabilities and developmental delays. Some children's services may be extended until age 5. Each state has a unique Early Intervention Program, so covered services may vary. Early intervention is important as motor delays have shown that as a child grows, their school performance, ability to perform activities of daily living, and participation in recreational sports are negatively affected (Dannemiller et al., 2020).

Fig. 18.1 Services provided by the pediatric PT

Developmental activities
Movement and mobility training
Strengthening
Motor learning
Balance and coordination
Recreation, play, and leisure
Daily care activities and routines
Equipment design, fabrication, and fitting
Tone management
Assistive technology
Posture, positioning, and lifting
Orthotics and prosthetics
Wound care
Cardiopulmonary endurance
Pain management
Safety, health promotion, and prevention programs

School Age

Under the Individuals with Disabilities Education Act (IDEA), students who meet criteria for one or more of 13 disability classifications, and whose disability impacts their educational performance, can qualify for school-based physical therapy (see Chap. 12) (Individuals With Disabilities Education Act, 2004). Physical therapy provided in the school setting is established for the child to fully participate in an educational program designed to allow the child to progress in school. Observation of the child within the school day is an essential part of a physical therapist's assessment. Ideally, an evaluation occurs in the classroom or natural environment. Goals specifically related to the child's education as well as parent concern and opinions should then be established as a team within the child's IEP (Individualized Education Program). A survey of school-based physical therapists found that only 80% of parent's concerns were "always" or "usually" addressed and 83% of physical therapists "always" or "usually" participate in the IEP development (Effgen & Kaminker, 2014). Functional activities such as being able to walk the hallways, climb school stairs, write, get on or off the bus, or carry their school bag are examples of school-based tasks that a physical therapist may address in the IEP. The therapist works with the child on skills to better navigate the school day. The therapist also works with the teacher or paraprofessional to teach strategies or identify adaptive equipment to maximize the child's success during school. A study found that children receive approximately 40 minutes of school-based physical therapy per week (Jeffries et al., 2019).

Physical therapy services for children are also offered through outpatient services using a medical model, where the child's medical diagnosis has accompanied sensory, motor, or developmental delays. In the outpatient setting, functional tasks and the ability to navigate their home and community are addressed. Children may work on non-school-related activities such as riding a bike, skipping, higher level balance, and coordination. Children may only qualify for clinic-based pediatric physical therapy if their deficits do not impact their learning within the school setting. It is also possible for children to receive both school- and clinic-based physical therapy in conjunction with one another.

Transition to Adult Care

Pediatric physical therapists typically transition patients to adult care providers between 18 and 21 years of age. Programs such as Early and Periodic Screening, Diagnosis, and Treatment (EPSDT), a federally mandated program that provides comprehensive coverage for children who receive Medicaid, stops at age 21 (Early and Periodic Screening, Diagnostic, and Treatment | Medicaid). A child that receives special education services in the school system is only guaranteed these services until they finish high school or are 21 years old. Transitioning care can be challenging for families and patients. The burden of obtaining services shifts to the adult who must seek the necessary accommodations for themselves. Many providers advocate early planning to help with the transition into adult services. Many therapists argue the need for a more student-centered approach during the transition process (Chandroo et al., 2018). A transition that encompasses students, parents, and teachers as a team to identify the student's strengths, weaknesses, and goals has been shown to have more positive outcomes (Laghi & Trimarco, 2020). For a more detailed review of addressing the needs of emerging adults and transitioning care, see Chap. 25 in this volume.

The Functional Approach to Diagnosis

A PT's functional approach focuses on the diagnosis of movement system impairments (Jiandani & Mhatre, 2018). While medical professionals may have intimate knowledge of the World Health Organization's International Classification of Diseases (WHO–ICD) for diagnosing and billing purposes, its' companion, the WHO

International Classification of Functioning, Disability and Health (WHO–ICF), may be less familiar. The WHO–ICF, which includes the WHO–ICF–Children and Youth classification, is used to characterize:

> what a person with a health condition can do in a standard environment (their level of capacity), as well as what they actually do in their usual environment (their level of performance). These domains are classified from body, individual and societal perspectives by means of two lists: a list of body functions and structure, and a list of domains of activity and participation. In ICF, the term functioning refers to all body functions, activities and participation. (WHO, 2002, p. 2)

Within pediatric physical therapy, the ICF–CY model works well to incorporate a broader understanding of the whole child to include not only their disability but also their abilities through function and participation levels. The framework uses a universal language that can help pediatric physical therapists move from a novice level using deductive reasoning to a more expert clinician who uses inductive reasoning and identifies "patterns" in health conditions (Atkinson & Nixon-Cave, 2011). When using the ICF–CY, the inclusion of both the child and caregiver information is important to characterize the complete picture of the child. Children with CP were found to focus on their abilities, while caregivers tended to discuss the limitations and disabilities of the child (Schiariti et al., 2014). Among the different domains within the ICF–CY, interpersonal interactions and relationships were addressed the least by healthcare professionals (Chien et al., 2014).

A limitation in using the ICF–CY model is the lack of consensus among clinicians on which tests or measures are the most appropriate to use and which outcome should be addressed. To address this, the Children and Youth with Disability in Society (CYDiS) research unit at the University of British Columbia (Canada), and the ICF Research Branch, the Classification, Terminology and Standards Team at WHO developed 5 ICF Cores Sets for Children and Youths (CY) with CP. These include a comprehensive set covering all the developmental stages from birth up to 18 years of age, as well as brief sets for CY with CP below 6 years of age, ≥6 and < 14 years of age, ≥14 and 18 years of age, and up to 18 years of age (ICF RESEARCH BRANCH – Home (icf-research-branch.org).

Treatment Planning

After a physical therapist completes the initial evaluation of the child, they will establish a treatment plan of care. This will include the frequency of therapy visits, short-term goals, and long-term goals for the child. Short-term goals may have a target achievement of a few weeks to 3 months, while long-term goals may be up to 6 months. Physical therapy goals should be specific, objective, measurable, and meaningful to the child or caregiver. Physical therapists will monitor a child's progress through reassessments at various intervals. Goals can then be adjusted based on the child's progress. Within school-based physical therapy, goal-setting is a collaborative process that involves the IEP team and includes factors such as the educational environment, level of support available, student strengths, and therapist participation (Wynarczuk et al., 2017).

Physical Therapy for Specific Disorders

Developmental Disability

Pediatric PTs are experts in the variations in motor development associated with developmental disability (DD) and implementing therapies to improve overall functionality. The prevalence of DD has increased over the past 25 years due to improved prenatal/perinatal care, increased access to care, and greater public awareness (Callaghan et al., 2017; Zablotsky et al., 2019); however, racial disparities remain a challenge(Howell et al., 2018). The National Health Interview Survey tracks the prevalence of ten developmental disabilities (attention-deficit/hyperactivity disorder, autism spectrum disorder, blindness, cerebral palsy, moderate to profound hearing loss, learning disability, intellectual disability, seizures, stuttering or stammering, and other developmental delays) in children 3–17 years of age. The 2015–2017 cycle of interviews found an increase in developmental delay compared to the 2009–2011 cycle, with prevalence of 17.8% and 16.2%, respectively (Zablotsky et al., 2019). Children with DD are at increased risk of functional motor impairment and developmental coordination disorder (Lachambre et al., 2021). The association between motor impairments and ADHD, ASD, epilepsy, cerebral palsy, ID, and numerous genetic disorders is well-documented (Busti Ceccarelli et al., 2020; Lee et al., 2021; van Hoorn et al., 2021). Preterm birth is also a risk factor for motor disability. The Extremely Preterm Infants in Sweden Study of children born at less than 27 weeks and without ID, cerebral palsy, hearing or visual impairment found a prevalence of developmental coordination disorder of 37.1% in those born preterm compared to 5.5% in controls at 6.5 years uncorrected age (Bolk et al., 2018). Specific learning disabilities have also been associated with motor impairments. Hussein et al. (2020) reported an association between reading disorders and balance, bilateral coordination deficits and mathematics learning disorder, and running speed and agility in children with more than one specific learning disorder. Fine motor integration and upper limb coordination impairment was also associated with learning disability.

Children with DD may be referred to a physical therapist as a newborn or when the pediatrician identifies a motor concern. The American Academy of Pediatrics offers guidance for the screening of newborns and ongoing developmental surveillance (Lipkin & Macias, 2020) as well as a clinical report on the evaluation of motor delays, which includes a review of genetic disorders with associated motor delay (Noritz & Murphy, 2013). Children with DD are best cared for in a medical home, where the coordinator can assist with specialty referrals such as PT and arrange for community supports such as Early Steps.

A Brief Review of General Movements

The science of fetal and infant movements has advanced dramatically since Prechtl first described general movements (GM) in the 1980s (Prechtl & Hopkins, 1986). Fetal movement is first detected at about 7 weeks postmenstrual age, with GM emerging in the fetus at 9–10 weeks (Lüchinger et al., 2008). GM are spontaneous and involve the whole body, providing a marker for the functionality of the developing nervous system (Prechtl et al., 1997). After delivery, between 0 and 9 weeks corrected age, normal GM are described as writhing movements and, at 6–20 weeks, as fidgety movements. After 20 weeks, GM fade as voluntary movements emerge (Prechtl, 1990). Fidgety movements are a motor reflection of the cortical and subcortical networks (Hadders-Algra, 2018). The assessment of GM considers the variation, complexity, and fluency of movements (Palisano et al., 1997; Prechtl et al., 1997). Absent or abnormal GM are associated with a poor neurological outcome (Hadders-Algra, 2018). Tools for assessing GM include the General Movement Assessment (Prechtl, 1997) and the Hammersmith Infant Neurological Examination (Haataja et al., 1999). Recently, assessment through artificial intelligence video analysis has proved promising (Silva et al., 2021). Preliminary studies have identified abnormal movement patterns which may serve as a biomarker for neurodevelopmental disorders including ASD (Caruso et al., 2020). While video analysis is becoming a useful tool, the evaluation by a pediatric PT provides the basis for therapy to improve functionality in children with DD.

PT Treatment for Children with Developmental Delay

PT treatment will vary based on the functional strengths and needs of the child. For the infant, the PT may assist with feeding strategies, environmental alterations to promote self-calming, parent education regarding expected motor milestones, and the importance of skin-to-skin contact, known as kangaroo care. PT also addresses the positioning that is necessary to avoid tight muscles due to poor alignment, as positioning can influence developing cerebral motor pathways (Wood et al., 2015), For toddlers, PT will support the development of motor milestones. For older children, the focus will be to increase functionality to allow participation in self-care, school, and the community. When necessary, the PT will assess for orthoses or other equipment. The best treatment plan to address child-specific needs is highly individualized, but improving overall fitness is a universal goal.

The American Physical Therapy Association (APTA) has provided guidance on the role of the pediatric PT in promoting fitness, with recommendations for strength training, aerobic exercise, and adapting sports programs for children with DD (O'Neil et al., 2012). Children with developmental disability are at increased risk for poor physical fitness and obesity (Neumeier et al., 2017). Overall strength, endurance, and cardiorespiratory fitness have been noted as concerns for children with IDD (Wouters et al., 2020). In addition, pain or discomfort may limit activity (Barney et al., 2020). Because of the motivation necessary to achieve and maintain fitness, PT for fitness addresses more than functionality. Personal goals, family, environment, stressors, and readiness for change must be considered to avoid derailing the fitness plan (Bloemen et al., 2017; Rowland et al., 2015). Rowland et al. (2015) summarizes over 30 measures available for assessing the fitness level and specific interventions for persons with DD. The Gross Motor Classification System (GMCS-E&R) is a foundational measure for children ages 2–18. The GMCS-E&R assigns one of five functional levels to a child's self-initiated movement, including sitting, transfers, mobility, and use of mobility devices to inform treatment planning, progress monitoring, and communication among professionals clinically and in research (Palisano et al., 2008). The PT will select additional measures of strength, endurance, aerobic capacity, or balance based on the GMCS-E&R (Rowland et al., 2015).

Autism Spectrum Disorder

Autism spectrum disorder (ASD) is a neurodevelopmental disorder characterized by social interaction and communication deficits with restricted and/or repetitive behaviors (American Psychiatric Association, American Psychiatric Association, & DSM-5 Task Force, 2013). While stereotyped or repetitive motor movements are a diagnostic feature of ASD, other motor impairments are common, including delays in fine and gross motor skills, postural delays, and imitation and praxis impairments (Bhat et al., 2011). In addition, motor development plays a role in other developmental domains.

Motor delays in children with ASD are common. The Western Australian Register for Children with ASD assessed over 2000 children at the time of ASD diagnosis with 79.1% scoring low or moderately low on the Vineland Motor Skills Domain. Only 24 of the children had been diagnosed with a motor disorder prior to the Vineland (Licari et al., 2020). The SPARK (Simons Foundation Powering Autism Research for Knowledge) autism research study queried nearly 16,000 families noting that 15.1% of children carried a developmental coordination disorder diagnosis, and 31.6% were receiving physical therapy. The Developmental Coordination Disorder Questionnaire was then administered with 86.9% showing a significant motor impairment (Bhat, 2020). These studies highlight that the PT needs of many children with ASD were not being met.

The possibility of early motor phenotype as a marker for ASD has been suggested, specifically, gait biomechanics, manual dexterity, and neuromuscular tone (Esposito & Pasca, 2013; Jequier Gygax et al., 2021; Licari et al., 2020; Lidstone et al., 2020; Paquet et al., 2017; Wilson et al., 2018). Moseley and Pulvermueller (2018) have proposed an action perception integration theory as a mechanism linking motor abnormalities to social communication and conceptual deficits in ASD centering on the transfer of information between motor and sensory regions. While research regarding action perception in persons with ASD is mixed, a meta-analysis suggests that perception and interpretation of motion lags behind TD peers with children with ASD showing a greater impairment than adolescents with ASD (Cusack et al., 2015; Todorova et al., 2019). With the increasing interest in the motor development of children with ASD, treating children with ASD is an essential competency for pediatric PTs and there has been a call for additional information on best practices for treating children with ASD (Cynthia et al., 2019; Rapport et al., 2014).

PT for children with ASD often addresses fundamental motor skills often through ECSS or in a school setting. Healy, Obusnikova, and Getchell (2019b) conducted a systematic review of fundamental motor skills (FMS) interventions targeting balance, locomotor, and object-control skills, with significant improvement noted in each category. Because PT goals can be advanced through games, physical education, and even applied behavioral analysis, PT supports cognitive and sensory development while increasing motor skills (Aniszewski et al., 2020; Bhat et al., 2015). A 10-year narrative review of motor skills intervention outcomes found improvements in fine motor, gross motor, and coordination skills with results regarding the impact of improved motor function on core social and communication features of ASD mixed (Busti Ceccarelli et al., 2020). While a more recent review suggests that interventions with a duration greater than 16 hours and greater intensity (frequency) are more effective (Case & Yun, 2019), this correlation was not noted by Colombo-Dougovito and Block (2019), who noted the need for more research to determine optimal interventions, duration, and frequency. Because gross motor delays have been associated with more problem behaviors with poorer quality of life for children with ASD, the need for assessment and treatment of motor delays should be considered in all children with ASD (Hedgecock et al., 2018).

In recognition of the contributions of physical activity to achieving developmental milestones, the U.S. Department of Health and Human Services recommends that preschool children have opportunities for physical activity throughout the day. One hour of moderate to vigorous activity is recommended for children and adolescents (Piercy et al., 2018). Children and teens with ASD are less likely than typically developing peers to meet this guideline, with adolescents lagging behind younger children (Healy, Aigner, et al., 2019a). The American Physical Therapy Association has published a clinical guideline for the *Prevention and Management of Obesity* (Orringer et al., 2020) addressing prevention, screening, diagnosis, and treatment for persons over the age of two; however, the needs of persons with ASD/IDD are not specifically addressed. There is not a general exercise program recommended for children with ASD. In addition to overall health benefits, aerobic exercise has been associated with a decrease in stereotypic and self-injurious behaviors and improved social skills and engagement (Andy, 2020; Aniszewski et al., 2020).

Cerebral Palsy

As movement specialists, physical therapists are critical to the health and development of children with cerebral palsy (CP) by assisting children to achieve greater strength, motor functionality, and independence through early intervention and ongoing targeted therapies. CP is the most prevalent motor disability in children, with the National Health Interview Survey for 2009–2016 reporting a weighted prevalence of CP as 3.2 per 1000 children ages 3–17 (McGuire et al., 2019). For children born before 32 weeks, this increases dramatically to 6.8% (Pascal et al., 2018). Cerebral palsy was defined by the International Workshop on Definition and Classification of Cerebral Palsy in 2006 as:

> a group of permanent disorders of the development of movement and posture, causing activity limitation, that are attributed to nonprogressive disturbances that occurred in the developing fetal or infant brain. (Rosenbaum et al., 2007, p. 9)

As with many descriptive diagnoses, the etiologies of CP are multifactorial and complex. Preconception, prenatal, perinatal, and postnatal risk factors for CP have been identified (Sadowska et al., 2020). In addition, an array of genetic and metabolic disorders have been identified as CP mimics (Leach et al., 2014; Pearson et al., 2019).

The American Academy of Neurology has published a practice parameter for the assessment for cerebral palsy (Ashwal et al., 2004). The American Academy of Pediatrics provides guidance to the medical home for children with CP (Liptak & Murphy, 2011), As the diagnosis of CP is beyond the scope of PT practice, here we will briefly review the assessment and treatment of CP provided by the PT.

The Evaluation of CP

The goal of the PT's evaluation is to synthesize physical, radiographic, and historical information in order to develop an individualized treatment plan detailing specific interventions and expected outcomes for a child with CP (Jewell, 2014). In addition to the general assessment procedure described above, there is careful attention to movement and postural control and tone. A Delphi study achieved a consensus definition of postural control as "control of the body's position in space for postural orientation and postural stability [balance]" (Dewar et al., 2017, p. 467). Postural control includes strength and stability within the core muscles to prepare for movement and the ability to hold oneself upright when sitting or standing. Good postural control allows the arms and legs to mover more freely. Postural tone refers to the tension of the muscles supporting the spine and is altered with spasticity and abnormal extensibility. Muscles are also assessed for shortening or overlengthening (Beaman et al., 2015). Movement of the spine, joints, and gait is assessed. As early intervention is critical to improve functionality, the assessment of the general movements of infants at risk for CP is critical. Absent fidgety movements at 3 months is strongly associated with a later diagnosis of CP, with abnormal movements also indicating risk (Ferrari et al., 2011; Øberg et al., 2015; Prechtl et al., 1997). Five classification systems are in common use by PTs for diagnosing a child's level of functionality and progress monitoring (Patel et al., 2020). See Table 18.2. The gold standard is the Gross Motor Function Classification System – Expanded and Revised categorizes self-initiated movement into five levels of functionality and is generally stable after the age of two (Palisano et al., 1997; Palisano et al., 2006; Palisano et al., 2008). The evaluation may also include assessment for orthotics, adaptive equipment, or assistive technology (Patel et al., 2020). If a PT orders durable medical equipment, the physician may receive an authorization request from the supplier in order to secure insurance payment. The evaluation of CP may be complicated by comorbid conditions including pain, epilepsy, intellectual disability, speech disorders, sleep disorders, and mental disorders, including ASD (Novak et al., 2012).

Table 18.2 Function assessment tools

Classification system	Domain
Gross motor function classification system – Expanded and Revised[1]	Classifies self-initiated movement, with emphasis on sitting, transfers, and mobility with a five-level classification system. Validated for 0–18-year-olds. Valid for use in persons with Down's syndrome.
Manual ability classification System[2]	Classifies how children 4–18 years of age use their hands to manipulate objects during activities of daily living in a five-level system.
Communication function classification System[3]	Classifies the effectiveness of daily communication using a five-level system. Validated for ages 4 and above.
Eating and drinking ability classification System[4]	Classifies functional eating and drinking ability using a five-level system. Validated for ages 3 and above.
Visual function classification System[5]	Classifies how visual ability is used in daily life in youth ages 1–19.

[1]Palisano et al. (2021), [2]Eliasson et al. (2006), [3]Hidecker et al. (2011), [4]Sellers et al. (2014); [5]Baranello et al. (2020)

The Treatment of CP

The treatment of CP is designed to maximize functional ability. The rapid development of the brain during the perinatal and first years of life offers an opportunity for early interventions to capitalize on neural plasticity. Early intervention is critical in order to promote the brain development that is the foundation for complex motor movements throughout life (Spittle et al., 2018). The International Clinical Practice Guideline for *Early Intervention for Children Aged 0–2 Years With or at High Risk of Cerebral Palsy* (Morgan et al., 2021) includes recommended interventions to improve motor skills, promote reduction in motor tone, and support parents. The Guideline specifically recommends against passive movement, sleep positioning systems, and a wait-and-see approach. Recommended interventions to improve motor skills include targeted motor training activities, constraint-induced movement therapy, bimanual movement therapy, comprehensive hypertonia management, regular use of standing equipment for positioning, and ankle-foot orthosis.

As children age, the goal of the PT remains maximizing functionality as youth engage with their environment in more complex ways. In addition to the interventions mentioned above, action observational training, treadmill training, function chewing training, context-focused therapy, and home programs using goal-directed training have shown efficacy (Novak et al., 2020). More recently, virtual reality-based therapies for improving upper limb functionality have shown promise (Demers et al., 2021; Rathinam et al., 2019). The National Longitudinal Transition Study 2, a study that examines the experience of students as they leave secondary school, documented a dramatic decrease in PT when a student left school, with 60% of students with CP receiving PT in school, but only 35% receiving PT as emerging adults (Liljenquist et al., 2018). The authors highlight the importance of care with a medical home and the provision of well-coordinated transition services for adolescents with CP.

Future Directions in PT

The role of the pediatric PT is critical to the development and quality of life of children living with developmental disability. The PT's focus on functionality and overall wellness not only addresses impaired musculoskeletal function and pain; PT interventions can prevent dysfunction, making the PT a critical medical home team member delivering care to promote maximum independence. With the pandemic, the PTs have deftly shifted to offering virtual therapy which many families find convenient and effective. Current research into developmental trajectory of motor skills in persons with ASD may lead to the identification of a biomarker, further increasing the role of PT in the diagnosis of ASD and the development of motor interventions for children with ASD (Parma & de Marchena, 2016).

References

Aertssen, W., Smulders, E., Smits-Engelsman, B., & Rameckers, E. (2019). Functional strength measurement in cerebral palsy: Feasibility, test-retest reliability, and construct validity. *Developmental Neurorehabilitation, 22*(7), 453–461. https://doi.org/10.1080/17518423.2018.1518963

American Psychiatric Association, & American Psychiatric Association. DSM-5 Task Force. (2013). *Diagnostic and statistical manual of mental disorders : DSM-5* (5th ed.). American Psychiatric Association. http://www.PsychiatryOnline.org

Andy, C. Y. (2020). Brief report: Impact of a physical exercise intervention on emotion regulation and behavioral functioning in children with autism spectrum disorder. *Journal of Autism and Developmental Disorders, 50*(11), 4191–4198.

Aniszewski, E., Almeida, H., & Alvernaz, A. (2020). Benefits of physical activity for the development of autistic children. *International Physical Medicine & Rehabilitation Journal, 5*(2), 79–80.

Ashwal, S., Russman, B. S., Blasco, P. A., Miller, G., Sandler, A., Shevell, M., & Stevenson, R. (2004). Practice parameter: Diagnostic assessment of the child with cerebral palsy. *Neurology, 62*(6), 851–862. https://doi.org/10.1212/01.WNL.0000117981.35364.1B

Atkinson, H. L., & Nixon-Cave, K. (2011). A tool for clinical reasoning and reflection using the international classification of functioning, disability and health (ICF) framework and patient management model. *Physical Therapy, 91*(3), 416–430. https://doi.org/10.2522/ptj.20090226

Baranello, G., Signorini, S., Tinelli, F., Guzzetta, A., Pagliano, E., Rossi, A., … Mercuri, E. (2020). Visual Function Classification System for children with cerebral palsy: development and validation. *Developmental Medicine & Child Neurology, 62*(1), 104–110.

Barney, C. C., Andersen, R. D., Defrin, R., Genik, L. M., McGuire, B. E., & Symons, F. J. (2020). Challenges in pain assessment and management among individuals with intellectual and developmental disabilities. *Pain Reports, 5*(4), e821. https://doi.org/10.1097/PR9.0000000000000822

Beaman, J., Kalisperis, F. R., & Miller-Skomorucha, K. (2015). *The infant and child with cerebral palsy. Pediatric Physical Therapy* (pp. 187–246). Lippincott Wolters Klumer.

Bhat, A. N. (2020). Is motor impairment in autism spectrum disorder distinct from developmental coordination disorder? A report from the SPARK study. *Physical Therapy, 100*(4), 633–644.

Bhat, A. N., Landa, R. J., & Galloway, J. C. (2011). Current perspectives on motor functioning in infants, children, and adults with autism spectrum disorders. *Physical Therapy, 91*(7), 1116–1129.

Bhat, A., Bubela, D., & Lande, R. (2015). Austim spectrum disorders and physical therapy. In J. Tecklin (Ed.), *Pediatric physical therapy* (5th ed.). Lippincott, Williams, and Wilkins.

Bloemen, M., Van Wely, L., Mollema, J., Dallmeijer, A., & de Groot, J. (2017). Evidence for increasing physical activity in children with physical disabilities: A systematic review. *Developmental Medicine & Child Neurology, 59*(10), 1004–1010.

Bolk, J., Farooqi, A., Hafström, M., Åden, U., & Serenius, F. (2018). Developmental coordination disorder and its association with developmental comorbidities at 6.5 years in apparently healthy children born extremely preterm. *JAMA Pediatrics, 172*(8), 765–774.

Boyce, W., Gowland, C., Rosenbaum, P., Lane, M., Plews, N., Goldsmith, C., Russell, D., Wright, V., Zdrobov, S., & Harding, D. (1992). Gross motor performance measure for children with cerebral palsy: study design and preliminary findings. *Canadian Journal of Public Health, 83*(Suppl 2), S34–S40. https://www.ncbi.nlm.nih.gov/pubmed/1468047

Bruininks, R. H., & Bruininks, B. D. (2005). *Bruininks-Oseretsky test of motor proficiency*. AGS Publishing.

Busti Ceccarelli, S., Ferrante, C., Gazzola, E., Marzocchi, G. M., Nobile, M., Molteni, M., & Crippa, A. (2020). Fundamental motor skills intervention for children with autism spectrum disorder: A 10-year narrative review. *Children, 7*(11), 250.

Callaghan, W. M., MacDorman, M. F., Shapiro-Mendoza, C. K., & Barfield, W. D. (2017). Explaining the recent decrease in US infant mortality rate, 2007–2013. *American Journal of Obstetrics and Gynecology, 216*(1), 73–e71.

Caruso, A., Gila, L., Fulceri, F., Salvitti, T., Micai, M., Baccinelli, W., Bulgheroni, M., & Scattoni, M. L. (2020). Early motor development predicts clinical outcomes of siblings at high-risk for autism: Insight from an innovative motion-tracking technology. *Brain Sciences, 10*(6), 379.

Case, L., & Yun, J. (2019). The effect of different intervention approaches on gross motor outcomes of children with autism spectrum disorder: A meta-analysis. *Adapted Physical Activity Quarterly, 36*(4), 501–526.

Chandroo, R., Strnadova, I., & Cumming, T. M. (2018). A systematic review of the involvement of students with autism spectrum disorder in the transition planning process: Need for voice and empowerment. *Research in Developmental Disabilities, 83*, 8–17. https://doi.org/10.1016/j.ridd.2018.07.011

Chien, C. W., Rodger, S., Copley, J., & Skorka, K. (2014). Comparative content review of children's participation measures using the international classification of functioning, disability and health-children and youth. *Archives of Physical Medicine and Rehabilitation, 95*(1), 141–152. https://doi.org/10.1016/j.apmr.2013.06.027

Colombo-Dougovito, A. M., & Block, M. E. (2019). Fundamental motor skill interventions for children and adolescents on the autism spectrum: A literature review. *Review Journal of Autism and Developmental Disorders, 6*(2), 159–171.

Coster, W. (1997). *Development of a functional outcome measure for children in an educational setting: Final report*. Unpublished manuscript.

Council on Children With, D., Section on Developmental Behavioral, P., Bright Futures Steering, C., & Medical Home Initiatives for Children With Special Needs Project Advisory, C. (2006). Identifying infants and young children with developmental disorders in the medical home: An algorithm for developmental surveillance and screening. *Pediatrics,* 118(1), 405–420. https://doi.org/10.1542/peds.2006-1231.

Cusack, J. P., Williams, J. H., & Neri, P. (2015). Action perception is intact in autism spectrum disorder. *The Journal of Neuroscience, 35*(5), 1849–1857. https://doi.org/10.1523/jneurosci.4133-13.2015

Cynthia, C., Duck, M., McQuillan, R., Brazill, L., Malik, S., Hartman, L., McPherson, A. C., Gibson, B. E., & Jachyra, P. (2019). Exploring the role of physiotherapists in the care of children with autism spectrum disorder. *Physical & Occupational Therapy in Pediatrics, 39*(6), 614–628.

Dannemiller, L., Mueller, M., Leitner, A., Iverson, E., & Kaplan, S. L. (2020). Physical therapy management of children with developmental coordination disorder: An evidence-based clinical practice guideline from the academy of pediatric physical therapy of the American Physical Therapy Association. *Pediatric Physical Therapy, 32*(4), 278–313. https://doi.org/10.1097/PEP.0000000000000753

Das, S. P., & Ganesh, G. S. (2019). Evidence-based approach to physical therapy in cerebral palsy. *Indian Journal of Orthopaedics, 53*(1), 20–34. https://doi.org/10.4103/ortho.IJOrtho_241_17

Demers, M., Fung, K., Subramanian, S. K., Lemay, M., & Robert, M. T. (2021). Integration of motor learning principles into virtual reality interventions for individuals with cerebral palsy: Systematic review. *JMIR Serious Games, 9*(2), e23822.

Dewar, R., Claus, A. P., Tucker, K., & Johnston, L. M. (2017). Perspectives on postural control dysfunction to inform future research: A Delphi study for children with cerebral palsy. *Archives of Physical Medicine and Rehabilitation, 98*(3), 463–479.

Effgen, S. K., & Kaminker, M. K. (2014). Nationwide survey of school-based physical therapy practice. *Pediatric Physical Therapy, 26*(4), 394–403. https://doi.org/10.1097/PEP.0000000000000075

Eliasson, A. C., Krumlinde-Sundholm, L., Rösblad, B., Beckung, E., Arner, M., Öhrvall, A. M., & Rosenbaum, P. (2006). The Manual Ability Classification System (MACS) for children with cerebral palsy: scale development and evidence of validity and reliability. *Developmental Medicine and Child Neurology, 48*(7), 549–554.

Esposito, G., & Pasca, S. (2013). Motor abnormalities as a putative endophenotype for autism spectrum disorders. *Frontiers in Integrative Neuroscience, 7*, 43.

Ferrari, F., Todeschini, A., Guidotti, I., Martinez-Biarge, M., Roversi, M. F., Berardi, A., Ranzi, A., Cowan, F. M., & Rutherford, M. A. (2011). General movements in full-term infants with perinatal asphyxia are related to basal ganglia and thalamic lesions. *The Journal of Pediatrics, 158*(6), 904–911.

Folio, M. R. (1983). Peabody developmental motor scales. *DLM Teaching Resources*.

Gorgon, E. J. R. (2018). Caregiver-provided physical therapy home programs for children with motor delay: A scoping review. *Physical Therapy, 98*(6), 480–493. https://doi.org/10.1093/ptj/pzy009

Jewell, D. V. (2014). Guide to evidence-based physical therapist practice. Jones & Bartlett Publishers.

Haataja, L., Mercuri, E., Regev, R., Cowan, F., Rutherford, M., Dubowitz, V., & Dubowitz, L. (1999). Optimality score for the neurologic examination of the infant at 12 and 18 months of age. *The Journal of Pediatrics, 135*(2), 153–161.

Hadders-Algra, M. (2018). Neural substrate and clinical significance of general movements: An update. *Developmental Medicine & Child Neurology, 60*(1), 39–46.

Healy, S., Aigner, C. J., Haegele, J. A., & Patterson, F. (2019a). Meeting the 24-hr movement guidelines: An update on US youth with autism spectrum disorder from the 2016 National Survey of Children's Health. *Autism Research, 12*(6), 941–951.

Healy, S., Obrusnikova, I., & Getchell, N. (2019b). Fundamental motor skill interventions in children with autism spectrum disorder: A systematic review of the literature including a methodological quality assessment. *Research in Autism Spectrum Disorders, 81*, 101717.

Hedgecock, J. B., Dannemiller, L. A., Shui, A. M., Rapport, M. J., & Katz, T. (2018). Associations of gross motor delay, behavior, and quality of life in young children with autism spectrum disorder. *Physical Therapy, 98*(4), 251–259.

Henderson, S. E. (1992). Movement assessment battery for children. *The Psychological Corporation*. https://ci.nii.ac.jp/naid/10010096709/en/

Henderson, S. E., Sugden, D. A., & Barnett, A. L. (2007). *Movement assessment battery for children* (Vol. 26). Harcourt Assessment.

Hidecker, M. J. C., Paneth, N., Rosenbaum, P. L., Kent, R. D., Lillie, J., Eulenberg, J. B., … Taylor, K. (2011). Developing and validating the Communication Function Classification System for individuals with cerebral palsy. *Developmental Medicine & Child Neurology, 53*(8), 704–710.

Houtrow, A., Murphy, N., Kuo, D. Z., et al. (2019). Prescribing physical, occupational, and speech therapy services for children with disabilities. *Pediatrics, 143*(4). e20190285. https://doi.org/10.1542/peds.2019-0285. PMID: 30910917.

Howell, E. A., Janevic, T., Hebert, P. L., Egorova, N. N., Balbierz, A., & Zeitlin, J. (2018). Differences in morbidity and mortality rates in black, white, and Hispanic very preterm infants among New York City hospitals. *JAMA Pediatrics, 172*(3), 269–277.

Hussein, Z. A., Abdel-Aty, S. A.-R., Elmeniawy, G. H., & Mahgoub, E. A.-M. (2020). Defects of motor performance in children with different types of specific learning disability. *Drug Invention Today, 14*(2), 302–307.

Individuals With Disabilities Education Act, 20 U.S.C. § 1400 (2004).

Jasmin, E., Tetreault, S., Lariviere, N., & Joly, J. (2018). Participation and needs of children with developmental coordination disorder at home and in the community: Perceptions of children and parents. *Research in Developmental Disabilities, 73*, 1–13. https://doi.org/10.1016/j.ridd.2017.12.011

Jeffries, L. M., McCoy, S. W., Effgen, S. K., Chiarello, L. A., & Villasante Tezanos, A. G. (2019). Description of the services, activities, and interventions within school-based physical therapist practices across the United States. *Physical Therapy, 99*(1), 98–108. https://doi.org/10.1093/ptj/pzy123

Jequier Gygax, M., Maillard, A. M., & Favre, J. (2021). Could gait biomechanics become a marker of atypical neuronal circuitry in human development?–The example of autism spectrum disorder. *Frontiers in Bioengineering and Biotechnology, 9*, 184.

Jiandani, M. P., & Mhatre, B. S. (2018). Physical therapy diagnosis: How is it different? *Journal of Postgraduate Medicine, 64*(2), 69.

Laboratories, A. T. S. C. o. P. S. f. C. P. F. (2002). ATS statement: Guidelines for the six-minute walk test. *American Journal of Respiratory and Critical Care Medicine, 166*(1), 111–117. https://doi.org/10.1164/ajrccm.166.1.at1102

Lachambre, C., Proteau-Lemieux, M., Lepage, J.-F., Bussières, E.-L., & Lippé, S. (2021). Attentional and executive functions in children and adolescents with developmental coordination disorder and the influence of comorbid disorders: A systematic review of the literature. *PLoS One, 16*(6), e0252043.

Laghi, F., & Trimarco, B. (2020). Individual planning starts at school. Tools and practices promoting autonomy and supporting transition to work for adolescents with autism spectrum disorder. *Ann Ist Super Sanita, 56*(2), 222–229. https://doi.org/10.4415/ANN_20_02_12

Leach, E. L., Shevell, M., Bowden, K., Stockler-Ipsiroglu, S., & van Karnebeek, C. D. M. (2014). Treatable inborn errors of metabolism presenting as cerebral palsy mimics: Systematic literature review. *Orphanet Journal of Rare Diseases, 9*(1), 1–14.

Lee, J., Mayall, L. A., Bates, K. E., Hill, E. L., Leonard, H. C., & Farran, E. K. (2021). The relationship between motor milestone achievement and childhood motor deficits in children with Attention Deficit Hyperactivity Disorder (ADHD) and children with Developmental Coordination Disorder. *Research in Developmental Disabilities, 113*, 103920.

Licari, M. K., Alvares, G. A., Varcin, K., Evans, K. L., Cleary, D., Reid, S. L., Glasson, E. J., Bebbington, K., Reynolds, J. E., & Wray, J. (2020). Prevalence of motor difficulties in autism spectrum disorder: Analysis of a population-based cohort. *Autism Research, 13*(2), 298–306.

Lidstone, D. E., Miah, F. Z., Poston, B., Beasley, J. F., & Dufek, J. S. (2020). Manual dexterity in children with autism spectrum disorder: A cross-syndrome approach. *Research in Autism Spectrum Disorders, 73*, 101546.

Liljenquist, K., O'Neil, M. E., & Bjornson, K. F. (2018). Utilization of physical therapy services during transition for young people with cerebral palsy: A call for improved care into adulthood. *Physical Therapy, 98*(9), 796–803.

Lipkin, P. H., & Macias, M. M. (2020). Promoting optimal development: Identifying infants and young children with developmental disorders through developmental surveillance and screening. *Pediatrics, 145*(1), e20193449. https://doi.org/10.1542/peds.2019-3449

Liptak, G. S., & Murphy, N. A. (2011). Providing a primary care medical home for children and youth with cerebral palsy. *Pediatrics, 128*(5), e1321. https://doi.org/10.1542/peds.2011-1468

Lüchinger, A. B., Hadders-Algra, M., Van Kan, C. M., & De Vries, J. I. P. (2008). Fetal onset of general movements. *Pediatric Research, 63*(2), 191–195.

Marks, K. P. (2020). AAP developmental surveillance and screening algorithm is improved but further discussion needed about 24 months and 4 years. *Pediatrics, 145*(6), e20200712A. https://doi.org/10.1542/peds.2020-0712A

McGuire, D. O., Tian, L. H., Yeargin-Allsopp, M., Dowling, N. F., & Christensen, D. L. (2019). Prevalence of cerebral palsy, intellectual disability, hearing loss, and blindness, National Health Interview Survey, 2009–2016. *Disability and Health Journal, 12*(3), 443–451.

Morgan, C., Fetters, L., Adde, L., Badawi, N., Bancale, A., Boyd, R. N., Chorna, O., Cioni, G., Damiano, D. L., Darrah, J., de Vries, L. S., Dusing, S., Einspieler, C., Eliasson, A. C., Ferriero, D., Fehlings, D., Forssberg, H., Gordon, A. M., Greaves, S., Guzzetta, A., Hadders-Algra, M., Harbourne, R., Karlsson, P., Krumlinde-Sundholm, L., Latal, B., Loughran-Fowlds, A., Mak, C., Maitre, N., McIntyre, S., Mei, C., Morgan, A., Kakooza-Mwesige, A., Romeo, D. M., Sanchez, K., Spittle, A., Shepherd, R., Thornton, M., Valentine, J., Ward, R., Whittingham, K., Zamany, A., & Novak, I. (2021). Early intervention for children aged 0 to 2 years with or at high risk of cerebral palsy: International clinical practice guideline based on systematic reviews. *JAMA Pediatrics, 175*(8), 846–858. https://doi.org/10.1001/jamapediatrics.2021.0878. **PMID: 33999106.**

Moseley, R. L., & Pulvermueller, F. (2018). What can autism teach us about the role of sensorimotor systems in higher cognition? New clues from studies on language, action semantics, and abstract emotional concept processing. *Cortex, 100*, 149–190.

Musumeci, M., & Chidambaram, P. (2018). Medicaid's role for children with special health care needs: A look at eligibility, services, and spending. *Henry J. Kaiser Family Foundation, February, 22*.

Neumeier, W. H., Grosso, C., & Rimmer, J. H. (2017). Obesity and individuals with intellectual and developmental disabilities: a research to policy brief.

Noritz, G. H., & Murphy, N. A. (2013). Motor delays: Early identification and evaluation. *Pediatrics, 131*(6), e2016. https://doi.org/10.1542/peds.2013-1056

Novak, I., Hines, M., Goldsmith, S., & Barclay, R. (2012). Clinical prognostic messages from a systematic review on cerebral palsy. *Pediatrics, 130*(5), e1285–e1312.

Novak, I., Morgan, C., Fahey, M., Finch-Edmondson, M., Galea, C., Hines, A., Langdon, K., Mc Namara, M., Paton, M. C. B., & Popat, H. (2020). State of the evidence traffic lights 2019: Systematic review of interventions for preventing and treating children with cerebral palsy. *Current Neurology and Neuroscience Reports, 20*(2), 1–21.

Øberg, G. K., Jacobsen, B. K., & Jørgensen, L. (2015). Predictive value of general movement assessment for cerebral palsy in routine clinical practice. *Physical Therapy, 95*(11), 1489–1495.

O'Neil, M. E., Fragala-Pinkham, M., Ideishi, R. I., & Ideishi, S. K. (2012). Community-based programs for children and youth: our experiences in design, implementation, and evaluation. *Physical & Occupational Therapy in Pediatrics, 32*(2), 111–119.

Orringer, K. A., Van Harrison, R., Nichani, S. S., Riley, M. A., Rothberg, A. E., Trudeau, L. E., & White, Y. (2020). Obesity prevention and management.

Palisano, R., Rosenbaum, P., Walter, S., Russell, D., Wood, E., & Galuppi, B. (1997). Gross motor function classification system for cerebral palsy. *Developmental Medicine and Child Neurology, 39*(4), 214–223.

Palisano, R. J., Cameron, D., Rosenbaum, P. L., Walter, S. D., & Russell, D. (2006). Stability of the gross motor function classification system. *Developmental Medicine and Child Neurology, 48*(6), 424–428.

Palisano, R. J., Rosenbaum, P., Bartlett, D., & Livingston, M. H. (2008). Content validity of the expanded and revised gross motor function classification system. *Developmental Medicine & Child Neurology, 50*(10), 744–750.

Palisano, R., Rosenbaum, P., Bartlett, D., & Livingston, M. (2021). Gross motor function classification system: expanded and revised. In: *CanChild Centre for Childhood Disability Research*. Accessed 4 July 2021.

Paquet, A., Olliac, B., Golse, B., & Vaivre-Douret, L. (2017). Evaluation of neuromuscular tone phenotypes in children with autism spectrum disorder: An exploratory study. *Neurophysiologie Clinique/Clinical Neurophysiology, 47*(4), 261–268.

Parma, V., & de Marchena, A. B. (2016). Motor signatures in autism spectrum disorder: The importance of variability. *Journal of Neurophysiology, 115*(3), 1081–1084. https://doi.org/10.1152/jn.00647.2015

Pascal, A., Govaert, P., Oostra, A., Naulaers, G., Ortibus, E., & Van den Broeck, C. (2018). Neurodevelopmental outcome in very preterm and very-low-birthweight infants born over the past decade: A meta-analytic review. *Developmental Medicine & Child Neurology, 60*(4), 342–355.

Patel, D. R., Neelakantan, M., Pandher, K., & Merrick, J. (2020). Cerebral palsy in children: A clinical overview. *Translational Pediatrics, 9*(Suppl 1), S125–S135. https://doi.org/10.21037/tp.2020.01.01

Pearson, T. S., Pons, R., Ghaoui, R., & Sue, C. M. (2019). Genetic mimics of cerebral palsy. *Movement Disorders, 34*(5), 625–636.

Piercy, K. L., Troiano, R. P., Ballard, R. M., Carlson, S. A., Fulton, J. E., Galuska, D. A., George, S. M., & Olson, R. D. (2018). The physical activity guidelines for Americans. *JAMA, 320*(19), 2020–2028.

Prechtl, H. F. (1990). Qualitative changes of spontaneous movements in fetus and preterm infant are a marker of neurological dysfunction. *Early Human Development, 23*, 151–158.

Prechtl, H. F. R. (1997). State of the art of a new functional assessment of the young nervous system. An early predictor of cerebral palsy. *Early Human Development, 50*(1), 1–11.

Prechtl, H. F. R., & Hopkins, B. (1986). Developmental transformations of spontaneous movements in early infancy. *Early Human Development, 14*(3–4), 233–238.

Prechtl, H. F. R., Einspieler, C., Cioni, G., Bos, A. F., Ferrari, F., & Sontheimer, D. (1997). An early marker for neurological deficits after perinatal brain lesions. *The Lancet, 349*(9062), 1361–1363.

Pritchard-Wiart, L., Thompson-Hodgetts, S., & McKillop, A. B. (2019). A review of goal setting theories relevant to goal setting in paediatric rehabilitation. *Clinical Rehabilitation, 33*(9), 1515–1526. https://doi.org/10.1177/0269215519846220

Rapport, M. J., Furze, J., Martin, K., Schreiber, J., Dannemiller, L. A., DiBiasio, P. A., & Moerchen, V. A. (2014). Essential competencies in entry-level pediatric physical therapy education. *Pediatric Physical Therapy, 26*(1), 7–18.

Rathinam, C., Mohan, V., Peirson, J., Skinner, J., Nethaji, K. S., & Kuhn, I. (2019). Effectiveness of virtual reality in the treatment of hand function in children with cerebral palsy: A systematic review. *Journal of Hand Therapy, 32*(4), 426–434.

Rosenbaum, P., Paneth, N., Leviton, A., Goldstein, M., Bax, M., Damiano, D., Dan, B., & Jacobsson, B. (2007). A report: The definition and classification of cerebral palsy April 2006. *Developmental Medicine and Child Neurology. Supplement, 109*(suppl 109), 8–14.

Rowland, J. L., Fragala-Pinkham, M., Miles, C., & O'Neil, M. E. (2015). The scope of pediatric physical therapy practice in health promotion and fitness for youth with disabilities. *Pediatric Physical Therapy, 27*(1), 2–15.

Sadowska, M., Sarecka-Hujar, B., & Kopyta, I. (2020). Cerebral palsy: Current opinions on definition, epidemiology, risk factors, classification and treatment options. *Neuropsychiatric Disease and Treatment, 16*, 1505–1518. https://doi.org/10.2147/NDT.S235165

Schiariti, V., Sauve, K., Klassen, A. F., O'Donnell, M., Cieza, A., & Masse, L. C. (2014). 'He does not see himself as being different': The perspectives of children and caregivers on relevant areas of functioning in cerebral palsy. *Developmental Medicine and Child Neurology, 56*(9), 853–861. https://doi.org/10.1111/dmcn.12472

Sellers, D., Mandy, A., Pennington, L., Hankins, M., & Morris, C. (2014). Development and reliability of a system to classify the eating and drinking ability of people with cerebral palsy. *Developmental Medicine & Child Neurology, 56*(3), 245–251.

Silva, N., Zhang, D., Kulvicius, T., Gail, A., Barreiros, C., Lindstaedt, S., Kraft, M., Bölte, S., Poustka, L., & Nielsen-Saines, K. (2021). The future of general movement assessment: The role of computer vision and machine learning–A scoping review. *Research in Developmental Disabilities, 110*, 103854.

Spittle, A. J., Morgan, C., Olsen, J. E., Novak, I., & Cheong, J. L. Y. (2018). Early diagnosis and treatment of cerebral palsy in children with a history of preterm birth. *Clinics in Perinatology, 45*(3), 409–420.

Todorova, G. K., Hatton, R. E. M., & Pollick, F. E. (2019). Biological motion perception in autism spectrum disorder: A meta-analysis. *Molecular Autism, 10*(1), 49. https://doi.org/10.1186/s13229-019-0299-8

van Hoorn, J. F., Schoemaker, M. M., Stuive, I., Dijkstra, P. U., Rodrigues Trigo Pereira, F., Van der Sluis, C. K., & Hadders-Algra, M. (2021). Risk factors in early life for developmental coordination disorder: A scoping review. *Developmental Medicine & Child Neurology, 63*(5), 511–519.

World Health Organization (WHO) (2002) *Towards a Common Language for Functioning, Disability and Health: ICF*. WHO, Geneva. http://www.who.int/classifications/icf/training/icfbeginnersguide.pdf.

Wilson, R. B., Enticott, P. G., & Rinehart, N. J. (2018). Motor development and delay: Advances in assessment of motor skills in autism spectrum disorders. *Current Opinion in Neurology, 31*(2), 134–139.

Wood, L., Burke, E., & Morrison, A. (2015). Individual cognitive behavioural therapy for psychosis (CBTp): A systematic review of qualitative literature. *Behavioural and Cognitive Psychotherapy, 43*(3), 285–297. https://doi.org/10.1017/S1352465813000970

Wouters, M., Evenhuis, H. M., & Hilgenkamp, T. I. M. (2020). Physical fitness of children and adolescents with moderate to severe intellectual disabilities. *Disability and Rehabilitation, 42*(18), 2542–2552.

Wynarczuk, K. D., Chiarello, L. A., & Gohrband, C. L. (2017). Goal development practices of physical therapists working in educational environments. *Physical & Occupational Therapy in Pediatrics, 37*(4), 425–443. https://doi.org/10.1080/01942638.2017.1290009

Zablotsky, B., Black, L. I., Maenner, M. J., Schieve, L. A., Danielson, M. L., Bitsko, R. H., Blumberg, S. J., Kogan, M. D., & Boyle, C. A. (2019). Prevalence and trends of developmental disabilities among children in the United States: 2009–2017. *Pediatrics, 144*(4), e20190811.

Planning for Treatment of Children with Autism and Other Developmental Disabilities with the Child and Adolescent Psychiatrist

19

Hannah Scott, Jennifer Creedon, and Joshua Sanderson

Abstract

Child and adolescent psychiatrists are physicians trained in brain structure and function with a focus on mental health disorders and their pharmacological treatment. Treatment planning for children with autism spectrum disorders benefits from the psychiatrist's perspectives on child development, parent–child relationships, therapy, and the pharmacological treatment of the disorders that present during infancy, childhood, and adolescence. This chapter reviews mental disorders and behavioral challenges common to children with autism with a focus on integrating pharmacological and behavioral interventions.

Keywords

Autism Spectrum Disorder · Psychiatry · Treatment planning · Pharmacology · Coordinating care · Child psychiatry · ASD · Therapy · Medication · Medication management

H. Scott (✉) · J. Creedon · J. Sanderson
Louisiana State University School of Medicine, New Orleans, LA, USA

Child and adolescent psychiatry is a subspecialty of psychiatry and neurology with a focus on development, behavior, and emotional well-being of infants, children, and adolescents. Psychiatrists are medical doctors that obtained a medical degree, M.D. or D.O., who then went on to train specifically in a psychiatry residency. Often, those who treat patients with autism or developmental disabilities pursued additional training in child and adolescent psychiatry. Most often, psychiatrists are boarded by the ABPN, the American Board of Psychiatry and Neurology; however, some are board certified by an alternate path through the American Society of Adolescent Psychiatry. Psychiatrists are trained in a number of different aspects of medicine, including pharmacology, neurology, substance abuse, relationship dynamics, communication, brain structure and function, as well as mental health disorders. A child psychiatrist obtains additional training in child motor development, cognitive development, parent–child relationships, play therapies, brain maturation, behavioral modification, and treatment of various disorders that present during infancy, childhood, and adolescence. Additionally, child psychiatrists receive experience working with school systems for specific accommodations and learning and behavioral plans that would benefit their patients. Some psy-

P. McPherson (ed.), *Handbook of Treatment Planning for Children with Autism and Other Neurodevelopmental Disorders*, Autism and Child Psychopathology Series,
https://doi.org/10.1007/978-3-031-06120-2_19

chiatrists work with the juvenile justice system as well to assist or treat those patients where the legal system and the mental healthcare system intersect.

A child and adolescent psychiatrist is educated in specific disorders that are relevant to children who may have an autism spectrum disorder and/or developmental disorder (ASD/DD) and who may suffer from other co-occurring mental illnesses. There is also a breadth of experience that builds during training and practice. For example, psychiatrists work alongside social workers who are aware of community coalitions that provide access and coordination to treatment to mental health resources in the community (Kataoka et al., 2018). The overall health of a person with ASD/DD may be enhanced through primary care and community-based health services (Saqr et al., 2017). A psychiatrist has access to a network of different organizations that offer a spectrum of services, understands how they operate and where they are located, and personally knows the people who work there. This can help the medical home team to match the patient with the appropriate services. Not every child is created equal nor is every organization. Assembling the proper medical home team is a key to successful treatment.

Some primary care practices directly employ the medical home model to develop a *shared plan of care* (SpoC), which will be discussed in more detail later in the chapter, with a goal to enhance the experience for the family and the practitioners. In some well-coordinated systems, pediatricians and family practitioners collaborate with the same psychiatrist in a multidisciplinary clinic where the patient and family can be referred back and forth for additional concerns. Alternatively, psychiatrists with their own private practice may have the primary care physicians refer the patient for treatment outside of the primary care setting. For complex cases, the psychiatrist may follow patients long term. Conversely, for straightforward cases, an opinion or recommendation may be offered to the PCP. The amount of involvement will depend on the complexity of the individual and must be determined on a case-by-case basis.

Preparing for the Appointment

Physicians rely on past records as part of the evaluation of a patient. Records from pediatricians, psychologists, and geneticists are just a few examples of the types of information needed by a psychiatrist. In order to receive these records, a Release of Information must be signed by the patient and/or guardian. Different hospitals, clinics, and schools have different policies regarding protected health information and may require their own Release of Information signed before releasing medical records. The process for gathering information can take time and thus requires patience on both the doctor and patient's part. To expedite the process, a family may maintain a copy of the information and can provide that to the doctor directly. The 21st Century Cures Act provides families direct access to medical records and is designed to promote care coordination.

A child diagnosed with ASD/DD does not automatically require a referral to a psychiatrist. Children and adolescents with autism are referred at different stages in their life for different reasons. Table 19.1 breaks down the ages and common reasons a child or adolescent is referred to a psychiatrist with or even prior to a diagnosis of autism or developmental disability.

Table 19.1 Common reasons for referrals to a child psychiatrist at different ages with or without an autism/developmental disability diagnosis

Age	Reason for referral
Birth to 2 years	Feeding issues, bonding concerns, developmental concerns, sleep issues, family adjustment to genetic disorder or other condition
2–5 years	Behavior problem, developmental delay, social-communication issues, self-abuse
5–12 years	Problems related to school and learning, behavior problems in school, attention and concentration issues, difficulty making friends, issues with the education system related to accommodations and learning plans, undiagnosed comorbid intellectual disability
13–18 years	Anxiety, depression, suicidality, development of psychosis, difficulty transitioning to adulthood, behavioral deterioration

Child and adolescent psychiatrists are often involved in the care of patients with ASD/DD, as these patients can be clinically complex and require an all-inclusive treatment plan with the aid of a multidisciplinary medical home team. This patient population often has a higher incidence of comorbid psychiatric conditions. The *Diagnostic and Statistical Manual of Mental Disorders (5th ed.; DSM-5; American Psychiatric Association,* 2013*)* now recognizes that persons with autism can have coexisting psychiatric disorders. A metanalysis done by Hossain et al. revealed a high rate of anxiety disorders; depression; psychotic disorders; disruptive, impulse-control, and conduct disorders; attention-deficit/hyperactivity disorder; and other psychiatric disorders in persons with autism (Hossain et al., 2020). Another study which looked at a population-derived cohort of 112 children aged 10–14 years noted that social anxiety disorder was the most common diagnosis, followed by ADHD and oppositional defiant disorder (ODD; Simonoff et al., 2008). Children with intellectual disability on average experience three to four times the rate of mental illness compared to the general population (Einfeld et al., 2006). Studies suggest that approximately 40% of children with ASD require psychiatric medication for issues related to anxiety, irritability, depression, sleep problems, and attention and concentration issues. However, access to general child psychiatric services can be difficult and finding services for children with ASD/DD can pose additional hardships (Marrus et al., 2014). There are a number of different mechanisms for incorporating psychiatry into the clinical treatment team, including a referral, telemedicine appointments, collaborative practice, or an independent referral to an outside psychiatrist. Some medical home primary care physicians may formally or informally "curbside" in a physician-to-physician consultation. In these circumstances, if the psychiatrist requires additional information related to behavior or family interaction, a one-time telemedicine or in-person evaluation may be warranted.

Coordination of care between specialty services and primary care services is one of the most important fundamental elements in the American Academy of Pediatrics' patient-and-family-centered medical home model (PFCMH). The medical home model places the family at the center and calls for partnership among specialties to optimize care for the patient and support for the family. Coordinating care among physicians may include discussing the patient's needs before the referral, followed by treatment recommendations after the visit. Communication can take place through different forms of telecommunication, secure fax, email, text messaging, or telephone. As technology becomes increasingly interconnected, team members of a patient's medical home may be able to access physician notes immediately. The medical home model benefits all families by identifying possible psychosocial and mental health needs such as linkage to state agencies coordinating respite care, personal care attendants, and other services. As shown in Table 19.2, when initiating a referral to a child and adolescent psychiatrist, the primary care physician can anticipate general questions about the referral.

If there are multiple child psychiatrists in your area who specialize in ASD/DD, there are a number of points to consider when choosing the "best fit." Prior to a referral, if ASD is in the differential diagnosis, the referring physician may want to consider genetic testing, IQ testing, or specific

Table 19.2 General questions posed from the child psychiatrist to the referring primary care physician

1. Reason for referral.
2. Identifying and demographic information: Age, sex, gender, race, ethnicity, etc.
3. The medical home shared plan of care, emergency plan, and other documentation related to medical diagnoses and associated treatments for the patient.
4. Prior mental health treatment or psychiatric medication usage.
5. Specific issues that may be relevant – Access to care, socioeconomic status, financial issues, housing issues, single-parent household, barriers to treatment.
6. Possible concern for genetic abnormalities or intellectual disability.
7. Any available reports already completed, i.e., sleep studies, GI or swallow studies related to poor feeding, EKGs, EEGs, genetic studies, and general lab work.
8. Sensitive information of concern for resistance or challenges to treatment.

psychological testing to diagnose an ASD prior to referral to a psychiatrist. Getting the ball rolling on multiple assessments simultaneously accomplishes one of the shared goals of the medical home. The results of the testing can be added to the medical home *shared plan of care* (SpoC) as previously mentioned and can guide treatment and recommendations for specialties outside of psychiatry as well. Lastly, a psychiatrist may be aware of ongoing clinical trials that are a good fit for the patient and can aid in coordination.

How to Prepare a Family for a Referral

Individual psychiatrists have different approaches to evaluating and treating patients. Many families experience anxiety when referred to a psychiatrist. The referring physician can ask the family if they have any questions about the referral and can anticipate questions regarding the referral as noted in Table 19.3. The family may gain comfort in knowing that their PCP, whom they know and trust, has initiated the care coordination by inviting the psychiatrist to partake in the medical home of the patient. Embracing a patient-centered approach is paramount to recognizing the unique challenges a family faces while also acknowledging that there will be shared decision-making concerning the treatment of the patient (Edison et al., 2020). It may be a good idea to contact the clinic before an initial appointment in order to inquire about the expectations of the first clinic visit. By discussing these questions even if the physician doesn't know the answer, the family will feel more informed and prepared for the referral. For instance, the potential for other issues or detection of a rare genetic disorder may be discovered, which can be devastating for a family.

Table 19.3 Common family concerns when a referral is being made to a child psychiatrist

Why is a referral being made?
What can a specialist provide that is currently not being provided?
What if they find something serious that is wrong?
Why would we need to see a geneticist?
What will other people think if we go see a psychiatrist?
What if I don't agree with their recommendations?
What if I do not want to put my child on medication?
What if they try to put my child in a psychiatric hospital?

What to Expect from the Initial Encounter with a Psychiatrist

Some clinicians may prefer to meet with caregivers independent of the child prior to having an appointment with others present. On the other hand, some clinicians may desire the family to bring records, including psychological, genetic, or laboratory testing that has been performed prior to the referral. The records can also be faxed or mailed beforehand for review, if possible, to make the most out of the time allotted for the appointment. If these documents are not in the family's possession, the family may request records or provide the clinician a signed Release of Information form, so they can obtain the information directly from other providers. It can often be beneficial for families to bring an additional caregiver in case the child gets unmanageable during the assessment, as the time it may take to complete an initial evaluation can vary. The average initial intake evaluation can last between 1 and 3 hours, depending on the clinic and the specific circumstances of the case.

Psychiatrists will often take a biopsychosocial approach in managing these complex cases. The assessment considers important aspects of the biological history, i.e., perinatal and postnatal history, genetics, overall health, and developmental history. The psychological aspect considers maladaptive and adaptive beliefs, behavior, emotional regulation, and level of skills, while the social component takes into consideration the family system, community support, education, environmental effects, and social support (Sperry & Sperry, 2012). The provider should educate the caregivers to this concept and encourage them not to be alarmed if questions seem intrusive or personal. Although some of these questions may be uncomfortable to answer, it is important to be honest so the patient can receive the best care. For example, being open and honest about substance use or a history of a sexually transmitted disease can help guide the clinician to providing the appropriate treatment to relieve suffering. The clinician is trained not to judge, and their aim is to help where help is needed. The clinician will attempt to learn as much as they can in order to accurately assess the resources that each specific family has at their disposal. For example, a single parent may be asked about involvement of other family members to determine if respite care would be beneficial for both the child and parent.

Components that Guide the Psychiatric Treatment Plan

The Assessment

The evaluation is composed of the clinical interview, review of records, possible lab work, and coordination of care with other providers. The assessment brings the individual parts to a whole. It involves incorporating science, medical education, and clinical experience in order to arrive at the appropriate plan with specific recommendations to meet the individual patient's need. Youngstrom and Van Meter outlined the assessment in three separate parts: the prediction, the prescription, and the process (Youngstrom & Van Meter, 2016). The prediction involves outlining symptoms and risk factors; the prescription involves matching treatment including therapy and pharmacological treatment with the diagnosis but taking into consideration the values of the patient and the family; lastly, the process is an ongoing work of monitoring symptoms and progress, defining and redefining goals, and measuring treatment (Youngstrom & Van Meter, 2016).

The assessment is an ongoing process. Understanding the past helps the psychiatrist plan and set treatment goals for the future. There is a difference in the assessment approach between a child who may have autism and a child that has already been diagnosed with autism; therefore, the more precise the referral question, the more beneficial the evaluation will be. Formal and very specific testing is often required in order for the child to receive government benefits. The psychiatrist may be able to address symptoms related to the diagnosis of an ASD while awaiting testing results. A psychiatrist will need to gather and review the reports, determine what testing is needed, and recommend or direct future referrals as needed.

There are key elements that guide treatment. The examination covers the child from birth to present, including the mother's pregnancy and delivery. Substance use, complications, including postpartum depression, illness, domestic abuse, and nutrition throughout pregnancy will be part of the initial evaluation visit. Developmental milestones, i.e., age when the child started walking, talking, and was potty trained, are objective measures that allow physicians to assess and compare the child to the norms of the general population. This helps determine if the child's development is delayed, normal, or advanced by comparing it to age-matched norms consistent with their peers. Some families may find it helpful to bring baby books documenting milestones or previous evaluations for reference. Physicians will be interested in a child's speech, language, executive functioning, and play habits as well as interactions with family members during the encounter.

A family history of mental illness, genetic conditions, medical illness, and substance use are also important topics – not only in predicting patterns of inheritance but also when prioritizing family support and psychosocial interventions. Autism alone has a heritability pattern of approximately 80% (Atbaşoğlu, 2020). This fact is important to discuss if the family is considering having additional children in the future. It is also important to screen family and caregivers for signs and symptoms of depression or anxiety given the nature of these conditions and the severity of symptoms in some cases. A child's exposure to chronic caregiver depression can be associated with worsening of cognitive and emotional problems, and developmental delays. A caregiver's depression can also affect their ability to respond to a child's needs with the patience required under these conditions (Öz et al., 2019). By uncovering issues presented in the examination, the physician can help guide treatment of the family as well. The primary caregivers may need therapy or psychiatric treatment themselves to lessen caregiver burnout, as the severity of autism along with caregiver depression results in the highest caregiver burden (Baykal et al., 2019).

The diagnostic system used by mental health professionals, *The Diagnostic and Statistical Manual 5* (DSM-5), classifies mental disorders based on symptom checklists. Under this nosology, a symptom may fall under an array of diagnoses. This requires the psychiatrist to develop a broad differential diagnosis list for a symptom constellation and then prioritize additional assessment and interventions to determine whether any comorbid conditions are present and to rule out other conditions that may be confused with autism. In short, to determine if symptoms are due to the ASD/DD or a comorbid mental disorder. Mood disorders such as major depressive disorder and bipolar disorder, personality disorders, anxiety disorders, and psychotic disorders, can lead to symptoms similar to the core symptoms of developmental disabilities and autism as well as present as comorbidities. The psychiatrist must determine if the symptom patterns warrant the additional diagnosis only if DSM-5 criteria were met and were not solely due to ASD/DD. Tic disorders, obsessive-compulsive disorder (OCD), attention-deficit/hyperactivity disorder, impulse-control disorders, intellectual disability, and other mental illnesses may also complicate the clinical picture (Atbaşoğlu, 2020) due to an overlap of symptoms. A history of physical, emotional, and sexual abuse as well as neglect are important factors during a patient's upbringing that can contribute to behavioral issues and emotional dysregulation. In fact, a trauma-informed approach is recommended for all patient encounters.

School performance will be assessed and descriptions of the child's interactions with family and other children will be obtained. Parents and families often note that school is an area of distress for a child with ASD/DD for a multitude of reasons. Children may already be receiving support through in-class interventions, 504 Accommodations, and/or an IEP (discussed in the next section of the chapter). A history of the child's current function in school and access to services is important. Other information will include current school, current grade level, history of academic testing, and history of services in school. If a child is having difficulty particularly in school, more information may be requested.

Other Medical Disorders

As a medical doctor, a psychiatrist is uniquely trained to assess for medical conditions presenting with mental health symptoms and address the reciprocal impact of physical and mental illnesses. When medical issues are appropriately dealt with and managed, mental health also improves and vice versa. A thorough medical history will likely entail a screening for frequently comorbid conditions, medical diseases that can be confused with mental illness, and medical illness that could complicate treatment with certain pharmacological interventions. Conditions such as fetal alcohol syndrome, neurofibromatosis, Wilson's disease, irritable bowel syndrome, seizure disorders, sleep disorders, and autoimmune conditions are frequently seen in patients with ASD/DD.

The evaluation may reveal genetic disorders or lab abnormalities that need to be treated. Medical diseases that present with symptoms similar to mood, anxiety, obsessive-compulsive, behavioral dysregulation, and psychotic disorders include hypothyroidism, hyperthyroidism, substance abuse, chemical poisonings (lead/mercury/carbon monoxide) and malignancies. Genetic disorders may be due to Rett syndrome, neurocutaneous disorders, phenylketonuria, Angelman syndrome, Smith-Lemli-Opitz syndrome, Fragile X, Williams syndrome, 5p-syndromes, Prader-Willi, Landau-Kleffner syndrome, trauma/head injury, adrenal hyperactivity/insufficiency, and infection (Dulcan, 2016; Johnson & Myers, 2007). Additionally, over-the-counter, herbal remedies, and caffeine can interact with prescription medications and can result in adverse reactions or conditions that mimic psychiatric illness. Medication interactions will need to be assessed to evaluate the possibility of increased medication levels leading to toxicity or decreased medication levels rendering them ineffective. A thorough review of medications is necessary to determine possible unwanted side effects.

Medical conditions that could complicate treatment usually involve conditions related to the heart, kidneys, liver, and gastrointestinal tract. Physicians may ask about a history of congenital cardiac malformations, arrhythmias, murmurs, syncope, and family history of early cardiac death prior to prescribing stimulants. Certain histories may warrant further evaluation and management. (This is discussed further in other chapters of this book). Common long-term side effects of psychiatric medications can include hypertension, elevated cholesterol, diabetes, and obesity. The initial evaluation sets the foundation and directs treatment. The psychiatrist may coordinate with the medical home for additional referrals as indicated.

The Evolving Evaluation

The psychiatric evaluation of a child with complex healthcare needs is a process, not an end point. The family will arrive at the physician's office and begin to build a relationship with the psychiatrist and mental health staff one step at a time. As in any relationship, trust is built and gained over time. The initial psychiatric evaluation will be the first stepping-stone to building a doctor–patient relationship that is family centered. The psychiatrist hopes to establish trust enabling a family to share sensitive information that is pertinent to a child's treatment.

The medical field has evolved over the past few decades to adopt a person-centered approach to evaluation and treatment of patients. During an appointment, the psychiatrist aims to understand the needs, expectations, and psychosocial context, reach an understanding of the problem with related treatment options that align with the family's values, and create a partnership between the doctor, the patient, and the caregiver to share in the decision-making responsibility (Małus et al., 2018). The first impression can set the tone for the doctor–patient relationship. A recent study revealed that patients formed their opinion and attitude toward their psychiatrist at the initial stage of treatment and maintained this attitude thereafter, which was usually positive (Małus et al., 2018). Furthermore, the longer a relationship between a psychiatrist and patient, the more positive the attitude regarding the interaction became (Małus et al., 2018). Building a trusting relationship with the psychiatrist as part of the treatment team is important. If a patient and their caregiver feel that they are not able to build a strong doctor–patient relationship after a few appointments, then it may be appropriate to seek help from another specialist.

The age at which a child is diagnosed with ASD or a development disability and subsequently referred to a psychiatrist will determine the direction of screening for other mental illnesses. Epidemiological data reveals the prevalence and incidence of specific mental illnesses occurring in different age groups. A psychiatrist is aware of this data and can screen a child appropriately. About 70% of people with ASD have another psychiatric disorder and about 40% will have two or more psychiatric disorders (DeFilippis, 2018). The frequency of co-occurring psychiatric disorders can compound existing challenges (Hossain et al., 2020). Continued evaluation and monitoring of those affected can help identify and address symptoms early for improved outcomes. The most common co-occurring disorders include anxiety disorders, depressive disorders, bipolar and mood disorders, schizophrenia spectrum disorder, and ADHD (Hossain et al., 2020). A recent meta-analysis from 2019 found that co-occurring mental illnesses occur at higher rates in those with ASD compared to the general population (Lai et al., 2019). About two out of three persons with ASD will have one comorbid mental disorder and almost half will have two or more that cannot be explained by the diagnosis of autism alone (DSM-5). Persons with intellectual disability can have up to four times the rate of other conditions including mental disorders or epilepsy. Other developmental disabilities can include global developmental delay, communication disorders, pervasive developmental disorder, cerebral palsy, fetal alcohol syndrome, and Down's syndrome. From ample clinical experience, a psychiatrist becomes an autism expert and can determine if behaviors are resulting from development disorders or they are from a comorbid diagnosis. This practice takes nuance and experience in evaluating symptoms. Prevention is key. Identifying symptoms early in childhood may decrease the risk and severity of psychiatric disorders in adulthood in those with ASD.

Along with screening and diagnosis comes testing. Testing can help delineate the problem and guide treatment. Testing is a broad term that encompasses psychological testing, blood samples for medical evaluation, blood or saliva samples for genetic testing, intelligence testing, adaptive skills testing, and achievement testing to name just a few. Tests can be problem-specific or broad in scope. Factors that must be accounted for are the age range the test covers, the time it takes to administer and score, the cost, the delivery method (i.e., via paper, electronic or person-to-person), and the available languages (Delahunty, 2015). The psychiatrist will determine what information is needed and order and prepare the patient and family accordingly. A psychiatrist has familiarity working with insurance companies to advocate and obtain prior approval for the tests. Lastly, a psychiatrist is able to determine the appropriate timing for the test to get the most accurate and beneficial data for treatment.

The doctor–patient–caregiver relationship and trust that is built over time lends to an undercurrent of support for the families that are having to deal with chronic and serious disorders. A psychiatrist is in a unique position to provide empathy because they understand the intricacies and difficulties that accompany the disorder. A caregiver of a child with ASD is more vulnerable when the diagnosis is new and the treatment and prognosis of the child is unknown. Psychiatrists understand that when a caregiver is suffering mentally there is a trickle-down effect to the people they are caring for, which increases the risk of delay in developmental progress (Osborne et al., 2008). Supporting and increasing resilience in the caregiver, identifying low confidence, feelings of hopelessness and isolation in a parent, and recognizing stigma associated with the diagnosis are all ways that a child psychiatrist may support a family and assist in identifying additional interventions that can benefit the child and caregivers (Papadopoulos et al., 2019).

Advocating and Navigating the School System

A child and adolescent psychiatrist works with children entering, navigating, and exiting the school system and is therefore intimately acquainted with the complexities inherent to the system. Chapter 12 in the volume addresses education in detail. The first intervention by a school may be a Response to Intervention (RTI) plan. The aim of an RTI is to identify students who are struggling in one or more areas and provide support. This support can take many different shapes and forms, ranging from class-wide instruction, to small group instruction, to individual intensive interventions. The intervention can vary depending on the need and the progress of the student. The RTI plan may or may not be shared with parents and does not reduce a student's workload or provide accommodations for students with a disability.

If a student continues to struggle in school, the next step or concurrent step may be 504 accommodations. A student can receive education in regular classrooms, in regular classes with supplementary services, and/or special education and related services. This allows for a wide range of accommodations from extended time for taking tests to access to physical and occupational therapy.

A child with ASD has the right to a Free Appropriate Public Education (FAPE) mandated by the federal government and delivered by the state government (US Department of Education, 2010). Autism is a specific disability that is eligible for special education and related services according to the Individuals with Disabilities Education Act (IDEA), through which an Individualized Education Plan (IEP) will be developed for qualifying children. Developing an IEP occurs with a team which includes certain educators in the public school, the child (when appropriate), the child's parents or guardians, and others whom the parents or guardians wish to be in attendance. This opens the door to have a psychiatrist or mental health advocate to promote your child's needs. Once assembled, an individualized plan is formulated with the goal to improve educational results. If a child does not have an IEP and is not meeting education goals, an IEP may be appropriate. Child and adolescent psychiatrists are familiar with the process of obtaining and advocating for a child in need of services and can help navigate this complex system.

Child and adolescent psychiatrists are well versed in federal laws that extend care and treatment to children with mental and physical disabilities in the school system and understand how their particular state executes those. The doctor can write letters to the school on behalf of the family to advocate for the child's specific needs. Some children may qualify for benefits through federally mandated state services established under federal developmental disability mandates. These programs vary in name and may have state-specific requirements for services. These services include programs for infants and toddlers age zero to three who exhibit developmental delay. The services provided can be in the form of therapy, monetary aid for families who care for children with profound or severe disabilities, and waiver programs to assist in providing living options in the least-restrictive environment possible. A Medicaid Waiver must be applied for and qualifications must be met in order for an individual to benefit from this service. A child may qualify for Medicaid to access insurance and cover costs of care. They may qualify for SSI (Supplemental Security Income) if they are determined to be disabled. A psychiatrist is in a unique position to aid a family through these hurdles.

Telemedicine

In our current ever-changing digital era, telemedicine has emerged as a means to deliver mental health services to those previously unreached. It has been defined many ways, but in general "telemedicine can be thought of as the application of medical expertise or services via a remote care delivery or support pathway" (Hilt, 2017). Telemedicine provides access and availability to mental health services. It is cost-effective and can be delivered securely to maintain confidentiality.

Autism spectrum disorder does not discriminate according to region. Those with ASD/DD can be found in urban and rural areas alike. Child and adolescent psychiatrists, however, are more likely to be located in urban areas. Telemedicine has increased the availability of mental health providers to these remote places. ASD can be accurately diagnosed using telemedicine (Juárez et al., 2018). In a meta-analysis, services delivered via telehealth were equivalent to those delivered in face-to-face encounters (Sutherland et al., 2018). Follow-up care was also accessible and the families who utilized telemedicine expressed satisfaction with the service (Juárez et al., 2018).

To treat children and family with ASD/DD via telehealth, certain accommodations should be considered. Logistically, a stable internet connection is needed to proceed through the evaluation along with a basic understanding of technology. The time allotted for the evaluation may need to be increased to allot for issues that arise. One study revealed that a telehealth visit required more time to complete the evaluation due to technical and cultural factors (Allen & Shane, 2014). In communities where mental health resources are scarce, telemedicine provides much needed access to services. Proper infrastructure, hardware, and software would need to be available and accessible by the treatment provider and the family. Family assessments or visits with multiple people could pose difficulties in assessing parent–child interactions. Telemedicine can overcome several barriers to treatment, but there continue to be limits to the utility of telemedicine.

Psychiatry Coordinated Services

Once a child has received a diagnosis of ASD/DD, the child psychiatrist will discuss potential treatment options for the symptoms related to autism since there is no cure for autism. There is no "one size fits all" treatment. Treatment modalities aim to reduce distressing symptoms while promoting development and enhancing learning opportunities as a child grows.

Referrals are likely to be made to a number of different experts and specialists. In the medical home model, the PCP coordinates care; however, in some instances, the child psychiatrist or another medical home team member may take on this role to serve as the leader. As autism impacts several different key areas of functioning, specialized therapies will be required to address deficits in language, social skills, motor skill, and comorbid psychiatric diagnoses. The types of referrals needed will depend upon the individual child and family's needs. Common specialized services involved in the treatment of children with autism include neurology, physical therapy, occupational therapy, speech therapy, and psychotherapy (primarily behavioral interventions, a brief review in this chapter, and a more in-depth view in the psychology chapter).

Secondary Psychiatric Diagnoses

Individuals with autism are more likely than the general population to have additional, or comorbid psychiatric diagnoses. The term "comorbid" is used to denote the presence of other disorders in addition to the primary diagnosis. It is estimated that approximately 70% of individuals with ASD have an additional comorbid psychiatric diagnosis, and that up to 40% of those with ASD have two or more additional diagnoses (Mansour et al., 2017). The most common comorbid conditions seen in children with autism are social anxiety disorder, attention-deficit/hyperactivity disorder (ADHD), and oppositional defiant disorder (Simonoff et al., 2008). Generalized anxiety disorder, panic disorder, and enuresis are also overrepresented in children with ASD. Additionally, irritability is common in children with ASD, and can often be misdiagnosed as hyperactivity, anxiety, or mania. Diagnosing comorbid psychiatric conditions can pose a particular challenge in individuals with ASD, given the complicated overlap of symptoms between ASD and other DSM-5 diagnoses. Because of the nuance required in teasing apart these symptoms and subsequently developing an effective treatment plan, it is critical to involve an expert in pediatric psychiatric illnesses, such as a child psychiatrist. Diagnosis and treatment of these comorbid conditions is a process in itself. Co-occurring diagnoses may be recognized prior to a formal diagnosis of ASD or Developmental Delay, they may be diagnosed during an ASD evaluation, and/or they may be identified later in the treatment process.

Irritability in Autism

A common presenting problem in patients with ASD is severe irritability, demonstrated through verbal or physical outbursts of anger or frustration. Caregivers of children with ASD may refer to these episodes as "meltdowns" or "temper tantrums." These symptoms may lead to disruption of school or daycare placement and can frequently be the presenting concern leading to inpatient hospitalization of children with ASD. One study estimated that approximately 20% of children with autism have symptoms of irritability, including aggression, tantrums, and deliberate self-injury (Lecavalier, 2006). Irritability can be a symptom of several other psychiatric diagnoses, including mood disorder, anxiety disorders and PTSD, a side effect from medication, or a symptom of ASD itself. Because of this, as well as the potential consequences of untreated irritability, it is important to obtain a thorough history with regards to irritability and aggressive behaviors in order to determine the most likely etiology and the best treatment plan. While irritability can be a particularly problematic symptom in ASD, it is also a symptom with effective pharmacological treatments, as will be discussed later in this chapter.

Anxiety Disorders in ASD

Multiple studies have demonstrated that anxiety disorders are the most common co-occurring disorders among children with ASD. A meta-analysis by van Steensel and colleagues of 2121 children and adolescents found the prevalence of anxiety disorders of approximately 35%. By comparison, in typically developing children, the lifetime prevalence of any anxiety disorder is estimated between 15% and 32% (Dulcan, 2016). Anxiety, particularly social anxiety, becomes more common as children with ASD age into adolescence. Studies also suggest that children and adolescents with higher IQs have a higher burden of anxiety disorders as compared to those with ASD who have lower IQs (van Steensel & Heeman, 2017). Identifying anxiety disorders in children with ASD can be challenging. There is significant overlap of symptoms, particularly with regards to social anxiety disorder. Additionally, individuals with ASD are more likely to display behavioral dysregulation in the setting of anxiety, perhaps due to deficits in expressive language and difficulty with emotional insight. A 2006 study compared adolescents with Asperger syndrome (which under DSM-5 would now be classified as ASD) and anxiety with peers with anxiety only. The findings revealed youth with an ASD diagnosis were more likely to experience negative automatic thoughts, behavioral problems, and overall impairment related to their anxiety symptoms as compared to the adolescents diagnosed with anxiety disorders but not ASD (Farrugia & Hudson, 2016).

Attention-Deficit/Hyperactivity Disorder (ADHD)

In earlier versions of the DSM, if two psychiatric disorders had significant overlap in presentation to the point that the symptoms of one could be sufficient to meet criteria of the other, exclusionary criteria in the manual would state that the two diagnoses could not both be made in one individual. Such was the case of ASD and ADHD prior to the DSM-5. The earlier iterations of the DSM included ADHD in its exclusion criteria for autism, meaning that if a child had a diagnosis of ADHD, they could not also receive a diagnosis of autism. However, as our understanding of the neurophysiology, etiology, and presentation of psychiatric illness evolves, so does our diagnostic manual. With the release of the DSM-5 in 2013, several changes were made to the diagnostic criteria for ASD, including the removal of ADHD as an exclusionary criterion.

While there is significant overlap in symptomatology between ASD and ADHD, studies have demonstrated consistent phenotypical differences between children with ASD only and those with ASD and ADHD. In otherwise neurotypical children, the prevalence of ADHD is estimated to be between 3% and 7% (Dulcan, 2016). ADHD is more common in children with ASD; a recent metanalysis reported the prevalence of ADHD in youth with ASD ranging from 25% to 65% (Hossain et al., 2020). In children who function on the higher end of the autism spectrum, ADHD symptoms may become apparent prior to ASD symptoms, especially once the child starts school. Careful screening for ASD should be considered in children diagnosed with ADHD who are not responding as robustly as expected to stimulant medications. It is estimated that between 15% and 25% of children with a diagnosis of ADHD also meet criteria for ASD (Leitner, 2014).

Disruptive Behavior Disorders

One in four children with ASD will also meet criteria for a disruptive behavior disorder, such as ODD (oppositional defiant disorder) or conduct disorder (Kaat & Lecavalier, 2013). Disruptive behavior disorders can significantly impact the clinical course and morbidity for children with ASD, as children with both ASD and disruptive behavior disorders are approximately five times more likely to be psychiatrically hospitalized than children with ASD alone (Mandell, 2008). Disruptive behavior disorders are also associated with increased use of psychotropic medications, more severe functional impairment, and increased parental stress (Storch et al., 2012; Witwer & Lecavalier, 2005; Lecavalier et al., 2006).

The symptoms of these disorders, particularly ODD, can sometimes be difficult to distinguish from behavioral difficulties related to autism, especially in children with lower verbal abilities. Children with non-existent or minimal verbal abilities may engage in disruptive or aggressive behaviors in an attempt to communicate their needs. Disruptive behavior may also be used to gain access to a desired object of perseverative interest, such as a video game or specific toy (Reese et al., 2005). A child with autism may also become emotionally dysregulated due to changes in routines or schedule. While such behaviors are common in this population, it is important to remember that disruptive behaviors such as impulsive aggression or emotional dysregulation are not sufficient for making a diagnosis of a disruptive behavior disorder. To receive a diagnosis of ODD on top of ASD, a child must also meet additional DSM-5 criteria, such as active defiance of adults, deliberate annoyance of others, and spiteful or vindictive behavior. Likewise, to receive a diagnosis of conduct disorder, the individual must have consistently demonstrated behaviors consistent with DSM-5 criteria, i.e., deliberate destruction of property, deceitfulness or theft, physical cruelty to people or animals, and/or school truancy to name a few.

Depression

Rates of depression in children and adolescents with ASD vary widely across studies, with estimated prevalence anywhere from 2% to 47% (Hossain et al., 2020). In comparison, rates of depression in otherwise typical-developing youth are estimated at 2% in children and 4–8% in adolescents (Dulcan, 2016). The variability in rates of depression detected in youth with ASD may reflect difficulty in making a diagnosis in children and adolescents who may have less insight into their own internal states. Further, depression can present atypically in individuals with ASD, with increased rates of irritability, mood lability, and aggression. Inquiring about classic DSM symptoms of depression, including anhedonia or loss of interest or pleasure in activities, changes in energy and sleep, suicidality, and changes in appetite can guide diagnosis, particularly regarding changes in these symptoms from a child's baseline. Youth with higher-functioning ASD experience increased rates of depression compared to those with more severe ASD. The theory behind this finding is interesting: A person with less severe ASD has a higher capacity for insight into their differences compared to neurotypical peers. This contributes to a decreased sense of self-worth which conversely increases the risk of depression (Vickerstaff et al., 2007).

Children with any psychiatric disorder are at higher risk of bullying from peers, and ASD is no exception. Rates of depression in children with ASD are also positively correlated with bullying from peers, suggesting the impact of the social environment on the development of mood symptoms (Magnuson & Constantino, 2011). Biologically, higher rates of depression are also found in family members of children with ASD, suggesting an underlying common genetic predisposition.

Obsessive-Compulsive Disorder (OCD)

As defined by the DSM-5, obsessive-compulsive disorder (OCD) is characterized by the presence of obsessions, compulsions, or both. Obsessions are defined as recurrent, intrusive, and distressing thoughts, urges, and impulses which the individual tries to suppress. Compulsions are repetitive behaviors (such as hand washing) or mental acts (such as counting) that the individual performs to reduce distress from the obsession or to prevent an imagined negative event or outcome. There is significant overlap in symptoms between ASD and OCD, making it yet another diagnosis that can be challenging to accurately make in children with ASD. Both disorders are marked by repetitive behaviors, as well as intrusive thoughts. It is estimated that anywhere between 9% and 22% of individuals with ASD also meet criteria for OCD (Hossain et al., 2020). There are some qualities of obsessions and compulsions that can be helpful in distinguishing the two diagnoses. Repetitive behaviors, or compulsions, in OCD are often described as unwanted and bothersome, while such behaviors in ASD are experienced as comforting. Additionally, the content of obsessions in individuals with OCD are more likely to revolve around themes of contamination, aggression, and religious or sexual preoccupations as compared to obsession in ASD individuals (Lewin et al., 2011). Differences in core symptoms of OCD and ASD may help to guide cognitive-behavioral therapy approaches to treatment in these patients.

Psychosis and Schizophrenia

Psychosis is a broad diagnostic term that can be defined as having a loss of contact with reality. Symptoms of psychosis include hallucinations, i.e., hearing, seeing, smelling, feeling, or tasting things that are not actually present, as well as delusions, e.g., paranoia, excessive religiosity, grandiosity, or believing one has special powers. Individuals who are psychotic can also demonstrate disorganized behavior and speech patterns. Psychosis can be present in a number of psychiatric disorders, including schizophrenia spectrum disorders, bipolar disorder, and major depressive disorder with psychotic features, as well as substance intoxication and some medical conditions. Schizophrenia is specifically defined in the DSM-5 as the presence of psychotic symptoms, with at least two of the following: hallucinations, delusions, disorganized speech, disorganized behavior, and negative symptoms (such as social isolation, apathy, decreased speech, and decreased affect). In order to meet criteria for a diagnosis of schizophrenia, symptoms must be present to some degree for at least 6 months. Other psychotic disorders include brief psychotic disorder, during which psychotic symptoms last between 1 day and 1 month, with subsequent return to baseline functioning; and schizophreniform disorder, in which symptoms have been present for between 1 and 6 months. Schizophreniform disorder can be a precursor to a schizophrenia diagnosis. Schizoaffective disorder is also classified as a psychotic disorder. The diagnosis of schizoaffective disorder requires an individual to experience symptoms consistent with schizophrenia, in addition to meeting the criteria for a mood disorder (major depression or mania) at least 50% of the time, with psychotic symptoms remaining present even with resolution of mood symptoms.

Individuals with ASD have a higher risk of schizophrenia and other psychotic disorders. A recent metanalysis found that individuals with ASD were more than three times more likely to have a diagnosis of schizophrenia than those without ASD (Zheng et al., 2018). A Swedish study looked at all individuals in the Stockholm Youth Cohort, which included all individuals aged 17 and younger living in Stockholm County, Sweden, over a decade, from January 1, 2001, to December 31, 2011. The study compared those individuals with an ASD diagnosis to their siblings and same-aged peers without ASD diagnosis. The researchers found that individuals with ASD were over six times more likely to meet criteria for a schizophrenia spectrum disorder compared to peers without ASD (Selten et al., 2015). Another study used the birth cohort of the Avon Longitudinal Study of Parents and Children (ALSPAC), which encompassed 14,062 live births in Avon, UK, between April 1, 1991, and December 31, 1992. Children from this cohort were interviewed at age 12 years to assess for psychotic symptoms or experiences. Investigators found that children with a documented ASD diagnosis were approximately three times more likely to have experienced psychotic symptoms by age 12. One interesting finding in this study was that deficits in pragmatic language skills (i.e., understanding sarcasm, irony, hints) were associated with increased likelihood of psychotic experiences. Pragmatic language skills require an understanding of the perspective of the other person, or "theory of mind," of which there are deficits in both ASD and schizophrenia (Sullivan et al., 2013). It should be noted that childhood-onset schizophrenia, which denotes onset of schizophrenia prior to age 13, is extremely rare, with an estimated prevalence of 1/40,000 (Gochman et al., 2011). That is to say that the experience of psychotic symptoms in childhood does not necessarily mean that a child has schizophrenia, even if they do have ASD. There are complex overlaps in symptoms, risk factors, and genetics of ASD and schizophrenia and other psychotic disorders. This makes for a complicated diagnostic picture when a child presents with symptoms of both disorders, further necessitating the involvement of an expert in the area of child and adolescent psychiatry.

Catatonia

Catatonia is a complex psychomotor syndrome that is estimated to occur in at least 10% of patients with acute psychiatric illness. It is becoming increasingly recognized in patients with ASD, with prevalence studies estimating rates of 4–17% in adolescents and adults with ASD (Dhossche 2014). Catatonia frequently goes unrecognized and is underdiagnosed due to low index of suspicion, as well as the overlap of symptoms between catatonia and autism. Prompt recognition and treatment of catatonia is important due to the risk of progression to malignant catatonia, which can be fatal. Symptoms that should prompt further evaluation of catatonia include increased psychomotor slowing, mutism, stereotypy, unusual voluntary movement, echolalia, or echopraxia. Because of the similarity between these symptoms and those of more severe autism, the use of structured catatonia rating scale, such as the Bush-Francis Catatonia Rating Scale (Bush et al., 1996). Improvement in symptoms following a dose of lorazepam can also be diagnostic of catatonia.

First-line treatment of catatonia in ASD, as in other cases, is benzodiazepines, most commonly lorazepam. High doses (up to 20–30 mg of lorazepam/day) are typically required (Dhossche & Wachtel, 2013). Symptoms should be tracked with an objective rating scale. If lorazepam does not prove effective, electroconvulsive therapy (ECT) is the definitive treatment for catatonia. Some patients may require maintenance treatment to prevent relapse of catatonic symptoms after resolution. Catatonia in ASD should be considered a medical emergency, and requires specialty, typically inpatient, care. Availability of care, particularly ECT, may limit treatment options in some communities.

Akathisia

Akathisia (from Greek meaning "inability to sit") is a subjective feeling of restlessness and need to be in constant motion. Akathisia can be extremely uncomfortable and distressing to patients, to the point that it is associated with increased rates of suicidal ideation. This syndrome is most commonly associated with antipsychotic medications; however, it can also be caused by antidepressants, and other non-psychiatric drugs (such as metoclopramide). While it is difficult to determine precise rates, patients with ASD clinically seem to be more susceptible to medication side effects generally, and akathisia specifically. Akathisia often goes unrecognized or is misdiagnosed as anxiety or agitation related to autism. Signs that should trigger investigation into possible akathisia include an increase in agitation and physical restlessness from baseline, particularly after initiation or increase in dose of a medication such as an antipsychotic or SSRI. Observed symptoms can include fidgeting of the legs, difficulty staying seated, rocking, or pacing. When able, patients may describe feeling a need to move, or a sense of having "ants in their pants."

Treatment options for akathisia include lowering the dose of the offending medication or changing to another agent. Effective adjunct treatment for akathisia includes beta-blockers (such as propranolol), 5HT2A receptor antagonists (such as mirtazapine or cyproheptadine), or in some cases, benzodiazepines (Tachere & Modirrousta, 2017).

Medication Management

No medication can cure autism, though there are medications that are very effective in the treatment of some of the debilitating symptoms, particularly irritability. The primary role of medication is to mitigate associated symptoms of ASD, such as irritability, aggression, hyperactivity, inattention, impulsivity, anxiety, insomnia, and self-injurious behavior. Because of the complexity and frequency of comorbidities in children with ASD, consultation with or treatment by a child psychiatrist is often warranted. Practice parameters published by the American Academy of Child and Adolescent Psychiatry (AACAP) help to guide both pharmacological and psychotherapeutic treatment recommendations.

Antipsychotics

Only two medications have FDA approval specifically for use in ASD: risperidone and aripiprazole. Numerous other medications are used off-label, with various levels of evidence.

Risperidone (Risperdal) is an atypical antipsychotic which has FDA approval for the treatment of irritability, self-injury, and aggressive behavior in children aged 5 and over with ASD. Two double-blind, placebo-controlled studies of children and adolescents with ASD found clinically significant improvement in irritability (DeFilippis & Wagner, 2016). A separate 6-month randomized control trial comparing risperidone to placebo demonstrated superiority of risperidone with regards to improvement in social and emotional responsiveness as measured by the Childhood Autism Rating Scale (or CARS) (Schopler et al., 1988), as well as reduction in severity of disturbance measured by Children's Global Assessment Scale (Shaffer et al., 1983). Children receiving risperidone also showed reduced hyperactivity and aggression (Nagaraj et al., 2006). Blinded discontinuation studies investigating long-term efficacy of risperidone found that individuals randomized to placebo had significantly higher rates of symptom relapse as compared to those remaining on risperidone. In these cases, relapse was defined as a 25% increase in irritability as measured by the irritability subscale of the Aberrant Behavior Checklist (Aman et al., 1986) and a score of much worse or very much worse on the Clinical Global Impression-Improvement (Guy, 1976) which measures illness severity as compared to symptoms prior to medication discontinuation (DeFilippis & Wagner, 2016).

The most common side effects of risperidone demonstrated by these studies included weight gain, increased appetite, fatigue/drowsiness, dizziness, and drooling. Risperidone can also be associated with elevated prolactin levels and related amenorrhea, breast enlargement, and galactorrhea.

Aripiprazole (Abilify) is the second medication that is FDA approved for irritability in children and adolescent with ASD ages 6–17. Efficacy of aripiprazole for the acute treatment of irritability in ASD was established with several randomized control trials comparing multiple doses (5, 10, and 15 mg) and placebo. In these studies, aripiprazole at all doses was superior to placebo with regards to improvement in irritability subscales of the Aberrant Behavior Checklist (DeFilippis & Wagner, 2016). Another double-blind, randomized, placebo-controlled trial investigated efficacy of long-term treatment with aripiprazole. This found no difference in time to the relapse of irritability symptoms between the treatment group and the placebo group; however, there were suggestions that some patients did benefit from maintenance treatment with aripiprazole (Findling et al., 2014). Another open-label 52-week study with aripiprazole found that the group that continued on aripiprazole maintained improvement in irritability as measured by the ABC irritability subscale and CGI improvement scores (Marcus et al., 2011). A trial comparing aripiprazole and risperidone found no difference between the two agents with regards to behavioral symptoms (Ghanizadeh et al., 2014).

The most common side effects of aripiprazole were weight gain, sedation, appetite increase, and insomnia. In one study, aripiprazole was also found to have higher rates of extra-pyramidal symptoms as compared to placebo.

Other antipsychotics are sometimes used for the management of irritability and aggression in ASD. This list includes second-generation antipsychotics, e.g., olanzapine (Zyprexa), lurasidone (Latuda), quetiapine (Seroquel), ziprasidone (Geodon), paliperidone (Invega), as well as the first-generation drug haloperidol (Haldol). These are used “off-label,” meaning they do not have an FDA indication for use in ASD. The efficacy data for these medications is mixed and based off of studies with small numbers of participants (DeFilippis & Wagner, 2016). The most common side effects for second-generation antipsychotics include weight gain, sedation, and increased appetite. First-generation antipsychotics, which are classified by more potent action at D2 dopamine receptors, are associated with dyskinesias, or abnormal, involuntary movements. One longitudinal study of 118 children with ASD treated with haloperidol found that 34% developed some form of dyskinesia with long-term use (Campbell et al., 1997).

Antidepressants

Selective-Serotonin Reuptake Inhibitors (SSRIs) are commonly prescribed to children and adolescents for the treatment of depressive and anxiety disorders, post-traumatic stress disorder, and OCD. Research into the efficacy of SSRIs in ASD has had mixed results. Most studies focus on the reduction of repetitive movements and behaviors, as well as irritability. Two studies of fluoxetine (Prozac) in this population had conflicting results. One demonstrated improvement in repetitive behaviors as assessed by the Children’s Yale-Brown Obsessive-Compulsion Scale, or C-YBOCS, (Goodman et al., 1989) for children on fluoxetine compared to placebo (Hollander et al., 2005). Another found no statistical difference in such symptoms compared to placebo (Herscu et al., 2019). It should be noted that in both of these studies, the mean dose of fluoxetine was 10 mg, which is relatively low when considering the dose required to treat OCD is typically between 40 and 80 mg. A randomized, placebo-controlled trial of youth with ASD compared citalopram (Celexa) to placebo and found no significant differences in repetitive movements as assessed with C-YBOCS or overall improvement as measured with CGI-I (King et al., 2009). A small open-label trial of escitalopram (Lexapro) did find significant improvement in irritability as assessed with the ABC irritability subscale and global improvements on CGI. It was noted in this study that a significant number of participants responded to low-dose escitalopram (<10 mg) and had adverse effects at doses at or above 10 mg (Owley et al., 2005). The most common side effects of SSRIs in this population were found to be restlessness, agitation, insomnia, and irritability, which appear in some cases to be dose related.

Other antidepressants have been investigated in children with ASD, including venlafaxine (Effexor), mirtazapine (Remeron), clomipramine, and desipramine. These studies have had small sample sizes but suggest possible efficacy of low-dose venlafaxine and clomipramine. However, more data is needed (DeFilippis & Wagner, 2016).

Tricyclic antidepressants (TCAs), an older class of antidepressants that block reuptake of serotonin and norepinephrine, were commonly used to treat behavioral problems in ASD prior to the development of SSRIs. Two small double-blind studies of clomipramine, desipramine, and placebo found that clomipramine was better than placebo and desipramine on ratings of anger and repetitive and compulsive behaviors (Gordon et al., 1992). TCAs fell out of favor due to frequency of side effects, particularly anticholinergic effects, as well as their potential to lower the seizure threshold. Clomipramine does remain an option for repetitive behaviors in ASD as well as for patients with comorbid OCD.

Stimulants and Alpha-2 Agonists

Studies on the use of stimulant medications in youth with ASD have typically looked at efficacy of these medications in treating co-occurring symptoms of ADHD, including hyperactivity, inattention, and impulsivity. A 2017 Cochrane Review of studies of methylphenidate in children and adolescents with ASD found only four adequate studies, which suggested short-term improvement in hyperactivity as assessed by parents and teachers (Sturman et al., 2017). In one double-blind study comparing methylphenidate with placebo, children on methylphenidate showed improvement in hyperactivity, however, with lower effect size or magnitude of change, compared to typical-developing children with ADHD (RUPP Autism Network, 2005). Children and adolescents with ASD were also more likely to experience adverse effects of the medication, including decreased appetite, insomnia, and increased irritability.

Alpha-2 agonists, such as guanfacine and clonidine, have also been investigated for the management of hyperactivity and impulsivity in children with ASD and ADHD symptoms. In one study of extended-release guanfacine vs. placebo, those receiving medication had statistically significant improvement in hyperactivity measured on the ABC-hyperactivity subscale and on global improvement measures based on CGI-I scores (Scahill et al., 2015). Additional smaller studies have shown improvement for children receiving guanfacine or clonidine vs. placebo on measures of hyperactivity. The most common adverse effects for alpha-2 agonist medications are sedation and a mild decrease in blood pressure (DeFilippis & Wagner, 2016).

Antiepileptics

Antiepileptic medications (AEDs) are commonly used in patients with ASD, particularly given the frequency of seizure disorders in this population. AEDs are also frequently used to treat behavioral symptoms in children and adolescents with ASD. However, there are limited studies to support the efficacy of these medications for the treatment of ASD symptoms. Studies of valproic acid have been mixed, with some demonstrating efficacy for valproic acid in reducing repetitive behaviors in children with autism when compared to placebo (Hollander et al., 2010), while others found no significant difference in irritability ratings between valproic acid and placebo. Lamotrigine has only one RCT investigating its efficacy in children with ASD, which did not show significant effect in the reduction of irritability or improvement in social behavior when compared to placebo (Perisco et al., 2021). Topiramate has typically been looked at as adjunct treatment to mitigate weight gain in patients on antipsychotic medications, in addition to its potential role in mood stabilization. RCTs and open-label studies have demonstrated modest reduction in neuroleptic-induced weight gain in patients on topiramate, with limited evidence for improvements in irritability or mood. Given limited evidence for efficacy in treating core symptoms of ASD, antiepileptic drugs may be best considered when there is another comorbid indication present.

Other Medications and Biological Treatments

Given the prevalence of ASD and significant morbidity caused by its symptoms, there is significant interest in identifying additional agents that may be efficacious for the treatment of ASD core symptoms. The following summary of treatments is by no means exhaustive.

Buspirone

Buspirone is a 5-HT(1A) receptor agonist that is used in the treatment of generalized anxiety disorder. An 8-week randomized double-blind placebo-controlled trial of 40 children and adolescents with ASD who were also taking risperidone found that low-dose buspirone in combination with risperidone resulted in significantly decreased scores on the irritability subscale of the Aberrant Behavior Checklist compared to those on risperidone and placebo. The average daily dose of buspirone was 6.7 mg per day (Ghanizadeh & Ayoobzadehshirazi, 2015). Given the favorable side effect profile of buspirone, it is a reasonable option for the treatment of irritability and anxiety in children with ASD, particularly as augmentation with other agents.

Amantadine

Amantadine acts as a noncompetitive NMDA receptor antagonist, nicotinic receptor antagonist, and dopamine reuptake inhibitor. With increased interest in the role of glutamate pathways in ASD and other psychiatric disorders, amantadine has been hypothesized as a potential agent to target symptoms of autism. An RCT comparing patients given amantadine as augmentation to risperidone, compared to those on risperidone and placebo, demonstrated improvement in irritability and hyperactivity in patients in the amantadine group (Mohammadi et al.). An earlier study of amantadine alone compared to placebo found no significant differences in hyperactivity or irritability based on parent rating scales, though there was significant improvement in hyperactivity and inappropriate speech based on clinician rating scales (King et al., 2001). While further study is needed, amantadine may be a viable option as augmentation for patients that do not experience a robust response to monotherapy.

Melatonin

As with other psychiatric disorders, sleep disturbances are common in children and adolescents with ASD. Melatonin is a naturally occurring hormone released by the pineal gland that helps to regulate the sleep–wake cycle. It is available as an over-the-counter supplement and is utilized for a variety of sleep disturbances. It is one of the best-studied complementary treatments used in ASD, with multiple double-blind, placebo-controlled studies. One study of 160 children with ASD compared melatonin, melatonin plus cognitive-behavioral therapy (CBT), CBT alone, and placebo. The combination of melatonin and CBT was found to be most effective after 12 weeks, although all active treatment groups (melatonin, CBT, and combination) showed better results than placebo (Cortesi et al., 2012). A smaller placebo-control study of 22 youth with ASD and sleep disturbances who had not responded to behavioral therapy found significant improvement in sleep latency and total sleep with melatonin as compared to placebo (Wright et al. 2011). Melatonin has consistently been shown to be well tolerated without significant adverse effects. It is generally considered to be a safe treatment option for sleep disturbances in ASD, though behavioral therapy should always be considered as well.

Behavioral Therapies for ASD

Behavioral therapies are commonly recommended for children with autism spectrum disorder. The type of therapy, location of therapy (home, office, school), and degree of parental involvement will vary based on a child's age, needs, and the severity of their symptoms. While it is rare for a child psychiatrist to provide behavioral therapies themselves, they will frequently provide recommendations and referrals for therapies, and will collaborate with therapy providers to help guide treatment.

Applied Behavioral Analysis (ABA), first applied to autism spectrum disorders in the 1960s, is one of the oldest forms of behavioral therapy used in the treatment of autism spectrum disorder, as well as other disorders. It is also referred to as early and intensive behavioral intervention when specifically referencing its use with young children (under 60 months) with ASD and is based on theories of operant conditioning. Operant conditioning is a method of learning occurring through rewards and punishments, with rewards increasing the frequency of a behavior while punishments decrease behavior frequency. This is in contrast with classical or Pavlovian conditioning, in which learning takes place through association of an unconditioned response with a neutral stimulus, producing a new, conditioned response (the classic example is Pavlov's dog conditioned to salivate at the sound of a bell). ABA therapy is the most researched of the behavioral therapies utilized in the treatment of ASD. The first study of long-term ABA therapy for children with autism was published in 1987 and compared a cohort of children who received 40 hours a week of ABA therapy, with a group that received 10 hours a week of behavioral intervention. Both groups received treatment for at least 2 years. Nearly 50% of the children in the ABA group achieved a normal IQ and were able to be maintained in regular first grade classrooms. By comparison, only 2% of children in the control group attained these milestones (Lovaas, 1987).

ABA looks at behavior as a three-step process. The first step is the antecedent, or cue, that occurs just before the targeted behavior. This can be a command, an object, or even a thought or feeling. Next comes the behavior, or the individual's response or lack of response to the antecedent. The behavior may be an action, a verbal response, or something else that is observed, or the lack of the observed behavior. The final step is the consequence, or the event that occurs after the behavior. The consequence may be a positive reinforcement for the desired behavior, or no reaction for undesired or incorrect behavior. The so-called ABCs of ABA therapy, namely, antecedent, behavior, and consequence previously described, can be utilized to reinforce and increase desired, adaptive behaviors and to reduce or extinguish unwanted or dangerous behaviors. Several procedures within the therapy are utilized to work toward these goals. "Shaping" refers to reinforcing successive approximations of a desired behavior. For example, if a child struggles to say "dog," the therapist may reward appropriate attempts at the word, starting with the "d" sound and then progressing until they can pronounce the full word. "Chaining" is used to break down a complicated behavior or series of behaviors that a child would not be able to learn all at once. For example, if attempting to teach a child to tie their shoes, they may be reinforced for first getting their shoes from their room, then putting them on, tightening the laces, and so on (Granpeesheh et al., 2009).

In these ways, ABA can help individuals with autism develop social and communication skills, as well as practical life skills, including grooming, motor dexterity, and self-care. An ABA therapist will help to prioritize certain behaviors that require change and set realistic, achievable behavioral goals to meet those challenges. These goals will be regularly reviewed with the therapist to ensure that progress is being made. Parental/caregiver involvement in ABA therapy is crucial for its success. ABA therapy can be a significant time commitment, as it is recommended that children participate in upwards of 35 hours a week of therapy, with some level of family involvement as well.

There are several types of interventions that fall under the umbrella of Applied Behavioral Analysis, including Discrete Trial Training (DTT) and Pivot Response Training (PRT), and Verbal Behavioral Therapy (VB). These are not separate forms of therapy, rather they represent different teaching formats within ABA that may be utilized at different times depending on the situation and the child's needs.

Discrete Trial Training (DTT) is a structured teaching strategy, in which the therapist will present multiple discrete opportunities, or "discrete trials" throughout the day. In DTT, the therapist will provide a prompt for a desired component of behavior and await the response. If a desired response is produced, the child will receive a reward that is motivating to them (candy, toy, preferred activity). However, if an undesired response is produced, the therapist will gently correct the child and repeat the prompt. For example, the therapist may ask the child to "point to the car" (cue). If the child points to the car (behavior), the therapist will provide the desired reinforcer, i.e., toy, candy, (consequence). If the child does not point to the car, the therapist will provide correction, and repeat the trial. DTT allows a child to have a large number of discrete trials during a session, each providing an opportunity for learning. A limitation of this type of teaching is difficulty of generalization of skills learned in DTT sessions to less structured settings.

Pivotal Response Training (PRT) is based on ABA therapy and targets areas considered "pivotal" to a child's development, such as communication and motivation, as opposed to one specific behavior. PRT emphasizes natural reinforcement. For example, if a child makes a good attempt to request a toy, they are rewarded with being handed the toy, not with an unrelated reward. PRT involves both structured and unstructured interactions, to help children generalize skills to everyday situations. Trials are also initiated by the child, instead of the therapist. For example, if a child points to a toy, the therapist may prompt the child to verbalize the request, the reward for which will be the toy. This differs from DTT, in which the therapist would start the interaction by instructing the child to name or point to the toy, regardless as to whether they had expressed interest in it.

Verbal Behavioral Therapy (VB) is also based on principles of ABA therapy and is used to teach communication and language. This technique teaches children to connect words with their purpose. Language is broken down into different types, with requests being the most basic form. Therapists first work with the child to learn that saying a word can produce a response, i.e., saying "cookie" can produce a cookie. This therapy works on the premise of approximation. In an individual with autism, the behavior is first reinforced in any way. For example, the child can point to something and be rewarded with the object they desired. Gradually, the therapist helps them develop their communication so that they are able to say or sign the appropriate word to communicate their request. The Picture Exchange Communication System (PECS) is also used to develop and reinforce language skills in non-verbal children. With this system, children are reinforced for utilizing specific picture cards to request preferred objects and activities (Roane et al., 2016). As technology has advanced, so have communication tools for non-verbal children. Devices such as the iPad can now be programmed with communication applications based on PECS, allowing a child to have easier access to a larger variety of words and pictures with which to communicate.

ABA therapists may implement a combination of DTT, PRT, VB, and other aspects of ABA therapy with an individual child, depending on the child's needs and learning style. Regardless of how ABA therapy is implemented, research makes it clear that the weekly number of hours and intensity of the therapy is crucial to its efficacy. The initial study done by Lovaas found that children receiving low-intensity ABA therapy (approximately 10 hours a week) had significantly smaller gains than those receiving high-intensity therapy (40 hours per week). Subsequent research has demonstrated that, for optimal efficacy, ABA therapy must be of adequate intensity, between 35 and 40 hours per week. It has also been found that duration of therapy has an impact on gains made. While studies have found that children do show benefit from shorter duration ABA therapy (9–14 months) as compared to control groups, longer-duration therapy (2 years or more) has consistently demonstrated superior outcomes. This is not to say that children who receive lower intensity or shorter duration of treatment do not benefit. A study of children receiving ABA therapy for 20 hours a week over 15 months found gains in IQ as compared to control groups; however, those gains were smaller than those who received therapy of higher intensity and longer duration (Sheinkopf & Siegel, 1998).

There has been extensive research into the efficacy of ABA therapy for young children with ASD. A metanalysis of 22 studies demonstrated medium-to-large effects of long-term comprehensive ABA therapy on intellectual functioning, language development, daily living skills, and social functioning (Virués-Ortega, 2010). A study of 332 children aged 2–7 years in Canada enrolled in community-based ABA programs demonstrated significant gains in cognitive, social, and adaptive developments, as well as reduction in autism symptom severity. This study was particularly reassuring, as it demonstrated efficacy of ABA interventions in a more "real world" setting (Perry et al., 2008).

There are a number of behavioral treatment approaches that integrate ABA procedures and theories into the treatment of children with ASD. One such approach is the Early Start Denver Model (ESDM), which is an interdisciplinary approach targeting young children with autism (ages 12–48 months). It utilizes PRT, in addition to other ABA techniques, and is based on developmental progression in toddlerhood. There are five essential developmental components targeted during treatment: imitation, non-verbal and verbal communication, social development, emotion sharing, and play (Roane et al., 2016). In a randomized-controlled trial of 48 toddlers with ASD, families were randomly assigned to intervention of 25 hours of ESDM one-on-one therapy over 2 years or a standard community-based treatment. Those in the intervention group showed an average gain of 19 IQ points, compared to a gain of 7 points in the control group (Dawson et al., 2010). A study of children aged 12–62 months with ASD enrolled in an autism-specific early learning program which utilized ESDM techniques with small groups of children found that, after 1 year, the younger cohort of children (aged 12–48 months) had made larger gains in verbal language development, as compared to the older cohort. Improvements in non-verbal and adaptive development did not differ between groups. This suggests that early intervention programs targeting toddlers and pre-school children may be particularly effective for children with verbal delays (Vivanti & Dissanayake, 2016). Additionally, it reinforces the importance of early recognition and diagnosis of ASD.

ABA therapy techniques can also be utilized in older children with ASD. In these cases, it is more common that therapy is targeting a particular challenging behavior, such as aggression, self-injury, food aversion, or destructive behavior. In these cases, the ABA therapist will typically start with a functional behavioral assessment (FBA) of the behavior, essentially identifying the function of the behavior and possible unintentional reinforcers of that behavior. For example, if a child throws a tantrum when he is asked to pick up his toys, his parents may give in and pick up the toys for him. The function of the behavior is to avoid the undesired chore, which is reinforced when the parents intercede and pick up the toys for the child. Once the FBA is completed, the therapist will help the parents/caregivers and others involved with the child to develop a behavioral intervention plan, which utilizes theories of ABA to extinguish unwanted behavior and increase desired behavior. This requires consistency among all caregivers in implementing the plan (Granpeesheh et al., 2009). There exists a body of evidence supporting the use of FBA and behavioral intervention plans to reduce atypical behavior in children with ASD (Roane et al., 2016). The use of FBAs has been recommended as a "best practice" by the American Academy of Pediatrics (Myers et al., 2007).

Limitations of ABA Therapy

While ABA has an extensive history and research base, it is not without controversy or limitations. Accessibility to quality therapists remains a limitation in many areas of the country. Additionally, the time commitment required to maintain fidelity to evidence-based implementation of ABA therapy is quite high, both with regards to intensity and duration. There has also been growing criticism from some within the autism community regarding the practices of ABA. Some of the concern stems from historical practices of early ABA therapy, which relied on punishment, including electric shocks, to help extinguish unwanted behaviors. This has largely fallen out of favor, both due to obvious ethical concerns as well as increasing evidence as to the superiority of positive reinforcement in behavioral therapies. There is also concern that skills developed in ABA therapy may not generalize outside of the therapeutic sessions, as well as criticism that ABA seeks to make children with ASD "normal" as opposed to embracing their individuality. All of these critiques should be taken into consideration when determining whether ABA-based therapy is appropriate for an individual child. As has been described, ABA encompasses a number of different behavioral approaches and is not a "one size fits all" therapy. Pediatricians and others working with children with ASD should become familiar with ABA providers in their area, in order to help families make the best treatment decisions for their child.

Other Psychotherapeutic Interventions for ASD

While ABA has the longest and richest research base, there are other interventions that have evidence for the treatment of children with ASD. Among these are Relationship Development Intervention (RDI) and Floortime.

Relationship Development Intervention (RDI) is a family-based intervention that focuses on the development of social and emotional skills. In RDI, parents or primary caregivers serve as the primary therapist for the child. Parents or caregivers first attend 6 days of intensive workshops about the theory and implementation of RDI, followed by weekly planning and coaching sessions with an RDI consultant. The focus of RDI is to help children with ASD develop dynamic intelligence, or the ability to think more flexibly. The six objectives of RDI are emotional referencing (the ability to learn from the experiences of others), social coordination (the ability to observe and control behavior in order to participate in social relationships), declarative language (the ability to communicate feelings and coordinate plans with others), flexible thinking (adapting and changing plans as circumstances change), relationship information processing (the ability to put things in context, and to solve problems that have no "right or wrong" answers), and foresight and hindsight (the ability to anticipate future experiences based on past experiences) . The RDI treatment plan is based upon an individual child's needs and goals. Plans are developed after an assessment of the child's interaction with parents and others. The treatment plan initially focuses on interactions between the child and parent but can progress to include other children or groups. While RDI still requires a significant time commitment, it can be more flexible and less costly as compared to ABA therapy.

There is limited research data on the efficacy of RDI. A study of 16 children aged 20–96 months at treatment onset demonstrated reduction in ADOS scores, particularly those related to communication and social interaction, as well as improvement in measures of cognitive flexibility and decreased need for special education at 30-month follow-up. The small sample size of this should be noted; additionally, all children in the study had an IQ of at least 70, limiting generalization to autistic children with more severe intellectual disabilities (Gutstein et al., 2007). More high-quality research is needed to further determine efficacy, as well as to help understand what children may benefit most from this intervention.

Floortime is another relationship-based intervention for children with autism. As its name implies, parents/caregivers get down on the floor with the child, with the goal to help the child expand their circle of communication. It can be used as an alternative to ABA, or in conjunction with ABA therapy. Floortime aims to help children with ASD reach six developmental milestone goals: self-regulation, engagement in relationships, two-way communication, complex communication, emotional ideas, and emotional thinking. Floortime sessions emphasize back-and-forth play. Parents/caregivers can build upon the development of shared attention and engagement to become more adept in complex communication and social situations. Floortime therapy sessions can take between 2 and 5 hours a day. They can also be used in school settings with typical-developing peers as the child progresses. A pilot study of children with ASD in Thailand randomized 32 children to either Floortime intervention or treatment-as-usual. Children were reassessed after 3 months. Those participating in Floortime showed significant improvement in emotional functioning as measured by the Functional Emotional Assessment Scale (FEAS) (Greenspan, et al., 2001) compared to the control group (Pajareya & Nopmaneejumruslers, 2011). Once again, this intervention requires further study. Some aspects make it an exciting option, including the potential for efficacy as a less time-intensive, less-costly intervention.

This review is not an exhaustive list of psychotherapeutic interventions for ASD, nor does it delve into therapies targeted at comorbid conditions. It is meant to provide information on some of the most common and most researched therapies currently available. Once again, it is important to consult with a specialist in child psychiatric disorders to help determine which interventions may be valuable for each individual child.

Engaging Outside Supports

Caring for a child with ASD can be all-consuming at times. Without appropriate support, caregiver burnout is quite common. Utilization of respite care, while not often discussed, can help to provide relief and improve well-being in families with a child with ASD. Respite care can take many forms. It may involve a respite caregiver staying with the child for a few hours, so that the primary caregivers can attend appointments, or have a date night. Given the unique needs of children with ASD, hiring trained caregivers to provide respite care in the home is generally recommended over hiring a regular babysitter. Having the same caregiver provide respite sessions can be useful, as it allows them and the child an opportunity to develop a relationship and serves to increase the child's comfort in staying with the caregiver. Qualified caregivers can be found through specific agencies that provide such services, as well as from inquiring at local autism centers, pediatric and child psychiatry clinics, and social services agencies. In some places, options also exist for respite day centers, at which an individual with autism can be dropped off and cared for by qualified staff, allowing the parent or caregiver to have time for themselves. These settings can also provide additional socialization with peers who also have ASD.

Respite care can also be provided for longer durations outside the home, either on a planned or emergent basis. Planned respite stays occur when a child with autism resides at a residential respite facility for a pre-determined amount of time (usually 3–4 days), to allow parents to have an extended break, to travel, or to prepare for a transition, such as the birth of a new baby. Emergent respite stays occur when a child cannot safely be managed at home and is referred to residential care or short-term foster care to allow for necessary interventions and support for the child and family. Within the US, residential respite care can be difficult to access due to limited availability and inconsistent insurance coverage. A review of a residential treatment home in Northamptonshire, England, found planned respite services to be effective, with high parental satisfaction. This study found that certain attributes of a residential respite program seemed to confer better outcomes, including the physical environment, staff's understanding of ASD, consistency in the use of ASD-oriented behavioral approaches, individualization of interventions, activities available, and opportunities to interact with other children (Preece, 2011).

Because of the variability in services, it is difficult to study the efficacy of respite care in the long-term outcomes for children with ASD. Respite care has been shown to reduce stress in families (Chan & Sigafoos, 2001), and high-quality respite care can also provide children with ASD with positive experiences (Preece & Jordan, 2010). It is recommended that families of children with ASD proactively discuss available options for respite services with their treatment providers, including pediatricians, child psychiatrists, and therapists.

References

Allen, A., & Shane, H. (2014). The evaluation of children with an autism spectrum disorder: Adaptations to accommodate a telepractice model of clinical care. *Perspectives on Telepractice, 4*(2), 42–51. https://doi.org/10.1044/teles4.2.42

Aman, M. G., Singh, N. N., Stewart, A. W., & Field, C. J. (1986). *Aberrant behavior checklist (ABC) [measurement instrument]*. Stoelting Publishers.

American Psychiatric Association. (2013). *Diagnostic and statistical manual of mental disorders* (5th ed.). https://doi.org/10.1176/appi.books.9780890425596

Atbaşoğlu, E. C. (2020). Autism spectrum disorder as an initial diagnosis in adults. *Noro Psikiyatri Arsivi, 57*(1), 1–2. Published 2020 Feb 13. https://doi.org/10.29399/npa.25050

Baykal, S., Karakurt, M. N., Çakır, M., & Karabekiroğlu, K. (2019). An examination of the relations between symptom distributions in children diagnosed with autism and caregiver burden, anxiety and depression levels. *Community Mental Health Journal, 55*(2), 311–317. https://doi.org/10.1007/s10597-018-0343-8

Bush, G., Fink, M., Petrides, G., Dowling, F., & Francis, A. (1996). Catatonia. I. Rating scale and standardized examination. *Acta Psychiatrica Scandinavica, 93*(2), 129–136.

Campbell, M., Armenteros, J., Malone, R., Adams, P., Eisenber, Z., & Overal, J. (1997). Neuroleptic-related dyskinesias in autistic children: A prospective, longitudinal study. *Journal of the American Academy of Child and Adolescent Psychiatry, 36*(6). https://doi.org/10.1097/00004583-199706000-00022

Chan, J., & Sigafoos, J. (2001). Does respite care reduce parental stress in families with developmentally disabled children? *Child and Youth Care Forum, 30*, 253–263. https://doi.org/10.1023/A:1014467226528

Cortesi, F., Giannotti, F., Sebastiani, T., Panunzi, S., & Valente, D. (2012). Controlled-release melatonin, singly and combined with cognitive behavioural therapy, for persistent insomnia in children with autism spectrum disorders: A randomized placebo-controlled trial. *Journal of Sleep Research, 21*(6). https://doi.org/10.1111/j.1365-2869.2012.01021.x

Dawson, G., Rogers, S., Munson, J., Smith, M., Winter, J., Greenson, J., Donaldson, A., & Varley, J. (2010). Randomized, controlled trial of an intervention for toddlers with autism: The early start Denver

model. *Pediatrics, 125*(1). https://doi.org/10.1542/peds.2009-0958
DeFilippis, M. (2018). Depression in children and adolescents with autism spectrum disorder. *Children, 5*, 112. Retrieved from https://doi.org/10.3390/children5090112
DeFilippis, M., & Wagner, K. D. (2016). Treatment of autism spectrum disorder in children and adolescents. *Psychopharmacology Bulletin, 46*(2), 18–41.
Delahunty, C. (2015). Developmental delays and autism: Screening and surveillance. *Cleveland Clinic Journal of Medicine, 82*(11 Suppl 1), 29–32. https://doi.org/10.3949/ccjm.82.s1.06
Dhossche, D. M. (2014). Decalogue of catatonia in autism spectrum disorders. *Frontiers in Psychiatry, 5*, 157.
Dhossche, D., & Wachtel, L. (2013). ECT for catatonia in autism. In M. Ghaziuddin & G. Walter (Eds.), *Electroconvulsive therapy in children and adolescents* (pp. 217–246). Oxford University Press.
Dulcan, M. (Ed.). (2016). *Dulcan's textbook of child and adolescent psychiatry* (2nd ed.). American Psychiatric Association Publishing.
Edison, T., Hess, A., Hess, T., & Kelly, A. (2020). Family engagement in the autism treatment and learning health networks. *Pediatrics, 145*(Supplement 1), S30–S34.
Einfeld, S., Piccinin, A., Mackinon, A., et al. (2006). Psychopathology in young people with intellectual disability. *JAMA: The Journal of the American Medical Association, 296*, 1981–1989. https://doi.org/10.1001/jama.296.16.1981
Farrugia, S., & Hudson, J. (2016). Anxiety in adolescents with Asperger syndrome: Negative thoughts, behavioral problems, and life interference. *Focus on Autism and Other Developmental Disabilities*. https://doi.org/10.1177/10883576060210010401
Findling, R. L., Mankoski, R., Timko, K., Lears, K., McCartney, T., McQuade, R. D., Eudicone, J., Amatniek, J., Marcus, R., & Sheehan, J. J. (2014). A randomized controlled trial investigating the safety and efficacy of aripiprazole in the long-term maintenance treatment of pediatric patients with irritability associated with autistic disorder. *The Journal of Clinical Psychiatry, 75*(1), 22–30. https://doi.org/10.4088/jcp.13m08500
Ghanizadeh, A., & Ayoobzadehshirazi, A. (2015). A randomized double-blind placebo-controlled clinical trial of adjuvant buspirone for irritability in autism. *Pediatric Neurology, 52*, 77–81.
Ghanizadeh, A., Sahraeizadeh, A., & Berk, M. (2014). A head-to-head comparison of aripiprazole and risperidone for safety and treating autistic disorders, a randomized double blind clinical trial. *Child Psychiatry and Human Development, 45*(2). https://doi.org/10.1007/s10578-013-0390-x
Gochman, P., Miller, R., & Rapoport, J. L. (2011). Childhood-onset schizophrenia: The challenge of diagnosis. *Current Psychiatry Reports, 13*(5), 321–322. https://doi.org/10.1007/s11920-011-0212-4
Goodman, W. K., Price, L. H., Rasmussen, A., Mazure, C., Fleischmann, R. L., Hill, C. L., Heninger, G. R., & Charney, D. S. (1989). Child Yale-Brown obsessive compulsive scale (CYBOCS). Retrieved from https://iocdf.org/wp-content/uploads/2016/04/05-CYBOCS-complete.pdf
Gordon, C. T., et al. (1992). Differential response of seven subjects with autistic disorder to clomipramine and desipramine. *The American Journal of Psychiatry, 149*(3), 363–366.
Granpeesheh, D., Tarbox, J., & Dixon, D. R. (2009). Applied behavior analytic interventions for children with autism: A description and review of treatment research. *Annals of Clinical Psychiatry: Official Journal of the American Academy of Clinical Psychiatrists, 21*(3), 162–173.
Greenspan, S. I., DeGangi, G., & Weider, S. (2001). *The functional emotional assessment scale (FEAS): For infancy & early childhood*. Interdisciplinary Council on Development & Learning Disorders.
Gutstein, S. E., Burgess, A. F., & Montfort, K. (2007). Evaluation of the relationship development intervention program. *Autism: The International Journal of Research and Practice, 11*(5), 397–411. https://doi.org/10.1177/1362361307079603
Guy, W. (1976). *Clinical global impression scale-improvement (CGI-I)*. Retrieved from https://assets.website-files.com/5cb8b10a48eebf8ee23d835b/5cd457135c25c41b6d8cb64f_CGI-GI.pdf
Herscu, P., Handen, B. L., Arnold, L. E., Snape, M. F., Bregman, J. D., Ginsberg, L., Hendren, R., Kolevzon, A., Melmed, R., Mintz, M., Minshew, N., Sikich, L., Attalla, A., King, B., Owley, T., Childress, A., Chugani, H., Frazier, J., Cartwright, C., … Autism Speaks Autism Clinical Trials Network. (2019). The SOFIA study: Negative multi-center study of low dose fluoxetine on repetitive behaviors in children and adolescents with autistic disorder. *Journal of Autism and Developmental Disorders*. https://doi.org/10.1007/s10803-019-04120-y
Hilt, R. J. (2017). Telemedicine for child collaborative or integrated care. *Child and Adolescent Psychiatric Clinics of North America, 26*(4), 637–645. https://doi.org/10.1016/j.chc.2017.05.001
Hollander, E., Phillips, A., Chaplin, W., Zagursky, K., Novotny, S., Wasserman, S., & Iyengar, R. (2005). A placebo controlled crossover trial of liquid fluoxetine on repetitive behaviors in childhood and adolescent autism. *Neuropsychopharmacology, 30*(3), 582–589. https://doi.org/10.1038/sj.npp.1300627
Hollander, E., et al. (2010). Divalproex sodium vs placebo for the treatment of irritability in children and adolescents with autism spectrum disorders. *Neuropsychopharmacology, 35*(4), 990–998.
Hossain, M. M., Khan, N., Sultana, A., Ma, P., McKyer, E., Ahmed, H. U., & Purohit, N. (2020). Prevalence of comorbid psychiatric disorders among people with autism spectrum disorder: An umbrella review of systematic reviews and meta-analyses. *Psychiatry*

Research, 287, 112922. https://doi.org/10.1016/j.psychres.2020.112922

Johnson, C., & Myers, S. (2007). Identification and evaluation of children with autism spectrum disorders. *Pediatrics, 120*, 1183–1215. https://doi.org/10.1542/peds.2007-2361

Juárez, A. P., Weitlauf, A. S., Nicholson, A., Pasternak, A., Broderick, N., Hine, J., Stainbrook, J. A., & Warren, Z. (2018). Early identification of ASD through telemedicine: Potential value for underserved populations. *Journal of Autism and Developmental Disorders, 48*(8), 2601–2610. https://doi.org/10.1007/s10803-018-3524-y

Kaat, A. J., & Lecavalier, L. (2013). Disruptive behavior disorders in children and adolescents with autism spectrum disorders: A review of the prevalence, presentation, and treatment. *Research in Autism Spectrum Disorders, 7*(12), 1579–1594. https://doi.org/10.1016/j.rasd.2013.08.012

Kataoka, S., Ijadi-Maghsoodi, R., Figueroa, C., Castillo, E., Bromley, E., Patel, H., & Wells, K. (2018). Stakeholder perspectives on the social determinants of mental health in community coalitions. *Ethnicity & Disease, 28*(Suppl 2). https://doi.org/10.18865/ed.28.S2.389

King, B. H., et al. (2001). Double-blind, placebo controlled study of amantadine hydrochloride in the treatment of children with autistic disorder. *JAACAP, 40*, 6.

King, B., Hollander, E., Sikich, L., McCracken, J., Scahill, L., Bregman, J., Donnelly, C., Anagnostou, E., Dukes, K., Sullivan, L., Hirtz, D., Wagner, A., Ritz, L., & STAART psychopharmacology Network. (2009). Lack of efficacy of citalopram in children with autism spectrum disorders and high levels of repetitive behavior: Citalopram ineffective in children with autism. *Archives of General Psychiatry, 66*(6), 583–590. https://doi.org/10.1001/archgenpsychiatry.2009.30

Lai, M., Kassee, C., Besney, R., Bonato, S., Hull, L., Mandy, W., Szatmari, P., & Ameis, S. H. (2019). Prevalence of co-occurring mental health diagnoses in the autism population: A systematic review and meta-analysis. *The Lancet Psychiatry, 6*(10), 819–829. https://doi.org/10.1016/S2215-0366(19)30289-5

Lecavalier, L. (2006). Behavioral and emotional problems in young people with pervasive developmental disorder: Relative prevalence, effect of subject characteristics, and empirical classification. *Journal of Autism and Developmental Disorders, 26*(8), 1101–1104.

Lecavalier, L., Leone, S., & Wiltz, J. (2006). The impact of behaviour problems on caregiver stress in young people with autism spectrum disorders. *Journal of Intellectual Disability Research: JIDR, 50*(Pt 3). https://doi.org/10.1111/j.1365-2788.2005.00732.x

Leitner, Y. (2014). The co-occurrence of autism and attention deficit hyperactivity disorder in children – What do we know? *Frontiers in Human Neuroscience, 8*. https://doi.org/10.3389/fnhum.2014.00268

Lewin, A., Wood, J., Gunderson, S., Murphy, T., & Storch, E. (2011). Phenomenology of comorbid autism spectrum and obsessive-compulsive disorders among children. *Journal of Developmental and Physical Disabilities, 23*. https://doi.org/10.1007/s10882-011-9247-z

Lovaas, O. I. (1987). Behavioral treatment and normal educational and intellectual functioning in young autistic children. *Journal of Consulting and Clinical Psychology, 55*(1), 3–9. https://doi.org/10.1037//0022-006x.55.1.3

Magnuson, K. M., & Constantino, J. N. (2011). Characterization of depression in children with autism spectrum disorders. *Journal of Developmental and Behavioral Pediatrics: JDBP, 32*(4), 332–340. https://doi.org/10.1097/DBP.0b013e318213f56c

Małus, A., Galińska-Skok, B., Konarzewska, B., & Szulc, A. (2018). How do patients perceive ambulatory psychiatric care and what are their needs? *Annals of Agricultural and Environmental Medicine: AAEM, 25*(1), 90–94. https://doi.org/10.5604/12321966.1233559

Mandell, D. S. (2008). Psychiatric hospitalization among children with autism spectrum disorders. *Journal of Autism and Developmental Disorders, 38*(6), 1059–1065. https://doi.org/10.1007/s10803-007-0481-2

Mansour, R., Dovi, A. T., Lane, D. M., Loveland, K. A., & Pearson, D. A. (2017). ADHD severity as it relates to comorbid psychiatric symptomatology in children with autism spectrum disorders (ASD). *Research in Developmental Disabilities, 60*, 52–64. https://doi.org/10.1016/j.ridd.2016.11.009

Marcus, R. N., Owen, R., Manos, G., Mankoski, R., Kamen, L., McQuade, R. D., Carson, W. H., Corey-Lisle, P. K., & Aman, M. G. (2011). Aripiprazole in the treatment of irritability in pediatric patients (aged 6-17 years) with autistic disorder: Results from a 52-week, open-label study. *Journal of Child and Adolescent Psychopharmacology, 21*(3), 229–236. https://doi.org/10.1089/cap.2009.0121

Marrus, N., Veenstra-Vanderweele, J., Hellings, J. A., et al. (2014). Training of child and adolescent psychiatry fellows in autism and intellectual disability. *Autism, 18*(4), 471–475. https://doi.org/10.1177/1362361313477247

Myers, S. M., Johnson, C. P., & American Academy of Pediatrics Council on Children With Disabilities. (2007). Management of children with autism spectrum disorders. *Pediatrics, 120*(5), 1162–1182. https://doi.org/10.1542/peds.2007-2362

Nagaraj, R., Singhi, P., & Malhi, P. (2006). Risperidone in children with autism: Randomized, placebo-controlled, double-blind study. *Journal of Child Neurology, 21*(6), 450–455. https://doi.org/10.1177/08830738060210060801

Osborne, L. A., McHugh, L., Saunders, J., & Reed, P. (2008). Parenting stress reduces the effectiveness of early teaching interventions for autistic spectrum disorders. *Journal of Autism and Developmental Disorders, 38*(6), 1092–1103. https://doi.org/10.1007/s10803-007-0497-7

Owley, T., Walton, L., Salt, J., Guter, S. J., Jr., Winnega, M., Leventhal, B. L., & Cook, E. H., Jr. (2005). An open-label trial of escitalopram in pervasive developmental disorders. *Journal of the American Academy of Child and Adolescent Psychiatry, 44*(4), 343–348. https://doi.org/10.1097/01.chi.0000153229.80215.a0

Öz, B., Yüksel, T., & Nasiroğlu, S. (2019). Depression-Anxiety Symptoms and Stigma Perception in Mothers of Children with Autism Spectrum Disorder. *Noro Psikiyatri Arsivi, 57*(1), 50–55. Published 2019 Nov 28. https://doi.org/10.29399/npa.23655

Pajareya, K., & Nopmaneejumruslers, K. (2011). A pilot randomized controlled trial of DIR/floortime™ parent training intervention for pre-school children with autistic spectrum disorders. *Autism: The International Journal of Research and Practice, 15*(5). https://doi.org/10.1177/1362361310386502

Papadopoulos, C., Lodder, A., Constantinou, G., & Randhawa, G. (2019). Systematic review of the relationship between autism stigma and informal caregiver mental health. *Journal of Autism and Developmental Disorders, 49*(4), 1665–1685. https://doi.org/10.1007/s10803-018-3835-z

Perry, A., Cummings, A., Geier, J. D., Freeman, N. L., Hughes, S., LaRose, L., Managhan, T., Reitzel, J.-A., & Williams, J. (2008). Effectiveness of intensive behavioral intervention in a large, community-based program. *Research in Autism Spectrum Disorders, 2*(4), 621–642. https://doi.org/10.1016/j.rasd.2008.01.002

Preece, D. (2011). Respite care for families with children with autism spectrum disorders. *Annales Universitatis Paedagogicae Cracoviensis, Studia Psychologica, IV*, 174.

Preece, D., & Jordan, R. (2010). Obtaining the views of children and young people with autism spectrum disorder about their experience of daily life and social care support. *British Journal of Learning Disabilities, 38*, 10–20. https://doi.org/10.1111/j.1468-3156.2009.00548.x

Reese, R. M., Richman, D. M., Belmont, J. M., & Morse, P. (2005). Functional characteristics of disruptive behavior in developmentally disabled children with and without autism. *Journal of Autism and Developmental Disorders, 35*(4), 419–428. https://doi.org/10.1007/s10803-005-5032-

Research Units on Pediatric Psychopharmacology Autism Network. (2005). Randomized, controlled, crossover trial of methylphenidate in pervasive developmental disorders with hyperactivity. *Archives of General Psychiatry, 62*(11), 1266–1274. https://doi.org/10.1001/archpsyc.62.11.1266

Roane, H., Fisher, W., & Carr, J. (2016). Applied behavior analysis as treatment for autism spectrum disorder. *The Journal of Pediatrics, 175*. https://doi.org/10.1016/j.jpeds.2016.04.023

Saqr, Y., Braun, E., Porter, K., Barnette, D., & Hanks, C. (2017). Addressing medical needs of adolescents and adults with autism spectrum disorders in a primary care setting. *Autism*. https://doi.org/10.1177/1362361317709970

Scahill, L., McCracken, J., King, B., Rockhill, C., Shah, B., Politte, L., Sanders, R., Minjarez, M., Cowen, J., Mullett, J., Page, C., Ward, D., Deng, Y., Loo, S., Dziura, J., McDougle, C., & Research Units on Pediatric Psychopharmacology Autism Network. (2015). Extended-release guanfacine for hyperactivity in children with autism spectrum disorder. *The American Journal of Psychiatry, 172*(12). https://doi.org/10.1176/appi.ajp.2015.15010055

Schopler, E., Reichler, R. J., DeVellis, R. F., & Daly, K. (1988). *Childhood autism rating scale (CARS)*. WPS Publishing.

Selten, J., Lundberg, M., Rai, D., & Magnusson, C. (2015). Risks for nonaffective psychotic disorder and bipolar disorder in young people with autism spectrum disorder: A population-based study. *JAMA Psychiatry, 72*(5), 483–489. https://doi.org/10.1001/jamapsychiatry.2014.3059

Shaffer, D., Gould, M. S., Brasic, J., Ambrosini, P., Fisher, P., Bird, H., & Aluwahlia, S. (1983). A children's global assessment scale (CGAS). *Archives of General Psychiatry, 40*(11), 1228–1231. https://doi.org/10.1001/archpsyc.1983.01790100074010

Sheinkopf, S. J., & Siegel, B. (1998). Home-based behavioral treatment of young children with autism. *Journal of Autism and Developmental Disorders, 28*(1), 15–23. https://doi.org/10.1023/a:1026054701472

Simonoff, E., Pickles, A., Charman, T., Chandler, S., Loucas, T., & Baird, G. (2008). Psychiatric disorders in children with autism spectrum disorders: Prevalence, comorbidity, and associated factors in a population-derived sample. *Journal of the American Academy of Child and Adolescent Psychiatry, 47*(8), 921–929. https://doi.org/10.1097/CHI.0b013e318179964f

Sperry, L., & Sperry, J. (2012). *Case conceptualization: Mastering this competency with ease and confidence*. Taylor & Francis.

Storch, E., Arnold, E., Jones, A., Ale, C., Wood, J., Ehrenreich-May, J., Lewin, A., Mutch, A. J., & Murphy, T. (2012). The role of co-occurring disruptive behavior in the clinical presentation of children and adolescents with anxiety in the context of autism spectrum disorders. *Child Psychiatry and Human Development, 43*(5). https://doi.org/10.1007/s10578-012-0294-1

Sturman, N., Deckx, L., & Driel, M. (2017). Methylphenidate for children and adolescents with autism spectrum disorder. *Cochrane Database of Systematic Reviews, 11*. https://doi.org/10.1002/14651858.CD011144.pub2

Sullivan, S., Rai, D., Golding, J., Zammit, S., & Steer, C. (2013). The association between autism spectrum disorder and psychotic experiences in the Avon longitudinal study of parents and children (ALSPAC) birth cohort. *Journal of the American Academy of Child and Adolescent Psychiatry, 52*(8), 806–814.e2. https://doi.org/10.1016/j.jaac.2013.05.010

Sutherland, R., Trembath, D., & Roberts, J. (2018). Telehealth and autism: A systematic search and review of the literature. *International Journal of Speech-Language Pathology, 20*(3), 324–336. https://doi.org/10.1080/17549507.2018.1465123

Tachere, R., & Modirrousta, M. (2017). Beyond anxiety and agitation: A clinical approach to akathisia. *Australian Family Physician, 45*(5), 296–298.

US Department of Education. (2010). Free appropriate public education under section 504.

van Steensel, F., & Heeman, E. J. (2017). Anxiety levels in children with autism spectrum disorder: A meta-analysis. *Journal of Child and Family Studies, 26*(7), 1753–1767. https://doi.org/10.1007/s10826-017-0687-7

Vickerstaff, S., Heriot, S., Wong, M., Lopes, A., & Dossetor, D. (2007). Intellectual ability, self-perceived social competence, and depressive symptomatology in children with high-functioning autistic spectrum disorders. *Journal of Autism and Developmental Disorders, 37*(9), 1647–1664. https://doi.org/10.1007/s10803-006-0292-x

Virués-Ortega, J. (2010). Applied behavior analytic intervention for autism in early childhood: Meta-analysis, meta-regression and dose-response meta-analysis of multiple outcomes. *Clinical Psychology Review, 30*(4), 387–399. https://doi.org/10.1016/j.cpr.2010.01.008

Vivanti, G., & Dissanayake, C. (2016). Outcome for children receiving the early start Denver model before and after 48 months. *Journal of Autism and Developmental Disorders, 46*(7), 2441–2449. https://doi.org/10.1007/s10803-016-2777-6

Witwer, A., & Lecavalier, L. (2005). Treatment incidence and patterns in children and adolescents with autism spectrum disorders. *Journal of Child and Adolescent Psychopharmacology, 15*(4). https://doi.org/10.1089/cap.2005.15.671

Wright, B., Sims, D., Smart, S., Alwazeer, A., Alderson-Day, B., Allgar, V., Whitton, C., Tomlinson, H., Bennett, S., Jardine, J., McCaffrey, N., Leyland, C., Jakeman, C., & Miles, J. (2011). Melatonin versus placebo in children with autism spectrum conditions and severe sleep problems not amenable to behaviour management strategies: A randomised controlled crossover trial. *Journal of Autism and Developmental Disorders, 41*(2). https://doi.org/10.1007/s10803-010-1036-5

Youngstrom, E. A., & Van Meter, A. (2016). Empirically supported assessment of children and adolescents. *Clinical Psychology: Science and Practice, 23*(4), 327–347. https://doi.org/10.1037/h0101738

Zheng, Z., Zheng, P., & Zou, X. (2018). Association between schizophrenia and autism spectrum disorder: A systematic review and meta-analysis. *Autism Research: Official Journal of the International Society for Autism Research, 11*, 1110–1119.

Autism Spectrum Disorders: Treatments and Psychological Interventions

20

Michelle M. Yetman, Adam T. Blancher, Nanette Massey, and Michael D. McGill

Abstract

Autism spectrum disorder (ASD) is a complex disorder with a wide range of symptoms, and thus treatment approaches are heterogeneous in nature. There is no single symptom that defines ASD, and therefore, no single treatment. While there is no cure for ASD, several interventions have been developed to help reduce symptoms, improve cognitive ability and daily living skills, and maximize the ability of the individual to ultimately live as independently as possible within the community. This chapter describes the psychologist's role in the identification of ASD and provides a review of a variety of treatments, including early intervention and specific forms of Applied Behavioral Analysis such as Pivot Response Training (PRT), Discrete Trial Training (DTT), and Verbal Behavioral Therapy. Other early interventions, including TEACCH, Early Start Denver Model, and Floortime, are also discussed. In addition, the importance of referring to specialists such as speech therapists, occupational therapists, and genetics on an as-needed basis is reviewed. The framework of treatment is presented in a series of vignettes, following a young child with ASD, from the initial diagnosis phase through adulthood.

Keywords

Autism · Autism spectrum disorder · Treatment · Intervention · Early intervention

Introduction

Psychologists commonly assess children identified as at risk for autism spectrum disorder (ASD) or other neurodevelopmental disorders and provide treatment when indicated. Psychology is a vast specialty. Similar to the specialization of physicians, psychologists train and practice in varied capacities. When pediatricians and primary care providers screen and conduct surveillance that determine a child is "at risk" for delay, a referral is typically made to a psychologist for diagnosis (see Table 20.1). The psychologist may practice independently or act as part of a team conducting a multidisciplinary diagnostic assessment. The psychologist's role is often broad and complex. The diagnosis of ASD or some other type of developmental delay is made following direct observation of the child and in-depth inter-

M. M. Yetman (✉) · A. T. Blancher · N. Massey
M. D. McGill
Louisiana State University – Shreveport, Shreveport, LA, USA
e-mail: michelle.yetman@lsuhs.edu

P. McPherson (ed.), *Handbook of Treatment Planning for Children with Autism and Other Neurodevelopmental Disorders*, Autism and Child Psychopathology Series,
https://doi.org/10.1007/978-3-031-06120-2_20

Table 20.1 Autism spectrum disorder referral process

Referrals for evaluation by a psychologist typically begin with the pediatrician/primary care physician. The following steps can be expected in the process:
1. Parent should discuss any developmental concerns with the pediatrician/primary care physician. During this appointment, the doctor will review concerns and ask parent to complete developmental and/or ASD-specific screening
2. If the results of the screening tools are negative for ASD and developmental delays, the pediatrician will provide anticipatory guidance and will continue to monitor during future well-visits
3. If the results of the screening tools are negative for ASD but place child at risk for developmental delays, the pediatrician will refer to local early intervention program (if child is not yet 3 years of age), local school system (if child is 3 years or older), or appropriate private agency
4. If the results of the screening tools place child at risk for ASD, pediatrician will write a prescription/referral to a local provider or agency. In addition, the pediatrician will refer to local early intervention program (if not yet 3 years of age), local school system (if child is 3 years or older), or appropriate private agency

views with the caregivers. In addition, psychological testing is often conducted to assess cognitive skills, adaptive functioning, language ability, and academic skills (Huerta & Lord, 2012; Volkmar et al., 2014). Following the diagnosis, psychologists may provide guidance to the primary care provider for the coordination of services and/or begin treatment.

A variety of specialized psychologists could participate in the treatment of a child with ASD or other special needs, including clinical psychologists, counseling psychologists, and/or school psychologists. American Psychological Association defines psychology as the study of both the mind and behavior, specifically focusing on how the mind affects behavior (VandenBos, 2007). While all of these psychological specialties focus on this core principle, the environments in which they practice differ. Clinical and counseling psychologists may work in a hospital, outpatient clinic, or private practice setting. They can assist with the diagnosis, treatment planning, and family support services. School psychologists can assist within the school setting, modifying the classroom setting, helping children engage with peers, establish behavior modification programs or decrease challenging behaviors, help children manage transitions, or manage bullying, if present. School psychologists may also work with parents and teachers to help the child with a developmental disability develop strategies for success. With the rise in popularity of Applied Behavioral Analysis (ABA) therapy – in part due to insurance coverage – behavioral psychologists have become increasingly in demand. Behavioral Psychologists examine the purpose and function of behaviors and develop ways to shape and change behavior in a desired way. Using a system of rewards (reinforcers), the behavioral psychologists will gradually work with a child and "shape" their behavior toward the desired outcome. For example, if the ultimate goal is to have a child sit down at a desk and do work, but initially the child is running around the room, the behavioral psychologist will develop a plan in which gradual approximations of the desired behavior are rewarded, e.g., stopping in response to name; coming when name is called; walking over to desk; standing at desk; working at puzzle for 30 seconds while standing at desk; working at puzzle for one minute while standing at desk; sitting in chair at desk; sitting in chair and working on puzzle at desk. Behavioral intervention can also be used to "extinguish" or treat serious problem behaviors that may be presenting safety concerns. Thus, the role of psychologists varies by specific training and the setting in which they work.

Psychologists play a significant and varied role in the diagnosis and treatment of autism and other neurodevelopmental disorders. Whether at a clinic or a school-based program, psychologists play a key role in both the evaluation and diagnosis of children with delays. Following the assessment, a treatment plan is developed and referrals are made. Goals within the treatment plan are communicated to both the family and the referring doctor.

Sam's Journey

Sam is a 24 month old toddler and has recently been diagnosed with ASD at a developmental clinic. He does not talk or make any speech sounds. When he is frustrated, he will scream, cry and bang his head. His mother noted that while he used to "eat everything", in recent months his diet has become very restricted, limited to milk and snack foods like chips and crackers. He refuses to try any new foods. He has begun waking up in the middle of the night and his mother is often unable to get him back to sleep. This makes for a long, challenging day. His parents took him to a Clinical Psychologist, and he received the diagnosis of autism. Now that she has the diagnosis of ASD, Sam's mother is feeling overwhelmed regarding treatment options and what paths she needs to explore.

Treatment Planning

Receiving the diagnosis of ASD is only the beginning of a long journey for parents and caregivers of children with ASD. Treatment is an important part of ASD, and while there is no cure for ASD, many aspects of the condition benefit from therapy and intervention. Similarly, when parents receive a diagnosis of a neurodevelopmental disorder, they may initially feel some relief, in that the diagnosis confirms their suspicion that something "was not quite right" with their child, they are often at a loss as to how to proceed after the diagnosis. Early intervention has been shown to be effective in improving the outcomes for children with ASD and other neurodevelopmental disorders (Roane et al., 2016). For this reason, intervention should begin as early as possible.

It is important to remember that therapy must be customized to a child's unique individual needs. Screening for and identifying delays in young children and then developing interventions based on identified delays are the goal of early intervention programs. Since IQ and language skills are still emerging at young ages, children are evaluated based on the development, typically in the following areas: physical skills, cognitive skills, communication skills, social skills, and adaptive/self-help skills. The role of the psychologist is to evaluate the child globally, based on their developmental profile, and refer to appropriate specialist, e.g., occupational therapy, speech therapy, for additional detailed treatment plan. As a child grows and develops, one form of therapy, which had been previously beneficial in the past, may no longer be effective. As a child's needs change, therapy/treatment must change to address the continuously evolving needs of the child.

Early Intervention Programs

After researching her options, Sam's mother decided to use a combined of early intervention home based therapy, which was provided by the state, and ABA therapy, which was covered by her insurance. She appreciated being involved in the home-based therapy, and included in the goal making process. She also feels encouraged by the strong research which supports ABA therapy and is optimistic that Sam will make progress. Targeted goals are listed on Sam's IFSP (Individualized Family Service Plan) and will be monitored on a regular basis.

The Early Intervention Programs for Infants and Toddlers with Disabilities provide for services and supports to children from birth through three years of age at risk for developmental delays or disabilities. These services can include speech–language therapy, occupational therapy, physical therapy, assistive technology, behavioral interventions, and more. The program offers free assessments as part of a federally funded program administered by each state under Part C of the Individuals with Disabilities Education Act (IDEA). Parents can request an evaluation if they are concerned about their child's development or a community Child Find provider, such as an early educator or pediatrician may suggest an evaluation. If the evaluation finds that a child qualifies for services, an Individualized Family Service Plan (IFSP) will be created detailing recommended services and goals. Treatment costs vary by state; there may be a no charge or a sliding scale for services. IDEA mandates that no child can be denied services because their family is unable to pay (Guralnick & Bruder, 2019).

Many therapies have overlapping features, but it is important to note that not every therapy is appropriate for every child. Variables such as intellectual ability and language ability, which vary greatly among individuals with ASD and developmental delays, will often determine which therapy is the best therapy for them at any given time. While numerous treatments exist for this population, it is important to note that no one intervention has been shown to be superior to others. The most important aspect to remember when it comes to treatment is that intervention can help, and the earlier the better (Dulcan, 2015).

There are currently many different types of early intensive behavioral intervention therapies for ASD in addition to supplemental therapies such as occupational therapy, speech–language therapy, mediation treatment, and social skills training. Regardless of intervention selected, research has shown that parents acting as interventionists for their young children with ASD and other developmental delays can be beneficial (National Autism Center, 2015a, b).

While the symptoms of both autism spectrum disorder and neurodevelopmental disorder can change with age, development, and in direct response to treatment, the condition itself is considered chronic and will impact a child's development throughout their lifespan, including their transition into adulthood. For this reason, treatment plans must be shared and modified based on the child's progress and in response to their developmental needs. For example, as a child begins to transition from the school system and into the workforce, compensatory skills that will allow them to participate in an employment setting to the best of their ability should be encouraged. Typically an Individualized Education Plan (IEP) dictates the goals for the child, which can include work-related goals as the child begins the process of transitioning of exiting the school system and entering the workforce. Reevaluation should occur periodically but no less than once every three years when a child has been formally enrolled in school. Reevaluations should occur on a more frequent basis when the child is young, as they can change greatly in their presentation and, in some cases, may no longer meet the criteria for the initial diagnosis. A small prospective longitudinal study found that 9% of children who initially met the criteria for the diagnosis of autism at age 2 no longer met the same criteria by age 19 (Anderson et al., 2014). Assessments monitor the child's progress, comparing their functioning level to same aged peers, and help determine to what degree the child is impacted by their disability. Therefore, this should be an ongoing process, as the degree to which the child is impacted at age 2 is not the same as the degree to which they are impacted at age 15 by their ASD diagnosis.

Speech–Language Pathology

> Sam's parents first voiced concerns about his developmental delay to his pediatrician regarding his lack of speech. Typically, parent's first concern related to developmental delays in their child is communication (Kozlowski et al., 2011). While the most overt sign of Sam's "speech" delay was that he was not using any words, there is much more to "communication" than just speech. Sam's parents were referred to a speech-language pathologist (SLP) for further evaluation and treatment.

While psychologists evaluate speech in terms of developmental milestones, e.g., is it grossly within normal limits for the child's age, an in-depth analysis of speech development is required – the specialized skill of a speech–language pathologist. Often the terms "speech" and "language" are used interchangeably, but ***speech*** refers to how humans sound, or say sounds and words. It includes articulation, voice, and fluency, all of which can have an effect on the sound of speech. ***Language***, however, refers to communication (including words, signs, gestures, augmentative and alternative communication) and how we use this code to express thoughts and ideas for a variety of purposes (expressive language) as well as an understanding of this code (receptive language) so that humans can decipher and understand others' communication (ASHA, n.d.). Speech–language pathologists are often included as core team members on the diagnostic team which evaluate young children suspected of having developmental delays. If a psychologist were to be operating in private practice, they would refer a child with communication delays to a speech–language pathologist for additional testing for concerns that were outside the scope of the psychologist. Psychologists, pediatricians, or other interventionalists often refer to an SLP whenever there is concern about a child's language, either the understanding of language or the expressive use of language.

Occupational Therapy

Based on the psychologist's explanation of Sam's needs, his mother elected to receive home based services through her states early intervention program. Sam qualified to receive both occupational therapy and speech therapy. Sam's mother worked with the Occupational Therapist to develop goals for Sam's Individual Family Service Plan (IFSP) which included learning to sit at the table during meal times with physical supports, learning to expand his diet to include a wider range of textures, and improving his fine motor skills. Additional goals may be related to toilet training.

Psychologists refer to occupational therapists (OTs) to provide services to a child who is having delays in a variety of areas, including sensory issues, feeding issues, and fine motor delays. Psychologists often identify delays in these areas during their assessment of the child, either through parent report or identifying delays during developmental testing. In-depth assessment and treatment would require a referral to an occupational therapist. Even without a formal diagnosis, developmental delays can lead to a referral to early intervention, which provides services including occupational therapy for infants and toddlers who are delayed in developing the baseline skills babies typically master in the first 3 years of life. When a child shows delays in mastering typical milestones or activities, or displays unusual or disruptive behavioral responses, the OT is in many instances the first professional to work with him/her (AOTA, 2019).

Occupational Therapy for Children from Birth to 3 Years: Early Intervention

Early intervention services aim to treat delays and developmental problems from the earliest time possible. The first 3 years of life are the most crucial in terms of the developing brain and body. Evidence-based research studies prove that the best window of opportunity for progress is from birth through age three (Roane et al., 2016). OT treatment with infants includes educating the parents/caregivers on activities they can do at home to encourage their infant to meet all of these developmental challenges. It is best done in the infant's home or other natural environments (i.e., daycare, grandparent's house, etc.). OT early intervention services are provided for children with suspected delays and identified delays such as autism spectrum disorders, fetal alcohol syndrome, intellectual disabilities, and other neurodevelopmental disorders. (Boyer & Thompson, 2014).

As a child grows and develops, repeat psychological assessments may prompt referrals to occupational therapy to address difficulties with their play skills, feeding skills, sensory processing problems, pre-writing and academic readiness skills, and/or self-help skills.

Sensory Processing Disorder

Among the many challenging behaviors that children with ASD can present with, a hypersensitivity to their environment is commonly seen. Sensory processing disorder or SPD (originally called sensory integration dysfunction) is a neurological disorder in which the sensory information that the individual perceives results in abnormal responses. A neurophysiologic condition in which sensory input either from the environment or from one's body is poorly detected, modulated, or interpreted and/or to which atypical responses are observed. Sensory processing refers to the way the nervous system receives messages from the senses and turns them into responses. For those with sensory processing disorder, sensory information goes into the brain but does not get organized into appropriate responses. Those with SPD perceive and/or respond to sensory information differently than most other people. Unlike people who have impaired sight or hearing, those with sensory processing disorder do detect the sensory information; however, the sensory information gets "disorganized" in their brain and therefore the responses are inappropriate in the context in which they find themselves. (Miller, 2014).

Parents often report challenging behavior in their children, and through careful questioning, the psychologist may determine that sensory triggers may be a factor in the child's challenging behavior. While a precise definition of sensory dysfunction has yet to be agreed upon, many children, not just those with ASD, experience an abnormal perception of sensory stimuli whether it is auditory, visual, or tactile and consequent behavioral responses (Noor et al., 2018).

Psychologists receive referrals for behavioral difficulties that appear to be, at least in part, reactionary to sensory input. For example, the child with ASD who has behavioral tantrums only during PE class or the lunchroom may be reacting to the loud noise in these environments. Psychologists may refer to an occupational therapist if they feel sensory oversensitivity or undersensitivity is contributing to a child's behavioral difficulties. Unfortunately, there have been conspicuously few research studies of sensory processing disorders and few theoretical models. Nevertheless, some of the most illuminating descriptions of this phenomenon are provided by individuals describing how they perceive the world (Kranowitz, 2016). It is often challenging to identify which specific environmental stimuli will trigger distress behavior in the individual with ASD or a developmental delay.

While a psychologist may be able to identify the pattern of triggers and responses, e.g., an extremely loud, sudden noise results in aggressive behavior, a referral to an occupational therapist for treatment would be typically done if it is believed sensory issues are contributing to behavioral challenges. Learning how to address the challenges imposed by sensory sensitivities can have a positive impact on a child's development. Working with an occupational therapist, one can develop approaches to address a child's specific concerns. The psychologist would encourage parents to keep a log of antecedents or triggers that appear to come before the behavior that may be contributing to the behavior, e.g., mother runs the vacuum cleaner or blender and the child throws a tantrum and becomes aggressive in response to noise. While some sensory experiences can be avoided, others may be reduced by introducing a barrier, e.g., earplugs or headphones. The psychologist often refers to an occupational therapist for assistance with possible environmental triggers, strategies to avoid/reduce such triggers, and coping methods to either block external sensory experiences or build up tolerance to these experiences. To reduce a reaction of fear or panic, the psychologist or occupational therapist could develop a program of desensitization to the fear-inducing object or situation (Fjeldsted & Xue, 2019).

The exact cause of sensory processing disorder has not yet been identified. Preliminary studies and research suggest that SPD is often inherited. Prenatal and birth complications have also been implicated as causal in SPD, as well as certain environmental factors. Sensory processing disorder can affect anyone. Current studies indicate that 5–16% of children exhibit symptoms of SPD (Case-Smith et al., 2015). Sensory processing disorder is not recognized as a formal medical dx at this time. Sensory differences are, however, included in Diagnostic and Statistical Manual of Mental Disorders (DSM-5) criteria for ASD.

Feeding Issues and Neurodevelopmental Disorders

Sam's feeding issues were present from early on in life. He would not transition from the bottle to rice cereal, depriving him of calories and needed nutrients for growth. His mother had been aware of potential delays, and decided to discuss the scope of her concerns with the pediatrician. Due to the parent reported concerns, the pediatrician referred them to early intervention, which provides services for infants and toddlers who are delayed in developing the basic skills babies typically develop in the first three years. After the initial evaluation from the early intervention team, while there was no formal diagnosis, everyone agreed that Sam had feeding issues due to oral sensitivity, in addition to other developmental delays. While Sam made some progress with tolerance for textures his feeding continues to be restricted in comparison to other children his age. A referral to an Occupational Therapist who specialized in feeding issues was included in his IFSP.

A large percentage of children with ASD demonstrate feeding difficulties from an early age and have a significantly less varied diet in comparison to non-ASD peers. It is estimated that 50% of children with ASD demonstrate limited food acceptance (Seiverling et al., 2018) in comparison to rate of 25% for neurotypical developing children (Hubbard et al., 2014). Despite such feeding difficulties, their overall caloric intake and growth does not appear to be significantly impacted (Emond, et al., 2010). This phenomenon, while not core to the diagnosis of ASD, is a very common experience for families who have children with ASD. Other types of feeding problems, such as Pica, are more common in individuals with both ASD and intellectual disabilities (ID). The consumption of non-nutrition food substances can be dangerous and, in some cases, life-threatening (Matson et al., 2013). Many families take it upon themselves to try diet manipulation with the hopes that it will reduce ASD symptomatology. One of the most popular diets with the fields of autism has been the gluten-free, casein-free diet (GFCF), with over half of families within the ASD community reporting having tried an elimination diet (Smith & Antolovich, 2000). While there are ample anecdotal stories and the lay press supporting the use of the GFCF diet, empirical research support for the diet is lacking. A 2013 review by Hurwitz of the literature found only five studies that could meet the inclusive criteria for "well-controlled" studies. Of these, three had no positive effects and the two remaining studies had quality concerns. The author concluded that parents would be best served by putting their time, energy, and resources into therapies that have proven outcomes rather than therapies which lack sufficient scientific support like GFCF diets.

Sam's mother believes feeding therapy has helped her son. Three years after voicing her initial concerns and choosing to pursue therapy for her son, he has not only transitioned from the bottle, he has expanded his diet to include crackers, fruits, some vegetables, and pasta. While he still does not like to combine textures he is willing to try new foods and slowly is incorporating them into his diet. His mother reports that like many children with an autistic spectrum disorder, her son continues to have a restricted diet but slowly, methodically he is accepting a wider variety of types and textures of foods. She has learned that having a mealtime routine and eating together is important. It is important to eat on a schedule. She has learned to space meals throughout the day and to eliminate snacks.

Feeding problems typically are multifactorial; many issues can contribute to the difficulty such as food aversion, sensory issues, behavioral problems, premature birth history, and feeding tube. For these reasons, often numerous professionals, including behavioral psychologists, occupational therapists, and other feeding specialists, work with the child and family. Children with feeding issues often have developed negative behaviors when the food is introduced. Behavioral psychologists can work with the child and the family to help change negative behaviors, such as throwing food, and make mealtime a more pleasant, positive experience. One of the first areas psychologists work on is helping teach parents to train the child's internal hunger signals to specific mealtimes. By reducing snacking, this helps the child's body expect and accept food at designated times. Oftentimes the role of either the psychologist or the occupational therapist with feeding therapy experience is to change the parents' expectation regarding mealtime as much as it is to change a child's feeding behaviors. The role of the parent is to provide healthy, nutritious food on a schedule, and the role of the child is to determine how much they will eat. Decreasing power struggles and expectations are an important part of feeding therapy. Reducing parent anxiety surrounding mealtime can in turn reduce the child's anxiety.

Many children with ASD have anxiety at mealtimes due to sensory aversions and fear of unfamiliar foods. Children with ASD are prone to anxiety, and caregivers can inadvertently make anxiety worse during mealtime by trying to force children to eat, thereby causing an association between mealtime and stress (Twachtman-Reilly et al., 2008). Fear and anxiety can supersede hunger by putting the child's body in a state of "fight or flight." Since many children with ASD experience this phenomenon of "fight or flight," a psychologist can assist parents in how to give positive reinforcement and how to help their child relax, such as before mealtime. Ways to do this include spending training the parents on modeling deep breathing or alternatively, blowing pinwheels, bubbles, or even a kazoo. Positive reinforcement during mealtime can include smiling, praise, singing, and other positive behaviors.

Intensive Interventions

Therapy can be delivered in either a school setting, a private setting, or a combination of the two. More and more, therapists are under increased pressure to show positive outcome results, particular if third party payers (insurance companies) are involved. However, as with all children, much of a child's educational experience is difficult to measure in objective data points. Similar to a teacher "teaching to the test" in order to obtain arbitrary benchmarks, if you focus on minor gains, you may lose the holistic educational experience. It is very challenging to measure teacher–child rapport, personal and emotional growth, as well as the child's ability to gain new experiences. Value can be still obtained in education even if it cannot be demonstrated in a dataset. For that reason, there is no one specific treatment for a child with autism or a developmental disability that is recommended over another. Parents are encouraged to review the programs available to them, the specific needs of their child, and determine the best "goodness of fit" approach. Many children often receive various forms of intervention throughout their lifetime. For example, some children will receive home-based early intervention through a state-sponsored program. Upon turning 3 years of age, they will transfer to the public school system and will receive special education services that may include supplemental supports such as speech therapy and occupational therapy. Along the way, their parents may have sought out private Applied Behavioral Analysis services. Later, when their child is older and no longer needs ABA therapy, they may enroll their child in social skills training. During adolescence, if their child develops mood symptoms, they may seek out the support of counselor for therapy and a psychiatrist for medication management. Thus, goals and plans change in response to the child's developmental needs to maintain the goodness of fit as the child matures.

Applied Behavioral Analysis (ABA)

Treatment planning often includes ABA with the goal of improving socially desirable behaviors, e.g., on-task behavior and communication, by using techniques that are based upon learning theory principles. In addition to the promotion of desirable behaviors, self-control and reducing negative behaviors which interfere with learning, e.g., stereotyped behaviors, self-injurious behaviors, are also targeted for intervention.

ABA has a long history of documented effectiveness as a treatment for children with ASD and other neurodevelopmental disorders (Makrygianni et al., 2018). Numerous studies have documented its effectiveness in increasing desired behaviors and teaching new skills (National Autism Center, 2015a, b; Wong et al., 2015). ABA has also been shown to be effective in decreasing, or in some cases eliminating, negative behavior that interferes with the process of learning (National Autism Center, 2015a, b). When ABA therapy is introduced at an early age (prior to age 4) and to an intensive degree (greater than 20 hours per week), preliminary studies suggest that significant gains are possible, which can reduce the need for therapy later in life (Reichow, 2012; Roane et al., 2016). Despite the need for additional research, in 1999, the United States Surgeon General took the strong position of concluding:

> Thirty years of research demonstrated the efficacy of applied behavioral methods in reducing inappropriate behavior and in increasing communication, learning and appropriate social behavior. (p. 178).

Applied Behavior Analysis is provided by a Board Certified Behavior Analyst (BCBA), who often has support staff working under their supervision. The goal of intensive ABA treatment is to address the core symptoms of ASD and the co-occurring behavior through a comprehensive curriculum. General goals for children in ABA with or without the diagnosis of ASD typically include improving daily living skills, e.g., feeding and mealtime skills, toileting; expressive and receptive language skills, e.g., following directions, responding to greetings; social skills, e.g., turn taking and reducing elopement risk, e.g., staying with therapist during outdoor excursions. The BCBA develops a customized program for each individual child. The support staff, often referred to as "line technicians," work directly with the child, monitoring the child's progress and collecting data. ABA can be offered through private centers, and some school systems. Children who show deficits in multiple areas and/or behavioral problems are often candidates for immersive ABA treatments that may involve 25 hours a week of therapy or more. Children who have a milder expression of ASD may receive time limited ABA therapy or therapy to a less frequent degree, e.g., 10 hours per week (Association for Professional Behavior Analysts, 2016).

Left to their own devices, children with ASD will often engage in non-productive, non-social, repetitive forms of play. They fail to learn vicariously through watching and observing others. The goal of specific, structured techniques such as Discrete Trial Training (DTT), which is a procedure used within ABA, is to help simplify and individualize instruction. As Lerman et al. (2016) explain, each discrete trial is broken down into five components:

1. *Cue*: the teacher cues the child, e.g., "what is this?" or "do this."
2. *Prompt*: either simultaneously or immediately following the cue, the teacher gives the child a prompt, e.g., models saying the word or moves the child's hand to imitate the desired behavior.
3. *Response*: The child responds either correctly or incorrectly to the teacher's cues. No response is an incorrect response.
4. *Consequence*: A correct response will immediately be reinforced with positive praise, hugs, small bites of food, or access to a toy. If an incorrect response is given, the teacher says "no," looks away, removes the material, or in some other way indicates the response was incorrect.
5. *Intertrial interval*: Following the consequence, the teacher pauses 1–5 seconds, then cues the next trial.

DTT was one of the very first interventions developed for children with ASD, and this technique has extensive research documenting its effectiveness (Eldevik et al., 2009). Less structured techniques used in ABA include Pivotal Response Training (PRT). PRT takes advantage of the child's natural interests and uses these for motivation. For example, if a child with autism shows an interest in cars, toy cars will be used as reinforcers, not random unrelated objects. The goal is to target "pivotal" areas of development rather than one specific skill. Pivotal areas include motivation, response to multiple cues, self-management, and initiation of social interactions (Lei & Ventola, 2017). By allowing the child to choose which activities to focus on, motivation is intrinsically increased. PRT reinforces not only correct responses but also any valid attempts made by the child in acquiring a skill, and this is also felt to increase motivation (Koegel et al., 1988). PRT has been shown to be effective as a behavioral intervention in the improvement of social-communication skills in children with ASD (Smith, 2013; Koegel et al., 2016). Finally, many ABA programs also use Verbal Behavioral Therapy (VBT). Conceptualized by B.F. Skinner in 1957, VBT attempts to teach communication and language use. Language is classified types, known as "operants" based on their function. For example, a "mand" is a request, such as when you ask a child to say "cookie" in order to receive a cookie. A "tact" is a comment used for social-communication purposes. For example, if a child points and says "dog" when they see a dog, they are not requesting the dog, they are drawing both their attention and other's attention to the dog. An "interverbal" is a word or response to a question. For example, if someone says "what is your name?" and the child correctly responds with Marcus, this is considered to be an interverbal. Finally, an "echoic" is a repeated word that a child may use as they are learning language. Communication is shaped initially by requesting mands or if the child cannot speak, any type of communication that indicates a request, such as pointing (DeSouza et al., 2017). VBT has been shown to be effective with young children who are just beginning to learn language and other children with delayed or disordered language (Akers et al., 2017).

The Early Start Denver Model (ESDM)

As the incidence of children with ASD and other neurodevelopmental disorders has increased and families have demanded effective treatments, there has been pressure on insurance companies to support evidence-based treatment. While ABA has a long-established treatment history, it is by no means the only treatment that has been shown to be effective in children with ASD and other neurodevelopmental disorders. The Early Start Denver Model (ESDM) is a comprehensive behavioral early intervention program designed for toddlers between the ages of 12 and 48 months. Building on the theory that targeted early intervention can have the most dramatic impact, ESDM utilizes a developmental curriculum that focuses on specific skills to be taught with specific strategies to be used to teach the skills. ESDM can be delivered in clinic settings, the child's home, and in individual or group settings. The ESDM is the early-age extension of the Denver Model. The Early Start Denver Model (ESDM) is a comprehensive, naturalistic developmental behavioral early intervention approach designed by Dr. Sally Rogers and Dr. Geraldine Dawson. The program can be used not only for children with a confirmed diagnosis of ASD but also for children who are "at risk" for ASD, such as children who have an older sibling with ASD. There is a genetic component to ASD, and if a family has one child with autism, there is a 1-in-5 chance that subsequent children will have ASD (Ozonoff et al., 2011). ESDM could be used to target the "at risk" sibling. ESDM starts from a position of "normal" development and learning, and uses the knowledge of how ASD and other neurodevelopmental disorders can affect early development, in order to improve outcomes for children who are neurologically atypical children. ESDM aims to reduce the severity of ASD symptomatology or negative behavior in children with neurodevelopmental disorder while promoting and accelerating their development in numerous areas such as language, cognitive, social-emotional, and adaptive skills. Therapists using an ESDM approach follow a manualized curriculum and use teaching practices that are based on naturalistic developmental-applied behavior analysis. ESDM programs typically involve 20–25 hours of scheduled therapy and families are encouraged to use the techniques outside of therapy (Canoy & Boholano, 2015). Intervention models such as ESDM are seen as effective since they have the potential to reshape the young child's developing brain to be receptive to social information by using a naturalistic approach that combines behavioral techniques and developmentally sensitive techniques (Bradshaw et al., 2015; Schreibman et al., 2015). While the ESDM has been shown to be effective with other populations, such as children with Fragile X syndrome, larger empirically supported research studies are needed to clearly establish its effectiveness among non-ASD populations (Vismara et al., 2019).

TEACCH

Developed by Dr. Eric Schopler and Dr. Robert Reichler in the 1960s at the University of North Carolina – Chapel Hill, the Treatment & Education of Autistic and Communication Related Handicapped Children (TEACCH) approach is both a clinical and a research-based intervention program which uses a variety of techniques based on the learning characteristics of individuals with ASD, such as strengths in visual information processing, difficulties with social-communication, and weakness in attention and executive functioning (D'Elia et al., 2014). The TEACCH program attempts to address both the neuropsychological weakness and the strengths of the individual with ASD by (1) using the physical organization of the environment by putting visual barriers and minimizing distractions, (2) customized visual work schedules so the student knows what to expect, (3) work systems that promote independent work, and (4) a visual structure of materials in tasks and activities. The use of this structured learning environment and use of clients' special interest to engage them as a reward for learning are designed to promote social engagement and self-initiated communication (Mesibov & Shea, 2009). Although an accepted practice today, TEACCH was groundbreaking in its approach to involving parents in the treatment of their children with ASD (Ciurlik et al., 2015). The use of a structured TEACCH approach is to encourage the child with ASD to be independent and flexible. This teaching approach can be used with other evidence-based intervention techniques, such as ABA. In terms of standalone effectiveness, some studies have shown that the effects of TEACCH are comparable to that of Applied Behavioral Analysis (ABA; Callahan et al., 2010). The TEACCH model, which has become a state-wide program in North Carolina, has become a model for many programs around the world. The TEACCH program has been shown to be effective in teaching functional skills to adults with ASD and intellectual disabilities (Siu et al., 2019). Despite positive findings, research gaps remain. In the one meta-analysis that has analyzed the effectiveness of TEACCH (Virues-Ortega et al., 2013), TEACCH was found to be effective in improving social behavior and reducing maladaptive behaviors, but dramatic benefits in the areas of cognition, perception, and motor area were not replicated from smaller studies. This limited support was found due to the small number of studies available, lack of randomized control trials, and small samples included in the studies. In addition, methodologies with different aged children and different severity of ASD no doubt impacted findings. Thus, as with most forms of interventions, additional studies, particularly those which look at the long-term impact of TEACCH across an individual's life, e.g., social, communication, daily living, are needed to better evaluate the program.

Floortime

Developed by Dr. Stanley Greenspan in the 1980s, Floortime is a specific therapeutic approach that utilizes the Developmental Individual Difference Relationship Model (DIR). In Floortime, the therapist or parent helps to expand the child's communication by meeting the child at their current developmental level and build on the child's strengths. Therapy is typically conducted in a play format on the floor – hence the term Floortime. Greenspan (1992) described six stages of development and the goal of Floortime is to help the child achieve these developmental milestones, which are critical to emotional and intellectual growth. The six stages are (1) birth to 3 months = self-regulation and interest in the world; (2) 2 to 7 months = Intimacy or a special love for the world of human relations; (3) 3 to 10 months = affective, reciprocal interactions which eventually lead to two-way communications; (4) 9 to 24 months = initiation of complex communication; (5) 1.5 to 4 years = development of emotional ideas; and (6) middle childhood through adolescence = progression of emotional thinking. Through the use of play, Floortime combines developmental tasks such as speech, motor, and cognitive development. By selecting a play activity that is interesting to the child, the therapist or parent attempts to engage the child initially at the level they are at and then gradually move the child toward more complex interactions (Cullinane, 2015). A review of the studies on DIR/Floortime generally supports the idea that it could be considered evidenced-based practice but does not qualify as evidenced-treatment (Mercer, 2015).

Behavioral Assessment

> The school psychologist was asked to conduct an FBA on Sam due to his tendency to run out of the classroom and attempt to leave the school building. After a couple of weeks of observations, data gathering and interviews with Sam, his teachers and his parents, the psychologist determined that Sam had become fixated on the construction site that was going on next to the school building. His attempts at leaving the school building occurred when he heard the heavy equipment and he want to leave to see the machinery in action. Numerous interventions were put in place such as placing large visual cues e.g., 'Stop' signs on exits to remind Sam it was not safe to leave. Sam was given a reinforcer of watching construction videos during his break time, which decreased his desire to leave the classroom.

Challenging behavior is often formally addressed using tools such as the Functional Behavioral Assessment (FBA) and the Behavior Intervention Plan (BIP), which allow a psychologist to work collaboratively with other school officials and determine the best way to intervene. The initial portion of a behavioral assessment a psychologist may conduct would be an FBA; students enrolled in regular education or special education may receive an FBA. While most often used in the school system, FBA are also created within ABA treatment programs and hospital settings. The purpose of an FBA is to identify behavior(s) that are directly interfering with the child's educational progress as well as to identify the specific target behavior, determine the purpose of the behavior, and what factors maintain the behavior (Reitzel et al., 2013). For example, if a child screams and yells when a teacher presents work to him, the identified behavior is screaming and yelling, the purpose of the behavior may be to get out of doing the work, and the behavior may be maintained because the work is too challenging for the child, and when he fusses, the teacher gives up and removes the challenging work. An FBA is part of a Behavior Intervention Plan (BIP). After the psychologist has identified the specific behaviors to be addressed in the FBA, as well as the precursors to the behaviors, staff will formulate a Behavior Intervention Plan (BIP). Typically it is necessary to complete an FBA when a specific behavior or emotional concern is disrupting the academic progress of a child and/or the other children within the classroom. In addition, if a behavior has escalated during the school year and has become a continuous concern, an FBA may be requested. A psychologist or other qualified school assessment personnel completes the FBA. The FBA consists of several key components. First, the severity of the behavior is determined. This is often determined through direction observation, data point gathering, as well as parent and teacher ratings. Information is gathered regarding the behavior such as when it occurs, what events have immediately occurred before the behavior or after the behavior, possible reasons for the behavior, and any recent changes in the student's life. This information is used to develop a hypothesis regarding what might be causing and/or supporting the behavior. Next, the BIP is developed and evaluated for effectiveness over time. The BIP is the actual plan created to help improve behavior. It is also the method through which interventions are formally documented and progress, or lack of response, is tracked over time.

If the child is receiving special education services, the FBA can be part of the Behavior Intervention Plan (BIP). Following observations and data collection, a written plan is developed with identified goals on how to change the challenging behavior. An FBA and BIP can be used for special education students or regular education students. When an FBA/BIP is conducted regarding students classified as a Special Education student, it becomes part of their Individualized Education Plan (IEP).

Wandering and Neurodevelopmental Disorders

Wandering, also called "elopement," occurs when a child or an individual leaves a caregiver or a safe area and wanders off; this is potentially dangerous behavior. Individuals with ASD and developmental disabilities are particularly prone to wandering. A survey of caregivers of children with ASD and/or ID found that 1 in 4 reported to have experienced an episode of wandering in the last year (Rice et al., 2016). In an online survey of families participating in the Interactive Autism Network (IAN), nearly half (49%) of respondents had attempted elopement at least once after 4 years of age. The impact of this behavior on the family was significant; 62% of parents reported their child's elopement risk prevented them from getting a good night's sleep and 56% reported it was one of the most stressful behaviors they had to deal with as caregivers of a child with ASD (Anderson et al., 2012).

The best strategy is a multitiered approach. Parents and caregivers should be given anticipatory guidance at the time of diagnosis that their child is "at risk" for wandering behaviors, and safety measures should be put in place even if the behavior is not currently a problem. Manuals such as "The Big Red Toolkit," available online by National Autism Association (nationalautismassociation.org) is designed for children with autism, although the safety strategies could apply to all children and adults with development disabilities who are at risk for wandering. Strategies include educating the child about safety and dangers using whatever means of communication works, including social stories, language and/or visual prompts such as placing stop signs on exits. Parents can make wandering prevention strategies part of a child's IEP and work with school officials to make sure that safety measures are put in place. Additional prevention methods include the use of locks, door and window alarms, and GPS tracking devices. These tracking devices can be registered with local law enforcement. Parents should specifically seek out tracking devices that are waterproof, as many children with ASD may be prone to seek out bodies of water. Parents of children or adults with developmental disabilities who have communication difficulties are encouraged to have that individual wear a medical alert ID bracelet. Keep in mind that communication skills often deteriorate in times of stress, such as during an interaction with a police officer. A child with a developmental disability may become stressed and may not be able to convey their safety information, such as name, address, and caregiver contact information. According to the National Institute for Elopement Prevention and Resolution (2015), a history of wandering may mean that this child or adult is at risk for future elopement and that additional supervision and other precautions may be necessary. If there is a history of wandering, questions to consider include:

1. When did the wandering behavior begin and how often does it occur?
2. Is it more frequent during daytime or nighttime hours?
3. Is the wandering behavior associated with any other factors, such as pain/discomfort or noise?
4. What type of travel pattern is exhibited (random, seeking out water, heavy machinery, etc.)
5. Does the wandering appear purposeful?

In recent years there has been several high-profile cases that have brought the danger of wandering to the public's attention (McIlwain & Fournier, 2010). Numerous children with ASD have reportedly died as a result of drowning, motor vehicle accidents, or wandering away from their caregivers. As public awareness increases, greater effort is being made to train parents, school officials, and law enforcement who are in the unique role to help manage and prevent the challenge of wandering behavior among ASD and neurodevelopmental delayed individuals.

Referral for Co-occurring Development Concerns

While communication deficits are part of the core definition of an autism spectrum disorder (APA, 2013), ASD and neurodevelopmental disorders have high frequencies of co-occurring conditions (Hyman et al., 2019). Since co-occurring conditions can influence the presentation of ASD symptomatology, additional evaluations are typically warranted. Since communication and cognitive ability tend to predict a child's ultimate level of functioning, valid assessment of both of these areas is considered to be a core part of evaluation (Hyman et al., 2019). Psychologists often focus on disorder-specific assessment, e.g., diagnosing autism with instruments designed for that purpose, such as the Autism Diagnosis Interview Revised (ADI-R) and the Autism Diagnosis Observation Scale, second Edition (ADOS-2). In addition, a psychologist may provide an assessment of developmental functioning or intellectual functioning as well as adaptive functioning. Assessing adaptive functioning means reviewing how well a child handles day-to-day demands and how independently they can function compared to other children their age. Areas specifically assessed included practical skills, e.g., self-care such as bathing and grooming; social skill, e.g., the child's ability to interact with others through play and/or conversation; and conceptual skills, e.g., emerging academic skills such as letter/number recognition, interest in books or being read to. Table 20.2 provides an overview of commonly requested assessments in a comprehensive developmental evaluation.

Table 20.2 Common assessments requested in an ASD or neurodevelopmental evaluation

Assessment type	Responsible clinician
Cognitive testing	Psychologist
Language testing	Speech language pathologist
Adaptive functioning	Psychologist
Motor assessment	Occupational therapist
Hearing assessment	Audiologist
Sensory assessment	Occupational therapist
Feeding assessment	Occupational therapist/Speech language pathologist
Genetic testing	Geneticist

Treatment for Psychological Comorbidities

Around age 7, Sam's mother noticed that his academic progress appeared to "plateau" and his peers were outpacing him in terms of academic skills. Other children his age were able to read and perform basic math such as addition and subtraction, while Sam was still struggling to consistently retain his letter sounds. It felt as if he would master the material one day but would forget it overnight. He often had trouble expressing his thoughts clearly and was hard to understand at times. In addition, deficits were noted in both gross and fine motor skills. Sam could not ride a bike and he struggled with buttons and zippers when getting dressed. He also began displaying mild aggression in the form of hitting when frustrated. Sam was referred for psychological testing, and he was found to have Mild Intellectual Disability in addition to his diagnosis of autism spectrum disorder. His IEP was updated to include both the diagnosis of ASD and ID and curriculum modifications were implemented. His list of services on his IEP includes special education services, speech therapy, occupational therapy, and adaptive Physical Education. His mother continues to take him to a private facility for 10 hours of ABA therapy a week.

Although the DSM-5 provides clinically relevant criteria in which clinicians can make accurate diagnoses, there is substantial overlap in symptoms with other neurodevelopmental disorders and DSM-5 diagnoses. In fact, approximately 70% of individuals with ASD may experience symptoms of one comorbid DSM-5 condition and up to 40% of individuals with ASD will experience comorbid symptoms of two or more DSM-5 conditions (APA, 2013). Children and adults with ASD may experience clinical symptoms of major depressive disorder and a number of anxiety-related disorders (e.g., generalized anxiety disorder, obsessive-compulsive disorder). With regard to neurodevelopmental disorders, symptoms of intellectual disability, attention-deficit/hyperactivity disorder, communication disorders, and learning disabilities commonly occur in children with ASD. The presence of co-occurring disorders is associated with lower quality of life, higher reliance on intervention services, and poorer overall prognosis (Vannucchi et al., 2014; Matson & Cervantes, 2014).

Therefore, a brief summary of the often co-occurring conditions and treatment recommendations in individuals with ASD is pertinent.

Among the comorbid conditions with ASD, depression and anxiety symptoms appear to be some of the most prevalent (Hofvander et al., 2009). Mood or depressive disorders are defined as the presence of sad or irritable mood accompanied by somatic (e.g., weight loss or gain, loss of energy) and cognitive (e.g., decreased ability to think, recurrent thoughts of hopelessness) changes that affect multiple areas of a person's functioning. All anxiety disorders share common features such as excessive fear (i.e., the emotional response to stimuli) and worry (i.e., the anticipation of future harm or threat). Prevalence rates for anxiety disorders in individuals with ASD range from 22% to 84% (Vasa & Mazurek, 2015). A recent meta-analysis suggested that approximately 7.7 of children (18 and under) experience depression and the lifetime prevalence for adults with ASD is approximately 40% (White et al., 2018). An important factor associated with the development of anxiety and depressive symptoms appears to be age. According to Lever and Geurts (2016), within the ASD population, anxiety disorders are more prevalent in childhood, whereas in adulthood there is a higher prevalence of depressive disorders. Diagnosis of these conditions in individuals with ASD is often challenging due to limitations in communication abilities, emotional expression, theory of mind, and general insight into social interaction. For clinicians, it may be important to understand and address the emergence of mood or anxiety symptoms when individuals with ASD exhibit changes in their functional level or an increase in maladaptive behavior (e.g., anhedonia, reduction in activities of daily living, increases in agitation and compulsive behaviors).

According to Individuals with Disabilities Education Act (IDEA), the school system must provide special education services to children who meet criteria for one or more identified disabilities, provided the disability significantly interferes with the child's educational achievement. Both autism and intellectual disability are identified as being in the disability designation under IDEA. It is essential that the student meet both criteria (the disability and its negative impact on school progress) in order to qualify for services. In the vignette with Sam, as he progresses through school and he is reevaluated on a regular basis, the global impact of his disability becomes more apparent with age.

Psychotherapy

Psychosocial interventions include, but are not limited to, Cognitive-Behavioral Therapy (CBT), relaxation, and mindfulness training. These treatments will assist with improving insight and the overall presentation of the patient. Secondary benefits may include improvement in other areas of functioning such as social, school, or work performance. In the early 1960s, Aaron Beck originally developed CBT to be a short-term (approximately 10 weeks) structured therapy for those with dysfunctional thinking that led to feelings of depression and anxiety. Beck's CBT has received significant empirical support within the literature (Yang et al., 2017; Li et al., 2018). The basic premise of the approach is that if a patient can become aware of their own maladaptive thinking patterns (i.e., cognitive distortions) and subsequently change these thinking patterns through rational evaluation, improvements in mood and anxiety can occur and have lasting effects (Beck, 2011). Although some of the core features of ASD (i.e., cognitive rigidity, difficulties with developing relationships, and adherence to routine, etc.) may impact the overall effectiveness, meta-analysis findings suggest that CBT can be effective with individuals with ASD (Weston et al., 2016). Some researchers recommend modifications to CBT to support the treatment's efficacy (Walters et al., 2016). For example, modifications suggested by the National Institute of Healthcare and Excellence (Crowe and Salt, 2015) in the United Kingdom include emotion recognition training (e.g., increasing a person's ability to attend to facial expressions, tone of voice, and body language and interpret and respond to these communication strategies) and using concrete and/or structured approaches such as visual and written activities that incorporate the special interests of the individual to enhance CBT treatment effectiveness.

Mindfulness and relaxation training has shown to be effective in the treatment of anxiety and depressive disorders in individuals with ASD. These therapeutic approaches encourage the individual to identify thoughts and feelings as they occur in the present moment without judgment in order to develop an increased awareness and compassionate stance toward experience. The goals of clinical mindfulness training are to enhance attention and awareness of the self-including thoughts, feelings, and bodily sensations and can reduce social-communication deficits and improve internalizing and externalizing behavior problems, attention, distress, and overall emotional well-being (Ridderinkhof et al., 2018).

In addition to individual therapy, children with ASD can benefit greatly from social skill training. Often these treatment approaches target the core deficits of social-communication (e.g., conversation, theory of mind, non-literal use of language) to improve reciprocity and help to develop and understand social relationships. Although there are many approaches to teaching social skills, common applications include group and individual therapy formats. Group treatment is most common and, according to recent meta-analyses (Gates et al., 2017; Wolstencroft et al., 2018), shows moderate efficacy. However, one limitation to group treatment in the clinic setting is generalization of learned skills to other environments (e.g., school, community). Individual treatment often utilizes role-playing and perspective-taking to teach children with ASD social skills. One popular approach is Social Stories™. Carol Gray (2010) developed and used stories to describe common social situations, skills, or concepts. The goal is to improve social responsiveness by teaching individuals the necessary social information and/or typical responses in a given situation. Social stories have received satisfactory empirical evidence (Test et al., 2011). It should be noted that social skills can be learned across the lifespan and social skill development can be an appropriate goal for older individuals with ASD as well. For example, programs such as Program for the Education and Enrichment of Relational Skills (PEERS), which was initially designed for adolescents (Laugeson & Frankel, 2011), was expanded to include preschoolers (Radley et al., 2018) and adults (Laugeson et al., 2015). Overall, targeted social skills training, in group or individual format, should be considered a priority during treatment planning for individuals with ASD.

Intellectual Disability and ASD

As noted above, a number of neurodevelopmental disorders have been found to be associated with ASD, including intellectual disability (ID), specific learning disability, and attention-deficit/hyperactivity disorder (ADHD). Intellectual disability is characterized by significant deficits in cognitive (Intellectual Quotient less than 70) and adaptive functioning (i.e., conceptual, practical, and social skills). Individuals with comorbid ID and ASD not only typically have difficulty with social-communication and restricted interests and/or behaviors but also struggle with general reasoning and often require enhanced care. Studies show that the co-occurrence of ASD and ID ranges from 30% (Polyak et al., 2015) to 70% (Fombonne, 2005) methodological research differences, severity level of ID symptoms and age of the individual may affect prevalence rates in the literature. Treatment typically involves establishing supports and services through various local (e.g., local speech/language and ABA therapy service providers) and state agencies (e.g., Office of Citizens with Developmental Disabilities) as well as coordination of academic interventions (e.g., special education). As these children progress through school-aged years, parents and other caregivers are encouraged to pay particular attention to limited basic cognitive and academic skills, as this will greatly impact the ability to obtain future employment and may limit independent functioning. In order to monitor a child's progress related to cognitive, social, academic, and overall development, routine psychological assessment should be made at regular intervals. Monitoring progress is useful for identification/acquisition of appropriate supports and services as the child ages.

Learning Disabilities and ASD

Just as a proportion of children with ASD or NDD will also have co-occurring ID, children with ASD or NDD may also struggle with achievement (i.e., acquisition of academic knowledge) and are susceptible to learning struggles. According to the DSM-5, a specific learning disability (SLD) is defined as significant deficits in an area of academics (e.g., basic reading kills, reading comprehension, fluency; math calculation, problem solving, and fluency; written language and fluency) as measured by standardized instruments in the absence of significant deficits in intellectual/cognitive functioning (e.g., verbal and nonverbal reasoning, working memory, processing speed). Therefore, it is important to note that a specific learning disability is not equivalent to ID. A child with ASD without significant cognitive deficits may struggle in school and may or may not meet DSM-5 criteria for SLD. In other words, ASD is not conceptualized as a learning disability but also does not preclude an SLD diagnosis. Due to the nature of ASD and the underlying neurocognitive deficits often seen in individuals with ASD, these individuals may struggle in school. In general, children with ASD may struggle within the school environment due to executive functioning deficits, sensory processing deficits, and challenging behaviors (e.g., tantrums in response to being overwhelmed). Furthermore, as children progress through school, academic material becomes more complex and abstract. Children with high-functioning forms of ASD (without ID) often struggle with the theory of mind (e.g., the ability to understand the desires, intentions, emotions, and beliefs of others) and conceptual reasoning (e.g., the ability to use previous experience to form new concepts; Williams et al., 2014). Some strategies to combat these deficits include using visual cues to represent the abstract concept, reducing the use of idioms, dual meaning words, and sarcasm to facilitate learning and understanding. In addition, teaching organizational skills can be an effective intervention to assist with lost materials, homework/classwork, or messy lockers or desks. In a situation in which a child with ASD also meets full DSM-5 criteria for specific learning disorder, strategies and interventions commonly used with individuals with these disorders can also be effective. Although multisensory models are preferred in most cases, some caution and modifications may be necessary when sensory defensiveness or seeking behavior interferes with effective teaching. Ultimately, any curriculum should be tailored to the specific strengths and weaknesses of each individual child.

ADHD and Autism Spectrum Disorder

Attention-deficit/hyperactivity disorder (ADHD) is a disorder characterized by inattention, hyperactivity/impulsivity, and poor executive functions which often manifests prior to age 12 years and significantly impairs multiple areas of functioning. In a review, Leitner (2014) suggested that the co-occurrence of ASD and ADHD symptoms ranged from 24% to up to 85%. Both ASD and ADHD are neurobiologically complex and there is significant overlap in symptoms related to these disorders. For example, the DSM-5 symptoms of "difficulty sustaining attention," "often does not seem to listen when spoken to," and "often has difficulty organizing tasks" (APA, 2013, p. 59) can all be seen in individuals with both ASD and ADHD. Therefore, comprehensive psychological and developmental evaluation often is necessary to provide diagnostic clarification. According to the DSM-5, abnormalities of attention (overly focused or easily distracted) are common in individuals with ASD, as is hyperactivity. A diagnosis of attention-deficit/hyperactivity disorder (ADHD) should be considered when attentional difficulties or hyperactivity exceeds that typically seen in individuals of comparable mental age (p. 58).

Children with co-occurring ASD and ADHD often have lower quality of life and poorer adaptive functioning as compared to those with only one of the disorders (Sikora et al., 2012). And therefore, it will be important to tailor treatment to address both sets of symptoms. According to the literature, ADHD is best treated using a three-pronged approach, which includes parent support and education, behavior modification, and medication. Parent education can provide insight into the nature of both disorders and will improve a parent's knowledge base for intervention and support strategies. Once parents have learned appropriate behavioral strategies, they are encouraged to implement these strategies in order to improve attention to detail and improve academic performance. Knowledge often empowers parents and increases their ability to advocate for their child. Equipped with this knowledge, parents can seek appropriate treatments and support in both the clinical and academic arenas. Parents are encouraged to seek formal academic support through specialized programs such as an Individualized Accommodation Plan (IAP; 504 plan) or an Individualized Education Program (IEP; special education). These are necessary because as children with ASD/ADHD progress through school they will likely find age-appropriate tasks difficult to complete given his difficulty with sustained attention, distractibility, mild impulsivity, and general social-communication deficits. Parents of children with ASD with comorbid ADHD should also provide their child with education about their condition and ways to manage their deficits (e.g., time management training, decision-making, behavioral strategies to improve attention and concentration). It will be important for these individuals to learn coping skills to manage attentional difficulties. To address behavioral concerns, it is recommended that individuals with ASD/ADHD consult with a psychologist or other qualified interventionist who specializes in behavior modification and/or those providing Applied Behavior Analysis therapy (ABA therapy). It is very common for these individuals to complete a functional behavior analysis (FBA). The behavior plan should include preventative and management strategies to address behavioral challenges such as tantrums, aggression, and elopement. Overall, a behavior plan will inform those working with the individual how to adequately and most effectively prevent and manage any behavioral challenges. Ideally, the interventions should be used in various settings to increase consistency across settings.

Parents are also encouraged to contact the child's pediatrician or other qualified healthcare provider (e.g., medical psychologist, psychiatrist) and discuss whether medication management is necessary to assist with treatment of symptoms related to ADHD. Medication may assist in reducing inattentive, distracted, impulsive, and hyperactive behavior. Some evidence suggests that although certain medications such as Methylphenidate reduce hyperactivity associated with ASD, the degree of improvement is less than in those typically developing children with ADHD (Research units, 2005). Although there is a risk of adverse side effects with the use of medication, the ASD with ADHD population does not appear to be at significantly more risk than children with ADHD alone, especially for symptoms such as irritability with use of stimulant medication (Ghanizadeh et al., 2019). A combination of medication and behavioral strategies will provide the optimal treatment approach for any individual with ADHD and ASD.

Autism spectrum disorder impacts three main areas in an individual's life: the social aspect, communication, and their behaviors. While individual signs and symptoms vary from person to person, regardless of the degree of impairment present (high functioning versus low functioning), all individuals present with the same core triad of deficits (Center for Disease Control, 2018). A child may look "more autistic" at a young age due to the presence of stereotyped behaviors, such as hand flapping and toe walking, as this child ages, if they continue to meet diagnostic criteria for autism the core deficits of impairments in all three areas will remain, although the degree to which symptoms may be observed can change. For example, some high-functioning individuals with ASD who have intact language may be mistaken for demonstrating signs of ADHD. A referral to a psychologist who specializes in ASD should be able to accurately determine the correct diagnosis. Using record reviews, parent and teacher interviews, and a variety of psychological tests, including some specific to ASD, the psychologist should be able to differentiate between the core features of ASD and other clinical conditions.

When Sam turned 15 years old, his IEP was updated to include a work transition plan. The psychologist was one of several professionals who participated in planning and developing goals that would help transition Sam from the school system to the work force. The goal was to have Sam remain in school until he was 21 years of age, but he would spend his later years learning job skills and begin the process of exiting the school system and into a supported work environment. Outside community agencies, such as Vocational Rehabilitation Services, began to attend his IEP meetings and supported Sam and his family during this transition process. Sam also began seeing a psychiatrist who prescribed anti-anxiety medication, which helped with Sam's difficulties with transitions and learning in new environments.

Genetics

The etiology of ASD is complex and it is a genetically heterogeneous disorder (Waye & Cheng, 2018). As our understanding of the architecture of the human genome improves, one area of increasing research interest is in the underlying genetic variations associated with ASD. At present, heritability estimates are varied, and some estimates are as low as 37% and some as high as 90%, and approximately 15% of ASD cases seem to be associated with a known genetic mutation (APA, 2013). These figures suggest that parents who have ASD are at an increased risk of a child also having ASD as well as parents without ASD who have a child with ASD have a significantly increased chance of having a second child with ASD. These are important factors due to the increase of ASD over the past three decades (CDC, 2017). Given the increase in genetic understanding as well as the value of genetic counseling for families who are considering future pregnancies, psychologists routinely referred newly diagnosed children and their families for genetic counseling. Psychologists follow the guidelines recommended by the American Academy of Pediatrics and the American Academy of Neurology and the Child Neurology Society, which state, among other things, "genetic testing in children with autism, specifically high-resolution chromosome studies (karyotype) and DNA analysis for Fragile X, should be performed in the presence of intellectual disability (or if intellectual disability cannot be excluded), if there is a family history of Fragile X or undiagnosed intellectual disability, or if dysmorphic features are present. However, there is little likelihood of positive karyotype or Fragile X testing in the presence of 'high-functioning autism'" (CDC, 2017). By understanding these and other genetic risks factors, researchers hope to ultimately find an accurate and reliable diagnostic marker to assist clinicians in the accurate diagnosis of ASD. In addition, the benefits of genetic testing can enable the parents to have a deeper understanding regarding the cause of ASD and may lead to earlier diagnosis and enhance treatment options (Rutz et al., 2019). Although additional research and understanding is necessary for genetic testing to become a more reliable diagnostic option, several organizations such as the American Academy of Pediatrics (AAP) and American College of Medical Genetics (ACMG) recommend that genetic testing be offered to all families affected by ASD.

Summary

This chapter summarizes the identification process of ASD as well as behavioral (e.g., ABA, PRT, DTT, TEACCH) and alternative (e.g., speech, occupational therapy) interventions for the disorder. Psychologists play an integral part in the identification and treatment of individuals with ASD and work closely with individuals with ASD, their families, and other treatment providers to help reduce symptoms, improve cognitive ability and daily living skills, and maximize the ability of the individual to ultimately live as independently as possible within the community. The treatment process is ever-evolving to the needs of the individual.

References

Akers, J., Fisher, W. W., Kaminskin, J., Greer, B. D., DeSouza, A., & Retzlaff, B. (2017). An evaluation of conditional manding using a four component multiple schedule. In V. Saini, *Recent advances in methods to improve functional communication training*. Symposium Conducted at the 43rd Annual Convention of the Association for Behavior Analysis International, Denver.

American Occupational Therapy Association (AOTA). (2019). *What is occupational therapy?* (Brochure). Author.

American Psychiatric Association (APA). (2013). *Diagnostic and statistical manual of mental disorder* (5th ed.). American Psychiatric Publishing.

American Speech-Language-Hearing Association (ASHA). (n.d.). *What is speech? What is language?* https://www.asha.org/public/speech/development/speech-and-language/

Anderson, C., Law, K., Daniels, A., Rice, C., Mandell, D., Hagopian, L., & Law, P. (2012). Occurrence and family impact of elopement in children with autism spectrum disorders. *Pediatrics, 130*(5), 870–877.

Anderson, D., Laing, J., & Lord, C. (2014). Predicting young adult outcome among more and less cognitively able individuals with autism spectrum disorders. *Journal of Child Psychology & Psychiatry, 55*(5), 485–494.

Association for Professional Behavior Analysts. (2016). *Licensure and other regulation of ABA practitioners*. Author. http://www.apbahome.net/APBALicensure.php. Accessed 2 Nov 2016. 59.

Beck, J. S. (2011). *Cognitive behavior therapy: Basics and beyond* (2nd ed.). The Guildford Press.

Boyer, V., & Thompson, S. (2014). Transdisciplinary model and early intervention: Building collaborative relationships. *Young Exceptional Children, 17*(3), 19–32.

Bradshaw, J., Steiner, A., Gengoux, G., & Koegel, L. (2015). Feasibility and effectiveness of very early intervention for infants at-risk for autism spectrum disorder: A systematic review. *Journal of Autism and Developmental Disorders, 45*(3), 778–794.

Callahan, K., Shukia-Mehta, S., Magee, S., & Wie, M. (2010). ABA versus TEACCH: The case for defining and validating comprehensive treatment models in autism. *Journal of Autism and Developmental Disorders, 40*(1), 74–88.

Canoy, J. P., & Boholano, B. (2015). Early start DENVER model: A meta-analysis. *Journal of Education and Learning, 9*(4), 314–327.

Case-Smith, J., Weaver, L., & Fristad, M. (2015). A systematic review of sensory processing interventions for children with autism spectrum disorders. *Autism, 19*(2), 133–148.

Centers for Disease Control. (2017, April 27). Autism rates increase slightly. *CDC. Morbidity and Mortality Weekly Report, 67*(6), 1–23.

Ciurlik, A., Tennenbaum, D., & Duer, J. (2015). Current studies on TEACCH & future directions. *Narrative Documents*, Book 3. http://scholarexchange.furman.edu/schopler-about/3

Crowe, B. H., & Salt, A. T. (2015). Autism: the management and support of children and young people on the autism spectrum (NICE Clinical Guideline 170). *Archives of Disease in Childhood-Education and Practice, 100*(1), 20–23.

Cullinane, D. (2015). *Evidence base for DIR/Floortime*. Retrieved from www.icdl.com/research

D'Elia, L., Valeri, G., Sonnino, F., Fontana, L., Mammone, A., & Vicari, S. (2014). A longitudinal study of the TEACCH program in different settings: The potential benefits of low intensity intervention in preschool children with autism spectrum disorder. *Journal of Autism and Developmental Disorders, 44*(3), 615–626.

DeSouza, A., Akers, J., & Fisher, W. (2017). Empirical application of skinner's verbal behavior to interventions for children with autism: A review. *Analysis of Verbal Behavior, 33*, 229–259.

Dulcan, M. (2015). *Dulcan's textbook of child & adolescent psychiatry*. American Psychiatric Publishing.

Eldevik, S., Hastings, R., Hughes, J. C., Jahr, E., Eikeseth, D., & Cross, S. (2009). Meta-analysis of early intensive behavioral intervention for children with autism. *Journal of Clinical and Adolescent Psychology, 38*, 439–450.

Emond, A., Emmett, P., Steer, C., & Golding, J. (2010). Feeding symptoms, dietary patterns, and growth in young children with autism spectrum disorders. *Pediatrics, 126*(2), 337–342.

Fjeldsted, B., & Xue, L. (2019). Sensory processing in young children with fetal alcohol spectrum disorders. *Physical & Occupational Therapy in Pediatrics, 39*(5), 553–565.

Fombonne, E. (2005). Epidemiology of autistic disorder and other pervasive developmental disorders. *Journal of Clinical Psychiatry, 66*(10), 3–8.

Gates, J., Kang, E., & Lerner, M. (2017). Efficacy of group social skills interventions for youth with autism spectrum disorder: A systematic review and meta-analysis. *Clinical Psychology Review, 52*, 164–181.

Ghanizadeh, A., Molla, M., & Olango, G. (2019). The effect of stimulants on irritability in autism comorbid with ADHD: A systematic review. *Neuropsychiatric Disease and Treatment, 15*, 1547–1555.

Gray, C. (2010). *The new social story book* (10th ed.). Future Horizons.

Greenspan, S. (1992). *Infancy and early childhood: The practice of clinical assessment and intervention with emotional and developmental challenges*. International Universities Press.

Guralnick, M., & Bruder, M. B. (2019). Early intervention. In J. Matson (Ed.), *Handbook of intellectual disabilities – Integrating theory, research and practice* (pp. 717–741). Springer International Publishing.

Hofvander, B., Delorme, R., Chaste, P., Nydén, A., Wentz, E., Ståhlberg, O., ... & Leboyer, M. (2009). Psychiatric and psychosocial problems in adults with normal-intelligence autism spectrum disorders. *BMC Psychiatry, 9*(1), 1–9.

Hubbard, K., Anderson, S. E., Curtin, C., Must, A., & Bandini, L. (2014). A comparison of food refusal related to characteristics of food in children with autism spectrum disorder and typically developing children. *Journal of the Academy of Nutrition and Dietetics, 114*(12), 1981–1987.

Huerta, M., & Lord, C. (2012). Diagnostic evaluation of autism spectrum disorders. *Pediatric Clinics of North America, 59*(1), 103–111.

Hurwitz, S. (2013). The gluten-free, casein-free diet and autism: Limited return on family investment. *Journal of Early Intervention, 15*(1), 3–19.

Hyman, S., Levy, S., & Myers, S. (2019). Identification, evaluation, and management of children with autism spectrum disorder. *Pediatrics, 145*(1), e20193447.

Koegel, R. L., O'Dell, M., & Dunlap, G. (1988). Producing speech use in nonverbal autistic children by reinforcing attempts. *Journal of Autism and Developmental Disorders, 18*(4), 525–538.

Koegel, L. K., Ashbaugh, K., & Koegel, R. L. (2016). Pivotal response treatment. In R. Lang, T. B. Hancock, & N. N. Singh (Eds.), *Early intervention for young children with autism spectrum disorder* (pp. 85–112). Springer International Publishing.

Kozlowski, A. M., Matson, J. L., Horovitz, M., Worley, J., & Neal, D. (2011). Parents' first concerns of their child's development in toddlers with autism spectrum disorders. *Developmental Neurorehabilitation, 14*(2), 72–78.

Kranowitz, C. (2016). *The out-of-sync child grows-up*. Penguin Random House.

Laugeson, E. A., & Frankel, F. (2011). *Social skills for teenagers with developmental and autism spectrum disorders: The PEERS treatment manual*. Routledge.

Laugeson, E. A., Gunman, A., Kapp, S. K., Poremski, K., & Ellingsen, R. (2015). A randomized controlled trial to improve social skills in young adults with autism spectrum disorder: The UCLA PEERS® program. *Journal of Autism and Developmental Disorders, 45*, 3978–3989.

Lei, J., & Ventola, P. (2017). Pivotal response treatment for autism spectrum disorder: Current perspectives. *Neuropsychiatric Disease and Treatment, 13*, 1613–1626.

Leitner, Y. (2014). The co-occurrence of autism and attention deficit hyperactivity disorder in children–what do we know? *Frontiers in Human Neuroscience, 8*, 268.

Lerman, D. C., Valentino, A. L., & LeBlanc, L. A. (2016). Discrete trial training. In *Early intervention for young children with autism spectrum disorder* (pp. 47–83). SpringerLink Publishing.

Lever, A., & Geurts, H. (2016). Psychiatric co-occurring symptoms and disorders in young, middle-aged, and older adults with autism spectrum disorder. *Journal of Autism and Developmental Disorders, 46*, 1916–1930.

Li, J., Zhang, Y., Su, W., Liu, L., Gong, H., Peng, W., & Jiang, C. (2018). Cognitive behavioral therapy for treatment-resistant depression: A systematic review and meta-analysis. *Psychiatry Research, 268*, 243–250.

Makrygianni, M. K., Gena, A., Katoudi, S., & Galanis, P. (2018). The effectiveness of applied behavior analytic interventions for children with autism Spectrum disorder: A meta-analytic study. *Research in Autism Spectrum Disorders, 51*, 18–31.

Matson, J. L., & Cervantes, P. E. (2014). Commonly studied comorbid psychopathologies among persons with autism spectrum disorder. *Research in Developmental Disabilities, 35*(5), 952–962.

Matson, J., Hattier, M., Belva, B., & Matson, M. (2013). Pica in persons with developmental disabilities: Approaches to treatment. *Research in Developmental Disabilities, 34*, 2564–2571.

Mcllwain, L., & Fournier, W. (2010, October 22). *Wandering & autism: The need for data and resources*. Interagency Autism Coordinating Committee Meeting. https://lacc.hhs.gov/events/2010/slides_fournier_mcllwain_102210.pdf

Mercer, J. (2015). Examiner DIR/Floortime as a treatment for children with autism spectrum disorders: A review of research and theory. *Research on Social Work Practice, 27*(5), 625–635.

Mesibov, G. B., & Shea, V. (2009). The TEACCH program in the era of evidence-based practice. *Journal of Autism and Developmental Disorders, 40*(5), 570–579.

Miller, L. J. (2014). *Sensational kids: Hope and help for children with sensory processing disorder* (2nd ed.). Perigee.

National Autism Center. (2015a). *Findings and conclusions: National standards project, phase 2*. Author.

National Autism Center. (2015b). *National standards report*. National Autism Center. http://www.nationalautismcenter.org/national-standards-project/phase-2/

National Institute for Elopement Prevention and Resolution. (2015). *The elopement issues workbook: Long term care, assisted living, & group homes*. National Institute for Elopement Prevention & Resolution.

Noor, I., Lawson, L., & Hartwell, J. (2018). Relationship between sensory processing and participation in daily occupations for children with autism spectrum disorders: A systematic review of studies that used Dunn's sensory processing framework. *American Journal of Occupational Therapy, 72*, 1–9.

Ozonoff, S., Young, G., Carter, A., Messinger, D., Yirmiya, N., Zwaigenbaum, L., Bryson, S., Carver, L., Constantino, J., Dobkins, K., Hutman, T., Iverson, J., Landa, R., Rogers, S., Sigman, M., & Stone, W. (2011). Recurrence risk for autism spectrum disorders: A baby siblings research consortium study. *Pediatrics, 128*(3), 488–495.

Polyak, A., Kubina, R., & Girirajan, S. (2015). Comorbidity of intellectual disability confounds ascertainment of autism: Implications for genetic diagnosis. *American Journal of Medical Genetics. Part B: Neuropsychiatric Genetics, 168*(7), 600–608.

Radley, K., Helbig, K., Murphy, A., McCargo, M., & Lown, E. (2018). The effects of a modified PEERS curriculum on accurate and novel responding of children with autism spectrum disorder. *Advances in Neurodevelopmental Disorders, 2*, 412–424.

Reichow, B. (2012). Overview of meta-analyses on early intensive behavioral intervention for young children with autism spectrum disorders. *Journal of Autism and Developmental Disorders, 42*, 512–520.

Reitzel, J., Summers, J., Lorv, B., Szatmari, P., Zwaigenbaum, L., Georgiades, S., & Duku, E. (2013). Pilot randomized controlled trial of a functional behavior skills training program for young children with autism spectrum disorder who have significant early learning skill impairments and their families. *Research in Autism Spectrum Disorders, 7*, 1418–1432.

Research Units on Pediatric Psychopharmacology Autism Network. (2005). Randomized, controlled, crossover trial of methylphenidate in pervasive developmental disorders with hyperactivity. *Archives of General Psychiatry, 62*(11), 1266–1274.

Rice, C. E., Zablotshy, B., Avila, R., Colpe, L., Schieve, L., Pringle, B., & Blumberg, S. (2016). Reported wandering behavior among children with autism spectrum disorder and/or intellectual disability. *The Journal of Pediatrics, 174*(2), 232–239.

Ridderinkhof, A., de Bruin, E., Blom, R., & Bögels, S. (2018). Mindfulness-based program for children with autism spectrum disorder and their parents: Direct and long-term improvements. *Mindfulness, 9*, 773–791.

Roane, H. S., Fisher, W. W., & Carr, J. E. (2016). Applied behavior analysis as treatment for autism spectrum disorder. *The Journal of Pediatrics, 175*, 27–32.

Rutz, A., Dent, K., Botto, L., Young, P., & Carbone, P. (2019). Brief report: Pediatrician perspectives regarding genetic evaluations of children with autism spectrum disorder. *Journal of Autism and Developmental Disorders, 49*, 794–808.

Schreibman, L., Dawson, G., Stahmer, A., Landa, R., Rogers, S., McGee, G., Kasari, C., Ingersoll, B., Kaiser, A., Bruinsma, Y., McNerney, E., Wetherby, A., & Halladay, A. (2015). Naturalistic developmental behavioral interventions: Empirically validated treatments for autism spectrum disorder. *Journal of Autism and Developmental Disorders, 45*(8), 2411–2428.

Seiverling, L., Towle, P., Hendy, H. M., & Pantelides, J. (2018). Prevalence of feeding problems in young children with and without autism spectrum disorder: A chart review study. *Journal of Early Intervention, 40*(4), 335–346.

Sikora, D., Vora, P., Coury, D., & Rosenburg, D. (2012). Attention-deficit/hyperactivity disorder symptoms, adaptive functioning, and quality of life in children with autism spectrum disorder. *Pediatrics, 130*(2), 91–97.

Siu, A., Lin, Z., & Chung, J. (2019). An evaluation of the TEACCH approach for teaching functional skills to adults with autism spectrum disorders and intellectual disabilities. *Research in Developmental Disabilities, 90*, 14–21.

Smith, T. (2013). What is evidence-based behavioral analysis? *The Behavior Analyst, 36*(1), 7–33.

Smith, T., & Antolovich, M. (2000). Parental perceptions of supplemental interventions received by children with autism in intensive behavior analytic treatment. *Behavioral Interventions, 15*, 83–97.

Test, D., Richter, S., Knight, V., & Spooner, F. (2011). A comprehensive review and meta-analysis of the social stories literature. *Focus on Autism and Other Developmental Disabilities, 26*(1), 49–62.

Twachtman-Reilly, J., Amaral, S. C., & Zebrowski, P. P. (2008). Addressing feeding disorders in children on the autism spectrum in school-based settings: Physiological and behavioral issues. *Language, Speech, and Hearing Services in Schools, 39*(2), 261–272.

United States Surgeon General. (1999). *Mental health: A report of the surgeon general*. Author.

VandenBos, G. R. (Ed.). (2007). *APA dictionary of psychology*. American Psychological Association.

Vannucchi, G., Masi, G., Toni, C., Dell'Osso, L., Marazziti, D., & Perugi, G. (2014). Clinical features, developmental course, and psychiatric comorbidity of adult autism spectrum disorders. *CNS Spectrums, 19*(2), 157–164.

Vasa, R. A., & Mazurek, M. O. (2015). An update on anxiety in youth with autism spectrum disorders. *Current Opinion in Psychiatry, 28*(2), 83–90.

Virues-Ortega, J., Julio, F. M., & Pastor-Barriuso, R. (2013). The TEACCH program for children and adults with autism: A meta-analysis of intervention studies. *Clinical Psychology Review, 33*(8), 940–953.

Vismara, L., McCormick, C. E., Shields, R., & Hessel, D. (2019). Extending the parent-delivered early start Denver model to young children with Fragile X syndrome. *Journal of Autism and Developmental Disorders, 49*, 1250–1266.

Volkmar, F., Siegel, M., Woodbury-Smith, M., King, B., McCracken, J., & State, M. (2014). Practice parameter for the assessment and treatment of children and adolescents with autism spectrum disorder. *Journal of the American Academy of Child and Adolescent Psychiatry, 53*(2), 237–257.

Walters, S., Loades, M., & Russell, A. (2016). A systematic review of effective modifications to cognitive behavioural therapy for young people with autism spectrum disorders. *Review Journal of Autism and Developmental Disorders, 3*, 137–153.

Waye, M. M., & Cheng, H. Y. (2018). Genetics and epigenetics of autism: A review. *Psychiatry and Clinical Neurosciences, 72*(4), 228–244.

Weston, L., Hodgekins, J., & Langdon, P. E. (2016). Effectiveness of cognitive behavioural therapy with people who have autistic spectrum disorders: A systematic review and meta-analysis. *Clinical Psychology Review, 49*, 41–54.

White, S. W., Simmons, G. L., Gotham, K. O., Conner, C., Smith, I. C., Beck, K., & Mazefsky, C. (2018). Psychosocial treatments targeting anxiety and depression in adolescents and adults on the autism spectrum:

Review of the latest research and recommended future directions. *Current Psychiatry Reports, 2*, 82–96.

Williams, D., Mazefsky, C., Walker, J., Minshew, N., & Goldstein, G. (2014). Associations between conceptual reasoning, and problem solving and adaptive ability in high-functioning autism. *Journal of Autism and Developmental Disorders, 44*(11), 2908–2920.

Wolstencroft, J., Robinson, L., Srinivasan, R., Kerry, E., Mandy, W., & Skuse, D. (2018). A systematic review of group social skills interventions, and meta-analysis of outcomes, for children with high functioning ASD. *Journal of Autism and Developmental Disorders, J48*(7), 2293–2307.

Wong, C., Odom, S., Hume, K., Cox, A., Fettig, A., Kucharczyk, S., Brock, M., Plavnick, J., Fleury, V., & Schultz, T. (2015). Evidence-based practices for children, youth, and young adults with autism spectrum disorder: A comprehensive review. *Journal of Autism and Developmental Disorders, 45*, 1951–1966.

Yang, L., Zhou, X., Zhou, C., Zhang, Y., Pu, J., Liu, L., Gong, X., & Xie, P. (2017). Efficacy and acceptability of cognitive behavioral therapy for depression in children: A systematical review and meta-analysis. *Academic Pediatrics, 17*(1), 9–16.

21

Occupational Therapy for Children with Autism Spectrum Disorder and Intellectual and Developmental Disability

Stefanie C. Bodison, Leah I. Stein Duker, Bonnie Nakasuji, Michaelann Gabriele, and Erna Imperatore Blanche

Abstract

Occupational therapy (OT) seeks to promote health and well-being by increasing the client's ability to engage in meaningful activities. Utilizing a strength-based approach, the occupational therapist focuses the evaluation on the ways the individual's personal strengths and social networks impact functioning in order to design interventions to improve those strengths and ultimately, the individual's participation in chosen occupations. The OT evaluation process considers a myriad of factors, including but not limited to: the individual's multiple roles in life; their values, beliefs, desires, and priorities; the demands of the task; specific performance skills required to engage in desired activities; and the context/environment in which these activities are performed. This chapter will explore the role of OT in working with individuals with autism spectrum disorder (ASD) and intellectual and developmental disability (IDD), including referral, assessment, intervention, and collaborative care.

Keywords

Occupational therapy · Autism · ASD · IDD · Disability · Medical home · Treatment planning

Introduction to Occupational Therapy

Occupations are "the everyday activities that people do as individuals, in families, and with communities to occupy time and bring meaning and purpose to life. Occupations include things people need to, want to and are expected to do" (WFOT, 2012, para. 2). The Occupational

S. C. Bodison (✉)
Department of Occupational Therapy, College of Public Health and Health Professions, University of Florida, Gainesville, FL, USA
e-mail: stefaniebodison@phhp.ufl.edu

L. I. Stein Duker
Chan Division of Occupational Science & Occupational Therapy, Ostrow School of Dentistry, University of Southern California, Los Angeles, CA, USA

B. Nakasuji
Therapy West, Los Angeles, CA, USA

M. Gabriele
Center for Developing Kids, Inc., Pasadena, CA, USA

E. I. Blanche
Chan Division of Occupational Science & Occupational Therapy, Ostrow School of Dentistry, University of Southern California, Los Angeles, CA, USA

Therapy West, Los Angeles, CA, USA

P. McPherson (ed.), *Handbook of Treatment Planning for Children with Autism and Other Neurodevelopmental Disorders*, Autism and Child Psychopathology Series,
https://doi.org/10.1007/978-3-031-06120-2_21

Therapy Practice Framework (AOTA, 2020) generally considers occupations in nine categories: activities of daily living (ADLs; e.g., bathing, toileting, dressing, eating, functional mobility), instrumental activities of daily living (IADLs; e.g., care of others, driving, health management), health management, rest and sleep, education, work, play, leisure, and social participation (AOTA, 2020).

The aim of occupational therapy (OT) is to promote health and well-being by increasing the client's ability to engage in these meaningful activities. Engagement in occupations is referred to as *participation*, which is broadly defined as "involvement in life situations" by the World Health Organization (WHO; 2008). Participation may be restricted due to any combination of impairments in motor or processing skills, environmental factors, or personal factors (e.g., demographic variables, personal values and beliefs) (WHO, 2008). Therefore, occupational therapy practitioners use a unique, whole-person approach to addressing these restrictions, often including interventions aimed to remediate (develop or restore a skill or ability) or compensate (adapt or modify an activity or environmental) in order to improve performance, prevent barriers to performance and participation, and/or promote health and wellness (AOTA, 2020).

OT is client-centered, focusing on improving client participation in those occupations individuals both *want* and *need* to do (WFOT, 2010), employing activities a person is motivated to perform in intervention. By utilizing these activities, the occupational therapist can capitalize on an individual's intrinsic motivation, allowing them to experience enjoyment and satisfaction throughout the intervention experience (thereby generating additional effort during intervention), not exclusively focusing on the final goal of the intervention(s) (Arnsten, 1990; McLaughlin Gray 1998). As a client-centered profession, goals in OT are often set collaboratively (Mroz et al., 2015), and treatment can include not only the client but also those individuals who may be involved in the client's care, support, treatment, or community re-integration (WFOT, 2010).

These tenets guide all occupational therapy practice, but it is important to note that OT may appear different depending on the setting (e.g., home, school, clinic, hospital, community) and the population (e.g., developmental disabilities, spinal cord injuries, stroke survivors, mental health diagnoses, etc.). This chapter will explore the role of OT in working with individuals with autism spectrum disorder (ASD) and intellectual and developmental disability (IDD), including referral, assessment, and intervention.

Referral and Funding for Occupational Therapy Services

Children with ASD and IDD present with impairments in motor performance, social interaction, and behavior regulation that impact participation; all of which could be addressed in OT. OT focuses on the relationship of impairments to participation; thus, the focus of the evaluation and intervention is on the link between impairments and engagement in occupations/activities or the link between impairments and decreased community participation.

The impairments addressed by occupational therapists vary along the developmental spectrum. Infants with ASD and IDD often present with motor and interactive delays that limit engagement in play and the acquisition of independence in daily activities (Bal et al., 2015; Kent et al., 2020; Lloyd et al., 2013). Preschoolers and school-age children are frequently referred to OT because of difficulties in social interaction, emotion regulation, behavioral organization, the development of sensory and motor skills, and impaired executive functions (Cibralic et al., 2019; Demetriou et al. 2019). Adolescents with ASD and ID are commonly referred to OT because of social isolation, mental health issues such as depression and anxiety, and decreased community participation (Jarrard et al., 2018; Ohl et al., 2020).

The referral process will depend on the age of the child, the funding source, and the type of difficulty the child presents. Infants are referred through health insurance or state agencies addressing early intervention. School-age children often receive services through the school systems, although they can also receive services through health insurance.

School-based services focus on academic performance; therefore, the occupational therapy referral needs to identify difficulties in accessing academic performance, such as attention difficulties, handwriting, participating in educational tasks, or socializing with peers. However, if the child is referred because of difficulties in getting dressed, eating/feeding, toileting, organizing daily tasks, and participating in family leisure activities, the services can be funded through health insurance. Table 21.1 details some of the most common sources of funding for occupational therapy services.

Once the individual is referred, the occupational therapist focuses on the underling impairment limiting engagement and participation. For example, a child who exhibits stereotypies may benefit from an occupational therapy evaluation of sensory processing. A child who is unable to feed themself may require an occupational therapy evaluation of motor planning. Occupational therapists are particularly proficient in the evaluation and intervention of sensory processing and feeding abilities. Many of these impairments can be related to sensory processing disorders. For example, emotion regulation could be linked to hyperresponsivity to sensory input in the environment.

Table 21.1 Funding sources for occupational therapy services

	Most salient impairments and participation issues	Funding sources
Infants (6–12 months)	Developmental delays (motor and engagement) Participation in play (peekaboo)	Private Health Maintenance Organizations (HMOs) Preferred Provider Organizations (PPOs) State and regional funding sources
Toddlers and Preschoolers	Behavior regulation Sensorimotor Emotion regulation Participation in daily tasks such as dressing, eating, toileting	Private Health Maintenance Organizations (HMOs) Preferred Provider Organizations (PPOs) School districts State and regional funding sources
School-age children 6–14 years	Behavior regulation Sensorimotor Emotion regulation Participation in academics, independent self-care, and social play	Private Health Maintenance Organizations (HMOs) Preferred Provider Organizations (PPOs) School districts
Adolescents 14+	Social skills Mental health issues Participation in school and leisure group activities	Private health insurance companies Preferred Provider Organizations (PPOs) School districts

Occupational Therapy Evaluation

An OT evaluation is unique and complex because of the profession's core value on meaningful occupations. The 'occupational' focus requires the evaluator utilize a global and holistic approach to understand what is 'meaningful' to the client and what is challenging or limiting engagement in those occupations. Including a consideration of the context in which the occupation is performed is also key. The evaluation is client-centered (Fisher & Short-DeGraff, 1993) and strength-based (Dunn, 2014; Tomchek et al., 2017; Watling & Spitzer, 2018) for which occupational therapists are uniquely trained. Utilizing this strength-based approach, the occupational therapist focuses the evaluation on the ways the individual's personal strengths and social networks assist in the individual's functioning, designing interventions to improve those strengths and ultimately, the individual's participation in chosen occupations.

The OT evaluation process considers a myriad of factors, including but not limited to: the individual's multiple roles in life; the individual's values, beliefs, desires, and priorities; the demands of the task (e.g., body functions, actions required to perform occupations, space and timing to successfully perform the task); specific performance skills (e.g., motor, sensory, cognitive abilities) required to engage in desired activities; and the context/environment in which these activities are performed (e.g., if the occupation is part of a routine, the physical space, social context, and the influence of factors such as culture and time).

A holistic occupational lens is then used to interpret the information for several purposes, including contributing to a diagnosis (Watling & Spitzer, 2018), determining the need for intervention, and creating an intervention plan with measurable goals. This holistic, client-centered and strengths-based approach is particularly well-suited for individuals with a diagnosis of ASD or other IDD as it values and respects the uniqueness of each person.

A comprehensive OT evaluation is multifaceted in the manner of gathering pertinent information. It begins with a review of records such as medical history, educational or school history, and reports from previous or currently provided services. The next step often gathers additional information through interview of the client or parent/caregiver of the child. This information begins the formation of the occupational profile, which provides a "summary of a client's occupational historyand experiences, patterns of daily living, interests, values, and need" (AOTA, 2020, p. 21). From this, the occupational therapist can plan the next steps in the evaluation process, deciding what additional information is needed and selecting the tools and means to gather this critical data. Using psychometric or standardized measures in addition to unstructured and structured clinical observations are helpful at this point. Table 21.2 provides a list of many of the assessment methods that may contribute to the understanding of occupational performance: narrative/interview methods, standardized and psychometrically sound objective testing measures, and structured and unstructured observations.

The final step in the evaluation process lies in the careful analysis and documentation of all the information gathered. This includes reporting test results, conclusions made about the client's performance that incorporate strengths and capacities, recommendations (for intervention, referrals, adaptations, accommodations, etc.), and intervention goals.

Vignette 1 – Matt – The Early Years – Mastering Feeding

After learning that their son would be born with Down syndrome (DS), Matt's parents worked with the perinatal team to prepare for his birth. They benefitted from the stories and expertise of other families after the perinatologist shared resources for a local support group and the National Association for Down Syndrome. After delivery, Matt's medical home team went into action. As the American Academy of Pediatrics health supervision guidelines for children with DS (Bull, 2011) were implemented, the hospital-based OT was front and center as part of the hospital's Infant Feeding Disorders team. The team assessed Matt for oral sensory issues and provided resources to help Matt's parents appreciate that children with DS may experience oral hypo- or hypersensitivity as well as impaired sucking due to limited motor control of the tongue and muscles involved in swallowing. They were cautioned that his neck may need extra support during feeding to avoid aspiration and that he might easily tire during feedings. Matt's parents were relieved to know that there would be ongoing support to promote Matt's healthy development and ability to feed as an infant and later develop independence with feeding.

Feeding challenges in infants and toddlers with DS are the result of multiple orosensorimotor issues and require ongoing monitoring across cross childhood (Anil et al., 2019). Food texture sensitivities impacting feeding have been identified in infants with DS as young as 4 months, with increased sensitivity noted in young children with DS compared to peers without DS (Ross et al., 2022). Children with DS have shown greater difficulty with chewy, firm, or juicy textures (Ross et al., 2019). Interestingly, allowing

Table 21.2 Examples of occupational therapy assessment methods

Assessment method	Occupational components	Examples
Narrative, interview	Available support systems Family/caregiver priorities, values, cultural considerations, etc. Interests Leisure Occupational history Patterns of daily living, interests, values, and needs Rest, work, and play balance Routines Social participation (family, friends, community)	Asking questions such as: Tell me about a favorite person/ best friend. Describe a typical day. Describe sleep routine. What is a favorite food? What do you like to do to have fun or relax? What do you like best about school or work? What is hard at school or work? Describe the support you find most helpful at work? at school?
Psychometric, standardized measures	ADL and IADL ASD symptoms Behavior (adaptive, emotional, social) Motor skills Play development and playfulness scales Preferences School function Sensory processing and sensory integration Self-perceptual abilities Social skills Visual perceptual/visual-motor skills Family concerns	Assessment tools: Bayley Scales of Infant & Toddler Development (2019) Beery-Buktenica Developmental Test of Visual Motor Integration (2010) Knox Preschool Play Scale (2008) Miller Function and Participation Scales (2006) Sensory Processing Mesure-2 (2021) Sensory Profile (2014) Social Skills Improvement System Rating Scales (2008) Test of Playfulness (2008)
Unstructured and structured observations	Adaptive abilities ADL/IADL Behavior and self-regulation Developmental and cognitive skills Environment (richness/limitations), objects and toys Fine motor skills (hand dominance, grasp patterns, manipulation skills, purposefulness, and quality of interactions with objects) Gross motor skills (balance, posture, coordination, energy, strength) Play choices, development, and playfulness Sensory processing (preferences, avoidances) Social interactions	Unstructured observations: Free choices made in environment Social interactions Independence level with self-care tasks (i.e., washing hands, toileting) Problem solving Control during manipulation Structured observations: Comprehensive Observations of Proprioception (2012) Prone extension/supine flexion Diadochokinesia Visual tracking

Adapted from AOTA (2020) and Watling (2018)

toddlers to play with food of different textures increases food acceptance (Nederkoorn et al., 2018).

Matt – The Early School Years – Self-Care Skills

At three, Matt graduated from the local Zero to Three program and entered school. An OT assessment was performed as part of his special education evaluation. His individualized education plan targeted self-care skills to promote independent toileting, feeding, and dressing. Under the direction of the OT, Matt's teacher encouraged use of toys that increased the fine motor skills needed to zip and fasten his sneakers. As he moved from the preschool to school setting, the OT addressed the graphomotor skills needed for writing, cutting with scissors, and helped Matt to participate and enjoy physical education activities with his classmates.

Mastering basic self-care skills is critical for a child's movement into the community and participation in school. Successful toilet training requires assessing readiness and supporting parents in implementing traditional toilet training strategies. Social stories, picture books, and picture charts can assist with toileting mastery. Children with DS master toileting more slowly than typically developing peers, with the majority mastering toileting by 60 months (Bruschini et al., 2003; Mrad et al., 2018) and dressing by the age of 18 (Krell et al., 2021). While handwriting skills are delayed in children with DS, finger dexterity exercises and activities supporting visual motor integration (puzzles, blocks, mazes, etc.) assist with writing skills (Moy et al., 2017). Adapted physical education for children with DS supports health and socialization (Villalvazo, 2017).

Matt – The Teenage Years – Increasing Independence

Matt's school and family supplemented his individualized educational plan with a transition plan when he entered high school. The transition plan included independent living skills and vocational training. Matt enjoyed greeting customers at a local grocery store. His OT assisted Matt with job skills, ensuring necessary workplace accommodations, and learning to be safe in the community.

A study of over 2600 persons with DS found that the majority could work independently by the age of 20, and 34% of those in the United States were living independently by the age of 31 (de Graaf et al., 2019). Trujillo et al. (2019) advocate for a biopsychosocial approach to transition planning with attention to *remediating deficits*, providing education about *strengths and differences*, training in compensatory strategies for *irremediable attributes*, and use of *meaningful activities* to achieve an occupation.

Occupational Therapy Intervention

Once the occupational therapy evaluation has been completed, the occupational therapist works with the individual with ASD or IDD and/or their caregivers to develop the most effective and meaningful intervention plan, including a series of goals to monitor progress during the intervention period. If requesting authorization for funding of OT services, consideration of the funding source perspective will be crucial to guide the specific focus of the occupational therapy intervention and goals (as exemplified in Table 21.3).

Occupational therapy intervention can be delivered in a variety of ways: (1) in a 1:1 capacity, where the OT works directly with the individual 1–2 hours per week toward the achievement of individualized goals over a series of months; (2) in a group capacity, where the individual might participate in 1–2 hour sessions at a time with the OT guiding a group of people working toward common goals and skill building; and/or (3) in a consultative fashion, where the OT works directly with the caregiver and/or other people in the individual's support system (teachers, other school personnel, community members, etc.). Depending on the individual's unique needs, the delivery of OT intervention might include one or more of the types described for a period anywhere from 6 months to multiple years.

During an OT intervention session for the individual with ASD or IDD, the OT will use a variety of methods to enhance the components of the individual's occupational performance (see Table 21.2) that were identified as needing support during the evaluation. For example, if a child with ASD is overly sensitive to touch and move-

Table 21.3 Relationship between funding source and intervention foci

Funding source	Perspective	Examples of intervention focus
Insurance Companies	Medical model	Self-care: grooming, eating, dressing, etc.
State Department of Developmental Services	Developmental model	Development of fine motor, gross motor, social-emotional maturity
School District	Educational model	Attention, readiness for learning, self-regulation, writing

ment such that classroom activities are causing emotional and physical outbursts, the OT might provide an intervention in a play-based environment to improve the ways the child's nervous system processes and integrates both tactile and vestibular sensory information. In this case, the OT will use their specialized knowledge and skills to grade the task and sensory characteristics of the activities during the intervention session so that the child's nervous system adapts over time and becomes more regulated. In another example, the OT may work with an individual with IDD who desires to develop increased independence in riding public transportation to get to their job. The OT might work with that individual in the community, developing the skills to understand the bus routes, successfully manage paying for each ride, and navigating the environment to and from work.

OT intervention for individuals with ASD and IDD is varied, complex, and tailored to the individual's unique needs. It is informed by the evaluation data and understanding of the funding source, and guided by a desire to enhance the individual's occupational performance in as meaningful a way to that individual as possible. No two interventions will look the same, making it often difficult for others to understand the complex nature of the ways occupational therapists can improve the lives of individuals with ASD and IDD.

Vignette 2 – Tena – A Toddler at Risk for ASD

At 4, Tena had mastered eating and toileting without assistance and had made a friend, Marci, at preschool. While professionals reassured her parents that Tena was meeting milestones, they were concerned because she exhibited some of the same behaviors as her six-year-old older brother, Ben, who was diagnosed with autism. At home, Tena would line up her toys and exhibited difficulty with changes in routine. She was particular about her clothes, preferring to wear the same soft cotton dress almost every day. Because Ben had made great progress with hypersensitivity to textures and noise with the help of his occupational therapist, Tena's parents requested an OT evaluation from a neurodevelopmental pediatrician. Sensory issues were identified, but a multidisciplinary assessment found that Tena did not meet criteria for ASD. A reevaluation was recommended should Tena's development regress or plateau.

While siblings of children with ASD are at increased risk for ASD or the broader autism phenotype, most siblings do not qualify for an ASD diagnosis (Pisula & Ziegart-Sadowska, 2015). In addition, younger siblings of children with ASD may adopt behaviors that they observe in the sibling with ASD (Bontinck et al., 2018). Hearing this, Tena's parents wondered if her sensory issues were a learned behavior from watching her brother.

The occupational therapist working with Tena and her family completed a comprehensive evaluation that included a caregiver-completed sensory responsivity questionnaire, observation of Tena interacting with sensory materials in a clinical setting, and a standardized assessment of sensory functions. The occupational therapist compiled all of the data gathered and determined that Tena, in fact, did have hypersensitivity to touch and motor planning problems, which caused her difficulty in learning new skills. Tena received occupational therapy for approximately 6-months, during which time her sensitivity to textures improved. She continued to have some difficulty with motor planning, but Tena's parents felt that she was actively engaging with her peers in gross motor activities and able to keep up so occupational therapy was discontinued.

Tena – The Early School Years – Increasing Challenges

Tena's academic progress in first and second grade was in the average range. She and Marci were still good friends, but Tena had difficulty making new friends and would become sullen or angry if Marci played with other girls. With the increasing demands of third grade, Tena's academic and social development plateaued. She was overwhelmed by the noise and activity in the cafeteria and stopped eating at school. As her PE class required increased motor control, Tena became frustrated and refused to participate. Tena's parents requested that the school evaluate Tena. At the age of 8, Tena was diagnosed with ASD. Her individualized educational program included OT to address sensory, motor, and social issues.

Decreased engagement is common among children with sensory processing issues (Chien et al., 2016) because often the sensory processing issues inhibit the development of age-appropriate social-emotional and motor skills. In fact, parents frequently identify social participation as an OT goal for their child with ASD (Schaaf et al., 2015). Noting that Tena has struggled learning new tasks since when she was younger because of motor planning issues, the occupational therapist gave her parents strategies to break tasks down into small steps and allow Tena the necessary time to practice each step to be successful. The occupational therapist also recommended another short period of direct occupational therapy services in a clinical setting where Tena could improve her sensorimotor skills.

Tena – The Teenage Years – Self Advocacy

Tena and her family were diligent at regularly attending therapy and implementing therapy strategies, with Tena developing mastery of strategies to address sensory issues by the time she completed middle school. This led to increased academic success and meant that she no longer required school-based OT services. However, Tena still struggled socially and needed additional support to navigate friendships and social situations. Thankfully, her insurance covered outpatient OT services so she could continue to get the support she needed. Social Stories were used to prepare Tena for high school and the nuances of adolescent relationships. Tena learned that running helped her manage her social anxiety and she joined the cross-country team. Tena credited her experiences with occupational therapy with giving her confidence to learn more about autism. Tena wrote her college admissions essay on her choice to use identity-first language.

The use of stories to illustrate life lessons is ubiquitous across cultures and time. Gray is credited with the development of Social Stories™ as a tool for exploring situations in a manner that is more accessible to teens with ASD (Gray & Garand, 1993). Social Stories allowed Tena to prepare for new situations through role playing in a non-threatening way. Tena learned about possible dilemmas she might face in social situations. Because she often has difficulty reading nuanced interactions and interpreting others emotional responses, Social Stories offer her a non-threatening way to build enhanced skills.

Anxiety disorders are five to six times more likely in teens with ASD compared to neurotypical peers, though presentation may vary (Cammuso & Vorderstrasse, 2020). Sensory issues are emerging as a key to explaining differences in presentation. A common neural relationship between sensory issues and anxiety has been proposed. Hypotheses highlight medial prefrontal cortex and limbic system involvement and call out the importance of considering sensory issues in the assessment and treatment of anxiety (Blank et al., 2019; Green & Ben-Sasson, 2010). In addition to sensory function, the ability to read emotions in others and the ability to tolerate ambiguity must be accounted for in the assessment and treatment of anxiety (Blank et al., 2019).

Tena views autism, her sensory issues, and her anxiety as a part of her core identity rather than a disorder. She appreciates that her OT doesn't want to "fix" her but help her flourish.

Occupational therapists are trained to work with their clients to help them identify what is meaningful to them and find the necessary strategies to support participation at whatever level is desired. Occupational therapists utilize strength-based approaches and teach their clients to advocate for their individualized needs. Tena's occupational therapist might help her find a support group of other teens her age and encourage her to develop a few ways to talk about her autism with others.

In summary, occupational therapists are uniquely qualified to evaluate the ways an individual's underlying factors might contribute to challenges or limitations to engagement in meaningful occupations. The client-centered, strength-based perspective supports individuals with ASD and IDD to develop the necessary skills and routines to participate and engage in the occupations that are most meaningful to them.

References

American Occupational Therapy Association. (2020). Occupational therapy practice framework: Domain and process (4th ed.). *American Journal of Occupational Therapy, 74*(Supplement 2), 1–87. https://doi.org/10.5014/ajot.2020.74S2001

Anil, M. A., Shabnam, S., & Narayanan, S. (2019). Feeding and swallowing difficulties in children with Down syndrome. *Journal of Intellectual Disability Research, 63*(8), 992–1014. https://doi.org/10.1111/jir.12617

Arnsten, S. M. (1990). The issue is: Intrinsic motivation. *American Journal of Occupational Therapy, 44*, 462–463.

Bal, V. H., Kim, S. H., Cheong, D., & Lord, C. (2015). Daily living skills in individuals with autism spectrum disorder from 2 to 21 years of age. *Autism, 19*(7), 774–784.

Bayley, N. (2019). *Bayley scales of infant and toddler development* (4th ed.). Harcourt.

Beery, K. E., Buktenica, N. A., & Beery, N. A. (2010). *Beery-Buktenica developmental test of visual-motor integration* (6th ed.). Pearson.

Blanche, E. I., Bodison, S., Chang, M. C., & Reinoso, G. (2012). Development of the comprehensive observations of proprioception (COP): Validity, reliability and factor analysis. *American Journal of Occupational Therapy, 66*(5), 691–598.

Blank, R., Barnett, A. L., Cairney, J., Green, D., Kirby, A., Polatajko, H., … Vinçon, S. (2019). International clinical practice recommendations on the definition, diagnosis, assessment, intervention, and psychosocial aspects of developmental coordination disorder. *Developmental Medicine and Child Neurology, 61*(3), 242–285. https://doi.org/10.1111/dmcn.14132

Bontinck, C., Warreyn, P., Van der Paelt, S., Demurie, E., & Roeyers, H. (2018). The early development of infant siblings of children with autism spectrum disorder: Characteristics of sibling interactions. *PLoS One, 13*(3), e0193367.

Bruschini, H., Faria, N., Garcez, E., & Srougi, M. (2003). Development of bladder control in mentally handicapped children. *International Brazilian Journal of Urology, 29*, 455–458.

Bull, M. J. (2011). Health supervision for children with Down syndrome. *Pediatrics, 128*(2), 393–406. https://doi.org/10.1542/peds.2011-1605

Cammuso, K., & Vorderstrasse, V. (2020). Anxiety in autism spectrum disorder: Diagnosis and therapy. *The Brown University Child and Adolescent Behavior Letter, 36*(11), 1–4.

Chien, C.-W., Rodger, S., Copley, J., Branjerdporn, G., & Taggart, C. (2016). Sensory processing and its relationship with children's daily life participation. *Physical & Occupational Therapy in Pediatrics, 36*(1), 73–87.

Cibralic, S., Kohlhoff, J., Wallace, N., McMahon, C., & Eapen, V. (2019). A systematic review of emotion regulation in children with autism spectrum disorder. *Research in Autism Spectrum Disorders, 68*, 101422.

de Graaf, G., Levine, S. P., Goldstein, R., & Skotko, B. G. (2019). Parents' perceptions of functional abilities in people with Down syndrome. *American Journal of Medical Genetics Part A, 179*(2), 161–176.

Demetriou, E. A., DeMayo, M. M., & Guastella, A. J. (2019). Executive function in autism spectrum disorder: History, theoretical models, empirical findings, and potential as an endophenotype. *Frontiers in Psychiatry, 10*, 753.

Dunn, W. (2014). *Sensory profile 2: Strengths-based approach to assessment and planning, user's manual*. Pearson.

Fisher, A. G., & Short-DeGraff, M. (1993). Improving functional assessment in occupational therapy: Recommendations and philosophy for change. *American Journal of Occupational Therapy, 47*(3), 199–201.

Gray, C. A., & Garand, J. D. (1993). Social stories: Improving responses of students with autism with accurate social information. *Focus on Autistic Behavior, 8*(1), 1–10.

Green, S. A., & Ben-Sasson, A. (2010). Anxiety disorders and sensory over-responsivity in children with autism spectrum disorders: Is there a causal relationship? *Journal of Autism and Developmental Disorders, 40*(12), 1495–1504. https://doi.org/10.1007/s10803-010-1007-x

Gresham, F. M., & Elliott, S. N. (2008). *Social skills improvement system rating scales*. Pearson Assessments.

Jarrard, P., Beckman, K., Boss, B., Catalano, B., McDaniel, A., & Nussbaumer, A. (2018). Reducing anxiety symptoms in adolescents with autism spectrum disorder. *American Journal of Occupational Therapy, 72*(4_Supplement_1), 7211515268p1.

Kent, C., Cordier, R., Joosten, A., Wilkes-Gillan, S., Bundy, A., & Speyer, R. (2020). A systematic review and meta-analysis of interventions to improve play skills in children with autism spectrum disorder. *Review Journal of Autism and Developmental Disorders, 7*(1), 91–118.

Knox, S. (2008). Development and current use of the Knox Preschool Play Scale. In D. Parham & L. Fazio (Eds.), *Play in occupational therapy for children* (2nd ed., pp. 55–70). Mosby/Elsevier.

Krell, K., Haugen, K., Torres, A., & Santoro, S. L. (2021). Description of daily living skills and independence: A cohort from a multidisciplinary Down syndrome clinic. *Brain Sciences, 11*(8), 1012.

Lloyd, M., MacDonald, M., & Lord, C. (2013). Motor skills of toddlers with autism spectrum disorders. *Autism, 17*(2), 133–146.

McLaughlin Gray, J. (1998). Putting occupation into practice: Occupation as ends, occupation as means. *American Journal of Occupational Therapy, 52*, 354–364.

Miller, L. J. (2006). *Miller function and participation scales manual*. Pearson.

Moy, E., Tardif, C., & Tsao, R. (2017). Predictors of handwriting in adolescents and adults with Down syndrome. *International Journal of Disability, Development and Education, 64*(2), 169–181.

Mrad, F. C. d. C., Figueiredo, A. A. d., Bessa, J. d., & Bastos Netto, J. M. (2018). Prolonged toilet training in children with Down syndrome: A case–control study. *Jornal de Pediatria, 94*(3), 286–292. https://doi.org/10.1016/j.jped.2017.06.011

Mroz, T. M., Pitonyak, J. S., Fogelberg, D., & Leland, N. E. (2015). Client centeredness and health reform: Key issues for occupational therapy. *American Journal of Occupational Therapy, 69*, 1–8.

Nederkoorn, C., Theiβen, J., Tummers, M., & Roefs, A. (2018). Taste the feeling or feel the tasting: Tactile exposure to food texture promotes food acceptance. *Appetite, 120*, 297–301.

Ohl, A., Schelly, D., Brown, D., Schulze, N., Smith, M., & Davies, B. (2020). A survey of occupational therapy services provided for adults with autism spectrum disorder. *The Open Journal of Occupational Therapy, 8*(2), 1–13.

Parham, L.D., Ecker, C. L., Kuhaneck, H., Henry, D. A., & Glennon, T. J. (2021). *Sensory Processing Measure, Second Edition*. Torrance, CA: Western Psychological Services.

Pisula, E., & Ziegart-Sadowska, K. (2015). Broader autism phenotype in siblings of children with ASD—A review. *International Journal of Molecular Sciences, 16*(6), 13217–13258.

Ross, C. F., Bernhard, C. B., & Smith-Simpson, S. (2019). Parent-reported ease of eating foods of different textures in young children with Down syndrome. *Journal of Texture Studies, 50*(5), 426–433.

Ross, C. F., Surette, V. A., Bernhard, C. B., Smith-Simpson, S., Lee, J., Russell, C. G., & Keast, R. (2022). Development and application of specific questions to classify a child as food texture sensitive. *Journal of Texture Studies, 53*(1), 3–17.

Schaaf, R. C., Cohn, E. S., Burke, J., Dumont, R., Miller, A., & Mailloux, Z. (2015). Linking sensory factors to participation: Establishing intervention goals with parents for children with autism spectrum disorder. *American Journal of Occupational Therapy, 69*(5), 6905185005p6905185001–6905185005p6905185008.

Skard, G., & Bundy, A. (2008). Test of playfulness. In D. Parham & L. Fazio (Eds.), *Play in occupational therapy for children* (2nd ed., pp. 71–94). Mosby/Elsevier.

Tomchek, S., Koenig, K. P., Arbesman, M., & Lieberman, D. (2017). Occupational therapy interventions for adolescents with autism spectrum disorder. *The American Journal of Occupational Therapy, 71*(1), 7101395010p1–7101395010p3.

Trujillo, C., Poach, M., & Carr, M. (2019). Exploring the applicability of occupational therapy transition assessments for students with disabilities. *World Federation of Occupational Therapists Bulletin, 75*(1), 71–77.

Villalvazo, O. (2017). *The effects of adapted physical education on students with Down syndrome: A Meta-Analysis*.

Watling, R. (2018). Tailoring evaluations for diverse individuals with ASD. In R. Watling & S. L. Spitzer (Eds.), *Autism across the lifespan: A comprehensive occupational therapy approach* (pp. 105–125). American Occupational Therapy Association.

Watling, R. & Spitzer, L.S . (Eds.), (2018). *Autism across the lifespan: A comprehensive occupational therapy approach* (4th ed.). Bethesda, MD: AOTA Press.

World Federation of Occupational Therapists. (2010). Position statement: Client-centeredness in occupational therapy. Retrieved from https://www.wfot.org/resources/client-centredness-in-occupational-therapy

World Federation of Occupational Therapists. (2012). About occupational therapy. Retrieved from https://www.wfot.org/about-occupational-therapy

World Health Organization. (2008). *International classification of functioning, disability, and health: ICF*. World Health Organization.

22 Treating Sleep Disturbances in Children with Developmental Disorders

Anjalee Galion, Monika Mathur, Thomas Megerian, and Christopher B. Min

Abstract

The treatment of sleep disturbances in children with developmental disorders requires a stepwise approach to understanding the nature of the sleep disturbance, the unique characteristics of underlying neurodevelopmental disorder (NDD), and familial and environmental factors that contribute to the child's disordered sleep. The aim of this chapter is to provide the reader with an understanding of the types of sleep disturbances associated with some of the most common neurodevelopmental disorders, as well as best practices for evaluation and treatment. The chapter presents autism spectrum disorders (ASDs) in depth and briefly examines sleep disorders that occur in children with other NDDs (many of which often overlap with ASD). This chapter highlights the ways that common mental health and medical comorbidities associated with ASDs and NDDs contribute to disordered sleep. The text provides an overview of evidence-based practices for evaluating and treating sleep disturbances in this population of children and reviews the pharmacological approaches that can be considered when nonpharmacological interventions fail. Finally, the authors provide clinical pearls gleaned from decades of combined experience in working with children with concomitant neurodevelopmental and sleep disorders.

Keywords

Sleep disorder · Epilepsy · Autism · Insomnia · Attention-deficit/hyperactivity disorder · Melatonin

Introduction

Difficulties with sleep onset and maintenance are experienced by 33% of toddlers and preschoolers and up to 80% of children with autism spectrum disorder (ASD) (Gail Williams et al., 2004) These challenges negatively affect the child as well as family members. Children with poor-quality sleep may display increased aggressive behaviors, anxiety, and developmental regression. The nature of these sleep difficulties varies across developmental stages. Typically, younger children may have sleep anxiety, bedtime resistance, difficulties with falling asleep in their own beds, nighttime awakenings, and early morning awakenings. Adolescents may exhibit poor sleep

A. Galion (✉) · M. Mathur · T. Megerian · C. B. Min
Department of Pediatrics, Division of Pediatric Neurology, University of California Irvine – Children's Hospital of Orange County, Orange, CA, USA
e-mail: agalion@hs.uci.edu

P. McPherson (ed.), *Handbook of Treatment Planning for Children with Autism and Other Neurodevelopmental Disorders*, Autism and Child Psychopathology Series,
https://doi.org/10.1007/978-3-031-06120-2_22

hygiene, anxiety related to sleep difficulties, circadian rhythm difficulties (generally delayed sleep phase), and daytime sleepiness (Goldman et al., 2012).

The vulnerability of patients with ASD to sleep difficulties is not surprising. Children with ASD often have difficulties with emotional regulation and anxiety, contributing to difficulties with initiating sleep and returning to sleep after nocturnal awakenings. These children may rely on certain stimuli (e.g., rocking) to fall asleep and return to sleep. Bedtime refusal behaviors may also be pronounced. Due to comorbid conditions such as anxiety and obsessive compulsive disorder (OCD), children with autism may take selective serotonin reuptake inhibitors (SSRI), which are known to affect the rapid eye movement (REM) stage of sleep. In some cases, treatment with SSRIs induces aggressive and violent motor dream enactment (Kotagal & Broomall, 2012). These alterations in the quality and quantity of sleep often lead to insufficient sleep and daytime sleepiness.

One challenge in the diagnosis and management of sleep issues in this population is the differentiation of night wakings associated with ASD from those associated with behavioral insomnia of childhood. Because of the complexity of these issues, the sleep specialist is a valuable member of the medical home team, with additional specialty training beyond that offered in residency and the capacity to translate recent advances in the field of sleep medicine into practice and clinical results. In this chapter, four sleep specialists review sleep disturbances that are common among children with ASD, with a focus on night wakings.

Sleep Disturbances in Children with ASD

In children with ASD, it may be difficult to differentiate night wakings from the behavioral insomnia of childhood. As defined by the International Classification of Sleep Disorders, third edition (ICSD -3), which is the official textbook of the American Academy of Sleep Medicine (AASM), insomnia in young children is "often the result of inappropriate sleep associations or inadequate limit setting" (p. 24). Specifically, if a child is dependent on a specific stimulus, such as being held or nursing, to initiate sleep, then sleep onset may be significantly delayed in the absence of said stimulus. In the case of limit-setting issues, stalling at bedtime or "bedtime refusal" occurs when a parent or caregiver has few or no limits that are inconsistently or unpredictably applied. Poorly set limits can lead to prolonged nocturnal awakenings.

The sleep problems observed in children with ASD are diverse in presentation and severity. Some studies note that particular subpopulations of children with ASD may be more prone to sleep problems than others. Giannotti et al. found that children with regressive ASD show significantly longer sleep latency, more difficulties with sleep maintenance, and more sleep disorders than children with nonregressive ASD (Giannotti et al., 2011). Others have found links between ASD symptom severity and sleep problems (Cortesi et al., 2010; Mayes & Calhoun, 2009; Polimeni et al., 2005). In a similar vein, ASD symptom severity and sleep problems may have a bidirectional relationship (Adams et al., 2014). Others suggest that sleep difficulties in children with ASD occur independently of ASD symptom severity (Gunes et al., 2019). Other studies speculate that ASD symptom severity may have a *negative* relationship with sleep problems, as children with Asperger disorder (i.e., less severe symptoms of ASD) had more severe sleep disturbance than children with ASD (Polimeni et al., 2005). To serve the patient best, the clinician should evaluate each child independently of such assumptions and explore ASD symptom severity in addition to other mental health factors.

Which features of ASD contribute most powerfully to difficulties with sleep? Sleep onset difficulties may especially be increased by hypersensitivity, particularly at bedtime, when lights are dim and sensitivity to touch and sound is heightened (Mazurek & Petroski, 2015; Tzischinsky et al., 2018). As children are in a supine position for sleep, their skin comes into increased contact with bedtime materials, such as blankets and bedtime clothing, exacerbating issues with sleep onset in a child with hypersensitivity to touch (Tzischinsky et al., 2018).

Children with ASD are also at higher risk of developing comorbid mental health disorders than typically developing counterparts (Simonoff et al., 2008; Smalley et al., 2007). Researchers have reported particularly high rates of anxiety and depression in those with ASD (Joshi et al., 2010). These mental health issues are commonly associated with sleep problems (American Psychiatric Association, 2013; Giannotti et al., 2011). It can be difficult to parse out which facet of a child's psychosocial profile is affecting sleep the most: concomitant mental health symptoms, parenting issues/behavioral mismanagement of the child's sleep patterns and routines, or challenges related to ASD itself. An awareness of common mental health symptoms that present in children with ASD may help the clinician to navigate treatment. It is therefore important to consider as many psychosocial factors as possible when assessing sleep problems in a child with ASD. The following sections detail some of the more common mental health symptoms related to ASD as well as issues common to ASD that may affect sleep.

Sleep Disturbances in Children with Internalizing Disorders

Studies consistently show that children with ASD are at increased risk for internalizing disorders, which include mood disorders, anxiety disorders, and trauma and stress-related disorders (Simonoff et al., 2008). Internalizing disorders are associated with sleep problems in children with ASD (Reynolds et al., 2017; Thomas et al., 2015) The following sections examine specific internalizing disorders and how they impact sleep in children with ASD.

Anxiety

Several studies also show that children with ASD tend to have higher levels of anxiety than typically developing children, and anxiety is related to poor sleep in children with ASD (Hollway et al., 2013; Mazurek & Petroski, 2015). Individuals with anxiety disorders often experience prolonged sleep onset latency (American Psychiatric Association, 2013). For children with ASD, nighttime can be a particularly anxiety-provoking part of daily living. At night, children have little to distract them from their thoughts as they lay in bed, which may lead to rumination on anxious thoughts and subsequent difficulty with sleep onset or maintenance. Fear of the dark, an upcoming test, or an impending new routine/experience the following day are all examples of things that might cause anxiety for a child. These fears and worries may be exacerbated in a child with ASD who tends to perseverate on their thoughts, particularly if the child's source of anxiety and rigid/repetitive interests converge (e.g., an adolescent with ASD who has an intense interest in his train collection and is anxious about a younger cousin coming to visit the following day who will want to play with his trains). The combination of anxiety and ASD in children seems to have a particularly strong influence on sleep.

Mood Disorders

Studies consistently link mood disorders and sleep problems in children with ASD (Malow et al., 2006; Richdale et al., 2014) For example, children with ASD who exhibit good sleep patterns display fewer mood issues than do those with poor sleep (Malow et al., 2006; Richdale et al., 2014). Adolescents with ASD have significantly depressed mood and increased pre-sleep arousal, compared to typically developing adolescents (Richdale et al., 2014). In major depressive disorder, insomnia or hypersomnia is listed as one of the nine primary symptoms (American Psychiatric Association, 2013). From a psychopathology perspective, sleep and mood disorders may have a reciprocal relationship (Dahl, 1995). Some suggest that, in children, sleep problems are more closely associated with depression than with anxiety (Gregory et al., 2006). Others suggest that the association between sleep problems and depression strengthens as children reach adolescence (Gregory & O'Connor, 2002).

Children with ASD who have mood disorders may have difficulty with sleep onset due to rumination about various stressors in their lives. It has been the observation of this writer that children with ASD often tend to present with comorbid anxiety at an earlier age, and as they develop into adolescence, the longstanding nature of their anxiety and lack of relief from chronic worry dampens their mood, leading often to depression. Moreover, this writer has observed that as children with ASD develop into adolescence, they become more cognizant of their social skills deficits and (oftentimes) resulting lack of peer relationships, further contributing to mood difficulties (MGM).

Sleep Disturbances in Children with Externalizing Disorders

Externalizing disorders involve difficulties with self-control and behavioral regulation, such as attention-deficit/hyperactivity disorder (ADHD) and/or delinquent and aggressive behaviors (Mayes & Calhoun, 2009; Smalley et al., 2007). The impact of externalizing problems on sleep can be profound, particularly during times of day that are close to sleep onset.

Attention-Deficit Hyperactivity Disorder

ADHD, which is defined by the *Diagnostic and Statistical Manual of Mental Disorders* (DSM–5) as "a persistent pattern of inattention and/or hyperactivity-impulsivity that interferes with functioning or development" (American Psychiatric Association, 2013, p. 59), is observed in 75% of children with ASD (Devnani & Hegde, 2015; Mayes & Calhoun, 2009; Smalley et al., 2007). Although sleep problems are not directly listed as a symptom of ADHD, difficulties with regulation can make bedtime routines and sleep onset difficult. Children who struggle with poor working memory may often find it difficult to adhere to the steps involved in a bedtime routine such as taking a bath, changing clothes, and brushing teeth. Moreover, children with ADHD may become easily distracted during bedtime routines, resulting in delayed sleep and decreased total sleep time. Hyperactivity may also play a part at bedtime as children with ADHD may struggle to regulate their bodies and lay still to bring on asleep (Sung et al., 2008). Executive function difficulties appear to persist across the trajectory of development for individuals with ASD (Demetriou et al., 2018).

Insufficient sleep in children with ASD may be the primary reason behind inattention and difficulty with regulation (Cremone-Caira et al., 2019; Gunes et al., 2019). One possible mechanism by which ADHD impacts sleep in children with ASD may involve co-occurring internalizing problems (Jensen et al., 1997) As previously noted, children with ASD are at increased risk for internalizing disorders such as anxiety and depression (Simonoff et al., 2008).

Although the relationship between multiple mental health disorders, sleep, and ASD in children is complex, it does appear that there is a cumulative effect of psychiatric symptoms on sleep problems in children with ASD (Chen et al., 2015).

Night Wakings in ASD and Developmental Disabilities (DD)

Roughly 65% of parents of children with ASD report sleep problems, including insomnia, difficulty falling asleep, bedtime resistance, prolonged night wakings, and short sleep duration. Children with ASD often have prolonged night wakings (1–3 hours), with a return to sleep in the early morning hours. These night wakings are strongly associated with problematic daytime behaviors, presenting a challenge when the child has to wake up for daily activities at a scheduled time.

As early as 1999, Diomedi et al. reported an increased number of arousals in children with ASD, compared to their typically developing controls. The most common tools used to evaluate night wakings include polysomnography, sleep diaries, and parental questionnaires. The evaluation of a child with autism and NDD should, at the minimum, include parental inquiries about the nature and duration of any night wakings. Parents typically report a prolonged night waking of multiple hours, during which time it is very difficult to have the child reinitiate sleep. The Childhood Sleep Habits Questionnaire (CSHQ) is one of the most widely used pediatric sleep questionnaires (Owens et al., 2000). It has two questions about night wakings: "Does your child wake up once during the night?" and "Does your child wake up multiple times throughout the night?" The questionnaire does not ask about the duration of each night waking. A study by Honomichl et al. evaluated 100 children with pervasive developmental disorder using the CSHQ and sleep diaries (Honomichl et al., 2002). The results revealed more night wakings in younger children, compared with older children. However, Malow et al. (2009) showed that the Family Inventory of Sleep Habits was significantly correlated with the CSHQ for night wakings for typically developing children but not in children with ASD.

Data from parental reports may not always match that obtained through diagnostic evaluation. In 2008, Sitnick et al. found that actigraphy had poor sensitivity for the detection of night wakings compared to concurrent polysomnography with video in looking at preschoolers with ASD, preschoolers with developmental delay but not ASD, and neuro-typical controls (Sitnick et al., 2008). The effects of these night wakings can be significant including negative effects on behavior, severity of autism, inattention, irritability, and effects on the family (Mazurek & Sohl, 2016). In 2014, Taylor et al. (2012) demonstrated a significant negative effect of night wakings on communication. This work fits with the findings of Kheirouri et al. that night wakings are correlated with ASD severity (Kheirouri et al., 2016). Mazurek & Sohl, (2016) showed that night wakings are significantly associated with physical aggression.

Another challenge in the identification of night wakings is that they may occur as part of other sleep disturbance. Parents may not think to differentiate them from difficulties with sleep initiation or early morning wakings. The high rate of chronic sleep disturbance in children with ASD may compel parents to stop differentiating the nature of the sleep disturbances, because they feel that these sleep issues are "normal" for their child. Parents with this mindset are even less likely to raise these issues during a routine healthcare visit, unless specifically questioned.

Disruptive and Aggressive Behaviors

Hyperactivity and aggression are particularly strong predictors of sleep difficulties in children with ASD (Mayes & Calhoun, 2009; Thomas et al., 2015) Children with ASD who have a tendency toward aggression may likely exhibit that aggression at bedtime, causing difficulties during the bedtime routine. While bedtime can often be a point of contention between typically developing children and their caregivers, children with ASD who become aggressive at bedtime have increased risk of sleep issues. Malhi et al. found that parent-reported daytime behavior difficulties were associated with more sleep problems in children with ASD and that children with ASD exhibited more bedtime resistance compared with typically developing children (Malhi et al., 2019). Children with ASD are also reported to have more variability in hours slept and less total sleep time, compared to typically developing children (Malhi et al., 2019).

Comorbid Conditions

The assessment of sleep disturbance in children with ASD or other DD includes identifying comorbid conditions that may make the child more prone to sleep disturbance. Common co-occurring medical conditions that impact sleep include gastrointestinal disorders (nighttime pain or discomfort from constipation and/or gastroesophageal reflux disease (GERD)), sleep apnea, epilepsy, asthma, and concomitant neuropsychiatric disorders such as anxiety and depression (Stores, 2014). In addition, medications used to treat co-occurring disorders, such as attention deficit disorders (ADD), seizure disorders, and psychiatric disorders, often have side effects that disrupt sleep (Coleman et al., 2019; Williams Buckley et al., 2019).

Respiratory Conditions

In the pediatric population, estimates for obstructive sleep apnea (OSA) range from 1% to 4%, but these may even be underestimated due to trends in pediatric obesity and the limited availability of pediatric sleep medicine providers and access to accredited pediatric sleep diagnostic centers (Bixler et al., 2009; Lumeng & Chervin, 2007). Typical symptoms include snoring, frequent nocturnal awakenings, restless sleep, witnessed apneas, and a preference for sleeping in a prone position with the mouth open and/or the neck hyperextended. Typical daytime sequelae include hyperactivity, difficulty with attention and focus, irritability, and excessive daytime sleepiness. Children may also complain of headaches. Diagnosis requires an overnight polysomnography study. In the case of adenotonsillar hypertrophy, evaluation by an otolaryngologist is recommended. When OSA persists despite surgical intervention or when surgery is not possible, as in the case of obese patients, continuous positive airway pressure (CPAP) is recommended. Although adherence to CPAP can often be challenging in this patient population, a specialized PAP desensitization program can increase the likelihood of PAP tolerance and compliance. This is typically achieved in a multidisciplinary setting with psychologists, child life specialists, and sleep medicine physicians who can closely monitor progress with therapy.

Other respiratory conditions linked to sleep disruption include allergic rhinitis and asthma. Children with allergic rhinitis may snore even in the absence of OSA. The antihistamine medications used to treat allergic rhinitis have known sedative side effects that may lead to daytime somnolence. Children with asthma may have nocturnal symptoms including cough, wheezing, and shortness of breath, which can lead to difficulties with sleep maintenance.

Vision Impairment

Children that have vision impairment, particularly those who are blind, often experience non–24-hour sleep–wake rhythm disorder, a type of circadian rhythm disturbance. Because daily retinal light exposure is necessary to synchronize circadian rhythms with the external 24-hour solar environment, vision disturbances often impair light entrainment of the body's internal pacemaker, the suprachiasmatic nucleus in the anterior hypothalamus. Affected individuals experience cyclical or periodic episodes of poor sleep and daytime dysfunction, which may severely interfere with social life and school. Strategies for management include behavioral modification and the use of melatonin or tasimelteon, a melatonin receptor 1 and 2 agonist (Quera Salva et al., 2017).

Epilepsy

Thirty percent of children with ASD have epilepsy (Trickett et al., 2018). Seizures and interictal discharges, the pathological brain activity observed between seizures, are often activated during sleep; some types of epilepsies occur only at night. Frequent nocturnal seizures and/or interictal discharges will interrupt sleep and lead to daytime sleepiness. In 33% of children with medically intractable epilepsy, OSA exacerbates seizure burden (St. Louis, 2011). The effective management of OSA provides therapeutic benefit comparable to that of adding another anti-epileptic medication for seizure treatment. One study found a 50% reduction in the frequency of seizures in 50–60% of patients whose OSA was managed effectively (Maris et al., 2016). Recognizing and treating co-existent sleep disorders in children with epilepsy may improve seizure burden and quality of life. Primary sleep disorders may even mimic epilepsy; for example, parasomnias manifest similarly to nocturnal seizures. By using video electroencephalography (EEG) in combination with polysomnography to localize seizures and to identify the sleep stages affected most severely, clinicians may be able to differentiate parasomnias from seizures. The prevalence of periodic limb movement disorder is noted to be as high as 47% in children with autism, compared to 8% in neuro-typical controls. Periodic limb movement disorder was also associated with decreased serum ferritin levels (Youssef et al., 2013). These nocturnal movements not only cause sleep fragmentation but also make it increasingly difficult for parents to identify seizure-like activity that is separate from sleep-related limb movement.

Sleep deprivation is a known trigger for seizures. Some children with nocturnal seizures may experience sleep anxiety if seizures occur at night, which can lead to an overall decreased quality of life for the parent as well as the child. The parent plagued with worry that their child may have a seizure at night is likely to have difficulty sleeping. Such parents may allow a co-sleeping arrangement, which can significantly decrease the quality and quantity of the parent's sleep and negatively impact a marriage (Larson, 2012).

Down Syndrome

Children with Down syndrome commonly report sleep disturbance, insomnia, excessive daytime sleepiness, parasomnias, and OSA. OSA occurs in more than 50% of children with Down syndrome. In a study by Maris et al. (2016), 66.4% of children with Down syndrome had OSA. This high prevalence was found across age groups. Even in those with a negative history of symptoms suggestive of OSA, the prevalence was 53.8%. Therefore, it is important to screen all children with Down syndrome with polysomnography, irrespective of age or parental reports of symptoms of sleep-disordered breathing. In 2011, the American Academy of Pediatrics began recommending polysomnography studies for all children with Down syndrome by 4 years of age, regardless of symptom history. The high prevalence of OSA in children with Down syndrome is likely due to anatomic factors (macroglossia, adenotonsillar hypertrophy, midface hypoplasia) and comorbid conditions, including obesity, hypothyroidism, hypotonia, and gastroesophageal reflux disease (GERD). Untreated OSA can lead to significant morbidity in patients with Down syndrome. Children with Down syndrome are at increased risk for cardiovascular complications (up to 56% have congenital heart disease), which, in turn, increases risk for pulmonary hypertension, especially in the setting of untreated OSA (Simpson et al., 2018). Children with Down syndrome also exhibit a wide spectrum of neurodevelopmental issues at baseline, including deficits in memory, language, visual perception, and executive function. These cognitive outcomes are often worse in children with comorbid OSA (Simpson et al., 2018).

Craniofacial Abnormalities

Children with certain craniofacial abnormalities are at increased risk for OSA. Airway obstruction may occur at different levels depending on the specific craniofacial abnormality. Children with craniosynostosis syndromes (e.g., Apert syndrome, Crouzon disease, Pfeiffer syndrome, Saethre–Chotzen syndrome) often have abnormalities of the skull base and accompanying maxillary hypoplasia, which can lead to nasopharyngeal obstruction. In children with syndromes that involve micrognathia (e.g., Treacher–Collins syndrome, Pierre Robin syndrome, Goldenhar syndrome), obstruction typically occurs at the hypopharyngeal level (Chan et al., 2004). The surgical management of OSA is typically more complex in this population. The patient should ideally be evaluated by a craniofacial team that includes pediatric specialists from plastic surgery, oral surgery, and otolaryngology.

Gastroesophageal Reflux Disease

The relationship between GERD and sleep disturbance is likely bidirectional. GERD is associated with many sleep disturbances, including decreased sleep duration, difficulty with falling asleep, arousals during sleep, poor sleep quality, and early morning wakings. Sleep deprivation has been found to induce esophageal hyperalgesia to acid perfusion. GERD patients with sleep disturbances report more severe symptoms and poorer quality of life than those without sleep disturbance (Fujiwara et al., 2012)

Neuromuscular Disorders

Children with neuromuscular disorders involving hypotonia are at increased risk for sleep-related breathing disorders such as OSA, hypoventilation, and central sleep apnea (due to diaphragmatic weakness). It is important to note that sleep-disordered breathing may be observed when muscle weakness is still mild and diurnal respiratory dysfunction is not yet apparent. The child may demonstrate other symptoms (excessive daytime sleepiness, insomnia, poor sleep quality, morning headaches, mood disturbance, attention deficit, and learning difficulties). Nocturnal symptoms such as air hunger, intermittent snoring or pauses in breathing, orthopnea, cyanosis, restlessness, and insomnia should prompt further diagnostic studies, including polysomnography. Children with neuromuscular disorders are at higher risk for developing complications of nocturnal hypoxemia, including pulmonary hypertension, cor pulmonale, and neurocognitive dysfunction (Alves et al., 2009). While OSA may be caused by fixed (e.g., adenotonsillar hypertrophy) or dynamic upper airway abnormalities (e.g., pharyngeal wall collapse, laryngomalacia), dynamic abnormalities are more common in children with hypotonia (Goldberg et al., 2005) Children with dynamic abnormalities are less likely to respond to surgical interventions and may require other treatment modalities, such as PAP or assisted ventilation.

Rett Syndrome

Approximately 87% of children with Rett syndrome have sleep problems, including daytime napping, nighttime laughter, teeth grinding, night screaming, and nocturnal seizures (Young et al., 2007). Irregular breathing patterns are observed during the day and at night, when hypoventilation and central and obstructive apnea manifest (Sarber et al., 2019). Polysomnography is a useful tool for evaluation. Typical treatments include adenotonsillectomy and PAP.

Angelman Syndrome

Sleep disturbances are observed in 48–70% of children with Angelman syndrome. Common symptoms include shorter sleep duration, longer sleep onset latency, more frequent nocturnal awakenings with difficulty falling back asleep, early morning waking, limb movements during sleep, sleep-disordered breathing, and daytime somnolence. Approximately 80% of children with Angelman syndrome have seizures, and, as previously discussed, poor sleep can impact seizure burden. Other common health conditions that impact sleep in this population include scoliosis and obesity, as children with Angelman syndrome may have impaired satiety and exhibit behaviors such as taking and storing food. Obesity increases the risk for OSA and the severity of GERD symptoms (Trickett et al., 2018).

Smith–Magenis Syndrome

It is estimated that 100% of children with Smith–Magenis syndrome (SMS) experience sleep disturbances, including decreased total sleep time, night waking, earlier final morning wake times, and inverted circadian rhythm. Seizures affect up to 18% of children with SMS. Other common comorbid health conditions that affect sleep in this population include scoliosis, chronic otitis media, and obesity. Many children with SMS have sleep-disordered breathing, likely related to obesity, but also likely due to midface hypoplasia, which affects over 90% of individuals with SMS. Symptoms indicative of GERD are also associated with sleep-disordered breathing in this population (Trickett et al., 2018).

Tuberous Sclerosis Complex

Approximately 74% of children with tuberous sclerosis complex (TSC) experience sleep disturbances, including decreased total sleep time, difficulties with sleep onset, and nocturnal awakenings. Seizures affect up to 70% of children with TSC, and a higher prevalence of night waking has been found in children with TSC who experienced seizures in the last 6 months or who had ≥3 seizures per night (Trickett et al., 2018).

Williams–Beuren Syndrome

Approximately 65% of children with Williams–Beuren syndrome (WBS) exhibit sleep disturbances, including sleep-disordered breathing, bedtime resistance, anxiety, frequent nocturnal awakenings, and daytime sleepiness. These children have lower nocturnal melatonin levels, which can lead to circadian rhythm dysfunction. Daytime sequelae include behavioral issues and memory problems (Santoro et al., 2016).

Prader–Willi Syndrome

Children with Prader–Willi syndrome (PWS) are at risk for OSA and sleep-related alveolar hypoventilation. The clinical features that contribute to these conditions include hypotonia, obesity, scoliosis, restrictive lung disease (and resulting alveolar hypoventilation), narrowed upper airway, facial dysmorphism, abnormalities of central respiratory control, and reduced ventilatory responses to hypoxia and hypercapnia. Central apnea is also frequently reported. Polysomnography should be considered for patients with symptoms suggestive of sleep-disordered breathing. The recommended treatment options include weight loss, adenotonsillectomy, and nocturnal ventilation (Nixon & Brouillette, 2002). Children with PWS often have hypersomnia, as indicated on multiple sleep latency tests (MSLT), and abnormalities in REM sleep (including sleep-onset REM periods, reduced latency to REM sleep, and reduced intervals between REM periods). These effects may be related to hypothalamic dysfunction. The hypothalamus is known to regulate NREM–REM cycling, which has been found to be abnormal in children with PWS (Camfferman et al., 2008).

Neoplasms of the Central Nervous System

Children with neoplasms of the central nervous system (CNS), especially the hypothalamus, thalamus, and brainstem, often have frequent, severe sleep problems. Approximately 80% of children with CNS neoplasms in these brain areas report excessive daytime sleepiness, and 46% have sleep-disordered breathing (Rosen & Brand, 2011). In these children, it is important to consider other comorbid health conditions and treatments, including seizures, adenotonsillar hypertrophy, side effects of medication, obesity, pain, anxiety, and drug use. Treatment may include sleep hygiene, behavioral counseling, PAP therapy, ventilation, sedative hypnotics, antidepressants, and/or stimulants (Rosen & Brand, 2011).

Delayed Sleep Phase

There are also normal variants in sleep behavior as children mature from infancy to adulthood. A common circadian disorder in teens and young adults is adolescent delayed sleep-phase syndrome. Crowley et al. established a "perfect storm" model that describes the multi-factorial nature of adolescent sleep behavior and includes both the circadian changes in adolescent maturation as well as psychosocial and societal pressures (Crowley et al., 2018). These effects are not well known in children with DD, who may be even more influenced by their frequent use of technology for both entertainment and communication.

Sleep disturbance is common among children with ASD or other DD. The nature of the sleep disturbance varies. The presence of a comorbid condition should dictate an individualized approach to assessment and treatment. A comprehensive, multidisciplinary approach to evaluation by pediatric subspecialists is often useful for the patient and their family members, as well as for the care team.

Molecular Basis of Sleep Disturbances in Children with ASD

ASD may cause abnormalities in the hypothalamic–pituitary–adrenal axis, which mediates the production of hormones and neurotransmitters, resulting in dysregulation of the child's circadian rhythm (Devnani & Hegde, 2015). A regular sleep–wake cycle requires a finely tuned balance in levels of serotonin, GABA, and melatonin; impaired levels of any of these hormones or neurotransmitters will directly impact sleep.

Melatonin is needed to maintain and synchronize the circadian rhythm. An enzyme called N-acetyl serotonin O-methyltransferase, which mediates the final step of melatonin synthesis, has impaired activity in children with ASD. Blood and urine samples from this population have low levels of melatonin; low levels of melatonin are associated with difficulty falling asleep and irregular sleep–wake patterns. Treatment with exogenous melatonin has been shown to improve sleep patterns in children with ASD.

GABA is secreted from the preoptic area in the hypothalamus, which regulates the sleep–wake cycle. The migration and maturation of interneurons that release GABA may be affected in autism. A region of genetic susceptibility has been identified on chromosome 15q, which contains GABA-related genes and is commonly implicated in ASD (Goldman et al., 2012).

Polysomnography studies have identified changes in sleep architecture that are common in children with ASD. These include decreased time in bed, decreased total sleep time, decreased quantity of REM sleep, increased latency to sleep onset, increased REM latency, and increased waking after sleep onset. These factors reduce sleep efficiency. Less total sleep time, lower proportion of REM sleep, and increased REM latency have been shown to correlate with increased severity on the childhood autism rating scale (Devnani & Hegde, 2015).

In children with ASD or other DD, sleep disturbance can lead to cognitive impairment, including difficulties with learning, memory, and academic achievement. Sleep disturbance is also associated with behavioral concerns such as increased aggression, irritability, hyperactivity, and social difficulties. The health concerns associated with sleep disturbance in these patients include poor appetite and impaired growth, perhaps caused by decreased secretion of growth hormone, which occurs during non-REM (NREM) stage 3 sleep (Kotagal & Broomall, 2012). When children with ASD or DD have poor sleep, their caregivers are also affected. The effects on caregivers include increased maternal stress, parental sleep disruption, and increased stress levels (Kotagal & Broomall, 2012).

Interventions for the Treatment of Sleep Disturbance

A recent Practice Guideline developed by the American Academy of Neurology (AAN) reviewed the literature for evidence-based treatments for insomnia and disrupted sleep in children and adolescents with ASD. The reviewers differentiated among bedtime resistance, sleep onset latency, sleep continuity (sleep efficiency and night wakings), total sleep time, and daytime behavior (Williams Buckley et al., 2020).

Behavioral management, including parent education and training on sleep hygiene and modification of a child/adolescent's behaviors are the most effective ways to treat sleep disorders in this population. Previous studies have also demonstrated that improving mood with traditional treatment methods such as behavioral activation can decrease sleep onset latency in children with ASD (Brand et al., 2015).

It is incumbent on the treating clinician to explore the presence of concomitant disorders and/or treatment medications for those disorders that may interfere with sleep. With the exception of melatonin, few pharmacologic interventions have been the subjects of high-quality Class I or Class II studies (Malow et al., 2012; Robinson-Shelton & Malow, 2016; Williams Buckley et al., 2019). Before considering pharmacologic treatment for sleep disorders, the following practices should be considered:

1. Assess for and treat medical comorbidities that can interfere with sleep (e.g., H2-blockers or proton pump inhibitors for GERD, CPAP for OSA, drugs for epilepsy).
2. Assess for and manage potential medication-induced insomnia or nighttime activation (certain antidepressants such as SSRIs can cause activation or restless legs; stimulants for ADHD cause prolonged latency).
3. Initiate sleep education program (e.g., sleep toolkit, cognitive behavioral therapy, parent training).

Behavioral Interventions

The simultaneous treatment of multiple comorbid symptoms may improve sleep in children with ASD (McCrae et al., 2020). A recent study by McCrae et al. found that cognitive behavioral therapy for childhood insomnia (CBT-CI) can improve sleep onset latency and decrease irritability, hyperactivity, and lethargy in children with ASD. As CBT-CI can involve relaxation skills and other techniques to improve not only sleep onset but also mood and anxiety, a shotgun approach that addresses both sleep and mental health symptoms may be most beneficial in improving the overall functioning of a child with ASD.

Aerobic exercise has been shown to improve sleep onset latency, efficiency, duration, and inhibitory control in children with ASD (Tse et al., 2019). Though the nature of the relationships between these variables is unknown, one might speculate that improved inhibitory control may facilitate bedtime routines and decrease bedtime resistance. Some sleep issues may require specialist intervention in the form of surgery or use of a CPAP device. To select the optimal approach to treatment, the clinician must understand the contributions of myriad medical, psychiatric, environmental, and social factors. In choosing the best treatment approach, be it educational, behavioral, or pharmacologic, it is critical to identify the type of sleep disturbance being addressed (Stores, 2016).

Many agree that adopting a biopsychosocial framework is best when conceptualizing and assessing mental health disorders, sleep, and ASD (Schreck & Richdale, 2020). Utilizing a biopsychosocial approach will encourage a broad-to-narrow approach, helping the clinician to uncover not only the mental health symptoms that might contribute to sleep problems but also to determine how caregivers and family members may unknowingly sustain sleep problems in their child.

Pharmacological Interventions

In general, evidence for the efficacy of pharmacologic treatment of sleep disorders specifically in children with ASD/DD is slim, often consisting of small randomized controlled trials that were short in duration, open-label trials lacking a control group, or retrospective studies and case series. Nearly all of the treatments to be discussed are off-label uses of approved medications or over-the-counter agents (Brown, 2019; Bruni et al., 2019; Cuomo et al., 2017; Robinson-Shelton & Malow, 2016).

Melatonin and Related Medications

Melatonin (N-actyl-5-methoxytryptamine) is a naturally occurring neurohormone derived from serotonin that is produced in the pineal gland. It is released in response to low light/darkness and plays a critical role in regulating circadian rhythm (Esposito et al., 2020; Kennaway, 2000). It is also noted to be an antioxidant, anti-inflammatory compound that plays a role in early neural development. Previous studies have shown that dysfunctional melatonin synthesis and release may play a role in sleep disorders in children and adolescents with ASD/DD (Kotagal & Broomall, 2012). However, other studies have reported normal levels of melatonin in children with concurrent developmental disabilities and sleep disruption. Several high-quality studies have demonstrated the effectiveness of short-acting and prolonged-release melatonin in treating multiple aspects of sleep disorders, over the short as well as long term (Coppola et al., 2004; Gringras et al., 2012, 2017). Doses of 0.5–12 mg given 3–60 minutes prior to bedtime have been reported to be effective. Decreases in sleep dysfunction are observed even in children with normal endogenous melatonin production (Goldman et al., 2014; Leu, 2019). While there were initial reports that long-term melatonin use may lead to tolerance, studies out to at least 4 years demonstrate continued effectiveness. Nonetheless, some children and adolescents may no longer experience benefit from melatonin after a period of effectiveness, and increased dosing does not overcome the issue, suggesting that tolerance is not at play in these instances (Andersen et al., 2008; Carr et al., 2007).

Melatonin has been shown to be effective when used alone or in combination with cognitive behavioral therapy (CBT) (Cortesi et al., 2012) Studies have also demonstrated melatonin's effectiveness in treating sleep-onset latency, sleep continuity, and total sleep time, both alone and in combination with CBT (Cortese et al., 2020; Cortesi et al., 2012; Cuomo et al., 2017; Malow et al., 2012; Parker et al., 2019; Williams Buckley et al., 2020; Wirojanan et al., 2009).

Melatonin's effectiveness in improving sleep appears to cross a wide range of DDs, though the effects on various subtypes of sleep dysfunction differ. For instance, while latency, total sleep time, and number of night wakings improved with melatonin in children with Angelman syndrome, ASD, and intellectual disability, the effectiveness of melatonin in Fragile X and tuberous sclerosis is based on improvements in latency and sleep time, but not on decreased night wakings. Moreover, studies evaluating the effectiveness of melatonin in treating sleep dysfunction in Rett syndrome have yielded inconsistent results, with some evidence suggesting positive effects on latency and sleep time (but not wakings) only in subjects with more severe sleep dysfunction at baseline (Leu, 2019; McArthur & Budden, 2008; Wirojanan et al., 2009).

In summary, melatonin is the most studied pharmacologic intervention for children with ASD/DD, with a significant weight of evidence favoring its use across multiple DDs for treating specific aspects of sleep disorders. The AAN practice guideline for Treatment for Insomnia and Disrupted Sleep Behaviors in Children and Adolescents with Autism Spectrum Disorder rates the evidence for melatonin as "probably effective" alone or in combination with CBT for treating most types of sleep disturbances in children with ASD (Williams Buckley et al., 2020).

Melatonin is commercially available in many forms, including liquid, tablet, rapid-dissolve tablets, chewables, and gummies, as well as concentrated liquid. There are several extended or prolonged-release formulations available at different strengths, though most (if not all) appear to require swallowing the dose whole, making them difficult to use in children who cannot swallow tablets. Most experts strongly recommend choosing pharmaceutical-grade melatonin to ensure consistent and reliable formulation, absorption, and purity (Cerezo et al., 2016; Erland & Saxena, 2017)

Randomized controlled trials have shown that the side effects of melatonin are mild. The most commonly reported adverse events are headaches, dizziness, nausea, diarrhea, enuresis, rash, daytime fatigue, and clouded thinking. Though not common, sleep terrors and vivid dreams have also been reported (Leu, 2019). Because melatonin is metabolized by CYP1A2 cytochrome enzymes, dosing adjustments may be necessary for use in conjunction with medications that inhibit this enzyme (e.g., fluvoxamine, mexiletine, cimetidine). Conversely, melatonin can inhibit CYP1A2 and CYP3A enzymes; adjustments may need to be made for concurrently administered medications metabolized by those enzymes (Horn & Hansten, 2007; Leu, 2019).

Ramelteon, a melatonin-1 (MT_1) receptor agonist, was approved for the treatment of prolonged sleep latency in adults. It has six-fold greater affinity for the MT_1 receptor than melatonin (Hatta et al., 2015; Kato et al., 2005). A limited number of case reports have reported its effectiveness in children with ASD (Asano et al., 2014; Hollway & Aman, 2011; Kawabe et al., 2014).

Alpha-2 (α2) Agonists

Clonidine, which is a noradrenergic *α2* agonist indicated for treating hypertension and ADD, is probably the second most widely used medication for treating sleep disorders in children with ASD/NDD. Several studies, mostly open label in nature, have demonstrated moderate effectiveness for reducing latency and night waking (Hollway & Aman, 2011; Ingrassia & Turk, 2005; Ming et al., 2008). Clonidine dosing typically starts at 0.05 mg and can go as high as 0.2–0.3 mg in adolescents with adult habitus. Formulations include oral tablet, weekly transdermal patch, and extended-release oral tablet. Tolerance is occasionally seen. The most frequent adverse events reported by respondents of a national survey of medications used to treat psychiatric and seizure disorders in children and adults who used clonidine included daytime fatigue/drowsiness, aggression/agitation/irritability, behavior problems, anxiety and dizziness, gastrointestinal symptoms, and paradoxical sleep problems (Coleman et al., 2019; Macleod & Keen, 2014).

Guanfacine is another *α2* agonist. Like clonidine, it was approved for treating hypertension and ADD. However, it is less sedating than clonidine and therefore less useful for promoting sleep in children with ASD/DD. In the parent/patient survey of psychiatric medications used in ASD referred to above, the rated benefit was small and far less than that of clonidine for reducing latency and improving sleep maintenance (Coleman et al., 2019). Indeed, other studies in children with ADD and ADD with ASD reported that extended-release guanfacine was found to have no effect or even to decrease total sleep time (Bruni et al., 2019; Politte et al., 2018; Rugino, 2018).

In summary, clonidine may have a moderate effect size in treating sleep disorders by improving sleep-onset latency, sleep continuity, and total sleep time, while decreasing the number of night wakings (Cuomo et al., 2017). The positive effects of these agents should be weighed against the known adverse effects, such as decreased REM-phase sleep. Nonetheless, they may be beneficial in children who do not respond to other behavioral therapies and/or melatonin.

Antipsychotic Medications

Risperidone is one of the two agents approved for the treatment of irritability in children and adolescents with ASD. In one study designed to examine the long-term effects of risperidone in children and adolescents with ASD, investigators noted improvement in sleep-onset latency, but not in sleep duration (Aman et al., 2005; Brown, 2019). However, other studies found a positive effect of risperidone on sleep duration (Cortese et al., 2020; Cuomo et al., 2017). Given the substantial metabolic and endocrinologic adverse events (e.g., weight gain, gynecomastia, dyslipidemia), as well as negative effects on REM sleep and the counterproductive potential to induce restless legs syndrome (and concomitant sleep disturbance), caution should be used in prescribing this medication solely for the purpose of treating sleep dysfunction in children with ASD/DD.

In one small study of adolescents with ASD, quetiapine was reported to be beneficial in improving sleep. The authors noted improvement on the CSHQ when quetiapine was tested in open-label fashion on children with concurrent aggressive behavior and elevated sleep scores (Golubchik et al., 2011). In a national survey ranking parent/patient-reported risk-benefit analyses of psychiatric medications in individuals with ASD, quetiapine was reported to have some benefit in treating sleep, though the risk:benefit ratio was reported as close to even (Coleman et al., 2019).

Antidepressants

Trazodone is an antidepressant (AD) that acts as a serotonin-2a ($5HT_{2a}$) and histamine-1 (H_1) receptor antagonist, and also modulates adrenoreceptor function. It has substantial sedating side effects, even at doses lower than used for their AD activity, a fact that has been capitalized on for use in childhood and adult sleep disorders. Although child psychiatrists frequently use trazodone in treating children and adolescents with

ASD/DD and other neuropsychiatric disorders, there is a paucity of literature supporting its use in children with ASD/DD (Brown, 2019; Hollway & Aman, 2011; Owens et al., 2010).

Similar to trazodone, mirtazapine is a sedating AD that is an H_1 and 5-HT_2 blocker. However, mirtazapine is a tetracyclic AD that also blocks 5-HT_3 and alpha-2 receptors, leading to increased release of norepinephrine and serotonin. It has the effect of reducing latency and increasing sleep duration without affecting REM. A small open-label study in children with ASD/DD showed improved sleep in a subset of children (Brown, 2019; Hollway & Aman, 2011; Owens et al., 2010; Posey et al., 2001). The side effects of both medicines include daytime drowsiness, dizziness, dry mouth, and constipation. Trazodone can also cause vision changes, and, rarely, hypotension, priapism, and cardiac arrhythmias (Brown, 2019; Hollway & Aman, 2011).

GABA Receptor Agonists

Clonazepam, a long-acting GABA receptor agonist, shares some properties with sleep agents approved for the treatment of insomnia in adults (Browne, 1976). Indicated for treating some seizure subtypes as well as panic disorder (in adults), it has been reported to be of some utility in treating children with ASD/DD. It can be particularly useful in children with concurrent seizure disorders and has been of benefit in helping with sleep and seizures in children with Angelman syndrome (Cuomo et al., 2017; Robinson-Shelton & Malow, 2016).

Anti-epileptic Medications

There is some evidence that gabapentin, dosed at 5 mg/kg up to 15 mg/kg at bedtime, may provide benefit for sleep latency and maintenance across several NDDs in children, including ASD, cerebral palsy, and ADHD. The reported side effects include paradoxical worsening of agitated awakening or prolonged sleep-onset latency due to activation in a subset of individuals. Gabapentin may be an especially good choice if co-occurring restless legs syndrome or seizure disorder is suspected in a child with ASD/DD (Accardo & Malow, 2014; Brown, 2019; Reynolds, 2019; Robinson & Malow, 2013).

Other anti-epileptics that have been associated with improvements in sleep latency and sleep efficiency including tiagabine, pregabalin, clobazam, and carbamazepine. (Accardo & Malow, 2014; Brown, 2019; Jain & Glauser, 2014; Reynolds, 2019; Robinson & Malow, 2013) However, it would be difficult to advise the use of these agents outside of a presentation of sleep dysfunction in the setting of epilepsy. Indeed, in the child with ASD/DD and co-occurring epilepsy, the clinician should be mindful of anti-epileptic drugs that may promote sleep rather than having no effect or even interfering (as can be seen with phenobarbital and phenytoin) (Nita & Weiss, 2019).

Cannabis

Barchel showed that cannabidiol may decrease bedtime resistance (Barchel et al., 2018). Administering cannabidiol to children with ASD helped with the child's motivation and ability to communicate with family and caregivers, ultimately decreasing disruptive behavior at night and bedtime resistance (Barchel et al., 2019).

In a review of cannabis in the treatment of ASD, the findings concerning the effects on sleep were inconsistent and mostly based on open-label or retrospective studies. In one study, parents reported improvement in 71% of children in whom sleep problems were present at the outset of the study. However, peer-reviewed studies found no significant improvement compared with "conventional" treatment methods (Barchel et al., 2019). Another study in children from Chile with ASD reported improved sleep after treatment with cannabis (Kuester et al., 2017). In contrast, a retrospective study that reported improvement in anxiety-related symptoms in children with ASD also reported significant side effects of hypervigilance, leading to exacerbated sleep concerns in 14% of children (Aran et al., 2019). The authors of the review confirmed the inconsistency of the effects of cannabis in treating sleep disorders in children with ASD, recognizing a need for more rigorous controlled studies to determine whether these treatments are of any benefit (Agarwal et al., 2019).

Antihistamines

Once a frequent go-to choice for inducing sleep in children with ASD/DD, studies have demonstrated that diphenhydramine is no better than placebo at improving sleep, despite its sedative effects. With the exception of an older study from 1976, which studied typically developing children and demonstrated improvements in latency and night wakings, there is no support for the use of this agent (Hollway & Aman, 2011; Russo et al., 1976). Hydroxyzine, indicated for anxiety, pruritis, and pre- and post-operative sedation, has also been reported to be a safer alternative for inducing sleep, but its use has not been well studied (Brown, 2019; Bruni et al., 2019).

While pharmacologic treatment may play a role in treating some aspects of sleep disorders in children with ASD/DD, its role should be secondary to parent education, training, and behavioral interventions. With the possible exception of melatonin, pharmacologic treatment should be used as an adjunct when parent training and behavioral interventions do not work, or in the rare instances where there is a need for more rapid temporizing measures to be employed while nonpharmacologic treatments can be optimized. The literature on melatonin is quite substantive, but also stands in stark contrast to the relative paucity of evidence for other pharmacologic options. As previously mentioned, the agents discussed in this section should be used with caution and in a parsimonious manner, based on the presence of other co-occurring conditions that might also benefit from treatment with a given agent.

Clinical Pearls

It is the opinion of this author (MGM) that the effective management of sleep disturbances requires careful attention to comorbid mental health disorders. The optimal approach will require close coordination among members of the medical home team, including a sleep specialist and mental healthcare provider. Healthcare providers with special training in sleep are essential because the average amount of sleep education received by pediatric residents is only 4 hours (Mindell et al., 2013). Despite the limited attention to sleep medicine in the conventional residency curriculum, sleep health plays a critical role in overall health and ongoing physician education and family education are vital.

In this writer's experience, misguided efforts to treat sleep without a proper evaluation and treatment of comorbid mental health symptoms can lead to exacerbation of the sleep problems *and* the mental health issues (MGM). A closely coordinated approach is even more warranted when a patient presents with severe psychiatric symptoms such as suicidal ideation or psychosis (symptoms that were not in the scope of this chapter). Thus, although a patient and family members may have initially presented at our multidisciplinary sleep clinic seeking treatment for sleep issues, we often educate the patient and family about the risks of prematurely addressing sleep problems without proper treatment of comorbid mental health symptoms. At our clinic, the treatment of sleep issues often begins with a referral for mental health care.

When investigating sleep disturbance in a child with DD, it is important to review the entire sleep history to fully characterize the sleep disturbance. Routine surveillance may allow night wakings to be overlooked, so the clinical evaluation should include attention to overall sleep habits, bedtime routine, behavioral sleep associations, and possible medical pathology contributing to sleep disturbance. The approach must encourage good sleep hygiene with consistent sleep and wake times, limited screen time prior to bedtime, limiting daytime naps, and daily physical activity. In 2014, Knight et al. demonstrated the significant effect of a behavioral treatment plan in the treatment of night wakings (Knight & Johnson, 2014). Similar results using bedtime fading and positive routines were reported by Delmere et al. (Delemere & Dounavi, 2018).

The overall sleep needs of the individual child must be considered carefully. A child with ASD/DD may have a different total sleep duration than their neuro-typical peers. If a child is consistently sleeping and waking at the same time with little to no night wakings and is able to participate in therapies and activities of daily living without somnolence or significant behavioral disturbance, then the total sleep time may be sufficient. Often, we find that parents have an expectation for children with ASD/DD to sleep the same amount as their neurotypical peers, which leads to inappropriate expectations for the child. There are significant demands upon the caregiver of a child with DD, and so they may put children to bed at 7–8 pm while trying to accomplish other work or household tasks – when the child's more physiologic sleep need may be sleeping from 11 pm to 7 am. The issue of misaligned parental expectations is often most notable for adolescent patients, who have a tendency for delayed sleep phase. Reviewing the individual sleep needs of the family and the child can realign the expectations for sleep and decrease parental anxiety.

If sleep hygiene strategies are not successful, sleep aids should be tried. As described above, endogenous melatonin is less effective in children with ASD than their neurotypical peers, and supplemental melatonin can be effective in the treatment of sleep-onset delay and night wakings. Some previous studies have indicated the effectiveness of nontraditional therapies. A 2017 study by Narasingharao et al. showed that a structured yoga intervention was effective in decreasing night wakings (Narasingharao et al., 2017). Despite the existence of few double-blind, placebo-controlled studies, there is significant popular demand for these therapies. Physicians may want to explore the utility of local resources, such as yoga for children and age-appropriate mindfulness training.

Night wakings in children with ASD/DD can have profound effects on patients and caregivers. Future work should focus on the incidence and prevalence of night wakings as well as treatment strategies. Even simple changes to the pediatric residence curriculum may offer significant opportunity to better understand night wakings in this population. This population of patients may have limited expressive language or atypical methods for communicating pain and discomfort.

These children may also have more difficulty than their neuro-typical peers in tolerating the multiple leads and diagnostic equipment required for in-lab polysomnography in addition to common fears associated with hospital- or medical facility-based diagnostics. Children with DD may not tolerate traditional polysomnography. If they are evaluated in a sleep lab used primarily to evaluate adult patients or undergo home testing, they may not be accurately diagnosed because of differences in the scoring systems used for adults vs. children. The 2017 AASM Pediatric Task Force published a position paper stating that home sleep apnea testing is not recommended for the pediatric population and cited references that the recommendation was in part due to a lack of available tests with sufficient sensitivity and specificity.

Another limitation in a typical visit with a child with DD is the number of concerns parents have that need to be addressed during a single visit. There is not always enough time for parents to address all concerns in 30–60 minutes, especially when the child is anxious or exhibiting behavioral issues, or the parent is chronically sleep deprived and managing multiple medical concerns. This highlights the importance of good referral information and preparing the family for the visit and perhaps collecting information with questionnaires ahead of time.

In terms of pharmacological interventions, the adverse event profile for antipsychotic medications should preclude their use solely for the purpose of treating sleep in children with ASD/DD. Assuming a parsimonious approach to medication choice, the practitioner should choose this class of medication to treat sleep if there is a need to target other symptoms susceptible to improvement with the same agent. We have had success with using low-dose quetiapine at night in children and/or adolescents who present with cyclic mood changes, especially when other interventions have been unsuccessful. We have also used it as a bridge while behavioral interventions are employed in cases where sleep disruption of the child has had significant negative effects on the entire family system (JTM).

We use trazodone (starting dose 25 mg, titrating up if needed to 150 mg) in the adolescent when family education and interventional techniques have failed due to uncontrollable environmental or psychosocial barriers and when melatonin, clonidine, and other options have failed. In our experience, there is a tendency to build tolerance with trazodone. Thus, its use as a long-term agent is questionable. As a result, we often ask families to reserve its use for evenings where latency has exceeded a pre-specified duration. Once a normal sleep pattern is established, we attempt to titrate the dose down to discontinuation or the lowest effective dose (JTM).

Conclusion

The relationship among sleep, psychopathology, and ASD in children is complex. One must consider the interplay between ASD and reported mental health symptoms and its effect on sleep (Bartlett et al., 1985; Johnson, 1996; Köse et al., 2017). The clinician is strongly encouraged to screen for mental health symptoms that are common to children with ASD, which may not be at the forefront of the reporting patient or family's concerns.

References

Accardo, J. A., & Malow, B. A. (2014). *Sleep, epilepsy, and autism*. https://doi.org/10.1016/j.yebeh.2014.09.081

Adams, H. L., Matson, J. L., Cervantes, P. E., & Goldin, R. L. (2014). The relationship between autism symptom severity and sleep problems: Should bidirectionality be considered? *Research in Autism Spectrum Disorder, 8*, 193–199.

Agarwal, R., Burke, S. L., & Maddux, M. (2019). Current state of evidence of cannabis utilization for treatment of autism spectrum disorders. *BMC Psychiatry, 19*(1), 328. https://doi.org/10.1186/s12888-019-2259-4

Alves, R., Resende, M., Skomro, R., Souza, F., & Reed, U. (2009). Sleep and neuromuscular disorders in children. *Sleep Medicine Reviews, 13*, 133–148.

Aman, M. G., Arnold, L. E., McDougle, C. J., Vitiello, B., Scahill, L., Davies, M., McCracken, J. T., Tierney, E., Nash, P. L., Posey, D. J., Chuang, S., Martin, A., Shah, B., Gonzalez, N. M., Swiezy, N. B., Ritz, L., Koenig, K., McGough, J., Ghuman, J. K., & Lindsay, R. L. (2005). Acute and long-term safety and tolerability of risperidone in children with autism. *Journal of Child and Adolescent Psychopharmacology, 15*(6), 869–884. https://doi.org/10.1089/cap.2005.15.869

Andersen, I. M., Kaczmarska, J. A., McGrew, S. G., & Malow, B. A. (2008). Melatonin for insomnia in children with autism spectrum disorders. *Journal of Child Neurology, 23*(5), 482–485. https://doi.org/10.1177/0883073807309783

Aran, A., Cassuto, H., Lubotzky, A., Wattad, N., & Hazan, E. (2019). Brief report: Cannabidiol-rich cannabis in children with autism spectrum disorder and severe behavioral problems – A retrospective feasibility study. *Journal of Autism and Developmental Disorders, 49*(3), 1284–1288. https://doi.org/10.1007/s10803-018-3808-2

Asano, M., Ishitobi, M., Kosaka, H., Hiratani, M., & Wada, Y. (2014). Ramelteon monotherapy for insomnia and impulsive behavior in high-functioning autistic disorder. *Journal of Clinical Psychopharmacology, 34*(3), 402–403. https://doi.org/10.1097/JCP.0000000000000137

American Psychiatric Association. (2013). *Diagnostic and statistical manual of mental disorders* (5th ed.). https://doi.org/10.1176/appi.books.9780890425596

Barchel, D., Stolar, O., De-Haan, T., Ziv-Baran, T., Saban, N., Or Fuchs, D., Koren, G., & Berkovitch, M. (2018). Oral cannabidiol use in children with autism spectrum disorder to treat related symptoms and co-morbidities. *Frontiers in Pharmacology, 9*, 1–5.

Barchel, D., Stolar, O., De-Haan, T., Ziv-Baran, T., Saban, N., Fuchs, D. O., Koren, G., & Berkovitch, M. (2019). Oral cannabidiol use in children with autism spectrum disorder to treat related symptoms and co-morbidities. *Frontiers in Pharmacology, 9*(January), 1521. https://doi.org/10.3389/fphar.2018.01521

Bartlett, L. B., Rooney, V., & Spedding, S. (1985). Nocturnal difficulties in a population of mentally handicapped children. *British Journal of Mental Subnormality, 31*, 54–59. https://doi.org/10.1179/bjms.1985.009

Bixler, E. O., Vgontzas, A. N., Lin, H. M., et al. (2009). Sleep disordered breathing in children in a general population sample: Prevalence and risk factors. *Sleep, 32*(6), 731–736.

Brand, S., Jossen, S., Holsboer-Trachsler, E., Puhse, U., & Gerber, M. (2015). Impact of aerobic exercise on sleep and motor skills in children with autism spectrum disorders – A pilot study. *Neuropsychiatric Disease and Treatment, 11*, 1911–1920.

Brown, L. W. (2019). In J. A. Accardo (Ed.), *Sleep in children with neurodevelopmental disabilities*. Springer International Publishing. https://doi.org/10.1007/978-3-319-98414-8

Browne, T. R. (1976). Clonazepam: A review of a new anticonvulsant drug. *Archives of Neurology, 33*(5), 326–332. https://doi.org/10.1001/archneur.1976.00500050012003

Bruni, O., Angriman, M., Melegari, M. G., & Ferri, R. (2019). Pharmacotherapeutic management of sleep disorders in children with neurodevelopmental disorders. *Expert Opinion on Pharmacotherapy*. https://doi.org/10.1080/14656566.2019.1674283

Camfferman, D., McEvoy, R., et al. (2008). Prader Willi syndrome and excessive daytime sleepiness. *Sleep Medicine Reviews, 12*(3), 65–75.

Carr, R., Wasdell, M. B., Hamilton, D., Weiss, M. D., Freeman, R. D., Tai, J., Rietveld, W. J., & Jan, J. E. (2007). Long-term effectiveness outcome of melatonin therapy in children with treatment-resistant circadian rhythm sleep disorders. *Journal of Pineal Research, 43*(4), 351–359. https://doi.org/10.1111/j.1600-079X.2007.00485

Cerezo, A. B., Leal, Á., Álvarez-Fernández, M. A., Hornedo-Ortega, R., Troncoso, A. M., & García-Parrilla, M. C. (2016). Quality control and determination of melatonin in food supplements. *Journal of Food Composition and Analysis, 45*, 80–86. https://doi.org/10.1016/j.jfca.2015.09.013

Chan, J., Edman, J. C., & Koltai, P. J. (2004). Obstructive sleep apnea in children. *American Family Physician, 69*(5), 1147–1154.

Chen, M. H., Wei, H. T., Chen, L. C., Su, T. P., Bai, Y. M., Hsu, J. W., Huang, K. L., Chang, W. H., Chen, T. J., & Chen, Y. S. (2015). Autistic spectrum disorder, attention deficit hyperactivity disorder, and psychiatric comorbidities: A nationwide study. *Research in Autism Spectrum Disorders, 10*, 1–6.

Coleman, D. M., Adams, J. B., Anderson, A. L., & Frye, R. E. (2019). Rating of the effectiveness of 26 psychiatric and seizure medications for autism spectrum disorder: Results of a national survey. *Journal of Child and Adolescent Psychopharmacology, 29*(2), 107–123. https://doi.org/10.1089/cap.2018.0121

Coppola, G., Iervolino, G., Mastrosimone, M., La Torre, G., Ruiu, F., & Pascotto, A. (2004). Melatonin in wake-sleep disorders in children, adolescents and young adults with mental retardation with or without epilepsy: A double-blind, cross-over, placebo-controlled trial. *Brain and Development, 26*(6), 373–376. https://doi.org/10.1016/j.braindev.2003.09.008

Cortese, S., Wang, F., Angriman, M., Masi, G., & Bruni, O. (2020). Sleep disorders in children and adolescents with autism spectrum disorder: Diagnosis, epidemiology, and management. *CNS Drugs, 34*(4), 415–423. https://doi.org/10.1007/s40263-020-00710-y

Cortesi, F., Giannotti, F., Ivanenko, A., & Johnson, K. (2010). Sleep in children with autistic spectrum disorder. *Sleep Medicine, 11*(7), 659–664.

Cortesi, F., Giannotti, F., Sebastiani, T., Panunzi, S., & Valente, D. (2012). Controlled-release melatonin, singly and combined with cognitive behavioural therapy, for persistent insomnia in children with autism spectrum disorders: a randomized placebo-controlled trial. *Journal of Sleep Research, 21*, 700–709.

Cremone-Caira, A., Buirkle, J., Gilbert, R., Nayudu, N., & Faja, S. (2019). Relations between caregiver-report of sleep and executive function problems in children with autism spectrum disorder and attention-deficit/hyperactivity disorder. *Research in Developmental Disabilities, 94*, 1034–1064.

Crowley, S. J., et al. (2018). An update on adolescent sleep: New evidence informing the perfect storm model. *Journal of Adolescence, 67*, 55–65. https://doi.org/10.1016/j.adolescence.2018.06.001

Cuomo, B. M., Vaz, S., Lee, E. A. L., Thompson, C., Rogerson, J. M., & Falkmer, T. (2017). Effectiveness of sleep-based interventions for children with autism spectrum disorder: A meta-synthesis. *Pharmacotherapy: The Journal of Human Pharmacology and Drug Therapy, 37*(5), 555–578. https://doi.org/10.1002/phar.1920

Dahl, R. E. (1995). Sleep in behavioral and emotional disorders. In M. K. R. Ferber (Ed.), *Principles and practice of sleep medicine in the child* (2nd ed., pp. 147–154). W.B. Saunders.

Delemere, E., & Dounavi, K. (2018). Parent-implemented bedtime fading and positive routines for children with autism spectrum disorders. *Journal of Autism and Developmental Disorders, 48*, 1002–1019.

Demetriou, E. A., Lampit, A., Quintana, D. S., Naismith, S. L., Song, Y. J. C., Pye, J. E., Hickie, I., & Guastella, A. J. (2018). Autism spectrum disorders: A meta-analysis of executive function. *Molecular Psychiatry, 23*, 1198–1204.

Devnani, P. A., & Hegde, A. U. (2015). Autism and sleep disorders. *Journal of Pediatric Neurosciences, 10*(4), 304–307. https://doi.org/10.4103/1817-1745.174438

Erland, L. A. E., & Saxena, P. K. (2017). Melatonin natural health products and supplements: Presence of serotonin and significant variability of melatonin content. *Journal of Clinical Sleep Medicine, 13*(02), 275–281. https://doi.org/10.5664/jcsm.6462

Esposito, D., Belli, A., Ferri, R., & Bruni, O. (2020). Sleeping without prescription: Management of sleep disorders in children with autism with non-pharmacological interventions and over-the-counter treatments. *Brain Sciences, 10*(7), 1–30. https://doi.org/10.3390/brainsci10070441

Fujiwara, Y., Arakawa, T., & Fass, R. (2012). Gastroesophageal reflux disease and sleep disturbances. *Journal of Gastroenterology, 47*, 760–769.

Gail Williams, P., Sears, L. L., & Allard, A. (2004). Sleep problems in children with autism. *Journal of Sleep Research, 13*, 265–268.

Giannotti, F., Cortesi, F., Cerquiglini, A., Vognoni, C., & Valente, D. (2011). Sleep in children with autism with and without autistic regression. *Journal of Sleep Research, 20*(2), 338–347. https://doi.org/10.1111/j.1365-2869.2010.00882.x. Epub 2010 Aug 16. PMID: 21518065.

Goldberg, S., Shatz, A., Picard, E., Wexler, I., Schwartz, S., Swed, E., Zilber, L., & Kerem, E. (2005). Endoscopic findings in children with obstructive sleep apnea: Effects of age and hypotonia. *Pediatric Pulmonology, 40*, 205–210.

Goldman, S. E., Richdale, A. L., Clemons, T., et al. (2012). Parental sleep concerns in autism spectrum disorders: Variations from childhood to adolescence. *Journal of Autism and Developmental Disorders, 42*, 531–538.

Goldman, S. E., Adkins, K. W., Calcutt, M. W., Carter, M. D., Goodpaste, R. L., Wang, L., Shi, Y., Burgess, H. J., Hachey, D. L., & Malow, B. A. (2014). Melatonin in children with autism Spectrum disorders: Endogenous and pharmacokinetic profiles in relation to sleep. *Journal of Autism and Developmental Disorders, 44*(10), 2525–2535. https://doi.org/10.1007/s10803-014-2123-9

Golubchik, P., Sever, J., & Weizman, A. (2011). Low-dose quetiapine for adolescents with autistic spectrum disorder and aggressive behavior. *Clinical Neuropharmacology, 34*(6), 216–219. https://doi.org/10.1097/WNF.0b013e31823349ac

Gregory, A. M., & O'Connor, T. G. (2002). Sleep problems in childhood: A longitudinal study of developmental change and association with behavioral problems. *Journal of the American Academy of Child and Adolescent Psychiatry, 41*, 964–971.

Gregory, A. M., Rijsdijk, F. V., Dahl, R. E., McGuffin, P., & Eley, T. C. (2006). Associations between sleep problems, anxiety, and depression in twins at 8 years of age. *Pediatrics, 118*, 1124–1132.

Gringras, P., Gamble, C., Jones, A. P., Wiggs, L., Williamson, P. R., Sutcliffe, A., Montgomery, P., Whitehouse, W. P., Choonara, I., Allport, T., Edmond, A., & Appleton, R. (2012). Melatonin for sleep problems in children with neurodevelopmental disorders: Randomised double masked placebo controlled trial. *BMJ (Online), 345*, 7882. https://doi.org/10.1136/bmj.e6664

Gringras, P., Nir, T., Breddy, J., Frydman-Marom, A., & Findling, R. L. (2017). Efficacy and safety of pediatric prolonged-release melatonin for insomnia in children with autism spectrum disorder. *Journal of the American Academy of Child and Adolescent Psychiatry, 56*(11), 948–957.e944. https://doi.org/10.1016/j.jaac.2017.09.414

Gunes, S., Ekinci, O., Feyzioglu, A., Ekinci, N., & Kalinli, M. (2019). Sleep problems in children with autism spectrum disorder: Clinical correlates and the impact of attention deficit hyperactivity disorder. *Neuropsychiatric Disease and Treatment, 15*, 763–771.

Hatta, K., Kishi, Y., & Wada, K. (2015). Ramelteon for delirium in hospitalized patients. *JAMA, 314*(10), 1071–1072.

Hollway, J. A., & Aman, M. G. (2011). *Pharmacological treatment of sleep disturbance in developmental disabilities: A review of the literature* (Vol. 32). Pergamon. https://doi.org/10.1016/j.ridd.2010.12.035

Hollway, J. A., Aman, M. G., & Butter, E. (2013). Correlates and risk markers for sleep disturbance in participants of the autism treatment network. *Journal of Autism and Developmental Disorders, 43*, 2830–2843.

Honomichl, R. D., Goodlin-Jones, B. L., Burnham, M., Gaylor, E., & Anders, T. F. (2002). Sleep patterns of children with pervasive developmental disorders. *Journal of Autism and Developmental Disorders, 32*(6), 553–561.

Horn, J. R., & Hansten, P. D. (2007). Get to know an enzyme: CYP1A2. *Pharmacy Times*. https://www.pharmacytimes.com/publications/issue/2007/2007-11/2007-11-8279

Ingrassia, A., & Turk, J. (2005). The use of clonidine for severe and intractable sleep problems in children with neurodevelopmental disorders: A case series. *European Child & Adolescent Psychiatry, 14*, 34–40. https://doi.org/10.1007/s00787-005-0424-4

Jain, S. V., & Glauser, T. A. (2014). Effects of epilepsy treatments on sleep architecture and daytime sleepiness: An evidence-based review of objective sleep metrics. *Epilepsia, 55*(1), 26–37. https://doi.org/10.1111/epi.12478

Jensen, P. S., Martin, D., Cantwell, D. P. (1997). Comorbidity in ADHD: implications for research, practice, and DSM-V. *Journal of the American Academy of Child and Adolescent Psychiatry, 36*(8), 1065–1079. https://doi.org/10.1097/00004583-199708000-00014. PMID: 9256586.

Johnson, C. R. (1996). *Sleep problems in children with mental retardation and autism* (Vol. 5). W.B. Saunders. https://doi.org/10.1016/s1056-4993(18)30355-9

Joshi, G., Petty, C., Wozniak, J., Henin, A., Fried, R., Galdo, M., Kotarski, M., Walls, S., & Biederman, J. (2010). The heavy burden of psychiatric comorbidity in youth with autism spectrum disorders: A large comparative study of a psychiatrically referred population. *Journal of Autism and Developmental Disorders, 40*(11), 1361–1370.

Kato, K., Hirai, K., Nishiyama, K., Uchikawa, O., Fukatsu, K., Ohkawa, S., Kawamata, Y., Hinuma, S., & Miyamoto, M. (2005). Neurochemical properties of ramelteon. *Neuropharmacology, 48*(2), 301–310.

Kawabe, K., Horiuchi, F., Oka, Y., & Ueno, S. (2014). The melatonin receptor agonist ramelteon effectively treats insomnia and behavioral symptoms in autistic disorder. *Case Reports in Psychiatry*, 1–5. https://doi.org/10.1155/2014/561071

Kennaway, D. J. (2000). Melatonin and development: Physiology and pharmacology. *Seminars in Perinatology, 24*(4), 258–266. https://doi.org/10.1053/sper.2000.8594

Kheirouri, S., Kalejahi, P., & Noorazar, S. (2016). Plasma levels of serotonin, gastrointestinal symptoms, and sleep problems in children with autism. *Turkish Journal of Medical Sciences, 46*(6), 1765–1772.

Knight, R., & Johnson, C. (2014). Using a behavioral treatment package for sleep problems in children with autism spectrum disorders. *Child & Family Behavior Therapy, 36*(3), 204–221.

Köse, S., Yılmaz, H., Ocakoğlu, F. T., & Özbaran, N. B. (2017). Sleep problems in children with autism spectrum disorder and intellectual disability without autism spectrum disorder. *Sleep Medicine, 40*, 69–77. https://doi.org/10.1016/j.sleep.2017.09.021

Kotagal, S., & Broomall, E. (2012). Sleep in children with ASD. *Pediatric Neurology, 47*, 242–251.

Kuester, G., Vergara, K., Ahumada, A., & Gazmuri, A. M. (2017). Oral cannabis extracts as a promising treatment for the core symptoms of autism spectrum disorder: Preliminary experience in Chilean patients. *Journal of the Neurological Sciences, 381*, 932–933. https://doi.org/10.1016/j.jns.2017.08.2623

Larson, A. M. (2012). Impact of pediatric epilepsy on sleep patterns and behaviors in children and parents. *Epilepsia, 53*(7), 1162–1169.

Leu, R. M. (2019). Melatonin. In *Sleep in children with neurodevelopmental disabilities* (p. 339). Springer International Publishing. https://doi.org/10.1007/978-3-319-98414-8_5

Lumeng, J. C., & Chervin, R. (2007). Epidemiology of pediatric obstructive sleep apnea. *Proceedings of the American Thoracic Society, 5*(2), 242–252.

Macleod, R. L., & Keen, D. V. (2014). Innovations in practice: "Off-label" clonidine: UK Paediatric and Child and Adolescent Psychiatry prescribing practice for sleep problems. *Child and Adolescent Mental Health, 19*(2), 147–150. https://doi.org/10.1111/camh.12032

Malhi, P., Kaur, A., Singhi, P., & Sankhyan, N. (2019). Sleep dysfunction and behavioral daytime problems in children with autism spectrum disorders: A comparative study. *Indian Journal of Pediatrics, 86*(1), 12–17.

Malow, B. A., Marzec, M. L., McGrew, S. G., Wang, L., Henderson, L. M., & Stone, W. (2006). Characterizing sleep in children with autism spectrum disorders: A multidimensional approach. *Sleep, 29*(12), 1563–1571.

Malow, B. A., Byars, K., Johnson, K., Weiss, S., Bernal, P., Goldman, S. E., Panzer, R., Coury, D. L., & Glaze, D. G. (2012). A practice pathway for the identification, evaluation, and management of insomnia in children and adolescents with autism spectrum disorders. *Pediatrics, 130*(2), S106–S124. https://doi.org/10.1542/peds.2012-0900I

Malow, B., Crowe, C., Henderson, L. et al. (2009). A sleep habits questionnaire for children with autism spectrum disorders. *Journal of Child Neurology, 24*(1), 19–24.

Maris, M., Verhulst, S., Wojciechowski, M., Van de Heyning, P., & Boudewyns, A. (2016). Prevalence of obstructive sleep apnea in children with down syndrome. *Sleep, 39*(3), 699–704.

Mayes, S. D., & Calhoun, S. L. (2009). Variables related to sleep problems in children with autism. *Research in Autism Spectrum Disorders, 3*, 931–941.

Mazurek, M. O., & Petroski, G. F. (2015). Sleep problems in children with autism spectrum disorder: Examining the contributions of sensory over-responsivity and anxiety. *Sleep Medicine, 16*(2), 270–279.

Mazurek, M., & Sohl, K. (2016). Sleep and behavioral problems in children with autism spectrum disorder. *Journal of Autism and Developmental Disorders, 46*, 1906–1915.

McArthur, A. J., & Budden, S. S. (2008). Sleep dysfunction in Rett syndrome: A trial of exogenous melatonin treatment. *Developmental Medicine & Child Neurology, 40*(3), 186–192. https://doi.org/10.1111/j.1469-8749.1998.tb15445

McCrae, C. S., Chan, W. S., Curtis, A. F., Deroche, C. B., Munoz, M., Takamatsu, S., Muckerman, J. E., Takahashi, N., McCann, D., McGovney, K., Sahota, P., & Mazurek, M. O. (2020). Cognitive behavioral treatment of insomnia in school-aged children with autism spectrum disorder: A pilot feasibility study. *Autism Research, 13*(1), 167–176.

Mindell, J. A., Bartle, A., Ahn, Y., et al. (2013). Sleep education in pediatric residency programs: A cross-cultural look. *BMC Research Notes, 6*, 130. https://doi.org/10.1186/1756-0500-6-130

Ming, X., Gordon, E., Kang, N., & Wagner, G. C. (2008). Use of clonidine in children with autism spectrum disorders. *Brain and Development, 30*(7), 454–460. https://doi.org/10.1016/j.braindev.2007.12.007

Narasingharao, K., Pradhan, B., & Navaneetham, J. (2017). Efficacy of structured yoga intervention for sleep, gastrointestinal and behaviour problems of ASD children: An exploratory study. *Journal of Clinical and Diagnostic Research, 11*(3), VC01–VC06.

Nita, D. A., & Weiss, S. K. (2019). Sleep and epilepsy. In *Sleep in children with neurodevelopmental disabilities* (pp. 227–240). Springer International Publishing. https://doi.org/10.1007/978-3-319-98414-8_20

Nixon, G. M., & Brouillette, R. T. (2002). Sleep and breathing in Prader-Willi syndrome. *Pediatric Pulmonology, 34*, 209–217.

Owens, J. A., Spirito, A., & McGuinn, M. (2000). The Children's Sleep Habits Questionnaire (CSHQ): Psychometric properties of a survey instrument for school-aged children. *Sleep-New York-, 23*(8), 1043–1052.

Owens, J. A., Rosen, C. L., Mindell, J. A., & Kirchner, H. L. (2010). Use of pharmacotherapy for insomnia in child psychiatry practice: A national survey. *Sleep Medicine, 11*(7), 692–700. https://doi.org/10.1016/j.sleep.2009.11.015

Parker, A., Beresford, B., Dawson, V., Elphick, H., Fairhurst, C., Hewitt, C., Scantlebury, A., Spiers, G., Thomas, M., Wright, K., & McDaid, C. (2019). Oral melatonin for non-respiratory sleep disturbance in children with neurodisabilities: Systematic review and meta-analyses. *Developmental Medicine & Child Neurology, 61*(8), 880–890. https://doi.org/10.1111/dmcn.14157

Polimeni, M. A., Richdale, A. L., & Francis, A. J. (2005). A survey of sleep problems in autism, Asperger's disorder and typically developing children. *Journal of Intellectual Disability Research, 49*(4), 260–268.

Politte, L. C., Scahill, L., Figueroa, J., McCracken, J. T., King, B., & McDougle, C. J. (2018). A randomized, placebo-controlled trial of extended-release guanfacine in children with autism spectrum disorder and ADHD symptoms: An analysis of secondary outcome measures. https://doi.org/10.1038/s41386-018-0039-3

Posey, D. J., Guenin, K. D., Kohn, A. E., Swiezy, N. B., & McDougle, C. J. (2001). A naturalistic open-label study of mirtazapine in autistic and other pervasive developmental disorders. *Journal of Child and Adolescent Psychopharmacology, 11*(3), 267–277. https://doi.org/10.1089/10445460152595586

Quera Salva, M. A., Hartley, S., Léger, D., & Dauvilliers, Y. A. (2017). Non-24-hour sleep-wake rhythm disorder in the totally blind: Diagnosis and management. *Frontiers in Neurology, 8*, 686.

Reynolds, A. (2019). Sleep-related movement disorders. In J. A. Accardo (Ed.), *Sleep in children with neurodevelopmental disabilities* (pp. 123–135). Springer.

Reynolds, K. C., Patriquin, M., Alfano, C. A., Loveland, K. A., & Pearson, D. A. (2017). Parent-reported problematic sleep behaviors in children with comorbid autism spectrum disorder and attention-deficit/hyperactivity disorder. *Research in Autism Spectrum Disorder, 39*, 20–32.

Richdale, A. L., Baker, E., Short, M., & Gradisar, M. (2014). The role of insomnia, pre-sleep arousal and psychopathology symptoms in daytime impairment in adolescents with high-functioning autism spectrum disorder. *Sleep Medicine, 15*(9), 1082–1088.

Robinson, A. A., & Malow, B. A. (2013). Gabapentin shows promise in treating refractory insomnia in children. *Journal of Child Neurology, 28*(12), 1618–1621. https://doi.org/10.1177/0883073812463069

Robinson-Shelton, A., & Malow, B. A. (2016). Sleep disturbances in neurodevelopmental disorders. In *Current psychiatry reports* (Vol. 18, pp. 1–8). Current Medicine Group LLC 1. https://doi.org/10.1007/s11920-015-0638-1

Rosen, G., & Brand, S. R. (2011). Sleep in children with cancer: Case review of 70 children evaluated in a comprehensive pediatric sleep center. *Support Care Cancer, 19*, 985–994.

Rugino, T. A. (2018). Effect on primary sleep disorders when children with ADHD are administered guanfacine extended release. *Journal of Attention Disorders, 22*(1), 14–24. https://doi.org/10.1177/1087054714554932

Russo, R. M., Gururaj, V. J., & Allen, J. E. (1976). The effectiveness of diphenhydramine HCI in pediatric sleep disorders. *The Journal of Clinical Pharmacology, 16*(5–6), 284–288. https://doi.org/10.1002/j.1552-4604.1976.tb02406

Santoro, S., Giacheti, C., et al. (2016). Correlations between behavior, memory, sleep-wake and melatonin in Williams Beuren syndrome. *Physiology & Behavior, 159*, 14–19.

Sarber, K. M., Howard, J., Dye, T. J., Pascoe, J. E., & Simakajornboon, N. (2019). Sleep-disordered breathing in pediatric patients with Rett syndrome. *Journal of Clinical Sleep Medicine, 15*(10), 1451–1457.

Schreck, K. A., & Richdale, A. L. (2020). Sleep problems, behavior, and psychopathology in autism: Interrelationships across the lifespan. *Current Opinion in Psychology, 34*, 105–111.

Simonoff, E., Pickles, A., Charman, T., Chandler, S., Loucas, T., & Baird, G. (2008). Psychiatric disorders in children with autism spectrum disorders: Prevalence, comorbidity, and associated factors in a population derived sample. *Journal of the American Academy of Child and Adolescent Psychiatry, 47*(8), 921–929.

Simpson, R., Oyekan, A., Ehsan, Z., & Ingram, D. G. (2018). Obstructive sleep apnea in patients with down syndrome: Current perspectives. *Nature and Science of Sleep, 10*, 287–293.

Sitnick, S. L., Goodlin-Jones, B., & Anders, T. F. (2008). The use of actigraphy to study sleep disorders in preschoolers: Some concerns about detection of nighttime awakenings. *Sleep, 31*(3), 395–401. https://doi.org/10.1093/sleep/31.3.395

Smalley, S. L., McGough, J., Moilanen, I. K., Loo, S. K., Taanila, A., Ebeling, H., Hurtig, T., Kaakinen, M., Humphrey, L. A., & McCracken, J. T. (2007). Prevalence and psychiatric comorbidity of attention-deficit/ hyperactivity disorder in an adolescent Finnish population. *Journal of the American Academy of Child and Adolescent Psychiatry, 46*(12), 1575–1583.

St. Louis, E. (2011). Sleep and epilepsy: Strange bedfellows no more. *Minerva Pneumologica, 50*(3), 159–176.

Stores G. (2014). Psychological disturbance associated with recurrent abnormal nocturnal events. *British Journal of Hospital Medicine (Lond), 75*(6), 337–341. https://doi.org/10.12968/hmed.2014.75.6.337. PMID: 25040410.

Stores, G. (2016). Multifactorial influences, including comorbidities, contributing to sleep disturbance in children with a neurodevelopmental disorder. *CNS Neuroscience & Therapeutics, 22*(11), 875–879. https://doi.org/10.1111/cns.12574

Sung, V., Hiscock, H., Sciberras, E., & Efron, D. (2008) Sleep problems in children with attention-deficit/hyperactivity disorder: prevalence and the effect on the child and family. *Archives of Pediatrics and Adolescent Medicine, 162*(4), 336–342. https://doi.org/10.1001/archpedi.162.4.336. PMID: 18391142

Taylor, M., Schreck, K., Mulick, J. (2012). Sleep disruption as a correlate to cognitive and adaptive behavior problems in autism spectrum disorders. *Research in Developmental Disabilities, 33*(5), 1408–1417.

Thomas, S., Lycett, K., Papadopoulos, N., Sciberras, E., & Rinehart, N. (2015). Exploring behavioral sleep problems in children with ADHD and comorbid autism spectrum disorder. *Journal of Attention Disorders*, 1–12.

Trickett, J., Heald, M., Oliver, C., et al. (2018). A cross-syndrome cohort comparison of sleep disturbance in children with Smith-Magenis syndrome, Angelman syndrome, autism spectrum disorder and tuberous sclerosis complex. *Journal of Neurodevelopmental Disorders, 10*(9), 1–14.

Tse, C. Y. A., Lee, H. P., Chan, K. S. K., Edgar, V. B., Wilkinson-Smith, A., & Lai, W. H. E. (2019). Examining the impact of physical activity on sleep quality and executive functions in children with autism spectrum disorder: A randomized controlled trial. *Autism, 23*(7), 1699–1710.

Tzischinsky, O., Meiri, G., Manelis, L., Bar-Sinai, A., Flusser, H., Michaelovski, A., Zivan, O., Ilan, M., Faroy, M., Menashe, I., & Dinstein, I. (2018). Sleep disturbances are associated with specific sensory sensitivities in children with autism. *Molecular Autism, 9*(22), 1–10.

Williams Buckley, A., Hirtz, D., Oskoui, M., Armstrong, M. J., Batra, A., Bridgemohan, C., Coury, D., Dawson, G., Donley, D., Findling, R. L., Gaughan, T., Gloss, D., Gronseth, G., Kessler, R., Merillat, S., Michelson, D., Owens, J., Pringsheim, T., Sikich, L., & Ashwal, S. (2019). Practice guideline: Treatment for insomnia and disrupted sleep behavior in children and adolescents with autism spectrum disorder. *Autism Speaks*. https://doi.org/10.1212/WNL.0000000000009033

Williams Buckley, A., Hirtz, D., Oskoui, M., Armstrong, M. J., Batra, A., Bridgemohan, C., Coury, D., Dawson, G., Donley, D., Findling, R. L., Gaughan, T., Gloss, D., Gronseth, G., Kessler, R., Merillat, S., Michelson, D., Owens, J., Pringsheim, T., Sikich, L., et al. (2020). Practice guideline: Treatment for insomnia and disrupted sleep behavior in children and adolescents with autism spectrum disorder: Report of the guideline development, dissemination, and implementation Subcommittee of the American Academy of Neurology. *Neurology, 94*(9), 392–404. https://doi.org/10.1212/WNL.0000000000009033

Wirojanan, J., Jacquemont, S., Diaz, R., Bacalman, S., Anders, T. F., Hagerman, R. J., & Goodlin-Jones, B. L. (2009). The efficacy of melatonin for sleep problems in children with autism, Fragile X syndrome, or autism and Fragile X syndrome. *Journal of Clinical Sleep Medicine, 5*(2), 145–150. https://doi.org/10.5664/jcsm.27443

Young, D., Nagarajan, L., de Klerk, N., Jacoby, P., Ellaway, C., & Leonard, H. (2007). Sleep problems in Rett syndrome. *Brain and Development, 29*(10), 609–616.

Youssef, J., Singh, K., Huntington, N., Becker, R., & Kothare, S. V. (2013). Relationship of serum ferritin levels to sleep fragmentation and periodic limb movements of sleep on polysomnography in autism spectrum disorders. *Pediatric Neurology, 49*(4), 274–278. https://doi.org/10.1016/j.pediatrneurol.2013.06.012

23 The Role of the SLP in the Evaluation and Treatment of Individuals with ASD and Other Neurodevelopmental Disorders

Sandra L. Hayes, Michael M. McGill, and Heather Anderson

Abstract

A delay in communication skills is often the first sign for autism spectrum disorder and other neurodevelopmental disorders in children. The speech-language pathologist is an integral part of the medical home model team and is often the initial provider to evaluate and identify these developmental delays. This chapter will discuss the role of the speech-language pathologist in the management of these children through young adulthood as illustrated by three case studies.

Keywords

Autism spectrum disorder · Neurodevelopmental disorders · Early intervention · Speech-language pathology · Medical home model · Developmental delay · Speech delay · Language delay

S. L. Hayes (✉) · M. M. McGill · H. Anderson
Louisiana State University Health Sciences Center, School of Allied Health, Shreveport, LA, USA
e-mail: sandra.hayes@lsuhs.edu

One week after Adam's second birthday, his parents took him for a wellness checkup. The pediatrician completed the checkup and asked about developmental milestones. Adam's parents expressed concern that he was not yet talking. Meanwhile, around the same time, Roderick's parents took him for a two-year-old wellness visit and had a similar conversation with his pediatrician. Both pediatricians made referrals for evaluation at an outpatient speech and hearing center. A third child, Billy, was also seen for a wellness visit. His parents noted that he seemed to be a picky eater. However, his weight was normal on the growth chart, and no overt signs of developmental delay were noted.

The speech-language pathologist (SLP) is a "first-line" provider, in that the SLP is tasked with identifying and characterizing delays in speech and language through an evaluation process that typically includes direct/formal evaluation, naturalistic observations of communication skills, parent interview, review of records, hearing screening, and screening of oral motor and feeding skills. SLPs provide evaluation and treatment in nine practice areas: articulation, language, fluency, voice, swallowing/feeding, hearing (screening/aural rehabilitation), modalities (non-speech modes of communication), social, and cognition. In cases where a young child is referred for an initial evaluation, the SLP will be particularly interested in language, cognition, pragmatics, articulation, and, if needed, swallowing/feeding and modalities. It is conventional for referring

P. McPherson (ed.), *Handbook of Treatment Planning for Children with Autism and Other Neurodevelopmental Disorders*, Autism and Child Psychopathology Series,
https://doi.org/10.1007/978-3-031-06120-2_23

providers, such as pediatricians, to characterize SLP services as "speech therapy;" however, from an SLP perspective, it is important to distinguish between speech and language. Language refers to auditory comprehension and verbal expression of spoken language. To evaluate language, the SLP will qualitatively and quantitatively assess the child's understanding and production of language in terms of form (syntax, morphology, phonology), content (semantics), and pragmatics (functional communication). Speech refers to the physical act of combining sounds into syllables, words, and sentences. Although a 2-year-old child is not expected to produce the full range of sounds, the SLP will evaluate speech by comparing his/her articulation (sound production) to expectations for his/her age. If a child is producing connected speech, the SLP will also evaluate fluency, which refers to rhythm and flow of speech. Depending on the case history, the SLP might also assess feeding and swallowing issues, such as food sensitivities, choking, or manipulation of food (ASHA, 2016).

The American Speech-Language-Hearing Association (ASHA) website lists major speech and language milestones from birth to age five at www.asha.org/public/speech/development/45/. Parents are also encouraged to refer to guidelines provided by the Centers for Disease Control (CDC) for more information concerning milestones for multiple aspects of development, including speech, social/emotional, motor, and cognition. These guidelines are available at www.cdc.gov/ncbddd/actearly/milestones.

The SLP will approach the evaluation with the knowledge that there are multiple possible etiologies for delayed communication skills development. In some cases, the child will be identified as having a delay in speech and/or language development without signs of delay in other developmental domains, such as cognition/play, social/emotional, fine and gross motor, and adaptive behavior. In other cases, speech/language delay may be the first and/or clearest presenting sign of an underlying condition, of which speech/language delay is one characteristic. Examples of underlying conditions include hearing loss, genetic syndromes, and neurodevelopmental disorders such as autism spectrum disorder (ASD). Upon identification of a speech/language delay or disorder, regardless of underlying etiology, the SLP will initiate treatment, which will involve some combination of referral to other professionals or agencies, direct therapy, family education, and coordination of care.

Based on the SLP scope of practice, ASHA has established guidelines for engagement by the SLP in a variety of assessment and treatment practices specific to individuals with autism spectrum disorder (ASD) (ASHA, n.d.). The Medical Home Model (MHM), developed by the US Department of Health and Human Services (USDHSS), Agency for Healthcare Research and Quality, focuses on five core elements of an ideal healthcare model (USDHSS, 2019), which also helps guide SLPs when providing comprehensive and continuous care to patients. These guidelines and elements are listed in Tables 23.1 and 23.2.

Early intervention by a speech-language pathologist is critical because communication difficulty is a central feature of ASD (ASHA, n.d.; Filipek et al., 1999), and the association of early intervention with positive long-term outcomes is well documented in the literature (Campbell & Ramey, 1994; Itzchak & Zachor, 2011; Majnemer, 1998). Although making a diagnosis of ASD is within the SLP's scope of practice (ASHA, 2016), best practice for a diagnosis of ASD is to involve assessment by an interdisciplinary team. Ideally, an interdisciplinary team includes SLP, psychologist, pediatrician, and depending on needs and availability, might also include geneticist, educational diagnostician, occupational therapist, neurologist, and teachers. The scope of practice in SLP does not include diagnosis of genetic/congenital syndromes, such as Down syndrome or fetal alcohol syndrome; however, SLPs are qualified to recognize potential signs of such disorders and refer the family to an appropriate specialist, such as a geneticist or neurologist, if indicated.

Using three illustrative cases, this chapter will explain the role of the SLP in evaluation and treatment services for individuals with ASD and other neurodevelopmental disorders. While the focus will be ASD, the information will also apply to

Table 23.1 Appropriate roles for speech language pathologists serving individuals with ASD

Providing information to individuals and groups known to be at risk for ASD, to their family members, and to individuals working with those at risk
Educating other professionals on the needs of persons with ASD and the role of SLPs in diagnosing and managing ASD
Screening individuals who present with language and communication difficulties and determining the need for further assessment and/or referral for other services
Conducting a culturally and linguistically relevant comprehensive assessment of language and communication, including social communication skills
Assessing the need and requirements for using augmentative and alternative communication (AAC) devices as a mode of communication
Assessing and treating feeding issues if present (e.g., patterns of food acceptance or rejection based on food texture; consumption of a limited variety of foods)
Diagnosing the presence or absence of ASD (typically as part of a diagnostic team or in other interdisciplinary collaborations)
Referring to other professionals to rule out other conditions, determine etiology, and facilitate access to comprehensive services
Making decisions about the management of ASD
Participating as a member of the school planning team (e.g., a team whose members include teachers, special educators, counselors, psychologists) to determine appropriate educational services
Developing speech and language goals focused on social language and literacy and assisting students with self-regulatory and social interactive functions so that they can participate in the mainstream curriculum as much as possible
Providing treatment, documenting progress, and determining appropriate dismissal criteria
Providing training in the use of AAC devices to persons with ASD as well as their families, caregivers, and educators
Counseling persons with ASD and their families regarding communication-related issues and providing education aimed at preventing further complications related to ASD (see ASHA's Practice Portal page on Counseling for Professional Service Delivery)
Consulting and collaborating with other professionals, family members, caregivers, and others to facilitate program development and to provide supervision, evaluation, and/or expert testimony, as appropriate (see ASHA's resources on interprofessional education/interprofessional practice [IPE/IPP], collaboration and teaming, and person- and family-centered care)

(continued)

Table 23.1 (continued)

Partnering with families in assessment and intervention with individuals who have ASD (see ASHA's resource on person- and family-centered care)
Providing parent education so that families may continue to provide intervention beyond the sessions
Remaining informed of research in the area of ASD and helping advance the knowledge base related to the nature and treatment of ASD
Advocating for individuals with ASD and their families at the local, state, and national levels
Serving as an integral member of an interdisciplinary team working with individuals with ASD and their families and, when appropriate, considering transition planning (see ASHA's resources on interprofessional education/interprofessional practice [IPE/IPP] and collaboration and teaming)
Providing quality control and risk management

American Speech-Language-Hearing Association (n.d.). *Autism* (Practice Portal). Retrieved month, day, year, from www.asha.org/Practice-Portal/Clinical-Topics/Autism/

other types of neurodevelopmental disorders such as Down syndrome, fetal alcohol syndrome (FAS), cerebral palsy (CP), global developmental delay (i.e., significant delays in multiple developmental domains), prematurity, hearing impairment, and environmental/social factors.

Evaluation and Treatment/Intervention Planning

Age Birth – Three

Evaluation and intervention services for children of age birth to three are offered through private or university-based clinics, private practitioners, or publicly funded programs. In addition to private agencies or clinics, early intervention services are offered in each state through a state- and federally funded program, under Part C of the Individuals with Disabilities Education Act (IDEA), for children of age birth–three with identified developmental and/or medical conditions. Each state has a different name for this program; for the purpose of discussion in this chapter, such programs will be referred to as Early Steps, the name used in Louisiana. Anyone concerned about a very young child's development can make a referral through a regional cen-

Table 23.2 Medical Home Model

Providing comprehensive and continuous medical care to patients
Comprehensive care: Meets patients' physical and mental health needs Includes prevention and wellness, and acute care and chronic care Team approach: physician, nurses, physician assistants, pharmacists, nutritionists, social workers, SLPs, occupational therapists, educators, etc.
Patient-centered: Relationship based Partnering with patients and families Respecting unique needs, cultures, values, and preferences
Coordinated care: Coordinated care across all elements of broad health care system Clear and open communication among patients and families, medical home, and all team members
Accessible services: Shorter wait times for urgent needs; enhanced in-person hours; 24-hour access to a team member; explore all options for service delivery (telehealth)
Quality and safety: Uses evidence-based practices and clinical decision-support tools to guide shared decision-making with patients and families Measures improvement and responsive to patient/family experiences and satisfaction Publicly shares robust quality and safety data and improvement activities

Adapted from US Department of Health and Human Services, AHRQ PCMH Patient-Centered Medical Home Resource Center Defining the PCMH https://pcmh.ahrq.gov/page/defining-pcmh for medical home model Retrieved 12/17/2019

tralized entry point known as the System Point of Entry (SPOE) for Early Steps. This referral initiates the process of evaluation to determine eligibility for early intervention services. The child's parents are then contacted by Early Steps personnel to begin the intake process and arrange initial screenings (Louisiana Department of Health, n.d.). The CDC website cdc.gov/ncbddd/actearly/parents/states.html provides links to each state's early intervention program.

Following Adam's and Roderick's referrals for outpatient evaluations, they were each seen by an SLP for an evaluation. Because Billy did not exhibit overt signs of developmental delay, he was not referred by his pediatrician for an evaluation at this point. The SLP who evaluated Adam administered the cognitive, social, and language subtests of the *Developmental Assessment of Young Children – 2nd Edition* (DAYC-2) (Voress & Maddox, 2012) and engaged with him in informal play scenarios. Adam was initially reluctant to participate and made limited eye contact, but eventually began smiling, making eye contact, and interacting with the SLP. Adam's mother reported that at home, he produced only a few words, and preferred to play while in his highchair, typically lining up objects in a row. During the evaluation, Adam initially lined up objects, but eventually demonstrated more appropriate use of objects. He also produced two words, occasionally gestured to make requests, and imitated several words such as "bubble" and "more."

The SLP who evaluated Roderick also administered subtests from the DAYC-2. Roderick showed limited eye contact with the SLP throughout the evaluation. However, he demonstrated strengths in some skills, such as stacking blocks and pointing to letters and numbers named by his mother. Roderick explored by dumping toys onto the floor. He kicked a ball and rolled a car but did not demonstrate any other diversity in his play skills, such as pretend play. Roderick imitatively counted to three by rote when prompted by his mother but did not produce any other imitative or spontaneous words or gestures. Roderick's mother also reported a history of feeding difficulties, which began with troubled breastfeeding. To address this concern, the SLP performed a clinical evaluation of swallowing. At the SLP's request, Roderick's mother brought a favorite snack and drink from home. No clinical signs of chewing or swallowing difficulty were observed. However, Roderick's mother reported that he refused many foods that she offered. For example, his fruit and vegetable intake was limited to soft-cooked or pureed apples, pears, squash and bananas, and he did not eat any meat except chicken nuggets. Roderick refused other snacks and beverages offered by the SLP. Scores on all three subtests for both children were approximately 1 to 1.5 standard deviations below the mean, falling in the mildly to moderately delayed range.

Roderick's history of feeding difficulty and the SLP's evaluation illustrate two important points concerning the SLP's role in treatment planning to address feeding problems. First, clinical and instrumental evaluation and treatment of pediatric dysphagia are part of the SLP scope of practice. Second, feeding or swallowing problems, such as failure to thrive, limited food intake or repertoire, or more overt chewing and swallowing problems, often occur in the presence of underlying conditions such as global delay, motor deficits, or neurodevelopmental disorders. When an SLP encounters feeding difficulties in a child who does not have a known medical diagnosis that would account for the difficulty, a key part of the SLP's treatment plan is referral and sharing of information (Arvedson & Brodsky, 2002; Ledford & Gast, 2006).

Another important finding that may arise during the speech-language evaluation is regression. Regression of skills at any age is great cause for concern and should be explored thoroughly. ASD is reported to occur in a bimodal distribution (Lord et al., 2004; Ozonoff et al., 2018). Parents of many children diagnosed with ASD report noticing subtle "signs" of ASD very early in their child's life such as failure to cuddle and bond, poor feeding skills, lack of eye contact, and overall flat affect. Other parents, however, report that their child was developing in a typical manner with an animated affect, reciprocal social smile, appropriate eye contact, easily socially engaged, and using several words along with simple gestures for a variety of communicative purposes, and suddenly all communication and social interactions stopped or decreased dramatically. This type of significant regression, which is different from the mild, temporary regression associated with life stressors such as a new sibling or a move, usually happens after the first 18 months to 2 years of life and has been inadvertently and falsely linked to childhood vaccinations. The most well known of these cases is the Wakefield article published in the *Lancet* 1998, which was eventually retracted (Eggertson, 2010). Other than ASD, neurological conditions such as childhood disintegrative disorder (CDD) or a lesion or tumor on the brain can be linked to regression in developmental skills; therefore, immediate referral to a neurologist is warranted.

Quantitative information (i.e., test scores) obtained from speech and language testing was similar for Roderick and Adam. However, the difference between their communication skills was found in the qualitative information. Both children showed some degree of "soft" signs of autism spectrum disorder, specifically in the atypical nature of their pragmatic language skills and play skills. When modeling and support were provided during the qualitative evaluation, important differences were noted. While Adam eventually began interacting and making eye contact with the examiner and imitating more varied play skills, Roderick's lack of eye contact and limited play skills remained consistent throughout the evaluation. Based on "soft signs" and parents' concerns, the initial treatment plan included a recommendation for further ASD screening for both children, along with the developmental evaluation necessary to qualify the boys for federally mandated early intervention programs. This scenario is an example of how the speech-language evaluation often has a dual purpose: characterizing communication skills and serving as part of a broader screening for neurodevelopmental and genetic disorders.

The route by which a child may receive further ASD screening may vary. The screening tool, *Baby and Infant Screen for Children with Autism Traits* (BISCUIT-Part 1) (Matson et al., 2007), is offered for all children enrolled in Early Steps in Louisiana (Louisiana Department of Health, n.d.). For children who are initially evaluated by SLPs in outpatient settings, the SLP may choose to refer the child back to the pediatrician, who may complete the screener *Modified Checklist for Autism in Toddlers, Revised with Follow-Up* (M-CHAT-RF) (Robins et al., 2009).

In the cases of Adam and Roderick, both failed the initial autism screening (BISCUIT and/or M-CHAT-RF) and were referred for completion of ASD interdisciplinary evaluation. Both were administered the *Autism Diagnostic Observation Schedule –2nd Edition* (ADOS-2) *(Module 1)* (Lord et al., 2012) along with the Autism Diagnostic Interview-Revised (ADI-R) (Rutter et al., 2008), during the formal interdisciplinary evaluation. Results of the interdisciplinary evaluation revealed that Roderick met criteria for ASD diagnosis, while Adam did not. This illustrates the significant differences in underlying etiology that can be present when a child exhibits late speech/language milestones. For the parents of both children, communication was the first concern identified; however, only one child was ultimately diagnosed with ASD. A similar process may occur in cases of other suspected neurodevelopmental conditions. While certain genetic syndromes, such as Down Syndrome, are more immediately identifiable at birth due to hallmark physical features, other conditions may not be suspected or identified until the child fails to meet milestones because the disability is "invisible," occurring without obvious signs or physical features. For example, a child with oral-verbal apraxia or mild static encephalopathy may present with more subtle "soft" neurological signs, such as slow oral motor development, slow or picky eating, or clumsy movements. While this may not initially raise a great deal of concern, when combined with subsequent failure to develop speech, concern increases and may lead to initiation of the referral process for a speech/language evaluation.

ASHA's Roles and Responsibilities of Speech Language Pathologists in Early Intervention: Guidelines (2008) serves as a roadmap for speech-language pathologists who are competent in providing services for infants and toddlers with communication needs. These guiding principles mirror MHM principles and state as follows: (1) services are family centered and culturally and linguistically responsive; (2) services are developmentally supportive and promote children's participation in their natural environments; (3) services are comprehensive, coordinated, and team-based; and (4) services are based on the highest quality evidence that is available.

Given the scenario of Roderick, Adam, and Billy, Roderick and Adam would qualify for speech language services during this time (birth to 3 years); however, Billy would likely not be identified yet since his parents were not expressing any communication concerns to his pediatrician and no notable delays in developmental milestones were identified by the pediatrician or his parents. The model of service delivery for Adam and Roderick may vary, but both should meet the standards and guidelines set forth by ASHA, as discussed earlier. While both toddlers should receive services delivered in the natural environment, Roderick's services may be delivered in a special education preschool classroom setting, such as an early head start classroom, while Adam's speech-language therapy services may be provided in his home environment. This is a shift from the traditional, clinical-based model of delivering speech/language services to a young child to a model that promotes speech language services delivered in the natural environment.

This model of providing services to young children in their natural environment also promotes inclusive practices and focuses on functional communication with the child and their various caregivers (parents, teachers, extended family members). While the SLP may initially be the primary agent delivering communication services, the communication goals are shared by all stakeholders, which increases the generalization of the various skills being taught and learned. This prioritizes the family as the primary stakeholder leading the way in setting goals, teaching, advocacy, etc.

Strategies for teaching and promoting increased communication skills in young children usually are considered to be (1) responsive interactions, (2) directive interactions, or (3) blended. (ASHA, 2008). Adam would be an ideal candidate for responsive interactions. This method involves following the child's lead and responding to the child's verbal and non-verbal communication with natural consequences.

> For example, if Adam were playing with a red truck and uttered "tuh," the adult would acknowledge Adam's verbal approximation, point to the truck and model the word "truck" and then may expand Adam's language by saying "Red truck. Adam is rolling the red truck" while playing with him using another red truck. At this point, the adult could also use parallel talk to "talk about" what he or she was doing with the truck, such as "I'm rolling the red truck over the blocks." The adult will also continue to try to expand language within the context of Adam's play interests or naturally occurring activities (i.e., eating lunch, playing in the sand, looking at a turtle, taking a bath, playing on the swing set, etc.).

Natural activities should not include screen time as this should be limited in this age group. The American Academy of Pediatrics (AAP) guidelines discourages screen time for children under age 2 and limits "screen time" to **2 hours a day** for children over age 2 (AAP, 2016).

Directive interaction strategies may benefit Roderick at this point in his communication journey. While directive interaction should still occur in the natural environment, the environment may be more structured with fewer choices offered to the child. Directive approaches are more behaviorally based and include strategies such as prompting, reinforcement, shaping, and fading. Depending on a myriad of factors, some children proceed along this path quickly and move to higher-level communication goals (and methods), while others do not and may never move beyond a certain level of communication.

For Roderick, one goal set by the parent may be for him to "ask for" or make a request. To start, a highly motivating reinforcer must be identified, such as a favorite snack, drink, or toy. In this case, Roderick loves pretzels but does not have an effective method of independently requesting them. Roderick will reportedly often cry or "throw a fit" in order to get pretzels and parents usually have to guess what he wants. After identifying a motivator, the SLP would then help Roderick's parents shape Roderick's "requesting" behavior from a cry to making a clear choice using a representative symbol (i.e., a photograph) to ideally using the words "I want pretzel" to make this request. This may start by Roderick making a clear choice between a food identified as a foil (something that he does not like) and a motivator (the pretzel that he likes). Once Roderick is making clear choices, this behavior would be faded and a new behavior would replace it. This new behavior may be Roderick choosing among a variety of photographs of food to choose the desired food and giving the photograph to his mother (request). Once this skill is mastered, Roderick's behavior may be shaped to include using verbal requests (words) to get his needs met.

Ideally, Roderick will move to a level of communication that involves a more directive approach for new targets and goals but allows a more responsive/natural approach for generalization of skills that he has mastered in a more structured setting. This blended approach will allow Roderick to move along the communication continuum toward more natural and functional speech and language. Young children often change rapidly and it is imperative that SLPs monitor progress and continually revise and modify communication intervention plans with the input from the family (ASHA, 2008).

Age 3–5

Age three is the time when children who are already receiving SLP services, whether through Early Steps or another agency, may enter the public school system after undergoing an evaluation to determine eligibility, often referred to as a "turning three" evaluation. Because Adam and Roderick were receiving Early Step services, they would undergo the "turning three" evaluation; however, other children may be initially evaluated for communication disorders or other developmental delays after entering preschool as the differences between their skills and those of same-age peers become more apparent.

An evaluation provided through the local school system will determine services based on identified exceptionalities, as defined by Department of Education (DOE) guidelines (Louisiana State Board of Elementary and Secondary Education, 2017), rather than medical diagnoses, so it is possible for a child to have both a medical diagnosis and an identified exceptionality. For example, in addition to the medical diagnosis of ASD, as identified by the DSM-5 and ICD-10, a child may also be identified with the exceptionality of autism according to Pupil Appraisal policies, but only if the child meets the Bulletin 1508 criteria for autism spectrum (Louisiana State Board of Elementary and Education, 2017). Based on the school system's evaluations, Adam met criteria for an exceptionality of developmental delay. Roderick met the Bulletin 1508 criteria for the exceptionality of Autism.

The notion of impact on educational performance is key in understanding how the evaluation process determines eligibility for services in the school system versus other environments. To receive services through the school system, evaluation results must show that the disability is both significant and likely to impact educational performance and depends on how the scope of education is interpreted by individual districts or states (Louisiana State Board of Elementary and Secondary Education, 2017). The fact that communication disorders, particularly ASD, may impact a child's educational experience will be considered by the evaluation team when determining if the child qualifies for speech/language therapy. Assessment of the impact of a child's disability on educational performance should not be "based only on discrepancies in age or grade performance in academic subject areas" (Posny, 2007, p. 1). The IDEA requires that children with disabilities receive services and supports that are appropriate for their individual needs, as determined by a wholistic assessment and is not limited to information about a child's academic performance (Posny, 2007). For example, the fact that a child is achieving passing grades when placed in the expected grade level for chronological age does not preclude eligibility. It is not necessary for a child to fail a grade in order to demonstrate impact of the disability on academic performance (Posny, 2007). The team also considers the presence of health conditions, such as Rett syndrome and CDD, which can mimic some of the signs and symptoms of ASD.

While eligibility criteria for private practitioners are different from those used by the school system, clinicians at private/hospital-based outpatient clinics also assess 3- to 5-year-old children to determine eligibility for services as well as to determine strengths and weaknesses for setting communication goals. The eligibility for services through an outpatient facility is determined by the SLP according to the policies of that facility and the clinician's knowledge of effective and ethical practice. In addition to eligibility for services through the local school system, as described above, Adam and Roderick may also be eligible for services through an outpatient facility. Additionally, because Roderick has an ASD diagnosis, he is also likely to qualify for and benefit from Applied Behavior Analysis (ABA) therapy (Makrygianni et al., 2018).

During this age range, Billy continues to meet developmental milestones, and although he rarely initiates communication with others, he completes his work (coloring, writing numbers, and letters). On the playground, Billy tends to gravitate toward activities where he can play alone, but he does not cause any problems. He is described as a quiet child but a hard worker and one that never causes problems. Neither Billy's parents nor his preschool teachers express concerns about Billy's "speech."

For Adam and Roderick, they continue receiving speech-language therapy, and their therapy sessions probably involve similar strategies. Although Roderick has a diagnosis of ASD, and Adam does not, language impairment in a young child may impact the child's ability to interact with others (peers and adults) and participate in normal routines (reading books, preparing for snack time). Both Adam and Roderick's goals may include appropriate interaction (initiation, turn-taking) with others. Again, goals such as these can be addressed using responsive, directive, or blended interaction strategies. The underlying goal for both children will be to facilitate development of relationships, participation in interactions with the family and in the community, and promote independence (Tables 23.1 and 23.2).

As Adam and Roderick progress in therapy and chronological age, goals for Adam are likely to focus on more traditional speech and language skills such as semantics, syntax, morphology, phonology, and early literacy/academic skills, while using a more directive interactional style. Examples of semantic skills to address with Adam may include building his vocabulary, specifically basic concepts (size, texture) and environmental words/phrases which are needed for functioning in the classroom (raise your hand). Syntactically, these words would be used to build longer, more age appropriate, complex sentences. If his articulation skills (phonology) require treatment, the SLP may work on individual sounds to increase intelligibility. Eventually, phonological awareness and preliteracy skills such as rhyming and familiarizing Adam with books (e.g., page turning, identifying letters on the page) may be necessary.

While goals for Roderick may include these areas, they will also continue to be determined by his strengths and weaknesses in pragmatic/social skills. While addressing social/pragmatic skills, the SLP may introduce new vocabulary words. Vocabulary will likely begin with words that meet functional needs but build into more basic and environmental words needed to function in the classroom. Syntactical goals for Roderick may first address using two to three words in a sentence and not focus on sentence complexity. Phonology and early literacy skills may be integrated into book reading activities but not specifically targeted within the treatment plan. By this point, Roderick's therapy sessions are primarily conducted using a blended approach.

The goals and treatment strategies (responsive, directive, or blended) for children with neurodevelopmental disorders will be similar to Adam's and Roderick's. However, it should be noted that for any child who exhibits a severe deficit in expressive communication, treatment planning may address the need for and potential use of augmentative-alternative communication (AAC). AAC refers to techniques, devices, and strategies used to supplement or replace speech for individuals who are unable to speak or lack sufficient speech to meet their communication needs (ASHA, 2020). AAC may be "high-tech," meaning computer-based, or "low tech/no tech" involving materials such as picture cards or communication boards, or use of gestures or pointing.

Severe impairment in speech and/or expressive language can occur due to a variety of causes. The most common medical diagnosis among children who benefit from AAC for communication is cerebral palsy (CP) (Beukelman & Mirenda, 2013), which is the most common pediatric physical disability in the United States (Edwards & Bacon, 2014; Yeargin-Allsopp et al., 2008). The condition often includes dysarthria, which may vary in terms of severity and nature of speech deficits, as well as impairments in language and cognition, and more global motor deficits that impact skills such as writing and gestures. Another example of a condition associated with potential AAC needs is apraxia of speech, a disorder of motor planning that impacts a child's ability to perform sequences of movements needed to produce and combine sounds for speech. Children with ASD may also benefit from some type of AAC as one component of treatment to improve overall functional communication (Beukelman & Mirenda, 2013).

Evaluation for use of AAC devices or methods is ideally performed by a multidisciplinary team that includes the family, SLP, occupational therapist, and other specialists depending on the child's needs and should address a range of potential approaches and materials. For example, a child who has a motor speech disorder may be evaluated to determine whether he or she will benefit from a high-tech device that produces synthesized speech, and/or a low-tech communication notebook. A child with ASD may be evaluated to determine whether he or she can use picture cards to make choices or requests, with potential progression to use of a communication board and/or speech. As with selection of an overall treatment approach, the SLP must tailor decisions concerning use of AAC to the child's needs, with input from parents and caregivers.

Providers and caregivers should be aware of two common misperceptions concerning AAC. First, while AAC can improve expressive communication for many children, it is not an instant replacement for speech. On the contrary, successful use of AAC requires a considerable investment of time, effort, and possibly money on the part of the family. Second, use of AAC devices or strategies will not prevent a child from developing speech or language skills. Providing a means of communication allows more opportunities for interaction with peers and adults, which enhances communication development (Beukelman & Mirenda, 2013).

Elementary School

During the elementary period, children with an earlier identified developmentally delayed exceptionality (by the school system) will undergo a new evaluation that includes testing for a more specific classification of exceptionality, such as intellectual disability, specific learning disability, emotional disturbance, or other health impairment (see Bulletin 1508 for a complete list of exceptionalities). This evaluation, often referred to as a "turning nine" evaluation, may not identify any exceptionality, resulting in the child no longer being eligible for special education services under Bulletin 1508 (Louisiana Board of Elementary and Secondary Education, 2017). However, the child may be eligible for accommodations and modifications under Section 504 if support is still needed. A child who received a specific classification of exceptionality, such as autism, at an earlier age, will not likely receive a "turning nine" evaluation.

All three boys are in elementary school, specifically third grade. Adam and Roderick are still enrolled in school-based services. Adam's "turning nine" evaluation was completed, since his exceptionality of developmentally delayed was no longer valid; his exceptionality changed to Speech and Language Impairment. For Roderick, his exceptionality of autism does not require a "turning nine" evaluation but he will undergo a reevaluation every three years (Louisiana State Board of Elementary and Secondary Education, 2017).

At this point, Billy has not been referred for a speech/language evaluation. When Billy entered first grade, he demonstrated mastery-level preacademic skills such as number/letter identification, phonemic awareness, decoding, and basic concepts such as colors and shapes. In second grade, he demonstrated a varied vocabulary and age-appropriate sentence length, although syntax has been noted to be peculiar at times. He also often becomes hyper focused on certain topics, such as favorite TV shows, dinosaurs, or Disney characters. An area that has posed challenges to Billy is the second to third grade transition from learning to read to "reading to learn" (Chall, 1983, as cited in Paul et al., 2018). He shows strengths in general knowledge of factual information, but the language demands of the curriculum are increasing. He is struggling with the increased use of nonliteral language and the unwritten social rules of the classroom and is not fitting in well with his peers. Overall, Billy has shown progressively more difficulty in school each year; however, he is maintaining A's and B's throughout elementary school and has shown no major behavior problems in the classroom.

For school-age children, communication skills may be addressed in the school setting, outpatient setting, or both. However, eligibility for services is interpreted differently in public versus private settings and there are advantages and disadvantages to each setting. As discussed previously, eligibility for school-based services requires documentation that the disorder impacts educational performance. Depending on how this policy is interpreted, children who benefit from services to the point where they perform well in the school environment may deemed ineligible for services.

By the end of elementary school, both Adam and Roderick might be considered for discharge from speech-language therapy. Adam may no longer meet criteria for eligibility for school-based services, or his outpatient provider may determine that he has made as much progress as can be expected at this time.
With appropriate early services, Roderick may show improvement in communication skills required for the academic environment, and it is possible that services might be discontinued if he is functioning well enough that services are not required.

Both should remain "on the radar" because as a child ages, the demands increase. The SLP should recommend to Adam's parents to have him rescreened if his academic performance is negatively affected. Meanwhile, Roderick may be functioning as expected in terms of academic performance, but is likely to struggle with social interactions and communicating with others, which can negatively impact performance in the classroom.

Without supportive services, the social impairment can have a major negative impact on education. It is important for the SLP, along with the parents and teacher, to be a strong advocate in this case for the child to remain "on the radar" to continue receiving services, even if the therapy model becomes consultative/indirect. As discussed earlier, SLPs and other professionals should be wary of limiting the notion of the "educational performance" to academic performance (Posny, 2007, p.1). If a child who is receiving services improves academic performance, he/she should not automatically have services discontinued. The impact of a neurodevelopmental disability such as ASD on educational performance is multifactorial and encompasses the child's social, cognitive, and language development in addition to progress in the formal academic curriculum. In the event that services are unavailable, decreased, or discontinued, the family may seek services through outside agencies, such as private or university-based clinics, which will require a referral by the primary care team (Table 23.1). Service eligibility through a private clinic is not dictated by specific criteria; however, services may be impacted by the insurance company's policies and approval of treatment services. Be aware that an insurance company may provide prior approval (PA) for therapy services at a private clinic but still not reimburse. Some insurance companies require medical necessity or that a neurologically based reason for the disorder is present. Children with a diagnosis of ASD, such as Roderick, may more readily qualify for private clinic services, whereas Adam may not qualify. Insurance companies will typically review the goals and objective, and these should clearly show how improvement will positively impact functional communication skills; however, the parents may still have to advocate with the insurance company to secure reimbursement for therapy services.

Since a child may engage in both available models, school-based and clinic-based, parents and professionals should be aware that both have advantages and disadvantages. School-based therapy is the most convenient and cost-effective for parents and families because treatment is provided onsite at school, at no cost to the parent. The costs are absorbed by the school system, with some supplemental third-party payments, if the school system bills Medicaid or other payors for the services. In this scenario, there is no discrepancy between services provided based on family resources, such as income, available transportation, scheduling flexibility, or insurance coverage. All children who meet eligibility criteria receive services regardless of insurance status or ability to pay. There is no need for the family to rearrange schedules or to procure transportation because therapy takes place at school and transportation is provided by the school system (Tables 23.1 and 23.2). Another advantage of school-based services is that the services are designed to directly support the child's educational participation. The SLP has direct access to the curriculum and the teachers. Regular contact with teachers allows the SLP to give and receive information and feedback regarding the child's needs, participation, and progress. Access to the curriculum allows the SLP to design intervention that includes, or is most relevant to, the specific materials, concepts, and activities that the child will be expected to engage within the classroom. A third advantage to school-based services is that services are typically provided in groups, which allows opportunities for social interaction with peers and allows the SLP to more directly address pragmatic language skills applied to actual social interactions. Some limits to school-based services include fewer opportunities for individual attention during speech-language therapy and, most typically, reduced frequency and intensity of services.

One advantage to clinic-based services is collaboration with the parent. In the private clinic, because the parent will be bringing the child to each therapy session, the therapist will see the parent on a regular basis. This provides many opportunities for discussion and demonstration of the child's progress and strategies for home carryover activities. The therapist in this setting can more easily gauge the impact of the treatment on the child's functioning in the home environment. School-based therapy ideally includes parent involvement as well; however, the frequency of parent contact is likely to be considerably less and is likely to be based on phone contact rather than live, face to face interactions. Another advantage of clinic-based services is that the setting is likely individual therapy rather than a group setting, which may allow for more customization of therapy to the child's individual needs. Sessions in private clinics are also typically longer in duration and can be scheduled more frequently if warranted compared to typical schedules for school-based therapy. Limitations of clinic-based therapy include potential scheduling difficulties and variability in families' ability to access services. Access to services may be impacted by location, scheduling, and insurance coverage or ability to pay. Although services are normally covered by Medicaid and private insurance plans, there is great variability in plans, including co-payments and treatment sessions allowed, and not all private clinics accept Medicaid. The scope of clinic-based services is also different from that of school-based services in that the focus is more likely to be on skills that are geared toward participation in the home or community activities. Clinic-based services may also include skills that are applicable to academic needs; however, they will not be as directly based on the school curriculum as school-based services typically are.

Therapy may also be conducted differently depending on the setting. As noted above, school-based services are more likely to be conducted in groups rather than individual sessions and will focus on supporting the child's progress in the school curriculum, requiring collaboration with teachers and others at school interacting with others in the classroom (Tables 23.1 and 23.2). Services provided in an outpatient setting may overlap with school-based services in terms of skills targeted and may incorporate elements of the child's schoolwork, but they are more likely to focus on activities and needs that are directly relevant to communication and social interactions with the family and in community activities. In both settings, the SLP role at this age is more likely to involve treatment of previously identified problems than initial identification of speech/language disorders or neurodevelopmental disorders. However, it is possible for children who have less severe disorders to be referred for evaluation and to be diagnosed with disorders at any time during elementary school, as communication and learning demands increase.

When Adam and Roderick were younger, speech-language procedures may have appeared similar since both would likely have occurred in the natural environment. Additionally, there may have been some overlap in objectives such as joint attending. However, for behaviors such as labeling and requesting, Roderick's objective may have been to increase nonverbal skills such as pointing, whereas Adam's objective was to verbally produce multiple words to show intent. As the two boys advance through elementary school, their needs and goals will likely become more differentiated, as the severity level and any coexisting disabilities, such as intellectual impairment, become clearer.

Roderick, whose functioning is toward the lower end of the autism spectrum, may benefit from goals related to making requests (nonverbal and verbal), engaging in reciprocal interactions with peers and teacher, and following verbal directions directly associated with school activities. A child functioning on the higher end of the spectrum might benefit more from goals related to expressing himself appropriately and understanding nonverbal cues and figurative language. Therapy for a child such as Adam, who is not on the autism spectrum, may focus more on skills such as narrative skills, articulation or comprehension of complex sentences or paragraphs, without the focus on pragmatic skills.

Beyond Elementary (Adolescent/Adult)

In fifth grade, Billy takes the test for entrance to a magnet middle school. He does not score high enough to attend the school. His parents are puzzled because he seems to have a large vocabulary and a great deal of knowledge about certain topics, such as his favorite historical era, World War I. Billy's parents thus make plans to enroll him at his district middle school. When he starts at the new school, Billy has trouble keeping up with classwork and homework, and his grades are considerably lower compared to elementary school. Now that Billy is changing classrooms six times per day, he is often late for class, and he forgets to retrieve needed books and papers from his locker. He also struggles to make new friends and does not seem to have anyone to sit with in the cafeteria for lunch. Billy's parents are very concerned that he seems miserable. They contact his pediatrician for a referral to a private counselor, who notes that Billy shows some signs of ASD and suggests a referral to a multidisciplinary evaluation center, where Billy is diagnosed with ASD. His parents then take the report from the evaluation center to the Pupil Appraisal department of the local school system. Although Billy now has a medical diagnosis of ASD, he does not meet criteria for an educational exceptionality of autism. Therefore, he does not receive services based on an IEP; however, he is eligible for Section 504 accommodations. His parents enroll him in speech-language therapy at an outpatient clinic.

Billy's case serves as an example of a scenario in which a child exhibits mild or subtle signs of a neurodevelopmental disorder at a young age but is not formally diagnosed until later. Although initial identification of ASD is much less common in adolescence than at younger ages, milder difficulties or differences in function that did not have a major handicapping effect at younger ages may become more evident or handicapping as the child matures and is expected to master more complex and advanced academic content. Like Billy, children on the higher end of the autism spectrum may function at a passing academic level without major behavior problems during elementary school. The transition to middle school involves an increase in academic and social demands that can sometimes trigger a referral for evaluation, which may take place through the school system or another route, ideally involving a multidisciplinary team.

For Billy, once he has been identified, treatment will be designed to improve functioning in areas most impacted by the communication deficits associated with ASD. One likely area of focus will be higher-level language skills, particularly nonliteral language. Individuals with ASD may have strengths in certain language skills. Billy, for example, may know literal meanings for a large number of words and produce sentences of age-appropriate length and complexity. However, difficulty with nonliteral language will impact both academic and social functioning. For example, middle school language arts curricula (Common Core Standards Initiative, 2020) typically include content that is dependent on nonliteral language skills such as figurative language and interpretation of poems and other types of literature. Billy is also likely to struggle with understanding sarcasm and idioms, which will impact interactions with others during class and social situations. As his peers develop more sophisticated and subtle humor during adolescence, Billy may not understand their jokes. Therefore, therapy focused on improving nonliteral language skills can impact functioning both academically and socially. Another area, closely related to nonliteral language, that is likely to be addressed in Billy's treatment is pragmatic skills. Billy may benefit from direct instruction in how to initiate, sustain, and close conversational interactions. He may need help learning to interpret and use nonverbal and paralinguistic aspects of language, such as facial expressions, eye contact, and intonation patterns, that provide nuance during conversational speech. Addressing these skills in therapy will not eliminate the underlying social-communication deficits but can give an individual like Billy needed tools to interact more appropriately with peers and adults.

In addition to ASD, another category of neurodevelopmental problem for which diagnosis and initiation of treatment may occur in the tween to teen years is language and learning disorders. Language is one of the broadest of the eight practice areas in the SLP scope of practice. While language difficulty is a component of the overall deficits experienced by individuals with underlying diagnoses or conditions such as ASD, cerebral palsy, or genetic syndromes, a language disorder can also occur as a primary disability in the absence of more global deficits in cognition, motor skills, or other domains. A language disorder impacts at least one subsystem of language (semantics, pragmatics, syntax) and at least one modality (written, spoken, signed). When the disorder impacts spoken language, as in specific language disorder, a child may be able to cope with the academic demands of elementary school, but may become overwhelmed as the complexity of material increases in late elementary and middle school, requiring increasing adeptness with auditory comprehension, metalinguistic skills, and deeper understanding and specialization of word meanings. A language disorder may also manifest as dyslexia. Dyslexia is classified as both a language disorder and a learning disability. While there are no universally accepted procedures for diagnosing dyslexia, characteristics include poor reading accuracy or fluency due to marked difficulties with reading, decoding, and spelling, in the absence of general intellectual impairment (Adlof & Hogan, 2018). Dyslexia is sometimes mistakenly attributed to visual problems because it impacts processing of written symbols. However, it is more accurately defined as a language disorder that primarily impacts the phonological and written language subsystems. Dyslexia may also be associated with late talking, poor semantic skills, or other signs of verbal language difficulty. Treatment of dyslexia may be provided by an SLP, reading specialist, and/or special education teacher (Adlof & Hogan, 2018) and will focus on improving and compensating for identified weaknesses. In addition to dyslexia, academic function may also be impacted by other learning or social/emotional disorders such as dysgraphia, dyscalculia, or attention deficit/hyperactivity disorder (ADHD) (US Department of Education, 2020). These disorders are less directly related to language skills and will most likely be treated by professionals other than SLPs.

While Billy was beginning the journey toward diagnosis and treatment, Roderick also enrolled at his district middle school. Roderick spends most of his day in a self-contained classroom, in which a special education teacher provides instruction in core academic subjects. Roderick is expected to make progress in all subjects, but the pace of learning is slower compared to a regular education classroom. He also attends home economics and choir with his regular-education peers. When Roderick enters high school at age 15, he enrolls in an alternative education track, which involves working toward a certificate of completion rather than a regular high school diploma. While his curriculum still includes content in math, reading, science, and social studies, there is increased focus on functional skills to foster independence in self-care, household tasks, and vocational readiness.

During childhood and adolescence, the overall goal of SLP treatment is to maximize functional communication in the home, school, and community. For a child with ASD or another neurodevelopmental disorder, the SLP will work to facilitate as much improvement as possible in speech and language skills, according to the child's needs and abilities, as the child progresses through developmental stages and academic levels. When the child begins the transition to adulthood, the overall goal will still be functional communication; however, the specific focus will shift or expand to address preparation for independent living. For example, for Roderick, the school-based SLP might target vocabulary skills related to classroom activities or community awareness; social negotiation skills; and functional auditory comprehension skills such as following and retaining directions to complete tasks. For Billy, therapy at the high school level is likely to continue focusing on the types of skills described above, but it is also likely to include goals specific to his needs at this age such as organizing information for written assignments, test taking strategies, and learning about job or college options. For a child with motor and speech impairments who uses an AAC device, the transition to a new school environment, such as high school, may be particularly challenging. In addition to the increased communication and learning demands, school personnel, including the school-based SLP, may be unfamiliar with the AAC device. The SLP will focus on adapting the device to include words and phrases for more age-appropriate communication, as well as vocabulary specific to new school subjects, in collaboration with parents and/or teachers. The academic status and level of independence that children with neurodevelopmental disabilities are expected to achieve will depend on the severity of disability; however, social interaction and employment are common concerns for most young adults with neurodevelopmental disabilities.

The K-12 setting offers built-in opportunities for interacting with peers on a regular basis, even if the way in which the interaction occurs is different or more challenging compared to neurotypical children. When the young adult moves on from the K-12 setting, whether to further education or to other activities, that daily opportunity and exposure is no longer in place. He or she may need support in developing and maintaining social connections and friendships. This may also be the point when the young adult begins dating, which can present an additional set of challenges and social rules that the individual must navigate. While the individual may have developed adequate functioning in the K-12 setting, the lifelong social and communication impairment associated with ASD will likely make the transition to less structured/supported environment difficult. SLP services at this stage are likely to focus on skills specific to social negotiation, conversation rules, and code switching, which means modifying communication style for different social situations and roles.

Neurodevelopmental disorders are lifelong conditions which continue to impact functional communication and social skills as the individual moves on to new roles, responsibilities, and adult developmental stages. It is important for providers and families to bear in mind that need for support services, including SLP, does not necessarily end when the child becomes an adult. While there is no single model of treatment that applies to all adults, the goals, frequency, and nature of treatment may change. For example, therapy may become more focused on specific, job-related or postsecondary education communication needs, or an individual may go without therapy services for a period of months or years, but then acquire a new AAC device or enter a new situation that challenges communication skills to the point where SLP services are needed.

After being discharged from speech and language therapy, Adam progressed through middle and high school, earning a regular high school diploma. After high school, he enrolls in a state university several hours from home. Adam joins a fraternity and enjoys making new friends, but has difficulty keeping up with his academic work. He attends class but often falls behind on assigned readings. He makes a C- on his first English paper and a D on his first exam in his biology class. He becomes increasingly anxious about his grades and seeks help at the university's center for academic assistance. The counselor at the center suggests that he undergo an evaluation to determine whether he qualifies for academic accommodations such as extended testing time and note-taking services. She also provides Adam with resources for academic tutoring.

Over the past several decades, many more students with disabilities, including neurodevelopmental and speech/language disorders, have attended college or university (Raue et al., 2011); such is the case for Billy and Adam. For the transition to postsecondary education, it is important for providers and families to understand how the process of qualifying for and obtaining support and rehabilitation services will change. During the K-12 years, the school system is responsible for providing services to address the potential impact of ASD on educational participation. In higher education settings, students with disabilities are still entitled to reasonable accommodations (US Department of Justice, n.d.). However, the responsibility for making contact with the disability services office at the college or university, providing documentation of a disability and needed accommodations, shifts to the individual. For students who need equipment, such as readers or AAC devices, they are often responsible for obtaining the equipment through other agencies. Depending on the policies at particular institutions, the disability service unit may notify instructors of needed accommodations, but it is likely that the student will be responsible for working with instructors to arrange certain accommodations such as extended test time or repeated directions. This type of negotiation is an example of the new communication demands in a higher education environment. The SLP may support students transitioning higher education in various ways. For example, the SLP may make recommendations to the family concerning the student's ability to manage the academic and communication demands of a higher education environment. For students who have language disorders, the SLP may provide documentation needed for accommodations during standardized testing (College Board, n.d.). Once the student enters the new setting, he or she may benefit from short-term direct instruction in communication skills such as negotiating accommodations and interacting appropriately with instructors, peers, and others.

Billy was diagnosed at a later age and began receiving therapy at an outpatient clinic. His therapy needs may address academic areas but will evolve over time to include more complex communication scenarios to prepare him for higher education and vocational needs. For educational purposes, he may need therapy to develop strategies to manage the new communication and interactional demands and executive function skills. However, after earning an associate or bachelor's degree, he may begin seeking full-time employment and need help with communication skills specific to interviewing or to performing his desired job such as interacting with customers or clients. An additional goal may be to help him function more independently, which may range from living independently to managing a checking account to buying groceries and paying bills. This would require collaboration with the family, other professionals, and support personnel (Tables 23.1 and 23.2).

Adam, who was not diagnosed with ASD but required treatment at a younger age, due to a language disorder, is less likely to need direct SLP services as an adult. Depending on the results of the evaluation recommended by the academic success center, he may or may not require accommodations at the educational level. However, if he does, he will most likely be able to negotiate and interact with instructors and peers without professional support services.

As illustrated by Roderick's case, many individuals with neurodevelopmental disorders have cognitive limitations that preclude participation in traditional higher education programs; however, communication needs and opportunities also change in various aspects of their life. The individual may obtain employment in a structured, supervised environment, which could include a sheltered workshop or through an employer such as a grocery store. The SLP may also need to coordinate services with other providers or agencies, consistent with ASHA guidelines (Table 23.1). For example, Roderick may receive assistance from Vocational Rehabilitation Services, a state-funded program designed to provide resources to support employment of people with disabilities. Vocational Rehabilitation Services may request an evaluation by an SLP so that Roderick's current level of functioning in communication skills can be considered in decision-making about job placement. An SLP evaluation in this scenario would focus on assessment of oral and written receptive and expressive language skills, speech intelligibility and fluency, pragmatic skills, and need for accommodations and supports, such as repeated directions, structured environment, or assistive technology (AAC). If communication skills are identified as a barrier in the workplace, Roderick might benefit from direct SLP services, with goals focused developing appropriate ways to communicate with supervisors and co-workers in that environment through activities such as role-play, verbal problem solving, or enrollment in a language-social skills group.

Adam, Roderick, and Billy are examples of the range of different circumstances and needs experienced by individuals with neurodevelopmental disorders from childhood to adulthood. Rehabilitation services, including speech-language pathology, are available and intended to allow the individual to function in his/her specific "world" (environment, community, family). For Adam, the improvements made in speech and language skills allowed him to achieve academic success and eventually attend college. For Billy, SLP services allowed him to more effectively interact with others and cope with academic demands at school. For Roderick, the functional communication skills fostered through SLP services allowed him to effectively communicate with friends and family, and work toward participation in a vocational setting. All three young men required and may still require services at and during different parts of their lives. Despite their differences, Adam, Billy and Roderick all became successful in their specific "worlds."

References

Adlof, S., & Hogan, T. (2018). Understanding dyslexia in the context of developmental language disorders. *Language, Speech, and Hearing Services in Schools, 49*(4), 762–773.

American Academy of Pediatrics. (2016). Media and young minds. *Pediatrics, 138*(5), e20162591. https://doi.org/10.1542/peds.2016-2591

American Speech-Language-Hearing Association. (n.d.). *Autism* (Practice Portal). Retrieved March 8, 2022 from www.asha.org/Practice-Portal/Clinical-Topics/Autism/.

American Speech-Language-Hearing Association. (2008). *Roles and responsibilities of speech-language pathologists in early intervention: Guidelines*. https://www.asha.org/policy/GL2008-00293/#sec1.7

American Speech-Language-Hearing Association. (2016). *Scope of practice in speech-language pathology*. https://www.asha.org/policy/SP2016-00343/

American Speech-Language-Hearing Association. (2020). *Augmentative and Alternative Communication*. https://www.asha.org/PRPSpecificTopic.aspx?folderid=8589942773&section=Overview

Arvedson, J. C., & Brodsky, L. (2002). *Pediatric swallowing and feeding: Assessment and management* (2nd ed.).

Beukelman, D. R., & Mirenda, P. (2013). *Augmentative & alternative communication: Supporting children and adults with complex communication needs* (4th ed.). Paul H. Brookes Publishing.

Campbell, F. A., & Ramey, C. T. (1994). Effects of early intervention on intellectual and academic achievement: A follow-up study of children from low-income families. *Child Development, 65*, 684–698.

College Board. (n.d.). *Communication disorders guidelines*. https://accommodations.collegeboard.org/documentation-guidelines/communication-disorders

Common Core Standards Initiative. (2020). *English language arts standards*. http://www.corestandards.org/ELA-Literacy/

Edwards, D., & Bacon, E. (2014). Optimizing function for patients diagnosed with cerebral palsy. *Perspectives on*

Swallowing and Swallowing Disorders (Dysphagia). https://doi.org/10.1044/sasd23.3.123

Eggertson, L. (2010). Lancet retracts 12-year-old article linking autism to MMR vaccines. *Canadian Medical Association Journal, 182*(4), E199.

Filipek, P. A., Accardo, P. J., Baranek, G. T., Cook, E. H., Dawson, G., Gordon, B., Gravel, J. S., Johnson, C. P., Kallen, R., & J., Levy, S.E., & Minshew, N. J. (1999). The screening and diagnosis of autistic spectrum disorders. *Journal of Autism and Developmental Disorders, 29*(6), 439–484.

Itzchak, E. B., & Zachor, D. A. (2011). Who benefits from early intervention in autism spectrum disorders? *Research in Autism Spectrum Disorders, 5*, 345–350.

Ledford, J. R., & Gast, D. L. (2006). Feeding problems in children with autism spectrum disorders: A review. *Focus on Autism and Other Developmental Disabilities, 21*(3), 153–166.

Lord, C., Shulman, C., & DiLavore, P. (2004). Regression and word loss in autistic spectrum disorders. *Journal of Child Psychology and Psychiatry, 45*(5), 1–41.

Louisiana Department of Health. (n.d.). *Early Steps system point of entry (SPOE) Referrals*. Retrieved 11 June 2020 from ldh.la.gov

Louisiana State Board of Elementary and Secondary Education. (2017). *Bulletin 1508 – Pupil appraisal handbook.* https://bese.louisiana.gov/policy

Majnemer, A. (1998). Benefits of early intervention for children with developmental disabilities. *Seminars in Pediatric Neurology, 5*(1), 62–69.

Makrygianni, M. K., Gena, A., Katoudi, S., & Galanis, P. (2018). The effectiveness of applied behavior analytic interventions for children with Autism Spectrum Disorder: A meta-analytic study. *Research in Autism Spectrum Disorders, 51*, 18–31.

Matson, J. L., Boisjoli, J. A., & Wilkins, J. (2007). *Baby and Infant Screen for Children with aUtism Traits – Part I (BISCUIT-Part1).* Disability Consultants, LLC.

Ozonoff, S., Young, G. S., Brian, J., Charman, T., Shephard, E., Solish, A., & Zwaigenbaum, L. (2018). Diagnosis of autism spectrum disorder after age 5 in children evaluated longitudinally since infancy. *Journal of the American Academy of Child & Adolescent Psychiatry, 57*(11), 849–857.

Paul, R., Norbury, C., & Gosse, C. (2018). *Language Disorders from Infancy through Adolescence*. Elsevier.

Posny, A. (2007). Personal communication. Retrieved June 11, 2020 from https://www2.ed.gov/policy/speced/guid/idea/letters/2007-1/Posnye030807disability1q2007.pdf

Raue, K., Lewis, L., & Coopersmith, J. (2011). Students with disabilities at degree-granting postsecondary institutions. *National Center for Education Statistics, US Department of Education, Statistical Analysis Report*, (2011-018).

Robins, D. L., Fein, D., & Barton, M. (2009). *Modified Checklist for Autism in Toddlers, Revised with Follow –Up.* https://mchatscreen.com/

Rutter, M., LeCouteur, A., & Lord, C. (2008). *Autism Diagnostic Interview – Revised (ADI-R).* Western Psychological Services.

U.S. Department of Education. (2020). *Learning disorders in children.* https://www.cdc.gov/ncbddd/childdevelopment/learning-disorder.html

U.S. Department of Health and Human Services. (2019). *Defining the PCMH.* https://pcmh.ahrq.gov/page/defining-pcmh

U. S. Department of Justice. (n.d.). *Americans with Disabilities Act of 1990*, as amended. https://www.ada.gov/pubs/adastatute08.htm

Voress, J., & Maddox, T. (2012). *Developmental Assessment of Young Children-2nd Edition (DAYC-2).* PRO-ED.

Yeargin-Allsopp, M., Braun, K. V. N., Doernberg, N. S., Benedict, R. E., Kirby, R. S., & Durkin, M. S. (2008). Prevalence of cerebral palsy in 8-year-old children in three areas of the United States in 2002: A multisite collaboration. *Pediatrics, 121*(3), 547–554.

Urologic Care of Children with Autism Spectrum Disorders or other Neurodevelopmental Disorders

24

Zachary M. Connelly, Tomas Paneque, and Alexander Gomelsky

Abstract

Autism spectrum disorder and neurodevelopmental disorders are understudied in the urologic field. This chapter aims to summarize the current knowledge in a comprehensive manner, in the context of the patient, caretaker, medical home, and healthcare team. This chapter will discuss the timing of a urologic consult, how a genitourinary physical exam should be conducted, and common tests that could be ordered. Moreover, clinical pearls at the end of the chapter describe the most important points from the text.

Keywords

Urology · Autism · Developmental disorder · ASD

Introduction

Limited data are available in the literature relating to urological symptoms and conditions associated with autism spectrum disorder (ASD) and neurodevelopmental disorders (NDDs). Delays in behavioral and communication development provide challenges to the management of these lifelong disorders. Urological consultation may be a part of a child's medical home care plan for many reasons. This chapter will provide a basic understanding of urologic evaluation, tests, common urologic manifestations, and treatment options. Throughout the text, several urologic conditions commonly associated with ASD and other NDDs will be discussed, including urinary incontinence, enuresis, urinary tract infections, cryptorchidism, hypospadias, Tuberous Sclerosis Complex, Fragile X syndrome, Down syndrome, Neurofibromatosis Type I, Prader–Willi Syndrome, CHARGE syndrome, Rett syndrome, and Spina Bidifa. At the completion of the chapter, the reader will have a solid understanding or urological manifestations in a patient with ASD and NDDs as well as a plan for proper lifelong management.

Initial Observation and Patient History (A Urologist's View)

Urology is a surgical subspeciality, and almost all patients will be received via a referral. A referral is necessary when the consulting physician deems that their patient's issues lie outside of their treatment capacity. A consult to urology may be needed if there is any concern for prob-

Z. M. Connelly (✉) · T. Paneque · A. Gomelsky
Department of Urology, Louisiana State University Health Shreveport, Shreveport, LA, USA
e-mail: zmc001@lsuhs.edu

P. McPherson (ed.), *Handbook of Treatment Planning for Children with Autism and Other Neurodevelopmental Disorders*, Autism and Child Psychopathology Series,
https://doi.org/10.1007/978-3-031-06120-2_24

lems relating to the kidneys, bladder, or genitalia. The reasons for urologic consultation include pre-natal hydronephrosis with ongoing hydronephrosis on post-natal ultrasound, vesicoureteral reflux demonstrated on voiding cystourethrogram, febrile/recurrent urinary tract infections (UTIs), neurogenic bladder (e.g., spina bifida), congenital anomalies (e.g., cystic kidneys, ureteropelvic junction obstruction, posterior urethral valves, ectopic ureter, ureterocele, duplex kidney, horseshoe kidney), hypospadias, disorders of sexual differentiation, and voiding dysfunction.

The first observation of the patient will be vital to proper urologic management. During the initial encounter, the patient's appearance, behavior, activity level, and interaction with parents or caregivers (Diaz-Saldano, 2017) are extremely important. It is critical that the treating physician remains observant of the interactions between the patient and his/her caregivers and looks for signs of potential abuse (Nicolaidis et al., 2014), as this population is at higher risk.

A history should include symptom onset, location, duration, characteristics, aggravation, and radiation. Additionally, voiding history and details (e.g., diapers, pads, indwelling catheters versus self-intermittent catheterization), urinary incontinence episodes, UTIs, and any episodes of urinary retention should be noted. A thorough history will dictate what tests need to be conducted. For example, a history of fever (especially $T > 101.5$ °F) and urine suspicious for UTI should prompt a review of past urine cultures (Diaz-Saldano, 2017). It is also important to make sure patients have reached their age benchmarks, such as achieving urinary and fecal continence as expected. Diurnal and nocturnal involuntary urine loss can have an impact not only on treatment options but also on the patients' psychological and academic development (Gor et al., 2012). Being cognizant of the patient and caregivers' education level will aid in effective communication between parties, including obtaining an accurate medical history. Some patients may have a slowed comprehension rate, so it is critical for the healthcare provider to talk slowly and make the patient feel at ease during their visit. This may involve the use of images and having the patient use non-verbal communication to indicate the pain or area of concern. This understanding of the patient's development may guide future management of their urologic conditions.

The initial workup should also include the birth history, past surgeries, family history, social history (i.e., living situation, abuse history), psychiatric history, past diagnosis of any genetic syndromes, and past and current treatment plans. All past and current medications and their adverse side effects should be reviewed, as these factors can contribute to the patient's urologic status and its treatment. For example, the use of some psychiatric medications in this patient population may be associated with somnolence and deep sleep leading to enuresis (Diaz-Saldano, 2017).

Review of Systems and Physical Exam

Several factors should be documented in a thorough review of symptoms. Recent weight changes should be elicited, as some medications can alter both appetite and body weight. A history of bleeding disorders, asthma, and reactive airway disease is important if surgical intervention is planned. A gastrointestinal review should include bowel habits. The psychiatric history, including medications, should be thoroughly documented, as well as a history of seizures, headaches, and hydrocephalus.

The physical exam is an extremely interactive event between the healthcare professional, the patient, and the parent/caretaker. There should be ongoing communication, where each move the physician makes is verbally explained and demonstrated visually with images or on a mannequin. The time expected for each part of the exam should be explained thoroughly. With each step, the clinician should try to gauge the patient's understanding prior to the actual movement. There should be a balance between a thorough physical exam and a quick exam to keep the amount of time where the patient is undressed to a minimum. If invasive testing is warranted, the use of numbing jelly or cream is highly encouraged. If available, a child life specialist may be able to communicate more effectively with the patient during the stressful office visit (Nicolaidis et al., 2014).

Referring physicians should make patients and their caregivers aware that a genitourinary exam is standard during urologic evaluation. Proper hygiene should be ensured to facilitate a quick and thorough examination. Patients should also be made aware that examination of intimate areas is only appropriate in the presence of a physician/clinician and/or caregivers. If the examination is challenging due to crying or positioning, caregivers should help to facilitate a proper examination.

Findings on examination of the head, such as a low set of ears, may suggest concomitant renal anomalies. The neurological exam should note any abnormal reflexes or muscle movement and any developmental delay. The cardiovascular exam should evaluate for heart murmurs and structural abnormalities. Failure to thrive and an abnormal thyroid examination should prompt an endocrine workup.

A urologic exam may include some evaluation of the abdomen, back/spine, genitalia, lower extremities, and neuromuscular function. Abdominal palpation may reveal constipation, while palpation of the kidneys and bladder may indicate flank tenderness or bladder distention, respectively. The patient should be asked to stand, and their gait should be observed for toe-walking, poor coordination, unsteady gait, and falling. Any of these uncoordinated exercises can be indicative of a neurological deficit which may have an impact on urination. During the exam, the patient should be turned on their side and the lumbosacral spine should be evaluated for dimples, hair tufts, sacral agenesis, subcutaneous lipomas, and hemangiomas. If any of these findings are present, a meningocele may be suspected. If the patient is an adolescent, their assent should be obtained for the caretaker to remain in the room (Diaz-Saldano, 2017). Following their decision, the patient should remove their clothing and the physician should examine the undergarments for any soiling. Lastly, the genitalia should be examined for rashes and dribbling of any urine with cough or Valsalva maneuver (Baskin & K. B., 2005).

Tanner staging guidelines can be found in various sources and are used to gauge puberty progression. Tanner staging provides a general scale for physical development and can help guide management of a variety of urologic issues. The five Tanner stages of puberty consist of prepubertal, three stages of development, and adult. During this time, pubic hair will develop and darken, testes and penis will enlarge, and breasts begin to develop in females. By the beginning of age 15, children can start to reach the adulthood plateau (Tanner, 1986).

Laboratory Testing

The history and physical examination will guide the selection of laboratory tests. A urinalysis (UA) of a mid-stream urine specimen may be collected and a urine dipstick allows rapid screening for infection, hematuria, level of hydration, and glycosuria. A urine culture and sensitivity testing may be performed when a bacterial infection is suspected. In cases of hematuria (blood in the urine) and dysuria (painful/difficult urination), random urine calcium to creatinine ratio may be used to screen for hypercalciuria.

Diagnostic Imaging

Diagnostic imaging can be used to gain further insight into the urologic anatomy and potential problems. An ultrasound of the bladder and kidneys is often the initial screening study owing to its noninvasive nature. Renal ultrasound can detect hydronephrosis, kidney stones, and ureteral duplications. A bladder ultrasound is useful for evaluating micturition by comparing pre- and post-voiding volume. A persistently elevated post-void residual (PVR) may place the patient at a greater risk for UTIs, bladder and renal dysfunction, and should prompt additional investigations.

Radiographic (X-rays) and nuclear medicine tests may also be useful, but can be more invasive. A voiding cystourethrogram (VCUG) is performed by instilling contrast (radio-opaque dye) into the bladder through a catheter and images are taken during bladder filling and emptying. A patient may benefit from numbing gel prior to this procedure. Vesicoureteral reflux may be seen if contrast travels up the ureters and into the kidneys. During the voiding phase, the bladder neck and urethra can be evaluated for obstruction or other anatomic abnormalities. A better understanding of kidney function may also be obtained with a nuclear medicine test where a tracer (mercaptoacetyltriglycine; MAG3) is injected into a vein. Images are taken at 1-minute intervals while the kidneys take up the tracer and then begin to excrete it. After a certain time, a diuretic is given to speed up excretion of the remaining tracer. This test provides valuable information regarding kidney shape, size, location, function, and excretion. Cross-sectional imaging, such as computed tomography (CT) and magnetic resonance imaging (MRI), may be indicated in some patients. While these tests provide an excellent description of anatomy, they require the patient to limit their movement which may be a challenge for some children.

Additional Urologic Tests

A basic, but extremely helpful, practice is having the patient or caretaker complete a voiding diary. See Fig. 24.1. Referring physicians may have patients and/or caregivers complete a voiding diary prior to urologic evaluation. Common presenting conditions that may prompt a voiding diary include recurrent UTIs, lower urinary tract symptoms (e.g., frequency, urgency, overactive bladder, incontinence), and bladder/bowel dysfunction. This diary is a continuous log of fluid intake and urinary output kept over a several-day period. This diary provides information regarding fluid consumption, timing and frequency of urination, incontinence episodes, and pain or discomfort. Patient and caretaker buy-in with a diary can be very helpful for the healthcare team to establish a baseline micturition profile and monitor for improvement after treatment. A child aged 1–3 years should void 500–600 cc/day, a 3–5 year-old should void 600–700 cc/day, a 5–8 year-old should void 650–1000 cc/day, and an 8–14 year-old should void 800–1400 cc/day (King et al., 2018). Although not commonly used, uroflowmetry measures the urine volume voided in a single urination and the rate at which urination occurs. Finally, multichannel urodynamics can evaluate bladder and urethral innervation and bladder detrusor function during bladder filling and emptying. This test may provide additional information regarding urinary incontinence, retention, and UTIs. While extremely useful in select patients, urodynamics requires catheterization of the bladder and rectum and, thus, requires patient and caregiver cooperation.

A Review of Normal Toilet Training

Toilet training is a complex process that involves the understanding and cooperation of caretakers and physician, as well as the patient's motor, physiological, and communicative readiness. Signs of readiness include dry periods greater than 2 hours (although dryness at night may not yet be present); predictable bowel movements; indications of sensory awareness of a full bladder; recognition that a soiled diaper is wet; ability to communicate about the toilet training process; having the motor skills to perform toilet tasks (e.g., remove clothes, sit, and flush); and being able to sit long enough to complete the task (King et al., 2018).

The general toilet training steps include: review pictures providing examples of teachable training steps necessary for successful toilet training; enter the bathroom and independently remove pants/undergarments (caretaker may initially help); sit on toilet; properly use toilet paper (both removal from roll and self-use); independently replace pants/undergarments; flush; wash hands; and, communicate/practice toilet training for use outside of home/training location. Praise should be given for each successful step and a pre-determined reward may be offered for independent completion (King et al., 2018).

Time	Number of Voids	Voided Volume	Leakage Episodes	What Caused Leakage?
12 AM – 2 AM				
2 AM – 4 AM				
4 AM – 6 AM				
6 AM – 8 AM				
8 AM – 10 AM				
10 AM – 12 PM				
12 PM – 2 PM				
2 PM – 4 PM				
4 PM – 6 PM				
6 PM – 8 PM				
8 PM – 10 PM				
10 PM – 12 AM				
Total				

Fig. 24.1 Voiding diary. This is an example of what a voiding diary may look like. *Voided Volume* is typically logged in mL, however, this can be changed to the easiest metric for the patient and caretaker

Treatments of Urologic Conditions

Urinary Incontinence

Urinary incontinence is defined as involuntary or uncontrollable leakage of urine occurring during the day (diurnal) or night (nocturnal; nocturnal enuresis) (Nevéus et al., 2006). An opportunity sampling of children in the United States by Bloom et al. found that 10% of children between the ages of 3 and 12 have diurnal urinary incontinence (Bloom et al., 1993). Enuresis is very common as well, with 6–9% of 7-year-old children and 3–7% of 10-year-old children affected (Byrd et al., 1996; Schaeffer & Diamond, 2013).

Patients with ASD, NDDs, and intellectual disabilities have a higher incidence of both enuresis and diurnal urinary incontinence (von Gontard et al., 2015). Overall, 40% of ASD children report at least one form of urinary incontinence, with 30–41% reporting enuresis and 25–41% having daytime incontinence (Kroeger & Sorensen-Burnworth, 2009; von Gontard et al., 2015). Another study found that 37% of patients had daytime urinary incontinence (Gubbiotti, Elisei, et al., 2019b). Twenty to thirty percent of the enuresis population has some form of a psychiatric diagnosis, which is approximately twice the prevalence of those without enuresis (Van Herzeele & Vande Walle, 2016). Through the first two years of life, 26% of the typically developing population is continent, while 98% and 100% are continent at 36 and 60 months, respectively (King et al., 2018). On average, children with ASD achieve continence later than their typically developing peers (von Gontard et al., 2015). As each case is unique by severity and type, there is no quantifiable data describing an exact time frame of obtaining continence in children with ASD. In addition to delayed continence, children with ASD have a higher prevalence of other lower urinary tract symptoms and behaviors, such as urinary urgency, postponement of voiding, and straining to void (De Gennaro et al., 2010; von Gontard et al., 2015). Some urologists have observed patients who would urinate into the toilet, but required other modalities for bowel passage (e.g., pull-ups, pads, diapers) (King et al., 2018). Urologic referral should be considered if symptoms are bothersome, negatively impact the patient's quality of life, or if symptoms worsen.

Anxiety is the trigger for incontinence in a subset of the ASD and NDDs population, instead of a toilet-learning deficit. When a patient has a daytime accident, it can further exacerbate their anxiety about the next urination event or lead to a fear of public humiliation. If a patient has a history of enuresis, they may be reluctant about attending evening events with friends, such as sleepovers or overnight camps, which can lead to self-quarantine. Lastly, the anxiety of incontinence can slow the development of maturity and independence (King et al., 2018).

First-line treatment for incontinence is caretaker support and education of the urinary tract function (Maternik et al., 2015). In most cases, toilet training can occur by taking small, achievable steps, albeit at a later period. The caretaker must remain tolerant while they aid the patient in understanding toilet training readiness. This will ensure the caretaker does not discourage the patient and delay continence. In patients with ASD, structural systematic approaches for teaching are important, as they allow the patient to gain independence for toilet training in the way that suits them most effectively.

Although urinary incontinence resolves in most cases, a recent study suggests that it can persist into adulthood with 81.8% of adults with ASD reporting any type of incontinence (Gubbiotti, Balboni, et al., 2019a). This may include fecal incontinence. Another study of adults with ASD demonstrated incontinence in 4.7% of those ages 18–29 and 22% in those older than 40 years of age (Fortuna et al., 2016). Given the small sample sizes of the studies, caution should be taken when considering the results and extrapolating to the patient's outlook on incontinence. In cases of incontinence persisting into adulthood, ASD patients would receive the same treatment plan as any other individual.

Enuresis

Enuresis is the involuntary release of urine during sleep in a child greater than 5 years old. It is considered primary enuresis if the child has never been dry for at least a 6-month period. Secondary enuresis is a recurrence of night-time wetting after a period of being dry for at least 6 months.

Similar to daytime incontinence, enuresis is a common childhood problem among those with and without NDDs. Treatment for enuresis should only be sought when the bed-wetting is inciting distress, anxiety, interfering with activities, or lowering self-esteem. This approach changes around the age of 7 where help is often encouraged. In this section, we will discuss behavior modifications as well as first- and second-line therapies to treat enuresis.

Primary care physicians may initiate management of nocturnal enuresis with either behavioral and/or pharmacologic therapies. There is no specific age or circumstance that should prompt urologic referral. If attempts at management fail to improve symptoms, urologic consultation should be considered. Also, if patients fail to show improvement with increasing age/maturity, or if symptoms become increasingly bothersome/distressing, urologic consultation should be considered. Other urologic issues such as UTIs or daytime lower urinary tract symptoms should also prompt urologic evaluation as they may be attributed to the enuresis.

Behavioral Therapy

Behavioral treatment for enuresis may involve several steps. First, more fluid intake should occur during the day and less at night. This is also termed "fluid shifting". Young children should consume 6–8 oz every 1–1.5 hours, and this fluid schedule should decrease during the 4–5 hours prior to bed. During this 4–5-hour block, liquid intake should be capped at 12 oz. It should also be noted that active individuals should attend activities well-hydrated and not consume large volumes of fluid post-event if it is later in the evening. Second, fiber consumption should be encouraged to promote at least one soft bowel movement a day. Third, caffeinated or carbonated beverages, as well as those with citric acid, should be discouraged. These ingredients can promote bladder irritation and increase urine production. Additionally, milk-based products should be avoided at night. Fourth, a regular voiding schedule should be encouraged to improve bladder health. Finally, an empty bladder at night may be achieved with double urination prior to bed. A bedtime routine with an initial urination and a second one just prior to getting into bed may be helpful (King et al., 2018). These steps are an excellent start to helping prevent enuresis; however, they may not be effective in every case. When these steps are insufficient, additional interventions such as alarms and pharmacologic agents can be utilized.

When deciding on treatment options for daytime incontinence or enuresis, comorbidities such as UTIs, obstructive sleep apnea, diabetes, and constipation should be assessed. A voiding diary (See Fig. 24.1) should be maintained to understand when the symptoms occur (e.g., in the middle of the sleep cycle, near waking, immediately after being put to sleep, or when dreams/nightmares occur). Additionally, a urine dipstick can be obtained to check for infection, proteinuria, or glycosuria. If the urine dipstick is normal, then further invasive testing is not warranted (Sharma & Paolinao, 2018).

First-Line Therapy

Enuresis alarms can be beneficial in some cases. These devices have sensors either in the bed or in the patient's clothing that make a sound when the presence of urine is detected. Recent developments have attempted to move away from an insertable device to textile moisture sensors woven into underwear to increase comfort and reduce the noticeability of the sensor to the wearer (Gaubert et al., 2020). Alarms have been successful in reducing the number of wet nights by three nights per week, compared to the control/no treatment group (Caldwell et al., 2020). Furthermore, wearing a sensor was associated with a seven times greater chance of achieving a complete response (14 nights dry) compared to no intervention (Caldwell et al., 2020). A cure as defined by two weeks of dryness was found in 80% and 85% of patients with alarm intervention of 16 and 20 weeks, respectively. A 3-month follow-up was conducted and 65% remained cured (Kosilov et al., 2018). It should be noted that patients with attention-deficit hyperactivity disorder can experience lower rates of success with this intervention (Nevéus et al., 2020).

Nevéus et al. (2020) provided guidelines regarding alarm therapy. The alarm should be used with well-motivated and informed patients. The apparatus should be demonstrated thoroughly, used nightly and, at minimum, the caretaker should awaken to the alarm. A follow-up appointment should be conducted in 1–3 weeks; if improvement is noted, continuous alarm therapy can be used until there are 14 consecutive dry days, or therapy should be stopped after 6 weeks of failure. Keten et al. looked at using an antidiuretic hormone in combination with the alarm. Although the combination treatment achieved success with preventing enuresis, the single arm of an enuresis alarm allowed for the best control when a patient relapsed (Keten et al., 2020).

Second-Line Therapy

Desmopressin is considered a borderline, first-line to strong, second-line therapy for enuresis. It is an analog to the antidiuretic human hormone vasopressin or antidiuretic hormone. It functions by binding to V2 receptors in the renal collecting duct, resulting in decreased aquaporin (water transport proteins) on the apical membrane (Bouby et al., 1984; Glazener & Evans, 2002). Thus, there is decreased water transport into the urine and, subsequently, slowed bladder filling. A common side effect of desmopressin is hyponatremia, and water intoxication syndrome can occur if a large amount of fluid is consumed following desmopressin therapy. Ideally, patients should not consume liquids two hours prior to medication administration and should continue a depleted fluid intake for the next eight hours. If possible, the patient is to be asleep for a majority, if not all, of the eight hours post-desmopressin administration (King et al., 2018).

If first-line therapy fails, second-line therapies consist of anti-cholinergic agents and tricyclic anti-depressants. Anti-cholinergic agents, such as oxybutynin and tolterodine, decrease bladder contractions and are often prescribed in cases of overactive bladder. These medicines are also effective in treating urinary urgency, frequency, and urgency urinary incontinence (Schröder & Thüroff, 2010). Anticholinergics may be given in combination with desmopressin for the treatment of enuresis, with long-acting agents used if the child sleeps more than eight hours. A 2021 study of children with primary enuresis revealed a 95% success rate after three months of combination therapy (anticholinergic and desmopressin) compared to 65% with monotherapy alone (Ghanavati et al., 2021). Ghanavati et al. advocate that this combination therapy should be used as a first-line therapy; however, constipation is a common side effect of anticholinergic agents and, in itself, can be associated with urinary incontinence.

Tricyclic anti-depressants (TCA), such as imipramine, may shorten the sleep cycle when enuresis occurs (Dean, 2012). TCAs may also possess some anticholinergic properties. As many patients with ASD and NDDs may already be on anti-depressants or similar therapies, it is important to monitor their medications closely. If a physician is considering prescribing a TCA for enuresis, it is prudent to consult the patient's other care providers to determine a treatment plan that is beneficial for the patient's overall well-being. Lastly, TCAs may be linked to prolongation of the QT interval and should be monitored closely (Okayasu et al., 2019).

Urinary Tract Infections

Approximately 7% of all febrile infants have a UTI, while 8% of children 2–19 years of age develop UTIs (potentially asymptomatic) (Shaikh et al., 2008). There is a dearth of data in the current literature regarding ASD and UTIs. One study demonstrated a greater than two-fold risk of UTI in children with ASD compared to the non-affected population; however, this study only focused on the first two years of life and not the entire pediatric spectrum (Rosen et al., 2007).

Diagnosis and treatment of UTIs should be done as early as possible to avoid adverse complications such as renal scarring, acute kidney injury, and urosepsis (Chen et al., 2003; Jahnukainen et al., 2005). As their ability to communicate may be impaired, the diagnosis of UTI in ASD patients can be challenging, and it is imperative for the clinician to have a high index of suspicion. Common presentations of UTIs can include worsening agitation, behavioral dysregulation, and urinary incontinence (Cetin et al., 2021; Sureshkumar et al., 2006).

Chronic constipation is associated with UTIs. By some estimates, 11% of chronically constipated and encopretic children have UTIs, with significantly more girls affected (Loening-Baucke, 1997). One proposed mechanism is that constipation may trigger bladder detrusor overactivity and subsequent urgency to urinate. Children may then use various posturing maneuvers to avoid urinary incontinence and attempts to obstruct urinary flow may lead to incomplete bladder emptying, bacterial colonization of the urine, and UTIs (Hellerstein & Linebarger, 2003; Jones et al., 2001). While the exact reason for the increased UTI incidence in ASD patients is unknown, their increased incidence of constipation is suspected to play a significant role (Pang & Croaker, 2011). In the Loening-Baucke study, all urologically intact patients had resolution of their UTIs once constipation was addressed, indicating a key UTI treatment target for the ASD population (Loening-Baucke, 1997).

In patients with a UTI, a mid-stream urine collection should be obtained to prevent contamination of the sample with normal skin or vaginal flora. Appropriate antibiotics can be prescribed for the patient following bacterial culture and empiric antibiotics may be appropriate in some patients while the bacterial culture is pending. Ideally, rapid screening methods could streamline bacterial testing and expedite diagnosis and treatment. One prospective study demonstrated a diaper-embedded urine test device as a safe and quick way to determine UTI presence (Paalanne et al., 2020). Similar tools may enable prompt diagnosis in a population that has a limited capacity to communicate.

Cryptorchidism and Hypospadias

Cryptorchidism, or undescended testes, is a gonadal malfunction that occurs in 2–8% of male infants (Virtanen & Toppari, 2008). Congenital cryptorchidism is hypothesized to occur from ineffective androgen/testosterone signaling (Hutson et al., 2010; Thorup et al., 2010; Virtanen & Toppari, 2008). Improper development of the reproductive tract is linked to several neurodevelopmental disorders. ASD and intellectual disabilities have associations with cryptorchidism, hypospadias, and micropenis (Rzhetsky et al., 2014). Additionally, data from the American

National Collaborative Perinatal Project suggests that patients with cryptorchidism may have an increased risk of cerebral palsy, lower IQ, poorer motor function, and mental retardation (Depue, 1988). Furthermore, a meta-analysis showed that cryptorchidism is associated with increased risk for intellectual disability, ASD, attention-deficit hyperactivity disorder, and anxiety (Chen et al., 2018). Although surgical intervention called orchiopexy is the common treatment for undescended testes, it did not show an improvement in associated risks. Therefore, patients presenting with cryptorchidism should be monitored for developmental delay, even following surgical intervention (Chen et al., 2018). Hypospadias, a condition where the urethral opening is on the underside (ventral side) of the penis, was likewise associated with an increased risk of intellectual disability, ASD, and attention-deficit hyperactivity disorder (Butwicka et al., 2015; Rzhetsky et al., 2014).

Common Urologic Conditions in Patients with Specific Neurodevelopmental Disorders

Tuberous Sclerosis Complex

Tuberous Sclerosis Complex (TSC) affects multiple organ systems with benign tumors found primarily in the brain, skin, and kidneys. TSC is inherited in an autosomal dominant fashion, in 30% of all TSC cases, with the remaining 70% attributed to spontaneous mutations. TSC has an incidence of 1 in 5800 without preference for racial or ethnic groups (Yohay, 2013). TSC results in mutations at the TSC1 gene locus (Chromosome 9q34) and TSC2 gene locus (Chromosome 16p13.3) encoding for the proteins hamartin and tuberin, respectively (Northrup et al., 1993; van Slegtenhorst et al., 1997). These proteins regulate molecular pathways for cell size, cell number, and, ultimately, organ size. The diagnosis of TSC is initiated with clinical suspicion and confirmed with genetic findings (Krueger & Northrup, 2013).

The most common finding with TSC is central nervous system involvement, which results in disorganized laminar structure, hypomyelination, epilepsy, tumorigenesis, and social behavior deficits (Carson et al., 2015). Cortical tubers found in the temporal lobe of the brain may be associated with a diagnosis of ASD (Franz et al., 2010). Up to 65% of TSC patients have ASD and TSC has a prevalence of 1–4% in the ASD population, even as high as 14% in ASD with seizure disorders (Zafeiriou et al., 2007). Renal disease is a common cause of early mortality among TSC patients. Eighty percent of patients with TSC can develop renal angiomyolipoma (AML), a tumor consisting of fat, smooth muscle, and vasculature (Dabora et al., 2001). It is estimated that 80% of children with TSC have an observable renal lesion by 10.5 years of age (Sharma & Paolinao, 2018). Renal cysts and AMLs have an incidence of 38–55% at less than six years of age, up to 80% in school-age children, and 86–100% in adults. In most cases, AMLs are asymptomatic; however, renal failure and/or hypertension may occasionally be present. Additional findings can include hematuria, intra-tumoral or retroperitoneal bleeding, and abdominal pain (Crino et al., 2006). TSC patients with AML will have multiple bilateral lesions, contrary to those with spontaneous AML. This occurs due to the tumor impairing normal kidney function. In some cases, these tumors can lead to hemorrhage from aneurysm formation (Yohay, 2013). Bilateral renal cysts commonly form in these patients; however, they present with a much smaller likelihood of hemorrhage. AMLs larger than 4 cm in diameter are associated with a higher chance of bleeding and consideration should be given to preemptive embolization (Pirson, 2013). If surgery is not an option, treatment with mammalian target of rapamycin (mTOR) inhibitors can be considered (Bissler et al., 2017). Once an AML is diagnosed, a renal ultrasound should be performed every 1–3 years, with MRI or CT imaging for any findings that are suspicious for malignancy (Yohay, 2013).

There are several other conditions associated with TSC that should be kept in mind by both care takers and physicians. Epithelial cysts, even more than AMLs, are associated with renal failure and hypertension. Glomerulocystic kidney disease is a rare case where glomerular cysts involve approximately 5% of nephrons without tubular dilatation (Murakami et al., 2012). Oncocytomas are benign epithelial tumors that are poor in fat and have polygonal cells with close-packed mitochondria. Oncocytomas may present similarly to TSC, as they can be seen on ultrasound, are bilateral, and present with hematuria. Furthermore, oncocytomas have a greater chance for a TSC patient to progress into renal cell carcinoma (RCC) than sporadic cases (Srinivas et al., 1985). Patients with TSC may develop RCC earlier and should be monitored at a younger age, with case reports demonstrating RCC as early as infancy (Washecka & Hanna, 1991). Lastly, there is weak evidence to support the association of TSC with parapelvic or perirenal lymphatic cysts in patients with extrapulmonary lymphangioleiomyomatosis, renal artery stenosis, aortic coarctation, and horseshoe kidney (Flynn et al., 1984; Matsui et al., 2000; Niemi et al., 2011).

Fragile X Syndrome

Fragile X Syndrome (FXS) is a genetic condition with a wide spectrum of developmental and cognitive delays. One-fourth to one-third of children with FXS concomitantly have ASD, and FXS makes up 2% of the ASD population (Zafeiriou et al., 2007). Patients with FXS may have anxiety, sensory processing deficits, and poor sequential learning. These factors may delay bathroom independence by up to seven years, compared to nonaffected patients (Riley et al., 2012). It is crucial that caretakers understand the delayed process, as rushing toilet training may be counterproductive. Anatomic findings of patients with FXS include enlarged testes (macroorchidism) after puberty in men and primary ovarian insufficiency in women (Zafeiriou et al., 2007). These patients may benefit from an endocrine consultation. Practitioners treating men with FXS should remain especially vigilant for the potential for the testes to twist around the spermatic cord (torsion) and patients having consistent testicular pain, as adults with macroorchidism remain more susceptible to such conditions (Flynn et al., 2002).

Down Syndrome

Trisomy 21, or Down syndrome (DS), is second only to trisomy 18 as a chromosomal abnormality resulting in congenital anomalies of kidney and urinary tract (CAKUT) (Stoll et al., 2014). In a study comparing almost 4,000 DS individuals to 3.5 million control patients, DS patients had a four to five times increased risk of CAKUT (Kupferman et al., 2009). DS and ASD coexist in up to 17% of cases (Zafeiriou et al., 2007). Ureteral and lower urinary tract abnormalities include hydroureteronephrosis, mega-bladder, ureteral stenosis, cystic dysplasia, obstructed megaureter, ureteral atresia, ureteropelvic junction abnormalities, and vesicoureteral reflux (Egli & Stalder, 1973; Mercer et al., 2004; Málaga et al., 2005). Posterior urethral valves also occur in this patient population and may lead to obstruction and reflux of urine into the bladder, ureters, and kidneys (Mercer et al., 2004).

DS also confers an increased risk for renal abnormalities. Fetal pyelectasis occurs at twice the rate of the unaffected population (Orzechowski & Berghella, 2013) and DS is associated with renal agenesis and horseshoe kidney (Limwongse, 1999). Twenty percent of DS patients have renal hypoplasia, which is the lack of a fully developed kidney, and 7% feature glomerular microcysts (Mercer et al., 2004; Subrahmanyam & Mehta, 1995). Additional renal conditions in DS include simple renal cysts, obstructive uropathy, and elevated serum creatinine (Mercer et al., 2004).

Males with DS have an increased incidence of hypospadias, cryptorchidism and hypogonadism, with the latter two associated with an increased risk of testicular cancer and infertility (Mercer et al., 2004).

The American Academy of Pediatrics has detailed the routine screenings required for persons with DS, including urological conditions (Bull, 2011). Males with Trisomy 21 should be closely monitored for testicular cancer with a routine testicular examination at each well-visit. If the testicles are not descended into the scrotum by one year of age, surgical intervention is indicated (Elseth & Hatley, 2020). All patients with DS should be routinely screened with renal ultrasound (Niamien-Attai et al., 2017) while the bladder should be imaged on ultrasound during the neonatal period (Niamien-Attai et al., 2017). Hematuria, incontinence, retention, or UTIs may be evaluated with renal and bladder ultrasound (King et al., 2018), while the need for VCUG and renal scans is dictated by the patient's clinical presentation (King et al., 2018).

Neurofibromatosis Type 1

Neurofibromatosis Type 1 (NF1) is an autosomal-dominant genetic condition occurring in 1/3000 live births (Evans et al., 2010). The NF1 gene encodes for neurofibromin, which is a tumor suppressor protein (Schröder et al., 2006). Patients who have been diagnosed with ASD have a 100-fold to 190-fold risk of an NF1 mutation, relative to the general population (Zafeiriou et al., 2007). When an NF1 mutation occurs, patients will present with café-au-lait spots, neurofibromas, optic pathway tumors, Lisch nodules, and dysplastic skeletal findings. Urological findings present in approximately 8% of NF1 mutation cases (Zafeiriou et al., 2007). Typically, this will consist of a retroperitoneal tumor in or around the urinary tract. If this is suspected due to the other clinical manifestations, an initial renal and bladder ultrasound is recommended (Schröder et al., 2006). Retroperitoneal tumors can displace the kidney or, if metastatic, can present in the kidney itself (Strauss et al., 2011). The bladder should also be imaged on ultrasound in case the retroperitoneal tumor is derived from the bladder (Purkayastha et al., 2017).

Prader–Willi Syndrome

Prader–Willi syndrome (PWS) is caused by a lack of paternally-derived imprinted genetic material on chromosome 15q11–13 and is present in 1 in 10,000 live births (Cassidy et al., 2000). Systematic reviews have revealed that 36% of PWS patients have ASD (Veltman et al., 2005). The most common findings include intellectual disability, developmental delay, and low muscle tone (Milner et al., 2005). Males with PWS and ASD often present with a small penis and scrotum and, possibly, small or undescended testicles (King et al., 2018). Hypogonadism as a result of underdeveloped gonads can lead to delayed or incomplete puberty and, ultimately, infertility (Bakker et al., 2015). Females may present with decreased clitoral and labial size and have a higher risk of delayed menses, even into their thirties. An endocrinologist should routinely see patients who present with PWS.

CHARGE Syndrome

CHARGE Syndrome (Coloboma, Heart defect, Atresia of the choanae, Retarded growth and development, Genitourinary anomalies, and Ear anomalies/deafness) occurs in 1 out of 10,000 live births (Hudson et al., 2017). ASD has been shown to be associated with CHARGE syndrome in up to 50% of cases (Lai et al., 2014; Richards et al., 2015; Zafeiriou et al., 2007). Genitourinary abnormalities involve both the urinary tract and the gonads. Forty percent of patients with CHARGE have urinary tract findings, including solitary kidney, renal hypoplasia, ectopic kidney, vesicoureteral reflux, and increased risk of UTIs (Ragan et al., 1999; Tellier et al., 1998). Renal and bladder ultrasound at birth is helpful to screen the genitourinary anatomy. In some cases, functional tests such as a VCUG may be performed (Williams, 2017). Both males and females with CHARGE syndrome frequently have gonadal developmental disorders. Eighty-five percent of males will have an underdeveloped penis or a micropenis, and 60% will have cryptorchidism. In females, 25% will have small labia and a small or absent uterus. Due to these gonadal deficits, 90% of these patients will need hormonal intervention or a lack of puberty may ensue (King et al., 2018; Kirk, 2017). As a result, these patients should be referred to an endocrinologist.

Rett Syndrome

Rett syndrome is a neurodevelopmental disorder occurring in 1 in 10,000 live female births (Amir et al., 1999; Hagberg, 1985). Almost all cases of Rett syndrome are due to a mutation in the x-linked methyl-CpG-binding-protein-2, which is heavily expressed in neuronal cells and regulates chromatin opening and condensing (Adkins & Georgel, 2011). As a result of this mutation, the loss of neuronal synapses leads to diminishing of previously acquired skills (Neul et al., 2010), including gait, purposeful hand skills, fine motor function, and spoken language. In addition, disrupted sleep, cardiac function abnormalities, breathing disturbances, cold extremities, and autonomic nervous system disruptions may be seen (Bissonnette et al., 2014; McCauley et al., 2011; Neul et al., 2010; Ogier et al., 2007; Viemari et al., 2005; Ward et al., 2011; Weese-Mayer et al., 2006). Urologic deficits associated with Rett syndrome are thought to occur from disruptions to the autonomic nervous system. In the only retrospective review of Rett syndrome patients, the most common urologic findings were urinary incontinence, kidney stones, UTIs, and urinary retention (Ward et al., 2016).

Spina Bifida

Spina bifida can be identified prenatally allowing for the establishment of a medical home prior to birth. The medical home will assist the family in coordinating subspeciality care, such as urology (Kerr & Hannon, 2019). Spina bifida is a spinal dysraphism where the spinal cord and membranes protrude out of the back through an unfused portion of the spinal column. Neural tube defects are the most common congenital central nervous system anomalies impacting between 70,000 and 100,000 people in the United States, with myelomeningocele (spinal bifida) being the most common (prevalence approximately 3.5 in 10,000 births) (Parker et al., 2010). Spina bifida is a dynamic neural tube defect that requires the continued care of a urologist to assist with management of various degrees of bowel and bladder dysfunction. Patients with spina bifida often present with varying degrees of weakness or absence of sensation affecting the lower extremities and bowel/bladder dysfunction, depending on the level of spinal lesion. The challenges evolve throughout the course of a patient's life; however, the ultimate goal of the urologist is to protect the upper urinary tracts (kidneys) from damage.

Folic acid supplementation has been shown to decrease the risk of neural tube defects prompting the CDC to recommend a 400 microgram daily supplement in women of childbearing age (Czeizel & Dudás, 1992). In the United States, fortified cereal grains have resulted in a 30–50% decreased in prevalence of spina bifida since compliance was made mandatory in 1998 (Williams et al., 2002). The level of spina bifida defect varies with lumbosacral being the most common at approximately 47% (Oakeshott et al., 2012). The level of lesion corresponds well with degree of ambulatory function, with most patients with sacral or lumbosacral lesions being ambulatory and those with high lumbar or thoracic lesions being non-ambulatory (Dicianno et al., 2015). The level of lesion does not correlate as well with bladder function; however, patients with lower level lesions and those able to ambulate are more likely to achieve continence (Liu et al., 2018).

Initial Evaluation: Management of Spina Bifida in Newborns and Infants

The urologic goals of management in patients with spina bifida include protecting the kidneys, achieving continence and independence, minimizing infections and urolithiasis, and avoiding reconstructive surgery, if possible. Typically, newborns with spina bifida undergo surgical closure of their spinal defect within the first few days of life. They may also have a ventriculoperitoneal shunt placed for management of hydrocephalus associated with Chiari malformation. A renal/bladder ultrasound is obtained during the first few days of life to evaluate for any hydronephrosis, renal parenchymal anomalies, and bladder appearance. A VCUG is often performed to further evaluate bladder appearance (smooth-walled or trabeculated) and the presence or absence of vesicoureteral reflux.

The initial management of newborns with spina bifida is highly variable and is primarily based on findings of the renal ultrasound, VCUG, and PVR (Lodwick et al., 2017). Some clinicians recommend starting infants on clean intermittent catheterization (CIC) with eventual cessation if bladder PVR volume remains low. Others will only start CIC if bladder residuals are high. The decision to proceed with CIC should be individualized, as instrumentation may increase the patient's risk for UTIs. However, consistently high PVRs, trabeculated bladders on VCUG, high-grade vesicoureteral reflux, and/or high-grade hydronephrosis should prompt initiation or continuation of CIC (Kaye et al., 2016; Timberlake et al., 2018).

A urodynamic study and repeat renal/bladder ultrasound are typically performed between two and four months of age. Urodynamic findings of bladder trabeculation, vesicoureteral reflux, high storage pressures evidenced by end fill pressure or detrusor leak point pressure > 40 cm H_2O and detrusor external sphincter dyssynergia have been associated with increased risk of UTIs and kidney scarring (Routh et al., 2016; Timberlake et al., 2018). These findings should prompt the initiation of CIC every four hours. In addition, patients may also be started on oxybutynin 0.2 mg/kg three times daily to help lower bladder filling pressure (Routh et al., 2016). Overnight continuous urethral catheter drainage can also be started to avoid high storage pressures by keeping the bladder empty throughout the night (Nguyen et al., 2005).

During the first year of life, patients with spina bifida should be assessed for any interim UTIs and renal/bladder ultrasound should be performed every 3–4 months. A urodynamic study is typically repeated at one year of age. While a DMSA (dimercaptosuccinic acid) scan may be considered in infancy, its utility is questionable in patients with no history of UTIs, vesicoureteral reflux, or bladder trabeculations (Timberlake et al., 2018).

Vesicostomy should be considered in patients with persistent UTIs or concern for renal damage despite regular CIC, anticholinergics, and overnight catheter drainage. A vesicostomy is a procedure in which a small opening is made just below the umbilicus to allow urine to drain from the bladder. These procedures are typically very successful in protecting the kidneys and may halt the need for further CIC and anticholinergics; however, they may be associated with risks such as bladder prolapse and stomal stenosis (Dönmez et al., 2017; Lee & Greenfield, 2005). The vesicostomy is typically reversed at the time of future reconstructive surgeries.

Management of Spina Bifida During Childhood

As patients age, the importance of achieving bowel and bladder continence becomes more important. Some children familiar with CIC, with or without the use of anticholinergics, may already be continent of urine and a smaller percentage may even void volitionally with the ability to remain continent between voids. CIC with anticholinergics for continence alone may be considered between the ages of 2 and 5, with most parents preferring to start sometime before school age (Timberlake et al., 2018). CIC is typically recommended every 3–4 hours. Anticholinergics such as oxybutynin may be added if the child experiences persistent leakage between voids. The frequency of CIC may be adjusted as needed by the patient and family to allow for fewer leakage episodes.

Another option for producing continence is onabotulinumtoxinA, which causes detrusor relaxation by preventing the release of acetylcholine at the neuromuscular junction. Typically, 200 units are injected in 20–30 sites throughout the bladder. In pediatric patients, the rate of continence varies widely from 30% to 100%, and repeat injections are needed approximately every 6–12 months (Hascoet et al., 2017). In 2021, the U.S. FDA approved intravesical onabotulinumtoxinA injection in pediatric patients five years of age or older who have inadequate response to, or are intolerant of, anticholinergics.

Bowel continence remains a challenging aspect of care. Spina bifida patients typically have both neurogenic bladder and bowel and a regimen for bowel continence should be established early. Most regimens use a combination of stool softeners/laxatives as well as a motility agent to promote bowel emptying (Ambartsumyan & Rodriguez, 2018; Freeman et al., 2017). Other methods such as timed toileting, suppositories, and enemas are supplemented to optimize bowel continence; however, finding a successful regimen often requires trial and error, as well as motivated patients and families.

Management of Spina Bifida During Adolescence and Adulthood

As children enter adolescence, approximately 25% of them will develop signs and symptoms concerning for symptomatic tethered cord and may require eventual cord release (Alsowayan et al., 2016; Phuong et al., 2002). Tethered cord may present with changes in bladder function as well as changes in lower extremity function and/or back pain (Alsowayan et al., 2016; Tarcan et al., 2006). Suspicion for tethered cord should be addressed with the patient's care team in order to facilitate tethered cord release surgery if needed.

The transition into adolescence and adulthood is marked by an emphasis on privacy, independence, and transition to a urologist who treats predominantly adults. The ability to achieve independence is a highly individualized process; however, attention to certain attendant conditions (i.e. obesity) can help to promote a more independent lifestyle (Polfuss et al., 2017). Attention to manual dexterity/fine motor function should also be considered when counseling patients on various catheter options.

With regard to transition to an adult urologist, insurance coverage and type represents a significant barrier to a successful transition. Often, social work and financial counseling services are implemented to help transition patients successfully with conversations regarding transition happening as early as age 16 (Grimsby et al., 2016). Transitional care should include an annual visit with renal ultrasound, serum creatinine, and assessment for any issues such as interim UTIs or new incontinence. A urodynamic study may be warranted if there is a change in urologic function or a reconstructive intervention is planned. Careful consideration for urolithiasis should be also given, as calculi are common surgical issues in adults.

Reconstructive Surgery in Patients with Spina Bifida

Reconstructive surgery may be considered in spina bifida patients with neurogenic bladder when there is a concern for kidney damage from a poorly compliant bladder and/or when continence and independence may be improved. The three main reconstructive surgeries function to increase bladder size and compliance, increase bladder outlet resistance, and improve bladder emptying by creating a catheterizable channel.

In patients with poorly compliant or small bladders, conservative management with CIC and medications may not be enough to prevent high bladder filling pressures, vesicoureteral reflux, and eventual renal deterioration. Surgical intervention should be offered at this time. Intravesical onabotulinumtoxinA has been shown to decrease overall bladder pressures as well as the frequency and amplitude of bladder contractions (Hascoet et al., 2017). The procedure is performed cystoscopically with direct injection into the detrusor muscle. The procedure is minimally invasive, and patients may return home the same day. However, the effects are limited, and repeat injections may be needed every 6–12 months. Bladder augmentation offers a more permanent option to increase bladder volume and compliance. Typically, bladder augmentation (augmentation cystoplasty) is performed by using detubularized and reconfigured small bowel or colon as a patch graft over a widely opened bladder. Bladder augmentation improves continence by enlarging bladder capacity, increasing the time interval between catheterizations, and protects the kidneys by improving compliance. However, the patient should be prepared for bladder irrigations and monitoring for some metabolic abnormalities may be necessary.

In patients with low outlet resistance and subsequent incontinence, several surgical options may be offered including formal urethral reconstruction, bladder neck slings, bladder neck bulking agents, and/or suburethral slings. These surgeries may be performed in the same setting as bladder augmentation to streamline care.

Since transferring and mobility are often difficult for patients with spina bifida, catheterization of the urethra can become challenging. Continent catheterizable channels allow for direct access to the bladder using a small abdominal opening. The two most common channels are the appendicovesicostomy (APV) and the tubularized intestinal catheterizable channel (Yang-Monti) (Mitrofanoff, 1980; Yang, 1993). Similar strategies may be used in the management of bowel continence and encopresis with an appendicocecostomy or a cecostomy tube which allows for antegrade continence enemas and improved bowel regimens (Malone et al., 1990).

Spina bifida is a complex condition that requires a large care team for the evolving problems faced throughout the patient's life. Urologists play a key role from birth until adulthood with the ultimate goal being to protect the kidneys. An understanding of the unique storage and emptying dynamics associated with neurogenic bladder allows providers to better care for patients with spina bifida and better understand the challenges relating to continence, independence, and transitions of care often faced by this population.

Clinical Pearls

- The recommended clinical urologic evaluation in patients with ASD does not differ significantly from the general population. Patient-specific strategies may be used to ease the anxiety of the patient and promote an easier and more productive office visit.
- Although an ASD diagnosis does not automatically categorize an individual with specific urologic conditions, some genetic syndromes linked to ASD or other NDDs may have predispositions for certain urologic findings.
- Referring physicians should make sure that the referred party is aware that a genitourinary exam is standard during urologic evaluation. The patients should know that examination of their intimate areas is only appropriate in the presence of a physician/clinician and/or caregivers.
- The major focus should be on behavioral training for incontinence as most patients will achieve continence; however, the age at which this occurs can vary. Including images, guides, step-by-step instructions, and providing designated rewards upon step completion can help. It is imperative not to rush this process, as regression or urinary anxiety may occur.
- Treatment of constipation is crucial to achieving urinary continence.
- Spina Bifida patients should have long-term care, with a plan in place to protect the kidneys. Understanding a neurogenic bladder will yield better care for patients who may present with continence, independence, and transitions of care.

Bibliography

Adkins, N. L., & Georgel, P. T. (2011). MeCP2: Structure and function. *Biochemistry and Cell Biology, 89*(1), 1–11. https://doi.org/10.1139/o10-112

Alsowayan, O., Alzahrani, A., Farmer, J. P., Capolicchio, J. P., Jednak, R., & El-Sherbiny, M. (2016). Comprehensive analysis of the clinical and urodynamic outcomes of primary tethered spinal cord before and after spinal cord untethering. *Journal of Pediatric Urology, 12*(5), 285.e281–285.e285. https://doi.org/10.1016/j.jpurol.2016.02.012

Ambartsumyan, L., & Rodriguez, L. (2018). Bowel management in children with spina bifida. *Journal of Pediatric Rehabilitation Medicine*. https://doi.org/10.3233/prm-180533

Amir, R. E., Van den Veyver, I. B., Wan, M., Tran, C. Q., Francke, U., & Zoghbi, H. Y. (1999). Rett syndrome is caused by mutations in X-linked MECP2, encoding methyl-CpG-binding protein 2. *Nature Genetics, 23*(2), 185–188. https://doi.org/10.1038/13810

Bakker, N. E., Wolffenbuttel, K. P., Looijenga, L. H., & Hokken-Koelega, A. C. (2015). Testes in infants with Prader-Willi syndrome: Human chorionic gonadotropin treatment, surgery and histology. *The Journal of Urology, 193*(1), 291–298. https://doi.org/10.1016/j.juro.2014.07.113

Baskin, L. S., & K. B. (2005). *Handbook of pediatric urology* (Illustrated ed.). Lippincott Williams & Wilkins.

Bissler, J. J., Kingswood, J. C., Radzikowska, E., Zonnenberg, B. A., Belousova, E., Frost, M. D., Sauter, M., Brakemeier, S., de Vries, P. J., Berkowitz, N., Voi, M., Peyrard, S., & Budde, K. (2017). Everolimus long-term use in patients with tuberous sclerosis complex: Four-year update of the EXIST-2 study. *PLoS One, 12*(8), e0180939. https://doi.org/10.1371/journal.pone.0180939

Bissonnette, J. M., Schaevitz, L. R., Knopp, S. J., & Zhou, Z. (2014). Respiratory phenotypes are distinctly affected in mice with common Rett syndrome mutations MeCP2 T158A and R168X. *Neuroscience, 267*, 166–176. https://doi.org/10.1016/j.neuroscience.2014.02.043

Bloom, D. A., Seeley, W. W., Ritchey, M. L., & McGuire, E. J. (1993). Toilet habits and continence in children: An opportunity sampling in search of normal parameters. *The Journal of Urology, 149*(5), 1087–1090. https://doi.org/10.1016/s0022-5347(17)36304-8

Bouby, N., Trinh-Trang-Tan, M. M., & Bankir, L. (1984). Stimulation of tubular reabsorption of magnesium and calcium by antidiuretic hormone in conscious rats. Study in Brattleboro rats with hereditary hypothalamic diabetes insipidus. *Pflügers Archiv, 402*(4), 458–464. https://doi.org/10.1007/bf00583948

Bull, M. J. (2011). Health supervision for children with Down syndrome. *Pediatrics, 128*(2), 393–406. https://doi.org/10.1542/peds.2011-1605

Butwicka, A., Lichtenstein, P., Landén, M., Nordenvall, A. S., Nordenström, A., Nordenskjöld, A., & Frisén, L. (2015). Hypospadias and increased risk for neurodevelopmental disorders. *Journal of Child Psychology and Psychiatry, 56*(2), 155–161. https://doi.org/10.1111/jcpp.12290

Byrd, R. S., Weitzman, M., Lanphear, N. E., & Auinger, P. (1996). Bed-wetting in US children: Epidemiology and related behavior problems. *Pediatrics, 98*(3 Pt 1), 414–419.

Caldwell, P. H., Codarini, M., Stewart, F., Hahn, D., & Sureshkumar, P. (2020). Alarm interventions for nocturnal enuresis in children. *Cochrane Database System Review, 5*(5), CD002911. https://doi.org/10.1002/14651858.CD002911.pub3

Carson, R. P., Kelm, N. D., West, K. L., Does, M. D., Fu, C., Weaver, G., McBrier, E., Parker, B., Grier, M. D., & Ess, K. C. (2015). Hypomyelination following deletion of Tsc2 in oligodendrocyte precursors. *Annals of Clinical Translational Neurology, 2*(12), 1041–1054. https://doi.org/10.1002/acn3.254

Cassidy, S. B., Dykens, E., & Williams, C. A. (2000). Prader-Willi and Angelman syndromes: sister imprinted disorders. *American Journal of Medical Genetics, 97*(2), 136–146. https://doi.org/10.1002/1096-8628(200022)97:2<136::aid-ajmg5>3.0.co;2-v

Cetin, N., Kavaz Tufan, A., & Tokar, B. (2021). The risk factors for recurrence of febrile urinary tract infection and renal scarring in children with functional urinary incontinence. *Lower Urinary Tract Symptoms, 13*(1), 160–167. https://doi.org/10.1111/luts.12349

Chen, J., Sørensen, H. T., Miao, M., Liang, H., Ehrenstein, V., Wang, Z., Yuan, W., & Li, J. (2018). Cryptorchidism and increased risk of neurodevelopmental disorders. *Journal of Psychiatric Research, 96*, 153–161. https://doi.org/10.1016/j.jpsychires.2017.10.006

Chen, M., Jahnukainen, T., Bao, W., Daré, E., Ceccatelli, S., & Celsi, G. (2003). Uropathogenic Escherichia coli toxins induce caspase-independent apoptosis in renal proximal tubular cells via ERK signaling. *American Journal of Nephrology, 23*(3), 140–151. https://doi.org/10.1159/000069853

Crino, P. B., Nathanson, K. L., & Henske, E. P. (2006). The tuberous sclerosis complex. *The New England Journal of Medicine, 355*(13), 1345–1356. https://doi.org/10.1056/NEJMra055323

Czeizel, A. E., & Dudás, I. (1992). Prevention of the first occurrence of neural-tube defects by periconceptional vitamin supplementation. *The New England Journal of Medicine, 327*(26), 1832–1835. https://doi.org/10.1056/nejm199212243272602

Dabora, S. L., Jozwiak, S., Franz, D. N., Roberts, P. S., Nieto, A., Chung, J., Choy, Y. S., Reeve, M. P., Thiele, E., Egelhoff, J. C., Kasprzyk-Obara, J., Domanska-Pakiela, D., & Kwiatkowski, D. J. (2001). Mutational analysis in a cohort of 224 tuberous sclerosis patients indicates increased severity of TSC2, compared with TSC1, disease in multiple organs. *American Journal of Human Genetics, 68*(1), 64–80. https://doi.org/10.1086/316951

De Gennaro, M., Niero, M., Capitanucci, M. L., von Gontard, A., Woodward, M., Tubaro, A., & Abrams, P. (2010). Validity of the international consultation on incontinence questionnaire-pediatric lower urinary tract symptoms: A screening questionnaire for children. *The Journal of Urology, 184*(4 Suppl), 1662–1667. https://doi.org/10.1016/j.juro.2010.03.075

Dean, L. (2012). Imipramine therapy and CYP2D6 and CYP2C19 genotype. In V. M. Pratt, S. A. Scott, M. Pirmohamed, B. Esquivel, M. S. Kane, B. L. Kattman, & A. J. Malheiro (Eds.), *Medical genetics summaries*. National Center for Biotechnology Information (US).

Depue, R. H. (1988). Cryptorchidism, and epidemiologic study with emphasis on the relationship to central nervous system dysfunction. *Teratology, 37*(4), 301–305. https://doi.org/10.1002/tera.1420370403

Diaz-Saldano, D. (2017). Assessment of children. In *Core curriculum for urologic nursing* (A. J. Pitman NJ, Ed. 1st ed.). Society for Urologic Nurses and Assoc.

Dicianno, B. E., Karmarkar, A., Houtrow, A., Crytzer, T. M., Cushanick, K. M., McCoy, A., Wilson, P., Chinarian, J., Neufeld, J., Smith, K., & Collins, D. M. (2015). Factors associated with mobility outcomes in a National Spina Bifida Patient Registry. *American Journal of Physical Medicine & Rehabilitation, 94*(12), 1015–1025. https://doi.org/10.1097/phm.0000000000000404

Dönmez, M., Carrasco, A., Jr., Saltzman, A. F., Vemulakonda, V., & Wilcox, D. T. (2017). Long-term outcomes of cutaneous vesicostomy in patients with neuropathic bladder caused by spina bifida. *Journal of Pediatric Urology, 13*(6), 622.e621–622.e624. https://doi.org/10.1016/j.jpurol.2017.05.015

Egli, F., & Stalder, G. (1973). Malformations of kidney and urinary tract in common chromosomal aberrations. I. Clinical studies. *Humangenetik, 18*(1), 1–15. https://doi.org/10.1007/bf00279026

Elseth, A., & Hatley, R. M. (2020). Orchiopexy. In *StatPearls*. StatPearls Publishing Copyright © 2020, StatPearls Publishing LLC.

Evans, D. G., Howard, E., Giblin, C., Clancy, T., Spencer, H., Huson, S. M., & Lalloo, F. (2010). Birth incidence and prevalence of tumor-prone syndromes: Estimates from a UK family genetic register service. *American Journal of Medical Genetics A, 152a*(2), 327–332. https://doi.org/10.1002/ajmg.a.33139

Flynn, B. J., Myers, S. M., Cera, P. J., & Mowad, J. J. (2002). Testicular torsion in an adolescent with Fragile X Syndrome. *Pediatrics, 109*(1), E16. https://doi.org/10.1542/peds.109.1.e16

Flynn, P. M., Robinson, M. B., Stapleton, F. B., Roy, S., 3rd, Koh, G., & Tonkin, I. L. (1984). Coarctation of the aorta and renal artery stenosis in tuberous sclerosis. *Pediatric Radiology, 14*(5), 337–339. https://doi.org/10.1007/bf01601889

Fortuna, R. J., Robinson, L., Smith, T. H., Meccarello, J., Bullen, B., Nobis, K., & Davidson, P. W. (2016). Health conditions and functional status in adults with autism: A cross-sectional evaluation. *Journal of General Internal Medicine, 31*(1), 77–84. https://doi.org/10.1007/s11606-015-3509-x

Franz, D. N., Bissler, J. J., & McCormack, F. X. (2010). Tuberous sclerosis complex: Neurological, renal and pulmonary manifestations. *Neuropediatrics, 41*(5), 199–208. https://doi.org/10.1055/s-0030-1269906

Freeman, K. A., Castillo, H., Castillo, J., Liu, T., Schechter, M., Wiener, J. S., Thibadeau, J., Ward, E., & Brei, T. (2017). Variation in bowel and bladder continence across US spina bifida programs: A descriptive study. *Journal of Pediatric Rehabilitation Medicine, 10*(3–4), 231–241. https://doi.org/10.3233/prm-170450

Gaubert, V., Gidik, H., & Koncar, V. (2020). Boxer underwear incorporating textile moisture sensor to prevent nocturnal enuresis. *Sensors (Basel), 20*(12). https://doi.org/10.3390/s20123546

Ghanavati, P. M., Khazaeli, D., & Amjadzadeh, M. (2021). A comparison of the efficacy and tolerability of treating primary nocturnal enuresis with Solifenacin Plus Desmopressin, Tolterodine Plus Desmopressin, and Desmopressin alone: A randomized controlled clinical trial. *International Braz J Urol, 47*(1), 73–81. https://doi.org/10.1590/s1677-5538.ibju.2019.0448

Glazener, C. M., & Evans, J. H. (2002). Desmopressin for nocturnal enuresis in children. *Cochrane Database Systematic Review*, (3), Cd002112. https://doi.org/10.1002/14651858.cd002112

Gor, R. A., Fuhrer, J., & Schober, J. M. (2012). A retrospective observational study of enuresis, daytime voiding symptoms, and response to medical therapy in children with attention deficit hyperactivity disorder and autism spectrum disorder. *Journal of Pediatric Urology, 8*(3), 314–317. https://doi.org/10.1016/j.jpurol.2010.10.009

Grimsby, G. M., Burgess, R., Culver, S., Schlomer, B. J., & Jacobs, M. A. (2016). Barriers to transition in young adults with neurogenic bladder. *Journal of Pediatric Urology, 12*(4), 258.e251–258.e255. https://doi.org/10.1016/j.jpurol.2016.04.015

Gubbiotti, M., Balboni, G., Bini, V., Elisei, S., Bedetti, C., Marchiafava, M., & Giannantoni, A. (2019a). Bladder and bowel dysfunction, adaptive behaviour and psychiatric profiles in adults affected by autism spectrum disorders. *Neurourology and Urodynamics, 38*(7), 1866–1873. https://doi.org/10.1002/nau.24081

Gubbiotti, M., Elisei, S., Bedetti, C., Marchiafava, M., & Giannantoni, A. (2019b). Urinary and bowel dysfunction in autism spectrum disorder: A prospective, observational study. *Psychiatria Danubina, 31*(Suppl 3), 475–478.

Hagberg, B. (1985). Rett syndrome: Swedish approach to analysis of prevalence and cause. *Brain Development, 7*(3), 276–280.

Hascoet, J., Manunta, A., Brochard, C., Arnaud, A., Damphousse, M., Menard, H., Kerdraon, J., Journel, H., Bonan, I., Odent, S., Fremond, B., Siproudhis, L., Gamé, X., & Peyronnet, B. (2017). Outcomes of intra-detrusor injections of botulinum toxin in patients with spina bifida: A systematic review. *Neurourology and Urodynamics, 36*(3), 557–564. https://doi.org/10.1002/nau.23025

Hellerstein, S., & Linebarger, J. S. (2003). Voiding dysfunction in pediatric patients. *Clinical Pediatrics (Phila), 42*(1), 43–49. https://doi.org/10.1177/000992280304200107

Hudson, A., Trider, C. L., & Blake, K. (2017). CHARGE syndrome. *Pediatr Review, 38*, 56–59. https://doi.org/10.1542/pir.2016-0050

Hutson, J. M., Balic, A., Nation, T., & Southwell, B. (2010). Cryptorchidism. *Seminars in Pediatric Surgery, 19*(3), 215–224. https://doi.org/10.1053/j.sempedsurg.2010.04.001

Jahnukainen, T., Chen, M., & Celsi, G. (2005). Mechanisms of renal damage owing to infection. *Pediatric Nephrology, 20*(8), 1043–1053. https://doi.org/10.1007/s00467-005-1898-5

Jones, E., Dimmock, P. W., & Spencer, S. A. (2001). A randomised controlled trial to compare methods of milk expression after preterm delivery. *Archives of*

Disease in Childhood. Fetal and Neonatal Edition, 85(2), F91–F95. https://doi.org/10.1136/fn.85.2.f91

Kaye, I. Y., Payan, M., & Vemulakonda, V. M. (2016). Association between clean intermittent catheterization and urinary tract infection in infants and toddlers with spina bifida. *Journal of Pediatric Urology, 12*(5), 284.e281–284.e286. https://doi.org/10.1016/j.jpurol.2016.02.010

Kerr, L. M., & Hannon, A. (2019). *Spina Bifida*. https://doi.org/https://www.medicalhomeportal.org/diagnoses-and-conditions/spina-bifida

Keten, T., Aslan, Y., Balci, M., Erkan, A., Senel, C., Oguz, U., Kayali, M., Guzel, O., Karabulut, E., & Tuncel, A. (2020). Comparison of the efficacy of desmopressin fast-melting formulation and enuretic alarm in the treatment of monosymptomatic nocturnal enuresis. *Journal of Pediatric Urology*. https://doi.org/10.1016/j.jpurol.2020.07.018

King, S., M., Y., H., R., & T., W. (2018). Urology. Humana Press.

Kirk, J. (2017). Genital abnormalities.

Kosilov, K. V., Geltser, B. I., Loparev, S. A., Kuzina, I. G., Shakirova, O. V., Zhuravskaya, N. S., & Lobodenko, A. (2018). The optimal duration of alarm therapy use in children with primary monosymptomatic nocturnal enuresis. *Journal of Pediatric Urology, 14*(5), 447.e441–447.e446. https://doi.org/10.1016/j.jpurol.2018.03.021

Kroeger, K., & Sorensen-Burnworth, R. (2009). Toilet training individuals with autism and other developmental disabilities: A critical review. *Research in Autism Spectrum Disorders, 3*(3), 607–618. https://doi.org/10.1016/j.rasd.2009.01.005

Krueger, D. A., & Northrup, H. (2013). Tuberous sclerosis complex surveillance and management: Recommendations of the 2012 international tuberous sclerosis complex consensus conference. *Pediatric Neurology, 49*(4), 255–265. https://doi.org/10.1016/j.pediatrneurol.2013.08.002

Kupferman, J. C., Druschel, C. M., & Kupchik, G. S. (2009). Increased prevalence of renal and urinary tract anomalies in children with Down syndrome. *Pediatrics, 124*(4), e615–e621. https://doi.org/10.1542/peds.2009-0181

Lai, M. C., Lombardo, M. V., & Baron-Cohen, S. (2014). Autism. *Lancet, 383*(9920), 896–910. https://doi.org/10.1016/s0140-6736(13)61539-1

Lee, M. W., & Greenfield, S. P. (2005). Intractable high-pressure bladder in female infants with spina bifida: Clinical characteristics and use of vesicostomy. *Urology, 65*(3), 568–571. https://doi.org/10.1016/j.urology.2004.11.034

Limwongse, C. (1999). Syndromes and malformations of the urinary tract. *Pediatric Nephrology*, 427–452.

Liu, T., Ouyang, L., Thibadeau, J., Wiener, J. S., Routh, J. C., Castillo, H., Castillo, J., Freeman, K. A., Sawin, K. J., Smith, K., Van Speybroeck, A., & Valdez, R. (2018). Longitudinal study of bladder continence in patients with spina bifida in the national spina bifida patient registry. *The Journal of Urology, 199*(3), 837–843. https://doi.org/10.1016/j.juro.2017.11.048

Lodwick, D., Asti, L., Deans, K., Minneci, P., & McLeod, D. (2017). Variation in practice patterns for the Management of Newborn Spina Bifida in the United States. *Urology, 100*, 207–212. https://doi.org/10.1016/j.urology.2016.08.003

Loening-Baucke, V. (1997). Urinary incontinence and urinary tract infection and their resolution with treatment of chronic constipation of childhood. *Pediatrics, 100*(2 Pt 1), 228–232. https://doi.org/10.1542/peds.100.2.228

Malone, P. S., Ransley, P. G., & Kiely, E. M. (1990). Preliminary report: The antegrade continence enema. *Lancet, 336*(8725), 1217–1218. https://doi.org/10.1016/0140-6736(90)92834-5.

Maternik, M., Krzeminska, K., & Zurowska, A. (2015). The management of childhood urinary incontinence. *Pediatric Nephrology, 30*(1), 41–50. https://doi.org/10.1007/s00467-014-2791-x

Matsui, K., Tatsuguchi, A., Valencia, J., Yu, Z., Bechtle, J., Beasley, M. B., Avila, N., Travis, W. D., Moss, J., & Ferrans, V. J. (2000). Extrapulmonary lymphangioleiomyomatosis (LAM): Clinicopathologic features in 22 cases. *Human Pathology, 31*(10), 1242–1248. https://doi.org/10.1053/hupa.2000.18500

McCauley, M. D., Wang, T., Mike, E., Herrera, J., Beavers, D. L., Huang, T. W., Ward, C. S., Skinner, S., Percy, A. K., Glaze, D. G., Wehrens, X. H., & Neul, J. L. (2011). Pathogenesis of lethal cardiac arrhythmias in Mecp2 mutant mice: Implication for therapy in Rett syndrome. *Science Translational Medicine, 3*(113), 113ra125. https://doi.org/10.1126/scitranslmed.3002982

Mercer, E. S., Broecker, B., Smith, E. A., Kirsch, A. J., Scherz, H. C., & C, A. M. (2004). Urological manifestations of Down syndrome. *The Journal of Urology, 171*(3), 1250–1253. https://doi.org/10.1097/01.ju.0000112915.69436.91

Milner, K. M., Craig, E. E., Thompson, R. J., Veltman, M. W., Thomas, N. S., Roberts, S., Bellamy, M., Curran, S. R., Sporikou, C. M., & Bolton, P. F. (2005). Prader-Willi syndrome: Intellectual abilities and behavioural features by genetic subtype. *Journal of Child Psychology and Psychiatry, 46*(10), 1089–1096. https://doi.org/10.1111/j.1469-7610.2005.01520.x

Mitrofanoff, P. (1980). Trans-appendicular continent cystostomy in the management of the neurogenic bladder. *Chiropractic Pediatrics, 21*(4), 297–305. (Cystostomie continente trans-appendiculaire dans le traitement des vessies neurologiques.).

Murakami, A., Gomi, K., Tanaka, M., Ohyama, M., Itani, Y., Ishikawa, H., Aida, N., Furuya, M., & Tanaka, Y. (2012). Unilateral glomerulocystic kidney disease associated with tuberous sclerosis complex in a neonate. *Pathology International, 62*(3), 209–215. https://doi.org/10.1111/j.1440-1827.2011.02777.x

Málaga, S., Pardo, R., Málaga, I., Orejas, G., & Fernández-Toral, J. (2005). Renal involvement in Down syndrome. *Pediatric Nephrology, 20*(5), 614–617. https://doi.org/10.1007/s00467-005-1825-9

Neul, J. L., Kaufmann, W. E., Glaze, D. G., Christodoulou, J., Clarke, A. J., Bahi-Buisson, N., Leonard, H., Bailey, M. E., Schanen, N. C., Zappella, M., Renieri, A., Huppke, P., & Percy, A. K. (2010). Rett syndrome: Revised diagnostic criteria and nomenclature. *Annals of Neurology, 68*(6), 944–950. https://doi.org/10.1002/ana.22124

Nevéus, T., Fonseca, E., Franco, I., Kawauchi, A., Kovacevic, L., Nieuwhof-Leppink, A., Raes, A., Tekgül, S., Yang, S. S., & Rittig, S. (2020). Management and treatment of nocturnal enuresis-an updated standardization document from the International Children's Continence Society. *Journal of Pediatric Urology, 16*(1), 10–19. https://doi.org/10.1016/j.jpurol.2019.12.020

Nevéus, T., von Gontard, A., Hoebeke, P., Hjälmås, K., Bauer, S., Bower, W., Jørgensen, T. M., Rittig, S., Walle, J. V., Yeung, C. K., & Djurhuus, J. C. (2006). The standardization of terminology of lower urinary tract function in children and adolescents: Report from the Standardisation Committee of the International Children's Continence Society. *The Journal of Urology, 176*(1), 314–324. https://doi.org/10.1016/s0022-5347(06)00305-3.

Nguyen, M. T., Pavlock, C. L., Zderic, S. A., Carr, M. C., & Canning, D. A. (2005). Overnight catheter drainage in children with poorly compliant bladders improves post-obstructive diuresis and urinary incontinence. *The Journal of Urology, 174*(4 Pt 2), 1633–1636.; ; discussion 1636. https://doi.org/10.1097/01.ju.0000179394.57859.9d

Niamien-Attai, C., Bacchetta, J., Ranchin, B., Sanlaville, D., & Cochat, P. (2017). Renal abnormalities in Down syndrome: A review. *Archives of Pediatrics*, 24(10), 1013-1018. https://doi.org/10.1016/j.arcped.2017.07.014 (Atteintes rénales de la trisomies 21.)

Nicolaidis, C., Kripke, C. C., & Raymaker, D. (2014). Primary care for adults on the autism spectrum. *The Medical Clinics of North America, 98*(5), 1169–1191. https://doi.org/10.1016/j.mcna.2014.06.011

Niemi, A. K., Northrup, H., Hudgins, L., & Bernstein, J. A. (2011). Horseshoe kidney and a rare TSC2 variant in two unrelated individuals with tuberous sclerosis complex. *American Journal of Medicine Genetics A, 155a*(10), 2534–2537. https://doi.org/10.1002/ajmg.a.34197

Northrup, H., Koenig, M. K., Pearson, D. A., & Au, K. S. (1993). Tuberous sclerosis complex. In M. P. Adam, H. H. Ardinger, R. A. Pagon, S. E. Wallace, L. J. H. Bean, K. Stephens, & A. Amemiya (Eds.), *GeneReviews(®)*. University of Washington. Copyright © 1993–2020, University of Washington, Seattle. GeneReviews is a registered trademark of the University of Washington, Seattle. All rights reserved.

Oakeshott, P., Hunt, G. M., Poulton, A., & Reid, F. (2012). Open spina bifida: Birth findings predict long-term outcome. *Archives of Disease in Childhood, 97*(5), 474–476. https://doi.org/10.1136/archdischild-2011-300624

Ogier, M., Wang, H., Hong, E., Wang, Q., Greenberg, M. E., & Katz, D. M. (2007). Brain-derived neurotrophic factor expression and respiratory function improve after ampakine treatment in a mouse model of Rett syndrome. *The Journal of Neuroscience, 27*(40), 10912–10917. https://doi.org/10.1523/jneurosci.1869-07.2007

Okayasu, H., Ozeki, Y., Fujii, K., Takano, Y., Shinozaki, T., Ohrui, M., & Shimoda, K. (2019). Investigation of the proarrhythmic effects of antidepressants according to QT interval, QT dispersion and T wave peak-to-end interval in the clinical setting. *Psychiatry Investigation, 16*(2), 159–166. https://doi.org/10.30773/pi.2018.12.11

Orzechowski, K. M., & Berghella, V. (2013). Isolated fetal pyelectasis and the risk of down syndrome: A meta-analysis. *Ultrasound in Obstetrics & Gynecology, 42*(6), 615–621. https://doi.org/10.1002/uog.12516

Paalanne, N., Wikstedt, L., Pokka, T., Salo, J., Uhari, M., Renko, M., & Tapiainen, T. (2020). Diaper-embedded urine test device for the screening of urinary tract infections in children: A cohort study. *BMC Pediatrics, 20*(1), 378. https://doi.org/10.1186/s12887-020-02277-5

Pang, K. H., & Croaker, G. D. (2011). Constipation in children with autism and autistic spectrum disorder. *Pediatric Surgery International, 27*(4), 353–358. https://doi.org/10.1007/s00383-010-2680-8

Parker, S. E., Mai, C. T., Canfield, M. A., Rickard, R., Wang, Y., Meyer, R. E., Anderson, P., Mason, C. A., Collins, J. S., Kirby, R. S., & Correa, A. (2010). Updated National Birth Prevalence estimates for selected birth defects in the United States, 2004-2006. *Birth Defects Research. Part A, Clinical and Molecular Teratology, 88*(12), 1008–1016. https://doi.org/10.1002/bdra.20735

Phuong, L. K., Schoeberl, K. A., & Raffel, C. (2002). Natural history of tethered cord in patients with meningomyelocele. *Neurosurgery, 50*(5), 989–993.; ; discussion 993–985. https://doi.org/10.1097/00006123-200205000-00011

Pirson, Y. (2013). Tuberous sclerosis complex-associated kidney angiomyolipoma: From contemplation to action. *Nephrology, Dialysis, Transplantation, 28*(7), 1680–1685. https://doi.org/10.1093/ndt/gft009

Polfuss, M., Bandini, L. G., & Sawin, K. J. (2017). Obesity prevention for individuals with spina bifida. *Current Obesity Reports, 6*(2), 116–126. https://doi.org/10.1007/s13679-017-0254-y

Purkayastha, A., Sharma, N., Vashisth, R., & Kishore, B. (2017). Transitional cell carcinoma of urinary bladder manifesting as extensive retroperitoneal and axillary lymph node metastasis: An extremely rare case scenario detected by (18)F-fluorodeoxyglucose positron emission tomography scan. *Indian Journal of Nuclear Medicine, 32*, 351–354. https://doi.org/10.4103/ijnm.IJNM_52_17

Ragan, D. C., Casale, A. J., Rink, R. C., Cain, M. P., & Weaver, D. D. (1999). Genitourinary anomalies in the CHARGE association. *The Journal of Urology, 161*(2), 622–625.

Richards, C., Jones, C., Groves, L., Moss, J., & Oliver, C. (2015). Prevalence of autism spectrum disorder phenomenology in genetic disorders: A systematic review and meta-analysis. *Lancet Psychiatry, 2*(10), 909–916. https://doi.org/10.1016/s2215-0366(15)00376-4

Riley, K., Picker, J., & Nasir, R. (2012). *Consensus of the fragile X clinical & research consortium on clinical practices*. National Fragile X Foundation. Retrieved August 2020. From https://fragilex.org/wp-content/uploads/2012/08/Toileting-Issues-in-Fragile-X-Syndrome2012-Oct.pdf

Rosen, N. J., Yoshida, C. K., & Croen, L. A. (2007). Infection in the first 2 years of life and autism spectrum disorders. *Pediatrics, 119*(1), e61–e69. https://doi.org/10.1542/peds.2006-1788

Routh, J. C., Cheng, E. Y., Austin, J. C., Baum, M. A., Gargollo, P. C., Grady, R. W., Herron, A. R., Kim, S. S., King, S. J., Koh, C. J., Paramsothy, P., Raman, L., Schechter, M. S., Smith, K. A., Tanaka, S. T., Thibadeau, J. K., Walker, W. O., Wallis, M. C., Wiener, J. S., & Joseph, D. B. (2016). Design and methodological considerations of the centers for disease control and prevention urologic and renal protocol for the newborn and young child with spina bifida. *The Journal of Urology, 196*(6), 1728–1734. https://doi.org/10.1016/j.juro.2016.07.081

Rzhetsky, A., Bagley, S. C., Wang, K., Lyttle, C. S., Cook, E. H., Jr., Altman, R. B., & Gibbons, R. D. (2014). Environmental and state-level regulatory factors affect the incidence of autism and intellectual disability. *PLoS Computational Biology, 10*(3), e1003518. https://doi.org/10.1371/journal.pcbi.1003518

Schaeffer, A. J., & Diamond, D. A. (2013). Pediatric urinary incontinence: Classification, evaluation, and management. *African Journal of Urology, 20*(1), 1–13. https://doi.org/10.1016/j.afju.2013.10.001

Schröder, A., Stein, R., Thüroff, J., & Gutjahr, P. (2006). Neurofibromatosis type 1 and its urological manifestation. *European Urology, 5*(2), 1–13.

Schröder, A., & Thüroff, J. W. (2010). New strategies for medical management of overactive bladder in children. *Current Opinion in Urology, 20*(4), 313–317. https://doi.org/10.1097/MOU.0b013e32833aa185

Shaikh, N., Morone, N. E., Bost, J. E., & Farrell, M. H. (2008). Prevalence of urinary tract infection in childhood: A meta-analysis. *The Pediatric Infectious Disease Journal, 27*(4), 302–308. https://doi.org/10.1097/INF.0b013e31815e4122

Sharma, A., & Paolinao, J. (2018). *Nephrology Chapter of The Massachusetts General Hospital Guide to Medical Care in Patients with Autism Spectrum Disorder* (E. P. Hazen, C. J. McDougle, & SpringerLink (Online service), Eds. 1st ed.). Humana Press. 169–179.

Srinivas, V., Herr, H. W., & Hajdu, E. O. (1985). Partial nephrectomy for a renal oncocytoma associated with tuberous sclerosis. *The Journal of Urology, 133*(2), 263–265. https://doi.org/10.1016/s0022-5347(17)48911-7

Stoll, C., Dott, B., Alembik, Y., & Roth, M. P. (2014). Associated nonurinary congenital anomalies among infants with congenital anomalies of kidney and urinary tract (CAKUT). *European Journal of Medical Genetics, 57*(7), 322–328. https://doi.org/10.1016/j.ejmg.2014.04.014

Strauss, D. C., Hayes, A. J., & Thomas, J. M. (2011). Retroperitoneal tumours: Review of management. *Annals of the Royal College of Surgeons of England, 93*(4), 275–280. https://doi.org/10.1308/003588411x571944

Subrahmanyam, A. B., & Mehta, A. V. (1995). Renal anomalies in Down syndrome. *Pediatric Nephrology, 9*(2), 253–254. https://doi.org/10.1007/bf00860766

Sureshkumar, P., Cumming, R. G., & Craig, J. C. (2006). Validity and reliability of parental report of frequency, severity and risk factors of urinary tract infection and urinary incontinence in children. *The Journal of Urology, 175*(6), 2254–2262. https://doi.org/10.1016/s0022-5347(06)00339-9

Tanner, J. M. (1986). Normal growth and techniques of growth assessment. *Clinics in Endocrinology and Metabolism, 15*(3), 411–451. https://doi.org/10.1016/s0300-595x(86)80005-6

Tarcan, T., Onol, F. F., Ilker, Y., Simsek, F., & Ozek, M. (2006). Does surgical release of secondary spinal cord tethering improve the prognosis of neurogenic bladder in children with myelomeningocele? *The Journal of Urology, 176*(4 Pt 1), 1601–1606.; ; discussion 1606. https://doi.org/10.1016/j.juro.2006.06.036

Tellier, A. L., Cormier-Daire, V., Abadie, V., Amiel, J., Sigaudy, S., Bonnet, D., de Lonlay-Debeney, P., Morrisseau-Durand, M. P., Hubert, P., Michel, J. L., Jan, D., Dollfus, H., Baumann, C., Labrune, P., Lacombe, D., Philip, N., LeMerrer, M., Briard, M. L., Munnich, A., & Lyonnet, S. (1998). CHARGE syndrome: Report of 47 cases and review. *American Journal of Medical Genetics, 76*(5), 402–409. https://doi.org/10.1002/(sici)1096-8628(19980413)76:5<402::aid-ajmg7>3.0.co;2-o

Thorup, J., McLachlan, R., Cortes, D., Nation, T. R., Balic, A., Southwell, B. R., & Hutson, J. M. (2010). What is new in cryptorchidism and hypospadias--a critical review on the testicular dysgenesis hypothesis. *Journal of Pediatric Surgery, 45*(10), 2074–2086. https://doi.org/10.1016/j.jpedsurg.2010.07.030

Timberlake, M. D., Jacobs, M. A., Kern, A. J., Adams, R., Walker, C., & Schlomer, B. J. (2018). Streamlining risk stratification in infants and young children with spinal dysraphism: Vesicoureteral reflux and/or bladder trabeculations outperforms other urodynamic findings for predicting adverse outcomes. *Journal of Pediatric Urology, 14*(4), 319.e311–319.e317. https://doi.org/10.1016/j.jpurol.2018.05.023

Van Herzeele, C., & Vande Walle, J. (2016). Incontinence and psychological problems in children: A common central nervous pathway? *Pediatric Nephrology, 31*, 689–692. https://doi.org/10.1007/s00467-016-3336-2

van Slegtenhorst, M., de Hoogt, R., Hermans, C., Nellist, M., Janssen, B., Verhoef, S., Lindhout, D., van den Ouweland, A., Halley, D., Young, J., Burley, M., Jeremiah, S., Woodward, K., Nahmias, J., Fox, M., Ekong, R., Osborne, J., Wolfe, J., Povey, S., … Kwiatkowski, D. J. (1997). Identification of the tuberous sclerosis gene TSC1 on chromosome 9q34. *Science, 277*(5327), 805–808. https://doi.org/10.1126/science.277.5327.805

Veltman, M. W., Craig, E. E., & Bolton, P. F. (2005). Autism spectrum disorders in Prader-Willi and Angelman syndromes: A systematic review. *Psychiatric Genetics, 15*(4), 243–254. https://doi.org/10.1097/00041444-200512000-00006

Viemari, J. C., Roux, J. C., Tryba, A. K., Saywell, V., Burnet, H., Peña, F., Zanella, S., Bévengut, M., Barthelemy-Requin, M., Herzing, L. B., Moncla, A., Mancini, J., Ramirez, J. M., Villard, L., & Hilaire, G. (2005). Mecp2 deficiency disrupts norepinephrine and respiratory systems in mice. *The Journal of Neuroscience, 25*(50), 11521–11530. https://doi.org/10.1523/jneurosci.4373-05.2005

Virtanen, H. E., & Toppari, J. (2008). Epidemiology and pathogenesis of cryptorchidism. *Human Reproduction Update, 14*(1), 49–58. https://doi.org/10.1093/humupd/dmm027

von Gontard, A., Pirrung, M., Niemczyk, J., & Equit, M. (2015). Incontinence in children with autism spectrum disorder. *Journal of Pediatric Urology, 11*(5), 264.e261–264.e267. https://doi.org/10.1016/j.jpurol.2015.04.015

Ward, C. S., Arvide, E. M., Huang, T. W., Yoo, J., Noebels, J. L., & Neul, J. L. (2011). MeCP2 is critical within HoxB1-derived tissues of mice for normal lifespan. *The Journal of Neuroscience, 31*(28), 10359–10370. https://doi.org/10.1523/jneurosci.0057-11.2011

Ward, C. S., Huang, T. W., Herrera, J. A., Samaco, R. C., Pitcher, M. R., Herron, A., Skinner, S. A., Kaufmann, W. E., Glaze, D. G., Percy, A. K., & Neul, J. L. (2016). Loss of MeCP2 causes urological dysfunction and contributes to death by kidney failure in mouse models of Rett syndrome. *PLoS One, 11*(11), e0165550. https://doi.org/10.1371/journal.pone.0165550

Washecka, R., & Hanna, M. (1991). Malignant renal tumors in tuberous sclerosis. *Urology, 37*(4), 340–343. https://doi.org/10.1016/0090-4295(91)80261-5

Weese-Mayer, D. E., Lieske, S. P., Boothby, C. M., Kenny, A. S., Bennett, H. L., Silvestri, J. M., & Ramirez, J. M. (2006). Autonomic nervous system dysregulation: Breathing and heart rate perturbation during wakefulness in young girls with Rett syndrome. *Pediatric Research, 60*(4), 443–449. https://doi.org/10.1203/01.pdr.0000238302.84552.d0

Williams, L. J., Mai, C. T., Edmonds, L. D., Shaw, G. M., Kirby, R. S., Hobbs, C. A., Sever, L. E., Miller, L. A., Meaney, F. J., & Levitt, M. (2002). Prevalence of spina bifida and anencephaly during the transition to mandatory folic acid fortification in the United States. *Teratology, 66*(1), 33–39. https://doi.org/10.1002/tera.10060

Williams, M. (2017). Urinary tract anomalies in charge: For the urologist.

Yang, W. H. (1993). Yang needle tunneling technique in creating antireflux and continent mechanisms. *The Journal of Urology, 150*(3), 830–834. https://doi.org/10.1016/s0022-5347(17)35625-2

Yohay K, K. S. (2013). *Tuberous Sclerosis*. Retrieved August 2020. From http://www.clinicaladvisor.com/pediatrics/tuberous-sclerosis

Zafeiriou, D. I., Ververi, A., & Vargiami, E. (2007). Childhood autism and associated comorbidities. *Brain and Development, 29*(5), 257–272. https://doi.org/10.1016/j.braindev.2006.09.003

25 Treatment Planning for Transition to Adult Services

Emma Simmons and Pamela McPherson

Abstract

Comprehensive transition planning is a critical element of treatment planning for youth with special needs and must begin during the early teen years to ensure that care is not disrupted. Medical home care coordinators can facilitate comprehensive transition planning which addresses healthcare transition (HCT) as well as legal, educational, housing, employment and self-care needs, among others. Adulthood is a biological, legal, and social construct with a myriad of implications for persons with developmental disorders. While the legal definition of adulthood will drive many aspects of transition planning, the successful launching of the adolescent into adulthood has more subtle nuances. Medical needs must be addressed as well as issues impacting the emerging adult including sexuality, decision-making, and social issues. For youth with special needs, emerging adulthood requires careful planning and supports to address any barriers to exploration, minimize risks posed by instability, and broaden the possibilities for the future.

Keywords

Transition · Medical home · Developmental disorder · Autism · Treatment planning

Introduction

The modern era of medical care for people living with disabilities (PLWD) dates back to the 1960s when President Kennedy established the National Institute of Child Health and Human Development within the National Institutes of Health and championed community-based care over institutionalization with the Community Mental Health Act of 1963. With this transition, the medical model of institution-based care has given way to the biopsychosocial model of community-based care. This later model informs the medical home model in promoting the empowerment and the accommodation of PLWD to participate fully the community. Participation has been fostered by additional legislation including Section 504 of the Rehabilitation Act of 1973, The Individuals with Disabilities Education Act, and the Americans with Disabilities Act (Ervin et al., 2014). In addition to the broad access guaranteed

E. Simmons
University of California Medical School, Riverside, CA, USA

P. McPherson (✉)
Department of Psychiatry and Behavioral Medicine, Louisiana State University Health Sciences Center, Shreveport, LA, USA
e-mail: Pam.mcpherson@lsuhs.edu

P. McPherson (ed.), *Handbook of Treatment Planning for Children with Autism and Other Neurodevelopmental Disorders*, Autism and Child Psychopathology Series,
https://doi.org/10.1007/978-3-031-06120-2_25

by these Acts, they mandate that healthcare providers accepting federal compensation provide effective accommodations, including communication options for PLWD. Through implementing a medical home model, the primary care doctor can serve an important role in improving the quality of life of PLWD, improving their access to quality health care, preventing unnecessary emergency room visits, and minimizing suffering, morbidity, and premature mortality. The medical home model also promotes improved communication and a healthy patient–caregiver relationship, intervening early to minimize disruptions, conflict, and other issues in the relationship.

With the estimates for ID ranging from 1.58% to 3% and developmental disability from 6.99% to 16.24%, healthcare providers must be prepared to accommodate PLWD in conceptual, social, and practical adaptive areas (Krahn, 2019). The increased prevalence of complex medical conditions among children, in addition to a concomitant decreased mortality means that primary care providers must develop the competencies necessary to treat PLWD (Perrin et al., 2014). The life expectancy of PLWD is increasing necessitating transition to an adult-oriented medical home team with a focus on lifelong health. Having a seamless transition from a pediatric to adult medical home is important for multiple reasons including improving or maintaining health care, reducing the anxiety and confusion of navigating healthcare systems, preventing complications requiring hospitalizations or relapses, and lowering healthcare costs (Sullivan et al., 2011). Still, fewer than 20% of adults with chronic medical conditions receive care in a medical home and this remains an area of needed improvement (Almalki et al., 2018). In recognition of this need and the systemic limitations in adult health care, the American College of Physicians and the American Academy of Family Physicians advocate for providing emerging adults with strategic transition planning to avoid fragmented and uncoordinated care, address the lack of time allotted for a typical office visit, and to fully address intersecting biopsychosocial factors impacting health care, as well as calling for educational and clinical training for healthcare providers treating PLWD (Iacono & Sutherland, 2006; White, Cooley, American Academy of,, & American Academy of Family, 2018a). Intellectual disability is a known risk factor for early death with the average age of death for persons with ID occurring decades earlier than peers without ID (Arvio et al., 2017). A seamless transition from the pediatric to the adult medical is necessary to maximize health in order to prolong and enhance the lives of PLWD.

Planning for Transition to Adult Heath Care

The promises of deinstitutionalization have not been fully realized for persons with intellectual and developmental disabilities (IDD). As people transitioned to the community from institutions, healthcare systems were unprepared to provide coordinated, personalized care (Lewis et al., 2016). In 2002, as a result of the poor medical outcomes of people with IDD, the US Surgeon General, Dr. David Satcher, issued, *Closing the Gap: A National Blueprint to Improve the Health of Persons with Mental Retardation*—a call to action to improve quality health care for people living with disabilities (Office of the Surgeon & Centers for Disease Control and, 2002). Five years later, the American Association on Intellectual and Developmental Disabilities noted “marked disparities” in the care provided to patients with IDD (Krahn et al., 2006). This led the 2006 United Nations Convention on Rights of Persons with Disabilities to advocate for the enjoyment of the highest attainable standard of health that is free of discrimination (Convention on the Rights of Persons with, 2006). These efforts have led to improved data collection regarding the healthcare needs of persons with IDD and the development of service delivery models including the patient-centered medical home. Research using the life course model of health inequities has identified emerging adulthood as a period of risk for with persons with IDD—requiring careful attention to transition planning (Acharya et al., 2017).

Comprehensive transition planning is a critical element of treatment planning for children with special needs and must begin prior to or during the early teen years to ensure that care is not disrupted. Medical home care coordinators facilitate comprehensive transition planning which addresses healthcare transition (HCT) as well as legal, educational, housing, employment, and self-care needs, among others. The American Academy of Pediatrics (AAP), the American Academy of Family Physicians, and the American College of Physicians have issued a joint clinical report affirming their 2011 tenets and algorithm for HCT, referencing the HCT needs of young adults with special needs to ensure optimal wellness (White, Cooley, et al., 2018a). See Fig. 25.1. Over 15 million persons aged 12–26 years of age experience a chronic healthcare condition, with children with special healthcare needs (SHCN) commonly experiencing medical complexity (Kuo & Houtrow, 2016; White, Schmidt, et al., 2018b). The 2016 National Survey of Children's Health found that many youth do not receive HCT services, 17% of youth ages 12–17 years with SHCN versus 14% of those without reported receiving HCT services (Lebrun-Harris et al., 2018). This chapter will review best practices and suggest resources for the transition of PLWD from pediatric to adult health care with attention to clinical care and the special challenges that may arise in the primary care medical home for emerging adults.

The Cornerstone of Care: The Physician–Patient Relationship

Health is vital to maintaining a quality of life. High-quality medical care is vital to the health and well-being of PLWD, yet basic preventative care, routine health monitoring, and chronic disease management are often neglected (Lewis et al., 2016). A cohort study of over 70,000 adults with ID/DD reported inpatient care and outpatient emergency care at three and two times benchmarks, respectively (Lauer et al., 2021). Arguably, many of these admissions and emergency visits might have been averted with preventative care and monitoring. Clearly, the primary care doctor can serve an important role in improving the quality of life of PLWD. Primary care doctors are the key to improving access to quality health care and preventing unnecessary emergency room visits and hospitalizations, as well as minimizing suffering, morbidity, and premature mortality. The primary care provider can also note early disruptions in care or difficulties in the patient and caregiver relationship; a relationship that is often vital to the health and community involvement of PLWD.

The relationship between the medical team and family or other caregivers must be executed delicately in order to provide the maximal level of privacy and confidentiality possible to the patient. While the patient is the primary person of record, the caregiver may be the only bridge to understanding the patient's pain, change in baseline status, or other symptoms. Similarly, patients may not be able to fully understand or process the information provided and thus are dependent on the caregiver for their health and wellness. This requires that the medical team simultaneously assess the patient's capacities for self-care and for medical decision making while maintaining respect for patient autonomy. If possible, the adult medical home team should try to engage with new PLWD and their families/caregivers prior to first "real" appointment to introduce them to the new practice setting. This can be facilitated by the pediatric medical home care coordinator. This "meet and greet" opportunity gives the patient time to establish trust without the need for a stethoscope or the typical angst associated with immunizations, physical examinations, and phlebotomy. It may be unusual for some providers of adult medical care to include family members or other caregivers during the office visit, but PWLD, especially older teens, may prefer others to be present or may be dependent on others for assisting with communication, shared decision making, or physical transitions.

The importance of the healthcare provider's relationship with the caregiver of the person with IDD cannot be overemphasized. Over 70% of persons with IDD live with a family member (Braddock et al., 2013). With caregivers at increased risk for depression, physical illness, got transition.

Health Care Transition Timeline
for Youth and Young Adults

Age 12-13

- Learn about your health condition, medications, and allergies.
- Ask your doctor questions about your health.
- Ask your doctor if and at what age they no longer care for young adults

Age 14-15

- Find out what you know about your health, health care, and family medical history. Both you and your parent/caregiver can take Got Transition's Transition Readiness Assessments* and discuss this together and with the doctor.
- Carry your own health insurance card.
- Learn more about your health and what to do in case of an emergency.
- Practice making a doctor's appointment and ordering prescription refills (either by phone, online, or through an app).
- Begin to see the doctor alone for part of the doctor's visit to help gain independence in managing your health and health care.

Age 16-17

- Make doctor's appointments, see the doctor alone, ask the doctor any questions you have, and refill medications.
- Ask the doctor to talk with you about your privacy rights when you turn 18.
- Work with your doctor to make a medical summary. Keep a copy for yourself.
- Before you turn 18 and become a legal adult, figure out if you will need help making health care decisions. If so, ask your Family Voices chapter for local resources.
- Talk with your parent/caregiver about the age you want to transfer to a new doctor for adult care.

Age 18-21

- You are a legal adult at age 18 and are legally responsible for your care. Parents/caregivers cannot access your medical information or be in the doctor's visit unless you agree.
- Work with your current doctor to find a new adult doctor, if needed. Make sure that the new doctor accepts your health insurance.
- Update your medical summary with your doctor. Have your doctor send this to your new adult doctor. Keep a copy for yourself.
- Call your new adult doctor to schedule the first appointment. Make sure the new office has your medical information, and learn if there are any charges at the visit.
- Learn if there are additional changes at 18 that affect you (e.g., health insurance, Social Security Income).

Age 22-25

- Continue to get care from your adult doctor, learn to manage your health and health care, and update your medical summary.
- Be sure to stay insured. If you change your health insurance, make sure your doctor takes your insurance, and learn if there are any charges at the visit.

*For a Transition Readiness Assessment for youth, visit https://gottransition.org/6ce/leaving-readiness-assessment-youth and for a version for parents/caregivers, visit https://gottransition.org/6ce/leaving-readiness-assessment-parent.

Got Transition (GotTransition.org) is supported by the Health Resources and Services Administration (HRSA) of the U.S. Department of Health and Human Services (HHS) (U1TMC31756). The contents are those of the author(s) and do not necessarily represent the official views of, nor an endorsement, by HRSA, HHS, or the U.S. Government.

Fig. 25.1 Resources for transition planning

social isolation, and economic hardship, the medical home model recognizes that attention should be given to the caregiver's need for supports (Williamson & Perkins, 2014). Caregivers report significant support needs in order to plan and care for PLWD (Gilson et al., 2017). The medical home team remains responsible for screening for the presence or absence of support systems and assisting in securing supports including personal care attendants and durable equipment (Burke et al., 2018). Novel approaches have been identified to support caregivers and young adults including mHealth apps that support independent patient healthcare management and online caregiver skill training (Bendixen et al., 2017; Guanxiu et al., 2018). Health education is a vital area of support for patients and caregivers. A scoping review found up to 52.5% of caregivers had low health literacy which was associated with poorer patient outcomes (Yuen et al., 2018).

Welcoming Emerging Adults to Your Practice

Establishing a medical practice to serve emerging adults with ASD/IDD must be approached intentionally. The American College of Physicians (https://www.acponline.org/) offers guidance on pediatric to adult care transition and the patient-centered medical home which can assist in the development of a practice that welcomes emerging adults living with disability. GOT Transition® maintains links to health professional educational materials. See Table 25.1. In addition to understanding population healthcare needs, providers must prepare and implement the necessary accommodations to help eliminate communication, physical, and fiscal barriers to care for their patients.

Understanding Emerging Adults

Adulthood is a biological, legal, and social construct. Historically, biology defined adulthood, with children becoming adults at puberty. Adolescence as a developmental stage is a relatively new idea, introduced at the turn of the twentieth century by psychologist G. Stanley Hall (1904). With recognition of the variation in biopsychosocial factors associated with adolescence, age has become the social and legal marker for adulthood, with the age of 18 generally signifying adulthood. The legal definitions of adulthood or majority often have critical significance for persons with developmental disabilities as services often have specific age requirements. For example, medical and educational decision-making passes to youth upon their 18th birthday, unless decision-making capacity is impaired and a court has appointed a guardian; youth receiving special education services may be entitled to accommodations until the age of 22, and a parent's insurance will have age requirements which may be waived for PLWD or enrolled in adult education. When assisting youth and families with transition planning, careful attention must be given to state and federal timelines. See Table 25.1 for links to state-specific timelines.

While the legal definition of adulthood will drive many aspects of transition planning, the successful launching of the adolescent into adulthood has subtle nuances. These subtle nuances are due to the plasticity of the teenage brain, which responds to experience and the environment with rapid neurological growth. Technological advances have spurred brain development research which supports the commonsense observation of every parent, namely, that the turning 18 holds no special magic, it is a sociolegal construct. The construct defines a time of increasing independence with multiple role transitions including accepting personal responsibility for one's health care. With brain maturation continuing well into a person's 20s, the transition is an evolving process celebrated at a symbolic point in time (Dosenbach et al., 2010). Arnett (2000) has proposed the developmental stage of *emerging adulthood* to capture the development unique to 18–29 year-olds with five cardinal features: self-focus, identity exploration, feeling in-between, embracing a sense of possibility for the future, and the appearance of instability as roles are explored and cast aside. The transition of the emerging adult into the role of personal healthcare administrator is determined by multiple factors, some unique to the individual and others due to structural and societal determinants of health (Schoon, 2015). Integrating the impact of social determinants of health on emerging adults into the shared plan of care (SPoC) is becoming a part of routine medical care. The American Medical Association has advanced the History and Physical 360 to capture social determinants of health or via screening instruments (Kirley et al., 2020). Screening instruments are also available for this purpose (Andermann, 2018).

Emerging adults with chronic healthcare conditions encounter significant challenges during the transition to adult health care with an overall negative impact on their health (Hart et al., 2019). Acharya, Meza, and Msall (2017) reported that youth with ASD, CP, and Down syndrome often have poor transition experiences with subsequent

Table 25.1 Resources for transition planning

Resource	Product description
AAP Transition ECHO https://www.aap.org/en-us/professional-resources/practice-transformation/echo/Pages/Transition.aspx	HCT-focused information including standardized provider curriculum, billing information, information for families, provider forum
ACP Pediatric to Adult Care Transitions https://www.acponline.org/clinical-information/high-value-care/resources-for-clinicians/pediatric-to-adult-care-transitions-initiative	A toolkit including a patient transition readiness assessment, model medical summary/transfer record, self-care assessment, and condition-specific tools.
Autism Speaks *Autism and the Transition to Adulthood Toolkit* https://www.autismspeaks.org/tool-kit/transition-tool-kit	Addresses family and youth needs regarding self-advocacy, independent living skills, education, health, legal issues, housing, employment options and more. Includes links for state-specific transition timelines.
Casey Life Skills https://www.casey.org/casey-life-skills-resources/	A competency assessment tool addressing domains – daily living, self-care, relationships/communication, housing/money management, work/study life, career/education planning, and self-efficacy.
Got Transition® https://www.gottransition.org/resources-and-research/policy-research.cfm	Comprehensive federally funded national resource center for HCT with information for young adults, caregivers, and professionals.
La Trobe Support for Decision Making Framework http://www.supportfordecisionmakingresource.com.au/	A seven-step process for assisting individuals with cognitive disability in making and communicating choices. The website includes provider education modules, a workbook, activity book, and other resources.
National Resource Center for Supported Decision Making http://www.supporteddecisionmaking.org/	Training and self-determination materials. State-specific links to SDM legislation and case law.
Self-Care Assessment for Young Adults with ID or DD https://www.acponline.org/system/files/documents/clinical_information/high_value_care/clinician_resources/pediatric_adult_care_transitions/gim_dd/idd_transitions_tools.pdf	Patient completed assessment of self-care knowledge and skills at a 4th grade reading level. Identifies education, support, and skill development needs.
Structural Vulnerability Assessment Tool https://sirenetwork.ucsf.edu/tools-resources/mmi/structural-vulnerability-assessment-tool	43-item questionnaire that guides a dialogue over 6 domains— – economic stability, education, social/community, health/clinical care, neighborhood/physical environment, and food.
Transition Readiness Assessment Questionnaire https://www.etsu.edu/com/pediatrics/traq/	TRAQ is a patient-centered 20 question instrument assessing five domains— managing medications, appointment keeping, tracking health issues, talking with providers, and managing daily activities.
Transition Readiness Assessment for Youth with Intellectual/Developmental Disabilities https://www.acponline.org/system/files/documents/clinical_information/high_value_care/clinician_resources/pediatric_adult_care_transitions/gim_dd/idd_transitions_tools.pdf	Youth completed tool used to inform education and skill development needs. Addresses legal choices for making healthcare decisions, personal care, health, healthcare, and self-care importance, and importance.
STARx Questionnaire https://www.med.unc.edu/transition/transition-tools/trxansition-scale/	A patient- or parent-rated instrument addressing three domains— – communication with medical provider, disease knowledge, and self-management. Patient-rated versions for pediatric and adult settings and a parent-rated version.

healthcare disparities. Similar concerns have been identified for youth with neurological conditions (Brown et al., 2016). Likewise, parents of young adults have identified the need for additional information, improved coordination of services, and family support during health care transition (Brown et al., 2020; Gauthier-Boudreault et al., 2018). Youth with ASD have expressed concern that they lack the skills necessary for managing their own health care (Cheak-Zamora et al., 2015). Youth with cerebral palsy noted that mastering a new healthcare system was a bigger challenge than self-care mastery in the journey to independence (Bagatell et al., 2017; Larivière-Bastien et al., 2013). The majority of HCT studies are qualitative, exploring parental perspectives and experiences (Betz et al., 2015). Additional research is needed to inform the experience of persons with autism/IDD.

During the emerging adult years, youth and their parents must master or relinquish roles in multiple domains including health care, education/vocation, housing, social and others, with each domain subject to its' own regulations and timelines. For youth with special healthcare needs, emerging adulthood requires careful planning and supports to address any barriers to exploration, minimize risks posed by role instability, and broaden the possibilities for the future (Cribb et al., 2019; Zukerman et al., 2011). Peer support programs offer advice based on first-hand experience with the transition to adult roles and are often well accepted by youth. HCT planning often requires coordination of many services and providers while negotiating changes in insurance coverage. Planning should begin early, with the AAP recommending that HCT be initiated by age 14 (White, Schmidt, et al., 2018b). The medical home plan of care should address transition planning, including the assessment of transition readiness and identification of supports for healthcare coordination as youth leave pediatric care. In addition to a general assessment of readiness (see Table 25.1), the pediatric medical home coordinator may need to arrange youth-specific assessments, keeping in mind that insurance coverage for assessments or reassessments may vary by age or if the youth is enrolled in school. For youth prescribed adaptive equipment or durable medical equipment, independence with use must be evaluated and the SPoC updated to provide education, accommodations, and/or supports as necessary. In addition to medical assessments, legal assessment of medical decision-making capacity may need to be evaluated.

Supported Decision Making and Medical Decision-Making Capacity with Emerging Adults

With adulthood comes medical decision-making capacity. The law considers adults competent to make medical decisions and consent to care unless this capacity is challenged, a judge rules the person incompetent, and, if necessary, a guardian is appointed in accordance with state law. Medical providers should be aware that treating a patient without consent is grounds for malpractice except in an emergency; therefore, it is the responsibility of providers to obtain consent after an assessment to ensure that medical decision-making capacity exists (Kolva & Rosenfeld, 2012). An evaluation of medical decision-making capacity assesses a patient's understanding of the relevant facts, application of those facts to their situation, and the quality of reasoning in arriving at a choice (Grisso & Appelbaum, 1998). Medical providers may be inclined to equate IDD or ASD with impaired decision making and are cautioned to consider the capacity to make decisions with regard to the specific decision at hand, respecting each patient's legal rights and autonomy.

The autonomy of the patient should be respected to the fullest extent possible. The United Nations' Convention on the Rights of Persons with Disabilities calls for freedom of expression and opinion which is operationalized in patient-centered approaches to medical decision making (Della Fina et al., 2017). Shared decision making, supported decision making, and advance directives are common approaches. Shared decision making refers to "a collaborative process between patients and providers that includes sharing information, eliciting informed preferences, and ensuring that these are integrated into the care plan" (Durand et al., 2015, p. 97). The Patient Protection and Affordable Care Act calls for the use of shared decision making through the use of evidence-based decision aids (Act, 2010). Shared decision making requires that the patient have the capacity to understand that a decision requires a choice, appreciates the potential consequences of each choice, and maintains the ability to communicate a reasoned choice (Werner & Chabany, 2016).

Should capacity for shared decision making be in question, supported decision making (SDM) is an alternative. SDM is a process that allows people to reach a decision through a discussion of the problem and making choices with friends and providers. Medical providers who assist in the process of SDM are acting in accordance with the principles of medical ethics and providing best care. SDM empowers self-determination by PLWD over important decisions, including health care. Improved health outcomes and better overall quality of life are associated with self-determination (Blanck & Martinis, 2015; Shogren et al., 2015). The United Nations' Convention on the Rights of Persons with Disabilities describes SDM as a process where

> the individual is the decision maker; the support person(s) explains(s) the issues, when necessary, and interpret(s) the signs and preferences of the individual. Even when an individual with a disability requires total support, the support person(s) should enable the individual to exercise his/her legal capacity to the greatest extent possible, according to the wishes of the individual. (Rights, 2007)

The La Trobe framework is an evidence-based SDM resource (Douglas & Bigby, 2020). See Table 25.1. The SDM process may be detailed in law or implemented in addition to substituted decision making.

Like guardianship, advance directives are a form of substituted decision making. Advance directives allow individuals with decision-making capacity to anticipate and memorialize wishes for future medical care should they become unable to make such decisions. An advance directive may be considered for persons with ASD/DD because of the increased risk for incapacity due to mental illness, IDD associated with regression, or early onset of Alzheimer's disease in some IDD including Down syndrome.

Preparing Medical Staff to Welcome Emerging Adults

In addition to the desire to work with patients with IDD, health caregivers need to have competencies in the management of IDD. Few medical schools or residencies offer a curriculum that focuses on the additional recommendations and modifications specific to people with IDD. Medical education regarding the presentation of mental illness is a specific deficit. In interviews of family and internal medicine residents, fewer than 35% reported exposure to any disability-focused education in medical school and 11.2% in residency (Stillman et al., 2021). The need for targeted education to care for persons with IDD is common across all medical professions (Auberry, 2018; Williamson et al., 2017). A systematic review of postgraduate medical education by Adirim et al. (2021) reported while educational experiences resulted in increased knowledge about IDD and patient needs, didactics alone rarely resulted in provider behavioral change or improved patient outcomes. Twenty-seven percent of the residents interviewed by Stillman et al. (2021) would consider a fellowship focusing on the care of people with IDD if available, indicating significant interest and opportunity. The National Council on Disability continues to advocate for minimal curricular content to be added to the elements required for graduation that are mandated by the Liaison Committee on Medical Education ("NCD LCME Second Response Letter," 2019).

Given the limited attention medical education pays to preparing physicians to care for PLWD, it is not surprising that mainstream health services rarely offer the necessary patient-centered medical home, multidisciplinary approach to health care. Family physicians readily acknowledge feeling overwhelmed when caring for persons with IDD due to a limited knowledge base, lack of confidence, and lack of familiarity with available resources (Wilkinson et al., 2012). In fact, less than half of the physicians surveyed by Iezzoni et al. (2021) reported that they were very confident that they could provide care to PLWD on par with patients without disability. To address this inequity, the Alliance for Disability Health Care (Alliance for Disability in Health Care, 2019) has published *Core Competencies on Disability for Healthcare Education* citing six competencies necessary for healthcare providers: knowledge regarding contextual and conceptual frameworks on disability, professionalism and patient-centered care, legal obligations and responsibilities for caring for patients with disabilities, teams and systems-based practice, clinical assessment, and clinical care over the lifespan and during transitions.

Honing Communication Skills

Identifying the required manner in which each patient communicates is key to providing quality health care, and it is the law under the Americans with Disabilities Act (PL 101–336,1990, amended 2008). Patients who are hard of hearing, have limited vision, speak a language other than English, or face other communication challenges may require additional support for their needs to be adequately met. In addition, support may be needed by family members or caregivers. The main goal is to deliver inclusive, patient-centered, quality health care, and it is the medical provider's ethical and legal responsibility to ensure that adequate accommodations are provided to meet this goal.

The healthcare provider must modify his/her approach to be a proficient communicator. This begins with reviewing transition documents from the pediatric medical home to identify communication abilities and accommodation needs. Patients should be addressed directly in an age-appropriate manner and the patient's consent should be obtained before information from caregivers is requested. Procedures should be explained and/or demonstrated to promote patient understanding, using visual aids when necessary. Attention should be paid to nonverbal cures. The patient should be informed that there is the right to privacy and that they may request that caregivers leave the examination room (Werner et al., 2017). Patients should be asked to share their understanding of what they have been told to check accuracy. The National Joint Committee on the Communication of Persons with Severe Disabilities has issued guidance for best practices in assessing and accommodating individuals with severe disabilities (Brady et al., 2016).

Persons with IDD or ASD are at risk for communication disorders (Fletcher, Barnhill, & McCarthy, 2016a). Attention must be paid to the patient's verbal and nonverbal communication, including receptive and expressive capacities, articulation, and comprehension. Even when a formal diagnosis has not been made, receptive and/or expressive issues are common with expressive ability typically exceeding that of the receptive capacity (Friedman & McNamara, 2018). In addition, providers should understand that experience and context may alter comprehension. For some patients, communication difficulties may solely be the result of problems with their muscle tone or overmedication. Some patients may have delayed responses or slow speech. Others may have difficulties with abstract language. Any combination of these issues may impact the ability of the patient to express their symptoms or wishes requiring providers to assess capacity and pay attention to communication style. It is important to note that PLWD may mask their symptoms or conform to the expectation of authority figures by agreeing or even saying they understand concepts they do not in order to circumvent embarrassment (Boardman et al., 2014). Because of the range of potential issues, it is imperative that each patient be considered uniquely and that baseline communication skills be established by explicitly asking and assessing, not assuming.

Providers should seek professional development to expand their knowledge and ability to care for people with IDD and recognize their own biases (Hemm et al., 2015). The Vanderbilt Kennedy Center for Excellence in Developmental

Disabilities' *Health Care Toolkit for Adults with Intellectual and Developmental Disabilities* is an excellent resource for professional development on communication and other issues (www.IDDtoolkit.org/). Bias that may impact health care includes diagnostic overshadowing. With diagnostic overshadowing, the provider attributes difficulties to IDD disregarding potential illness; for example, attributing slurred speech or poor coordination to IDD without taking a history or fully exploring the presentation (Geiss et al., 2017). A related concept is diagnostic masking which occurs when characteristics of IDD obscure features of illness (Manohar et al., 2016).

Preparing the Office to Welcome Emerging Adults

Preparing the office begins when the appointment is scheduled and continues with inquiry regarding necessary accommodations. Ideally, the pediatric care coordinator will act as a bridge to facilitate the transition from the pediatric to adult medical home where the skilled nurse manager or adult care coordinator will facilitate a successful transition. Attention to reducing the physical barriers includes making sure that the size of the room is large enough to accommodate the equipment (i.e. wheelchair or gurney) of the patient as well as their caregivers. Frequently, multiple people, in addition to the patient, are in the room during the office visit and must also be accommodated. Guidance is provided by the US Departments of Justice and Health and Human Services (Access to Medical Care for Individuals with Mobility Disabilities, 2010, Last updated 2020). The number of rooms in the practice as well as any items included in the room and the decoration of the room should be carefully planned. Similar consideration to the decorum and set up or placement of the waiting area is needed. Once inside the individual examination rooms, the type of lighting should be considered with preferential installation of non-fluorescent lights in order to avoid potential sensory discomfort. Examination tables should allow for height adjustments in order to facilitate transition from wheelchairs or gurneys. A variety of scale types may be necessary, with grab bars strategically placed to facilitate transfer on to the scale (Ervin et al., 2014).

Some patients may have difficulty waiting for even short periods of time. If necessary, bypassing the waiting areas by offering alternate entrances, immediate access, or offering the first or last visit of the day can be helpful. Longer appointment times will be necessary to establish rapport with the patient and caregivers, to accommodate patient needs, explain and verify understanding, and obtain consent for treatment. Adding an electronic health record field specific for accommodations in the scheduling and direct care tabs has been shown to improve accessibility for patients (Mudrick et al., 2020).

Treatment Planning and Coordinating Care

It is a well-established fact that the delivery of medical care should be integrated. Integrated healthcare can easily become a sound bite, but the difficulty lies in implementing this framework with the promotion of overall health as the key goal. Patients, care providers, and the medical home team must educate each other and embrace the SPoC. (See Chap. 1 for details on the SPoC.) The medical home team coordinates care with multiple consultants, ideally, providers who are knowledgeable, receptive, and responsive to the unique needs of the patient. It may be prudent to call the specialist prior to their visit to provide essential information about the patient and discuss the SPoC in order to maximize the outcomes of the specialist appointment and make the visit go more smoothly for your patient. If at all possible, having co-visitation with a specialist, virtually or in person, may also be a way to make the visits with the patient more efficient and productive. Similarly, coordination of the timing of imaging and blood work conserves valuable resources, including the patient's time and energy.

Treatment planning and coordination of care are where much of the complex medical decision making happens for PLWD. At the initial appointment, the emerging adult with ASD/IDD often arrives with a list of diagnoses and years of experience with healthcare professionals. The level of medical complexity is frequently not comparable to the typical presentation of a young adult without IDD, and adequate time must be allowed for the appointment. While evaluation and testing are necessary to assess the status of a disease process, this part of the patient visit is typically straightforward. What an assessment means for the patient and the subsequent planning for care of the patient in light of that person's unique life circumstances is where much of the medical home's time is spent. Deciding how to best promote health for the patient occurs during treatment planning and the creation of a SPoC. Implementing the SPoC requires a knowledgeable and skilled care coordinator.

The importance of a skilled person to handle care coordination cannot be overstated. Physicians are rarely taught best practices for the care of PLWD during residency and acquiring this skill poses the challenge of climbing a steep learning curve on the constantly shifting terrain of regulations and community supports (Ervin et al., 2014). Stillman et al. (2021) found that 78% of internal medicine and family practice residents did not know how to link PLWD to care coordination or community-based services. The prevalence of complex medical conditions has increased in children with neurodevelopmental and mental health disorders representing a significant portion of the medically complex conditions (Perrin et al., 2014). McManus and White (2017) estimated that one million teens with chronic medical conditions become adults each year. While preventive care needs to remain front and center, the medical home team must also attend to care coordination. Past records and specialist reports must be interpreted and included in the SPoC. Medication lists must be frequently reconciled. Polypharmacy and side effects of the medications must be assessed. Assessment for the durable medical equipment is necessary as it is essential to patients' reaching their maximal independence and improving their satisfaction with their life. The correct hospital bed, wheelchair, or Hoyer lift may translate to a patient's ability to socialize and more fully participate in their community. Care coordination frequently includes the timely completion of paperwork such as forms for nursing agencies and other in-home support services, annual visits with group homes, as well as forms or letters for the department of motor vehicle, jury duty accommodations or exemptions, Special Olympics physicals and needs, housing, work or accommodation, and other periodic forms. Timely assistance with these tasks can greatly reduce the stress of patients and their families.

Anticipating Challenges to Care from Outside the Office

Flexibility and planning for the unexpected are keys to successful care coordination as barriers to care will inevitably arise. The medical home team must maintain focus on a broad preventive approach to health care delivery despite the multiple compelling issues that may present at each visit and between visits. With this team approach, improved patient outcomes are documented; still, there are barriers outside the medical office that confound even the most skilled clinicians and must be addressed systemically (Hand et al., 2021). For adult patients, characteristics associated with inclusion in a medical home include private insurance, employment, family economic status, and living in the north or east (Almalki et al., 2018). These characteristics reflect social determinants of health and highlight how these determinants contribute to healthcare inequity. In 2021, the Centers for Disease Control called out racism as a "serious threat to the public's heath" (para. 1). Structural racism has been identified as a significant factor in access to diagnostic and treatment services for people living with ASD and IDD (Broder-Fingert et al., 2020; Jones et al., 2020; Kelleher & Wheeler, 2020). Increasing, the availability of a medical home intervention has been identified as one approach to addressing inequity in health care through attention to social determinants of health (Sheth et al., 2017). And increasingly, physicians and healthcare systems are recognizing that addressing social determinants of health as a responsibility that healthcare providers share with the community at large.

Social determinants of health can impact health care in some fundamental ways including transportation, employment opportunities, and community participation. There is a growing awareness that transportation may represent a significant issue for patients and their caregivers (Solomon et al., 2020). The prohibitive cost and maintenance of adaptive vehicles is a barrier that is too high for many PLWD. Patients must therefore heavily rely on public transportation, including city buses or smaller ride share vans, which are not known for timeliness or reliability. Additional time to account for transportation issues should be factored into most, if not all, visits. Transportation is one of many factors that restrict work and social opportunities for PLWD (Bascom & Christensen, 2017). Unemployment negatively impacts the health, mental health, and quality of life of emerging adults (Vancea & Utzet, 2017). Employment and other forms of community participation are necessary for young adults to develop social skills and a sense of self-efficacy (Perez & Crowe, 2021). In addition, community participation is necessary to address loneliness and a sedentary lifestyle that can contribute to poor mental and physical health. Community participation by PLWD is increasingly including volunteering and advocacy (Carter & Bumble, 2018). In summary, the medical home team must look beyond the clinic walls and anticipate challenges to promote health and wellness for emerging adults as they transition to adult healthcare systems.

Special Issues in the Treatment of Emerging Adults

Emerging adulthood is a time when issues related to substance use and sexuality may arise. As PLWD have asserted their autonomy and educated the medical and broader community, barriers to addressing sexuality and substance use have been challenged. These are complex and highly personal matters that must be approached with respect for patient autonomy and addressed in the way that meets the individual's needs.

Substance-Related and Addictive Disorders

Persons living in the community have access to alcohol, cigarettes, and a broad range of substances as well as gambling. While substance use disorders (SUD) among persons living with disability were traditionally considered to be low, recent studies indicate that it is a significant concern. Furthermore, the functional implications are often more serious due to multimorbidity, drug–medication interactions, and the impact of substances on cognitive functioning and health (Fletcher, Barnhill, & McCarthy, 2016a). The risk for addiction is multifactorial, with environment, genetics, temperament, epigenetics, and social determinants of health including adverse childhood experiences each contributing to risk (Butwicka et al., 2017; Cecil et al., 2016; Ressel et al., 2020).

Estimates for SUD co-occurring with ASD and IDD vary widely due to methodologies and subject selection, for example, whether the sample was from the community, a clinical, forensic, or hospital setting. A range from 0.7% (Abdallah et al., 2011) to a high of 36% (Mandell et al., 2012) has been reported. A Swedish longitudinal study of 26,986 persons with ASD reported an odds ratio of 5.2 for persons with ASD compared to controls, with increased risk for ASD and comorbid ADHD (OR 8.3) and ASD with ADHD and ID (OR 4.6) but not with ASD with ID (Butwicka et al., 2017). This is not surprising as ADHD is associated with SUD (Groenman et al., 2017; Wimberley et al., 2020). Among persons with IDD, the prevalence of SUD ranges from 0.5% (Cooper et al., 2007) to 6.4% (Lin et al., 2016). Sparse data are available regarding specific DD. The Canadian National FASD Database, including 726 adolescents and emerging adults with prenatal alcohol exposure, found 38% alcohol misuse and 46% other substance misuse (McLachlan et al., 2020). Whitney et al. (2019) reported that men with cerebral palsy without ID,

ASD, or epilepsy experienced at greater prevalence of alcohol and/or opioid disorders compared to controls, 4.7% versus 3%. The prevalence of SUDs among PLWD is an area in need of additional study as most studies report on SUD as one group with limited data available for specific categories of SUD.

The Diagnostic and Statistical Manual of the American Psychiatric Association, 5th edition, classifies substance-related disorders (SRD) as substance use disorders or substance-induced disorders, addressing 10 classes of drugs (American Psychiatric Association & American Psychiatric Association DSM-5 Task Force., 2013). The Diagnostic Manual- Intellectual Disability (DM-ID) provides a cross reference for the diagnosis mental disorders for persons with ID, noting the importance of seeking out accurate and reliable information prior to rendering a diagnosis of SUD as patients and carers may each have only a part of the overall picture (Fletcher, Barnhill, & McCarthy, 2016a). Due to the complex association of SUD with motivation and reward neural pathways, motivational interviewing techniques have been created to promote patient engagement in the treatment relationship and have been adapted to accommodate the needs of person with ID (Frielink & Embregts, 2013). Few screening instruments have been adapted for persons with IDD or ASD. Two Dutch instruments, the Substance Use Risk Profile Scale (Pieterse et al., 2020) and Substance Use and Misuse in Intellectual Disability–Questionnaire (Van der Nagel et al., 2011), are validated for persons with ID. For persons with ASD without ID, the ASSIST, which screens for use, and CRAFFT, which screens for use and substance-related behaviors, are commonly used. The CRAFFT may require language modification to accommodate patient language use (Kunreuther, 2021). For example, the CRAFFT asks about others suggesting a need to decrease substance use, but some persons with ASD report others notice they exhibit improved social skills when they consume alcohol, which in turn may decrease concerns by family or friends (Lalanne et al., 2015; Tinsley & Hendrickx, 2008).

Treating SUD in persons living with disability requires an individualized approach which considers neurocognitive ability in addition to social and environmental factors and is informed by best practices. Psychosocial and pharmacological interventions should be considered on a case-by-case basis. Psychosocial interventions include psychoeducation, cognitive behavioral therapies, cognitive remediation, and group interventions (Didden et al., 2020; Lalanne et al., 2017). Individualizing treatment requires considering language and communication skills, vocabulary, attention span, memory, executive functions, and the ability to conceptualize recovery (Helverschou et al., 2019; Roberts & Kwan, 2018). Treatment providers should be prepared to offer shorter, more frequent sessions that hone in on specific concepts or skills. Treatment planning should be holistic and address comorbid health and mental health conditions as well as the role of community supports and services.

Nicotine Products

The US Surgeon General has identified nicotine products as the "chief preventable killer in America" (Courtney, 2015, p. 6). A review of tobacco use prevalence among persons with ID between 1980 and 2016 found 22 studies with reported prevalence ranging from 0% to 62.9% (Huxley et al., 2019). A study of self-reported smoking found male teens with mild to moderate ID at greater risk for cigarette smoking than peers without ID, 28% versus 22%, and females at less risk, 22% versus 30% (Robertson et al., 2020). Among US Special Olympic athletes, tobacco use was reported by 8.1% of the men and 3.8% of the women (Eisenbaum, 2016). A cross-sectional study found less tobacco use among adults with ASD compared to adults without, 5.2% versus 31.9% (Fortuna et al., 2016). Croen et al. (2015) reviewed records of adults enrolled in the Kaiser Permanente in Northern California which documented that 11.9% of adults with autism used tobacco products versus 28.9% of adults without. The limited information regarding smoking cessation programs for persons living with disability cites psychoeducation, mindfulness, individualized health discussions, group interventions, role-play as possible interventions; however, additional study is needed (Didden et al., 2016).

Gambling Disorder

The DSM 5 recognizes gambling disorder as *persistent and recurrent problematic gambling behavior* (p. 585) if there is *clinically significant impairment or distress* (American Psychiatric Association. & American Psychiatric Association DSM-5 Task Force, 2013, p. 585). Gambling can take many forms and has been associated with significant harms to individual gamblers and others including detrimental effects on health, financial stability, and mental health (Price et al., 2021). The DSM 5 cites a lifetime prevalence rate for gambling disorder as 0.4–1% (American Psychiatric Association. & American Psychiatric Association. DSM-5 Task Force., 2013) although recent studies suggest the rate is higher (Abbott, 2020). As there has been limited study of gambling in persons with IDD, a prevalence rate for gambling disorder is unknown (Scheidemantel et al., 2019). A qualitative study in Australia found that persons with ID often misunderstand gambling but gamble in ways that similar to persons without ID, perhaps because they typically gamble with family or carers (Pitt et al., 2021). The DM-ID advises that the diagnosis of a gambling disorder in persons with ID should consider the individual's access to money, treatment, and control over gambling options as these diagnostic criteria may require modification (Fletcher, Barnhill, McCarthy, & Strydom, 2016b). An association between autistic traits and problem gambling has been suggested (Grant & Chamberlain, 2020). A literature search for gambling disorder treatment for persons with ID or ASD did not yield results, indicating that this is a topic ripe for additional study.

Gaming Disorder

While the DSM 5 classifies internet gaming disorder (IGD) as a condition for further study, the World Health Organization recognized gaming disorder (GD) as an official diagnosis in 2020 (American Psychiatric Association. & American Psychiatric Association. DSM-5 Task Force., 2013; Gaming Disorder, 2022). Darvesh et al. conducted a scoping review of the prevalence of GD and IGD, reporting large variation in prevalence 0.21–57.5% in the general population, with variation largely due to methodological variability. They could not identify prevalence studies for GD. Stevens, Dorstyn, Delfabbro, and King (2021) reported a GD prevalence of 1.96–3.05%. in the general population. Compared to non-gamers, depression, conduct disorder, hyperactivity/inattention, and suicidal and self-injurious behaviors have been associated with gaming (Strittmatter et al., 2015). Problem gaming has been associated with poorer psychosocial well-being, particularly for young men (Teng et al., 2020). Adult gamers in the general population report positive outcomes related to gaming, including enhanced relaxation and thinking skills, improved dexterity. and better pain management (Pontes et al., 2020). Proposed treatments have included cognitive behavioral therapy, pharmacotherapy, family therapy, and self-help websites, without an evidence-based intervention emerging to date (Zajac et al., 2017). Two systematic reviews concluded that persons with ASD are at increased risk for problematic gaming compared to the peers without ASD (Craig et al., 2021; Murray et al., 2021). Additional study is needed to better understand GD and IGD in PLWD.

Addressing Sexuality with Emerging Adults

Addressing sexual health and wellbeing is never an easy conversation in part because it can be clouded by the values or beliefs or embarrassment of the patient, family, or provider. Two scenarios are presented to highlight the diverse situations that may arise. Sexuality and safety are important components of health, and it is incumbent on medical professionals to initiate the conversations that will educate and protect patients.

Scenario A Elena is a 23-year-old female who has been diagnosed with Cerebral Palsy. She has very protective parents who did not let her out of their sight until she graduated high school at 22. Elena is now able to work part time in her day program. She started her menses at age 13. She is able to communicate but due to her dysarthria, her speech is largely unintelligible. She really wants a relationship and would like to one day get married. She meets Johnny, a colleague in the program who has Down syndrome. They like to hold hands and identify themselves as boyfriend and girlfriend. Everyone thinks that they are so cute as a couple. Johnny's mom has recently had a few conversations with him which led her to believe that they may be engaging in a consensual sexual relationship. She asks you how approach this with Johnny, Elena, and Elena's parents.

Scenario B Elena is a 23-year-old female who has been diagnosed with Cerebral Palsy. She has very protective parents who did not let her out of their sight until she graduated high school at 22. Elena is now able to work part time in her day program. She started her menses at age 13. She is able to communicate but due to her dysarthria, her speech is largely unintelligible. She really wants a relationship and would like to one day get married. She tells you that she has met Jason, her typically developed new manager, and he told her that he will "marry her" but only if she will meet him in the stock room.

Healthcare providers must be sensitive to and aware of the sexuality of patients who are living with disabilities in order to honor their basic human rights. Sexual feelings are normal for everyone and certainly no less thought or consideration should be given for patients with I/DD and physical disabilities. For some patients with I/DD, healthy sexual relationships may be possible (M. Brown & McCann, 2018; McDaniels & Fleming, 2016). Higher functioning patients should be supported in exploring their sexuality with a partner if there is consent and adequate knowledge among both parties. Even patients with severe and profound developmental delay and physical limitations may continue to have sexual desires that need to be explored and addressed. Continued awareness of and reassurance that these desires and feelings are part of normal development and helping patients and their caregivers to navigate these desires in socially appropriate ways are needed. Open and developmentally tailored conversations, questions, and feedback should be encouraged with the patient. Resources provided in a manner that is understandable to the patient are essential to help enhance knowledge, as well as to maintain a sense of privacy and independence. Supplemental conversations with various stakeholders, including by not limited to parents, caregivers, and /or social services and legal agencies about the limits of capacity may also be necessary to help best support your patients (Wilkinson et al., 2012). Importantly, this is an area that needs to be explicitly addressed with the patients and their caregivers with further guidance and direction based on their individual and collective preferences and unique situations.

While it is important to educate and support normal sexual activity among patients with I/DD, it is equally important to recognize the real and ever-present potential for abuse and mistreatment. In the United States, and beyond, there is an unconscionable history of sexual abuse and misuse in the institutions housing those with disabilities. There continues to be a culture of benign neglect or misinformation around the sexuality of young adults with IDD (Walters & Gray, 2018). Due to their inability to verbally or physically protect themselves or understand consent, patients with I/DD may be vulnerable to abuse. Department of Justice estimates suggest that persons with I/DD experience sexual assault more than seven times that of their peers without disabilities (Shapiro & O'Neil, 2018).

Evidence also indicates that PLWDD are less likely to receive sex education at home or school so the healthcare provider can play an important role in teaching patients about the full spectrum of reproductive health, sexuality, and appropriate and inappropriate relationships to reduce poor outcomes including but not limited to pregnancy, sexually transmitted infections, and abuse (Schmidt et al., 2021). Parents and caregivers may understandably feel protective of their adolescents/young adults and try to shield them from public humiliation and sexual abuse. This may unfortunately leave the youth with little to no information to protect themselves from abuse, if it presents itself, or the ability to engage in a mutually consensual relationship (Roden et al., 2020).

It is well documented that many healthcare providers are uncomfortable or do not feel adequately trained to have a meaningful discussion about sexuality with their patients (Fennell & Grant, 2019; Greenwood, 2019; Haboubi & Lincoln, 2003; Schaafsma et al., 2014; Thompson et al., 2016). PLWDD experience puberty and fertility during similar time frames as their peers, with some exceptions (Walters & Gray, 2018). They are also exposed to peers, television, books, radio, and social media that are increasingly flooded with images of sexuality. This can lead to confusion, frustration, fear, and depression if not properly addressed. Patients may not understand that masturbation should only be done in private spaces or that there are inappropriate 'touches' that need to immediately be stopped and then communicated to a trusted ally. Providers can provide a great service to their patients by simply starting the conversation and then delivering and adapting the information that they would typically present to patients who are not disabled. These conversations should be confidential, respectful of the families' religious and cultural, tailored to the developmental level of the patient, and use appropriate tools for optimal communication.

It is more important than ever to prepare patients who are living with disabilities with the tools and resources to live a developmentally appropriate, painless, healthy, and fulfilled sexuality (M. Brown & McCann, 2018). Given the reticence of parents, schools, and healthcare providers to discuss this important topic, the primary care provider is provided an opportune responsibility to broach this subject with the patients and their caregivers. PCPs do not have to be the 'experts' on all things sexual, but rather primarily be identified as knowledgeable and approachable partners who begin and advance the conversation in a safe and respectable manner that educates all involved to the best of their abilities. Resources are available (Got Transition, Mass gov) to assist with preparing the PCP to talk about anatomy, puberty, menstruation, sexual experiences and desires, contraception, sexually transmitted infections, and sexual abuse, as well as how to prepare for healthy sexual experiences as a basic human right.

Future Directions for the Transition of PLWD to Adult Medical Care

Emerging adults with developmental disabilities represent an expanding percentage of primary care practices. The medical home team must be prepared to accommodate a wide range of needs as primary care providers can play a major role in providing quality care and quality of life for this diverse and sometimes vulnerable population. Although there are many factors and barriers to consider when providing care to PLWD, the rewards for participating in the transition, care, and subsequent management of these young patients are well worth it. It is the hope of the authors that the information provided within this chapter encourages more physicians to become more familiar and comfortable with the approach to this special population.

References

Abbott, M. W. (2020). The changing epidemiology of gambling disorder and gambling-related harm: Public health implications. *Public Health, 184*, 41–45. https://doi.org/10.1016/j.puhe.2020.04.003

Abdallah, M. W., Greaves-Lord, K., Grove, J., Nørgaard-Pedersen, B., Hougaard, D. M., & Mortensen, E. L. (2011). Psychiatric comorbidities in autism spectrum disorders: Findings from a Danish Historic Birth Cohort. *European Child & Adolescent Psychiatry, 20*(11), 599–601.

Access to Medical Care for Individuals with Mobility Disabilities. (2010, Last updated 2020). US Department of Justice and US Department of Health and Human Services Retrieved from https://www.ada.gov/medcare_mobility_ta/medcare_ta.htm

Acharya, K., Meza, R., & Msall, M. E. (2017). Disparities in life course outcomes for transition-aged youth with disabilities. *Pediatric Annals, 46*(10), e371–e376.

Act, P. P. a. A. C. (2010). Patient protection and affordable care act. *Public law, 111*(48), 759–762.

Adirim, Z., Sockalingam, S., & Thakur, A. (2021). Postgraduate medical training in intellectual and developmental disabilities: a systematic review. *Academic Psychiatry, 45*(3), 1–11.

Alliance for Disability in Health Care, E. (2019). *Core competencies on disability for health care education*. Alliance for Disability in Health Care Education.

Almalki, Z. S., Karami, N. A., Almsoudi, I. A., Alhasoun, R. K., Mahdi, A. T., Alabsi, E. A., … Alotaib, T. M. (2018). Patient-centered medical home care access among adults with chronic conditions: National Estimates from the medical expenditure panel survey. *BMC Health Services Research, 18*(1), 744. https://doi.org/10.1186/s12913-018-3554-3

American Psychiatric Association., & American Psychiatric Association. DSM-5 Task Force. (2013). *Diagnostic and statistical manual of mental disorders : DSM-5* (5th ed.). American Psychiatric Association.

Andermann, A. (2018). Screening for social determinants of health in clinical care: Moving from the margins to the mainstream. *Public Health Reviews, 39*(1), 19.

Arnett, J. J. (2000). Emerging adulthood: A theory of development from the late teens through the twenties. *American Psychologist, 55*(5), 469.

Arvio, M., Salokivi, T., & Bjelogrlic-Laakso, N. (2017). Age at death in individuals with intellectual disabilities. *Journal of Applied Research in Intellectual Disabilities, 30*(4), 782–785. https://doi.org/10.1111/jar.12269

Auberry, K. (2018). Intellectual and developmental disability nursing: Current challenges in the USA. *Nursing: Research and Reviews, 8*, 23–28.

Bagatell, N., Chan, D., Rauch, K. K., & Thorpe, D. (2017). "Thrust into adulthood": Transition experiences of young adults with cerebral palsy. *Disability and Health Journal, 10*(1), 80–86. https://doi.org/10.1016/j.dhjo.2016.09.008

Bascom, G. W., & Christensen, K. M. (2017). The impacts of limited transportation access on persons with disabilities' social participation. *Journal of Transport & Health, 7*, 227–234.

Bendixen, R. M., Fairman, A. D., Karavolis, M., Sullivan, C., & Parmanto, B. (2017). A user-centered approach: Understanding client and caregiver needs and preferences in the development of mHealth apps for self-management. *JMIR mHealth and uHealth, 5*(9), e141.

Betz, C. L., Nehring, W. M., & Lobo, M. L. (2015). Transition needs of parents of adolescents and emerging adults with special health care needs and disabilities. *Journal of Family Nursing, 21*(3), 362–412.

Blanck, P., & Martinis, J. G. (2015). "The right to make choices": The national resource center for supported decision-making. *Inclusion, 3*(1), 24–33.

Boardman, L., Bernal, J., & Hollins, S. (2014). Communicating with people with intellectual disabilities: a guide for general psychiatrists. *Advances in Psychiatric Treatment, 20*(1), 27–36.

Braddock, D., Hemp, R., Rizzolo, M. C., Tanis, E. S., Haffer, L., Lulinski, A., & Wu, J. (2013). *State of the states in developmental disabilities 2013: The great recession and its aftermath*. American Association on Intellectual and Developmental Disabilities.

Brady, N. C., Bruce, S., Goldman, A., Erickson, K., Mineo, B., Ogletree, B. T., … Siegel, E. (2016). Communication services and supports for individuals with severe disabilities: Guidance for assessment and intervention. *American Journal on Intellectual and Developmental Disabilities, 121*(2), 121–138.

Broder-Fingert, S., Mateo, C. M., & Zuckerman, K. E. (2020). Structural racism and autism. *Pediatrics, 146*(3), e2020015420. https://doi.org/10.1542/peds.2020-015420.

Brown, L. W., Camfield, P., Capers, M., Cascino, G., Ciccarelli, M., de Gusmao, C. M., … Zupanc, M. (2016). The neurologist's role in supporting transition to adult health care: A consensus statement. *Neurology, 87*(8), 835–840. https://doi.org/10.1212/wnl.0000000000002965

Brown, M., Higgins, A., & MacArthur, J. (2020). Transition from child to adult health services: A qualitative study of the views and experiences of families of young adults with intellectual disabilities. *Journal of Clinical Nursing, 29*(1–2), 195–207. https://doi.org/10.1111/jocn.15077

Brown, M., & McCann, E. (2018). Sexuality issues and the voices of adults with intellectual disabilities: A systematic review of the literature. *Research in Developmental Disabilities, 74*, 124–138. https://doi.org/10.1016/j.ridd.2018.01.009

Burke, M., Arnold, C., & Owen, A. (2018). Identifying the correlates and barriers of future planning among parents of individuals with intellectual and developmental disabilities. *Intellectual and Developmental Disabilities, 56*(2), 90–100.

Butwicka, A., Långström, N., Larsson, H., Lundström, S., Serlachius, E., Almqvist, C., … Lichtenstein, P. (2017). Increased risk for substance use-related problems in autism spectrum disorders: A population-based cohort study. *Journal of Autism and Developmental Disorders, 47*(1), 80–89. https://doi.org/10.1007/s10803-016-2914-2

Carter, E. W., & Bumble, J. L. (2018). The promise and possibilities of community conversations: Expanding opportunities for people with disabilities. *Journal of Disability Policy Studies, 28*(4), 195–202.

Cecil, C. A. M., Walton, E., & Viding, E. (2016). Epigenetics of addiction: Current knowledge, challenges, and future directions. *Journal of Studies on Alcohol and Drugs, 77*(5), 688–691.

Cheak-Zamora, N. C., Teti, M., & First, J. (2015). 'Transitions are scary for our kids, and they're scary for us': Family member and youth perspectives on the challenges of transitioning to adulthood with autism. *Journal of Applied Research in Intellectual Disabilities, 28*(6), 548–560.

Convention on the Rights of Persons with, D. (2006). *Convention on the rights of persons with disabilities*. Department of Economic and Social Affairs.

Cooper, S.-A., Smiley, E., Morrison, J., Williamson, A., & Allan, L. (2007). Mental ill-health in adults with intellectual disabilities: Prevalence and associated factors. *The British Journal of Psychiatry, 190*(1), 27–35.

Courtney, R. (2015). The Health Consequences of Smoking—50 Years of Progress: A Report of the Surgeon General, 2014 Us Department of Health and Human Services Atlanta, GA: Department of Health and Human Services, Centers for Disease Control and Prevention, National Center for Chronic Disease Prevention and Health Promotion, Office on Smoking and Health, 2014 1081 pp. Online (grey literature): http://www.surgeongeneral.gov/library/reports/50-years-of-progress. Wiley Online Library.

Craig, F., Tenuta, F., De Giacomo, A., Trabacca, A., & Costabile, A. (2021). A systematic review of problematic video-game use in people with autism spectrum disorders. *Research in Autism Spectrum Disorders, 82*, 101726.

Cribb, S., Kenny, L., & Pellicano, E. (2019). 'I definitely feel more in control of my life': The perspectives of young autistic people and their parents on emerging adulthood. *Autism, 23*(7), 1765–1781.

Croen, L. A., Zerbo, O., Qian, Y., Massolo, M. L., Rich, S., Sidney, S., & Kripke, C. (2015). The health status of adults on the autism spectrum. *Autism, 19*(7), 814–823.

Della Fina, V., Cera, R., & Palmisano, G. (2017). *The United Nations convention on the rights of persons with disabilities: A commentary*. Springer.

Didden, R., VanDerNagel, J., & van Duijvenbode, N. (2016). Substance use and health related issues. In *Behavioral health promotion and intervention in intellectual and developmental disabilities* (pp. 197–209). Springer.

Didden, R., VanDerNagel, J., van Duijvenbode, N., Delforterie, M., Otten, R., & Poelen, E. (2020). Treatment of substance abuse in dual diagnosis. *Handbook of Dual Diagnosis*, Matson (editor) Handbook of Dual Diagnosis: Autism and Child Psychopathology Series. Springer. 549–563.

Dosenbach, N. U. F., Nardos, B., Cohen, A. L., Fair, D. A., Power, J. D., Church, J. A., … Lessov-Schlaggar, C. N. (2010). Prediction of individual brain maturity using fMRI. *Science, 329*(5997), 1358–1361.

Douglas, J., & Bigby, C. (2020). Development of an evidence-based practice framework to guide decision making support for people with cognitive impairment due to acquired brain injury or intellectual disability. *Disability and Rehabilitation, 42*(3), 434–441. https://doi.org/10.1080/09638288.2018.1498546

Durand, M. A., Barr, P. J., Walsh, T., & Elwyn, G. (2015). Incentivizing shared decision making in the USA--where are we now? *Healthc (Amst), 3*(2), 97–101. https://doi.org/10.1016/j.hjdsi.2014.10.008

Eisenbaum, E. N. (2016). Tobacco use among people with intellectual and developmental disabilities.

Ervin, D. A., Hennen, B., Merrick, J., & Morad, M. (2014). Healthcare for persons with intellectual and developmental disability in the community. *Frontiers in Public Health, 2*, 83.

Fennell, R., & Grant, B. (2019). Discussing sexuality in health care: A systematic review. *Journal of Clinical Nursing, 28*(17–18), 3065–3076.

Fletcher, R., Barnhill, J., & McCarthy, S.-A. (2016a). *Diagnostic manual-intellectual disability (DM-ID): A clinical guide for diagnosis of mental disorders in persons with intellectual disability in* (2nd ed.). NADD.

Fletcher, R. J., Barnhill, J., McCarthy, J., & Strydom, A. (2016b). From DSM to DM-ID. *Journal of Mental Health Research in Intellectual Disabilities, 9*(3), 189–204. https://doi.org/10.1080/19315864.2016.1185324

Fortuna, R. J., Robinson, L., Smith, T. H., Meccarello, J., Bullen, B., Nobis, K., & Davidson, P. W. (2016). Health conditions and functional status in adults with autism: a cross-sectional evaluation. *Journal of General Internal Medicine, 31*(1), 77–84.

Friedman, C., & McNamara, E. (2018). Home-and community-based speech, language, and hearing services for people with intellectual and developmental disabilities. *Research and Practice for Persons with Severe Disabilities, 43*(2), 111–125.

Frielink, N., & Embregts, P. (2013). Modification of motivational interviewing for use with people with mild intellectual disability and challenging behaviour. *Journal of Intellectual and Developmental Disability, 38*(4), 279–291.

Gaming Disorder (2022). http://www.who.int/newsroom/questions-and-answers/item/addictive-behavioursgaming-disorder. Accessed 21 June 2022.

Gauthier-Boudreault, C., Couture, M., & Gallagher, F. (2018). How to facilitate transition to adulthood? Innovative solutions from parents of young adults with profound intellectual disability. *Journal of Applied Research Intellectual Disabilities, 31*(Suppl 2), 215–223. https://doi.org/10.1111/jar.12394

Geiss, M., Chamberlain, J., Weaver, T., McCormick, C., Raufer, A., Scoggins, L., … Edmonson, D. (2017). Diagnostic overshadowing of the psychiatric population in the emergency department: Physiological factors identified for an early warning system. *Journal of the American Psychiatric Nurses Association, 24*(4), 327–331. https://doi.org/10.1177/1078390317728775

Gilson, C. B., Bethune, L. K., Carter, E. W., & McMillan, E. D. (2017). Informing and equipping parents of people with intellectual and developmental disabilities. *Intellectual and Developmental Disabilities, 55*(5), 347–360.

Grant, J. E., & Chamberlain, S. R. (2020). Autistic traits in young adults who gamble. *CNS Spectrums*, 1–6. https://doi.org/10.1017/S1092852920001571

Greenwood, J. A. (2019). *Sexuality and Romance in Individuals with Down Syndrome: Assessing the Relationship Between Parental Attitudes, Sexual Knowledge, and Experiences with Romance*. University of California, Irvine. https://escholarship.org/content/qt8635h55r/qt8635h55r_noSplash_bd5c544d7b6d19ea2695e14be51c5e52.pdf

Grisso, T., & Appelbaum, P. S. (1998). *Assessing competence to consent to treatment: A guide for physicians and other health professionals*. Oxford University Press.

Groenman, A. P., Janssen, T. W. P., & Oosterlaan, J. (2017). Childhood psychiatric disorders as risk factor for subsequent substance abuse: a meta-analysis. *Journal of the American Academy of Child & Adolescent Psychiatry, 56*(7), 556–569.

Guanxiu, T., Hang, Y., Shuping, Y., Meili, X., Pingping, Y., Wei, L., … Jun, L. (2018). Development and evaluation of an online training system for home caregivers. *Studies in Health Technology and Informatics, 250*, 7–10.

Haboubi, N. H. J., & Lincoln, N. (2003). Views of health professionals on discussing sexual issues with patients. *Disability and Rehabilitation, 25*(6), 291–296.

Hall, G. S. (1904). Adolescence: Its psychology and its relations to physiology. *Anthropology, Sociology, Sex, Crime, Religion and Education, 2*, 40–95. D. Appleton

Hand, B. N., Coury, D. L., White, S., Darragh, A. R., Moffatt-Bruce, S., Harris, L., … Garvin, J. H. (2021). Specialized primary care medical home: A positive impact on continuity of care among autistic adults. *Autism, 25*(1), 258–265.

Hart, L. C., Patel-Nguyen, S. V., Merkley, M. G., & Jonas, D. E. (2019). An evidence map for interventions addressing transition from pediatric to adult care: a systematic review of systematic reviews. *Journal of Pediatric Nursing, 48*, 18–34.

Helverschou, S. B., Brunvold, A. R., & Arnevik, E. A. (2019). Treating patients with co-occurring autism spectrum disorder and substance use disorder: a clinical explorative study. *Substance Abuse: Research and Treatment, 13*, 1178221819843291.

Hemm, C., Dagnan, D., & Meyer, T. D. (2015). Identifying training needs for mainstream healthcare professionals, to prepare them for working with individuals with intellectual disabilities: A systematic review. *Journal of Applied Research in Intellectual Disabilities, 28*(2), 98–110. https://doi.org/10.1111/jar.12117

Huxley, A., Dalton, M., Tsui, Y. Y. Y., & Hayhurst, K. P. (2019). Prevalence of alcohol, smoking, and illicit drug use amongst people with intellectual disabilities. *Drugs: Education, Prevention and Policy, 26*(5), 365–384.

Iacono, T., & Sutherland, G. (2006). Health screening and developmental disabilities. *Journal of Policy and Practice in Intellectual Disabilities, 3*(3), 155–163.

Iezzoni, L. I., Rao, S. R., Ressalam, J., Bolcic-Jankovic, D., Agaronnik, N. D., Donelan, K., … Campbell, E. G. (2021). Physicians' perceptions of people with disability and their health care: Study reports the results of a survey of physicians' perceptions of people with disability. *Health Affairs, 40*(2), 297–306.

Jones, D. R., Nicolaidis, C., Ellwood, L. J., Garcia, A., Johnson, K. R., Lopez, K., & Waisman, T. C. (2020). An expert discussion on structural racism in autism research and practice. *Autism in Adulthood, 2*(4), 273–281.

Kelleher, B. L., & Wheeler, A. C. (2020). Introduction to special issue on outcome measures for IDD: Where we have been, where we are now, and where we are heading. *American Journal on Intellectual and Developmental Disabilities, 125*(6), 413–417.

Kirley, K., Hayer, R., Khan, T., Johnson, E., Sanchez, E. S., Kosowicz, L., … Park, Y. S. (2020). Expanding the traditional history and physical examination to address chronic diseases and social needs: A multisite randomized control trial of 4 medical schools. *Academic Medicine, 95*(11S), S44–S50.

Kolva, E. A., & Rosenfeld, B. (2012). *Legal perspectives on civil capacity and competence* in *Civil Capacities in Clinical Neuropsychology: Research Findings and Practical Applications*, 17–38. Demakis (Editor) Oxford Press.

Krahn, G. L. (2019). A call for better data on prevalence and health surveillance of people with intellectual and developmental disabilities. *Intellectual and Developmental Disabilities, 57*(5), 357–375.

Krahn, G. L., Hammond, L., & Turner, A. (2006). A cascade of disparities: Health and health care access for people with intellectual disabilities. *Mental Retardation and Developmental Disabilities Research Reviews, 12*(1), 70–82.

Kunreuther, E. (2021). Autism spectrum disorder and substance use disorder: a dual diagnosis hiding in plain sight. *Psychiatric Clinics, 44*(1), 35–49.

Kuo, D. Z., & Houtrow, A. J. (2016). Recognition and management of medical complexity. *Pediatrics, 138*(6), e20163021.

Lalanne, L., Weiner, L., & Bertschy, G. (2017). Treatment of addiction in adults with autism spectrum disorder. In *Handbook of treatments for autism spectrum disorder* (pp. 377–395). Springer.

Lalanne, L., Weiner, L., Trojak, B., Berna, F., & Bertschy, G. (2015). Substance-use disorder in high-functioning autism: Clinical and neurocognitive insights from two case reports. *BMC Psychiatry, 15*(1), 1–5.

Larivière-Bastien, D., Bell, E., Majnemer, A., Shevell, M., & Racine, E. (2013). Perspectives of young adults with cerebral palsy on transitioning from pediatric to adult healthcare systems.

Lauer, E., Lindgren, S., Momany, E., Cope, T., Royer, J., Cogan, L., … Armour, B. (2021). Health service utilization patterns among medicaid-insured adults with intellectual and developmental disabilities: Implications for access needs in outpatient community-based medical services. *The Journal of Ambulatory Care Management, 44*(2), 138–147.

Lebrun-Harris, L. A., McManus, M. A., Ilango, S. M., Cyr, M., McLellan, S. B., Mann, M. Y., & White, P. H. (2018). Transition planning among US youth with and without special health care needs. *Pediatrics, 142*(4), e20180194.

Lewis, M. A., Lewis, C. E., Leake, B., King, B. H., & Lindemanne, R. (2016). The quality of health care for adults with developmental disabilities. *Public Health Reports, 117*(2), 174–184.

Lin, E., Balogh, R., McGarry, C., Selick, A., Dobranowski, K., Wilton, A. S., & Lunsky, Y. (2016). Substance-related and addictive disorders among adults with intellectual and developmental disabilities (IDD): An Ontario population cohort study. *BMJ Open, 6*(9), e011638.

Mandell, D. S., Lawer, L. J., Branch, K., Brodkin, E. S., Healey, K., Witalec, R., … Gur, R. E. (2012). Prevalence and correlates of autism in a state psychiatric hospital. *Autism, 16*(6), 557–567.

Manohar, H., Subramanian, K., Kandasamy, P., Penchilaiya, V., & Arun, A. (2016). Diagnostic masking and overshadowing in intellectual disability—How structured evaluation helps. *Journal of Child and Adolescent Psychiatric Nursing, 29*(4), 171–176. https://doi.org/10.1111/jcap.12160

McDaniels, B., & Fleming, A. (2016). Sexuality education and intellectual disability: Time to address the challenge. *Sexuality and Disability, 34*(2), 215–225.

McLachlan, K., Flannigan, K., Temple, V., Unsworth, K., & Cook, J. L. (2020). Difficulties in daily living experienced by adolescents, transition-aged youth, and adults with fetal alcohol spectrum disorder. *Alcoholism: Clinical and Experimental Research, 44*(8), 1609–1624.

McManus, M., & White, P. (2017). Transition to adult health care services for young adults with chronic medical illness and psychiatric comorbidity. *Child and Adolescent Psychiatric Clinics, 26*(2), 367–380.

Mudrick, N. R., Breslin, M. L., Nielsen, K. A., & Swager, L. C. (2020). Can disability accommodation needs stored in electronic health records help providers prepare for patient visits? A qualitative study. *BMC Health Services Research, 20*(1), 1–18.

Murray, A., Koronczai, B., Király, O., Griffiths, M. D., Mannion, A., Leader, G., & Demetrovics, Z. (2021). Autism, problematic internet use and gaming disorder: A systematic review. *Review Journal of Autism and Developmental Disorders, 9*, 120–140.

NCD LCME Second Response Letter. (2019). Retrieved from https://ncd.gov/publications/2019/ncd-lcme-2nd-response-letter

Office of the Surgeon General (US); National Institute of Child Health and Human Development (US); Centers for Disease Control and Prevention (US). (2002) Closing the Gap: A National Blueprint to Improve the Health of Persons with Mental Retardation: Report of the Surgeon General's Conference on Health Disparities and Mental Retardation. Washington (DC): US Department of Health and Human Services. Available from https://www.ncbi.nlm.nih.gov/sites/books/NBK44346/

Perez, T. S., & Crowe, B. M. (2021). Community participation for transition-aged youth with intellectual and developmental disabilities: A systematic review. *Therapeutic Recreation Journal, 55*(1), 19–41.

Perrin, J. M., Anderson, L. E., & Van Cleave, J. (2014). The rise in chronic conditions among infants, children, and youth can be met with continued health system innovations. *Health Affairs, 33*(12), 2099–2105. https://doi.org/10.1377/hlthaff.2014.0832

Pieterse, M. E., VanDerNagel, J. E. L., Ten Klooster, P. M., Turhan, A., & Didden, R. (2020). Psychometric qualities of the Dutch version of the substance use risk profile scale adapted for individuals with mild intellectual disabilities and borderline intellectual functioning. *Journal of Mental Health Research in Intellectual Disabilities, 13*(3), 250–266.

Pitt, H., Thomas, S. L., Watson, J., Shuttleworth, R., Murfitt, K., & Balandin, S. (2021). How do people with intellectual disability engage with and understand gambling? A qualitative study of adults in Victoria, Australia. *Frontiers in Public Health, 8*, 979.

Pontes, H. M., Stavropoulos, V., & Griffiths, M. D. (2020). Emerging insights on internet gaming disorder: Conceptual and measurement issues. *Addictive Behaviors Reports, 11*, 100242. 11 Dec. 2019. https://doi.org/10.1016/j.abrep.2019.100242

Price, A., Hilbrecht, M., & Billi, R. (2021). Charting a path towards a public health approach for gambling harm prevention. *Journal of Public Health, 29*(1), 37–53. https://doi.org/10.1007/s10389-020-01437-2

Ressel, M., Thompson, B., Poulin, M.-H., Normand, C. L., Fisher, M. H., Couture, G., & Iarocci, G. (2020). Systematic review of risk and protective factors associated with substance use and abuse in individuals with autism spectrum disorders. *Autism, 24*(4), 899–918.

Rights, U. N. O. o. t. H. C. o. H. (2007). From provisions to practice: Implementing the convention – Legal capacity and supported decision-making. *United Nations handbook for parliamentarians on the convention on the Rights of persons with disabilities*. Retrieved from https://www.un.org/development/desa/disabilities/resources/handbook-for-parliamentarians-on-the-convention-on-the-rights-of-persons-with-disabilities/chapter-six-from-provisions-to-practice-implementing-the-convention-5.html

Roberts, L., & Kwan, S. (2018). Putting the C into CBT: Cognitive challenging with adults with mild to moderate intellectual disabilities and anxiety disorders. *Clinical Psychology & Psychotherapy, 25*(5), 662–671. https://doi.org/10.1002/cpp.2196

Robertson, J., Emerson, E., Baines, S., & Hatton, C. (2020). Self-reported smoking, alcohol and drug use among adolescents and young adults with and without mild to moderate intellectual disability. *Journal of Intellectual & Developmental Disability, 45*(1), 35–45.

Roden, R. C., Schmidt, E. K., & Holland-Hall, C. (2020). Sexual health education for adolescents and young adults with intellectual and developmental disabilities: Recommendations for accessible sexual and reproductive health information. *The Lancet Child & Adolescent Health, 4*(9), 699–708.

Schaafsma, D., Kok, G., Stoffelen, J. M. T., Van Doorn, P., & Curfs, L. M. G. (2014). Identifying the important factors associated with teaching sex education to people with intellectual disability: A cross-sectional survey among paid care staff. *Journal of Intellectual and Developmental Disability, 39*(2), 157–166.

Scheidemantel, T., Braun-Gabelman, A., Stefanac, K., Ruedrich, S., & Kotz, M. (2019). Playing with a stacked deck: Literature review and case series of problem gambling in adults with intellectual and developmental disabilities. *Journal of Gambling Studies, 35*(3), 987–995.

Schmidt, E. K., Beining, A., Hand, B. N., Havercamp, S., & Darragh, A. (2021). Healthcare providers' role in providing sexual and reproductive health information to people with intellectual and developmental disabilities: A qualitative study. *Journal of Applied Research in Intellectual Disabilities, 35*, 1019–1027. https://doi.org/10.1111/jar.12861

Schoon, I. (2015). Diverse pathways: Rethinking the transition to adulthood. In *Families in an era of increasing inequality* (pp. 115–136). Springer.

Shapiro, B. K., & O'Neill, M. (2018). Intellectual disability. In Nelson Textbook of Pediatrics. 21st ed. Elsevier.

Sheth, S., Mleczko, V., Choi, A. L., & Sharma, A. E. (2017). The impact of the patient-centered medical home on health disparities in adults: a systematic review of the evidence. *Journal of Health Disparities Research and Practice, 10*(1), 5.

Shogren, K. A., Wehmeyer, M. L., Palmer, S. B., Rifenbark, G. G., & Little, T. D. (2015). Relationships between self-determination and postschool outcomes for youth with disabilities. *The Journal of Special Education, 48*(4), 256–267.

Solomon, E. M., Wing, H., Steiner, J. F., & Gottlieb, L. M. (2020). Impact of transportation interventions on health care outcomes: a systematic review. *Medical Care, 58*(4), 384–391.

Stevens, M. W., Dorstyn, D., Delfabbro, P. H., & King, D. L. (2021). Global prevalence of gaming disorder: A systematic review and meta-analysis. *Australian & New Zealand Journal of Psychiatry, 55*(6), 553–568.

Stillman, M. D., Ankam, N., Mallow, M., Capron, M., & Williams, S. (2021). A survey of internal and family medicine residents: Assessment of disability-specific education and knowledge. *Disability and Health Journal, 14*(2), 101011.

Strittmatter, E., Kaess, M., Parzer, P., Fischer, G., Carli, V., Hoven, C. W., … Apter, A. (2015). Pathological internet use among adolescents: Comparing gamers and non-gamers. *Psychiatry Research, 228*(1), 128–135.

Sullivan, W. F., Berg, J. M., Bradley, E., Cheetham, T., Denton, R., Heng, J., … Korossy, M. (2011). Primary care of adults with developmental disabilities: Canadian consensus guidelines. *Canadian Family Physician, 57*(5), 541–553.

Teng, Z., Pontes, H. M., Nie, Q., Xiang, G., Griffiths, M. D., & Guo, C. (2020). Internet gaming disorder and psychosocial well-being: A longitudinal study of older-aged adolescents and emerging adults. *Addictive Behaviors, 110*, 106530.

Thompson, V. R., Stancliffe, R. J., Wilson, N. J., & Broom, A. (2016). The content, usefulness and usability of sexual knowledge assessment tools for people with intellectual disability. *Sexuality and Disability, 34*(4), 495–512.

Tinsley, M., & Hendrickx, S. (2008). *Asperger syndrome and alcohol: Drinking to cope?* Jessica Kingsley Publishers.

Van der Nagel, J. E. L., Kiewik, M., van Dijk, M. K., De Jong, C. A. J., & Didden, H. C. M. (2011). Handleiding SumID-Q: Meetinstrument voor het in kaart brengen van middelengebruik bij mensen met een Lichte Verstandelijke Beperking.

Vancea, M., & Utzet, M. (2017). How unemployment and precarious employment affect the health of young people: A scoping study on social determinants. *Scandinavian Journal of Public Health, 45*(1), 73–84.

Walters, F. P., & Gray, S. H. (2018). Addressing sexual and reproductive health in adolescents and young adults with intellectual and developmental disabilities. *Current Opinion in Pediatrics, 30*(4), 451–458.

Werner, S., & Chabany, R. (2016). Guardianship law versus supported decision-making policies: Perceptions of persons with intellectual or psychiatric disabilities and parents. *American Journal of Orthopsychiatry, 86*(5), 486–499. https://doi.org/10.1037/ort0000125

Werner, S., Yalon-Chamovitz, S., Rinde, M. T., & Heymann, A. D. (2017). Principles of effective communication with patients who have intellectual disability among primary care physicians. *Patient Education and Counseling, 100*(7), 1314–1321.

White, P. H., Cooley, W. C., & American Academy of, P., & American Academy of Family, P. (2018a). Supporting the health care transition from adolescence to adulthood in the medical home. *Pediatrics, 142*(5), 182–200.

White, P. H., Schmidt, A., McManus, M., & Irwin, C. E. (2018b). Readying youth and young adults for transition to adult care during preventive care visits: New clinician toolkit. *Journal of Adolescent Health, 63*(6), 673–674.

Whitney, D. G., Warschausky, S. A., Ng, S., Hurvitz, E. A., Kamdar, N. S., & Peterson, M. D. (2019). Prevalence of mental health disorders among adults with cerebral palsy. *Annals of Internal Medicine, 171*(5), 328–333. https://doi.org/10.7326/M18-3420

Wilkinson, J., Dreyfus, D., Cerreto, M., & Bokhour, B. (2012). "Sometimes I feel overwhelmed": Educational needs of family physicians caring for people with intellectual disability. *Intellectual and Developmental Disabilities, 50*(3), 243–250.

Williamson, H. J., Contreras, G. M., Rodriguez, E. S., Smith, J. M., & Perkins, E. A. (2017). Health care access for adults with intellectual and developmental disabilities: A scoping review. *OTJR: Occupation, Participation and Health, 37*(4), 227–236.

Williamson, H. J., & Perkins, E. A. (2014). Family caregivers of adults with intellectual and developmental disabilities: Outcomes associated with US services and supports. *Mental Retardation, 52*(2), 147–159.

Wimberley, T., Agerbo, E., Horsdal, H. T., Ottosen, C., Brikell, I., Als, T. D., … Mors, O. (2020). Genetic liability to ADHD and substance use disorders in individuals with ADHD. *Addiction, 115*(7), 1368–1377.

Yuen, E. Y. N., Knight, T., Ricciardelli, L. A., & Burney, S. (2018). Health literacy of caregivers of adult care recipients: A systematic scoping review. *Health & Social Care in the Community, 26*(2), e191–e206.

Zajac, K., Ginley, M. K., Chang, R., & Petry, N. M. (2017). Treatments for internet gaming disorder and internet addiction: A systematic review. *Psychology of Addictive Behaviors, 31*(8), 979.

Zukerman, J. M., Devine, K. A., & Holmbeck, G. N. (2011). Adolescent predictors of emerging adulthood milestones in youth with spina bifida*. *Journal of Pediatric Psychology, 36*(3), 265–276. https://doi.org/10.1093/jpepsy/jsq075

Index

P. McPherson (ed.), *Handbook of Treatment Planning for Children with Autism and Other Neurodevelopmental Disorders*, Autism and Child Psychopathology Series,
https://doi.org/10.1007/978-3-031-06120-2